W0235445

Economic Synthesis of Heterocycles
Zinc, Iron, Copper, Cobalt, Manganese and Nickel Catalysts

RSC Catalysis Series

Series Editor:
Professor James J Spivey, *Louisiana State University, Baton Rouge, USA*

Advisory Board:
Krijn P de Jong, *University of Utrecht, The Netherlands*
James A Dumesic, *University of Wisconsin-Madison, USA*
Chris Hardacre, *Queen's University Belfast, Northern Ireland*
Enrique Iglesia, *University of California at Berkeley, USA*
Zinfer Ismagilov, *Boreskov Institute of Catalysis, Novosibirsk, Russia*
Johannes Lercher, *TU München, Germany*
Umit Ozkan, *Ohio State University, USA*
Chunshan Song, *Penn State University, USA*

Titles in the Series:
 1: Carbons and Carbon Supported Catalysts in Hydroprocessing
 2: Chiral Sulfur Ligands: Asymmetric Catalysis
 3: Recent Developments in Asymmetric Organocatalysis
 4: Catalysis in the Refining of Fischer–Tropsch Syncrude
 5: Organocatalytic Enantioselective Conjugate Addition Reactions:
    A Powerful Tool for the Stereocontrolled Synthesis of Complex Molecules
 6: *N*-Heterocyclic Carbenes: From Laboratory Curiosities to Efficient
    Synthetic Tools
 7: *P*-Stereogenic Ligands in Enantioselective Catalysis
 8: Chemistry of the Morita–Baylis–Hillman Reaction
 9: Proton-Coupled Electron Transfer: A Carrefour of Chemical Reactivity
    Traditions
10: Asymmetric Domino Reactions
11: C-H and C-X Bond Functionalization: Transition Metal Mediation
12: Metal Organic Frameworks as Heterogeneous Catalysts
13: Environmental Catalysis Over Gold-Based Materials
14: Computational Catalysis
15: Catalysis in Ionic Liquids: From Catalyst Synthesis to Application
16: Economic Synthesis of Heterocycles: Zinc, Iron, Copper, Cobalt,
    Manganese and Nickel Catalysts

How to obtain future titles on publication:
A standing order plan is available for this series. A standing order will bring
delivery of each new volume immediately on publication.

For further information please contact:
Book Sales Department, Royal Society of Chemistry, Thomas Graham House,
Science Park, Milton Road, Cambridge, CB4 0WF, UK
Telephone: +44 (0)1223 420066, Fax: +44 (0)1223 420247
Email: booksales@rsc.org
Visit our website at www.rsc.org/books

Economic Synthesis of Heterocycles
Zinc, Iron, Copper, Cobalt, Manganese and Nickel Catalysts

Xiao-Feng Wu and Matthias Beller
Leibniz Institute for Catalysis, Rostock, Germany
Email: xiao-feng.wu@catalysis.de; matthias.beller@catalysis.de

THE QUEEN'S AWARDS
FOR ENTERPRISE:
INTERNATIONAL TRADE
2013

RSC Catalysis Series No. 16

Print ISBN: 978-1-84973-935-1
PDF eISBN: 978-1-78262-083-9
ISSN: 1757-6725

A catalogue record for this book is available from the British Library

© Wu and Beller, 2014

All rights reserved

Apart from fair dealing for the purposes of research for non-commercial purposes or for private study, criticism or review, as permitted under the Copyright, Designs and Patents Act 1988 and the Copyright and Related Rights Regulations 2003, this publication may not be reproduced, stored or transmitted, in any form or by any means, without the prior permission in writing of The Royal Society of Chemistry or the copyright owner, or in the case of reproduction in accordance with the terms of licences issued by the Copyright Licensing Agency in the UK, or in accordance with the terms of the licences issued by the appropriate Reproduction Rights Organization outside the UK. Enquiries concerning reproduction outside the terms stated here should be sent to The Royal Society of Chemistry at the address printed on this page.

The RSC is not responsible for individual opinions expressed in this work.

Published by The Royal Society of Chemistry,
Thomas Graham House, Science Park, Milton Road,
Cambridge CB4 0WF, UK

Registered Charity Number 207890

For further information see our web site at www.rsc.org

Printed in the United Kingdom by CPI Group (UK) Ltd, Croydon, CR0 4YY, UK

Preface

Around 90% of naturally occurring molecules have heterocycles as their core structure, and heterocyclics have broad applications in pharmaceuticals, agrochemicals, dyes, and many others. With this background, syntheses of heterocyclic compounds have become one of the largest branches in modern organic chemistry. After decades developing experience, numerous synthetic methodologies have been successfully introduced. Nowadays, with the advent of sustainable development, the application of cheap metal salts as catalysts in heterocyclic synthesis has become important and attractive. Among the family of cheap metals, zinc, iron, copper, cobalt, manganese, and nickel are representative examples, as they are inexpensive, have low toxicity, are biocompatible, and are environmentally benign.

This book outlines the main contributions in this are. The contents are organized according to the catalyst applied and then subdivided by the size of the ring formed. The text starts with a short introduction, followed by chapters on the use of salts of the above metals in the synthesis of heterocyclic compounds.

The book took around 10 months to complete, and every attempt was made to ensure that the literature cited is as up-to-date as possible and that out descriptions are accurate. However, it is always possible that some literature might have been missed and there may be errors, for which we apologize. We sincerely hope that overall the content of this book will be useful to all those working in the field of heterocyclic synthesis.

Xiao-Feng Wu

RSC Catalysis Series No. 16
Economic Synthesis of Heterocycles: Zinc, Iron, Copper, Cobalt, Manganese and Nickel Catalysts
By Xiao-Feng Wu and Matthias Beller
© Wu and Beller, 2014
Published by the Royal Society of Chemistry, www.rsc.org

This book is dedicated to my wife and children, Qing-Yuan Wei, Nuo-Yu Wu and Nuo-Lin Wu, who are gratefully thanked for their understanding and support.

Contents

RSC Catalysis Series No. 16
Economic Synthesis of Heterocycles: Zinc, Iron, Copper, Cobalt, Manganese and Nickel Catalysts
By Xiao-Feng Wu and Matthias Beller
© Wu and Beller, 2014
Published by the Royal Society of Chemistry, www.rsc.org

CHAPTER 1

Introduction

Around 90% of naturally occurring molecules have heterocycles as their core structure, and heterocyclics have broad applications in pharmaceuticals, agrochemicals, dyes, and many other areas. With this background, syntheses of heterocyclic compounds have become one of the largest branches of modern organic chemistry.[1] After decades developing experience, numerous synthetic methodologies have been successfully introduced. Among these procedures, methods involving transition metal catalysts constitute a large percentage.

In recent years, sustainable development has been accepted by the wider social community. With the combination of transition metal catalysts and the concept of sustainable development, the use of cheap metal salts as catalysts has attracted the interest of synthetic chemists. Although the catalytic abilities of noble metals in coupling reactions are impressive, as demonstrated by the award of the 2010 Nobel Prize in Chemistry jointly to Richard F. Heck, Ei-ichi Negishi and Akira Suzuki for palladium-catalyzed cross-couplings in organic synthesis,[2] their high cost and toxicity are clear disadvantages.

Of the available cheap metals, zinc (Zn), iron (Fe), copper (Cu), cobalt (Co), manganese (Mn) and nickel (Ni) are representative examples: they are inexpensive, have low toxicity, are biocompatible and are environmentally benign.

Zinc is the 24th most abundant element in the Earth's crust. It was first discovered as a pure metal in 1746 by the German chemist Andreas Sigismund Marggraf by heating a mixture of calamine and carbon in a closed vessel without copper, and this process had become commercially practical by 1752. As a material, the major application of zinc is in the corrosion-resistant zinc plating of steel, with other applications in batteries and alloys. Biologically, zinc is an essential mineral with great biological and public health importance, and it is an essential component of thousands of

RSC Catalysis Series No. 16
Economic Synthesis of Heterocycles: Zinc, Iron, Copper, Cobalt, Manganese and Nickel Catalysts
By Xiao-Feng Wu and Matthias Beller
© Wu and Beller, 2014
Published by the Royal Society of Chemistry, www.rsc.org

proteins in plants, although it is toxic in excess. Zinc deficiency may cause many diseases in adults and also lead to growth retardation, delayed sexual maturation, infection susceptibility and diarrhea in children.[3] Chemically, zinc metal has the electron configuration $[Ar]3d^{10}4s^2$ and is a strong reducing agent. Zinc tends to form bonds with a greater degree of covalency and it forms much more stable complexes with N- and S-donors. Complexes of zinc are mostly 4- or 6-coordinated, but 5-coordinated complexes are also known. The applications of zinc in organic chemistry are mainly used in the preparation of organozinc compounds for organic synthesis, such as the applications of zinc reagents in the Reformatsky reaction, Frankland–Duppa reaction, Negishi reaction, Fukuyama reaction and so on.[4] The application of zinc complexes as catalysts for organic transformations has also been explored, but they are mainly limited by their 'Lewis acid' properties.[5] They are still undeveloped in the area of coupling reactions.

Iron is the fourth most common element in the Earth's crust, with a wide range of oxidation states (-2 to $+6$). Based on this property, iron catalysts have been widely applied in redox reactions.[6] Concerning cross-coupling reactions, the catalytic abilities of iron catalysts have also been explored, especially the iron-catalyzed cross-coupling of organohalides with Grignard reagents, which have even been applied in the total synthesis of biologically active molecules.[7] Iron is also abundant biologically. Iron-containing proteins are found in all living organisms, ranging from the evolutionarily primitive Archaea to humans. The color of blood is due to hemoglobin, an iron-containing protein.

The catalytic activities of copper salts are more remarkable, even comparable to those of noble metals such as palladium catalysts in cross-coupling reactions.[8] Numerous catalytic systems have been developed for C–O, C–N, C–S and C–C bond formation. Biologically, copper is an essential trace element in plants and animals and copper proteins have diverse roles in biological electron transport and oxygen transportation.

The word 'cobalt' is derived from the German kobalt, from kobold meaning 'goblin,' a superstitious term used for the ore of cobalt by miners. As an element, cobalt has the electron configuration as $[Ar]4s^23d^7$ and has oxidation states $2+$ and $3+$. Cobalt has many applications in a wide range of areas. In materials, cobalt is primarily used as the metal in the preparation of magnetic, wear-resistant and high-strength alloys. In biology, cobalt is the active center of coenzymes called cobalamins, the most common example of which is vitamin B_{12}. As such it is an essential trace dietary mineral for all animals. Cobalt in inorganic form is also an active nutrient for bacteria, algae and fungi. In chemistry, various cobalt compounds are used in chemical reactions as oxidation catalysts. Cobalt acetate is used for the conversion of xylene to terephthalic acid, the precursor to the bulk polymer polyethylene terephthalate. Typical catalysts are the cobalt carboxylates (known as cobalt soaps). They are also used in paints, varnishes and inks as 'drying agents' through the oxidation of drying oils. The same carboxylates are used to improve the adhesion of steel to rubber in steel-belted radial tires. Cobalt-based catalysts are also important in reactions involving carbon monoxide.[9] Steam reforming, useful in hydrogen production, uses cobalt oxide-based catalysts. Cobalt is a catalyst in the Fischer–Tropsch process, used in the hydrogenation of carbon monoxide to give liquid fuels. The hydroformylation of alkenes often relies on cobalt octacarbonyl as the catalyst, although such processes have been partially displaced by more efficient iridium- and rhodium-based catalysts, *e.g.*, in the Cativa process. The hydrodesulfurization of petroleum uses a catalyst derived from cobalt and molybdenum. This process helps to rid petroleum of sulfur impurities that interfere with the refining of liquid fuels.

Manganese is a silvery gray metal that resembles iron. It is hard and very brittle, difficult to fuse, but easy to oxidize. The most common oxidation states of manganese are +2, +3, +4, +6 and +7, although oxidation states from −3 to +7 are observed. In biology, manganese is an essential trace nutrient in all known forms of life. The classes of enzymes that have manganese cofactors are very broad. The reverse transcriptases of many retroviruses (although not lentiviruses such as HIV) contain manganese. There is about 12 mg of manganese present in the human body, which is stored mainly in the bones; in the tissues, it is mostly concentrated in the liver and kidneys. In the human brain, manganese is bound to manganese metalloproteins, most notably glutamine synthetase in astrocytes. Manganese is also important in photosynthetic oxygen evolution in chloroplasts in plants. In chemistry, manganese is mainly used as oxidant for organic substrates.[10] Recently, its catalytic properties have also been explored.

Nickel was first isolated and classified as a chemical element in 1751 by Axel Fredrik Cronstedt, and has two electron configurations, $[Ar]4s^2 3d^8$ and $[Ar]4s^1 3d^9$, with very close energies. From the application point of view, nickel plays important roles in the biology of microorganisms and plants. The plant enzyme urease (an enzyme that assists in the hydrolysis of urea) contains nickel. The [NiFe] hydrogenases contain nickel in addition to iron–sulfur clusters. Such [NiFe] hydrogenases characteristically oxidize H_2. In synthetic chemistry, nickel catalysts have been explored in carbonylation reactions, coupling reactions and many other types of catalytic transformations. More recently, nickel-catalyzed C–O bond activation has made important achievements and showed superior activity to palladium.[11] In $C(sp^3)$–X transformations, nickel gave excellent activities with boronic acids and organozinc reagents as coupling partner. Nickel catalysts have also been studied in the area of heterocycle synthesis.

With this background, it is extremely interesting and important to develop methodologies that involve Zn, Fe, Cu, Co, Mn and Ni as catalysts in the heterocycle syntheses. In the following chapters, we present detailed discussions on this topic. The chapters are subdivided according to the sizes of the rings formed. The book concludes with a personal outlook.

References

1. (a) J. A. Joule and K. Mills *Heterocyclic Chemistry*, 5th edn, Wiley-Blackwell, Oxford, 2010; (b) A. R. Katritzky, C. A. Ramsden, E. F. V. Scriven and R. J. K. Taylor (eds), *Comprehensive Heterocyclic Chemistry III*, Elsevier, Oxford, 2008; (c) R. V. A. Orru and E. Ruijter (eds), *Synthesis of Heterocycles via Multicomponent Reaction I*, Springer, Berlin, 2010; (d) E. Eycken and C. O. Kappe (eds), *Microwave-Assisted Synthesis of Heterocycles*, Springer, Berlin, 2006.
2. X.-F. Wu, P. Anbarasan, H. Neumann and M. Beller, *Angew. Chem. Int. Ed.*, 2010, **49**, 9047–9050.
3. (a) K. M. Hambidge and N. F. Krebs, *J. Nutr.*, 2007, **137**, 1101–1105; (b) A. S. Prasad, *BMJ*, 2003, **326**, 409–410; (c) W. Maret and H. H. Sandstead, *J. Trace Elem. Med. Biol.*, 2006, **20**, 3–18; (d) A. S. Prasad, F. W. Beck, S. M. Grabowski, J. Kaplan and R. H. Mathog, *Proc. Assoc. Am. Physicians*, 1997, **109**, 68–77; (e) L. Rink and R. Gabriel, *Proc. Nutr. Soc.*, 2000, **59**, 541–552.
4. (a) P. Knochel and P. Jones, *Organozinc Reagents: a Practical Approach*, Oxford University Press, Oxford, 1999; (b) W. A. Herrmann, *Synthetic Methods of Organometallic and Inorganic Chemistry: Catalysis*, Georg Thieme, Stuttgart, 2002.
5. (a) X.-F. Wu and H. Neumann, *Adv. Synth. Catal.*, 2012, **354**, 3141–3160; (b) X.-F. Wu, *Chem. Asian J.*, 2012, **7**, 2502–2509.
6. (a) D. Bézier, J.-B. Sortais and C. Darcel, *Adv. Synth. Catal.*, 2013, **355**, 19–33; (b) P. Stavropoulos, R. Celenligil-Cetin and A. E. Tapper, *Acc. Chem. Res.*, 2001, **34**, 745–752; (c) W. Nam, *Acc. Chem. Res.*, 2007, **40**, 522–531; (d) S. Gladiali and E. Alberico, *Chem. Soc. Rev.*, 2006, **35**, 226–236; (e) M. Costas, M. P. Mehn, M. P. Jensen and L. Que Jr, *Chem. Rev.*, 2004, **104**, 939–986; (f) S. V. Kryatov, E. V. Rybak-Akimova and S. Schindler, *Chem. Rev.*, 2005, **105**, 2175–2226; (g) M. Costas, K. Chen and L. Que Jr, *Coord. Chem. Rev.*, 2000, **200–202**, 517–544.
7. (a) B. D. Sherry and A. Fürstner, *Acc. Chem. Res.*, 2008, **41**, 1500–1511; (b) C. Wang and B. Wan, *Chin. Sci. Bull.*, 2012, **57**, 2338–2351; (c) C. Bolm, J. Legros, J. Le Paih and L. Zani, *Chem. Rev.*, 2004, **104**, 6217–6254; (d) A. Correa, O. Garcia Mancheno and C. Bolm, *Chem. Soc. Rev.*, 2008, **37**, 1108–1117; (e) C.-L. Sun, B.-J. Li and Z.-J. Shi, *Chem. Rev.*, 2011, **111**, 1293–1314.
8. (a) S. V. Ley and A. W. Thomas, *Angew. Chem. Int. Ed.*, 2003, **42**, 5400–5449; (b) F. Monnier and M. Taillefer, *Angew. Chem. Int. Ed.*, 2009, **48**, 6954–6971; (c) D. Ma and Q. Cai, *Acc. Chem. Res.*, 2008, **41**, 1450–1460; (d) S. R. Chemler and P. H. Fuller, *Chem. Soc. Rev.*, 2007, **36**, 1153–1160; (e) J. E. Moses and A. D. Moorhouse, *Chem. Soc. Rev.*, 2007, **36**, 1249–1262; (f) T. Jerphagnon, M. G. Pizzuti, A. J. Minnaard and B. L. Feringa, *Chem. Soc. Rev.*, 2009, **38**, 1039–1075; (g) D. S. Surry and S. L. Buchwald, *Chem. Sci.*, 2010, **1**, 13–31; (h) J. Hassan, M. Sévignon, C. Gozzi, E. Schulz and M. Lemaire, *Chem. Rev.*, 2002, **102**, 1359–1469;

(i) S. Reymond and J. Cossy, *Chem. Rev.*, 2008, **108**, 5359–5406; (j) K.-i. Yamada and K. Tomioka, *Chem. Rev.*, 2008, **108**, 2874–2886; (k) L. M. Stanley and M. P. Sibi, *Chem. Rev.*, 2008, **108**, 2887–2902; (l) C. Deutsch, N. Krause and B. H. Lipshutz, *Chem. Rev.*, 2008, **108**, 2916–2927; (m) M. Meldal and C. W. Tornoe, *Chem. Rev.*, 2008, **108**, 2952–3015; (n) G. Evano, N. Blanchard and M. Toumi, *Chem. Rev.*, 2008, **108**, 3054–3131; (o) I. P. Beletskaya and A. V. Cheprakov, *Coord. Chem. Rev.*, 2004, **248**, 2337–2364; (p) G. Lefevre, G. Franc, A. Tlili, C. Adamo, M. Taillefer, I. Ciofini and A. Jutand, *Organometallics*, 2012, **31**, 7694–7707.

9. (a) A. Y. Khodakov, W. Chu and P. Fongarland, *Chem. Rev.*, 2007, **107**, 1692–1744; (b) F. Hebrard and P. Kalck, *Chem. Rev.*, 2009, **109**, 4272–4282.

10. (a) G. G. Melikyan, *Synthesis*, 1993, 833–850; (b) B. B. Snider, *Chem. Rev.*, 1996, **96**, 339–363; (c) G. Cahiez, C. Duplais and J. Buendia, *Chem. Rev.*, 2009, **109**, 1434–1476; (d) J. M. Concellón, H. Rodríguez-Solla and V. del Amo, *Chem. Eur. J.*, 2008, **14**, 10184–10191; (e) M. Mondal and U. Bora, *RSC Adv.*, 2013, 3, 18716–18754.

11. (a) S. Gu, P. Ni and W. Chen, *Chin. J. Catal.*, 2010, **31**, 875–886; (b) J. Montgomery, *Angew. Chem. Int. Ed.*, 2004, **43**, 3890–3908; (c) J. Montgomery, *Acc. Chem. Res.*, 2000, **33**, 467–473; (d) X. Hu, *Chem. Sci.*, 2011, 2, 1867–1886; (e) F. S. Han, *Chem. Soc. Rev.*, 2013, **42**, 5270–5298; (f) B. M. Rosen, K. W. Quasdorf, D. A. Wilson, N. Zhang, A. M. Resmerita, N. K. Garg and V. Percec, *Chem. Rev.*, 2011, **111**, 1346–1416; (g) Z. X. Wang and N. Liu, *Eur. J. Inorg. Chem.*, 2012, 901–911; (h) J. Yamaguchi, K. Muto and K. Itami, *Eur. J. Org. Chem.*, 2013, 19–30.

CHAPTER 2

Zinc-Catalyzed Heterocycle Synthesis

Zinc is elementally essential in our daily lives, with a wide range of applications in materials, and in addition the adult body contains about 2–3 g of elemental zinc. Zinc salts have also been used in plant fertilizers. In organic synthesis, zinc salts are mainly used as Lewis acids. With the accepted importance of heterocyclic compounds and the environmentally benign properties of zinc salts, it is of interest to explore the applications of zinc catalysts in heterocycle syntheses.[1]

2.1 Five-Membered Heterocycles

2.1.1 Zinc-Catalyzed Synthesis of Carbonates

The application of zinc catalysts in the polymerization of epoxides and CO_2 has been known for many years,[2] and the alternative production of carbonates from epoxides and CO_2 by changing the reaction conditions and using zinc catalyst is also of interest. Carbonates are aprotic polar solvents (and nowadays considered as 'green' solvents) and are used as intermediates for pharmaceuticals and fine chemicals.[3] Various procedures have been developed, and the systems are becoming more well-defined or heterogenized.

Based on previous reports on zinc-catalyzed cyclization and copolymerization of CO_2 and peroxides, Kim and colleagues carried out a detailed mechanistic study.[4] A pyridinium alkoxy ion-bridged dimeric zinc complex was isolated and characterized. Subsequently, they carried out a systematic study. The reactions of CO_2 and epoxides to produce cyclic carbonates were performed in the presence of a catalyst $[L_2ZnX_2]$ (L = pyridine or substituted pyridine; X = Cl, Br, I). The effects of pyridine and halide ligands on the

RSC Catalysis Series No. 16
Economic Synthesis of Heterocycles: Zinc, Iron, Copper, Cobalt, Manganese and Nickel Catalysts
By Xiao-Feng Wu and Matthias Beller
© Wu and Beller, 2014
Published by the Royal Society of Chemistry, www.rsc.org

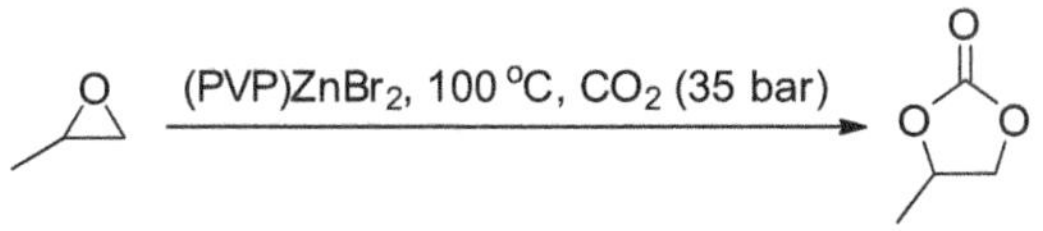

Scheme 2.1 Heterogeneous $ZnBr_2$-catalyzed reaction of propylene oxide and CO_2.

catalytic activity and the formation of active species were investigated. Catalysts with electron-donating substituents on the pyridine ligands exhibited higher activity than those with unsubstituted pyridine ligands. On the other hand, catalysts with electron-withdrawing substituents on the pyridine ligands showed no activity; this demonstrates the importance of the nucleophilicity of the pyridine ligands. A zinc complex containing a strongly chelating 2,2'-dipyridyl ligand was found to be totally inactive, indicating that ligand dissociation is also an important factor in the catalysis process. They also prepared a well-defined heterogeneous catalyst by supporting the zinc catalyst on PVP [poly(4-vinylpyridine)].[5] This catalyst shown high selectivity and activity for the reaction of CO_2 with ethylene oxide or propylene oxide (Scheme 2.1). Solid NMR characterization of the PVP-supported $ZnBr_2$ catalyst and its reaction product with propylene oxide and/or CO_2 showed that a pyridinium alkoxy ion-bridged zinc bromide complex is functioning as an active species, such as in the homogeneous catalysis with L_2ZnBr_2 (L = pyridine or methyl-substituted pyridine). Since then, they have developed several other procedures, such as imidazolium zinc tetrahalide systems[6] and phosphine-bound zinc halide complexes.[7]

Xia and co-workers found the combination of $ZnCl_2$ and [BMIm]Br (1-butyl-3-methylimidazolium bromine salt) to be an efficient system for the reaction of CO_2 with terminal epoxides, and several carbonates were synthesized in good yields (Scheme 2.2).[8] Interestingly, only a trace of product was observed with [BMIm]BF_4 and [BMIm]PF_6 and low activity with [BMIm]Cl. Later, they immobilized the catalytic system with a biopolymer (chitosan).[9] Under the same reaction conditions, using chitosan-supported zinc chloride, good yields of cyclic carbonates were produced and the catalyst system can be recycled five times without a significant decrease in activity.

Darensbourg and co-workers studied the effects of pyridine derivatives on the reaction of CO_2 and cyclohexene oxide.[10] Good to excellent yields (80–91%) of carbonate can be produced with 2,6-dimethoxypyridine, 2-phenylpyridine or 3-trifluoromethylpyridine as ligand and with $ZnBr_2$, ZnI_2 or $ZnCl_2$ as the catalyst. Zinc(II) benzoate complexes for the reaction of carbon dioxide with epoxide were also studied. Good activities were found under 41.3 bar of CO_2 at 80 °C. Additionally, catalytic systems based on $ZnBr_2$/n-Bu_4NI,[11] $ZnBr_2$/Ph_4PI,[12] Zn/ionic liquid,[13] Zn/SiO_2[14] and Zn/hydroxyapatite[15] were developed. A highly active homogeneous system based on Zn(salphen) was established recently.[16] In combination with Bu_4NI, under solvent-free conditions, excellent conversion and selectivity towards cyclic carbonate product can be obtained (Scheme 2.3).

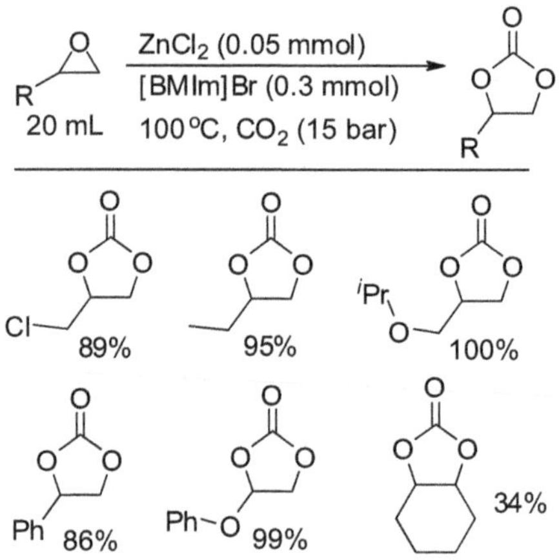

Scheme 2.2   ZnCl$_2$/[BMIm]Br-catalyzed carbonate synthesis.

Reaction Procedure (Scheme 2.2): A 100 mL stainless-steel pressure reactor was charged with propylene oxide (20 mL, 16.6 g, 0.285 mol), ZnCl$_2$ (6.8 mg, 0.05 mmol), [BMIm]Br (65 mg, 0.3 mmol) and the reaction vessel was heated at 100 °C for1h under a constant pressure (1.5 MPa) of carbon dioxide. Once the reaction was completed, the vessel was cooled to ambient temperature and the pressure released, then the contents were transferred into a round-bottomed flask. Unreacted substrate was removed in a vacuum first, and then the product was obtained. In the case of chitosan-supported ZnCl$_2$, it was prepared by refluxing ZnCl$_2$ (1.36 g) and chitosan (1.9 g) in 30 mL of ethanol for 24 h before use.

In addition to the reaction of epoxides with CO$_2$, carbonates can also be produced from 1,2-propylene glycol and urea.[17] By using ZnO as catalyst supported on NaY, propylene carbonate was produced in good yield.

2.1.2   Zinc-Catalyzed Synthesis of Tetrazoles

Tetrazoles are a class of heterocycles with a wide range of pharmaceutical and industrial activities. The most straightforward reaction pathway is the addition of azide ion to nitriles. In 2001, Demko and Sharpless reported the first example of the zinc bromide-mediated reaction of sodium azide with nitriles in water.[18] The reaction showed a broad substrates scope and functional group tolerance. A variety of aromatic nitriles, activated and unactivated alkyl nitriles, substituted vinyl nitriles, thiocyanates and cyanamides were converted into the corresponding tetrazoles in good to excellent yields (Scheme 2.4). Considering the recognized importance of the tetrazole moiety in peptidomimetic chemistry, they succeeded in extending

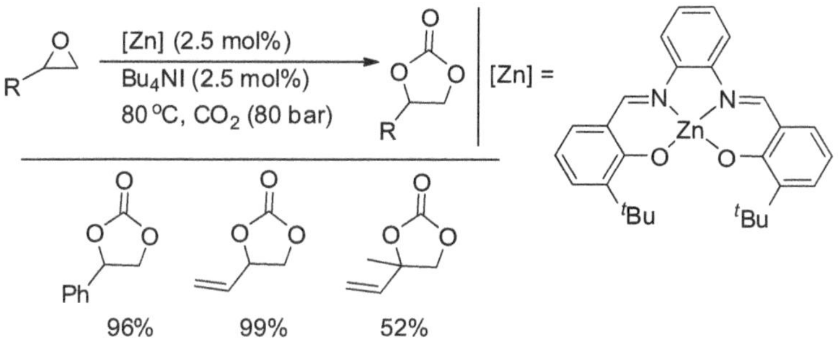

Scheme 2.3 Zn(salphen)-catalyzed synthesis of carbonates.

Reaction Procedure (Scheme 2.3): Zn(salphen) (0.025 g, 5.0×10^{-5} mol), Bu$_4$NI (5.0×10^{-5} mol) and the internal standard, mesitylene (0.27 mL, 2.0×10^{-3} mol), were weighed into a glass vial. To this mixture, the epoxide (2.0×10^{-3} mol) was added and the solution was stirred until complete dissolution occurred. The vial containing this solution and a magnetic stirring bar was transferred into a stainless-steel reactor in a high-throughput unit. Air was removed from the lines and from the reactor by purging with N$_2$ for 5–10 min. Next, the system was purged with CO$_2$ at 15 bar for 15 min before increasing the CO$_2$ pressure to 80 bar. First, the unit was pressurized to 60 bar (1 h), while keeping the block at room temperature. Next, the temperature was increased to 80 °C and the pressure was adjusted to the final value of 80 bar. As soon as the desired conditions were reached, the entrance valve of the reactor was closed and stirring at 900 rpm was started. The sample was allowed to react for the required time (3 or 5 h). Depressurization was started when the reactor had cooled to room temperature.

their methodology to α-aminonitriles by slightly changing the reaction conditions.[19] Tetrazole analogs of α-amino acids were prepared in good yields by refluxing the starting materials in water-2-propanol at 80 °C (Scheme 2.5a). It was later demonstrated that this reaction can be performed under solvent-free conditions.[20] Based on density functional theory (DFT) calculations, the coordination of the nitrile to the zinc ion is the dominant factor affecting the catalysis, which substantially lowers the barrier for nucleophilic attack by azide.[21] Interestingly, the tetrazole can be fragmented into the corresponding triazoles with the assistance of Zn(OTf)$_2$ and under microwave irradiation (Scheme 2.5b).[22] BF$_3 \cdot$ OEt$_2$, Al(OTf)$_3$, AgOTf, Cu(OTf)$_2$, In(OTf)$_3$ and ZnCl$_2$ were also tested as Lewis acids, and lower yields of triazole were observed; no product was formed with FeCl$_3$ as catalyst.

In 2007, Bolm's group extended this methodology to the synthesis of *N*-(1*H*)-tetrazole sulfoximines (Scheme 2.6).[23] In the presence of NaN$_3$ and ZnBr$_2$, various *N*-cyano compounds which were prepared from readily

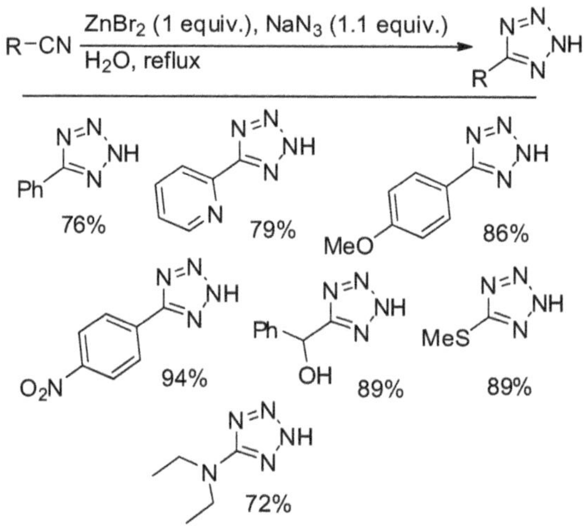

Scheme 2.4 ZnBr₂-mediated synthesis of tetrazoles.

Reaction Procedure (Scheme 2.4): To a 250 mL round-bottomed flask was added the nitrile (20 mmol), sodium azide (1.43 g, 22 mmol), zinc bromide (4.50 g, 20 mmol) and 40 mL of water. The reaction mixture was refluxed for 24 h; vigorous stirring is essential. HCl (3 M, 30 mL) and ethyl acetate (100 mL) were added, and vigorous stirring was continued until no solid was present and the aqueous layer had a pH of 1. If necessary, additional ethyl acetate was added. The organic layer was isolated and the aqueous layer extracted with 2 × 100 mL of ethyl acetate. The combined organic layers were evaporated, 200 mL of 0.25 M NaOH were added, and the mixture was stirred for 30 min, until the original precipitate had dissolved and a suspension of zinc hydroxide was formed. The suspension was filtered, and the solid washed with 20 mL of 1 M NaOH. To the filtrate were added 40 mL of 3 M HCl with vigorous stirring, causing the tetrazole to precipitate. The tetrazole was filtered and washed with 2 × 20 mL of 3 M HCl and dried in a drying oven to furnish the tetrazole as a white or slightly colored powder.

available sulfides and cyanogen amine or by direct cyanation of the corresponding *N*-(*H*)-sulfoximine were transformed. This reaction is stereospecific, allowing easy access to novel enantiopure sulfoximines from the corresponding optically active *N*-cyano derivatives. The transformation of these compounds into *N*-benzylated tetrazoles and 5-substituted 1,3,4-oxadiazole sulfoximines was also accomplished. Later, arylaminotetrazoles were prepared from arylcyanamides under similar reaction conditions.[24] Generally, isomers of 5-arylamino-1*H*-tetrazole can be obtained from arylcyanamides carrying an electron-withdrawing substituent on the aryl ring and, as the electropositivity of substituent was increased, the product

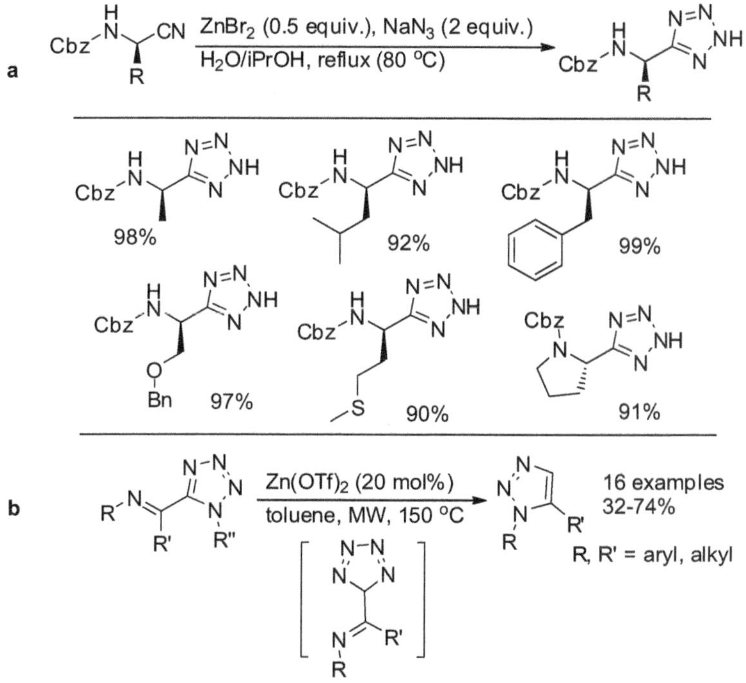

Scheme 2.5   ZnBr$_2$-mediated synthesis of tetrazole analogs of α-amino acids.

was shifted toward the isomer of 1-aryl-5-amino-1*H*-tetrazole. Since the first report from Sharpless, several heterogeneous zinc catalysts were developed and applied in the reaction of nitriles with azides. The typical catalysts included, ZnO,[25] ZnS[26] and others.[27] One main advantage of these methodologies is the reusability of the catalysts, but a high reaction temperature was still needed and the reactions should be carried out in an organic solvent.

Wu and co-workers reported a domino reaction of 2-alkynylbenzonitriles with sodium azide.[28] Under microwave irradiation (75 W), treatment of 2-alkynylbenzonitriles with 1.5 equiv. of sodium azide in DMSO at 140 °C gave 4,5-disubstituted-2*H*-1,2,3-triazoles in 60–99% yields. Additionally, adding 8 equiv. of ZnBr$_2$ and using 8 equiv. of sodium azide in DMF at 100 °C led to the formation of tetrazolo[5,1-*a*]isoquinolines in yields of up to 87% (Scheme 2.7).

A versatile and highly efficient protocol for the synthesis of 1,5-disubstituted tetrazoles was developed in 2007 by Hajra *et al.*[29] Zn(OTf)$_2$ was found to be the best catalyst for this one-pot reaction of alkenes, NBS, nitriles and TMSN$_3$. La(OTf)$_3$, Yb(OTf)$_3$, Y(OTf)$_3$, Sm(OTf)$_3$ and In(OTf)$_3$ were also effective for this transformation, but longer reaction times were required. Use of different combinations of alkenes and nitriles generated a variety of 1,5-disubstituted tetrazoles containing an additional α-bromo functionality of the N^1-alkyl substituent (Scheme 2.8).

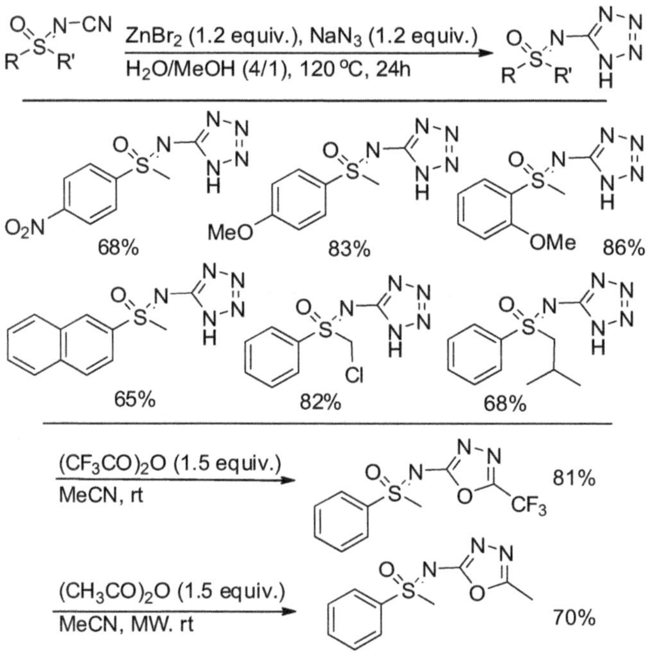

Scheme 2.6 ZnBr$_2$-mediated synthesis of *N*-(1*H*)-tetrazole sulfoximines.

Reaction Procedure (Scheme 2.6): A mixture of sulfoximine (1.67 mmol), NaN$_3$ (2.00 mmol) and ZnBr$_2$ (2.00 mmol) in a 4 : 1 mixture of H$_2$O-MeOH (4 mL) was stirred vigorously at 120 °C in a sealed tube for 8–24 h until the sulfoximine was consumed. The solvent was concentrated under reduce pressure, 1 M HCl (3 mL) and AcOEt (8 mL) were added, the aqueous solution was extracted with AcOEt (3 × 2 mL) and the solvent was evaporated. The residue was treated with 1 M NaOH (12 mL) for 20 min and the white suspension formed was filtered. The filtrate was acidified with 1 M HCl to pH 1–2 and the aqueous solution extracted with AcOEt (4 × 5 mL). The organic layer was dried over anhydrous MgSO$_4$, filtered and evaporated. The residue was purified by triturating with 1 : 1 Et$_2$O–pentane.

2.1.3   Zinc-Catalyzed Synthesis of Pyrroles

The importance and prevalence of pyrroles have been demonstrated by the wide range of applications in the areas of natural products, bioactive molecules and electrically conducting materials. As a development of the Paal–Knorr reaction, Zn(BF$_4$)$_2$ was applied as catalyst at room temperature under solvent-free conditions.[30] Various penta- and trisubstituted pyrroles were prepared in excellent yields in a clean reaction (Scheme 2.9).

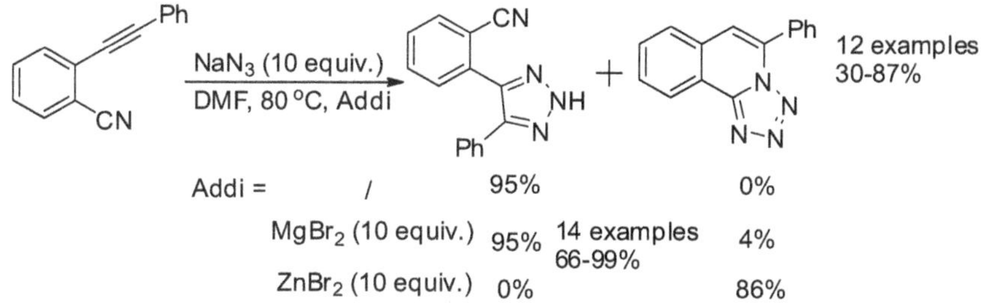

Scheme 2.7 ZnBr$_2$-mediated reaction of 2-alkynylbenzonitriles with sodium azide.

Reaction Procedure (Scheme 2.7): An appropriate 2-alkynylbenzonitrile (0.4 mmol), zinc bromide (3.2 mmol), and sodium azide (3.2 mmol) were suspended in DMF (3 mL) in a 10 mL glass vial. The mixture was irradiated using an irradiation power of 75 W. After completion of the reaction, the vial was cooled to 25 °C. The mixture was then diluted with water (20 mL), extracted with EtOAc (3 × 10 mL) and dried over anhydrous MgSO$_4$. After filtration and removal of solvent under reduced pressure, the residue was purified by column chromatography on silica gel to furnish the product.

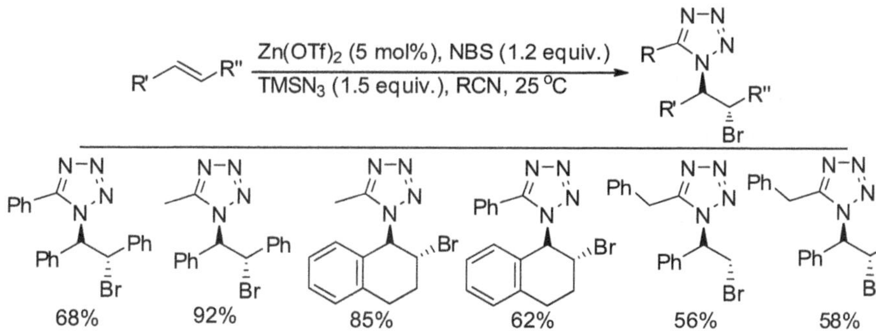

Scheme 2.8 Zn(OTf)$_2$-catalyzed synthesis of tetrazoles.

Reaction Procedure (Scheme 2.8): To a well-stirred suspension of 4 Å molecular sieve (0.100 g) and Zn(OTf)$_2$ (0.009 g, 0.025 mmol) in dry RCN (1.0 mL) were successively added alkene (0.50 mmol), TMSN$_3$ (0.1 mL, 0.75 mmol), and NBS (0.107 g, 0.60 mmol) under an argon atmosphere at rt (25 °C). The reaction was monitored by TLC. On completion, it was quenched with saturated aqueous NaHCO$_3$ and extracted with CH$_2$Cl$_2$ (3 × 30 mL). The combined organic layers were washed with water (25 mL) and brine (25 mL), dried over Na$_2$SO$_4$ and concentrated under vacuum. Flash column chromatographic purification of the crude material using petroleum ether (60–80 °C)-ethyl acetate as eluent afforded the pure tetrazole.

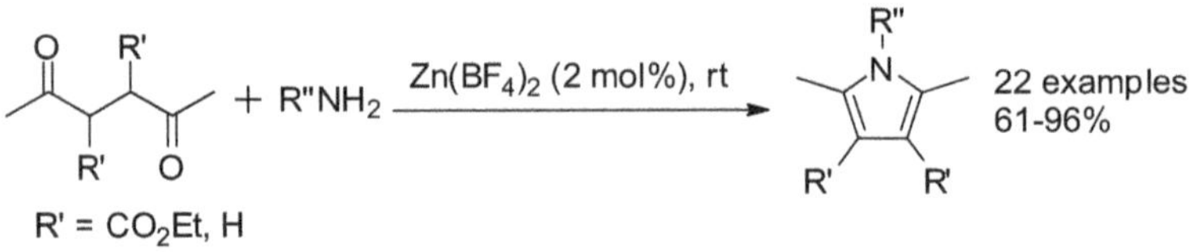

R' = CO₂Et, H

Scheme 2.9   Zn(BF₄)₂-catalyzed synthesis of pyrroles.

Reaction Procedure (Scheme 2.9): A mixture of a primary amine (1.2 mmol) and 1,4-diketone (1 mmol) was stirred at room temperature in the presence of a catalytic amount (0.01 mL, 2 mol%) of an aqueous solution (40%) of zinc tetrafluoroborate for a few minutes as required to complete the reaction (TLC). The reaction mixture was then extracted with diethyl ether (3 × 10 mL) and the extract was washed successively with aqueous HCl (1.3 M) and water. It was then dried (Na₂SO₄) and evaporated to leave the crude product, which was purified by column chromatography on silica gel (hexane–ethyl acetate, 95 : 5) to provide the pure product.

In 2007, Driver and co-workers reported a zinc-catalyzed synthesis of pyrroles from dienyl azides.[31] A range of 2,5-disubstituted and 2,4,5-trisubstituted pyrroles can be synthesized by this procedure at room temperature (Scheme 2.10a). Cu(OTf)₂ and Rh₂(O₂CC₃F₇)₄ were also effective for this transformation. This procedure was even possible without any catalyst if the reaction was carried out at 145 °C. Here, the starting materials were prepared from the corresponding methyl 2-azidoacetate and vinylaldehydes.

Alternatively, the cyclization of homopropargyl azides offers another procedure (Scheme 2.10b).[32] Moderate to high yields of pyrroles can be produced from this methodology, but the preparation of the starting materials is more tedious. This catalyst system proved to be more effective for aryl than for alkyl substituents. The relatively short reaction times *via* a microwave-assisted protocol provide an alternative to currently available methods. This approach allows for the facile preparation of *N*-unprotected highly substituted pyrroles with aryl, (fused) cycloalkyl and alkyl groups, and additionally substituents that are not readily introduced by other methods, such as cycloalkyls, can be easily inserted here.

Langer's group reported a regioselective synthesis of functionalized 1-aminopyrroles by ZnCl₂-catalyzed one-pot 'conjugate addition/cyclization' reactions of 1,3-bis(silyl enol ethers) with 1,2-diaza-1,3-butadienes.[33] These reactions were easy to carry out and proceeded under mild conditions and with high yields (Scheme 2.11a). It is noteworthy that the products were not directly available from the β-dicarbonyl compounds. Later, another approach using 1,2-diaza-1,3-butadienes as substrates was developed (Scheme 2.11b).[34] The sequence involved the preliminary preparation of aminohydrazones by Michael addition of primary amines to 1,2-diaza-1,3-dienes. The treatment of these intermediates with dialkyl acetylenedicarboxylates produced

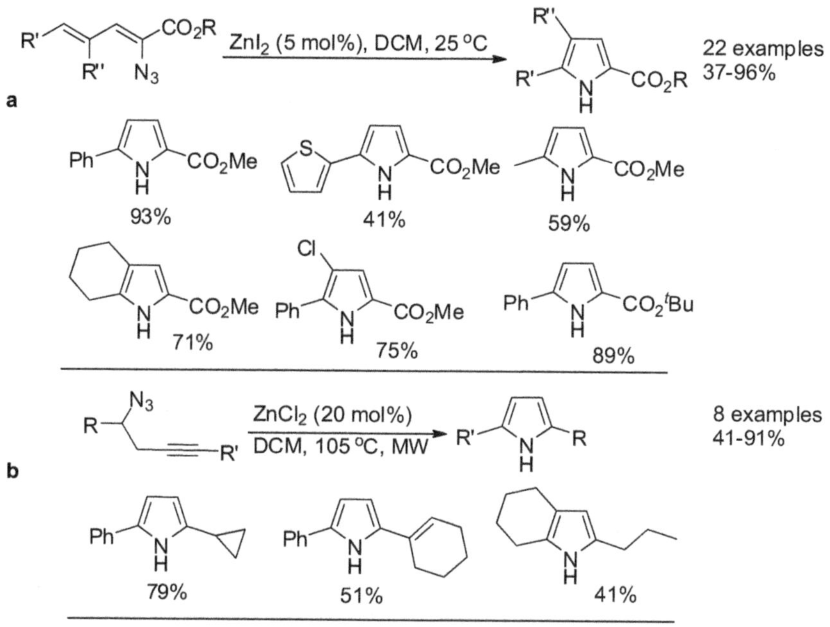

Scheme 2.10 Zinc-catalyzed synthesis of pyrroles from azides.

Reaction Procedure (Scheme 2.10): To a mixture of 0.100 g of dienyl azide (0.436 mmol) and 0.007 g of ZnI_2 (5% mol) was added 0.300 mL of solvent (1.5 M). The resulting mixture was stirred at room temperature for 15 h. The heterogeneous mixture was filtered through SiO_2. The filtrate was concentrated *in vacuo*. Purification via MPLC (0:100–30:70 EtOAc–hexanes) provided the pyrrole.

α-(*N*-enamino)hydrazones that were converted into the corresponding pyrroles. The substituents on the carbon at position 4 of 1,2-diaza-1,3-dienes drive the regioselectivity of the ring closure process. Starting from 4-aminocarbonyl-1,2-diaza-1,3-dienes, only dialkyl 1-substituted 5-aminocarbonyl-1*H*-pyrrole-2,3-dicarboxylates were obtained by Lewis acid-catalyzed ring closure. Screening of several Lewis/Brønsted acid catalysts was performed. Zinc(II) triflate was the most efficient catalyst. Under similar reaction conditions, employing 4-alkoxycarbonyl-1,2-diaza-1,3-dienes, only 4-hydroxy-1*H*-pyrrole-2,3-dicarboxylates were synthesized. The latter reactions can be accomplished regioselectively also in one pot. Using 4-aminocarbonyl-1,2-diaza-1,3-dienes, diamines and dialkyl acetylenedicarboxylates, the sequence provided the corresponding α,ω-di(*N*-pyrrolyl)alkanes.

Zinc salts [$Zn(OAc)_2$, $Zn(NO_3)_2$, $ZnCl_2$] have also been shown to be efficient and regioselective catalysts facilitating the hydroamination of functionalized *C*-propargyl vinylogous carbamates into pyrroles (Scheme 2.12).[35] Advantages are that these metal salts are commonly available, inexpensive

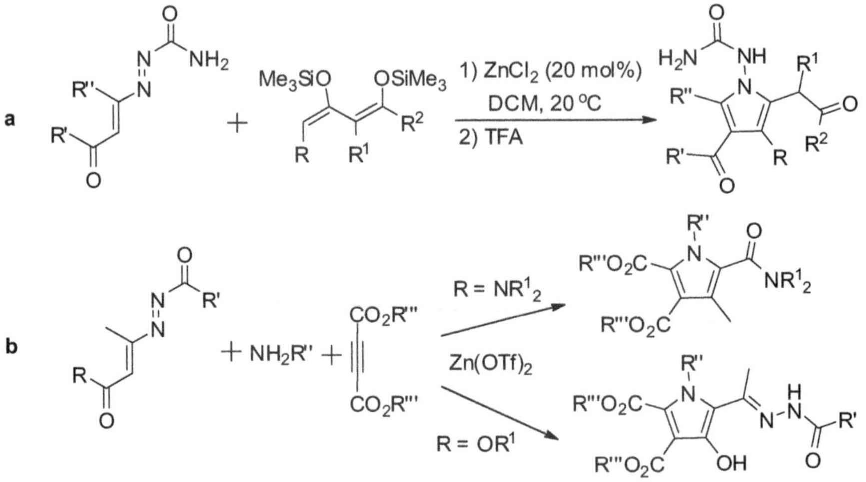

Scheme 2.11   Zinc-catalyzed synthesis of 1-aminopyrroles.

Reaction Procedure (Scheme 2.11): A 5mL microwave reactor vial, equipped with a magnetic stirring bar and fitted with a rubber septum, was charged with azide (1.00 mmol) and 1,2-dichloroethane (2 mL). Zinc chloride (1.0 M in diethyl ether; 0.20–0.40 mL) was added dropwise with a syringe and the vial was capped. The reaction was carried out with the following microwave reactor parameters: temperature 105 °C, run time 60 min (130 °C), power high (pressure ~2 bar). Silica gel column chromatography (hexanes–ethyl acetate, 100 : 0–20 : 1) gave the pure product.

and stable to air and moisture. The zinc catalysts with the exception of ZnO gave the highest yields under mild conditions; however, all of the group 12 metals afforded excellent yields of product when longer reaction times, higher catalytic loadings and increased microwave energy were utilized. The stability of the catalysts in solution, and also the Lewis acidity of the catalysts, play an important role in the efficiency of hydroamination reactions. Changing the counterion had a noticeable effect on hydroamination and it has been suggested that less coordinating anions result in higher rates up to a point at which the catalyst stability in solution becomes compromised as a result of anions becoming too non-coordinating.

   Zhan and co-workers developed an air- and moisture-tolerant zinc-catalyzed regioselective proppargylation/amination/cycloisomerization reaction of propargylic acetates, enoxysilanes and primary amines in a single pot.[36] Various aromatic and aliphatic propargylic acetates participate well in the reaction, providing the propargylation/amination/cycloisomerization products in good yields with complete regioselectivity (Scheme 2.13). The one-pot multicomponent coupling reaction furnishes substituted pyrroles in high yields by circumventing the isolation of intermediates. Zinc(II) chloride, working as a multifunctional catalyst, catalyzes three

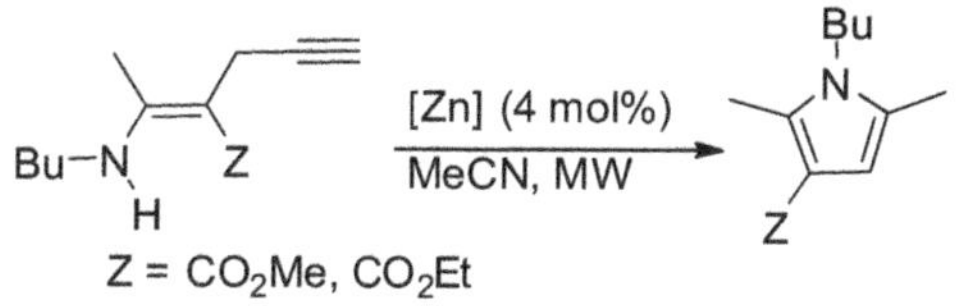

Scheme 2.12 Zinc-catalyzed synthesis of pyrroles from *C*-propargyl vinylogous carbamates.

Reaction Procedure (Scheme 2.12): Zinc salt (0.0358 mmol, 0.04 equiv.) was added to a solution of crude *C*-propargyl vinylogous amide (200 mg, 0.896 mmol) and acetonitrile (2 mL) in a microwave reaction vessel, and subjected to 40 W of microwave irradiation for 20 s whilst being stirred and force cooled.

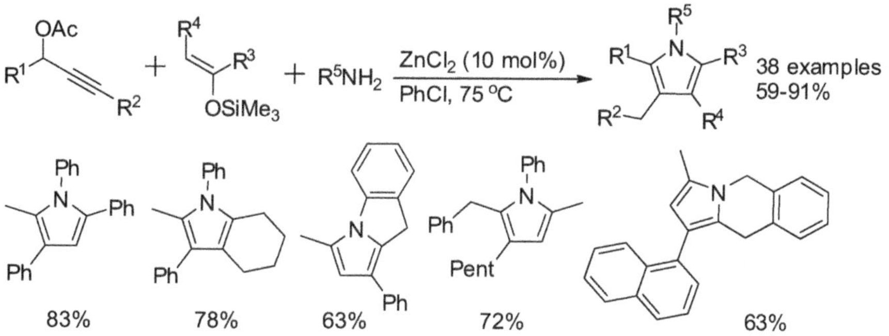

Scheme 2.13 Zinc-catalyzed multicomponent synthesis of pyrroles.

Reaction Procedure (Scheme 2.13): To a 10 mL flask, propargylic acetates (0.5 mmol), enoxysilanes (1.0 mmol), chlorobenzene (2.0 mL) and $ZnCl_2$ (0.05 mmol) were successively added. The reaction was stirred at 75 °C for 0.3 h, followed by the addition of primary amines (1.0 mmol). The reaction mixture was heated at reflux temperature for an additional 1.5–12 h until completion by TLC. Upon cooling to room temperature, the reaction mixture was quenched with 1M HCl (2 mL), the organic and aqueous layers were separated and the aqueous layer was extracted with Et_2O (3×5 mL). The combined organic layers were dried over $MgSO_4$ and filtered. The filtrate was concentrated *in vacuo*, then the residue was purified by silica gel column chromatography (EtOAc–hexane, 1:100) to afford the corresponding substituted pyrroles.

mechanistically distinct processes in a single pot under the same conditions. A wide range of aromatic propargylic acetates bearing terminal or internal alkyne groups are readily available and a number of functionalities are tolerated. Additionally, the secondary aliphatic propargylic acetates could also be efficiently incorporated into the pyrrole framework. The broad scope, mild reaction conditions and experimental ease of this transformation have made it a valuable alternative to currently available transformations. The protocol developed has been extended to the synthesis of *N*-bridgehead pyrroles containing polycyclic fragments.

2.1.4   Zinc-Catalyzed Synthesis of Furans

Furans are found in biologically active compounds and natural products and are useful synthetic intermediates. Vicente, López and co-workers reported a novel approach to the catalytic generation of synthetically valuable zinc carbenes by using alkynes and zinc salts (Scheme 2.14).[37] These

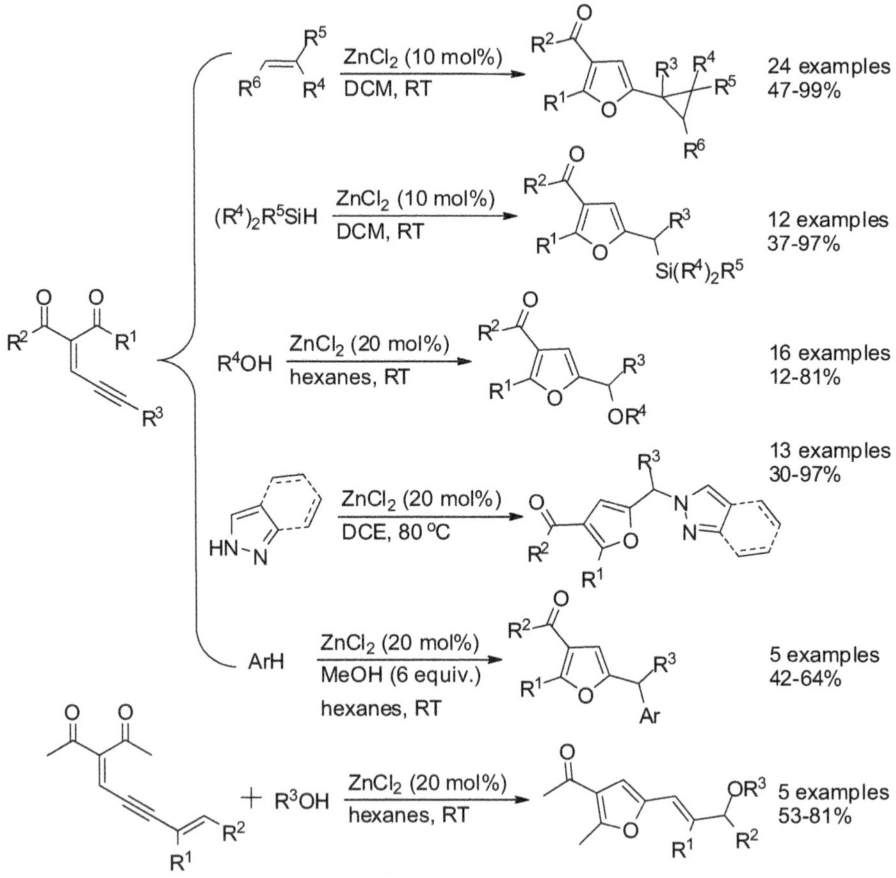

Scheme 2.14   Zinc-catalyzed synthesis of furans from alkynes.

Reaction Procedure (Scheme 2.14): ZnCl$_2$ (2.7 mg, 0.02 mmol, 10 mol%) was added to a solution of the enyne (42 mg, 0.2 mmol) and styrene (125 mg, 1.2 mmol, 6.0 equiv.) in CH$_2$Cl$_2$ (2 mL). The mixture was stirred at room temperature until the disappearance of enyne (monitored by TLC; 1.5 h). The solvent was removed under reduced pressure and the resulting residue was purified by flash chromatography (SiO$_2$, *n*-hexane–EtOAc, 20 : 1) to yield the product.

new zinc carbenes have been employed in cyclopropanation and Si–H bond insertion reactions, which can now be accomplished in a catalytic manner. The overall process allowed the efficient synthesis of highly substituted furans, thus making use of readily available materials and inexpensive ZnCl$_2$, having low toxicity, as the catalyst. Moreover, they also combined this methodology with known Knoevenagel condensations to develop convenient selective multicomponent or intramolecular reactions. A computational study indicates the participation of a 2-furyl–zinc(II) Fischer-type carbene complex, with a structure different from those previously proposed for Simmons–Smith reactions. Later, they reported another methodology that allowed the synthesis of valuable functionalized furfuryl ether derivatives when using alcohols. Analogously, the use of azoles permitted a straightforward preparation of relevant unsymmetrical triarylmethane derivatives. Further, zinc chloride proved capable of catalyzing a cascade comprising cyclization/C–C bond formation, to afford different triarylmethanes. Computational studies indicated a mechanism involving activation of the alcohol (or the azole) by zinc, which mimics the function of zinc metalloenzymes.

In 2007, a zinc-catalyzed cycloisomerization of alk-3-yn-1-ones to furans was developed by Dembinski and co-workers.[38] This methodology was later applied in the preparation of furopyrimidine nucleosides. The results showed that ZnCl$_2$ is an efficient, non-transition metal catalyst for the quantitative cycloisomerization of the alk-3-yn-1-one unit at room temperature and in the absence of base. The simple isolation protocol facilitates high yields. The relatively short reaction time and an easy-to-handle catalyst provide an appealing alternative to currently available methods. This approach allows the facile preparation of highly substituted furans (Scheme 2.15).

The combination of Zn(OTf)$_2$ with Pd(OAc)$_2$ for the synthesis of tetrasubstituted furans from aromatic alkynes were developed by Jiang and co-workers.[39] The reactions were carried out with 2 mol% of Pd(OAc)$_2$ and 30 mol% of Zn(OTf)$_2$, in MeOH under 7.6 bar of O$_2$ at 100 °C; six examples of furans were prepared in 56–82% yields. ZnCl$_2$ was applied in the synthesis of 2,3-disubstituted furans by Paquette's group in 2002.[40] Starting from α,β-unsaturated enones, the conjugate addition of organocopper reagents was performed through an aldol reaction with (tetrahydropyranyloxy)acetaldehyde under zinc chloride catalysis, then treatment of the resulting product with *p*-toluenesulfonic acid in THF afforded the target 2,3-disubstituted furans in moderate yields (Scheme 2.16).

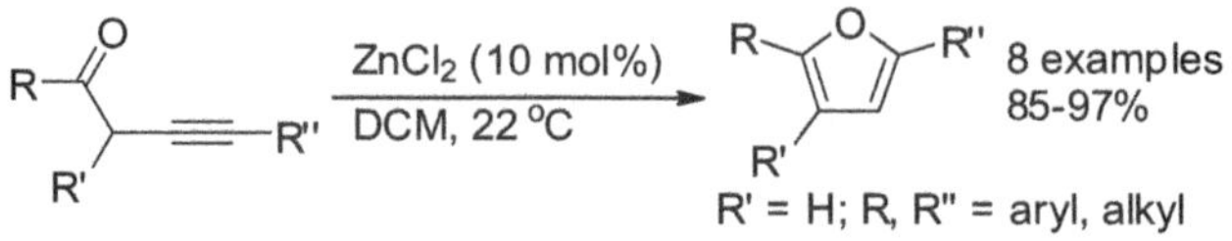

Scheme 2.15   Zinc-catalyzed cycloisomerization of alk-3-yn-1-ones to furans.

Reaction Procedure (Scheme 2.15): A round-bottomed flask was charged with substrate (0.234 g, 1.00 mmol) and CH_2Cl_2 (10 mL). Zinc chloride (1.0 M in diethyl ether, 0.10 mL, 0.10 mmol) was added dropwise with a syringe. The solution was stirred at room temperature (22 °C) for 2 h.

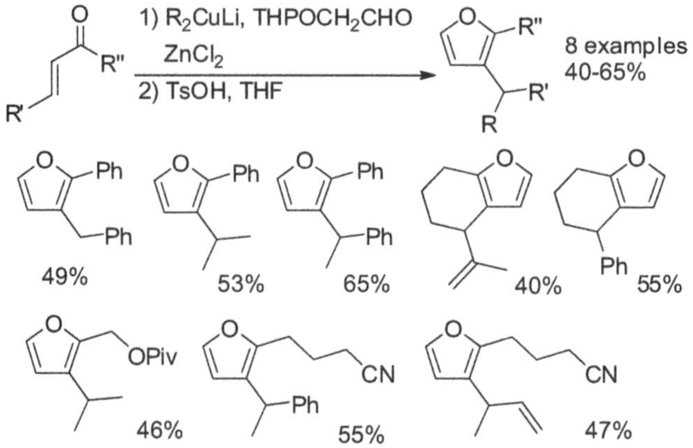

Scheme 2.16   Zinc-catalyzed synthesis of 2,3-disubstituted furans.

2.1.5   Zinc-Catalyzed Synthesis of Pyrrolidines

Pyrrolidine as a five-membered nitrogen-containing heterocycle can be easily prepared by zinc-catalyzed intramolecular hydroamination of alkenes and alkynes. A breakthrough was achieved by Blechert, Roesky and co-workers with their aminotroponate zinc complex.[41] They introduced various aminotroponiminate zinc alkyl and amido complexes as catalysts for the intramolecular hydroamination. The influence of steric modifications of the alkyl groups at the nitrogen atoms of the aminotroponiminate ligand and electronic modifications of the seven-membered ring were systematically studied. The influence of the leaving group was also studied. The major influence of the steric and electronic environment around the zinc atom on both the reactivity and stability of the corresponding complexes was demonstrated. The resulting complexes possess interesting advantages compared with other metal catalysts as they show a very high tolerance towards polar functional groups including ethers, thioethers and amides.

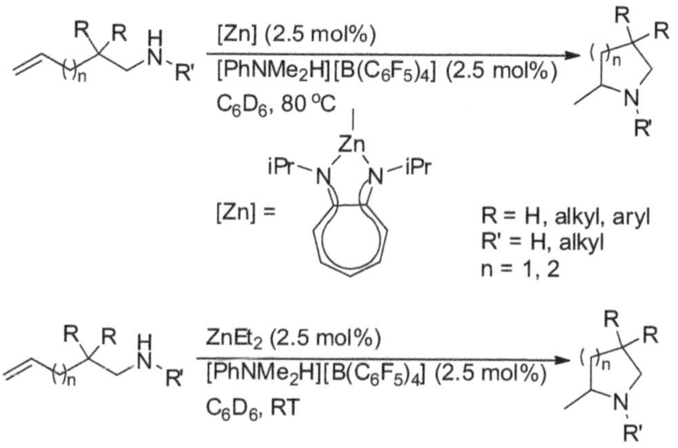

Scheme 2.17 Zinc-catalyzed hydroamination reactions.

The catalysts also show good activity in the catalytic conversion of non-activated C–C multiple bonds and a relatively high stability towards moisture and air (Scheme 2.17). Further, it was found that the combination of ZnEt$_2$ and [PhNHMe$_2$][B(C$_6$F$_5$)$_4$] can catalyze the hydroamination of aminoalkenes and aminoalkynes even at room temperature.[42] The products were obtained in a quantitative manner.

In order to combine the stability of heterogeneous systems with the high functional group tolerance of aminotroponiminate zinc compounds, Asefa and co-workers immobilized the aminotroponiminate zinc complex on mesoporous silica.[43] This non-toxic, inexpensive compound was expected to serve as a good recyclable heterogeneous catalyst for intramolecular hydroamination. [*N*-Propyl-2-(propylamino)troponiminato]methylzinc-functionalized SBA-15 (ZnDPRT) was synthesized in a multistep process. Highly ordered mesoporous silica SBA-15 was treated with 3-aminopropyltrimethoxysilane and the remaining silanol groups were then capped by reaction with an excess of hexamethyldisilazane to form capped aminopropyl-functionalized SBA = 15 (CAPI). Then the reaction of CAPI with ethylated *N*-(propylamino)-tropone gave immobilized *N*-propyl-2-(propylamino)troponimine (DPRT). Ethylated *N*-(propylamino)tropone was synthesized by treating *N*-(propylamino)-tropone with Meerwein's salt (Et$_3$OBF$_4$). Finally, DPRT was reacted with ZnMe$_2$ to form the target compound ZnDPRT. In order to investigate the catalytic activity and recyclability of ZnDPRT in intramolecular hydroamination, it was used as a catalyst for the cyclization of (2,2-diphenyl-4-pentyl)-(4-nitrobenzyl)amine. [PhNMe$_2$H][B(C$_6$F$_5$)$_4$] was used as a co-catalyst to increase the reactivity of ZnDPRT. The catalyst was fully recyclable over five runs without losing catalytic activity. After each run, the catalyst was recovered by vacuum filtration, washed with toluene and used directly for the next catalytic reaction by adding a fresh equivalent of the co-catalyst.

Following the results from Blechert, Roesky and co-workers, several other novel ligands and complexes were discovered and synthesized

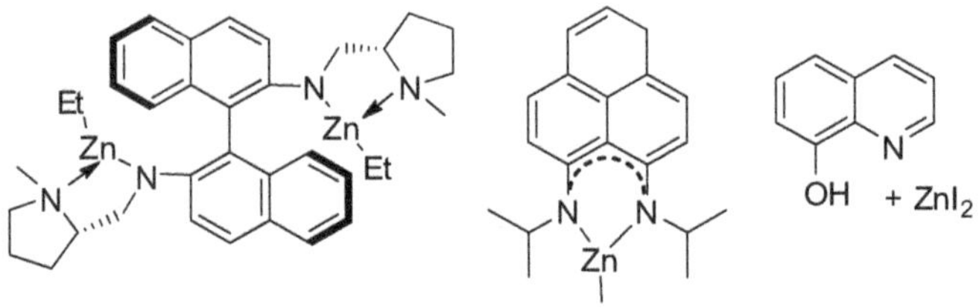

Scheme 2.18   Zinc complexes for hydroamination reactions.

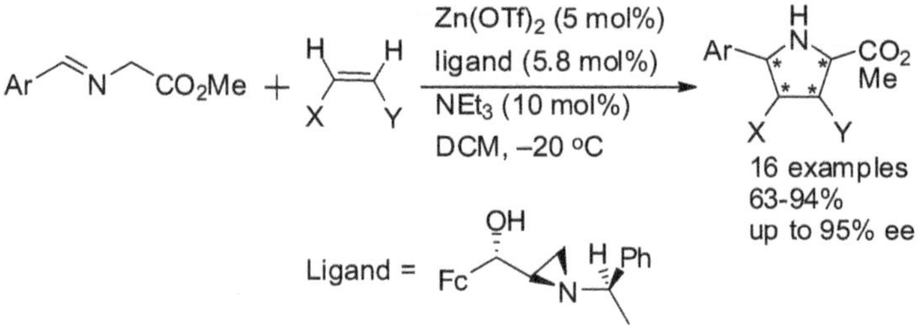

Scheme 2.19   Zinc-catalyzed asymmetric synthesis of substituted pyrrolidines.

(Scheme 2.18).[44] Based on DFT calculations, the alkene activation was found to be the preferred reaction pathway to the alternative amine activation pathway. The substitution on the amine was found have a significant influence on catalytic activity. The complex with *i*Pr on the nitrogen showed much higher activity than that substituted with Me.

Dogan, Garner and co-workers developed a zinc-based catalyst for asymmetric azomethine ylide cycloaddition to pyrrolidines.[45] With a chiral aziridino alcohol as ligand, *N*-arylideneglycine methyl esters react with a variety of dipolarophiles to give substituted pyrrolidines in very good to excellent yields and up to 95% *ee* (Scheme 2.19). A key feature of this ligand system is its ease of preparation, which is facilitated by the judicious incorporation of a ferrocenyl group into its structure. The synthesis of ferrocene-substituted pyrrolidine derivatives was also successfully achieved by a similar strategy using Et_2Zn as catalyst.

2.1.6   Zinc-Catalyzed Synthesis of Indoles, Pyrazoles and Other Heterocycles

The importance of indoles has been demonstrated by the prevalence of this moiety in biologically active molecules, and numerous methodologies have been developed for their preparation.[46] Among the methodologies, intramolecular hydroamination of alkynylamine derivatives is a representative example, usually catalyzed by Pd or Cu. Zinc has also been applied as a cheap catalyst in this type of transformation. Zhao and co-workers developed an Et_2Zn-catalyzed intramolecular hydroamination of alkynylsulfonamides.[47]

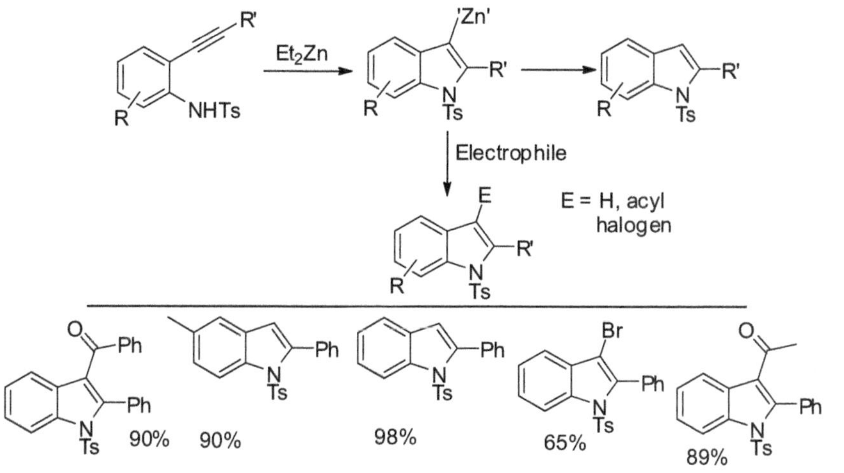

Scheme 2.20 Et$_2$Zn-catalyzed indole synthesis.

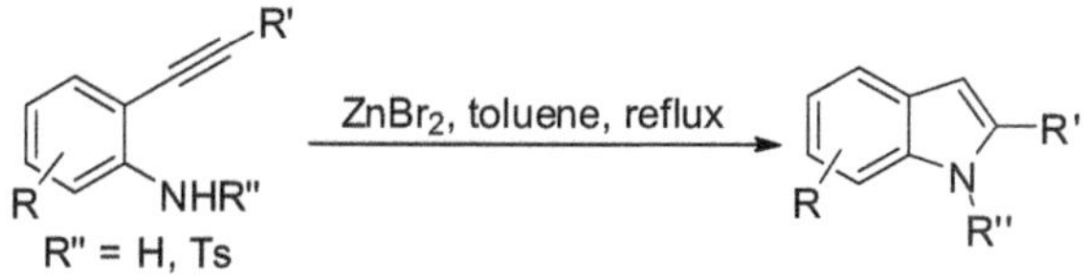

Scheme 2.21 ZnBr$_2$-catalyzed indole synthesis.

With a catalytic amount of Et$_2$Zn (20 mol%), indole derivatives were produced in excellent yield (Scheme 2.20). A tandem cyclization/nucleophilic addition procedure involving reaction of the indole zinc salt intermediate with acid chlorides or halides was also developed to provide an efficient approach to C3-substituted indole derivatives when an excess of Et$_2$Zn (120 mol%) was used.

Et$_2$Zn is air and moisture sensitive and Okuma and co-workers found that ZnBr$_2$ can be used as an alternative catalyst.[48] By treating 2-phenylethynyl *N*-tosylanilide with ZnBr$_2$ (3 equiv.) in refluxing toluene, *N*-tosyl-2-phenylindole was produced in 93% yield. With a catalytic amount of ZnBr$_2$ (0.05 equiv.), only 10% of *N*-tosyl-2-phenylindole was formed whereas 2-alkynylanilines reacted effectively and afforded the corresponding 2-substituted indoles in high yields (Scheme 2.21).

Beller and co-workers found a novel zinc-mediated system for indoles synthesis.[49] Starting from hydrazines and terminal alkynes, indoles were produced in moderate to excellent yields in the presence of an equivalent amount of ZnCl$_2$ or Zn(OTf)$_2$ in THF at 100 °C (Scheme 2.22). A numbers of functional groups are tolerated, but this methodology has difficulty in the preparation of 3-amidoindoles. In 2011, they found that ZnBr$_2$ could effectively promote the reaction between propargylamines and phenylhydrazines to give 3-amidoindoles.[49b] Either by thermal heating or

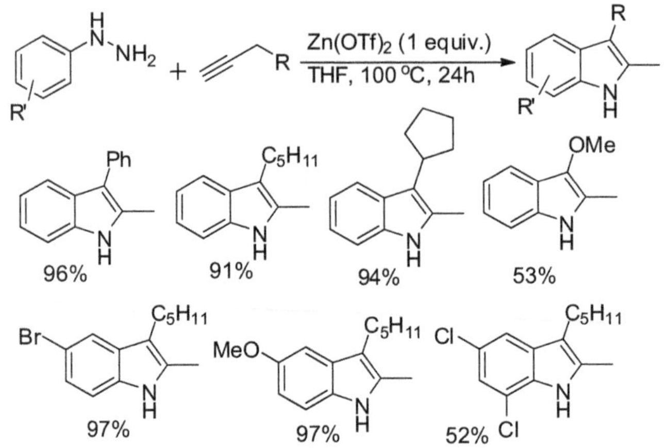

Scheme 2.22   Zn(OTf)$_2$-mediated Fischer indole synthesis.

Reaction Procedure (Scheme 2.22): ZnCl$_2$ (4.5 mmol, 545.3 mg) or Zn(OTf)$_2$ (1.5 mmol, 613.3 mg) was dissolved in THF in an ACE pressure tube under an argon atmosphere. Arylhydrazine (1.95 mmol) and alkyne (1.5 mmol) were then added to this solution. The pressure tube was sealed and the reaction mixture was heated at 100 °C for 24 h. After removal of the solvent *in vacuo*, the indole product was purified by column chromatography.

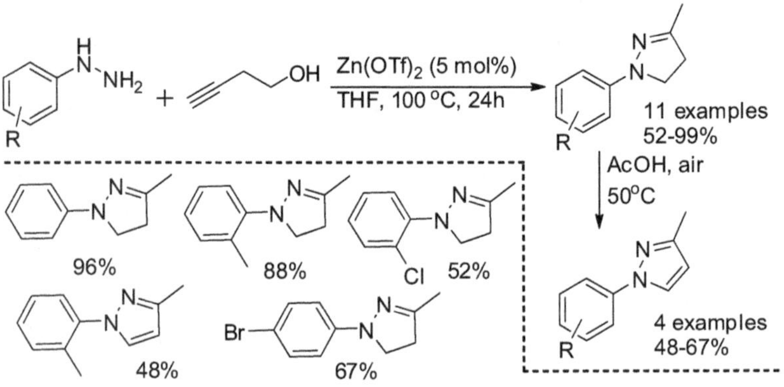

Scheme 2.23   Zn(OTf)$_2$-catalyzed pyrazoline synthesis.

by using microwave irradiation, biogenic 3-amidoindoles were produced in good yields with broad functional group tolerance.

Interestingly, when 3-butynol was treated with phenylhydrazines in the presence of zinc catalyst, pyrazolines were produced instead of their expected indole derivatives (Scheme 2.23).[50] Subsequent one-pot oxidation with air led to the corresponding pyrazoles in moderate yields. In this methodology, $Zn(OTf)_2$ is superior to the other zinc salts.

In 2011, Patil and Singh reported a zinc-mediated double hydroamination of 1,3-enynes with phenylhydrazines to give pyrazolines.[51] Compared with the above-discussed case, the advantage of this method is certainly its relatively broad substrate scope (Scheme 2.24). However, the disadvantage is the availability of starting materials, and the enynes have to be pre-prepared. The oxidation of pyrazolines to pyrazoles was also carried out.

Above we discussed a one-pot oxidation of pyrazolines to pyrazoles using air as oxidant; some methodologies intend for the preparation of pyrazoles have also been developed. A typical procedure is the condensation of 1,3-dicarbonyl compounds with hydrazine in the presence of a Lewis acid.[52] Interestingly, $Zn(L\text{-proline})_2$ was found to be an effective catalyst to drive the condensation of 1,3-dicarbonyl compounds with hydrazine in water at room temperature. Various pyrazoles were prepared in good yields. The catalyst can be recycled and reused five times without loss of activity. Another novel zinc-catalyzed procedure for pyrazole synthesis is based on diazoacetate compounds and terminal alkynes.[53] A series of pyrazoles were prepared

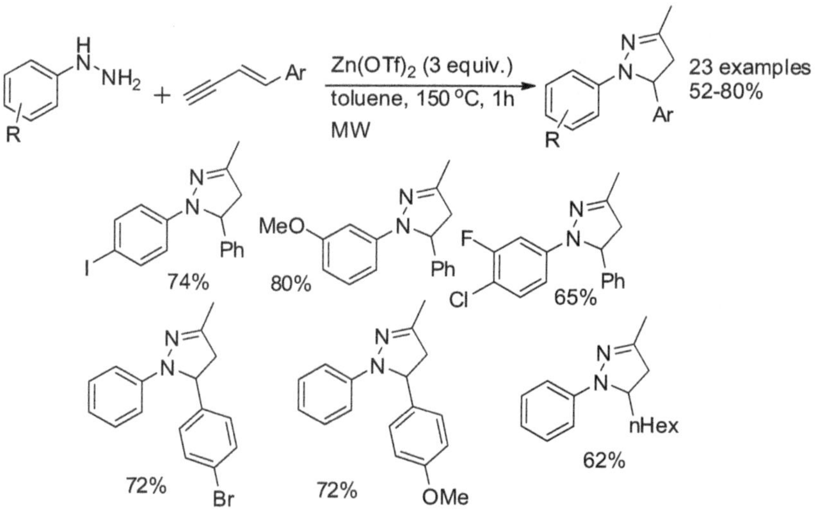

Scheme 2.24 $Zn(OTf)_2$-mediated synthesis of pyrazolines from enynes.

Reaction Procedure (Scheme 2.24): A solution of enyne (0.391 mmol), arylhydrazine (1.172 mmol) and $Zn(OTf)_2$ (3 equiv.) in toluene (2 mL) was sealed under nitrogen in a reaction vial and irradiated in a microwave reactor (Biotage, initiator 8, single-mode reactor) at 150 °C for 1 h. On cooling of the reaction mixture to ambient temperature the solvent was removed under reduced pressure and the residue was purified by column chromatography using hexane–EtOAc as eluent to afford pyrazoline.

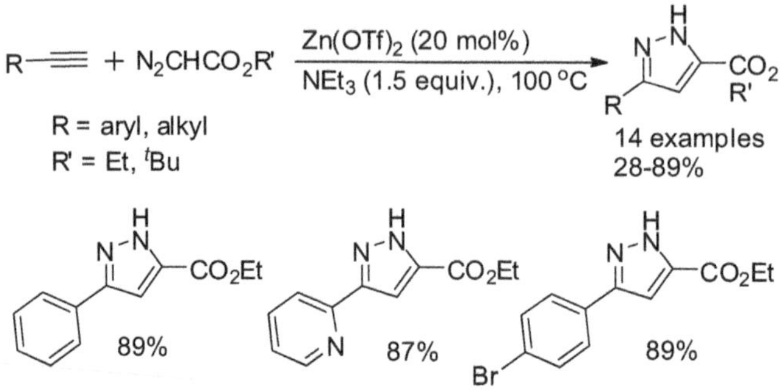

Scheme 2.25   Zinc-catalyzed pyrazole synthesis.

Reaction Procedure (Scheme 2.25): Anhydrous $Zn(OTf)_2$ (73 mg, 20 mol%), phenylacetylene (102 mg, 1 mmol), and ethyl diazoacetate (137 mg, 1.2 mmol) were added to a flame-dried bottle, then Et_3N (165 mg, 1.5 mmol) was added. The mixture was stirred for 8 h at 100 °C. The reaction mixture was quenched with a saturated aqueous solution of ammonium chloride and extracted with CH_2Cl_2. The combined organic extracts were washed with water and saturated brine. The organic layer was dried (Na_2SO_4) and concentrated *in vacuo*. The crude product was purified by chromatography on silica gel (hexane–AcOEt, 8 : 1).

---

in good yields *via* 1,3-dipolar cycloaddition of diazoacetate compounds to terminal alkynes promoted by $Zn(OTf)_2$ under mild conditions (Scheme 2.25). No solvent was needed for this transformation. $Cu(OTf)_2$ and $Cu(PPh_3)Br$ were also tested as catalysts, but resulted in the decomposition of the starting materials. Zn alkynilide was proposed as an intermediate for this transformation.

Dialkylzinc compounds are known to have the ability to initiate a radical process.[54] This type of activity was successfully applied in the cyclization of alkynes to give lactams (Scheme 2.26).[55] $ZnCl_2$ was applied in the cyclization of alkenes to give 2-iodomethyl-2,3-dihydrobenzofuran derivatives in MeOH with I_2.[56]

Oxazolines constitute a class of heterocycles that are naturally occurring and also have broad applications in both organic chemistry and coordination chemistry.[57] The synthesis, properties and valuable applications of chiral oxazolines and their metal complexes in asymmetric synthesis and catalysis have been extensively studied.[58] With this background, a two-step procedure for the synthesis of oxazolines from nitriles and amino alcohols based on Pt(IV) and Zn(II) catalysts was reported.[59] The first step is a Pt(IV)-mediated addition of amino to nitrile and gave amidines as intermediates. Then, in the presence of catalytic amounts of $ZnCl_2$, oxazolines were produced. The cyclization step also works without catalyst, but a much longer reaction time was needed. In 2006, a zinc-catalyzed conversion of

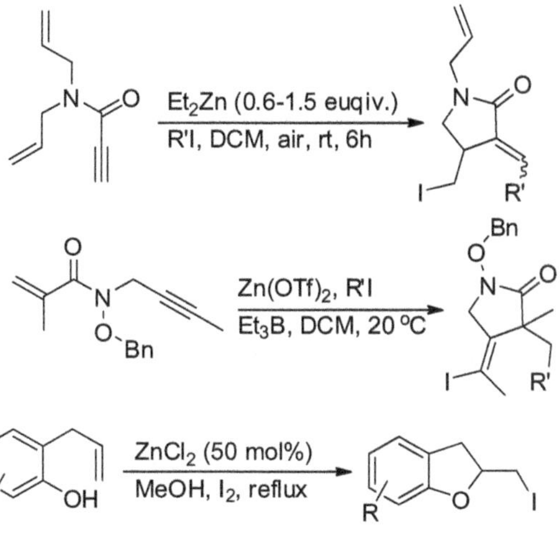

Scheme 2.26 Zinc-initiated radical reactions.

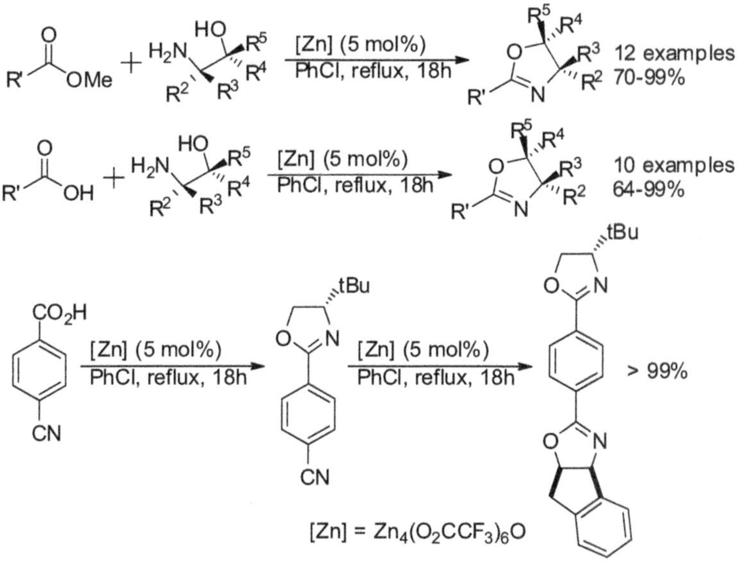

Scheme 2.27 Zinc-catalyzed oxazoline synthesis.

esters, lactones and carboxylic acids to oxazolines was developed.[60] A prepared tetranuclear zinc cluster was essential for this reaction; not only for condensation reactions but also cyclodehydration reactions. In the presence of this zinc cluster, carboxylic acids reacted much faster than nitriles or esters, allowing easy access to heterochiral bisoxazolines that contains different oxazoline moieties (Scheme 2.27).

Liu and co-workers developed a $Zn(OTf)_2$-catalyzed cyclization of propargyl alcohols with anilines, phenols and amides.[61] Various indoles, benzofurans and oxazoles were prepared in good yields in toluene *via* different annulation mechanisms (Scheme 2.28). The α-carbonyl intermediates were isolated as reaction intermediates. The 1,2-nitrogen shift in the formation of indole is catalyzed by Zn(OTf)$_2$ and its mechanism has been elucidated. In the case of

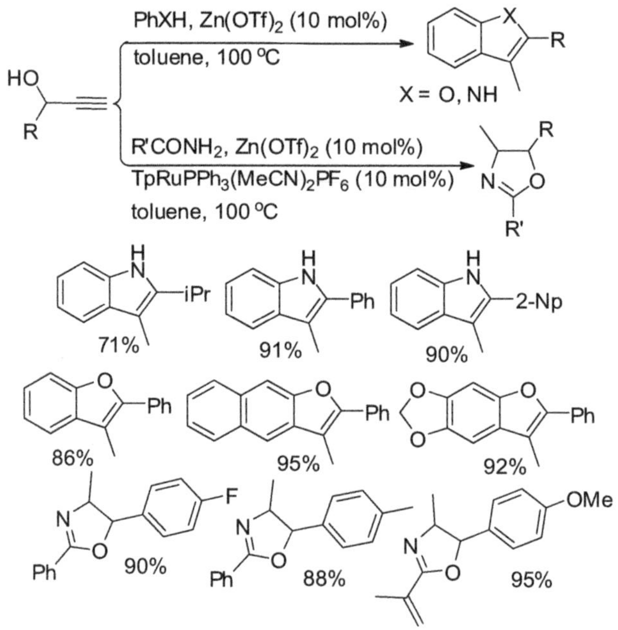

Scheme 2.28   Zn(OTf)$_2$-catalyzed indole, benzofuran and oxazole synthesis.

Reaction Procedure (Scheme 2.28): **Indole:** To a toluene solution (0.80 mL) of aniline (132 mg, 1.4 mmol) were added pent-1-yn-3-ol (100 mg, 1.1 mmol) and zinc triflate (43 mg, 0.01 mmol) and the reaction mixture was heated at 100 °C for 8 h. The solution was filtered over a short silica bed, washed with diethyl ether (5 mL), and concentrated under reduced pressure to afford the pure product.

   Benzofuran: To a toluene solution (0.80 mL) of phenol (87 mg, 0.92 mmol) were added 1-phenylprop-2-yn-1-ol (100 mg, 0.75 mmol) and zinc triflate (27 mg, 0.075 mmol) and the reaction mixture was heated at 100 °C. The reaction was monitored by TLC. After completion, the reaction mixture was filtered over a short silica bed and concentrated under reduced pressure.

   Oxazole: To a toluene solution (0.80 mL) of benzamide (110 mg, 0.90 mmol) were added 1-phenylprop-2-yn-1-ol) (100 mg, 0.75 mmol), zinc triflate (27 mg, 0.075 mmol), and TpRuPPh$_3$(CH$_3$CN)$_2$PF$_6$ (57 mg, 0.075 mmol), and the reaction mixture was heated at 100 °C for 5 h. The solution was filtered over a short silica bed and concentrated under reduced pressure.

the preparation of oxazoles, a ruthenium catalyst was needed as an additive. Oxazoles can be produced in 28% yield with only 10 mol% of $Zn(OTf)_2$ as catalyst; no activity was detected with only [TpRu] in the absence of $Zn(OTf)_2$.

Xia and Ganem reported a metal-promoted variant of the Passerini reaction.[62] They studied the effect of replacing the Brønsted acid in classic Passerini reactions with a mild Lewis acid. Triggered by metal-promoted silylation, condensations of aliphatic or aromatic carbonyl compounds with appropriately substituted isonitriles capable of neighboring group donation afford α-hydroxyamides, substituted oxazoles and other useful heterocycles (Scheme 2.29).

A highly regioselective S_N2-type ring opening of 2-aryl-*N*-tosylaziridines with carbonyl compounds was reported in 2007.[63] In the presence of a Lewis acid, various 1,3-oxazolidines were produced in excellent yields and moderate to high enantioselectivity (Scheme 2.30). The 1,3-oxazolidines can be further converted to 1,2-amino alcohols in the presence of PTSA at room temperature in MeOH. Not only $Zn(OTf)_2$ but also $Cu(OTf)_2$ and $BF_3 \cdot OEt_2$ were used as catalysts for this transformation. The formation of non-racemic products provides convincing evidence for the S_N2-type ring-opening mechanism.

Wang and co-workers developed an interesting Lewis acid-catalyzed regio-controlled synthesis of oxazolines and oxazoles.[64] ZnI_2 and $FeCl_3$ promoted the cyclization of acetylenic amides to achieve selectively oxazolines and oxazoles via C–O bond formation (Scheme 2.31). This Lewis acid-promoted

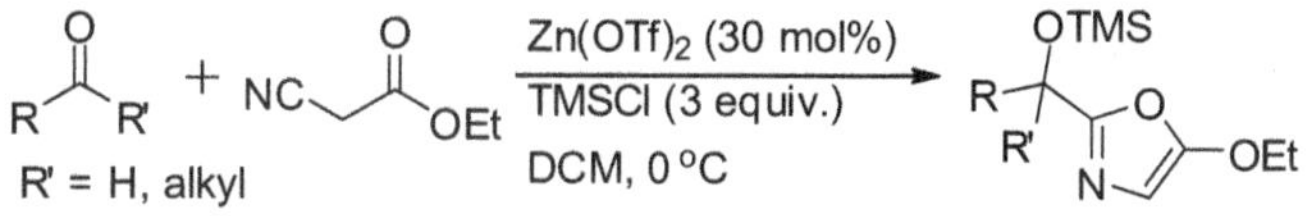

Scheme 2.29 Metal-promoted Passerini reaction.

Reaction Procedure (Scheme 2.29): To $Zn(OTf)_2$ (30 mol%) and benz-aldehyde (1 mmol) in DCM (5 mL) was added TMSCl (3 equiv.) at 0° C under argon. The heterogeneous mixture was stirred for 5 min at 0 °C, then isonitrile (1.05 equiv.) was added. The reaction mixture was warmed to rt and stirred for 24 h. Saturated aqueous $NaHCO_3$ was added and the mixture was stirred for another 20 min. The layers were separated and the aqueous phase was extracted twice with DCM. The combined organic layers were washed with brine, dried over $MgSO_4$ and filtered. Pure product was isolated by silica gel flash column chromatography.

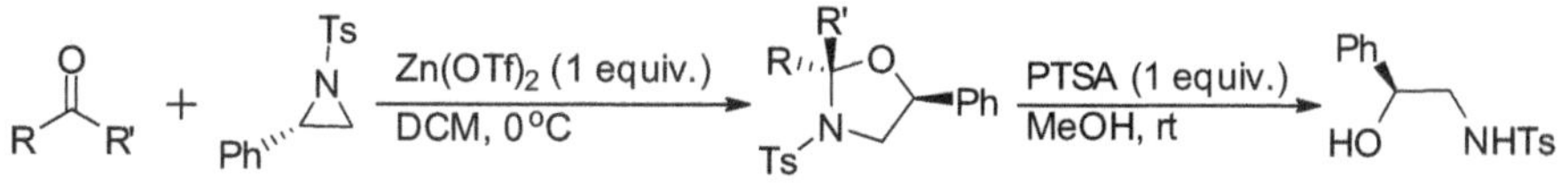

Scheme 2.30 Ring opening of 2-aryl-*N*-tosylaziridines with carbonyl compounds.

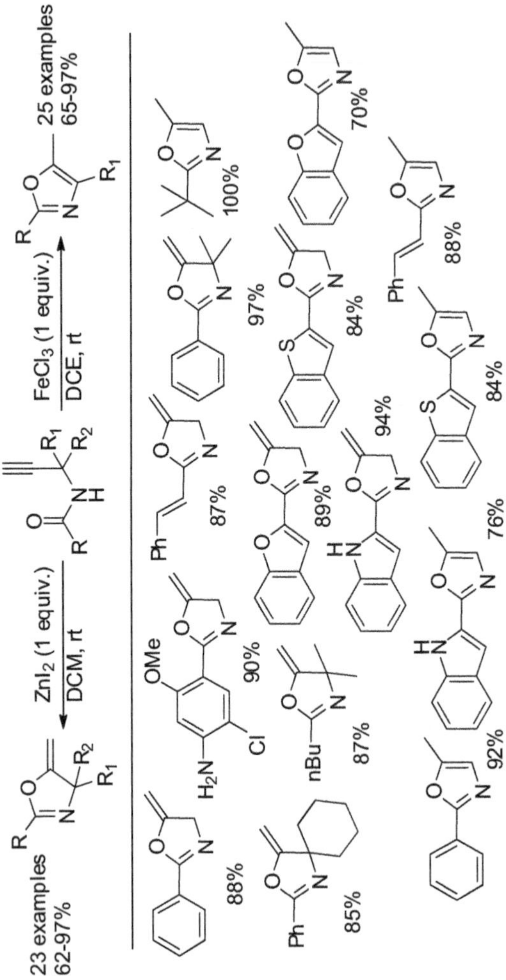

Scheme 2.31 Synthesis of oxazolines and oxazoles.

Reaction Procedure (Scheme 2.31): The substrate was dissolved in CH_2Cl_2 followed by the addition of ZnI_2 (1 equiv.) and stirred at room temperature $3 \sim 16$ h until the completion of reaction by TLC. The reaction mixture was diluted with water. The aqueous layer was extracted with CH_2Cl_2 (3×10 mL), and the combined organic layers were washed with water and brine, dried over Na_2SO_4 and concentrated under reduced pressure. Silica gel chromatography (ethyl acetate–hexane) gave the desired oxazoline derivatives.

cyclization is a practical route to oxaza heterocycles. On the basis of the results obtained, several features should be noted: (1) an inexpensive, regioselective, alternative and efficient approach in a 5-*exo-dig* cyclization mode is used; (2) the feasibility of the reaction was studied with a wide range of functionality with good to excellent yields; (3) the feasibility of one-pot synthesis *via* sequential addition gave the desired oxazolines and oxazoles in moderate yields; (4) no special precautions were required such as N_2 or argon to carry out the reaction, and this approach could find applicability in further cyclizations leading to the formation of new heterocycles; and (5) the Lewis acid used here is abundant, affordable and environmentally benign.

Carreira and co-workers developed a novel cyclization reaction of propargylic *N*-hydroxylamines to give 2,3-dihydroisoxazoles under mild reaction conditions in the presence of catalytic amounts (10 mol%) of ZnI_2 and DMAP (Scheme 2.32).[65] The method provides a new approach to this useful class of substituted heterocycles that complements extant methods. The intriguing reactivity of the substrates in the presence of Zn(II) and DMAP may have additional applications in related alkene or alkyne cyclization reactions.

In 2004, a metal-catalyzed tandem 1,4-addition/cyclization between propargyl alcohol and a Michael acceptor, such as an alkylidene malonate, was developed.[66] In the presence of catalytic amounts of zinc triflate [$Zn(OTf)_2$] and triethylamine (Et_3N), various 2-alkylidene-1,3-dicarbonyl compounds reacted with propargyl alcohol to give 3- or 4-methylenetetrahydrofurans in excellent yields (Scheme 2.33). Concerning the reaction mechanism, first zinc alkoxide is formed by the reaction of propargyl alcohol with Et_3N and $Zn(OTf)_2$, and subsequently adds to the Michael acceptor. The zinc enolate intermediate must be reactive enough to undergo cyclization quickly, since the initial 1,4-adduct was never observed. On the other hand, the reverse 1,4-addition must be even faster that the cyclization and overall the equilibrium of the first stage favors the starting material side (see above). The intramolecular carbozincation of the zinc enolate intermediate followed by protonation of the alkenylzinc intermediate furnishes the tetrahydrofuran product and regenerates the zinc alkoxide.

An improved method for the synthesis of 1,2-disubstituted benzimidazoles was reported in 2009 by Jacob *et al.*[67] $SiO_2/ZnCl_2$ was used as a heterogeneous catalyst in solvent-free conditions, and 1,2-disubstituted benzimidazoles were prepared in good yields from *o*-phenylenediamine and

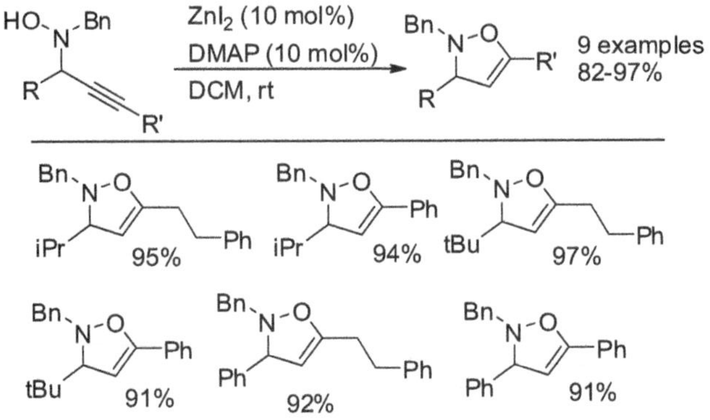

Scheme 2.32   Zinc-catalyzed synthesis of isoxazoles.

Reaction Procedure (Scheme 2.32): A flask was charged with 0.1 equiv. of zinc salt and purged with nitrogen for 15 min. To the flask was added a solution of 1 equiv. of hydroxylamine in 2 mL of CH_2Cl_2. The resulting mixture was stirred at 23 °C for 10 min, then 0.1 equiv. of DMAP was dissolved in 1 mL of CH_2Cl_2 and added to the flask in one portion. Upon completion, the reaction was quenched by the addition of 2 mL of saturated aqueous NH_4Cl solution. The reaction mixture was poured into a separating funnel containing 10 mL of CH_2Cl_2. The layers were separated and the aqueous layer was extracted with CH_2Cl_2 (3×10 mL). The combined organic layers were washed with brine, dried over $MgSO_4$ and concentrated *in vacuo*.

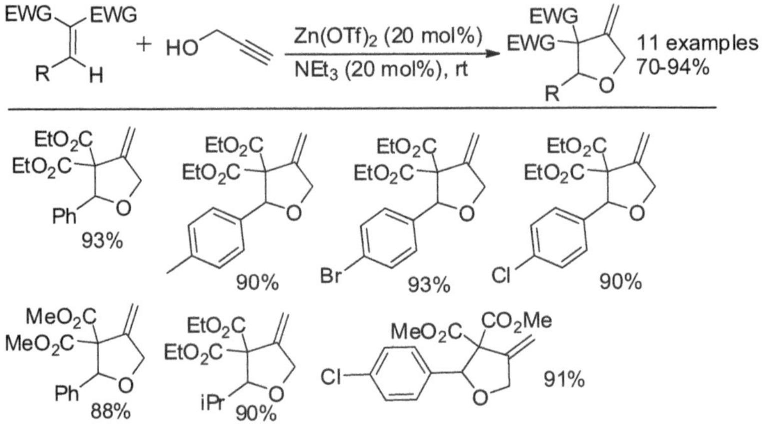

Scheme 2.33   Zinc-catalyzed synthesis of furans.

Reaction Procedure (Scheme 2.33): $Zn(OTf)_2$ (73 mg, 0.2 mmol) was placed in a dry Schlenk tube and dried *in vacuo* (0.05 mmHg) for 1 h with gentle heating (90–100 °C). The reaction vessel was allowed to cool to room temperature and filled with dry nitrogen. Propargyl alcohol (0.30 mL, 5 mmol) and triethylamine (0.2 mmol) were added sequentially. After stirring at 25 °C for 10 min, alkene (1 mmol) was added to the clear, colorless solution. After 12 h the alkene was consumed completely. The reaction mixture was filtered through a plug of silica gel with Et$_2$O as eluent. After removal of the solvent, purification with silica gel chromatography gave pure product.

aldehydes. This general, simple, fast and clean protocol minimizes the organic solvent and energy demands, and also the reaction time could be reduced from hours to a few minutes using MW irradiation. Another heterogeneous zinc catalyst, ZnO, was prepared and applied in the synthesis of indole derivatives.[68] Nanorods ZnO was obtained by employing Zn(OAc)$_2 \cdot$2H$_2$O, NH$_3$ and polyethylene glycol (PEG, M_w = 2000) and characterized in terms of its structural aspects. This catalyst is very active heterogeneous catalyst in the preparation of 3-indolyl-3-hydroxyoxindoles. Various isatins and indoles reacted at 80 °C to yield the corresponding 3-indolyl-3-hydroxyoxindoles in good to high yields. The catalyst can be readily recovered and reused.

Yang, Wang and co-workers developed an efficient zinc-catalyzed tandem synthesis of maleimides and carbazoles from isonitriles and allenic esters.[69] By this process, maleimides and carbazoles were produced in moderate to good yields (Scheme 2.34). The advantages of this method are that the substrates employed in these processes are readily available; the reaction conditions are exceptionally mild; and the compounds generated are amenable to further structural modifications.

2.2 Six-Membered Heterocycles

Zinc-catalyzed intramolecular hydroamination of 6-aminohex-1-yne to 2-methyl-1,2-dehydropiperidine was reported by Müller and co-workers.[70] Zn(OTf)$_2$ was shown to be an effective catalyst for this transformation, and zinc zeolite as a heterogeneous catalyst was found to be even more active. A direct addition of NH across a C–C multiple bond (hydroamination) was developed by the same group in a liquid–liquid two-phase system. The latter comprised a polar catalyst phase of Zn(CF$_3$SO$_3$)$_2$ in the ionic liquid 1-ethyl-3-methylimidazolium trifluoromethanesulfonate and a substrate mixture in heptane.

Roesky, Blechert and co-workers demonstrated that their newly prepared zinc complex is an efficient catalyst for homogeneous intramolecular hydroamination reactions.[71] The advantages, such as particularly high functional group tolerance, good activity in the catalytic conversion of non-activated C–C multiple bonds and a relatively high stability towards moisture and air, made this catalyst more practical. The reaction conditions allowed the manipulation of a multitude of polar functional groups, including

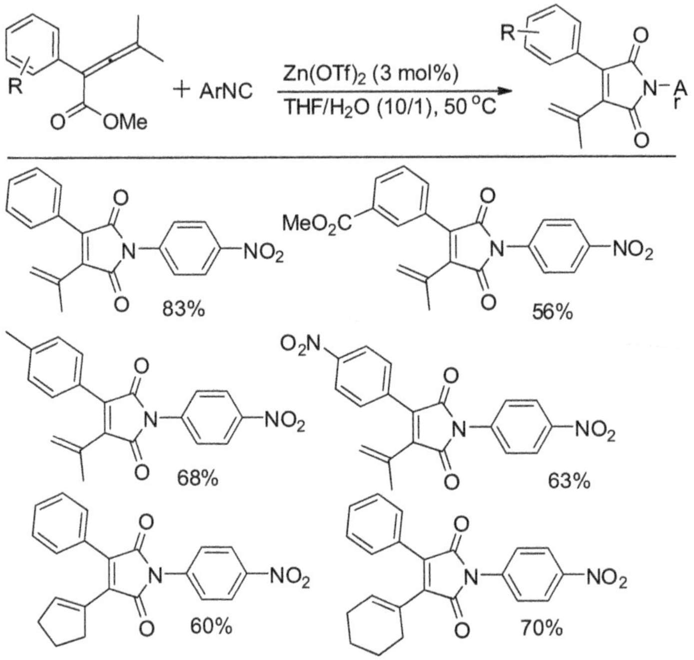

Scheme 2.34   Zinc-catalyzed synthesis of maleimides and carbazoles.

Reaction Procedure (Scheme 2.34): To a solution of an allene (0.6 mmol) in THF (8 mL) and H_2O (0.8 mL) were added isocyanide (0.5 mmol) and $Zn(OTf)_2$ (0.015 mmol). The mixture was stirred at 50 °C for 24 h. After removal of the solvent under vacuum, the residue was extracted with dichloromethane (3×20 mL) and the combined extract was concentrated to give a crude product for purification.

---

ethers, thioethers and amides. All the desired products were produced in good yields (Scheme 2.35).

Ding and co-workers prepared an interesting 3,3'-Br$_2$-BINOL–Zn complex which was applied in hetero-Diels–Alder reactions.[72] The complex can be made *in situ* by mixing Et$_2$Zn and 3,3'-dibromo-1,1'-bi-2-naphthol (3,3'-Br$_2$-BINOL) in the reaction solution. 2-Substituted 2,3-dihydro-4*H*-pyran-4-ones were produced in an enantioselective manner from Danishefsky's diene and aldehydes (Scheme 2.36). By the addition of diimine as additional ligand, dialdehyde can be selectively converted in two distinct reaction pathways. Later, a zinc-catalyzed aza-Diels–Alder reaction of *N*-arylimines with Danishefsky's diene was developed,[73] employing zinc(II)–BINOL as Lewis acid catalyst (10 or 100 mol%) for aza-Diels–Alder reactions, comparing the ester-substituted system with phenyl-, naphthyl-, furyl- and dimethylacetal-substituted imines. The phenyl and naphthyl systems are oddly unreactive

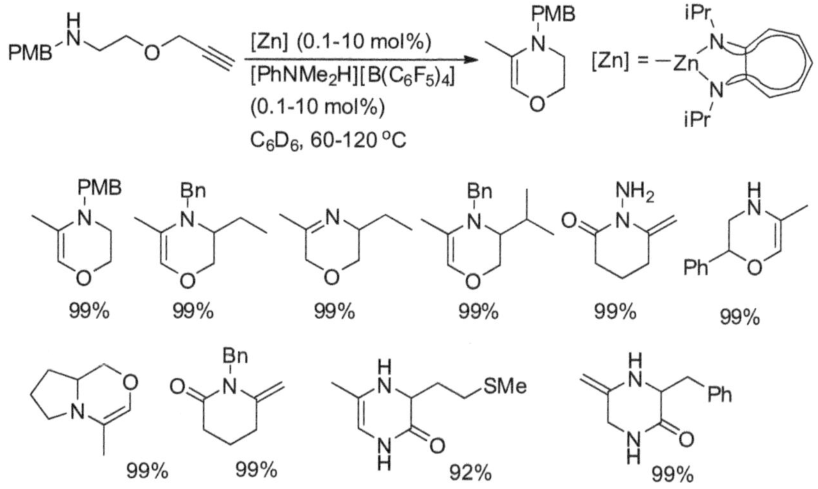

Scheme 2.35 Zinc-catalyzed hydroamination of alkynes.

Reaction Procedure (Scheme 2.35): This was an NMR tube-scale intramolecular hydroamination. All NMR tube-scale reactions were performed in an N_2-filled glove-box. The aminoalkynes (430 mmol) were dissolved in [$^2H_6$]benzene (0.5 mL) and then added to the catalyst (e.g., 4.30 mmol for 1 mol %). The mixture was injected into an NMR tube, which was removed from the glove-box and flame-sealed under vacuum. The reaction mixture was then heated to the appropriate temperature for the stated duration of time.

with zinc(II)–BINOL, the furylimine has low reactivity, which is highly solvent dependent, and the dimethylacetal-substituted system is surprisingly reactive. The asymmetric induction in these types of reactions with Danishefsky's diene varies from poor to good depending on the catalyst loading, solvent, temperature and substrate. However, the major enantiomer obtained was *S* when the (*S*)-BINOL complex was employed in each case. Both efficient aza-Diels–Alder reaction and asymmetric induction are proposed to be dependent upon the formation of bidentate zinc–imine complexes.

A general method for the catalytic formation of substituted pyridines from a variety of unactivated nitriles and α,ω-diynes was reported by Okamoto and co-workers. The reactions were catalyzed by 5 mol % of dppe/CoCl$_2 \cdot$ 6H$_2$O in the presence of Zn powder (10 mol%) at rt to 50 °C. 23 Examples of pyridines were produced in good yields with high functional compatibility and regioselectivity.[74] In 2009, another zinc-catalyzed procedure for pyridine synthesis was developed.[75] The reaction can be promoted using either microwave or thermal heating. In the presence of ZnCl$_2$ in ethanol, 2-amino-3,5-dicarbonitrile-6-thiopyridines were formed in moderate yields (Scheme 2.37).

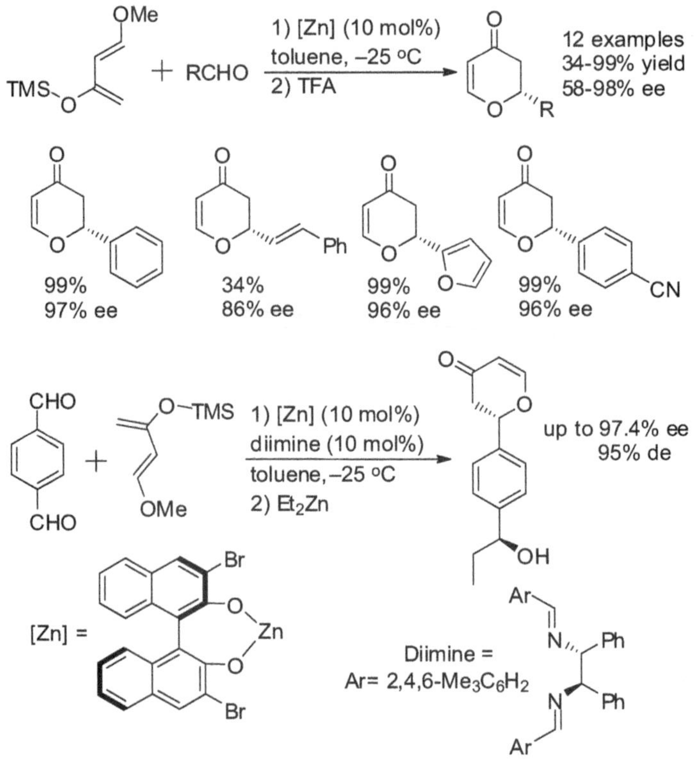

Scheme 2.36 Zinc-catalyzed synthesis of furanpyranones.

Reaction Procedure (Scheme 2.36): To a 1.5 mL polypropylene microtube was added 0.001 mmol of zinc catalyst prepared by mixing ligand and ZnEt$_2$ in a 1:1.2 molar ratio in toluene (40 L, 0.25 M). Freshly distilled benzaldehyde (10.6 mg, 0.10 mmol) was added and Danishefsky's diene (17.2 mg, 0.1 mmol) was charged after the reaction mixture had been kept at 0 °C for 30 min. The reaction was quenched by introducing 5 drops of trifluoroacetic acid after 24 h. Internal standard biphenyl (10 mg) in toluene (0.1 mL) and sodium bicarbonate saturated aqueous solution (0.5 mL) were added to the quenched mixture. The organic layer was separated and subjected to HPLC analysis for the determination of yields and enantiomeric excesses (*ee*).

ZnO nanoparticles as an efficient and recyclable heterogeneous catalyst were applied in the preparation of benzopyran-substituted pyridines in 2013.[76] Starting from β-enaminones, ammonium acetate and different active methylene compounds, the desired products were produced in excellent yields *via* a Michael addition, cyclodehydration and elimination sequence.

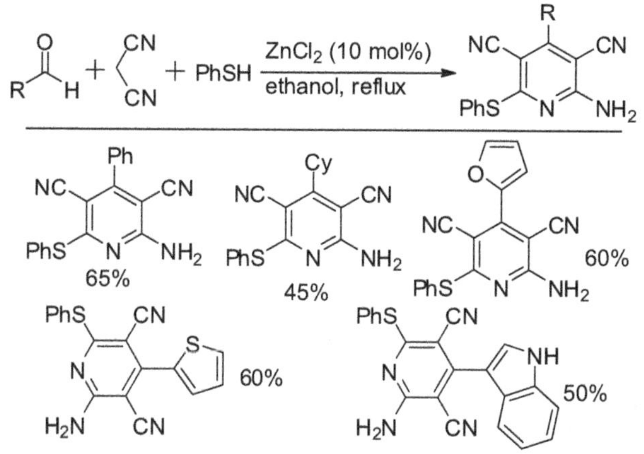

Scheme 2.37 Zinc-catalyzed synthesis of pyridines.

Reaction Procedure (Scheme 2.37): *p*-Fluorobenzaldehyde (0.5 g, 4.0 mmol), malononitrile (0.53 g, 8.0 mmol), thiophenol (0.45 g, 4.0 mmol), zinc chloride (0.1 g, 20 mol%), and ethanol (1 mL) were placed in a 10 mL pressure tube and subjected to microwave heating (CEM Discover, 180 W, 250 psi, 100 °C) for 2 min. The mixture was extracted with ethyl acetate (3 × 10 mL) and the combined extract was washed with water (1 × 5 mL) and brine (1 × 5 mL) and dried over anhydrous $MgSO_4$. The organic layer was concentrated under reduced pressure and the crude product was purified by column chromatography (silica gel, 100–200 mesh; ethyl acetate–hexane, 1:7) to afford the corresponding product.

In 2002, Jiang and Si reported a zinc-mediated alkynylation–cyclization of *o*-trifluoroacetylanilines to give quinolines.[77] In the presence of 1.2 equiv. of $ZnCl_2$ or $Zn(OTf)_2$ and NEt_3 at 25 °C, 10 examples of 4-trifluoromethylated quinolines were produced in moderate to good yields (Scheme 2.38). Notable features of this methodology are the mild conditions, simple manipulation and wide functional group tolerance. However, the loading of zinc salts still needs to be decreased.

Miller's group developed an $SnCl_2$- and $ZnCl_2$-mediated synthesis of quinolines from *o*-nitrobenzaldehyde and ketones.[78] The reactions were carried out in ethanol at 70 °C. As a development of the Friedländer reaction, the preparation and isolation of *o*-aminobenzaldehyde was avoided. Some interesting functional groups are tolerated and the desired products were isolated in good to excellent yields (Scheme 2.39).

Pyrimidine derivatives were found as the core structure in biologically active molecules such as voriconazole and avitriptan. In 2009, a zinc-catalyzed three-component synthesis of pyrimidine was developed.[79] Using 10 mol% of $ZnCl_2$ as catalyst, 4,5-disubstituted pyrimidine derivatives were produced from

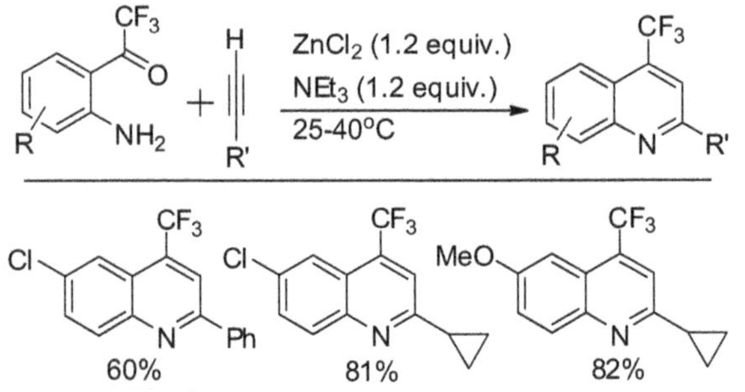

Scheme 2.38   Zinc-catalyzed synthesis of quinolines.

Reaction Procedure (Scheme 2.38): Under anargon atmosphere, a flame-dried flask was charged with phenylacetylene (122 mg, 1.2 mmol), anhydrous $Zn(OTf)_2$ (436 mg, 1.2 mmol), triethylamine (1.2 mmol), and toluene (2.5 mL). The mixture was stirred at 50 °C for 2 h and cooled to 25 °C, then 2-trifluoroacetyl-4-chloroaniline (223 mg, 1 mmol) was added. The reaction mixture was stirred at 25 °C for 4 h. Water (10 mL) was added and the mixture wasextracted with ethyl acetate (3 × 10 mL). The combined organic phase was washed with brine and dried over an-hydrous Na_2SO_4. After removal of the solvent, the residue was purified by flash chromatography on silica gel.

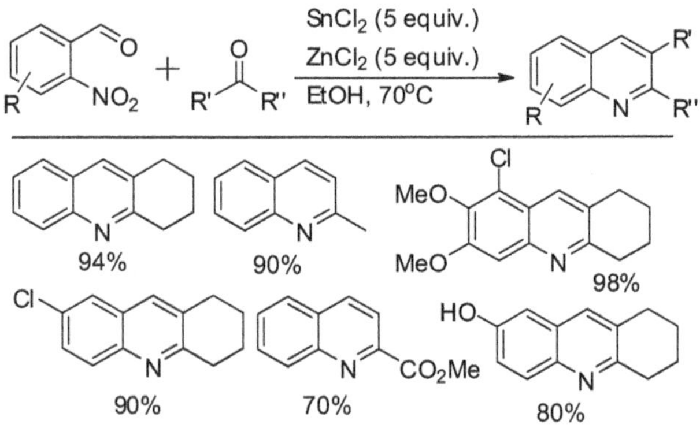

Scheme 2.39   Zinc-mediated Friedländer synthesis of quinolones.

Reaction Procedure (Scheme 2.39): A 150 mL round-bottomed flask equipped with a stir bar and reflux condenser was flame dried under an atmosphere of N_2. *o*-Nitrobenzaldehyde (0.5 g, 3.3 mmol) and cyclo-hexanone (0.325 g, 3.3 mmol) were added, followed by 20 mL of an-hydrous ethanol. $SnCl_2$ (3.138 g, 16.55 mmol, 5 equiv.), $ZnCl_2$ (2.257 g, 16.55 mmol, 5 equiv.), and ∼0.5 g of 4 Å molecular sieves were added to the solution. This mixture was then heated at 70 °C under an atmosphere of nitrogen for 3 h. The reaction was cooled to room temperature and rendered basic (pH 8) with 50 mL of 10% sodium bicarbonate solution. The mixture was transferred to a separating funnel and extracted with 3×20 mL of ethyl acetate. The organic phases were combined and washed thoroughly with saturated NaCl solution, dried over Na_2SO_4 and filtered through Celite. Following removal of the solvent *in vacuo*, the material remaining was subjected to chromatography on silica gel (20% ethyl acetate in hexane).

enamines, triethyl orthoformate and ammonium acetate in a single step (Table 2.1). The procedure was successfully applied to the efficient synthesis of mono- and disubstituted pyrimidine derivatives using methyl ketone derivatives instead of enamines. Interestingly, a 61% yield was achieved without catalyst, and $InCl_3$ and $Cu(OTf)_2$ did not improve the yield. Other zinc salts [$ZnBr_2$, $Zn(OTf)_2$] could also give quantitative yields.

Quinazolinone derivatives have many important applications in the drugs and agrochemicals areas. Many procedures have been developed for their preparation, including the use of zinc catalysts. Owing to the reducing ability of zinc metal,[80] it was applied in the reduction of nitro groups for the synthesis of quinazolinones,[81] but more than the stoichiometric amount of zinc metal was required.

Later, a $Zn(OTf)_2$-promoted synthesis of quinazolinone derivatives was reported by Tseng and Chu.[82] With the assistance of zinc triflate, the direct one-pot double cyclodehydration of linear tripeptides to the total synthesis of pyrazino[2,1-*b*]quinazoline-3,6-diones on a solid support was achieved with good overall yields and a short reaction time (Scheme 2.40). These syntheses of the pyrazino[2,1-*b*]quinazoline-3,6-diones were conveniently achieved in only three steps, starting from the amino acid-bound Wang resin. Remarkably, no desired product was formed when using $Al(OTf)_3$, $Cu(OTf)_2$, $Mg(OTf)_2$ or $La(OTf)_2$ as catalyst. A decreased yield resulted with $Sn(OTf)_2$ as promoter.

Wang *et al.* prepared a new type of combined Lewis acid–surfactant catalyst, zinc(II) perfluorooctanoate [$Zn(PFO)_2$].[83] This material was applied as a favorable surface-active catalyst in the three-component one-pot cyclocondensation reaction of isatoic anhydride with amines and aldehydes to afford the corresponding quinazolinone derivatives in good yields (Scheme 2.41). The reactions occurred in aqueous micellar media with high atom economy. It was found that this catalyst could be easily quantitatively recovered after the reaction was completed and could be reused for at least three cycles without any loss of activity.

A zinc catalyst has also been applied in the preparation of fluorescent sensors.[84] A monoboronic acid fluorescent sensor was conveniently synthesized from 3-nitronaphthalic anhydride and 3-aminophenylboronic acid in pyridine using $Zn(OAc)_2$ (0.61 equiv.) as catalyst. This novel

Reaction Procedure (Table 2.1): To a PhMe (1 mL) solution of ZnCl$_2$ (6.8 mg, 0.050 mmol) and acetal (220 mg, 1.5 mmol) in a screw-capped vial were added enamine (0.50 mmol) (or ketone) and ammonium acetate (77 mg, 1.0 mmol), and the vial was sealed with a cap containing a PTFE septum. The mixture was heated at 100 °C and monitored by TLC or GC analysis until the enamine had been consumed. To quench the reaction, a saturated aqueous solution of NaHCO$_3$ (5 mL) was added to the mixture. The mixture was extracted several times with CHCl$_3$ and the combined organic extracts were dried over Na$_2$SO$_4$, filtered and then concentrated under reduced pressure. The crude product was purified by silica gel chromatography.

Table 2.1   Zinc-catalyzed synthesis of pyrimidines.

Entry	Enamine	Product	Yield (%)
1			99
2			80
3			97
4			80
5			82

Table 2.1 (*Continued*)

Entry	Enamine	Product	Yield (%)
6			91
7			90
8			94
9			77
10			65
11			77

Entry	Ketone	Product	Yield (%)
12			86
13			70

Table 2.1 (*Continued*)

Entry	Ketone	Product	Yield (%)
14	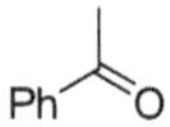	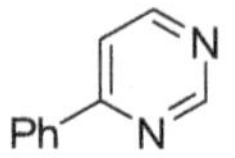	70
15	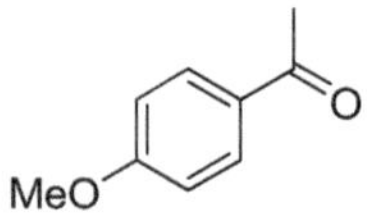	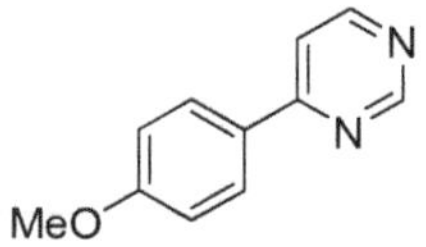	54
16	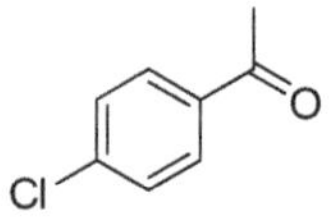	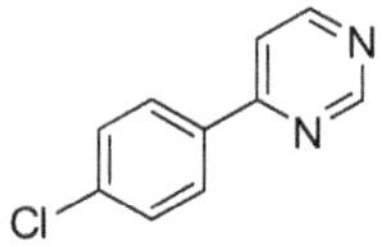	61

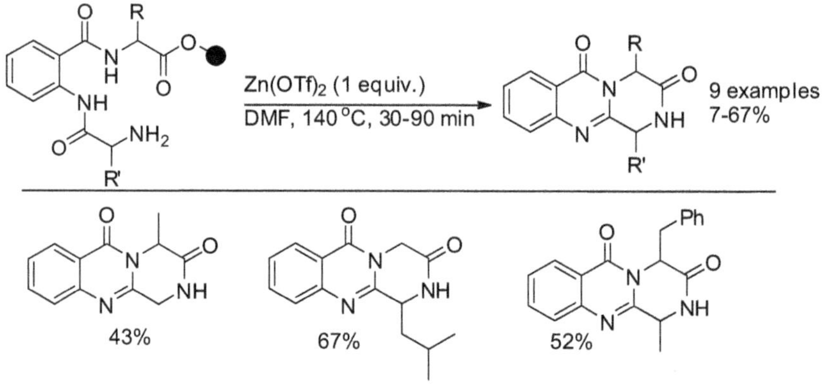

Scheme 2.40 Zn(OTf)$_2$-promoted synthesis of quinazolinone derivatives.

saccharide probe exhibits dual emission suitable for ratiometric sensing and displays a remarkable sensitivity for glucose relative to fructose and galactose

In 2008, Beller and co-workers developed a novel method for the synthesis of aryl-substituted 4,5-dihydro-3(2*H*)-pyridazinones based on domino hydrohydrazination and condensation reactions.[85] Eight substituted aryl-hydrazines reacted with 4-pentynoic acid in the presence of ZnCl$_2$ to give the corresponding pyridazinone derivatives in a one-pot process in moderate to good yields (Scheme 2.42). Notably, this convenient and practical procedure does not require any special handling or unusual reagents and proceeds without the exclusion of air or water.

A zinc-catalyzed three-component domino reaction for the synthesis of pyrano[3,2-*c*]quinolin-5(6*H*)-ones was developed in 2013.[86] Starting from 4-hydroxy-1-methylquinolin-2(1*H*)-one, aromatic aldehydes and

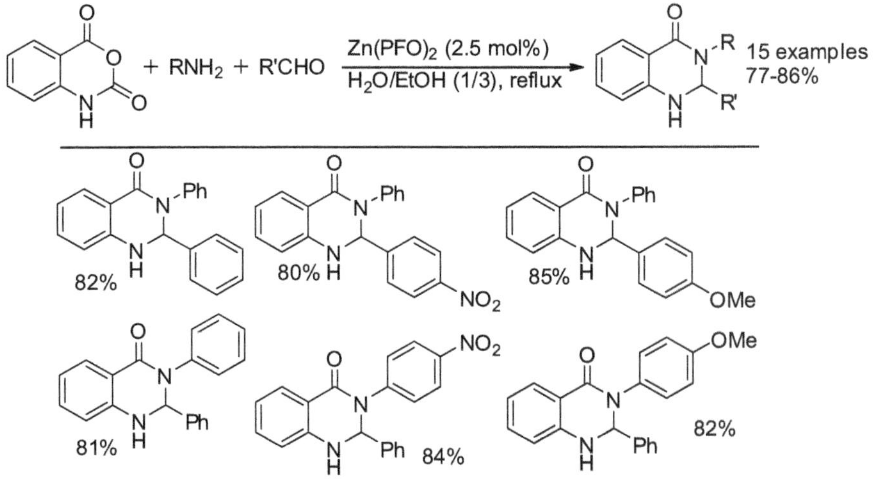

Scheme 2.41 Zn(PFO)$_2$-catalyzed synthesis of quinazolinone derivatives.

Reaction Procedure (Scheme 2.41): A stirred mixture of isatoic anhydride (1 mmol), amine (1.1 mmol), aldehyde (1 mmol) and Zn(PFO)$_2$ (0.027 g, 0.03 mmol) in H$_2$O–EtOH (1:3) (5 mL) was refluxed for 6 h. When the reaction was completed, as indicated by TLC, the reaction mixture was cooled to room temperature. The resulting solid residue was filtered, washed with 3 × 5 mL of water and recrystallized from EtOH to give the pure product.

nitroketene *N,S*-methylacetal, using ZnCl$_2$ as an inexpensive catalyst, 4*H*-pyrano[3,2-*c*]quinolin-5(6*H*)-ones were produced in good yields in refluxing ethanol (Scheme 2.43). Notably, no extraction or chromatography was needed for the purification. Only the reaction of 4-hydroxy-1-methylquinolin-2(1*H*)-one with aldehyde occurred in the absence of catalyst. A significantly decreased yield of quinolin-5(6*H*)-one was obtained when using CuBr$_2$ or CuCl$_2$ as the catalyst under the same conditions.

Nair *et al.* developed a Lewis acid-promoted Diels–Alder reaction of *o*-quinone dibenzenesulfonimides with allylstannane.[87] Potentially biologically active quinoxaline derivatives were easily prepared in good to excellent yields (Scheme 2.44). An ionic mechanism was proposed to illustrate this transformation. First the Lewis acid coordinates with the quinoneimine. The initial attack of allylstannane depends mainly on the basicity of the quinoneimine nitrogen. The sulfonyl substituent on nitrogen is more electrophilic and allylstannane attacks it to form an intermediate tin-coordinated carbocation that is stabilized by hyperconjugative interaction with the tin moiety. The thus formed carbocation is quenched by the N-terminus of the metal-coordinated nitrogen to furnish tetrahydroquinoxaline as the terminal product. When the substituent is benzoyl (imide

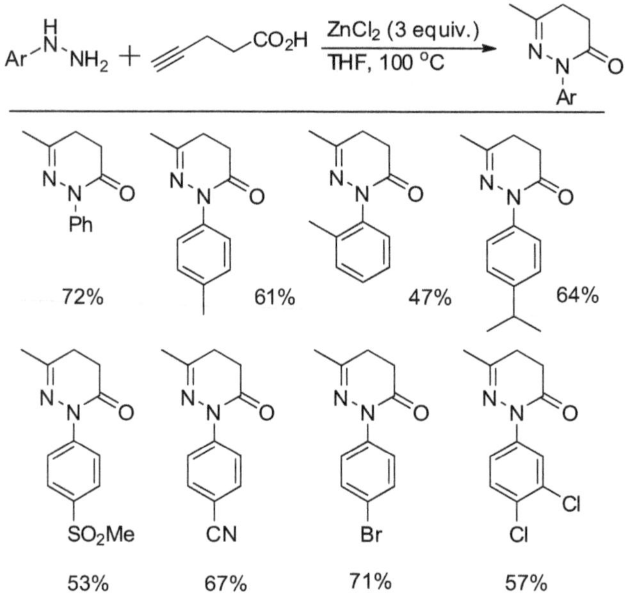

Scheme 2.42   ZnCl$_2$-promoted pyridazinones synthesis.

Reaction Procedure (Scheme 2.42): In an ACE pressure tube under an argon atmosphere, ZnCl$_2$ (613 mg, 4.5 mmol) was dissolved in 4 mL of dry THF. To this solution, 1.95 mmol arylhydrazine and 1.5 mmol alkyne were added. The reaction mixture was heated at 100 °C for 24 h. After cooling to room temperature, the solvent was removed *in vacuo* and the pyridazinone product was purified by column chromatography (eluent: chloroform).

nitrogen is more basic), the initial nucleophilic attack of the allylstannane occurs in a 1,4- or 1,6-manner depending on the substituents on the aromatic ring. The resulting carbocation suffers destannylation by the nucleophile to furnish the by-product.

Yan and co-workers recently presented an anhydrous ZnCl$_2$-catalyzed one-pot three-component reaction of tryptamines, propiolates and α,β-unsaturated aldehydes and also arylideneacetones.[88] Several functionalized 1,2,6,7,12,12b-hexahydroindolo[2,3-a]quinolizines were produced in moderate to high yields and with high diastereoselectivity (Scheme 2.45). The reaction mechanism involved the sequential Michael addition and Pictet–Spengler reactions of β-enamino ester generated *in situ*.

Du's group developed a convenient catalytic asymmetric tandem Friedel–Crafts alkylation/Michael addition reaction of nitroalkene enoates with indoles using a tridentate bis(oxazoline) I-Zn(OTf)$_2$ complex as catalyst.[89] Moderate to high stereoselectivities (up to 95:5 *dr*, up to 99% *ee*) and good to excellent yields of the functionalized chiral chromans were obtained (Table 2.2).

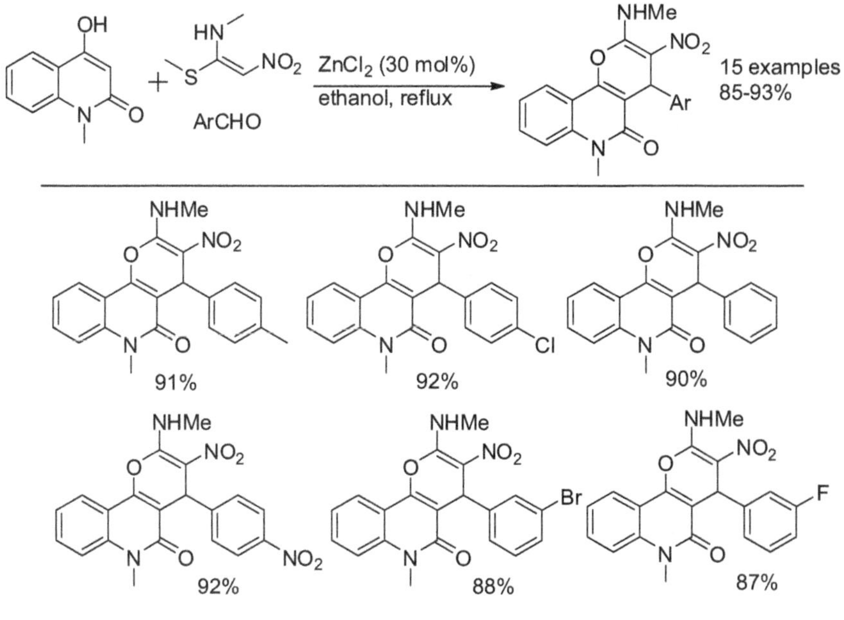

Scheme 2.43 ZnCl$_2$-catalyzed synthesis of quinolin-5(6H)-ones.

Reaction Procedure (Scheme 2.43): A mixture of 4-hydroxy-1-methylquin-olin-2(1H)-one (1 mmol), aromatic aldehydes (1 mmol), and nitroketene N,S-methylacetal (1 mmol) in the presence of ZnCl$_2$ (30 mol %) in ethanol (15 mL) was stirred at reflux for the time required. After completion of the reaction (TLC), the reaction mixture was cooled to room temperature and the resulting solid was filtered off and recrystallized from dichloromethane to obtain pure products.

More recently, a zinc bromide-promoted domino sequence involving 5-*endo-dig* hydroamination and intramolecular cyclization between the in-dole nitrogen with an amide group was reported.[90] The approach features mild conditions, benign functional group tolerance and non-indole starting materials, which is appropriate for the preparation of an array of indolo[1,2-c]quinazoline derivatives (Scheme 2.46). In this methodology, InBr$_3$ can be applied as an alternative catalyst, whereas only trace amounts of product were formed using FeCl$_3$, AlCl$_3$, CuI or PdCl$_2$.

Additionally, zinc salts have even been applied in four-component reactions.[91] 1,4-Dihydropyridines and octahydroquinazolinones were se-lectively synthesized in good yields. In addition to the use of the mentioned homogeneous zinc catalysts in the synthesis of heterocycles, heterogeneous zinc catalysts have also been developed and applied in organic synthesis. For example, the above-mentioned zinc-catalyzed Diels–Alder reaction can also be catalyzed by a solid-supported zinc catalyst (SiO$_2$/ZnCl$_2$) under

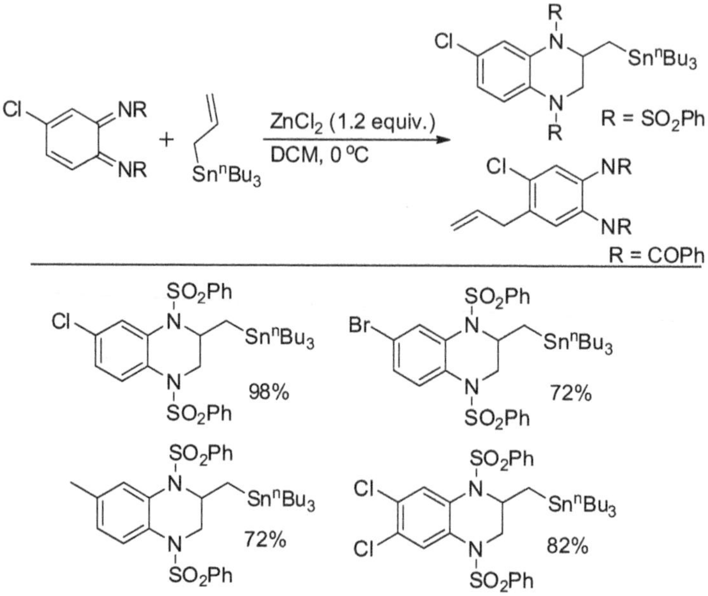

Scheme 2.44   ZnCl$_2$-promoted synthesis of tetrahydroquinoxaline.

Reaction Procedure (Scheme 2.44): A solution of *o*-quinonediimide (0.2 mmol) and allyltributyltin (0.24 mmol) in dry dichloromethane was cooled to 0° C. To this well-cooled solution, zinc chloride (0.24 mmol) was added and the resulting mixture was stirred at the same temperature for 30 min and then quenched with water. The reaction mixture was extracted with dichloromethane (3 × 10 mL), washed with water and saturated brine solution, then concentrated and the residue was subjected to silica gel column chromatography using hexane-ethyl acetate to afford the products, which were recrystallized from hexane–dichloromethane.

MW irradiation without a solvent.[92] Octahydroacridines were produced from (+)-citronellal and *N*-arylamines in good yields (six examples; 75–92%). In 2007, a zinc-catalyzed condensation of *o*-phenylenediamine and 1,2-dicarbonyl compounds was reported. The reaction works at room temperature and excellent yields of the products were isolated. A few years later, a ZnO-beta zeolite was prepared and applied in this reaction. Quinoxalines were produced in quantitative yields and the catalyst could be reused four times without losing any activity.[93] In 2009, a ZnO-catalyzed domino Knoevenagel–intramolecular hetero-Diels–Alder reaction was reported, which provides an efficient route for the formation of polycyclic indole derivatives in a single step (Scheme 2.47).[94] The major advantage of this reaction is the ease of the work-up, during which the products can be isolated without chromatography. This method also offers other advantages, such as clean reactions,

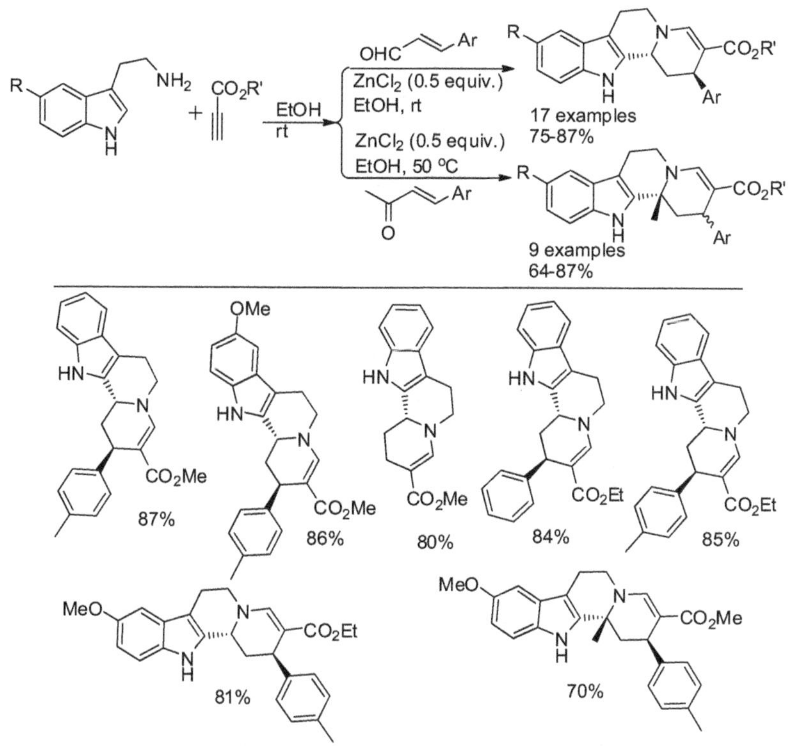

Scheme 2.45 ZnCl$_2$-promoted synthesis of quinolizines.

Reaction Procedure (Scheme 2.45): A mixture of tryptamines (2.0 mmol) and methyl or ethyl propiolate (2.0 mmol) in 10.0 mL of ethanol was stirred at room temperature for 20 min. Then arylideneacetones (2.0 mmol) and anhydrous zinc chloride (1.0 mmol, 0.136 g) were added and the solution was stirred at about 50 °C for an additional 36 h. The solvent was removed by rotary evaporation and the residue was titrated with cold alcohol to give the pure product.

low loading of catalyst, high yields of products, short reaction times and the use of ZnO as a non-toxic, non-corrosive, commercially available and inexpensive heterogeneous catalyst, which make it a useful and attractive strategy for the synthesis of pentacyclic indole derivatives.

In 2013, an ecofriendly, one-pot, three-component ZnO nanoparticle-mediated synthesis of 4*H*-chromene in water under thermal conditions was described.[95] The highly product-selective three-component electrophilic reaction of 2-hydroxybenzaldehyde with an active methylene compound and another carbon-based nucleophile of varied nature was also developed, involving a reversible alkylation procedure using a greener 'NOSE' approach. Greenness of the process was well adapted, as water was used both as the

Table 2.2 Zinc-catalyzed synthesis of chromans.

Entry	R^1	R^2	R^3	R^4	Yield (%)	dr	ee
1	H	H	H	H	88	98:2	98/88
2	H	H	H	Me	88	44:56	62/98
3	H	H	H	OMe	96	28:72	73/99
4	H	H	H	Cl	58	76:24	82/53
5	H	H	Me	H	89	82:18	65/79
6	H	Cl	H	H	86	74:26	95/95
7	H	Br	H	H	100	60:40	91/93
8	H	NO$_2$	H	H	89	49:51	95/95
9	OMe	H	H	H	85	58:42	39/83
10	OEt	H	H	H	94	54:46	30/97
11	Br	Br	H	H	75	38:62	53/90

reaction medium and as a medium for the synthesis of the catalyst. In these reactions, the use of nano-ZnO as a catalyst was found to be crucial for rendering the reactions possible in a water medium, and replacing nano-ZnO with other acids or bases resulted in the generation of too many side products. The catalyst can be efficiently recycled in up to six runs, an essential point in the area of green chemistry. The methodology provides cleaner conversion, shorter reaction times and high selectivity, which make the protocol globally putative. The crystal structure of 4*H*-chromene, easily produced by a chromatography-free, highly product-selective reaction, was explored by means of single-crystal X-ray diffraction analysis and H-bonding arrangements, and results for one typical compound prepared are presented. Under the optimized mild conditions, the isolated yields were 86–93% (Scheme 2.48).

Zinc oxide has also been applied in several other types of organic reactions,[96] such as the application of ZnO in the Biginelli reaction for the synthesis of dihydropyrimidinones, the synthesis of *N*-arylhomophthalimides and benzannelated isoquinolinones from homophthalic acids and amines and the production of tetrahydrobenzo[*b*]pyrans from aromatic aldehydes, activated methylenes and 1,3-diketones.

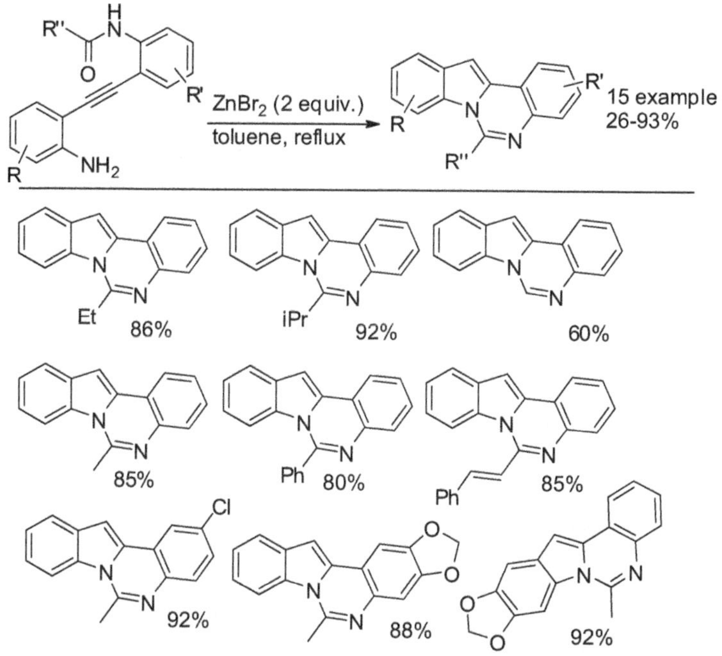

Scheme 2.46 ZnBr$_2$-promoted synthesis of indolo[1,2-*c*]quinazoline derivatives.

Reaction Procedure (Scheme 2.46): Zinc bromide (0.225 g, 1.0 mmol) was added in one portion to a stirred solution containing *N*-{2-[(2-amino-phenyl)ethynyl]phenyl}amide (0.5 mmol) and anhydrous toluene (4.0 mL) at room temperature. The reaction mixture was heated at 110 °C for 16 h. After cooling, the mixture was transferred directly to the top of z chromatographic column and purified by flash chromatography on silica gel with EtOAc–PE to afford the pure product.

2.3 Other Heterocycles

Epoxides are valuable compounds in organic synthesis and polymerizations. Many catalyst systems have been developed for epoxidation reactions. Zinc salt as a class of environmentally benign catalysts have also been applied in epoxide synthesis. In 2004, Minatti and Dötz carried out systematic studies on the asymmetric epoxidation of electron-deficient α,β-enones with a simple zinc–BINOL catalyst in the presence of *tert*-butyl hydroperoxide and cumene hydroperoxide (Scheme 2.49).[97] The epoxidation proceeded in moderate to excellent yields with complete diastereoselectivity and high enantiomeric excesses by using enantiomerically pure BINOL. Mechanistic investigations were also carried out. Electrophilic activation of the substrates

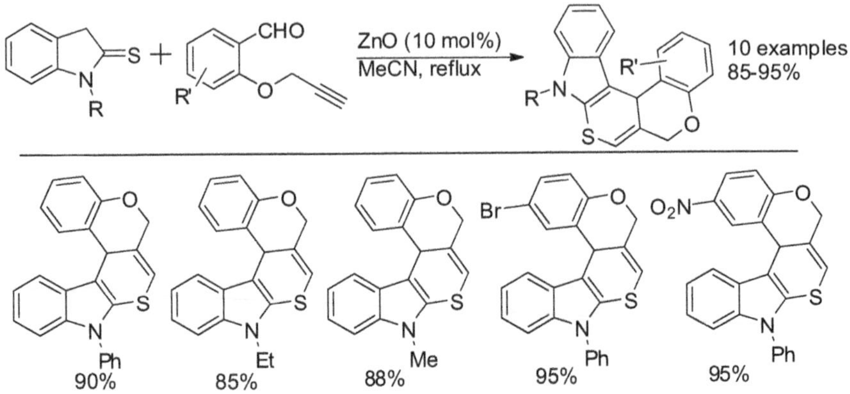

Scheme 2.47 ZnO-catalyzed synthesis of indole derivatives.

Reaction Procedure (Scheme 2.47): To a stirred suspension of ZnO (10 mol%) in CH_3CN (10 mL) were added indolin-2-thione (0.5 mmol) and O-propargylated salicylaldehyde derivative (0.5 mmol). The reaction mixture was stirred at reflux for3h and the progress of the reaction was monitored by TLC. After completion of the reaction, the mixture was poured into ice-cold water and stirred for 10 min, which resulted in precipitation of the product. The solid precipitate was filtered, dried, washed with petroleum ether to remove any residual starting material and, after drying, recrystallized from ethanol.

by the chiral zinc–BINOL catalyst was followed by a subsequent nucleophilic attack of the oxidant. This mechanistic proposal was supported additionally by a non-linear effect, the absolute product configuration and NMR studies. This catalyst system was later modified and self-supported zinc–BINOL catalysts were developed for this transformation.[98] Good yields and enantioselectivity were observed, and the catalyst could be easily recovered and reused several times.

A zinc triflate-catalyzed cyclization of ethenetricarboxylate derivatives with 2-ethynylanilines was developed in 2009.[99] Reaction of 1,1-diethyl 2-*tert*-butyl ethenetricarboxylate with 2-(trimethylsilylethynyl)aniline substrate in the presence of Zn(OTf)$_2$ gave bridged quinoline derivatives in 43–85% yield. The reaction of 1,1-diethyl 2-*tert*-butyl ethenetricarboxylate with 2'-amino-acetophenone also gave the bridged quinoline derivative in 41% yield. Thermal reaction of bridged quinolines (180–190 °C) afforded indole derivatives in moderate to good yields.

Zn(L-proline)$_2$ as a reusable catalyst was applied in a one-pot synthesis of 1,5-benzodiazepine derivatives from 1,2-diamines and ketones under solvent-free conditions. The efficiency of the catalyst was examined by a conventional method and a microwave irradiation technique. 1,5-Benzo-diazepines were obtained in moderate to good yields (up to 93%) in all the

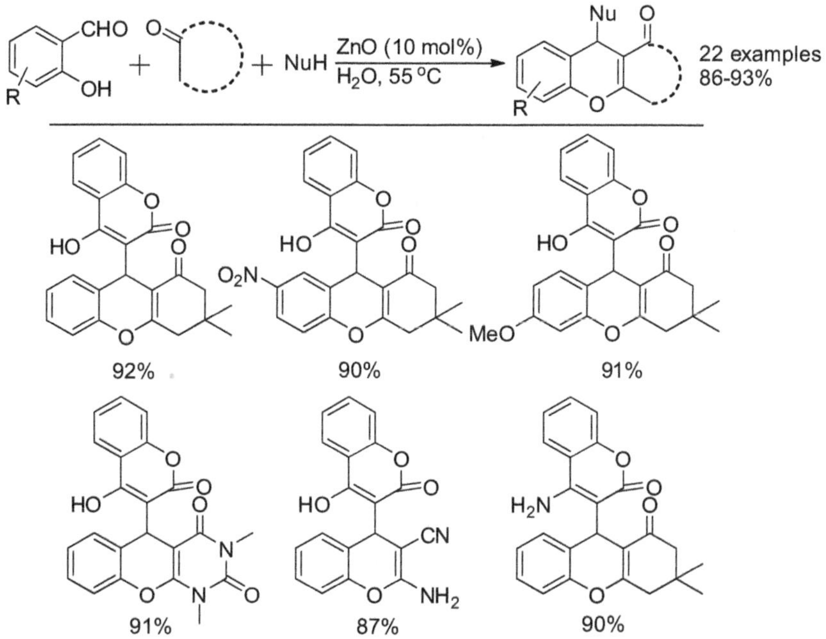

Scheme 2.48 ZnO-catalyzed synthesis of chromene derivatives.

Reaction Procedure (Scheme 2.48): A mixture of 2-hydroxybenzaldehyde (1 mmol), an active methylene compound (1 mmol), carbon-based nucleophile (1 mmol) and 10 mol% ZnO nanoparticles in 5 mL of water was stirred at 55 °C for the stipulated time. After completion of the reaction (monitored by TLC), the free-flowing solid was filtered and washed with water (20 mL) to afford the desired products as pale yellow solids, which were recrystallized from ethanol.

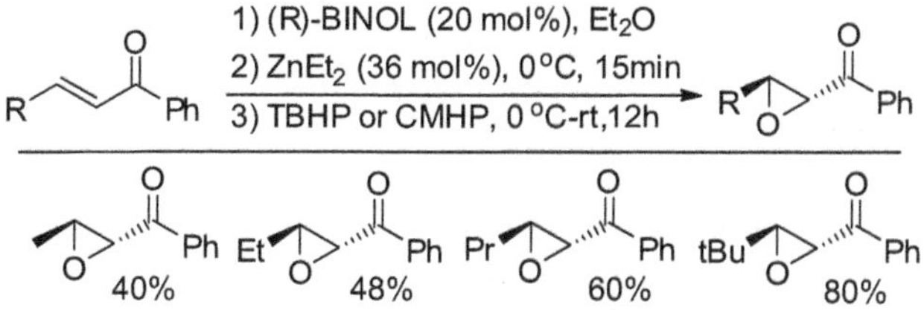

Scheme 2.49 Zinc-BINOL-catalyzed epoxide reaction.

reactions with a shorter reaction time under microwave irradiation. The recycling ability of the catalyst was also evaluated. The catalyst is reusable up to five times without appreciable loss of its catalytic activity. Later, an SiO$_2$/ZnCl$_2$ system was applied in this reaction and also showed high efficiency.[100]

2.4 Summary

The main developments in zinc-catalyzed heterocycle synthesis have been summarized and discussed. The results indicate that the main role of zinc catalysts is still as Lewis acids. The potential of zinc in coupling chemistry is still unexplored. In the future, more efficient catalyst systems are needed. Well-defined complexes as catalysts can help in the elucidation of reaction mechanisms and further assist the design of catalytically active complexes. With the clear advantages of zinc salts, more effort should be devoted to exploring the applications of zinc catalysts in organic chemistry.

References

1. X.-F. Wu and H. Neumann, *Adv. Synth. Catal.*, 2012, **354**, 3141–3160.
2. (a) D. J. Darensbourg and M. W. Holteamp, *Macromolecules*, 1995, **28**, 7577–7579; (b) M. Super, E. Berluche, C. Costello and E. Beckman, *Macromolecules*, 1997, **30**, 368–372; (c) D. J. Darensbourg, S. A. Niezgoda, J. D. Draper and J. H. Reibenspies, *J. Am. Chem. Soc.*, 1998, **120**, 4690–4698; (d) M. Cheng, E. B. Lobkovsky and G. W. Coates, *J. Am. Chem. Soc.*, 1998, **120**, 11018–11019; (f) T. Sârbu and E. J. Beckman, *Macromolecules*, 1999, **32**, 6904–6912; (g) D. J. Darensbourg and M. S. Zimmer, *Macromolecules*, 1999, **32**, 2137–2140.
3. (a) A.-A. G. Shaikh and S. Sivaram, *Chem. Rev.*, 1996, **96**, 951–976; (b) B. Schäffner, F. Schäffner, S. P. Verevkin and A. Börner, *Chem. Rev.*, 2010, **110**, 4554–4581.
4. (a) H. S. Kim, J. J. Kim, S. D. Lee, M. S. Lah, D. Moon and H. G. Jang, *Chem. Eur. J.*, 2003, **9**, 678–686; (b) H. S. Kim, J. J. Kim, S. D. Lee, O. S. Jung, H. G. Jang and S. O. Kang, *Angew. Chem. Int. Ed.*, 2000, **39**, 4096–4098.
5. H. S. Kim, J. J. Kim, H. N. Kwon, M. J. Chung, B. G. Lee and H. G. Jang, *J. Catal.*, 2002, **205**, 226–229.
6. (a) H. S. Lim, J. J. Kim, H. Kim and H. G. Jang, *J. Catal.*, 2003, **220**, 44–46; (b) J. Palgunadi, O.-S. Kwon, H. Lee, J. Y. Bae and B. S. Ahn, *Catal. Today*, 2004, **98**, 511–514; (c) H. S. Kim, P. Jelliarko, J. S. Lee, S. Y. Lee, H. Kim, S. D. Lee and B. S. Ahn, *Appl. Catal. A: Gen.*, 2005, **288**, 48–52; (d) J. K. Lee, Y. J. Kim, Y.-S. Choi, H. Lee, J. S. Lee, J. Hong, E.-K. Jeong, H. S. Kim and M. Cheong, *Appl. Catal. B: Environ.*, 2012, **111–112**, 621–627; (e) B. Lee, N. H. Ko, B. S. Ahn, M. Cheong, H. S. Kim and J. S. Lee, *Bull. Korean Chem. Soc.*, 2007, **28**, 2025–2028.
7. H. S. Kim, J. Y. Bae, J. S. Lee, O-S. Kwon, P. Jelliarko, S. D. Lee and S.-H. Lee, *J. Catal.*, 2005, **232**, 80–84.
8. F. Li, L. Xiao, C. Xia and B. Hu, *Tetrahedron Lett.*, 2004, **45**, 8307–8310.
9. L. Xiao, F. Li and C. Xia, *Appl. Catal. A: Gen.*, 2005, **279**, 125–129.
10. (a) D. J. Darensbourg, S. J. Lewis, J. L. Rodgers and J. C. Yarbrough, *Inorg. Chem.*, 2003, **42**, 581–589; (b) D. J. Darensbourg, J. R. Wideson and J. C. Yabrough, *Inorg. Chem.*, 2002, **41**, 973–980.

11. (a) J. Sun, S.-I. Fujita, F. Zhao and M. Arai, *Appl. Catal. A: Gen.*, 2005, **287**, 221–226; (b) A. Ion, V. Parvulescu, P. Jacbos and D. de Vos, *Appl. Catal. A: Gen.*, 2009, **363**, 40–44; (c) H. Xie, S. Li and S. Zhang, *J. Mol. Catal. A: Chem.*, 2006, **250**, 30–34; (d) Y. Wang, J. Sun, D. Xiang, L. Wang, J. Sun and F.-S. Xiao, *Catal. Lett.*, 2009, **129**, 437–443.
12. (a) S.-S. Wu, X.-W. Zhang, W.-L. Dai, S.-F. Yin, W.-S. Li, Y.-Q. Ren and C.-T. Au, *Appl. Catal. A: Gen.*, 2008, **341**, 106–111; (b) J. Sun, L. Wang, S. Zhang, Z. Li, X. Zhang, W. Dai and R. Mori, *J. Mol. Catal. A: Chem.*, 2006, **256**, 295–300.
13. (a) J. Sun, S.-i. Fujita, F. Zhao and M. Arai, *Green Chem.*, 2004, **6**, 613–616; (b) T. A. Zevaco, A. Janssen and E. Dinjus, *Arkivoc*, 2007, (iii), 151–163; (c) W. Cheng, Z. Fu, J. Wang, J. Sun and S. Zhang, *Synth. Commun.*, 2012, **42**, 2564–2573; (d) S.-i. Fujita, M. Nishiura and M. Arai, *Catal. Lett.*, 2010, **135**, 263–268.
14. (a) M. Ramin, N. van Vegten, J.-D. Grunwaldt and A. Baiker, *J. Mol. Catal. A: Chem.*, 2006, **258**, 165–171; (b) M. Ramin, J.-D. Grunwaldt and A. Baiker, *J. Catal.*, 2005, **234**, 256–267.
15. K. Mori, Y. Mitani, T. Hara, T. Mizugaki, K. Ebitani and K. Kaneda, *Chem. Commun.*, 2005, 3331–3333.
16. M. Taherimehr, A. Decortes, S. M. Al-Amsyar, W. Lueangchaichaweng, C. J. Whiteoak, E. C. Escudero-Adán, A. W. Kleij and P. P. Pescarmona, *Catal. Sci. Technol.*, 2012, **2**, 2231–2237.
17. G.-L. Yu, X.-R. Chen and C.-L. Chen, *React. Kinet. Catal. Lett.*, 2009, **97**, 69–75.
18. Z. P. Demko and K. B. Sharpless, *J. Org. Chem.*, 2001, **66**, 7945–7950.
19. Z. P. Demko and K. B. Sharpless, *Org. Lett.*, 2002, **4**, 2525–2527.
20. S. Rostamizadeh, H. Ghaieni, R. Aryan and A. Amani, *Chin. Chem. Lett.*, 2009, **20**, 1311–1314.
21. F. Himo, Z. P. Demko, L. Noodleman and K. B. Sharpless, *J. Am. Chem. Soc.*, 2003, **125**, 9983–9987.
22. L. E. Kaim, L. Grimaud and P. Pravin, *Eur. J. Org. Chem.*, 2013, 4752–4755.
23. O. G. Mancheño and C. Bolm, *Org. Lett.*, 2007, **9**, 2951–2954.
24. D. Habibi, M. Nasrollahzadeh, A. R. Faraji and Y. Bayat, *Tetrahedron*, 2010, **66**, 3866–3870.
25. (a) M. L. Kantam, K. B. S. Kumar and C. Sridhar, *Adv. Synth. Catal.*, 2005, **347**, 1212–1214; (b) L. V. Myznikov, Y. A. Efimova, T. V. Artamonova and G. I. Koldobskii, *Russ. J. Org. Chem.*, 2011, **47**, 728–730; (c) D. Habibi and M. Nasrollahzadeh, *Synth. Commun.*, 2012, **42**, 2023–2032; (d) A. Sinhamahapatra, A. K. Giri, P. Pal, S. K. Pahari, H. C. Bajaj and A. B. Panda, *J. Mater. Chem.*, 2012, **22**, 17227–17235; (e) S. M. Agawane and J. M. Nagarkar, *Catal. Sci. Technol.*, 2012, **2**, 1324–1327; (f) G. Zhu, Y. Liu, Z. Ji, S. Bai, X. Shen and Z. Xu, *Mater. Chem. Phys.*, 2012, **132**, 1065–1070; (g) A. K. Giri, A. Sinhamahapatra, S. Prakash, J. Chaudhar, V. K. Shahi and A. B. Panda, *J. Mater. Chem. A*, 2013, **1**, 814–822.

26. (a) G. Qi, W. Liu and Z. Bei, *Chin. J. Chem.*, 2011, **29**, 131–134;
    (b) L. Lang, B. Li, W. Liu, L. Jiang, Z. Xu and G. Yin, *Chem. Commun.*,
    2010, **46**, 448–450; (c) L. Lang, H. Zhou, M. Xue, X. Wang and Z. Xu,
    Mater. Lett., 2013, **106**, 443–446.

27. (a) X. Meng, X. Xu, T. Gao and B. Chen, *Eur. J. Org. Chem.*, 2010, 5409–
    5414; (b) G. Aridoss and K. K. Laali, *Eur. J. Org. Chem.*, 2011, 6343–6355;
    (c) D. Habibi and M. Nasrollahzadeh, *Monatsh. Chem.*, 2012, **143**, 925–
    930; (d) M. L. Kantam, V. Balasubrahmanyam and K. B. S. Kumar,
    Synth. Commun., 2006, **36**, 1809–1814.

28. C.-W. Tsai, S.-C. Yang, Y.-M. Liu and M.-J. Wu, *Tetrahedron*, 2009, **65**,
    8367–8372.

29. S. Hajra, D. Sinha and M. Bhowmick, *J. Org. Chem.*, 2007, **72**, 1852–
    1855.

30. B. C. Ranu, S. Ghosh and A. Das, *Mendeleev Commun.*, 2006, **16**, 220–
    221.

31. H. Dong, M. Shen, J. E. Redford, B. J. Stokes, A. L. Pumphrey and
    T. G. Driver, *Org. Lett.*, 2007, **9**, 5191–5194.

32. P. Wyrebek, A. Sniady, N. Bewick, Y. Li, A. Mikus, K. A. Wheeler and
    R. Dembinski, *Tetrahedron*, 2009, **65**, 1268–1275.

33. (a) A. Schmidt, V. Karapetyan, O. A. Attanasi, G. Favi, H. Görls,
    F. Mantellini and P. Langer, *Synlett*, 2007, 2965–2968; (b) V. Karapetyan,
    S. Mkrtchyan, A. Schmidt, O. A. Attanasi, G. Favi, F. Mantellini,
    A. Villinger, C. Fischer and P. Langer, *Adv. Synth. Catal.*, 2008, **350**,
    1331–1336.

34. (a) O. A. Attanasi, S. Berretta, L. De Crescentini, G. Favi, G. Giorgi and
    F. Mantellini, *Adv. Synth. Catal.*, 2009, **351**, 715–719; (b) O. A. Attanasi,
    S. Berretta, L. De Crescentini, G. Favi, G. Giorgi, F. Mantellini and
    S. Nicolini, *Adv. Synth. Catal.*, 2011, **353**, 595–605.

35. A. M. Prior and R. S. Robinson, *Tetrahedron Lett.*, 2008, **49**, 411–414.

36. X.-T. Liu, L. Hao, M. Lin, L. Chen and Z.-P. Zhan, *Org. Biomol. Chem.*,
    2010, **8**, 3064–3072.

37. (a) R. Vicente, J. González, L. Riesgo, J. González and L. A. López, *Angew.
    Chem. Int. Ed.*, 2012, **51**, 8063–8067; (b) J. González, J. González,
    C. Pérez-Calleja, L. A. López and R. Vicente, *Angew. Chem. Int. Ed.*, 2013,
    52, 5853–5857.

38. (a) A. Sniady, A. Durham, M. S. Morreale, K. A. Wheeler and
    R. Dembinski, *Org. Lett.*, 2007, **9**, 1175–1178; (b) A. Sniady, A. Durham,
    M. S. Morreale, A. Marcinek, S. Szafert, T. Lis, K. R. Brzezinska,
    T. Iwasaki, T. Ohshima, K. Mashima and R. Dembinski, *J. Org. Chem.*,
    2008, **73**, 5881–5889.

39. A. Wang, H. Jiang and Q. Xu, *Synlett*, 2009, 929–932.

40. J. Méndez-Andino and L. A. Paquette, *Org. Lett.*, 2000, **2**, 4095–4097.

41. (a) M. Dochnahl, J.-W. Pissarek, S. Blechert, K. Löhnwitz and
    P. W. Roesky, *Chem. Commun.*, 2006, 3405–3407; (b) N. Meyer,
    K. Löhnwitz, A. Zulys, P. W. Roesky, M. Dochnahl and S. Blechert,
    Organometallics, 2006, **25**, 3730–3734; (c) M. Dochnahl, K. Löhnwitz,

J.-W. Pissarek, M. Biyikal, S. R. Schulz, S. Schön, N. Meyer, P. W. Roesky and S. Blechert, *Chem. Eur. J.*, 2007, **13**, 6654–6666; (d) M. Dochnahl, K. Löhnwitz, J.-W. Pissarek, P. W. Roesky and S. Blechert, *Dalton Trans.*, 2008, 2844–2848; (e) K. Löhnwitz, M. J. Molski, A. Lühl, P. W. Roesky, M. Dochnahl and S. Blechert, *Eur. J. Inorg. Chem.*, 2009, 1369–1375; (f) J. Jenter, A. Lühl, P. W. Roesky and S. Blechert, *J. Organomet. Chem.*, 2011, **696**, 406–418; (g) H. P. Nayek, A. Lühl, S. Schulz, R. Köppe and P. W. Roesky, *Chem. Eur. J.*, 2011, **17**, 1773–1777.

42. J.-W. Pissarek, D. Schlesiger, P. W. Roesky and S. Blechert, *Adv. Synth. Catal.*, 2009, **351**, 2081–2085.

43. (a) C. T. Duncan, S. Flitsch and T. Asefa, *ChemCatChem*, 2009, **1**, 365–368; (b) C. T. Duncan, S. Flitsch and T. Asefa, *ACS Catal.*, 2011, **1**, 736–750.

44. (a) P. Horrillo-Martinez and K. C. Hultzsch, *Tetrahedron Lett.*, 2009, **50**, 2054–2056; (b) G.-Q. Liu, W. Li, Y.-M. Wang, Z.-Y. Ding and Y.-M. Li, *Tetrahedron Lett.*, 2012, **53**, 4393–4396; (c) A. Mukherjee, T. K. Sen, P. Kr. Ghorai, P. P. Samuel, C. Schulzke and S. K. Mandal, *Chem. Eur. J.*, 2012, **18**, 10530–10545.

45. (a) Ö. Dogan, H. Koyuncu, P. Garner, A. Bulut, W. J. Youngs and M. Panzner, *Org. Lett.*, 2006, **8**, 4687–4690; (b) Ö. Dogan and H. Koyuncu, *J. Organomet. Chem.*, 2001, **631**, 135–138.

46. (a) S. Cacchi and G. Fabrizi, *Chem. Rev.*, 2011, **111**, PR215–PR283; (b) M. Shiri, *Chem. Rev.*, 2012, **112**, 3508–3549; (c) G. R. Humphrey and J. T. Kuethe, *Chem. Rev.*, 2006, **106**, 2875–2911; (d) A. J. Kochanowska-Karamyan and M. T. Hamann, *Chem. Rev.*, 2010, **110**, 4489–4497.

47. (a) Y. Yin, W. Ma, Z. Chai and G. Zhao, *J. Org. Chem.*, 2007, **72**, 5731–5736; (b) Y. Yin, Z. Chai, W.-Y. Ma and G. Zhao, *Synthesis*, 2008, 4036–4040.

48. K. Okuma, J.-i. Seto, K.-i. Sakaguchi, S. Ozaki, N. Nagahora and K. Shioji, *Tetrahedron Lett.*, 2009, **50**, 2943–2945.

49. (a) K. Alex, A. Tillack, N. Schwarz and M. Beller, *Angew. Chem. Int. Ed.*, 2008, **47**, 2304–2307; (b) A. Pews-Davtyan and M. Beller, *Org. Biomol. Chem.*, 2011, **9**, 6331–6334.

50. K. Alex, A. Tillack, N. Schwarz and M. Beller, *Org. Lett.*, 2008, **10**, 2377–2379.

51. N. T. Patil and V. Singh, *Chem. Commun.*, 2011, **47**, 11116–11118.

52. M. Kidwai, A. Jain and R. Poddar, *J. Organomet. Chem.*, 2011, **696**, 1939–1944.

53. S. He, L. Chen, Y.-N. Niu, L.-Y. Wu and Y.-M. Liang, *Tetrahedron Lett.*, 2009, **50**, 2443–2445.

54. T. Akindele, K.-i. Yamada and K. Tomioka, *Acc. Chem. Res.*, 2009, **42**, 345–355.

55. (a) L. Feray and M. P. Bertrand, *Eur. J. Org. Chem.*, 2008, 3164–3170; (b) H. Miyabe, R. Asada and Y. Takemoto, *Beilstein J. Org. Chem.*, 2013, **9**, 1148–1155.

56. V. A. Mahajan, P. D. Shinde, A. S. Gajare, M. Karthikeyan and R. D. Wakharkar, *Green Chem.*, 2002, **4**, 325–327.

57. For a review on the coordination chemistry of oxazolines, see M. Gómez, G. Muller and M. Rocamora, *Coord. Chem. Rev.*, 1999, **193–195**, 769–835.

58. (a) G. Helmchen and A. Pfaltz, *Acc. Chem. Res.*, 2000, **33**, 336–345; (b) J. S. Johnson and D. A. Evans, *Acc. Chem. Res.*, 2000, **33**, 325–335.

59. A. V. Makarycheva-Mikhailova, V. Y. Kukushkin, A. A. Nazarov, D. A. Garnovskii, A. J. L. Pombeiro, M. Hauka, B. K. Keppler and M. Galanski, *Inorg. Chem.*, 2003, **42**, 2805–2813.

60. T. Ohshima, T. Iwasaki and K. Mashima, *Chem. Commun.*, 2006, 2711–2713.

61. M. P. Kumar and R.-S. Liu, *J. Org. Chem.*, 2006, **71**, 4951–4955.

62. Q. Xia and B. Ganem, *Org. Lett.*, 2002, **4**, 1631–1634.

63. M. K. Ghorai and K. Ghosh, *Tetrahedron Lett.*, 2007, **48**, 3191–3195.

64. G. C. Senadi, W.-P. Hu, J.-S. Hsiao, J. K. Vandavasi, C.-Y. Chen and J.-J. Wang, *Org. Lett.*, 2012, **14**, 4478–4481.

65. P. Aschwanden, D. E. Frantz and E. M. Carreira, *Org. Lett.*, 2000, **2**, 2331–2333.

66. M. Nakamura, C. Liang and E. Nakamura, *Org. Lett.*, 2004, **6**, 2015–2017.

67. R. G. Jacob, L. G. Dutra, C. S. Radatz, S. R. Mendes, G. Perin and E. J. Lenardao, *Tetrahedron Lett.*, 2009, **50**, 1495–1497.

68. M. Hosseini-Sarvari and M. Tavakolian, *Appl. Catal. A: Gen.*, 2012, **441–442**, 65–71.

69. Y. Li, H. Zou, J. Gong, J. Xiang, T. Luo, J. Quan, G. Wang and Z. Yang, *Org. Lett.*, 2007, **9**, 4057–4060.

70. (a) J. Penzien, T. E. Müller and J. A. Lercher, *Micropor. Mesopor. Mater.*, 2001, **48**, 285–291; (b) V. Neff, T. E. Müller and J. A. Lercher, *Chem. Commun.*, 2002, 906–907; (c) T. E. Müller, J. A. Lercher and N. V. Nhu, *AIChE J.*, 2003, **49**, 217–224; (d) J. Penzien, C. Haessner, A. Jentys, K. Köhler, T. E. Müller and J. A. Lercher, *J. Catal.*, 2004, **221**, 302–312.

71. (a) A. Zulys, M. Dochnhal, D. Hollmann, K. Löhnwitz, J.-S. Herrmann, P. W. Roesky and S. Blechert, *Angew. Chem. Int. Ed.*, 2005, **44**, 7794–7798; (b) J. Jenter, A. Lühl, P. W. Roesky and S. Blechert, *J. Organomet. Chem.*, 2011, **696**, 406–418.

72. (a) H. Du and K. Ding, *Org. Lett.*, 2003, **5**, 1091–1093; (b) H. Du, J. Long, J. Hu, X. Li and K. Ding, *Org. Lett.*, 2002, **4**, 4349–4352; (c) H. Du, X. Zhang, Z. Wang and K. Ding, *Tetrahedron*, 2005, **61**, 9465–9477.

73. (a) L. D. Bari, S. Guillatme, J. Hanan, A. P. Henderson, J. A. K. Howard, G. Pescitelli, M. R. Probert, P. Salvadori and A. Whiting, *Eur. J. Org. Chem.*, 2007, 5771–5779; (b) P. R. Girling, T. Kiyoi and A. Whiting, *Org. Biomol. Chem.*, 2011, **9**, 3105–3121.

74. K. Kase, A. Goswami, K. Ohtaki, E. Tanabe, N. Saino and S. Okamoto, *Org. Lett.*, 2007, **9**, 931–934.

75. M. Sridhar, B. C. Ramanaiah, C. Narsaiah, B. Mahesh, M. Kumaraswamy, K. K. R. Mallu, V. M. Ankathi and P. S. Rao, *Tetrahedron Lett.*, 2009, **50**, 3897–3900.

76. Z. N. Siddiqui, N. Ahmed, F. Farooq and K. Khan, *Tetrahedron Lett.*, 2013, **54**, 3599–3604.
77. B. Jiang and Y.-G. Si, *J. Org. Chem.*, 2002, **67**, 9449–9451.
78. B. R. McNaughton and B. L. Miller, *Org. Lett.*, 2003, **5**, 4257–4259.
79. T. Sasada, F. Kobayashi, N. Sakai and T. Konakahara, *Org. Lett.*, 2009, **11**, 2161–2164.
80. X.-F. Wu, *Chem. Asian J.*, 2012, **7**, 2502–2509.
81. P. Zhichkin, E. Kesicki, J. Treiberg, L. Bourdon, M. Ronsheim, H. C. Ooi, S. White, A. Judkins and D. Fairfax, *Org. Lett.*, 2007, **9**, 1415–1418.
82. M.-C. Tseng and Y.-H. Chu, *Tetrahedron*, 2008, **64**, 9515–9520.
83. L.-M. Wang, L. Hu, J.-H. Shao, J. Yu and L. Zhang, *J. Fluorine Chem.*, 2008, **129**, 1139–1145.
84. H. Cao, D. I. Diaz, N. DiCesare, J. R. Lakowicz and M. D. Heagy, *Org. Lett.*, 2002, **4**, 1503–1505.
85. K. Alex, A. Tillack, N. Schwarz and M. Beller, *Tetrahedron Lett.*, 2008, **49**, 4607–4609.
86. P. Gunasekaran, P. Prasanna, S. Perumal and A. I. Almansour, *Tetrahedron Lett.*, 2013, **54**, 3248–3252.
87. V. Nair, D. C. Rajesh, M. M. Bhadbhade and K. Manoj, *Org. Lett.*, 2004, **6**, 4743–4745.
88. J. Sun, L. Zhang and C. Yan, *Tetrahedron*, 2013, **69**, 5451–5459.
89. J. Peng and D.-M. Du, *Beilstein J. Org. Chem.*, 2013, **9**, 1210–1216.
90. M. Xu, K. Xu, S. Wang and Z.-J. Yao, *Tetrahedron Lett.*, 2013, **54**, 4675–4678.
91. (a) V. Sivamurugan, R. S. Kumar, M. Palanichamy and V. Murugesan, *J. Heterocycl. Chem.*, 2005, **42**, 969–974; (b) P. M. Shah and M. P. Patel, *Med. Chem. Res.*, 2012, **21**, 1188–1198.
92. R. G. Jacob, G. Perin, G. V. Botteselle and E. J. Lenardão, *Tetrahedron Lett.*, 2003, **44**, 6809–6812.
93. (a) M. M. Heravi, M. H. Tehrani, K. Bakhtiari and H. A. Oskooie, *Catal. Commun.*, 2007, **8**, 1341–1344; (b) S. S. Katar, P. H. Mohite, L. S. Gadekar, B. R. Arbad and M. K. Lande, *Cent. Eur. J. Chem.*, 2010, **8**, 320–325.
94. M. Kiamehr and F. M. Moghaddam, *Tetrahedron Lett.*, 2009, **50**, 6723–6727.
95. P. P. Ghosh and A. R. Das, *J. Org. Chem.*, 2013, **78**, 6170–6181.
96. (a) M. Hosseini-Sarvari and S. Shafiee-Haghighi, *Chem. Heterocycl. Compd.*, 2012, **48**, 1307–1313; (b) K. Bahrami, M. M. Khodaei and A. Farrokhi, *Synth. Commun.*, 2009, **39**, 1801–1808; (c) V. Krishnakumar, K. M. Kumar, B. K. Mandal and F. N. Khan, *Res. Chem. Intermed.*, 2012, **38**, 1881–1892.
97. (a) A. Minatti and K. H. Dötz, *Synlett*, 2004, 1634–1636; (b) A. Minatti and K. H. Dötz, *Eur. J. Org. Chem.*, 2006, 268–276.
98. H. Wang, Z. Wang and K. Ding, *Tetrahedron Lett.*, 2009, **50**, 2200–2203.

99. S. Yamazaki, S. Morikawa, K. Miyazaki, M. Takebayashi, Y. Yamamoto, T. Morimoto, K. Kakiuchi and Y. Mikata, *Org. Lett.*, 2009, **11**, 2796–2799.

100. (a) V. Sivamurugan, K. Deepa, M. Palanichamy and V. Murugesan, *Synth. Commun.*, 2004, **34**, 3833–3846; (b) R. G. Jacob, C. S. Radatz, M. B. Rodrigues, D. Alves, G. Perin, E. J. Lenardão and L. Savegnago, *Heterocycl. Chem.*, 2011, **22**, 180–185.

CHAPTER 3

Iron-Catalyzed Heterocycle Synthesis

In Chapter 2, a detailed discussion of the zinc-catalyzed synthesis of heterocycles was presented. The main or even the only role of a zinc catalyst is as a Lewis acid. This chapter discusses the iron-catalyzed synthesis of heterocyclic compounds in detail.

3.1 Five-Membered Heterocycles

3.1.1 Iron-Catalyzed Synthesis of Pyrrolidine Derivatives

In 2012, an efficient and convenient method for the synthesis of (Z)-4-(arylchloromethylene)-substituted spiropyrrolidines from cyclic 8-aryl-5-tosyl-5-aza-2-en-7-yn-1-ols using $FeCl_3$ as a non-toxic and inexpensive promoter was developed.[1] The reaction has many advantages: it is virtually instantaneous, even in the presence of air, the required $FeCl_3$ loading is 1.2 molar equivalent, no extra ligand is necessary and the yields are good to excellent (Scheme 3.1a). Under the same reaction conditions, $ZnCl_2$ and $Fe(NO_3)_3 \cdot 9H_2O$ did not give any product. This $FeCl_3$-promoted cyclization/chlorination can be applied to the formation of carbospirocycles. Under similar conditions, cyclopropanation of a cyclic enynol was involved and the ring was easily expanded. Interestingly, iodo-substituted products can be formed when the cyclopropanation and cyclization reaction was carried out in the same reaction pot. The desired products were observed in moderate to good yields (Scheme 3.1b).[2]

A simple iron salt-catalyzed intramolecular hydroamination of unactivated alkenes was demonstrated in 2006.[3] The reaction proceeds under mild conditions and it is not necessary to exclude air and moisture. Of special importance is the tolerance of aminoalkenes containing halide moieties,

RSC Catalysis Series No. 16
Economic Synthesis of Heterocycles: Zinc, Iron, Copper, Cobalt, Manganese and Nickel Catalysts
By Xiao-Feng Wu and Matthias Beller
© Wu and Beller, 2014
Published by the Royal Society of Chemistry, www.rsc.org

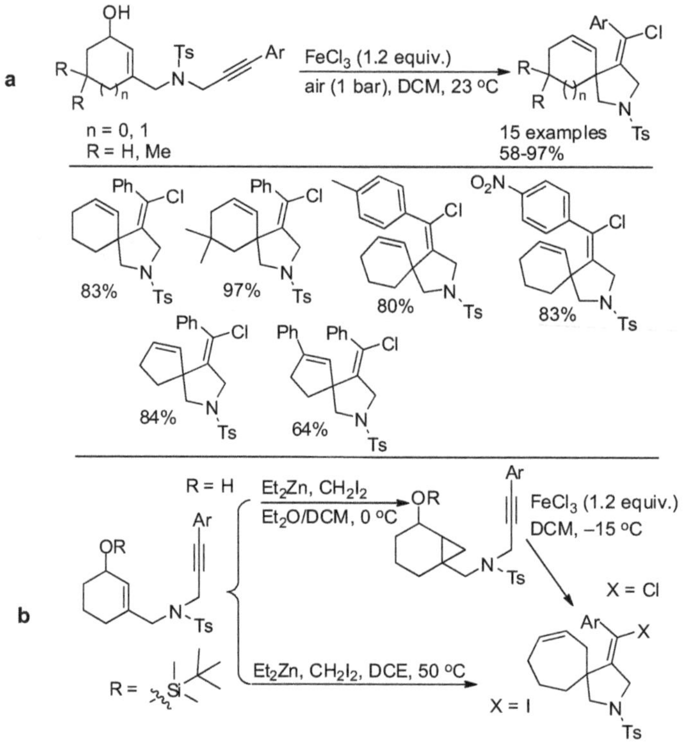

Scheme 3.1   FeCl₃-mediated synthesis of pyrrolidines.

Reaction Procedure (Scheme 3.1): (a) To a 10 mL round-bottomed flask equipped with a stirrer bar was added substrate (0.5 mmol), DCM (5.0 mL) and FeCl₃ (97 mg, 0.6 mmol) under air. The reaction mixture was stirred at room temperature for 1 min. The mixture was added to 10 mL of water, the resulting solution was extracted with diethyl ether (10 × 3 mL) and the combined extracts were washed with brine and dried (MgSO₄). The filtrate was concentrated *in vacuo* and purified by flash column chromatography on silica gel (10:1 hexanes–ethyl acetate) to give the pure product.

   Reaction Procedure: (b) To a dried DCE solution (5.0 mL) of substrate (0.50 mmol) were added Et₂Zn (0.66 mL, 1.0 mmol) and CH₂I₂ (0.54 g, 2.0 mmol) at 0 °C under 1 atm of nitrogen for 1 h. The resulting mixture was then stirred at 50 °C until no starting substrate was detected on TLC (∼9–12 h). The reaction mixture was added to 10 mL of water and extracted with diethyl ether (10 × 3 mL). The combined organic layer was washed with brine, dried over MgSO₄, filtered through a bed of Celite, and concentrated to give a crude mixture, which was purified by flash column chromatography to give pure product.

which is rarely observed with group 9 and 10 metals. Moreover, 1,2-disubstituted aminoalkenes readily react in this system. With regard to this FeCl₃-catalyzed reaction, a simple acid catalyzed pathway can be excluded. It was also demonstrated that the activity of the catalyst could be much improved with a silver salt additive, which allows the reaction of substrate to be completed within 1 h. Notably, only little or even no product was observed when the reaction was carried out in coordinative solvents such as benzene, 1,4-dioxane, THF, 2-propanol, DMSO and DMF. All the products were isolated in excellent yields (Scheme 3.2).

In 2010, Ishibashi's group developed an interesting procedure for the redox radical cyclization of 1,6-dienes and enynes.[4] By using iron phthalocyanine [Fe(Pc)] or FeCl₃ as catalyst, five-membered carbo- or heterocyclic compounds were produced in moderate yields. With different in radical traps, hydroxyl groups, halogen atoms and nitro groups can be introduced (Scheme 3.3). In the case when NaBH₄ is needed, LiBH₄ can be applied as an alternative reagent, but Et₃SiH does not give any product under the same conditions. Fe(NO₃)₃·9H₂O is known to be able under thermal heating to generate nitrogen dioxide (NO₂), which is a radical species. On attempting to use LiNO₃ as an NO₂ source together with LiCl, no reaction was observed.

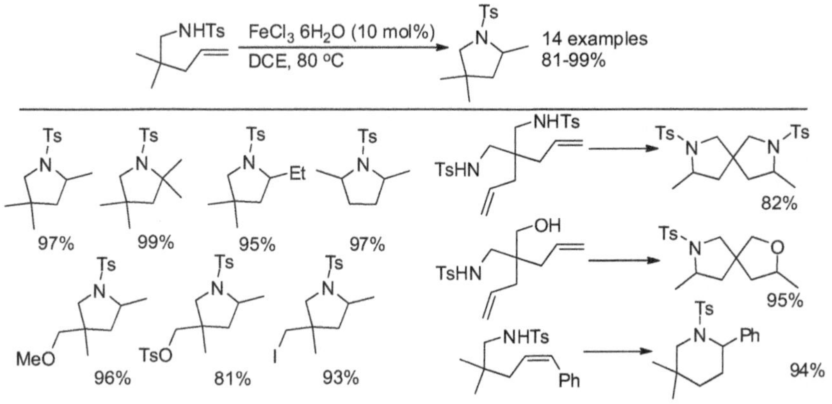

Scheme 3.2   FeCl₃-catalyzed hydroamination.

Reaction Procedure (Scheme 3.2): A mixture of substrate (1 mmol) and FeCl₃·6H₂O (27 mg, 0.1 mmol) in DCE (10 mL) was heated at 80 °C for 2 h while being monitored by silica gel TLC. The reaction mixture was allowed to cool and quenched with water (10 mL). The aqueous phase was extracted with diethyl ether (20 mL). The combined organic layer was washed with brine (10 mL), dried over MgSO₄, filtered and evaporated. The crude product was purified by column chromatography on silica gel (60–230 mesh) with hexane–ethyl acetate (3:1) as eluent to provide the pure product.

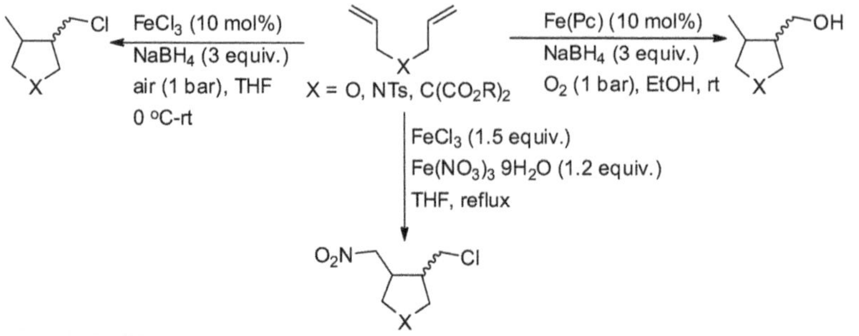

Scheme 3.3 Iron-catalyzed cyclization of 1,6-dienes.

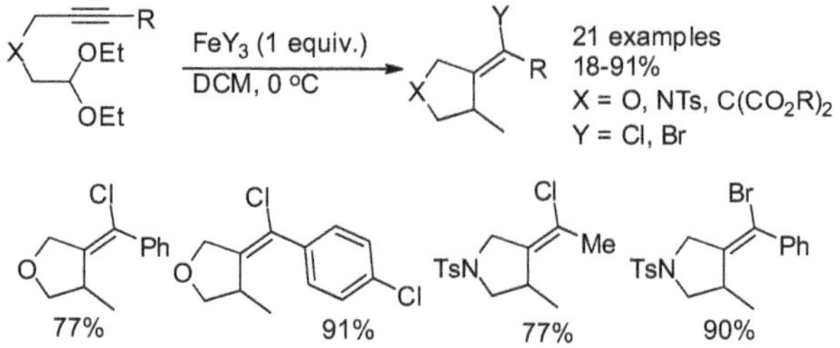

Scheme 3.4 Iron-promoted cyclization of alkynyl diethyl acetals.

> **Reaction Procedure** (Scheme 3.4): Under a nitrogen atmosphere, a solution of substrate (0.5 mmol) in CH_2Cl_2 (5 mL) was cooled to 0 °C and then transferred to a flask containing $FeCl_3$ (81 mg, 0.5 mmol) by a cannula. The mixture was stirred at 0 °C for 30 min, then quenched with water (10 mL). The organic layer was separated and the aqueous phase was extracted with CH_2Cl_2 (3×10 mL). The combined organic phase was dried over anhydrous Na_2SO_4, filtered and concentrated under reduced pressure. The resultant residue was purified by flash silica gel column chromatography [eluent: petroleum ether (b.p. 60–90 °C)-diethyl ether (20 : 1)] to afford the product.

Yu and co-workers described an iron-promoted cyclization/halogenation of alkynyl diethyl acetals (Scheme 3.4).[5] In the presence of an equal amount of $FeCl_3$ or $FeBr_3$, alkynyl diethyl acetals were converted into the corresponding (E)-2-(1-halobenzylidene or alkylidene)-substituted five-membered carbo- and heterocycles, which were then efficiently transformed to vinylarenes by palladium-catalyzed Suzuki coupling. $CuCl_2 \cdot 2H_2O$ and $FeCl_2$ were tested instead of $FeCl_3$, but neither gave any trace of the desired product.

Hennessy and Betley developed an dipyrrinatoiron catalyst that increases the reactivity of iron-bearing metal–ligand multiple bonds to promote the direct amination of aliphatic C–H bonds (Scheme 3.5).[6] Exposure of organic azides to the dipyrrinatoiron catalyst furnishes saturated cyclic amine

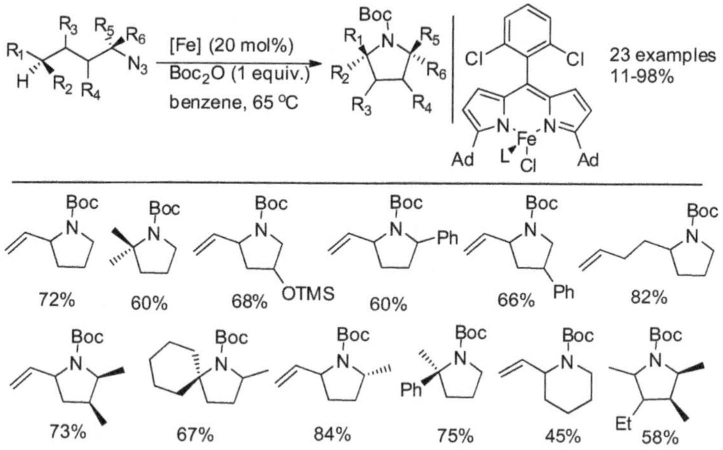

Scheme 3.5 Iron-catalyzed C–H activation.

Reaction Procedure (Scheme 3.5): Under an inert N_2 atmosphere, $(^{Ad}L_{Cl2})FeCl(OEt_2)$ (40.0 mg, 0.0556 mmol) and di-*tert*-butyl dicarbonate (1–10 equiv., depending on substrate) were added to a stirred solution of the desired azide (1–10 equiv, depending on substrate) in 3 mL of benzene in a 5 mL scintillation vial. The resultant inky, dark-red solution was then transferred to a pressure vessel and heated at 65 °C for 12 h. The reaction mixture was then flash chromatographed through a short pipette of deactivated silica gel using a 20:1 mixture of CH_2Cl_2 and MeOH as eluent to give a clear, bright-orange solution in order to remove paramagnetic materials. The solution was concentrated *via* rotary evaporation. The internal standard trimethoxybenzene or ferrocene (0.0556 mmol, 1 equiv.) was added to the reaction mixture, followed by 1 mL of $CDCl_3$ in order to dissolve the contents of the entire flask into a homogeneous solution. The crude products were purified *via* silica gel chromatography using hexanes–ethyl acetate as eluent.

products (*N*-heterocycles) bearing complex core-substitution patterns. This study highlighted the development of C–H bond functionalization chemistry for the formation of saturated cyclic amine products and should find broad application in the context of the synthesis of both pharmaceuticals and natural products.

Chirik's group applied a bis(imino)pyridine–iron catalyst for the cycloaddition of α,ω-dienes.[7] Later, they carried out a systematic study, in which a family of bis(imino)pyridine–iron metallacycles were synthesized, isolated and structurally characterized, allowing the elucidation of the electronic structure of important intermediates in iron-catalyzed hydrogenative cyclization and cycloaddition reactions. All of the experimental and computational data support a monoreduced bis(imino)pyridine radical anion throughout the catalytic turnover and that an Fe(I)–Fe(III) cycle is operative for C–C bond formation. These features enable a sufficiently high oxidation state iron complex to permit facile sp^3–sp^3 C–C reductive elimination and avoid reduced iron compounds that would result in catalyst decomposition owing to metal deposition. In 2008, Fürstner *et al.* prepared a series of Fe(0)-ate complexes by the reaction of ferrocene with lithium in the presence of either ethylene or cyclooctadiene (COD).[8] As a result, these complexes effect a variety of cycloaddition and cycloisomerization reactions of the Alder–ene [4 + 2], [5 + 2] and [2 + 2 + 2] types. These transformations are compatible with a variety of substitution patterns, potentially reducible functional groups, basic sites and even fairly CH-acidic terminal alkyne units. Labeling experiments together with the isolation of two novel Fe(I) complexes, which incorporate different primary cycloadducts, provide compelling evidence for the proposed mechanism.

Wang and co-workers developed a novel protocol for the convenient and effective construction of functionalized 2-pyrrolines from readily available substrates under mild conditions.[9] The use of a simple iron salt as catalyst renders the protocol suitable for large-scale synthesis, providing a valuable route to the assembly of biologically active molecules (Scheme 3.6). A plausible mechanism was proposed. Aziridine was activated by FeCl$_3$ to form a zwitterionic intermediate, which was then attacked by the phenyl-acetylene to generate an aryl-substituted alkenyl cation. An intramolecular cyclization gives the desired product and regenerates the iron catalyst.

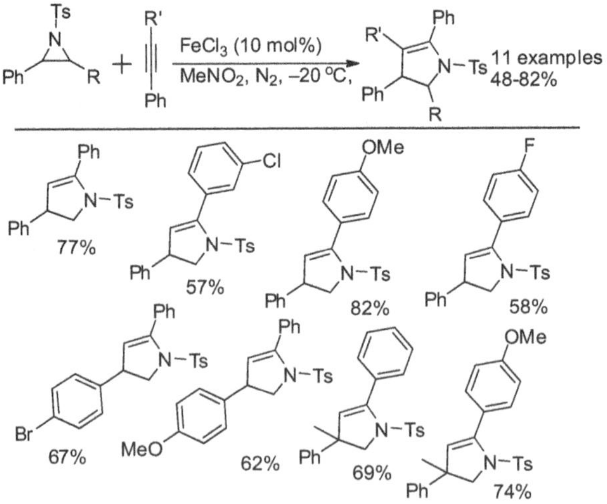

Scheme 3.6   Iron-catalyzed synthesis of 2-pyrrolines.

Reaction Procedure (Scheme 3.6): FeCl$_3$ (8.1 mg, 0.05 mmol) and aziridine (137.0 mg, 0.5 mmol) were loaded into a 10 mL oven-dried flask and the system placed under vacuum and charged with N$_2$ three times.

The system was cooled to −20 °C and a solution of alkyne (1.5 mmol) in freshly distilled CH_3NO_2 (2.0 mL) was added over 20 min using a syringe. The resulting mixture was maintained at −20 °C for a further 10 min, and then the solvent was evaporated under reduced pressure. The residue was purified by column chromatography on silica gel.

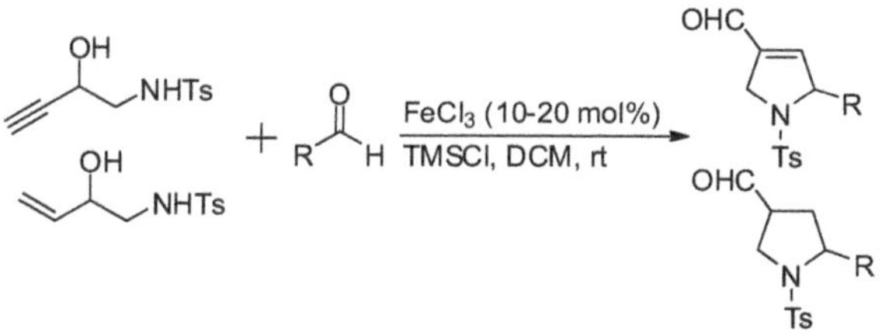

Scheme 3.7   Iron-catalyzed synthesis of pyrrolidines.

Reaction Procedure (Scheme 3.7): To a solution of *N*-(2-hydroxybut-3-enyl)-4-methylbenzenesulfonamide [2-hydroxyhomoallyltosylamine] (1.0 equiv.) in dry CH_2Cl_2 (0.1 M) were added anhydrous $FeCl_3$ (0.1 equiv.) and TMSCl (1.0 equiv.). Aldehyde (1.0 equiv.) was then added to the reaction mixture, which was stirred at room temperature until TLC analysis showed complete formation of product. The reaction was then quenched by addition of water with stirring and extracted with CH_2Cl_2. The combined organic layers were dried over $MgSO_4$ and the solvent was removed under reduced pressure. The crude reaction mixture was purified by flash silica gel column chromatography.

An efficient alkene aza-Cope–Mannich cyclization between 2-hydroxyhomoallyltosylamine and aldehydes in the presence of iron(III) salts to obtain 3-alkyl-1-tosylpyrrolidines in good yields was described in 2010 (Scheme 3.7).[10] This process is based on the consecutive generation of a *γ*-unsaturated iminium ion, 2-azonia-[3,3]-sigmatropic rearrangement and further intramolecular Mannich reaction. Iron(III) salts were also shown to be excellent catalysts for the new aza-Cope–Mannich cyclization using 2-hydroxyhomopropargyltosylamine.

3.1.2   Iron-Catalyzed Synthesis of Pyrrole Derivatives

In 1985, an $Mo(CO)_6$- or $Fe_2(CO)_9$-mediated conversion of phenyl-substituted 1,3-oxazepines to pyridines and pyrroles was reported.[11] The reaction pathway involves C2–O and C7–O bond cleavage are and a coordinated pyridine-2,3-oxide is the intermediate. In 1990, an $Fe_3(CO)_{12}$-mediated reductive de-oxygenation of 5,6-dihydro-4*H*-1,2-oxazines to pyrroles was reported.[12] Iron carbonyl complexes other than $Fe_3(CO)_{12}$ are also effective for this reaction.

The efficiency of the complexes for this reaction decreases in the order $Fe_3(CO)_{12} \gg Et_3NH[HFe_3(CO)_{11}] > Fe_2(CO)_9 \gg Fe(CO)_5$. Various substituted pyrroles were produced in moderate to good yields (Scheme 3.8).

An operationally simple, practical and economical protocol for iron(III) chloride-catalyzed Paal–Knorr pyrrole synthesis in water in good to excellent yields was reported in 2009 (Scheme 3.9).[13] Several *N*-substituted pyrroles

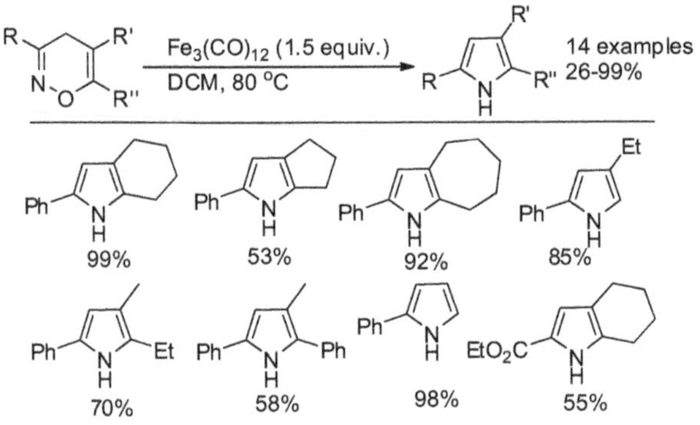

Scheme 3.8   Iron-mediated synthesis of pyrroles.

Reaction Procedure (Scheme 3.8): A mixture of substrate (1.0 mmol) and $Fe_3(CO)_{12}$ (1.5 mmol) in DCE (15 ml) was stirred under argon at 80 °C for 3–20 h and filtered. The solvent was removed and the residue was chromatographed on silica gel with hexane–benzene (1:1) as eluent to give the pure product.

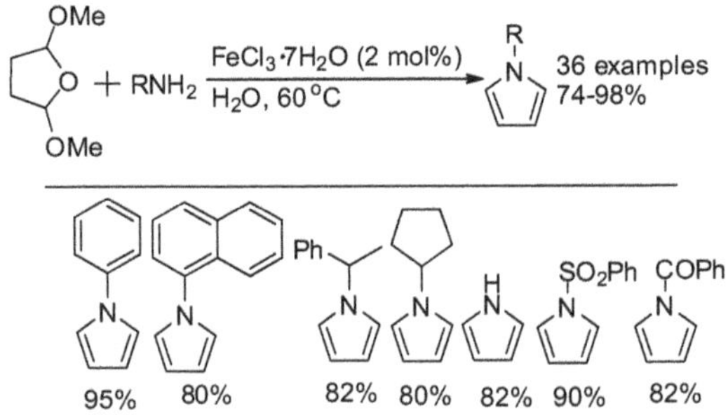

Scheme 3.9   Iron-catalyzed synthesis of pyrrole derivatives.

Reaction Procedure (Scheme 3.9): To a mixture of the amine (5 mmol) and 2,5-dimethoxytetrahydrofuran (6 mmol) in H_2O (4 mL) at 60 °C $FeCl_3 \cdot 7H_2O$ (2 mol%) was added. The mixture was stirred at this temperature for 1–4 h, then diluted with EtOAc and filtered. The organic solution was evaporated under vacuum affording the pyrrole derivative with good analytical purity.

were readily prepared from the reaction of 2,5-dimethoxytetrahydrofuran and aryl/alkyl, sulfonyl and acyl amines under mild reaction conditions. Interestingly, poor yields of product were formed in organic solvents such as DCM, THF, benzene, MeCN and EtOH.

An efficient method for easy access to 1,2,3,5-tetrasubstituted pyrroles was developed in 2010.[14] Using iron(III) chloride as the catalyst, pyrrole derivatives were produced in good yields from phenacyl bromide or its derivatives, amine and dialkyl acetylenedicarboxylate (Scheme 3.10). The mild reaction conditions, application of a readily available and less expensive catalyst,

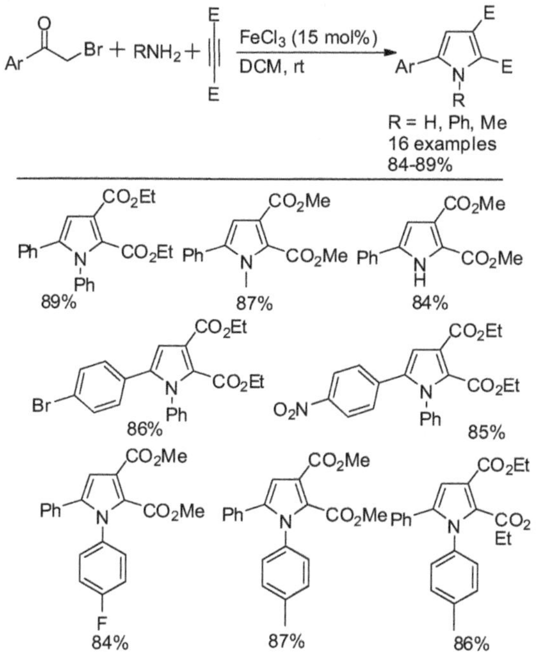

Scheme 3.10 Iron-catalyzed synthesis of 1,2,3,5-tetrasubstituted pyrrole derivatives.

Reaction Procedure (Scheme 3.10): To a stirred solution of phenacyl bromide or its derivatives (1 mmol) and amine (1 mmol) in CH_2Cl_2 (6 mL) was added $FeCl_3$ (15 mol%). After 5 min, dialkyl acetylenedicarboxylate (1 mmol) was added dropwise. The mixture was stirred at r.t. for 14–16 h (monitored by TLC). After completion, the solvent was evaporated; the residue was washed with cold water (2 × 5 mL) and subsequently extracted with EtOAc (2 × 10 mL). The extract was dried (anhydrous Na_2SO_4) and concentrated under vacuum. The residue was subjected to column chromatography (silica gel, 3–8% EtOAc–hexane as eluent) to give the pure pyrrole derivative.

operational simplicity and impressive yields are the advantages of the method. ZnCl$_2$ and CuI were also tested as catalysts; 45–63% yields of the desired product were obtained.

A simple, convenient and multicomponent coupling strategy for the synthesis of highly functionalized pyrroles catalyzed by iron(III) salts was developed in 2010.[15] This strategy demonstrated four-component coupling reactions of 1,3-dicarbonyl compounds, amines, aromatic aldehydes and nitroalkanes without an inert atmosphere. This methodology provides an alternative approach for easy access to highly substituted pyrroles in moderate to very good yields using four simple and readily available building blocks *via* a one-pot tandem reaction (Scheme 3.11). This method has several advantages: (1) inexpensive and environmentally friendly FeCl$_3$ (10 mol%) was used as catalyst; (2) all components are readily available and inexpensive; (3) it allows the direct introduction of carbonyl functionalities on to the pyrrole skeleton; (4) a wide variety of functional groups can survive; (5) it is

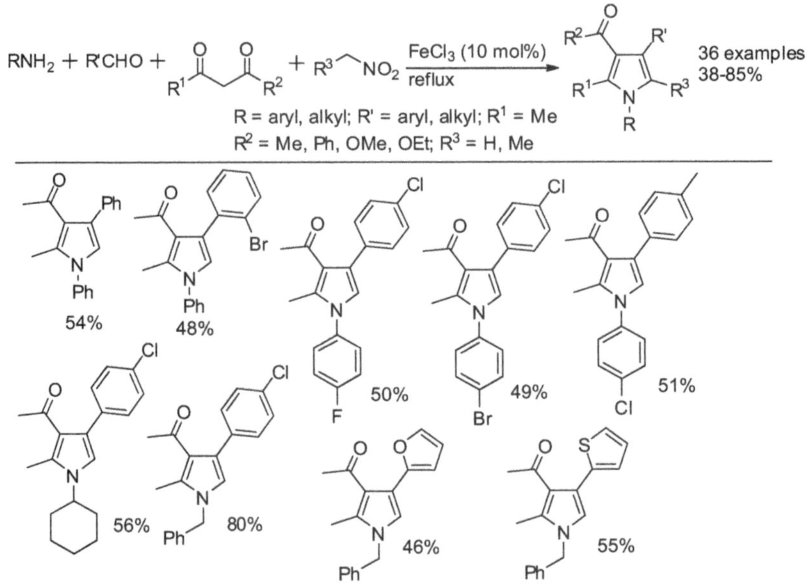

Scheme 3.11   Iron-catalyzed four-component synthesis of pyrrole derivatives.

Reaction Procedure (Scheme 3.11): To a stirred solution of *p*-anisdine (185 mg, 1.5 mmol), *p*-chlorobenzaldehyde (140 mg, 1 mmol), and acetylacetone (100 mg, 1 mmol) in nitromethane (1 mL) was added anhydrous FeCl$_3$ (16 mg, 0.1 mmol). The mixture was heated at reflux slowly for 7 h and cooled to room temperature. The excess solvent was removed under vacuum and the residue was directly purified by silica gel column chromatography to afford the product.

highly selective and allows the isolation of the desired pyrroles in moderate to high yields (38–85%); and (6) the reactions are applicable to a combination of substrates such as aromatic and aliphatic amines, aromatic and aliphatic aldehydes, β-diketones and β-keto esters with nitroalkanes.

Another highly efficient iron-catalyzed approach to polysubstituted pyrroles was developed in 2011.[16] Through the [4C + 1N] cyclization of 4-acetylenic ketones with primary amines, a variety of tetra- and fully substituted pyrroles and also fused pyrrole derivatives were produced in good to excellent yields (Scheme 3.12). Excellent yields of pyrroles were obtained in 89 and 91% yields with $FeCl_3$ and $FeBr_3$ as catalyst, respectively. $Fe(OTf)_3$ gave a similarly high yield of pyrroles in a short time, whereas 10 mol% CF_3SO_3H produced the target product in 78% yield in 18 h. $Fe(acac)_3$ afforded a mixture of the pyrrole and unreacted substrate in a ratio of 43 : 57 determined by ^1H NMR spectroscopic analysis of the reaction mixture. Interestingly, the oxidation state of iron appears to have no influence on the

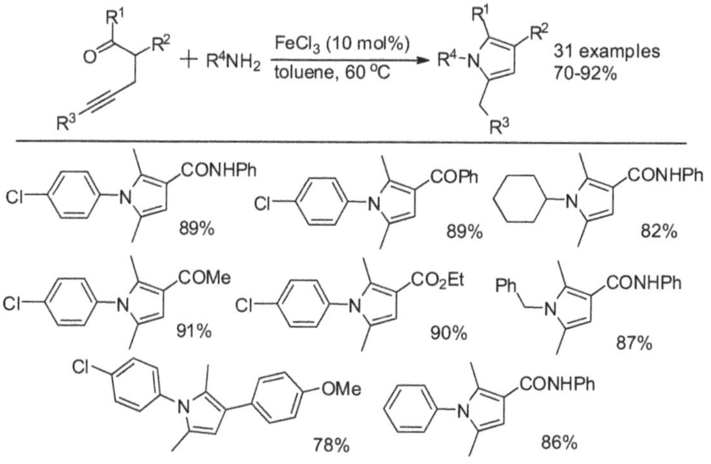

Scheme 3.12   Iron-catalyzed synthesis of pyrroles from 4-acetylenic ketones and amines.

Reaction Procedure (Scheme 3.12): To a solution of 2-acetyl-*N*-phenylpent-4-ynamide (215 mg, 1 mmol) and 4-chloroaniline (152 mg, 1.2 mmol) in toluene (1.0 mL), $FeCl_3$ (17 mg, 0.1 mmol) was added. The reaction mixture was warmed to 60 °C and stirred until the staring material was consumed (monitored by TLC). After cooling to room temperature, the reaction mixture was quenched with 1 M HCl (2 mL). The organic and aqueous layers were separated and the aqueous layer was extracted with dichloromethane (3×10 mL). The combined organic layers were dried over $MgSO_4$ and filtered. The filtrate was concentrated *in vacuo* and the residue was purified by silica gel column chromatography.

catalysis, because the FeCl$_2$ afforded a similar yield of the product to that under a nitrogen atmosphere, which could to a great extent protect FeCl$_2$ from oxidizing to FeCl$_3$. Other iron sources such as Fe$_2$O$_3$ and Fe powder all gave poor yields of pyrroles and most of the substrate was recovered.

Guan and co-workers developed an efficient tandem one-pot protocol for the synthesis of highly functionalized NH pyrroles based on FeCl$_3$-catalyzed addition and cyclization of Blaise reaction intermediates and nitroalkenes.[17] The reaction showed good functional group tolerance and afforded a series of substituted NH pyrroles in good yields (Scheme 3.13). The diverse and readily available nitriles and nitroalkenes make this method versatile and powerful. ZnBr$_2$, Cu(OTf)$_2$ and Yb(OTf)$_2$ were also tested as catalysts and gave slightly lower yields of the desired product.

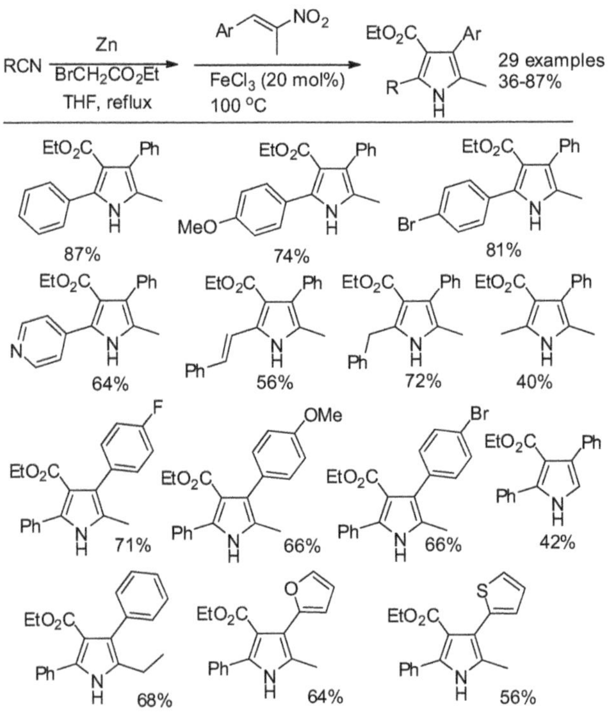

Scheme 3.13 Iron-catalyzed synthesis of pyrrole derivatives from nitriles.

Reaction Procedure (Scheme 3.13): In a 25-mL round-bottomed flask, a solution of methanesulfonic acid (5 mol%) in anhydrous THF (1 mL) was added to the stirred suspension of commercial zinc dust (1.2 mmol, 0.0785 g). After refluxing for 10 min (90 °C oil bath), nitrile (0.6 mmol) and ethyl bromoacetate (0.9 mmol, 0.1503 g) were added successively to the flask. The mixture was refluxed for 2–12 h, then FeCl$_3$ (20 mol%,

9.7 mg), nitroolefin (0.3 mmol) and THF (2 mL) were added and heating under reflux was continued until the completion of the reaction (detected by TLC). The mixture was cooled to room temperature, quenched with saturated aqueous NH_4Cl (20 mL) and extracted with ethyl acetate (20 mL). The organic layer was washed with brine (20 mL), dried over by anhydrous Na_2SO_4 and evaporated under vacuum. The desired pyrrole was obtained after purification by flash chromatography on silica gel with hexane–ethyl acetate as eluent.

3.1.3 Iron-Catalyzed Synthesis of Other Five-Membered Nitrogen-Containing Heterocycles

Iron as a 'green' and abundant catalyst has also been applied in the carbonylative synthesis of heterocycles. As early as 1970, an $Fe(CO)_5$-mediated carbonylative reaction of isocyanates and carbodiimide with alkynes was reported.[18] In the reaction of phenyl isocyanates with phenylacetylene in the presence of $Fe(CO)_5$, 4-benzylidene-1,3-diphenylhydantoin was obtained in 85% yield by addition and hydrogen shift reactions. The reaction of diphenylcarbodiimide with phenylacetylene in the presence of $Fe(CO)_5$ gave 4-benzylidene-1,3-diphenyl-2,5-bis(phenylimino)imidazolidine and 4-benzylidene-1,3-diphenyl-2-phenyliminoimidazolidin-5-one in 78 and 17% yields, respectively. 1,3,4-Triphenylpyrroline-2,5-dione and 1,3,4-triphenyl-5-phenyliminopyrrolin-2-one were obtained in 42 and 15% yields, respectively, in the reaction using phenyl isocyanate, diphenylacetylene and $Fe(CO)_5$. Later, the reaction of phenylbromoacetylene with heterocumulenes in the presence of iron pentacarbonyl was investigated by the same group. A catalytic version and systematic study was performed in 2012 by Mathur *et al.*[19] In the presence of a catalytic amount of $Fe(CO)_5$, terminal acetylenes, isocyanates and CO undergo $[2+2+1]$ cyclization to form substituted maleimides and hydantoins; when internal alkynes are used, maleimide formation is exclusively observed. While the maleimides can be obtained as the major products, in up to 90% yield, when the reaction is carried out in a CO atmosphere, in the absence of CO the hydantoins are formed in up to 87% yield. Formation of maleimides has been shown to occur *via* the formation of a ferrole intermediate, whereas the hydantoins are proposed to form through successive insertion of isocyanate into the iron–acetylide bond.

Imhof's group carried out a systematic study on the iron-catalyzed cycloaddition reaction of 1,4-diazabutadienes, carbon monoxide and ethylene by experimental investigation and density functional theory (DFT) calculation (Scheme 3.14).[20]

Based on experimental experience, the reaction with ethylene is considered to take place before any interaction with carbon monoxide (Figure 3.1). According to the computational results, the reaction does not proceed by ligand dissociation followed by addition of ethylene and

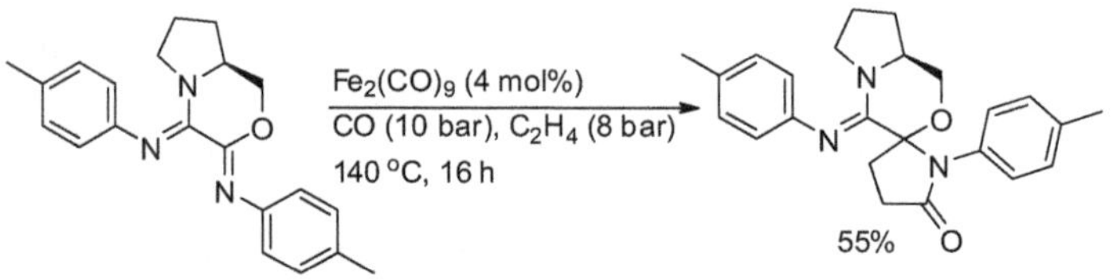

Scheme 3.14   Iron-catalyzed carbonylation of ketimines.

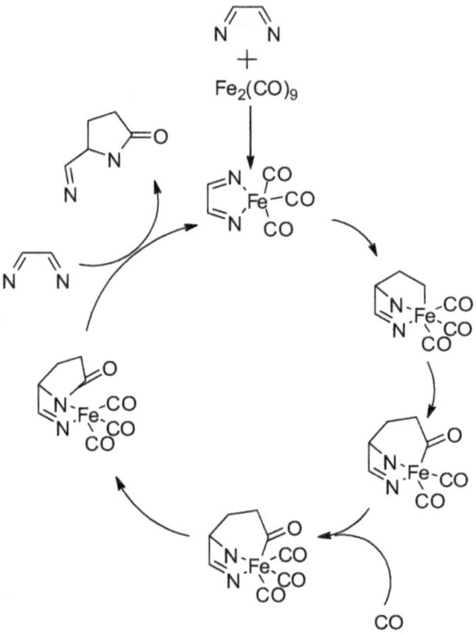

Figure 3.1   Reaction mechanism of iron-catalyzed cycloaddition.

subsequent intramolecular activation steps but by the approach of an ethylene molecule from the base of the square-pyramidal complex. This reaction yields an intermediate in which ethylene is coordinated to the iron center and a new C–C bond between ethylene and one of the imine groups is formed. The insertion of a terminal carbon monoxide ligand into the metal–carbon bond between ethylene and iron produces the key intermediate. The reaction proceeds by metal-assisted formation of a lactam. The catalytic cycle is closed by a ligand-exchange reaction in which the diazabutadiene ligand substitutes with re-formation of starting complex. This reaction pathway is found to be energetically favored over a reductive elimination, and leads to the experimentally observed heterocyclic product and a reactive $[Fe(CO)_3]$ fragment. Calculations on the substituents on the dienes were also carried out to explain the high regioselectivity. Analogous transition structures and reaction products starting from 1,4-diazabutadienes with a 2-fluoro, 2-hydroxo or 2-amino substituent revealed that the regioselectivity is not determined by the electronegativity of the heteroatom and thus by the differences in the NPA charges or the resulting Coulombic interactions in the transition structures. The main reason for the observed regioselectivities is the π-donor ability of the substituent to contribute to a delocalized π system incorporating the adjacent imine moiety. The increasing π-donor capability results in decreased reactivity of this moiety and increases the (relative) reactivity of the second imine group. This effect can even over-compensate for strong intramolecular Coulombic attractions in the transition structures.

An iron-catalyzed conversion of allenylimines to 3-alkylidene-4-pyrrolin-2-ones was reported in 1994.[21] Catalytic carbon–nitrogen bond formation was achieved by iron carbonyls in the [4 + 1] cycloaddition of allenylimines with CO. The $Fe(CO)_5$ photochemically catalyzed reaction of allenylimines and CO gives preparatively useful yields of 3-alkylidene-4-pyrrolin-2-ones (Scheme 3.15). These reactions take place under mild conditions and only require fluorescent light. Good control of the alkylidene bond stereochemistry is achieved when the terminal allene groups are *tert*-butyl and methyl. Experiments in the dark showed that stoichiometric $Fe_2(CO)_9$ can mediate [4 + 1] assembly by a purely thermal reaction to give good yields of the pyrrolinone products. These new methods for the construction of 3-alkylidene-4-pyrrolin-2-ones are complementary to existing procedures and allow for a greater variety of alkylidene substituents.

An iron-catalyzed double aminocarbonylation of alkynes with amines was developed by Beller and co-workers in 2009.[22] This provides a convenient one-pot method for the synthesis of various substituted succinimides. By starting from commercially available amines (or ammonia) and alkynes, a range of interesting succinimides were obtained selectively in the presence of catalytic amounts of either $[Fe(CO)_5]$ or $[Fe_3(CO)_{12}]$ (Scheme 3.16). For this novel environmentally friendly reaction, no expensive catalyst was required.

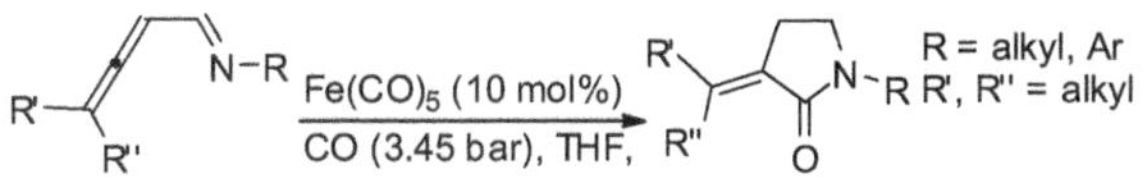

Scheme 3.15 Iron-catalyzed carbonylation of allenylimines.

Reaction Procedure (Scheme 3.15): 7 mL of freshly distilled (from Na/K benzophenone) THF was placed in a thick-walled glass bomb equipped with a Teflon vacuum valve, followed by 12.4 mg (10 mol%) of $Fe(CO)_5$. The bomb was pressurized to give 80 mM CO in solution. The bomb was irradiated for 48 h with fluorescent light. After demetalation (1 g flash silica gel, 1% triethylamine in CH_2Cl_2) and filtration, column chromatography of the residue on flash silica gel (CH_2Cl_2) yielded the pure product.

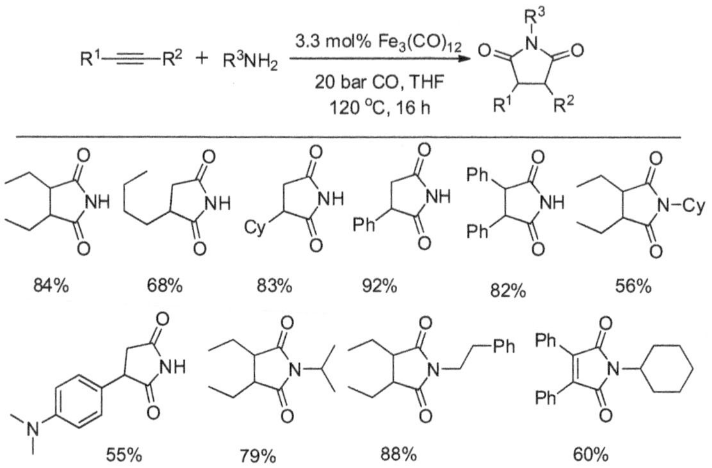

$$R^1{-}{\equiv}{-}R^2 + R^3NH_2 \xrightarrow[\substack{20 \text{ bar CO, THF} \\ 120\,°C,\ 16\,h}]{3.3 \text{ mol\% Fe}_3(CO)_{12}}$$

84%　　68%　　83%　　92%　　82%　　56%

55%　　79%　　88%　　60%

Scheme 3.16　Iron-catalyzed carbonylation of alkynes with ammonia or amines.

Reaction Procedure (Scheme 3.16): $[Fe(CO)_5]$ or $[Fe_3(CO)_{12}]$ was dissolved in THF under an argon atmosphere in a 50 mL Schlenk flask. The alkyne and amine were added to this solution before being transferred into an autoclave. When ammonia was required, it was condensed from a small bomb into the autoclave. Subsequently, the autoclave was pressurized with carbon monoxide and heated to 120 °C. The reaction was carried out for 16 h before the reaction mixture was cooled to room temperature. The pressure was then released and isooctane (internal standard) was added to the mixture. After removal of the solvent *in vacuo*, the crude succinimide product was purified by column chromatography on silica gel (10 : 1–1 : 1 heptane–ethyl acetate).

Soon afterwards, they applied this methodology in the synthesis of himanimide A and B. They proposed a reaction mechanism for this novel transformation (Figure 3.2). In the first stage, the amine reacts with $[Fe_3(CO)_{12}]$ to form an 'amine-$[Fe(CO)_4]$' and an $[Fe_2(CO)_8]$ species, which on further reaction with alkynes leads to five-membered rings. In the presence of an excess amount of amine, the corresponding cyclic imides are obtained *via* C–N bond formation. The order of adding the reagents is significant. Notably, when the amine is added to the catalyst solution before the alkyne the yield increases by about 10%. The authors assumed that double carbonylation and not a stepwise reaction *via* acrylamide is responsible because the product yield decreases on addition of acrylamide to the model system. Obviously, iron pentacarbonyl shows a completely different reaction behavior to $[Co_2(CO)_8]$, which reacts with acrylamides to give the corresponding succinimides.[23]

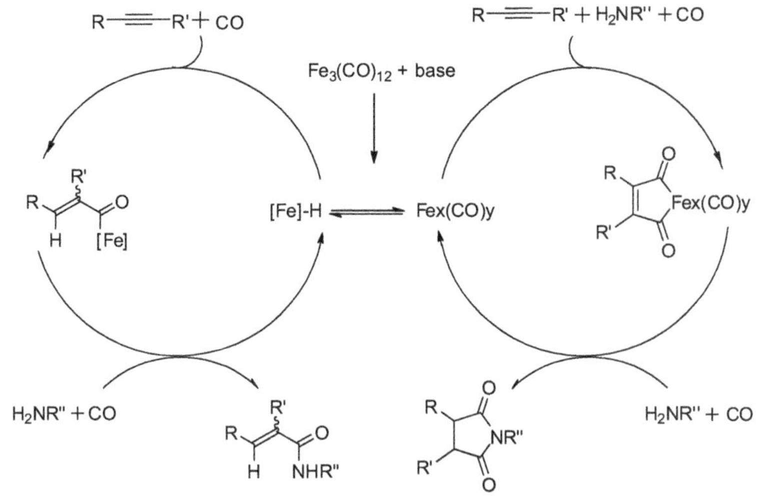

Figure 3.2 Proposed mechanism for iron-catalyzed carbonylation of alkynes.

3.1.4 Iron-Catalyzed Synthesis of Furan Derivatives

Zhan and co-workers developed an efficient method for the synthesis of substituted furans using $FeCl_3$ as catalyst.[24] Iron(III) chloride, operating as a bifunctional catalyst, catalyzes two mechanistically distinct processes in a single pot under the same reaction conditions, namely propargylation followed by cycloisomerization. First, the ionization of propargylic alcohol would lead to a propargylic cation and the subsequent nucleophilic attack of the enol would give a γ-alkynyl ketone. Coordination of iron(III) to the alkyne would form the π-alkyne–iron complex and would enhance the electrophilicity of the alkyne. Subsequent 5-*exo-dig* nucleophilic attack of the hydroxyl group on the β-carbon of the Fe(III)–alkyne complex would generate the alkenyliron derivative. Protonolysis of this complex would afford a dihydrofuran, which would then undergo isomerization and desilylation to give the furan as the final product. This facile methodology allows rapid access to a variety of tetrasubstituted furans (Scheme 3.17). $FeCl_3$ as the catalyst offers several advantages including being inexpensive and commercially available and allowing mild reaction conditions.

Jiang *et al.* described an iron-catalyzed domino reaction for the synthesis of α-carbonylfurans from readily available electron-deficient alkynes and 2-yn-1-ols (Scheme 3.18).[25] The first step of this transformation, the DABCO- or PBu_3-promoted nucleophilic addition of propargyl alcohol to an electron-deficient alkyne, afforded the enyne adduct. Subsequently, a 6-*endo-dig* addition of the enol ether to the iron(III)–alkyne complex resulted in the formation of a six-membered intermediate, which collapsed into the β-allenic ketone. Finally, this β-allenic ketone underwent sequential cyclization, carbene oxidation and dehydration–oxidation to form the furan.

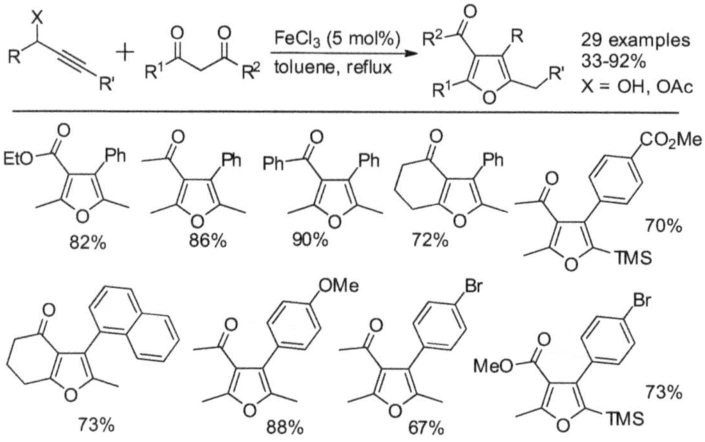

Scheme 3.17 Iron-catalyzed synthesis of furans.

> **Reaction Procedure** (Scheme 3.17): Propargylic alcohols or propargylic acetates (0.5 mmol), 1,3-dicarbonyl compounds (2.0 mmol), toluene (2.0 mL) and FeCl$_3$ (0.025 mmol, 4 mg) were successively added to a 5-mL flask. The reaction mixture was stirred at reflux and monitored periodically by TLC. Upon completion, toluene was removed under reduced pressure by an aspirator, then the residue was purified by silica gel column chromatography (EtOAc–hexane) to afford the corresponding tetrasubstituted furans.

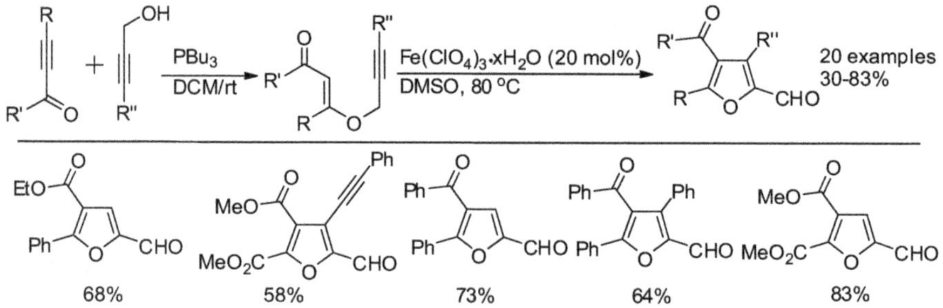

Scheme 3.18 Iron-catalyzed synthesis of α-carbonylfurans.

> **Reaction Procedure** (Scheme 3.18): A solution of 1,3-diphenylprop-2-yn-1-one (103 mg, 0.5 mmol), prop-2-yn-1-ol (28 mg, 0.5 mmol) and PBu$_3$ (0.1 mmol) in CH$_2$Cl$_2$ was stirred for 30 min at room temperature. The solvent was evaporated under reduced pressure. Fe(ClO$_4$)$_3$ and DMSO were added at 80 °C under atmospheric pressure. After completion of the reaction (monitored by TLC), the solvent was evaporated under reduced

pressure. Water (8 mL) was added, and the aqueous solution was extracted with diethyl ether (3×8 mL). The combined extracts were dried with anhydrous MgSO$_4$. The solvent was removed, and the crude product was separated by column chromatography.

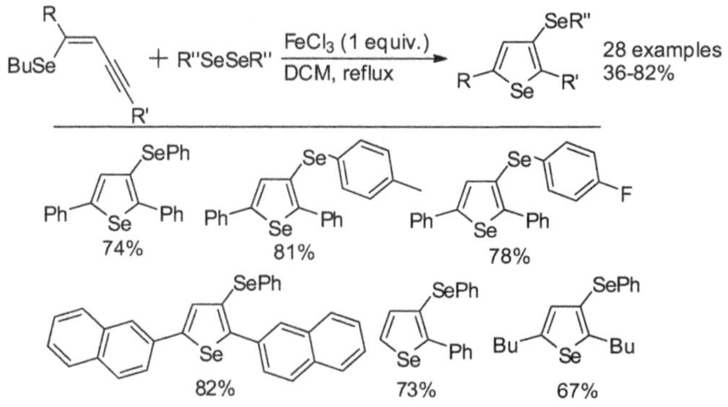

Scheme 3.19 Iron-mediated synthesis of selenophenes.

Reaction Procedure (Scheme 3.19): To a Schlenk tube, under argon, containing a mixture of FeCl$_3$ (0.25 mmol) in dry CH$_2$Cl$_2$ (2 mL) was added the appropriate diorganoyl dichalcogenide (0.55 equiv.). The resulting solution was stirred for 15 min at room temperature. (Z)-Chalcogenoenyne (0.25 mmol) in CH$_2$Cl$_2$ (1 mL) was then added and the resulting solution was stirred under reflux for 3 h. The solution was cooled to room temperature, diluted with dichloromethane (10 mL) and washed with saturated aqueous NH$_4$Cl (3×10 mL). The organic phase was separated, dried over MgSO$_4$ and concentrated under vacuum. The residue was purified by flash chromatography on silica gel using hexane as the eluent.

Zeni and co-workers reported the synthesis of a series of 2,5-disubstituted-3-(organoseleno)selenophenes by the FeCl$_3$- and diorganyl dichalcogenide-mediated intramolecular cyclization of (Z)-chalcogenoenynes.[26] The cyclized products were obtained in good yields (Scheme 3.19). The results showed that some of the products, evaluated in a mouse forced-swimming test, elicited an antidepressant-like activity. The studies clearly show that the phenyl group at the 2-position and an organoselenium group at the 3-position of the selenophene ring are essential for the antidepressant-like activity of selenophenes. Close inspection of the results also revealed that the fluorophenyl portion in the organoselenium group is fundamental for the antidepressant-like action of this class of organochalcogens.

3.1.5   Iron-Catalyzed Synthesis of Tetrahydrofuran Derivatives

Based on the known ability of iron catalysts in hydroalkoxylation reactions, an intramolecular hydroalkoxylation of alkenes was reported in 2007.[27] Komeyama *et al.* demonstrated the catalytic activity of cationic iron complexes in the intramolecular hydroalkoxylation of unactivated alkenes under mild conditions. The reaction system is compatible with many functional groups, giving rise to various types of cyclic ethers (Scheme 3.20). The role of the cationic iron catalysts might be dual activation of both the hydroxy group and the alkene moiety. The most striking feature is that the environmentally friendly catalysis enables us readily to induce hydroalkoxylation of unactivated alkenes, which is hardly attainable under mild conditions.

The cyclization of triynes offers a promising route for the preparation of different sized rings. In 2005, an $FeCl_3$/Zn-catalyzed intramolecular cyclotrimerization of triynes was reported (Scheme 3.21).[28] 1,3-Bis(2,6-diisopropylphenyl)imidazolium chloride (IPr · HCl) was applied as a ligand in this transformation. $CoCl_2$/Zn can be applied as alternative catalyst system with an *N*-heterocyclic carbene as ligand.

An Fe(III)-mediated ring opening of cyclopropyl ethers was reported in 2003.[29a] Cyclopropyl ethers bearing a phenyl-substituted butenyl side chain lead to the generation of β-keto radicals that undergo 5-*exo*-cyclization followed by a novel cascade sequence resulting in the formation of tricyclic ethers. Later, an iron-mediated oligomerization of stilbenes was developed by Velu *et al.*[29b] Oligostilbenoid dimers were prepared in low yields by a one-electron oxidation pathway.

An iron-catalyzed cyclization of alkynyl aldehyde acetals was reported by Li and co-workers in 2010.[30] By using acetyl chloride or bromide as a halogen source in dichloromethane, this interesting $FeCl_3 \cdot 6H_2O$- and $FeBr_3$-catalyzed Prins cyclization/halogenation of alkynyl aldehyde acetals was realized and afforded 2-(1-halobenzylidene or alkylidene)-substituted five-membered carbo- and heterocycles as the products (Scheme 3.22). This methodology provides an alternative route for vinylic C–Cl and C–Br bond formation. Five- to eight-membered cyclic enones were efficiently synthesized by $FeCl_3 \cdot 6H_2O$-catalyzed intramolecular cyclization of alkynyl aldehyde acetals in acetone under mild conditions. An oxocarbonium species generated *in situ* was proposed to initiate the reaction and the target

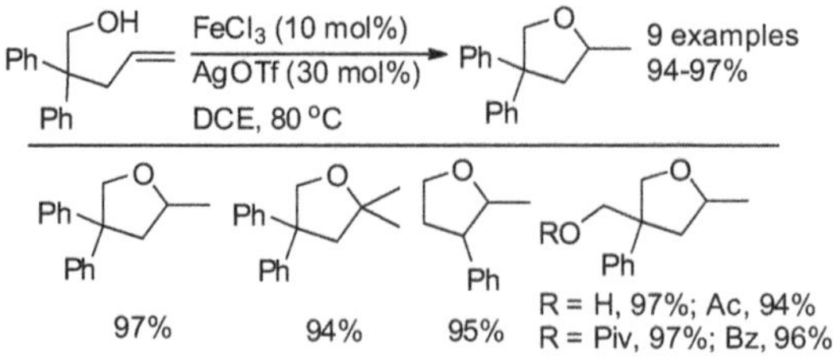

Scheme 3.20   Iron-catalyzed intramolecular hydroalkoxylation of alkenes.

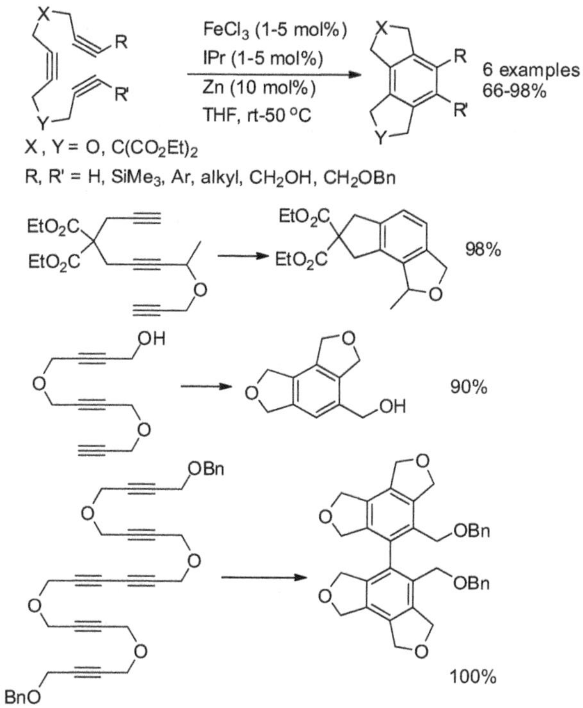

Scheme 3.21 FeCl$_3$/Zn-IPr·HCl-catalyzed cyclization of triynes.

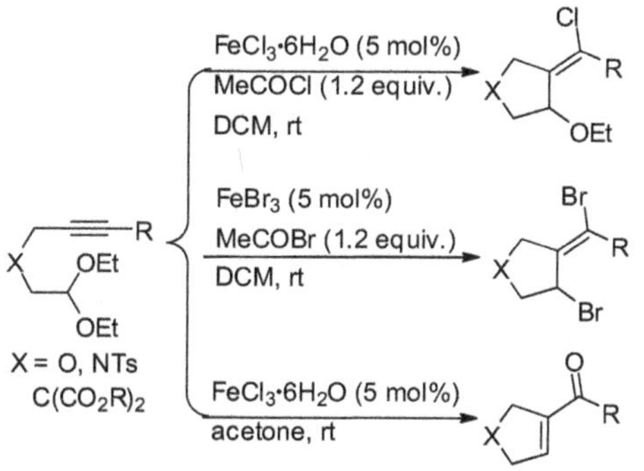

Scheme 3.22 Iron-catalyzed cyclization of alkynyl aldehyde acetals.

Reaction Procedure (Scheme 3.22): Substrate (0.5 mmol) and CH$_3$COCl (47 mg, 0.6 mmol) were added to a suspension of FeCl$_3$·6H$_2$O (6.8 mg, 5 mol%) in CH$_2$Cl$_2$ (5 mL). The mixture was stirred at ambient temperature and monitored by TLC on silica gel. When the starting acetal substrate was completely consumed, the resulting mixture was concentrated under reduced pressure. The resulting residue was purified by flash column chromatography on silica gel [eluent: petroleum ether (b.p. 60–90 °C)–diethyl ether (20:1)] to afford the target product.

Reaction Procedure: $FeCl_3 \cdot 6H_2O$ (6.8 mg, 5 mol%) was added to a solution of substrate (0.5 mmol) in acetone (5 mL) in air. The mixture was stirred at room temperature and monitored by TLC on silica gel. When the substrate was completely consumed (0.5 h), all volatile substances were evaporated from the resulting mixture under reduced pressure. The resulting residue was purified by flash column chromatography on silica gel [eluent: petroleum ether (b.p. 60–90 °C)–diethyl ether (10:1) to afford the target enone product.

products are formed *via* a vinylogous carbenium cation and oxete intermediates according to DFT calculations. Intermolecular reactions of alkynes and aldehyde acetals were also investigated with 20–40 mol% $FeCl_3 \cdot 6H_2O$ catalyst and produced α,β-unsaturated enones and chlorinated indene derivatives. The authors demonstrated the application of the protocol in the synthesis of carbo-, oxa- and azacycles.

An iron-catalyzed intermolecular ring expansion reaction of an aryl epoxide with several dienes, acrylates, enynes or styrenes was reported in 2005.[31] Tetrahydrofuran derivatives were generated in a highly chemo- and regioselective fashion (Table 3.1). The process could be used in an unprecedented way for the one-step synthesis of racemic calyxolane A and calyxolane B with acceptable diastereoselectivity. Here, the catalytic system can be either $FeCl_2$–PPh_3–Zn–1,3-bis(2,4,6-trimethylphenyl)imidazolium chloride or $FeCl_2$(dppe)–Zn. Later, the same group developed another catalytic system for this transformation.[32] Optimization of the catalyst system revealed that a preformed [Fe(salen)] complex minimizes the formation of polymerization side products so that increased yields of intermolecular reactions were obtained. However, more importantly, the scope of the reaction could also be enlarged considerably. The iron-catalyzed ring-expansion reaction can now be applied to some styrene oxide derivatives, acting as radical donors, and also to a wide variety of acceptor-substituted acyclic alkenes and cyclic dienes that act as radical acceptors. The use of unsymmetrical radical acceptors led to interesting questions concerning the regiochemistry of the reactions. The conservation of the stereochemistry of the starting materials in the products was investigated through a study of the reactions of *E*- and *Z*-configured acceptor-substituted double bonds. The reactions of fumaric and maleic esters were performed and the ratios of diastereomeric and regioisomeric products were determined.

Tang and co-workers developed a catalytic formal [4 + 1] annulation *via* a nitrogen ylide route, providing an easy route to dihydrofurans and dihydropyrroles in good to excellent yields with high diastereoselectivities.[33] In the presence of a catalytic amount of pyridine and Fe(Tcpp)Cl [tetra(*p*-chlorophenyl)porphyrin iron chloride], α-ylidene-β-diketones and α,β-unsaturated imines react with diazoacetates to provide dihydrofurans and

> **Reaction Procedure** (Table 3.1): In a Schlenk tube under a nitrogen atmosphere, $FeCl_2$ (0.2 mmol), PPh_3 (0.1 mmol), NHC-Ligand (0.1 mmol) [or the preformed $FeCl_2$(dppe) complex], zinc dust (1.4 mmol) and NEt_3 were suspended in CH_3CN (1 mL) and heated until boiling. After 5 min of agitation, the alkene (5 mmol) and styrene epoxide (1 mmol) were added. The mixture was stirred at 60 °C for 4 h and filtered through a pad of silica using Et_2O (100 mL) as eluent. After evaporation under reduced pressure, the crude product was purified by flash chromatography.

Table 3.1   Iron-catalyzed ring expansion of epoxides with alkenes.

[Fe] A = $FeCl_2$ (20 mol%), PPh_3 (20 mol%), NHC (20 mol%)
[Fe] B = $FeCl_2$(dppe) (20 mol%)
[Fe] C = Fe(Salen) (20 mol%)

Entry	Alkene	Ring expansion (%)		
		Conditions A	*Conditions B*	*Conditions C*
1	Ph alkene	42	41	71
2	alkyne alkene	24	37	72
3	EtO_2C alkene	25	43	65
4	EtO_2C alkene	–	–	81
5	MeO_2C alkene	40	37	–
6	Ph alkene	–	–	79
7	Ph, CO_2Me alkene	–	–	54
8	EtO_2C, CO_2Et alkene	–	–	46

dihydropyrroles, respectively, in up to 96% yield. The loading of the catalyst pyridine could be reduced to 1 mol%.

A facile and mild protocol for the synthesis of substituted tetra-hydrofurans, 2-deoxy *C*-aryl glycosides and *C*-aryl glycosides by use of 20 mol% FeCl$_3$ was described in 2002.[34] In the reaction, the benzylic car-bocation generated undergoes intramolecular substitution by the oxygen nucleophile to result in the tetrahydrofurans and *C*-aryl glycosides. Yb(OTf)$_3$ can also be used as an acid catalyst. Ten examples of tetrahydrofurans were produced in 63–81% yields at temperatures from room temperature to 40 °C. An FeBr$_2$-mediated fragmentation of 1,4-diaryl-2,3-dioxabicyclo[2.2.2]octanes could also give tetrahydrofuran derivatives.[35]

3.1.6 Iron-Catalyzed Synthesis of Lactones and Some Other Compounds

In 2007, a cationic iron-catalyzed addition of carboxylic acids to alkenes was reported.[36] In the case of intramolecular addition, lactones were produced in excellent yields by the combination of FeCl$_3$ with AgOTf in DCE (Scheme 3.23). Moderate yields were achieved with only FeCl$_3$ as catalyst under the same conditions, and no product was formed with Zn(OTf)$_2$ and low yields with AgOTf or Cu(OTf)$_2$.

A highly efficient oxidative coupling of 2-naphthols and a rearrangement tandem reaction to afford unique spiro compounds in the presence of FeCl$_3 \cdot 6H_2O$ in up to 88% yield was developed by Tsubaki and co-workers in 2010 (Scheme 3.24).[37] As the starting materials are readily available and FeCl$_3 \cdot 6H_2O$ is inexpensive, this tandem reaction should be very valuable in synthetic chemistry. It is especially worthwhile to note that this rearrange-ment reaction is associated with two types of natural products, biaryl com-pounds such as blestriarene C5 and spiro compounds such as dendrochrysanene, which are isolated from distinct origins.

Pappo and co-workers developed a novel cross dehydrogenative coupling (CDC) method for the coupling of cyclic and acyclic β-keto esters, based on an inexpensive, non-toxic iron catalyst, which leads to the formation of ei-ther polyaromatic spirolactones or polyaromatic hemiacetal architectures.[38] The ligand effect was identified, with 1,10-phenanthroline or 2,2'-bipyridine having a significant influence on the efficiency of the reaction and on the

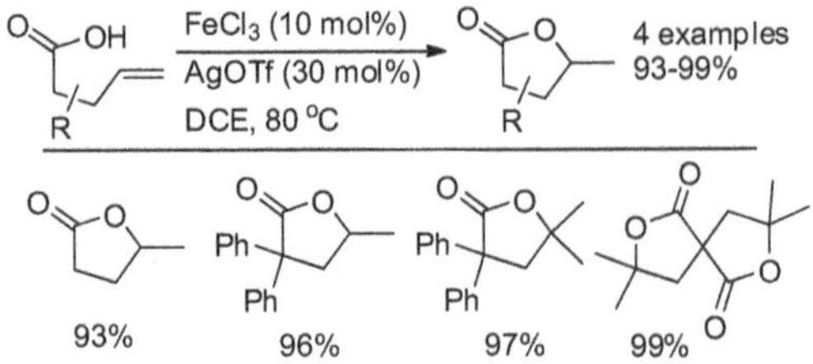

Scheme 3.23 Iron-catalyzed synthesis of lactones.

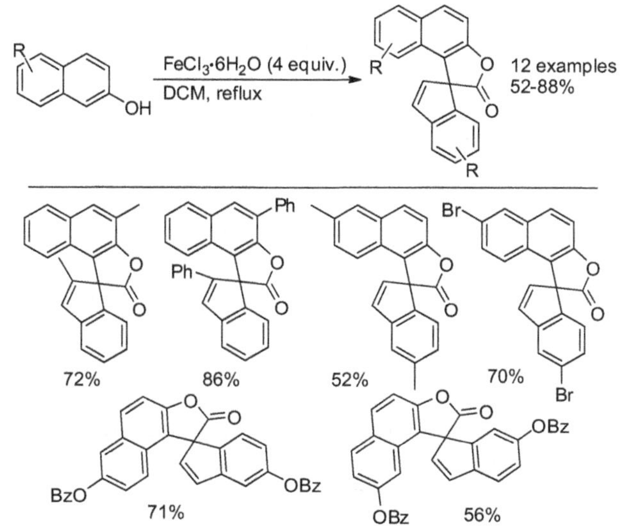

Scheme 3.24 Iron-mediated reaction of 2-naphthols.

Reaction Procedure (Scheme 3.24): A mixture of 2-naphthol (40 mg, 0.277 mmol) and $FeCl_3 \cdot 6H_2O$ (300 mg, 1.11 mmol) in CH_2Cl_2 (3.0 mL) was stirred at reflux temperature for 5 h. The reaction mixture was poured into water–ethyl acetate mixed solvent. The aqueous layer was extracted twice with ethyl acetate. The organic layer was combined, washed with brine and dried over sodium sulfate and evaporated to give a residue. The residue was purified by PTLC (20 : 1 hexane–ethyl acetate) to afford the pure product.

deceleration of side reactions. The wide applicability of this transformation was demonstrated by the various successful reactions between β-keto esters and different phenol derivatives. The coupling was shown to be chemo-, regio- and stereoselective and was successfully applied to the synthesis of the lachnanthospirone central core *via* a possible biomimetic approach. This method can be applied to the synthesis of many valuable phenolic compounds and is likely to find wide utility in the synthesis of natural products.

Eaton and co-workers reported as early as 1993 that allenyl ketones and aldehydes can undergo iron-catalyzed [4 + 1] cycloaddition with carbon monoxide to give α-alkylidenebutenolides (Scheme 3.25).[39] Good control of alkylidene bond stereochemistry was achieved when the terminal allene substituents were methyl and *tert*-butyl groups. Reactions could be performed in a wide variety of solvents, lending to flexibility of the method. The fastest reaction rates were observed in benzene and relative quantum yield experiments suggest a photochemically initiated catalytic [4 + 1] cycloaddition reaction in this solvent. To avoid the formation of the catalytically

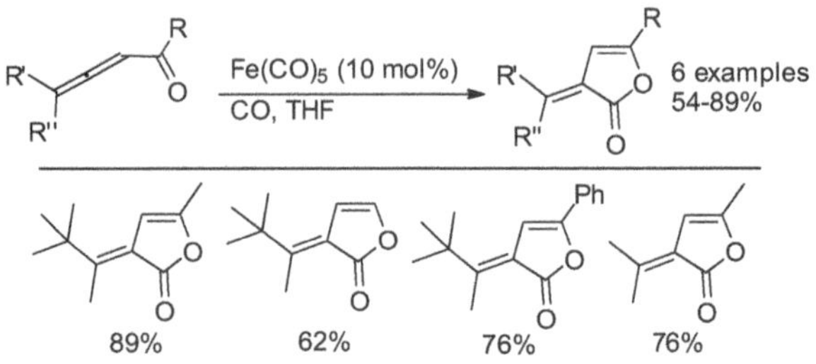

Scheme 3.25 Iron-catalyzed synthesis of α-alkylidenebutenolides.

inactive dinuclear cluster, lower energy irradiation was preferred, giving significantly higher yields of cycloaddition products.

A one pot synthesis of α,β-vinyl esters and alkoxy-substituted γ-lactones was reported by Mathur *et al.*[40] These products were produced by the photochemical reaction of a terminal acetylene (ferrocenyl, phenyl, trimethylsilyl, hexyl and cyclohexyl) with an alcohol (methanol, ethanol and 2-propanol) and carbon monoxide in the presence of iron pentacarbonyl as a catalyst. The selectivity of the compounds depends on the time of photolysis of the reaction and the solvent used. A stable ferrole reaction intermediate was isolated and further photolysis with alcohols resulted in the formation of α,β-vinyl esters.

A mild and efficient method for the preparation of acetonides from epoxides catalyzed by iron(III) chloride was developed in 2008 by Roy and co-workers.[41] In the presence of $FeCl_3$ in acetone at room temperature, epoxides react with acetone to give acetonides in good to excellent yields.

In 2000, an iron-catalyzed intramolecular aminochlorination of alkenes was reported by Bach *et al.*[42] In the presence of $FeCl_2$ with TMSCl in EtOH, 2-alkenyloxycarbonyl azides were converted into the corresponding 4-(chloromethyl)oxazolidinones in good yields (60–84%). The reaction presumably went through a stepwise single-electron transfer (SET) pathway. Soon afterwards, they reported a more systematic study on this topic.[43] They found that the facial diastereoselectivity of the ring-closing C–N bond-forming step is good in both cyclic and acyclic substrates (>90% *ds*). The subsequent chlorine atom transfer occurs selectively in cyclic systems and in systems that exhibit a conformational bias in the postulated radical intermediate. The lifetime τ of this elusive intermediate was estimated from the loss of stereochemical information in conformationally unrestricted systems and from the data obtained with a radical clock. 2-Alkynyloxycarbonyl azides also yield chloroamination products that are obtained exclusively as the *Z*-isomers (81–99% yield). The products of the *tert*-butyl-substituted substrates underwent an immediate rearrangement/solvolysis reaction in the reaction mixture and gave the 5-alkoxyoxazolidinones (93–99% yield). The scope of this methodology was further explored and extended to 2-alkynyloxycarbonyl azides. One-pot reactions starting from the corresponding acid derivatives were also realized. All the reactions gave good to excellent yields (Scheme 3.26).[44]

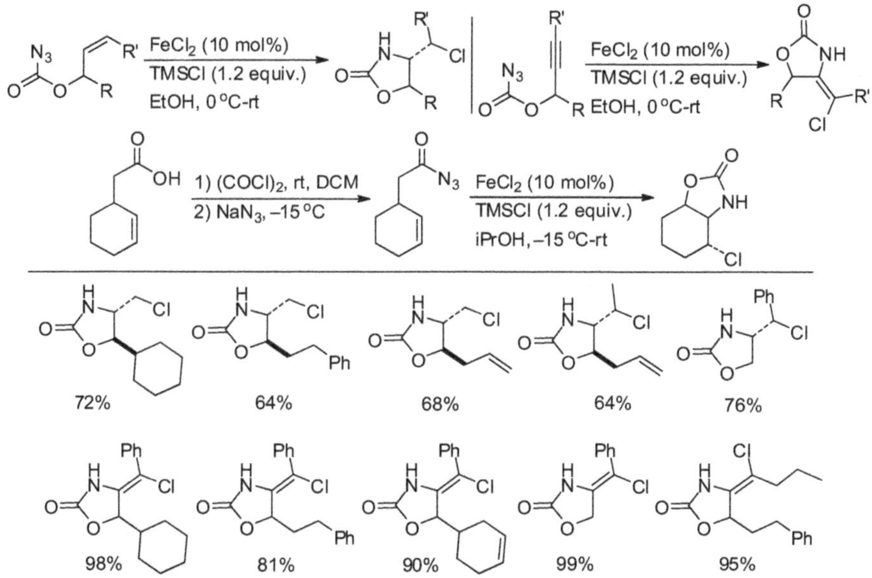

Scheme 3.26   Iron-catalyzed aminochlorination of alkenes and alkynes.

Reaction Procedure (Scheme 3.26): 1-Azidocarbonyloxy-1-cyclohexylprop-2-ene (209 mg, 1.00 mmol) was dissolved in anhydrous EtOH (5 mL) and the solution was degassed with a stream of argon for 15 min at 0 °C. Trimethylsilyl chloride (163 mg, 0.19 mL, 1.50 mmol) was added to the stirred solution *via* a syringe. Solid anhydrous FeCl$_2$ (13 mg, 0.10 mmol) was subsequently added in one portion. The solution was allowed to warm to r.t. during 21 h. EtOAc (10 mL) was added and the resulting solution was washed with water (10 mL) and brine (2×10 mL). The organic layer was dried (MgSO$_4$) and the solvent was removed *in vacuo*. After purification by flash chromatography, the pure product was obtained.

In 2002, an iron-promoted amidochlorination and amidoglycosylation of allal C3-azidoformates was described by Rojas's group.[45] In the presence of alcohols, one-pot β-glycosylation followed. Without added alcohol, amidochlorination occurred using FeCl$_2$, providing an anomeric mixture of glycosyl chlorides that could be used in subsequent silver ion-mediated couplings. The glycosylating agent for the *in situ* FeCl$_2$-promoted reactions may be a 2-amidoglycosyl chloride with a nitrogen-bound iron, which loses its glycosylating ability upon N–O migration of the iron center. On replacing FeCl$_2$ with FeBr$_2$ or FeI$_2$, aminobromination or aminoiodination occurred. The latter reaction was applied in the total synthesis of (−)-agelastatin A.[46]

Yoshimitsu and co-workers developed an intramolecular aminobromination of allyl *N*-tosyloxycarbamates in 2012.[47] β-Brominated oxazolidinones were produced in good yields with FeBr$_2$–n-Bu$_4$NBr as catalyst system in *tert*-BuOH (Scheme 3.27).

More recently, a new iron system was developed for the aminohy-droxylation of alkenes.[48] By using K$_4$Fe(CN)$_6$ as an unusual iron precursor, oxazolidinones were produced in moderate to excellent yields from the corresponding substrates (Scheme 3.28). Based on a preliminary mechanistic

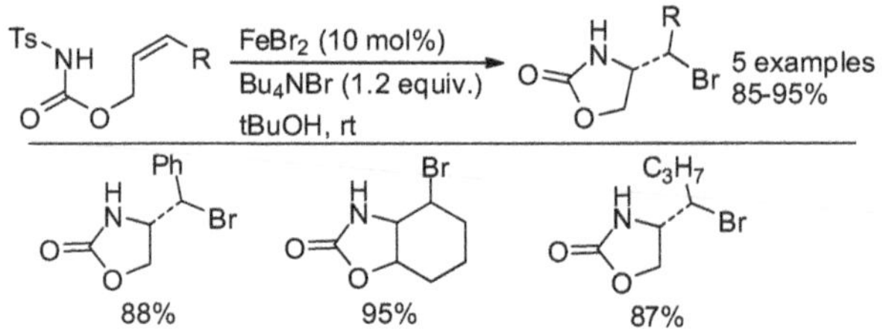

Scheme 3.27   Iron-catalyzed aminobromination of allyl *N*-tosyloxycarbamates.

Reaction Procedure (Scheme 3.27): n-Bu$_4$NBr (109 mg, 0.34 mmol) and FeBr$_2$ (6.2 mg, 0.028 mmol) were added to a stirred solution of *N*-tosy-loxycarbamate (0.28 mmol) in t-BuOH (5 mL) at room temperature. After sonication for 3 min, the mixture was stirred at room temperature for a further 4 h. The mixture was transferred to a separating funnel, where it was partitioned between H$_2$O and EtOAc. The organic phase was washed with brine, dried over MgSO$_4$, filtered and concentrated under reduced pressure. The residue was purified by silica gel flash column chroma-tography (2 : 5–2 : 3 EtOAc–n-hexane) to give the pure product.

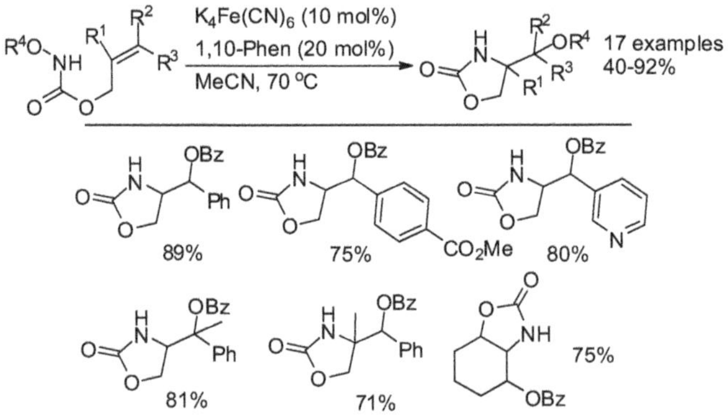

Scheme 3.28   Iron-catalyzed synthesis of oxazolidinones.

Reaction Procedure (Scheme 3.28): To a flame-dried Schlenk tube were added K$_4$Fe(CN)$_6$ (0.02 mmol) and phenanthroline (0.04 mmol). After evacuation with an oil pump and refilling with Ar gas three times, anhydrous and degassed CH$_3$CN (3 mL) was added. The suspension

> obtained was stirred vigorously at room temperature for 30 min, after which the substrate (0.2 mmol) was added. The reaction mixture was heated at 70 °C until all the starting material was fully consumed. The reaction mixture was cooled and concentrated to a residue, which was chromatographed using 1 : 1 hexane–ethyl acetate as the eluent to give the designed product.

study, the authors proposed an iron nitrenoid as the possible intermediate, which can undergo either aminohydroxylation or aziridination by careful selection of the counteranion and ligand combination.

Che and co-workers prepared a non-heme iron complex and applied it in amination reactions.[49] They found that the newly prepared non-heme iron complex [Fe(qpy)(MeCN)$_2$](ClO$_4$)$_2$ is an active catalyst for the intra- and intermolecular amination of various C(sp^3)–H bonds, including the amination of cyclic alkanes and of cycloalkane/linear alkane moieties in saturated steroids (Table 3.2). The substrate scope spans cycloalkanes, benzylic/allylic hydrocarbons, natural products α-pinene and β-pinene and a wide variety of sulfamate esters including those bearing methylcyclohexane, isononane, menthane, spirost-5-ene, cholane and androstane backbones. The amination reactions can be performed by employing the cyclic alkane substrates as limiting reagents with PhI = NR or 'PhI(OAc)$_2$ + H$_2$NR' (R = Ts, Ns) as nitrogen source. The interesting catalytic activity of the '[Fe] + PhI = NR' system likely arises from the generation of reactive cationic seven-coordinate iron–imide/nitrene intermediate(s) [Fe(qpy)(NR)(X)]$_n$ + (CX, X = NR, solvent or anion) as proposed on the basis of experimental studies (including ESI-MS analysis) and DFT calculations. The DFT-optimized ground states of these proposed reactive iron–imide/nitrene species, such as CNTs, feature imide ligand(s) with radical nitrene characteristics. This work provided unique examples of non-heme iron-catalyzed amination of cyclic alkanes and steroid compounds and lends credence to the possible use of non-heme iron complexes as practical catalysts for C(sp^3)–H amination reactions.

Additionally, an iron-catalyzed cycloaddition of aziridine with heterocumulenes using water as solvent was reported in 2013.[50] Isoselenocyanates and isocyanates were applied as reaction partners together with aziridines. The products were formed in good to excellent yields (Scheme 3.29). Organic solvents, such as toluene, DCM and DCE, were also tested as media for this transformation, but no desired product was formed.

Oxazole and oxazoline derivatives have important applications in biologically active compounds and are also applied as useful reagent/intermediates in organic synthesis. In 2012, an iron-mediated synthesis of oxazoles from the corresponding acetylenic amides was reported by Wang and co-workers (Scheme 3.30).[51] Here, iron acts as a Lewis acid that promotes the tautomerization of acetylenic amide by 5-*exo-dig* cyclization to produce the key intermediate. An internal proton transfer from the key intermediate resulted in the formation of the required oxazole derivatives.

Reaction Procedure (Table 3.2): PhI(OAc)$_2$ (90 mg, 0.28 mmol) and MgO (18.4 mg, 0.46 mmol) were added to a solution of sulfamate ester (0.2 mmol) and [Fe] (7.2 mg, 0.01 mmol) in anhydrous MeCN(2 mL) at 40 or 80 °C under argon. The mixture was stirred for 12 h, then diluted with CH$_2$Cl$_2$ (10 mL) and filtered through Celite. The residue on the Celite was washed with CH$_2$Cl$_2$ (2×5 mL). The filtrate was evaporated to dryness under reduced pressure. The residue was purified by flash column chromatography on silica gel with hexane–EtOAc as eluent.

Table 3.2   Iron-catalyzed amination of C(sp^3)–H.

Entry	Product	Yield (%)
1		84
2		87
3		87
4		87
5		81
6		85

Table 3.2 (*Continued*)

Entry	Product	Yield (%)
7	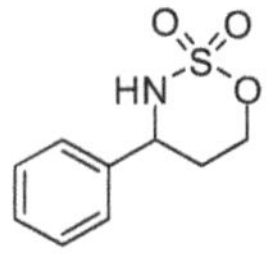	82
8	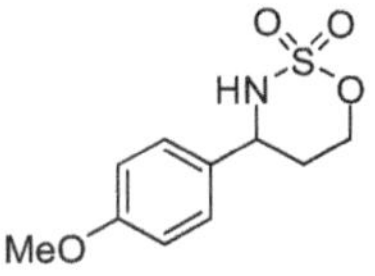	92
9	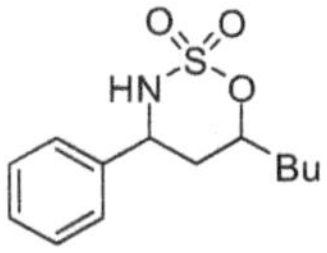	89
10	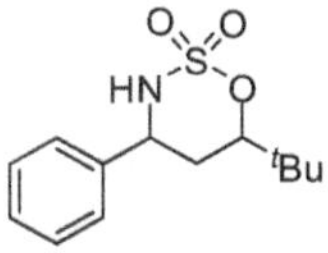	85
11	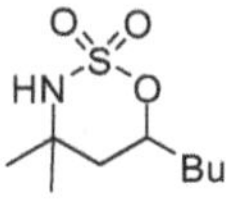	86
12	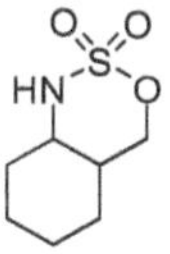	79

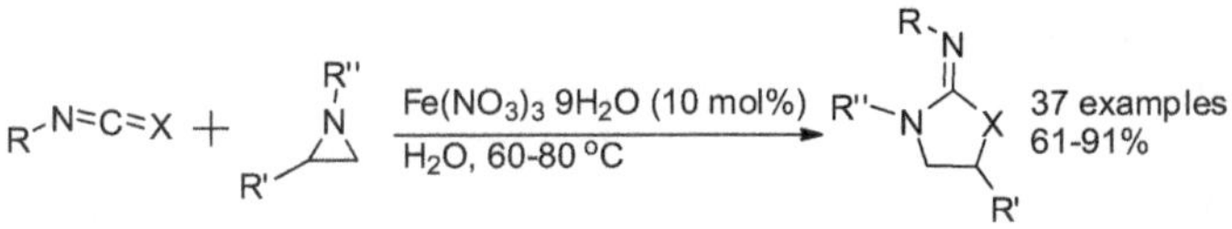

Scheme 3.29 Iron-catalyzed reaction of aziridines with heterocumulenes.

An iron-catalyzed synthesis of isoxazoles and isoxazolines from propargylic alcohols was developed in 2010.[52] Gold was applied as co-catalyst for this transformation. By using *N*-sulfonyl-protected hydroxylamines as binucleophiles, good yields of the desired products were obtained. This concept was also applied in the synthesis of thiazoles.[53] Starting with the reaction of propargylic alcohols and primary amides, followed by the addition of Lawesson's reagent, three substituted thiazoles were produced in

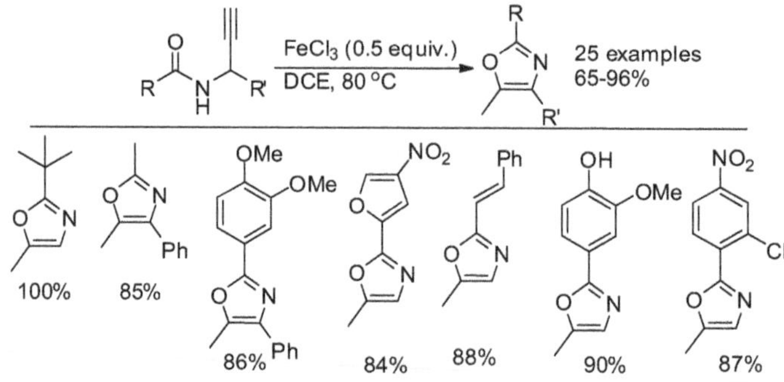

Scheme 3.30   Iron-mediated synthesis of oxazoles.

Reaction Procedure (Scheme 3.30): To a cooled solution of an appropriate propargylamine (1.3 equiv.) in CH_2Cl_2 (5 mL) were added triethylamine (3 equiv.) and an appropriate acid chloride (1 mmol) and the resulting solution was allowed to reach room temperature. Completion of the reaction was monitored by the disappearance of acid chloride by TLC. The mixture was successively diluted with water. The aqueous layer was extracted with CH_2Cl_2 (3×10 mL) and the combined organic layers were washed with saturated $NaHCO_3$ followed by water and brine, dried over Na_2SO_4 and concentrated under reduced pressure to obtain the crude alkynylamides. The resulting crude compound was dissolved in 1,2-DCE followed by the addition of $FeCl_3$ (0.5 equiv.) and stirred at 80 °C for 3–12 h until the complet consumption of starting material by TLC. The reaction mixture was diluted with water. The aqueous layer was extracted with CH_2Cl_2 (3×10 mL) and the combined organic layers were washed with water and brine, dried over Na_2SO_4 and concentrated under reduced pressure. Silica gel chromatography (ethyl acetate–hexane) gave the desired oxazole derivatives.

---

moderate yields. Alternatively, isoxazole derivatives can also be produced from alkynone O-methyloximes using iron as catalyst.[54] This procedure offers a reaction pathway to 4-organoselenylisoxazoles from alkynone O-methyloximes *via* $FeCl_3$-mediated intramolecular cyclization in the presence of substituted diorganyl diselenides (Scheme 3.31). The cyclization protocol is straightforward and allows the construction of highly functionalized isoxazole derivatives in moderate to good yields. The authors demonstrated three main advantages of this methodology: a short reaction time was required and the reactions were carried out under ambient atmosphere and had atom economy (the two PhSe groups from PhSeSePh are incorporated into the isoxazole ring). In addition, the 4-organoselenylisoxazole obtained in the course of this work proved to be convenient as a substrate for the preparation of more functionalized isoxazole derivatives, becoming a promising alternative to the construction of heterocycle libraries.

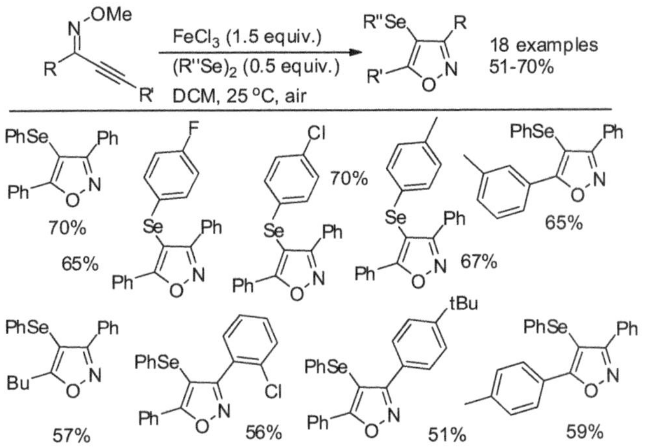

Scheme 3.31 Iron-mediated synthesis of isoxazoles.

Reaction Procedure (Scheme 3.31): In a Schlenk flask, under ambient atmosphere, containing CH_2Cl_2 (1.5 mL) were added $FeCl_3$ (0.061 g, 1.5 equiv.) and the appropriate diorganyl diselenide (0.5 equiv.). The reaction mixture was stirred for 20 min at room temperature. The corresponding alkynone O-methyloxime (0.25 mmol) was then added, diluted in CH_2Cl_2 (1 mL), and the reaction mixture was stirred at room temperature for 5 min, then diluted with CH_2Cl_2 (20 mL) and washed with a saturated aqueous solution of NH_4Cl (3×10 mL). The organic phase was separated, dried over $MgSO_4$ and concentrated under vacuum. The residue was purified by flash chromatography using hexane–ethyl acetate (95:5) as eluent.

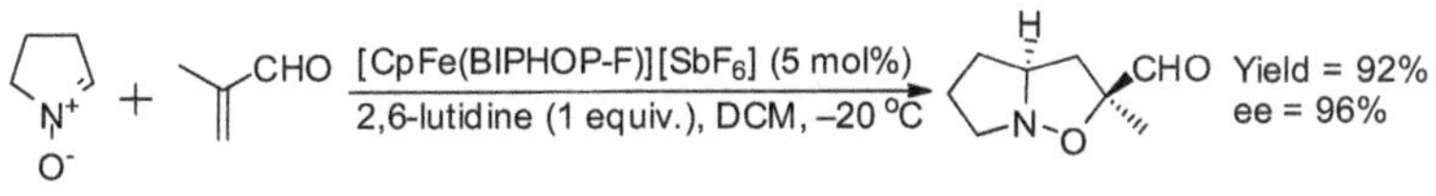

Scheme 3.32 Iron-catalyzed 1,3-cycloaddition.

Isoxazolidines are precursors of 1,3-amino alcohols; enantiomerically pure ones are even more important. In 2002, an iron-catalyzed asymmetric 1,3-dipolar cycloaddition between nitrones and enals was reported by Kündig and co-workers (Scheme 3.32).[55] The corresponding ruthenium complex is also active for this transformation. With [CpFe(BIPHOP-F)][SbF₆] as catalyst, isoxazolidines were produced in good yields and high enantioselectivity.

Additionally, oxazolidines as precursors for 1,2-amino alcohols were also prepared using an iron catalyst. In 2010, Yoon's group reported an Fe(acac)₃-catalyzed aminohydroxylation of alkenes to give oxazolidines.[56] By using the simple iron salt as catalyst, oxazolidines were produced in good yields and

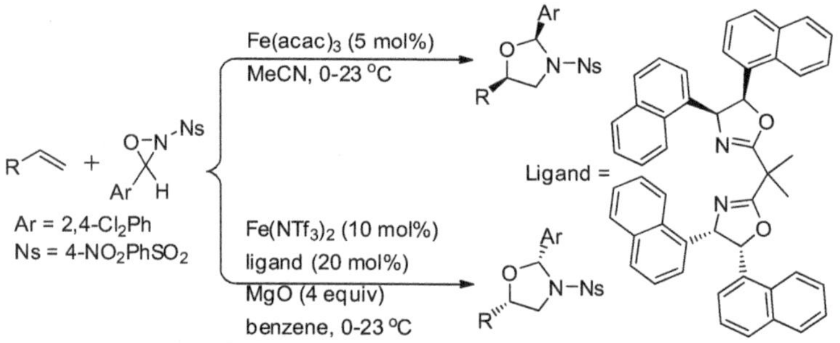

Scheme 3.33   Iron-catalyzed aminohydroxylation of alkenes.

were subsequently transformed into the corresponding 1,2-amino alcohols. (±)-Octopamine as a natural product was also prepared by this methodology. Later, the same group reported another iron catalyst system for this transformation with opposite regioselectivity.[57] With the assistance of a bis(oxazoline) ligand, the reactions proceeded in a highly enantioselective manner (Scheme 3.33).

Tetrazoles are important heterocycles for explosives, photographic agents and pharmaceuticals. The reaction between nitriles and azides constitutes a straightforward approach for the preparation of 5-substituted 1H-tetrazoles. In 2009, Bolm's group reported an iron-catalyzed version of the reaction between nitriles and azides.[58] In the presence of Fe(OAc)$_2$ (10 mol%) in DMF–H$_2$O at 80 °C, 5-substituted 1H-tetrazoles were produced in good yields using TMSN$_3$ as azide source (Scheme 3.34). Later, several heterogeneous iron catalysts were developed and sodium azide was applied instead of TMSN$_3$.[59]

A novel iron-catalyzed aminolysis of β-carbonyl 1,3-dithianes with various amines was reported in 2011 by Bi and co-workers.[60] This methodology can lead to the synthesis of stereodefined β-enaminones and 3,4-disubstituted pyrazoles (Scheme 3.35). This catalytic procedure is striking in terms of the wide range of applicable substrates, mild reaction conditions and excellent stereoselectivity.

3.1.7   Synthesis of Benzo-Fused Five-Membered Heterocycles

Benzofuran derivatives are important heterocycles because of their wide occurrence in natural products, broad range of biological activities and significant pharmaceutical potential. In 2009, a novel synthetic approach to 3-functionalized benzo[b]furan derivatives using FeCl$_3$ as promoter was demonstrated.[61] Starting from readily available α-aryl ketones, various benzofurans were produced in moderate to excellent yields (Scheme 3.36). This interesting FeCl$_3$-mediated ring closure of the electron-rich α-aryl ketones realized the construction of benzo[b]furan rings by joining the O-atom on the side chain to the benzene ring *via* direct oxidative aromatic C–O bond formation.

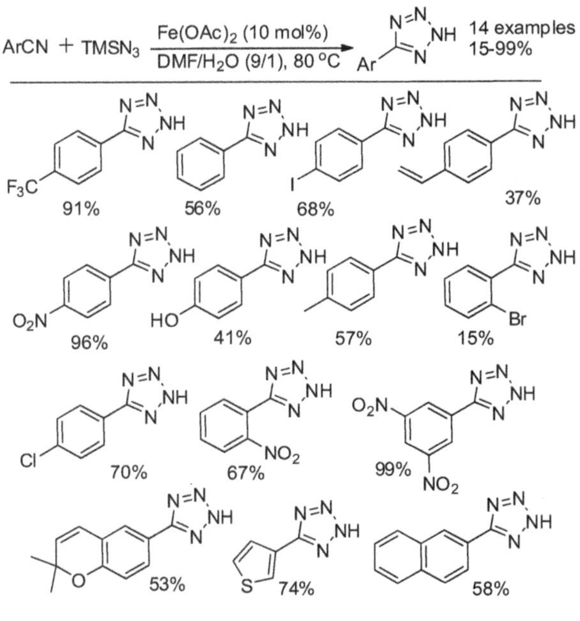

Scheme 3.34 Iron-catalyzed reaction of nitriles with azides.

Reaction Procedure (Scheme 3.34): A sealable tube equipped with a magnetic stir bar was charged with the aryl nitrile (1.0 equiv.) and Fe(OAc)$_2$ (0.1 equiv.). A rubber septum was used to cover the aperture of the tube, an argon atmosphere was established, and trimethylsilyl azide (1.5 equiv.) and 9:1 DMF–MeOH solution (1 mL) were added by using a syringe. The rubber septum was then replaced with a Teflon-coated screw cap and the reaction vessel was heated at 80 °C. After stirring at this temperature for 24 h, the mixture was cooled to room temperature and diluted with ethyl acetate. The resulting solution was washed with 1 M HCl, dried over anhydrous Na$_2$SO$_4$ and concentrated. An aqueous solution of NaOH (0.25 M) was added to the residue and the mixture was stirred for 30 min at room temperature. The resulting solution was washed with ethyl acetate, then 1 M HCl was added until the pH of the water layer became 1. The aqueous layer was extracted three times with ethyl acetate and the combined organic layers were washed with 1 M HCl. The organic layer was dried over anhydrous Na$_2$SO$_4$ and concentrated.

Li and co-workers demonstrated a novel and regioselective method for the construction of polysubstituted benzofurans in 2009 (Scheme 3.37).[62] The combination of FeCl$_3$ · 6H$_2$O and (*t*-BuO)$_2$ is an effective system in the reactions of simple phenol derivatives and β-keto esters. The effect of water in the iron-catalyzed oxidative reaction is interesting. Control experiments showed that not only water but also various alcohols and protic acids could accelerate the

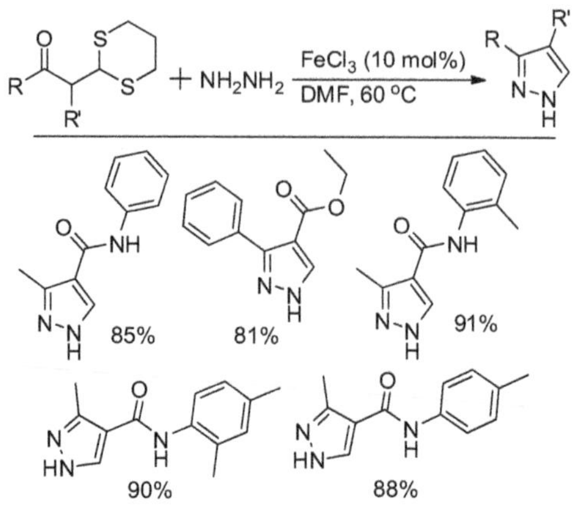

Scheme 3.35   Iron-catalyzed synthesis of pyrazoles.

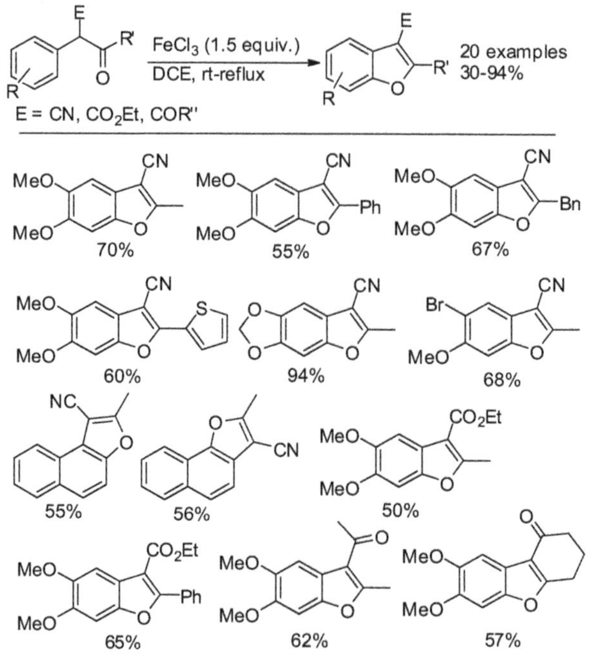

Scheme 3.36   Iron-mediated synthesis of benzofurans.

Reaction Procedure (Scheme 3.36): To a solution of α-aryl ketones (2.5 mmol) in dried 1,2-dichloroethane (30 mL) was added iron(III) chloride (6.25 mmol) in one portion with stirring at the designated temperature, and the reaction progress was monitored by TLC. The reaction mixture was then evaporated *in vacuo* to remove the solvent. The residue was purified by silica gel chromatography using petroleum ether–ethyl acetate as eluent to give the desired products.

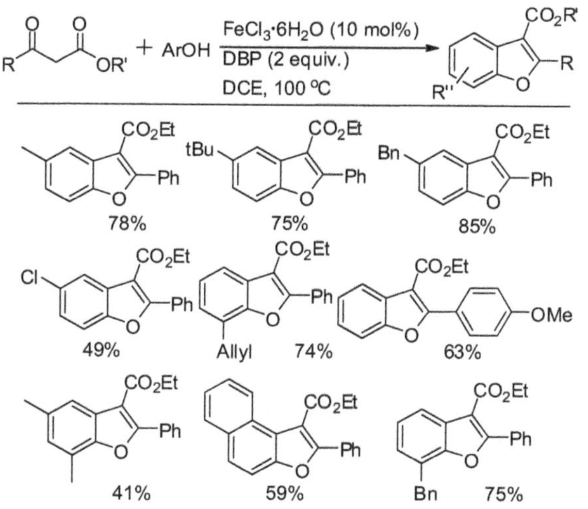

Scheme 3.37 Iron-mediated synthesis of polysubstituted benzofurans.

Reaction Procedure (Scheme 3.37): To a mixture of ethyl benzoylacetate (0.5 mmol), phenol (1.5 mmol), and $FeCl_3 \cdot 6H_2O$ (0.05 mmol) was added 1,2-dichloroethane (1.0 mL) under nitrogen at room temperature. Then di-*tert*-butyl peroxide (1.0 mmol) was dropped into the mixture under nitrogen. The reaction temperature was raised to 100 °C for 1 h, then decreased to room temperature. The resulting reaction solution was quenched with 2 mL of saturated $NaHCO_3$ and extracted four times with 15 mL of diethyl ether. The extract was washed twice with 10 mL of saturated $NaHCO_3$ and twice with 10 mL of deionized water. The extract was dried over $MgSO_4$. The solvent was evaporated *in vacuo* to afford the crude products and the residue was purified by flash column chromatography on silica gel with ethyl acetate–petroleum ether (1 : 100) as eluent.

reactions. These oxidative reactions are chemo- and regiospecific, which was also confirmed by X-ray diffraction. Kinetic isotopic effect (KIE) experiments gave $k_H/k_D = 1.0 \pm 0.1$. The isotopic effect indicates that aromatic C–H bond cleavage is a fast step and not involved in the rate-determining steps in this transformation. In addition, two possible intermediates were synthesized, which were transformed into the desired benzofuran under standard reaction conditions. This is a novel oxidative Pechmann-type condensation, in which benzofuran is generated instead of coumarin. Iron catalysts together with organic peroxides have been demonstrated to be very efficient in oxidative C–C bond formation. These results clearly demonstrate the dichotomous catalytic behavior of the iron catalysts, which are transition metal catalysts in the oxidative coupling step and Lewis acids in the condensation step.

A general synthesis of 3-chalcogen benzo[*b*]furans from the readily available 2-alkynylanisoles, *via* $FeCl_3$–diorganyl dichalcogenide intramolecular cyclization, was developed by Zeni and co-workers in 2010.[63] Aryl and alkyl groups directly bonded to the chalcogen atom were used as cyclization agents. The results revealed that the reaction depends significantly on the electronic effects of substituents in the aromatic ring bonded to the selenium atom of the diselenide species. The pathway of reaction was not sensitive to the nature of the substituents in the aromatic ring of anisole since both the electron-donating and the electron-withdrawing groups delivered the products in similar yields (Scheme 3.38). In addition, the heterocycles obtained were readily converted to more complex products by using a chalcogen–lithium exchange reaction with n-BuLi followed by trapping of the lithium intermediate with aldehydes, furnishing the desired secondary alcohols in good yields.

A concise and efficient approach to the syntheses of coumestan analogs was developed in 2011.[64] The strategy involves an FeCl$_3$-mediated direct intramolecular oxidative annellation of 4-hydroxy-3-phenyl-2H-chromen-2-one derivatives. Utilizing this synthetic protocol, a variety of coumestan derivatives were conveniently obtained from readily available reagents (Scheme 3.39).

A radical initiated cascade reaction to dihydrobenzofurans was reported in 2013 (Scheme 3.40).[65] Starting with a 5- or 6-exo-type cyclization of an aryl

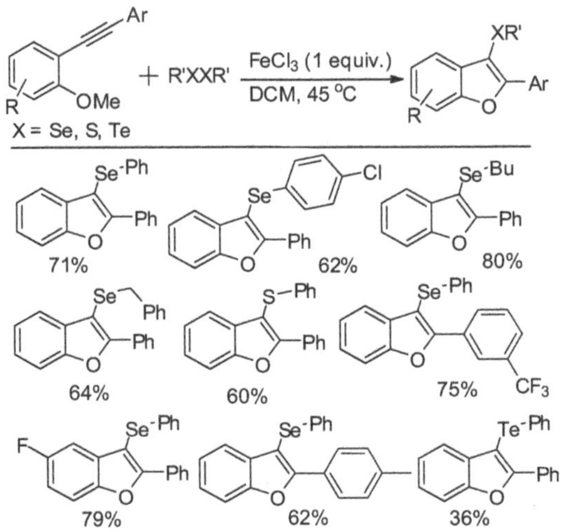

Scheme 3.38 Iron-mediated synthesis of 3-chalcogen benzo[b]furans.

Reaction Procedure (Scheme 3.38): To a two-necked round-bottomed flask equipped with a reflux condenser, under argon, containing a solution of FeCl$_3$ (0.081 g, 0.5 mmol) in CH$_2$Cl$_2$ (3 mL) was added 2-alkynylanisole (0.5 mmol) in CH$_2$Cl$_2$ (2 mL). The reaction mixture was stirred for the desired time at 45 °C. The mixture was then diluted with CH$_2$Cl$_2$ (20 mL) and washed with a saturated solution of NH$_4$Cl (20 mL). The organic phase was separated, dried over MgSO$_4$ and concentrated under vacuum. The residue was purified by flash chromatography with hexane as eluent.

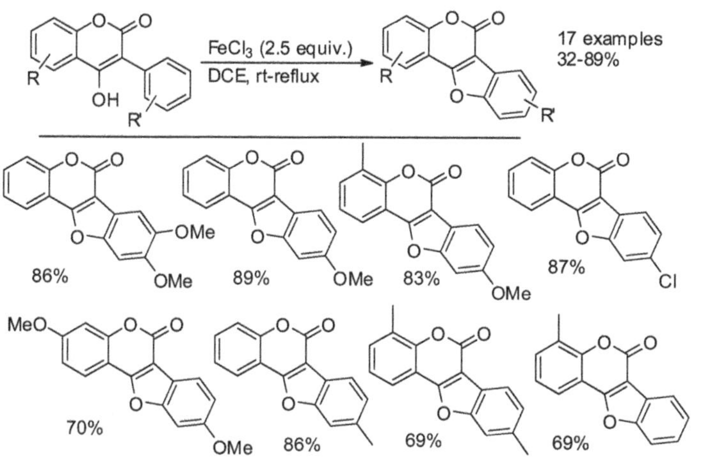

Scheme 3.39 Iron-mediated synthesis of coumestans.

Reaction Procedure (Scheme 3.39): To a solution of substituted 4-hydroxy-coumarin (2.5 mmol) in dried 1,2-CH$_2$Cl$_2$ (30 mL) was added FeCl$_3$–SiO$_2$ (50:50 w/w, 6.25 mmol) in one portion with stirring at ambient or reflux temperature, and the reaction process was monitored by TLC. The reaction mixture was then evaporated under vacuum to remove the solvent. The residue was purified by silica gel chromatography to give the desired products.

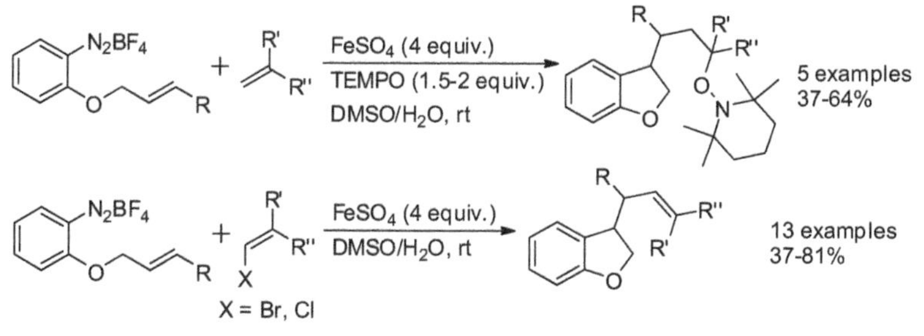

Scheme 3.40 Iron-mediated synthesis of dihydrobenzofurans.

Reaction Procedure (Scheme 3.40): A suspension of FeSO$_4$·7H$_2$O (4.0 equiv., 2.00 mmol), alkene (15–25 equiv.), and TEMPO (1.5–2.0 equiv.) in DMSO–H$_2$O (95:5, 1.0 mL) was degassed with argon for 15 min. Then aryldiazonium tetrafluoroborate (1.0 equiv., 0.50 mmol) dissolved in de-gassed DMSO–H$_2$O (95:5, 0.6 mL) was added by syringe pump over a period of 30 min. The mixture was stirred for a further 30 min and then ascorbic acid (3.0 equiv.) was added and stirring was continued for 15 min.

Water (30 mL) was added followed by extraction with Et$_2$O (3 × 30 mL). The combined organic phases were washed with brine and dried over Na$_2$SO$_4$. Removal of the solvents under reduced pressure and purification by column chromatography on silica gel gave the desired products.

Reaction Procedure: A suspension of FeSO$_4$ · 7H$_2$O (556 mg, 2.00 mmol) and the alkene (5.00 mmol) in DMSO–H$_2$O (95 : 5, 2 mL) was degassed with argon for 15 min. Then aryldiazonium tetrafluoroborate (0.50 mmol) dissolved in degassed DMSO–H$_2$O (95 : 5, 0.6 mL) was added by syringe pump over a period of 60 min. The mixture was stirred for a further 30 min and then water (30 mL) was added followed by extraction with Et$_2$O (3 × 30 mL). The combined organic phases were washed with brine, dried over Na$_2$SO$_4$, and the solvent was carefully evaporated (**Caution:** the product may be volatile). The crude products were purified by column chromatography.

radical on to a non-activated alkene gives versatile tools for the fast assembly of diverse products from readily available precursors. Such cyclizations can easily be combined with carbohydroxylation, vinylation and allylation reactions when diazonium salts are used as aryl radical precursors in the presence of iron(II) sulfate as the reductant. The products, accessible through a cascade reaction in which two carbon–carbon bonds are formed, offer many further options for modification and thus represent valuable synthetic intermediates.

Benzoxazoles have been reported with important biological and therapeutic activities that have applications in drug synthesis and have wide occurrence in natural products. In 2008, a practical iron-catalyzed intramolecular *O*-arylation for the preparation of 2-arylbenzoxazoles was developed.[66] Starting from readily available 2-haloanilines, various benzoxazoles were prepared in moderate to excellent yields (Scheme 3.41). The combination of the cheap and environmentally friendly FeCl$_3$ and 2,2,6,6-tetramethyl-3,5 heptanedione (TMHD) as the catalyst system increases the potential for large scale application of this methodology.

2-Substituted benzoxazoles can be prepared from 2-aminophenols and aldehydes using an iron catalyst. In 2010, a practical FeCl$_3$-catalyzed aerobic oxidation for the synthesis of benzoxazoles, benzothiazoles and benzimidazoles was developed by Zhang and co-workers.[67] The desired products were produced in good yields (Scheme 3.42). JTP-426467, as a selective antagonist for peroxisome proliferator-activated receptor, was synthesized by this methodology from ready available starting materials.

Alternatively, another iron-catalyzed system starting from *O*-nitrophenols and benzylic alcohols was developed in 2012,[68] and various 2-arylbenzoxazoles were selectively obtained in good to excellent yields (Scheme 3.43). Functional groups such as methyl, methoxy, fluoro, chloro and bromo were all well tolerated under the optimized reaction conditions. 2-Aminophenols

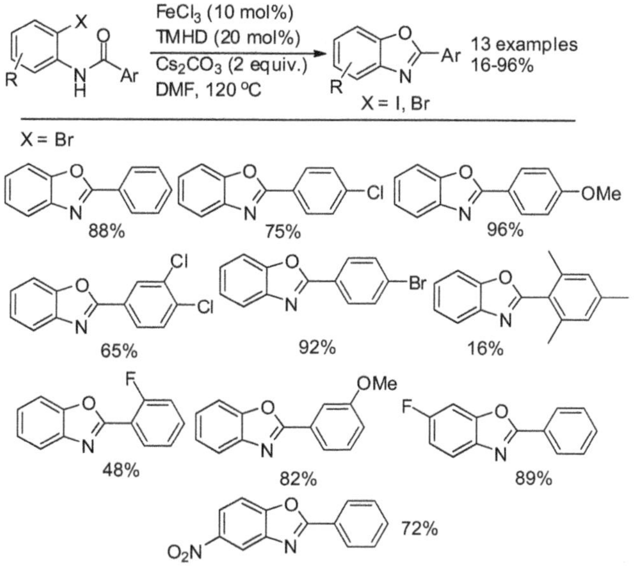

Scheme 3.41 Iron-catalyzed synthesis of 2-arylbenzoxazoles.

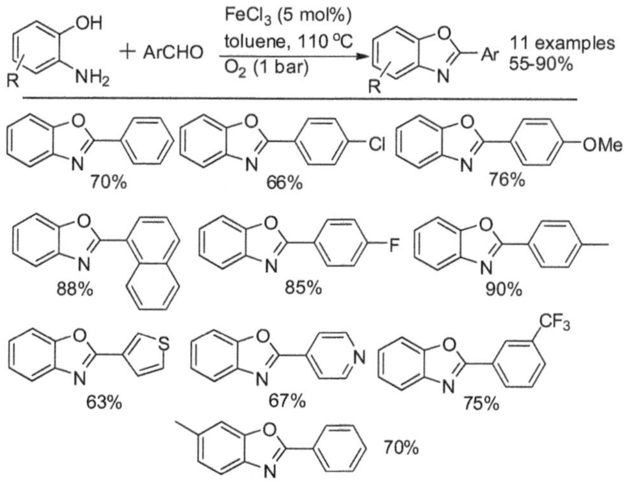

Scheme 3.42 Iron-catalyzed synthesis of benzoxazoles from 2-aminophenols.

and benzaldehydes were produced *in situ* and the alcohol acted as both coupling reagent and reductant. Alcohol oxidation, nitro reduction, heterocycle formation and heterocycle oxidation were realized in a cascade manner.

A ferric perchlorate-mediated reaction of C_{60} with various nitriles in *o*-dichlorobenzene under a nitrogen atmosphere affording the rare fullerooxazoles was developed by Wang and co-workers.[69] In the presence of

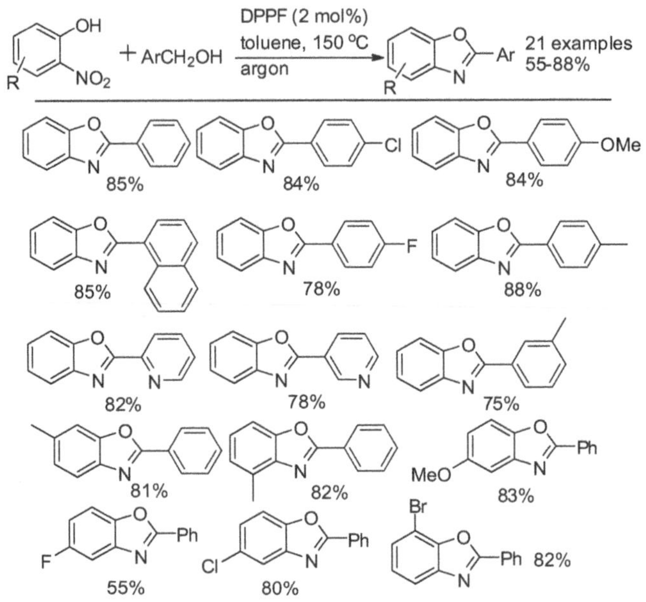

Scheme 3.43   Iron-catalyzed synthesis of benzoxazoles from O-nitrophenols.

an equimolar amount of $Fe(ClO_4)_3 \cdot 6H_2O$, eight examples of fullerooxazoles were produced in 41–84% yields.

A novel and efficient synthesis of pyrido[1,2-a]benzimidazoles through direct intramolecular aromatic C–H amination of N-aryl-2-aminopyridines was reported in 2010.[70] The reaction, co-catalyzed by $Cu(OAc)_2$ and $Fe(NO_3)_3 \cdot 9H_2O$, was carried out in DMF under a dioxygen atmosphere. Diversified pyrido[1,2-a]benzimidazoles containing various substitution patterns were obtained in moderate to excellent yields by using this procedure (Scheme 3.44). However, electron-withdrawing substituents in the *meta* position of the aniline ring and any position of the pyridine ring are unfavorable. The results of mechanistic studies suggested that a Cu(III)-catalyzed electrophilic aromatic substitution (S_EAr) pathway is operating in this process. The unique role of iron(III) is believed to lie in its ability to facilitate the formation of the more electrophilic Cu(III) species. In the absence of iron(III), a much less efficient and reversible Cu(II)-mediated S_EAr process takes place. In this process, the pyridinyl nitrogen in the substrates acts as both a directing group and nucleophile.

A novel iron(III)-catalyzed one-pot cascade reaction between nitroalkenes and 2-aminopyridines to give imidazo[1,2-a]pyridines was demonstrated in 2013.[71] The bielectrophilic nature of nitroalkenes was explored by this methodology and the desired products were produced in moderate to good yields (Scheme 3.45). This procedure could be successfully applied for the synthesis of zolimidine, a useful drug for the treatment of peptic ulcers. The reaction proceeds through Michael addition followed by intramolecular

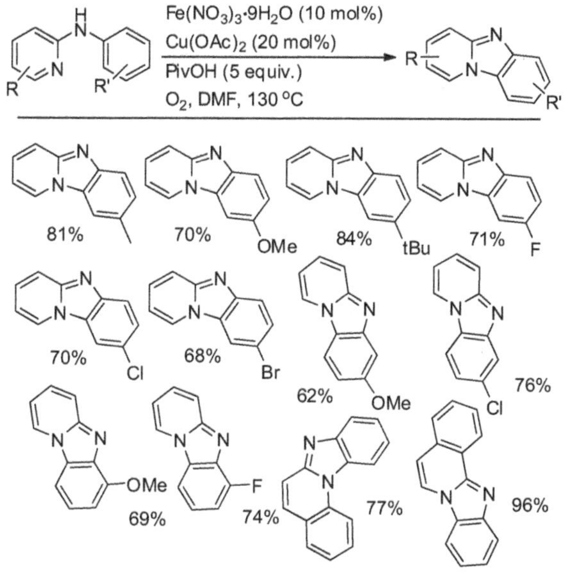

Scheme 3.44 Iron-catalyzed synthesis of pyrido[1,2-*a*]benzimidazoles.

Reaction Procedure (Scheme 3.44): A mixture of *N*-aryl-2-aminopyridine (0.5 mmol), Cu(OAc)$_2$ (20 or 100 mol%), Fe(NO$_3$)$_3$·9H$_2$O (10 mol%) and PivOH (2.5 mmol) in DMF (1.0 mL) was stirred at 130 °C under a balloon pressure of O$_2$. The reaction was cooled to room temperature after complete consumption of starting material as monitored by TLC. Water (10 mL), triethylamine (1.0 mL), and EtOAc (10 mL) were added successively to the reaction mixture. The organic phase was separated and the aqueous phase was further extracted with EtOAc (3×10 mL). The combined organic layers were dried over anhydrous Na$_2$SO$_4$, concentrated and purified by flash chromatography.

cyclization and *in situ* denitration. Some other common Lewis acids, such as In(OTf)$_3$, CuI, Cu(OAc)$_2$, Cu(OTf)$_2$, AlCl$_3$, ZnCl$_2$, LaCl$_3$ and BF$_3$·Et$_2$O, were also tested for this conversion, but low or no yields of the desired products were obtained.

Liu and co-workers reported a Cu(II) and Fe(III) co-catalyzed diamination of 2-aminopyridines and 2-aminoisoquinolines with readily available alkynes.[72] The strategy allows the direct synthesis of imidazo[1,2-*a*]pyridines and imidazo[1,2-*a*]isoquinolines in yields of up to 92% (Scheme 3.46). These structures are ubiquitously found in a large variety of compounds which possess important pharmacological properties that are essential to the biopharmaceutical and chemical sectors. This methodology provides simple, facile and straightforward access to an extensive array of compounds and synergizes the well-explored intramolecular diamination of alkenes to

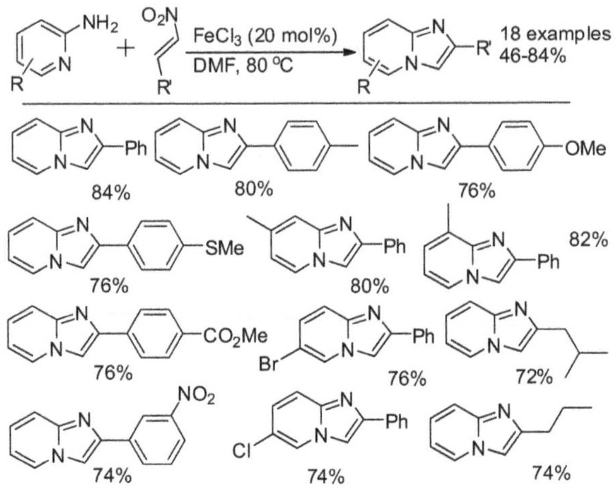

Scheme 3.45 Iron-catalyzed synthesis of pyrido[1,2-*a*]benzimidazoles from nitroalkenes.

Reaction Procedure (Scheme 3.45): A mixture of 2-aminopyridine (112 mg, 1.2 mmol) and nitroalkene (149 mg, 1 mmol) was stirred in the presence of anhydrous $FeCl_3$ (20 mol%) in DMF (2 mL) at 80 °C for 2 h (monitored by TLC). After completion, the reaction mixture was cooled to room temperature and extracted with CH_2Cl_2 (10 mL) followed by washing with brine (5 mL) and drying over Na_2SO_4. After evaporation of the solvent, the crude product was purified by column chromatography on silica gel using petroleum ether–ethyl acetate (3 : 1–2 : 1) as eluent.

alkynes in an intermolecular manner. For aminopyridines and aminoquinolines, high levels of chemo- and regioselectivity of the two nitrogens were demonstrated for a range of compounds, demonstrating the flexibility and good reactivity of this strategy. Concerning the reaction mechanism, the coordination of the endocyclic nitrogen atom of the 2-aminopyridine to the copper center was proposed for the first stage.

Maes *et al.* developed a sustainable direct amination protocol for the synthesis of substituted pyrido[1,2-*e*]purines.[73] Preactivation of nitrogen was not necessary and O_2 was used as the oxidant with iron as the catalyst; all of the desired products were isolated in moderate to excellent yields (Scheme 3.47). This method has excellent functional-group compatibility and the chemoselectivity towards halogens will allow post-functionalization in the annulated ring. A radical mechanism was proposed for this interesting transformation.

Indoles are ubiquitous substructures of natural or synthetic biologically active products. In 2007, a new $FeCl_3$–$PdCl_2$ catalytic combination for the preparation of indoles by annulation of the parent alkynylanilines was developed (Scheme 3.48).[74] High yields were obtained by using low loadings of

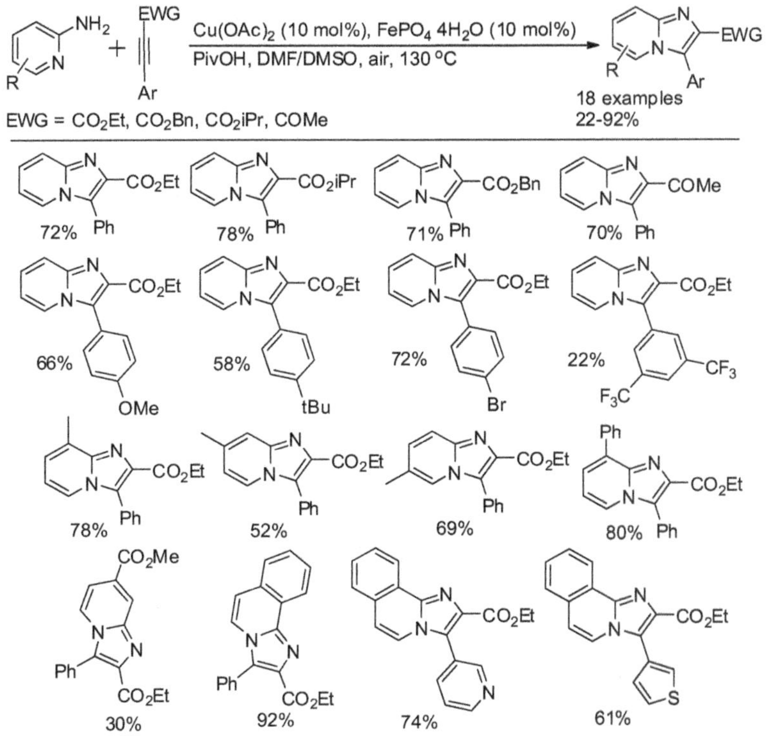

Scheme 3.46 Iron-catalyzed synthesis of pyrido[1,2-*a*]benzimidazoles from alkynes.

the transition metal complex (FeCl$_3$–PdCl$_2$, 2 and 1 mol%, respectively). Remarkably, only a trace of product was formed with palladium alone and no product was obtained with FeCl$_3$ alone as the catalyst. One-pot routes to bis(indolyl)methanes and trisubstituted indoles through annulation–Friedel–Crafts alkylation and annulation–1,4-Michael addition sequences, in which FeCl$_3$ acts as both a co-oxidant and a Lewis acid, were described. Cesium carbonate-promoted FeCl$_3$-assisted cyclization of *o*-alkynylanilides to indoles was also reported.[75] Increased yields of indoles can be produced by adding 10 mol% of FeCl$_3 \cdot$ 6H$_2$O to the reaction mixture.

An iron trichloride-promoted cyclization of *o*-alkynylaryl isocyanates was developed by Cossy and co-workers in 2009.[76] This reaction follows a cationic reaction pathway (Scheme 3.49). 2-(Arylethynyl)aryl isocyanates were cyclized to 3-(arylchloromethylene)oxindoles, which can subsequently be converted stereoselectively to (*Z*)-3-(aminoarylmethylene)oxindoles. For 2-(n-alkylethynyl)aryl isocyanates, the FeCl$_3$-promoted cationic cyclization was followed by C–H bond functionalization leading to 3-(1,2-dichloroalkylidene)oxindoles.

Zeni and co-workers reported an iron-promoted cyclization of *o*-alkynylanilines in 2013.[77] This methodology offers an alternative synthetic protocol for the preparation of 3-organoselenylindole derivatives *via* FeCl$_3$–diorganyl

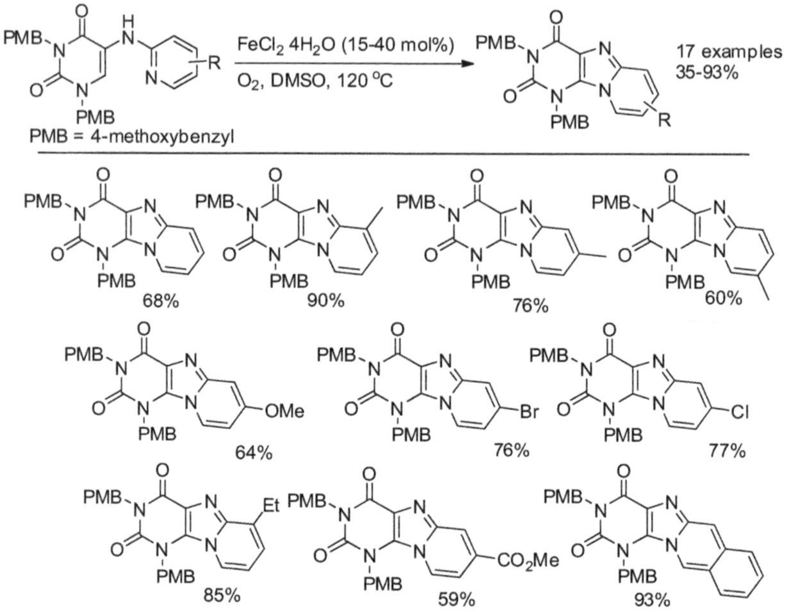

Scheme 3.47 Iron-catalyzed synthesis of annulated purines.

Reaction Procedure (Scheme 3.47): A microwave vessel was flushed with oxygen and loaded with 1,3-bis(4-methoxybenzyl)-5-(pyridine-2-ylamino)-pyrimidine-2,4(1H,3H)-dione (0.222 g, 0.5 mmol) and FeCl$_2$ · 4H$_2$O (15 mg, 0.075 mmol). Subsequently, 1 mL of DMSO was added and the vessel was sealed with a pressure cap and equipped with a balloon of oxygen. The reaction mixture was placed in a heating block at 120 °C and stirred for 18 h. After cooling to room temperature, the combined reaction mixtures was transferred to a separating funnel and 100 mL of CH$_2$Cl$_2$ was added. The organic phase was washed with 20 mL of saturated NaHCO$_3$ followed by a wash with 20 mL of NH$_4$OH–brine mixture (2:3). The combined water fractions were back-extracted with 30 mL of CH$_2$Cl$_2$. The combined organic fractions were dried on MgSO$_4$, filtered and evaporated to dryness *in vacuo*. The crude reaction product was purified *via* an automated chromatography system.

diselenide-promoted intramolecular cyclization of *o*-alkynylanilines. Through this cyclization method, the preparation of highly functionalized indole heterocyclic units in moderate to good yields became achievable (Scheme 3.50). This synthetic approach presents important economic and environmental advantages, including atom and energy economy. In addition, the cyclization reactions were carried out at room temperature in the presence of air (open flask) and both RSe moieties from diorganyl diselenides (RSeSeR) were incorporated in the final product. The

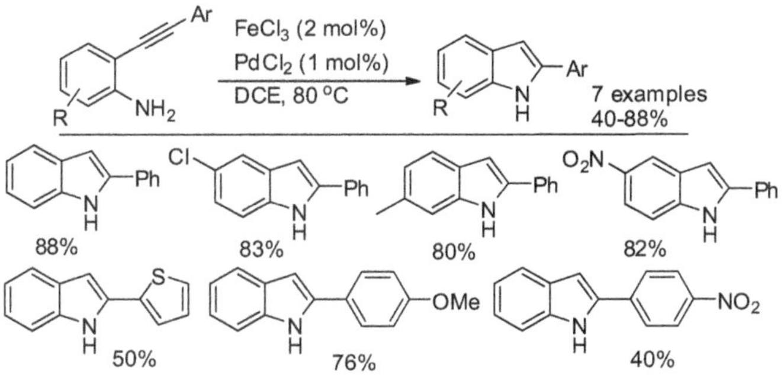

Scheme 3.48 Fe–Pd-co-catalyzed synthesis of indoles.

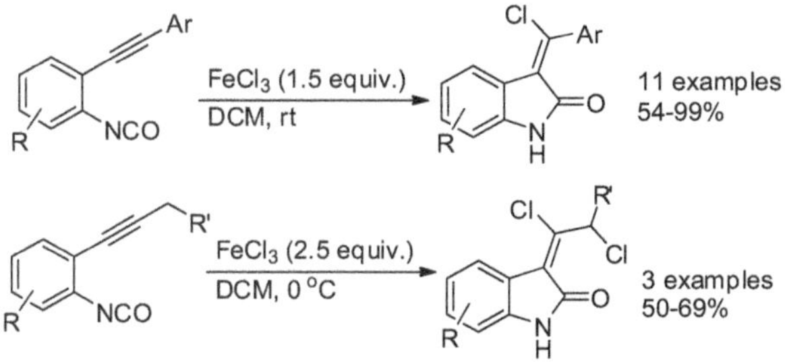

Scheme 3.49 Iron-promoted cyclization of *o*-alkynylaryl isocyanates.

3-organoselenylindoles obtained proved to be convenient synthetic intermediates for the synthesis of a more substituted indole nucleus in particular, the 3-phenylselenylindole furnished the desired 3-bromo- and 3-iodoindole derivatives *via* a selenium–lithium exchange reaction followed by trapping of the indolyllithium intermediate by bromine and iodine.

Subsequently, the same group modified their methodology and applied it in the synthesis of chalcogenophene[2,3-*b*]thiophenes.[78] By using FeCl$_3$–diorganoyl dichalcogenide as a simple and efficient system for the cyclization of alkynylthiophenes, substituted heterocycles were obtained in good yields (Scheme 3.51). The cyclization was carried out at room temperature under an air atmosphere, affording 4-(organochalcogen) heterocycles exclusively *via* a 5-*endo-dig* cyclization process. The best conditions worked well with a broad range of 2-organochalcogen-3-alkynylthiophenes and diorganoyl dichalcogenides. This approach to chalcogenophenes should prove useful in synthesis, particularly when one considers that there are many ways to transform the resulting tellurium and selenium functionalities into other substituents.

A novel and direct synthesis of indoles from nitroaromatics and alkynes was developed by Penoni and Nicholas in 2002.[79] The reaction used [(η^5-C$_5$H$_5$)Fe(CO)$_2$]$_2$ or [(η^5-C$_5$Me$_5$)Fe(CO)$_2$]$_2$ as catalyst, which could be replaced by [(η^5-C$_5$Me$_5$)Ru(CO)$_2$]$_2$; 52.5 bar of carbon monoxide was required for this transformation. Based on their mechanistic study, CO acts as a reductant for reducing nitroarenes to the corresponding nitrosoarenes or hydroxylamines.

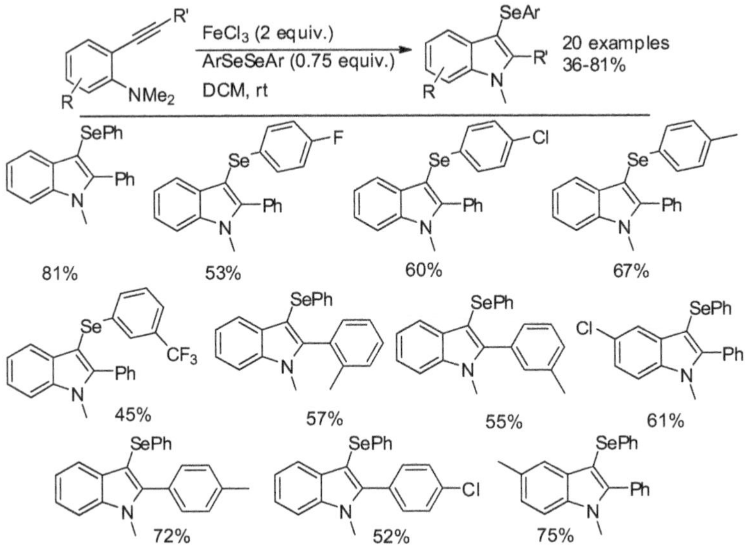

Scheme 3.50   Iron-mediated synthesis of 3-organoselenyl indoles.

Reaction Procedure (Scheme 3.50): To a Schlenk tube, open to the air, containing CH_2Cl_2 (3 mL) were added $FeCl_3$ (0.081 g, 2 equiv.) and diorganyl diselenide (0.061 g, 0.75 equiv.). The reaction mixture was stirred for 20 min at r.t., then *o*-alkynylaniline (0.25 mmol) was added, diluted in CH_2Cl_2 (2 mL). The reaction mixture was stirred at r.t. for the required time, then diluted with CH_2Cl_2 (20 mL) and washed with a saturated aqueous solution of $NaHCO_3$ (3×10 mL). The organic phase was separated, dried over $MgSO_4$ and concentrated under vacuum. The residue was purified by flash chromatography using hexane as eluent.

This was proved by the production of indoles from nitrosoarenes or hydroxylamines with alkynes in the absence of CO using the same catalyst (Scheme 3.52).

An $FeCl_3$-mediated intramolecular cyclization of 3-alkoxyimino-2-arylalkylnitriles was developed in 2008.[80] *N*-Alkoxyindoles were produced in good to excellent yields (Scheme 3.53). Later, a cyclization of 2-aryl-3-substituted hydrazonoalkylnitriles to *N*-aminoindoles was also described.[81] A mechanism for this intramolecular oxidative C–N bond formation process was proposed: the authors believe that the reaction starts with abstraction of the benzylic hydrogen atom from substrate by a SET process to give a carbon-based radical, with an *N*-radical resonance structure. Then, mediated by iron(III) bromide, a second SET process occurs to convert the *N*-radical to the nitrenium ion. Finally, nucleophilic attack on the nitrenium ion by the benzene ring results in a carbocation, which undergoes re-aromatization *via* the loss of a proton to afford the desired product.

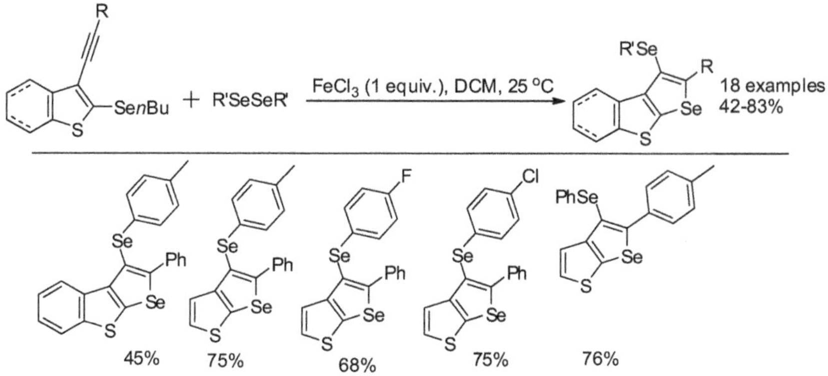

Scheme 3.51 Iron-mediated cyclization of alkynylthiophenes.

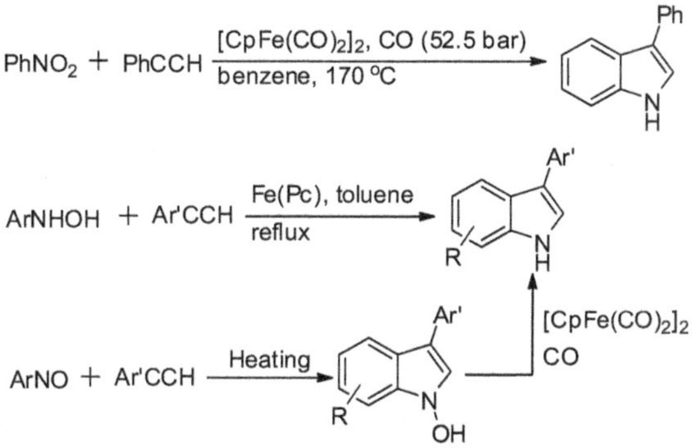

Scheme 3.52 Iron-catalyzed synthesis of indoles from alkynes.

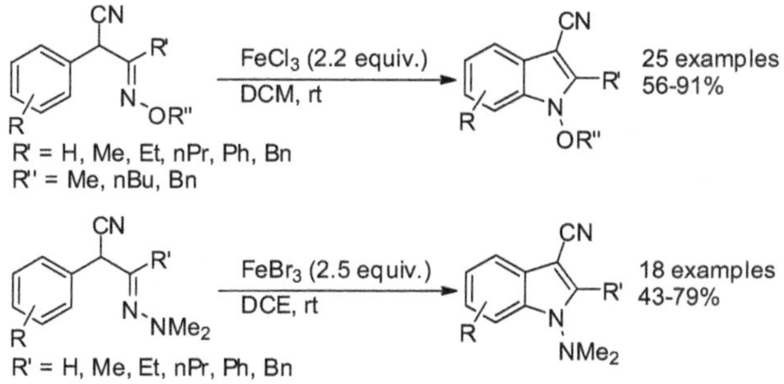

Scheme 3.53 Iron-mediated synthesis of indoles.

Reaction Procedure (Scheme 3.53): To a stirred solution of the 2-aryl-3-dimethylhydrazonoalkylnitrile (2.0 mmol) in DCE (20 mL) was added one portion of FeBr$_3$ powder (5.0 mmol) at room temperature under a nitrogen atmosphere. TLC was used to monitor the reaction process until total consumption of the starting material. To the solution water (20 mL) was added and stirring was continued for an additional 5 min. The reaction mixture was extracted with CH_2Cl_2 (3 × 30 mL) and the organic layer was dried over anhydrous sodium sulfate. The solvent was removed under reduced pressure and the residue was purified by column chromatography using petroleum ether–EtOAc as eluent to afford the pure product.

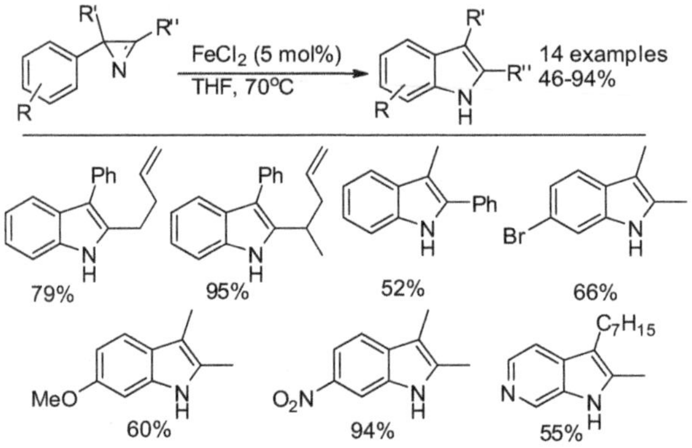

Scheme 3.54 Iron-catalyzed reaction of 2*H*-azirines.

Reaction Procedure (Scheme 3.54): Azirine (1 mmol) was placed in a disposable test-tube with a stirer bar and carefully dried under high vacuum for ∼10 min (until no air bubbles were seen escaping from the azirine). The test-tube was filled with nitrogen and then placed in a nitrogen-filled glove-box where $FeCl_2$ (0.1 or 0.2 mmol) was added. After removal from the glove-box, a nitrogen balloon was placed on the top of the test-tube and THF (1 mL) was added. The nitrogen balloon was detached from the test-tube and the nitrogen-filled tube was then placed in a preheated oil-bath (70 °C) for 24 h. After removal from the oil-bath, the mixture was diluted with CH_2Cl_2 or EtOAc. The diluted reaction mixture was washed with water (2 mL) and brine (2 mL) and then dried over Na_2SO_4. The solvent was evaporated under reduced pressure to obtain a crude mass that was purified by column chromatography on silica gel.

A general method for the synthesis of 2,3-disubstituted indoles using an iron catalyst was described by Zheng and co-workers.[82] The key feature of this method is the amination of aromatic C–H bonds *via* $FeCl_2$-catalyzed ring opening of 2*H*-azirines. The method tolerates a variety of functional groups such as Br, F, NO_2, OMe, CF_3, OTBS, alkenes and OPiv (Scheme 3.54). The method can also be extended to the synthesis of azaindoles. Concerning the reaction mechanism, the reaction starts with the coordination of Fe(II) to the imine nitrogen atom of 2*H*-azirine and formation of an iron–azirine complex; subsequent cleavage of the C–N bond would provide an iron–vinylnitrene complex; and finally, an indole could be formed by a five-centered 6π electrocyclization of the iron–vinylnitrene complex.

Che and co-workers developed an iron-catalyzed cyclization of aryl azides to indoles.[83] By using a commercially available and air-stable [Fe(F$_{20}$TPP)Cl] complex [H$_2$F$_{20}$TPP = *meso*-tetrakis(pentafluorophenyl)porphyrin] as an effective catalyst, indoles, indolines, tetrahydroquinolines, dihydroquinazolinones and quinazolinones were produced in good yields *via* intramolecular amination of sp^2 and sp^3 C–H bonds. In this methodology, aryl azides were used as the nitrogen source. Bolm's group reported a similar methodology with a cheaper and more convenient iron salt [Fe(OTf)$_2$] as the catalyst.[84] This procedure is more practical and the indoles were produced in moderate to excellent yields (Scheme 3.55).

Alternatively, another iron-catalyzed method for the preparation indoles from other types of azide compounds was developed by Driver and co-workers in 2013.[85] Iron(II) bromide was found to be active in promoting these tandem C–H bond amination–1,2-migration reactions. *Ortho*-substituted aryl azides were converted into the corresponding 2,3-disubstituted indoles in moderate to good yields (Scheme 3.56). The 1,2-shift component of this tandem reaction is very selective and enabled the migration aptitude to be predicted as Me < primary carbon < secondary carbon < Ph.

In addition to the above-mentioned procedures, indoles can also be prepared by iron-catalyzed oxidative coupling. In 2010, Liang and co-workers reported an iron-catalyzed oxidative coupling of enamines to the corresponding indoles.[86] Cu(OAc)$_2$ · CuCl$_2$ was used as the oxidant together with 10 mol% of FeCl$_3$ as catalyst. Interestingly, only 30–40% of indole was produced when Cu(OAc)$_2$ or CuCl$_2$ alone was applied as the oxidant. The products were obtained in moderate to good yields (Scheme 3.57).

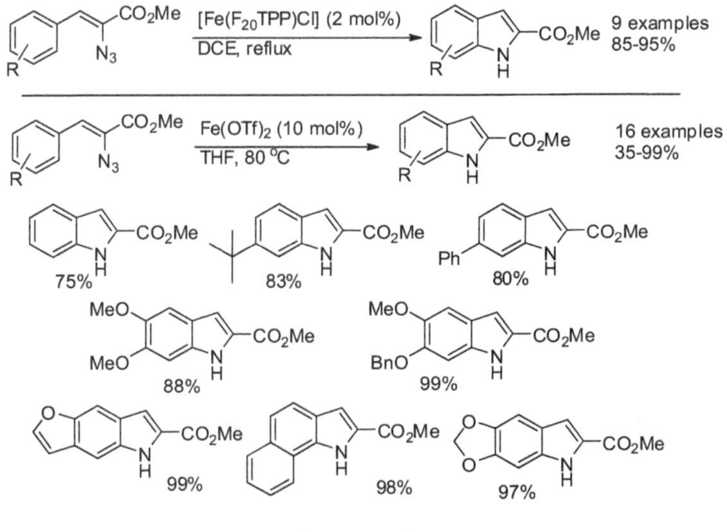

Scheme 3.55 Iron-catalyzed reaction of aryl azides.

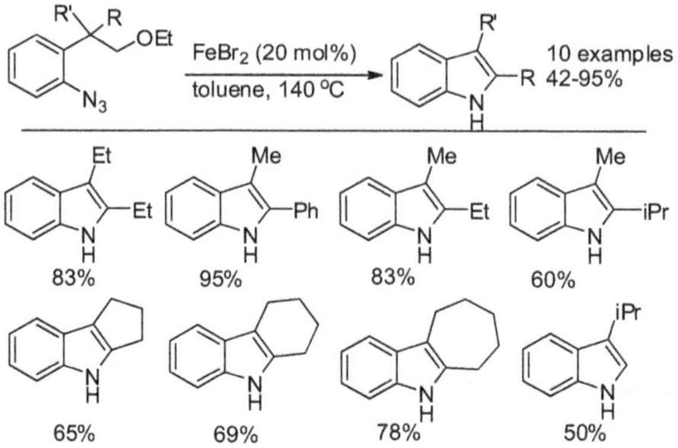

Scheme 3.56   Iron-catalyzed synthesis of indoles from aryl azides.

Reaction Procedure (Scheme 3.56): To a mixture of aryl azide (0.10 mmol) and FeBr$_2$ (20 mol%) in a Schlenk tube was added 1.20 mL of PhMe. The resulting mixture was heated at 140 °C. After 16 h, the heterogeneous mixture was cooled to room temperature. Purification of the reaction mixture by MPLC with a pad of Al$_2$O$_3$ afforded the pure indole.

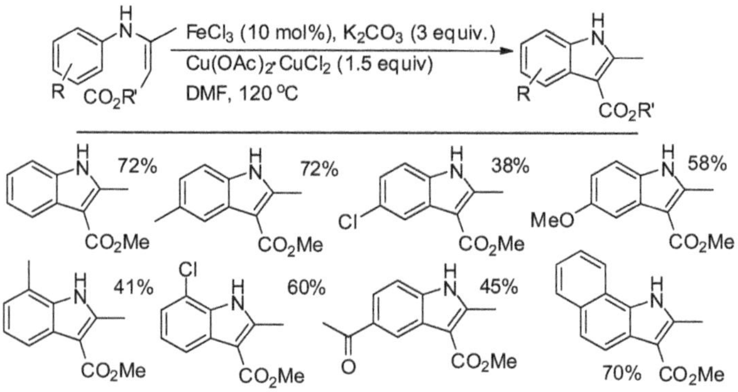

Scheme 3.57   Iron-catalyzed oxidative synthesis of indoles.

Reaction Procedure (Scheme 3.57): A mixture of methyl (Z)-3-(phenyl-amino)but-2-enoate (57 mg, 0.3 mmol), FeCl$_3$ (4.8 mg, 10 mol%), Cu(OAc)$_2$ · CuCl$_2$ (142 mg, 0.45 mmol) and K$_2$CO$_3$ (124 mg, 0.9 mmol) was stirred in DMF (3 mL) at 120 °C for 2 h. After completion of the reaction (monitored by TLC), the reaction mixture was cooled to room temperature, diluted with EtOAc (15 mL) and washed with ammonia solution (10%) (20 mL). The organic layers were dried over anhydrous Na$_2$SO$_4$ and evaporated *in vacuo*. The desired indole (41 mg) was obtained in 72% yield after purification by flash chromatography on silica gel with hexane–ethyl acetate–triethylamine (100 : 20 : 3) as eluent.

Shen and Driver reported an iron-catalyzed benzimidazoles synthesis using aryl azides as substrates.[87] They used 2-azidoaniline as substrate and condensation with aldehydes gave the corresponding 2-azidoarylimines as intermediates. In the presence of 30 mol% of FeBr$_2$, these 2-azidoarylimines were transformed into the corresponding benzimidazoles generally in good yields (Scheme 3.58). The reaction was proposed to start with the coordination of iron(II) bromide to the imine nitrogen to increases its electrophilicity and trigger nucleophilic attack by the pendant azide. After expulsion of N$_2$ and dissociation of the iron catalyst, 2*H*-benzimidazoles were produced. A slow dissociation step would account for the diminished catalytic activity of more Lewis acidic reagents, such as iron(III) bromide. Tautomerization of 2*H*-benzimidazoles forms 1*H*-benzimidazoles. The intermediacy of 2*H*-benzimidazoles would render this transformation non-stereospecific.

An Fe(NO$_3$)$_3$-catalyzed oxidative cyclization of 1,2-phenylenediamine with aldehydes was reported in 2009.[88] H$_2$O$_2$ was applied as the oxidant and

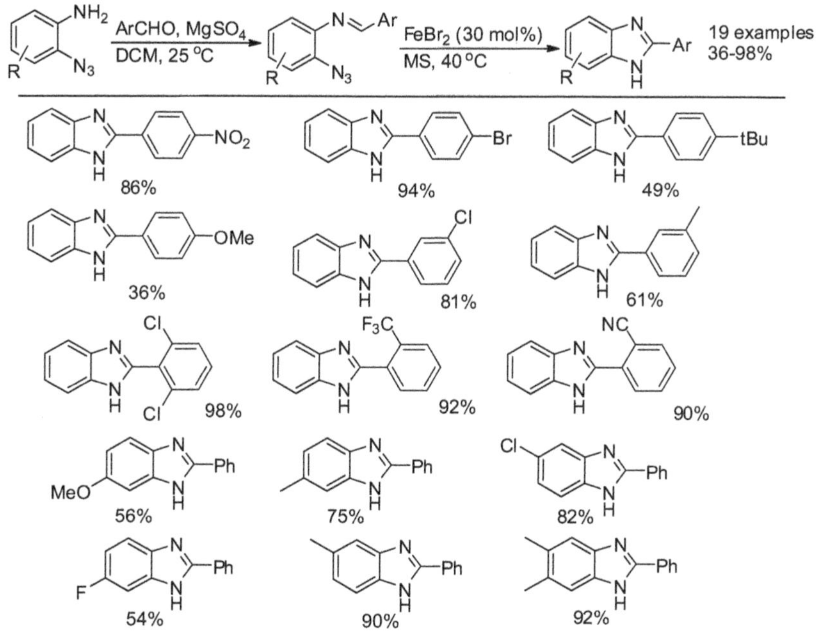

Scheme 3.58 Iron-catalyzed synthesis of benzimidazoles.

Reaction Procedure (Scheme 3.58): To a mixture of 2-azidoarylimine, FeBr$_2$ (30 mol%) and 0.120 g of crushed 4 Å molecular sieves (150 wt%) was added CH$_2$Cl$_2$ to make a 1.0 M solution. The resulting mixture was heated at 40 °C. After 12 h, the reaction mixture was cooled to room temperature and the heterogeneous mixture was filtered through SiO$_2$. The filtrate was concentrated *in vacuo*. Purification using MPLC (0:100–2:98 MeOH–CH$_2$Cl$_2$) provided the benzimidazole as a solid.

benzimidazoles were produced in good yields at 50 °C (Scheme 3.59).
2-Aminobenzenethiols can also be used as substrates, in which case 2-aryl-
benzothiazoles were produced in excellent yields. Additionally, a one-pot
procedure for the conversion of aromatic and heteroaromatic 2-nitroamines
into 2*H*-benzimidazoles was described by Hanan *et al.*[89] The procedure
employs formic acid, iron powder and an additive such as NH$_4$Cl to reduce
the nitro group and effect the imidazole cyclization with high-yielding conversions generally within 1–2 h. The compatibility with a wide range of
functionality demonstrates the general utility of this procedure.

An *in situ*-generated iron sulfide-catalyzed redox–condensation cascade
reaction between *o*-nitroaniline and methylhetarene was developed by
Nguyen *et al.* in 2013.[90] 2-Substituted benzimidazoles and benzoxazoles were
prepared in good yields (Scheme 3.60). Interestingly, only 30% of the desired
product was produced using already prepared FeS as the catalyst. Traces of

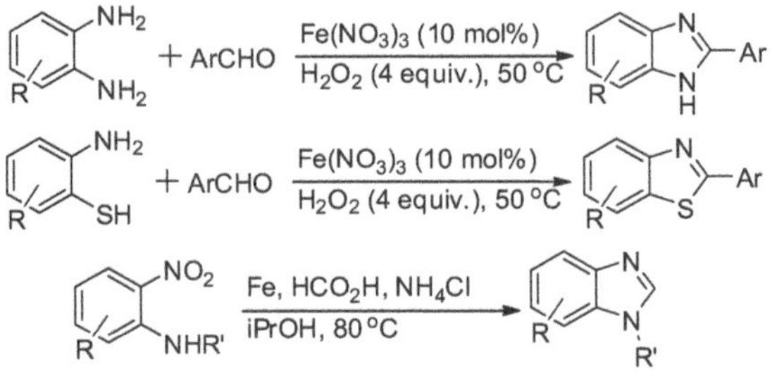

Scheme 3.59   Iron-catalyzed synthesis of benzimidazole derivatives.

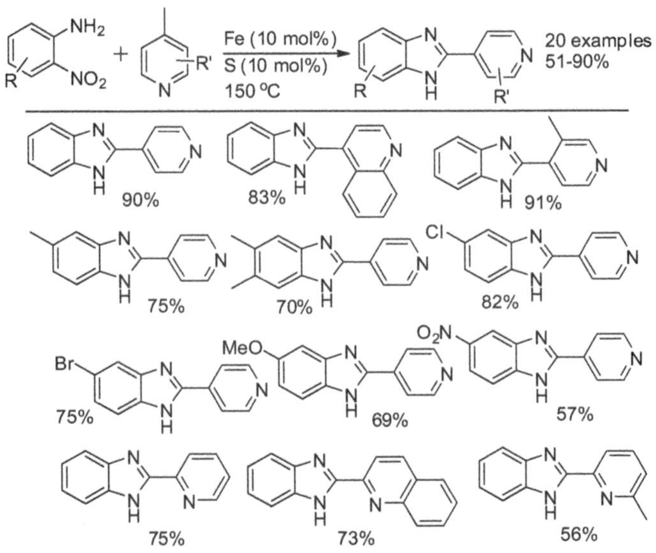

Scheme 3.60   FeS-catalyzed benzimidazoles synthesis.

product was obtained with the other iron salts [$FeCl_3$, $FeSO_4$, $Fe(NO_3)_3$, $Fe(acac)_3$; yields <5 mol%].

In 2008, Bolm and co-workers reported an iron-catalyzed *N*-arylation of amides.[91] Intermolecular reaction of aryl iodides with primary amides gave various secondary amides in good yields. In the case of intramolecular *N*-arylation, heterocycles can be produced (Scheme 3.61).

Li and co-workers developed an iron-catalyzed tandem reaction method for the synthesis of 2-aminobenzothiazoles in 2009.[92] In the presence of FeF_3 and 1,10-phenanthroline, a variety of 2-halobenzenamines and iso-thiocyanates underwent the tandem reactions successfully to afford the corresponding 2-aminobenzothiazoles in moderate to excellent yields (Scheme 3.62). It is noteworthy that the scope can be extended to 2-iodo-benzenamines and 2-bromobenzenamines by using an inexpensive and

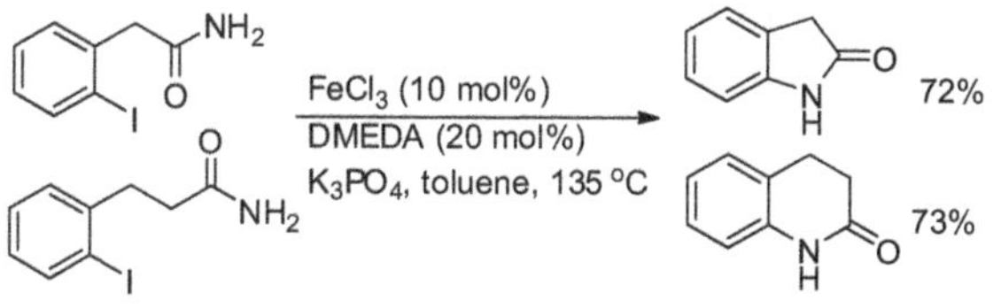

Scheme 3.61 Iron-catalyzed intramolecular *N*-arylation reactions.

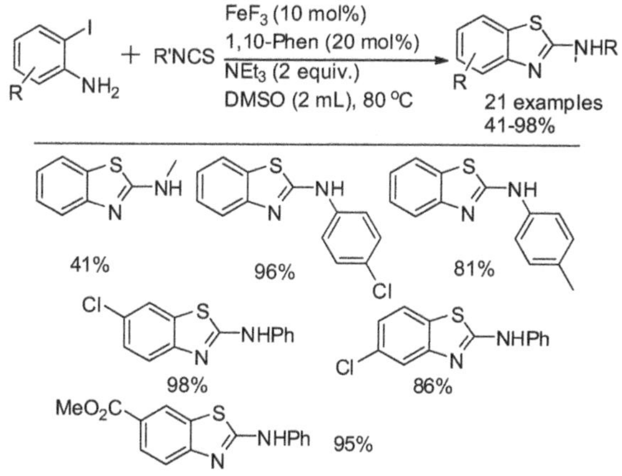

Scheme 3.62 FeF_3-catalyzed synthesis of benzothiazoles.

Reaction Procedure (Scheme 3.62): A mixture of 2-iodoaniline (0.2 mmol), isothiocyanate (0.22 mmol), FeF_3 (10 mol%), 1,10-phenanthroline (0.2 equiv.), Et_3N (2 equiv.) and DMSO (2 mL) was stirred at 80 °C until complete consumption of starting material. When the reaction was complete, the mixture was poured into ethyl acetate and washed with saturated NaCl (10 × 3 mL). After the aqueous layer had been extracted with ethyl acetate, the combined organic layers were dried over anhydrous Na_2SO_4, evaporated under vacuum and purified by flash column chromatography.

environmentally benign iron catalytic system. The effect of copper was also studied; low yields of the desired product resulted. One year later, this reaction was modified and water was applied as solvent.[93] Alternatively, 2-aminobenzothiazoles could be straightforwardly prepared by the reaction of *o*-aminobenzenethiols and isothiocyanates in the presence of a catalytic amount of iron catalyst.[94] The reaction gave 2-aminobenzothiazoles in good selectivity by using DABCO or Na_2CO_3 as base at 80 °C.

More recently, an efficient and versatile iron phthalocyanine-catalyzed method has been developed for the *N*-alkylation of various amines with alcohols.[95] Readily available alcohols were used as alkylating agents for direct *N*-alkylation of aminobenzothiazoles, aminopyridines and aminopyrimidines. *N*-Alkylation of *ortho*-substituted anilines (–NH$_2$, –SH and –OH) led to the synthesis of 2-substituted benzimidazoles, benzothiazoles and benzoxazoles in one pot in good to excellent yields (Scheme 3.63).

Lei and co-workers developed an efficient iron-catalyzed C–H functionalization/C–S bond formation reaction under mild conditions.[96] This transformation could be conveniently carried out to give various benzothiazoles in moderate to excellent yields (Scheme 3.64). Preliminary mechanistic studies revealed that the reaction required the coexistence of substrate,

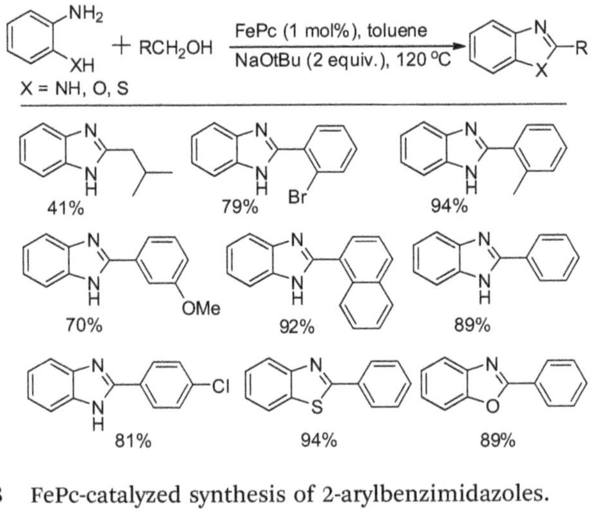

Scheme 3.63   FePc-catalyzed synthesis of 2-arylbenzimidazoles.

Reaction Procedure (Scheme 3.63): To a stirred suspension of FePc (1 mol%) and NaO*t*Bu (2 mmol) in toluene (5 mL) were added the alcohol (1.5 mmol) and *ortho*-substituted aniline (1.0 mmol) at room temperature and then the temperature was raised to 120 °C for 36 h. On completion of the reaction (monitored by TLC), the reaction mixture was filtered and passed through anhydrous Na_2SO_4 and dried under vacuum. The crude product was analyzed by GC–MS and purified by column chromatography over silica-gel (60–120 mesh) with *n*-hexane–ethyl acetate as eluent.

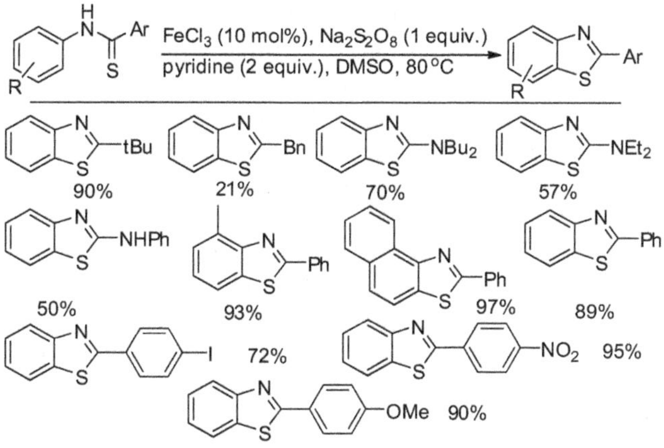

Scheme 3.64 FeCl$_3$-catalyzed oxidative synthesis of benzothiazoles.

Reaction Procedure (Scheme 3.64): A Schlenk tube equipped with a stir-bar was charged with FeCl$_3$ (8.3 mg, 0.01 mmol), substrate (0.50 mmol) and Na$_2$S$_2$O$_8$ (122.5 mg, 1.0 mmol). The reaction tube was purged with nitrogen, then pyridine (79.0 mg, 1.0 mmol) and 2 mL of DMSO were added to the reaction tube *via* a syringe. The mixture was stirred at 80 °C for 3 h. After cooling to room temperature, the mixture was quenched with water and extracted with ethyl acetate (2×20 mL). The organic layers were combined, dried over Na$_2$SO$_4$ and concentrated under reduced pressure, then purified by silica gel chromatography (1 : 20 ethyl acetate–petroleum ether as eluent) to yield the pure product.

oxidant, FeCl$_3$ and pyridine. Kinetic studies indicated that pyridine was crucial for the high selectivity of this transformation and the reaction was first order in the substrate and zeroth order in oxidant Na$_2$S$_2$O$_8$. Accordingly, for the reaction mechanism the authors proposed that the substrate *N*-phenylbenzothioamide was oxidized by Fe(III) to form a thioyl radical intermediate, and Fe(III) was reduced to Fe(II). The Fe(II) species was re-oxidized by Na$_2$S$_2$O$_8$ to regenerate Fe(III). Then, the cyclization of the thioyl radical intermediate followed by oxidation in the presence of Na$_2$S$_2$O$_8$ gave the product 2-phenylbenzothiazole.

3.2 Six-Membered Heterocycles

Zenneck and co-workers described two routes for the synthesis of (η^4-1,5-cyclooctadiene)(η^6-phosphinine)iron(0) complexes, which can be applied in the synthesis of pyridines from nitriles and alkynes at room temperature.[97] A three-component reaction of iron vapor with COD and 2-(trimethylsilyl)-4,5-dimethylphosphinine at low temperature yields 20% (COD)[2-(trimethylsilyl)-4,5-dimethylphosphinine]Fe(0), whereas ligand exchange of one COD ligand of the *in situ*-prepared metal vapor product $(COD)_2Fe$ by 2-(trimethylsilyl)-4,5-dimethylphosphinine or $(\eta^1$-2-chloro-4,5-dimethylphosphinine)$Cr(CO)_5$ gave the corresponding (COD)(phosphinine)Fe(0) complexes in more than 80% isolated yield. As the phosphinine derivatives are prochiral, complexation leads to the racemate of two enantiomers. Both enantiomers are found in the unit cell of a single crystal and are related by an inversion center. This iron complex is a novel room-temperature catalyst for the $[2+2+2]$ cyclic addition reaction of one molecule of butyronitrile with two molecules of methyl propargyl ether giving up to 160 mol of pyridine derivatives per mole. A chemically robust species, the iron complex is an air-stable crystalline material, but exposure to oxygen in solution causes slow decomposition.

In 2002, Guerchais and co-workers reported $[Fe(C_5Me_5)(CH_3CN)_2(PMe_3)][PF_6]$ and $[Fe(C_5Me_5)(CH_3CN)_3][PF_6]$ as two new iron complexes for the $[2+2+2]$ cyclic addition reaction of nitriles and alkynes to pyridine derivatives.[98] They proposed the formation of a metallacyclopentadiene intermediate in the first stage, which resulted from coupling of the two coordinated alkyne molecules. The coordination of the CC bond of the third molecule of the alkyne could be inhibited by the presence of the heteroatom. Only a molecule of acetonitrile could interact and then insert into the Fe–C bond to afford the observed pyridine product.

Since 2011, Louie and co-workers have developed several interesting iron-catalyzed cycloadditions of nitriles and alkynes to pyridines. First, they reported the cycloaddition of alkynenitriles and alkynes.[99] In the presence of $Fe(OAc)_2$ and pyridylbisimine ligand, 22 examples of pyridine derivatives were prepared in moderate to good yields (Scheme 3.65). Notably, this reaction system is very sensitive to the substituents on the imine ligand.

Alternatively, this system is also active in the cycloaddition of diynes and cyanamides,[100] and 2-aminopyridines were produced in good yields (Scheme 3.66). This system was also used to cyclize two terminal alkynes and a cyanamide to afford a 2,4,6-trisubstituted pyridine product regioselectively. The reaction pathway is believed to involve *in situ* ligand coordination and reduction by zinc, then a reduced Fe catalyst binds the diyne and facilitates oxidative coupling of the two alkyne units to form a ferracyclopentadiene. Insertion of cyanamide and reductive elimination subsequently afford the pyridine product.

Subsequently, the same group reported the first iron-catalyzed catalytic $[2+2+2]$ cycloaddition to produce aromatic diazaheterocycles.[101] Remarkably, iron, which traditionally has been an inefficient cycloaddition catalyst for nitrile incorporation, can now incorporate multiple nitriles into aromatic products (Scheme 3.67). Furthermore, traditionally more efficient catalysts were ineffective in this strategy of 2-aminopyrimidine synthesis, such as nickel, gold and silver salts.

Around the same time, Wan and co-workers developed a simple and highly efficient method for the iron-catalyzed $[2+2+2]$ cycloaddition of diynes and unactivated nitriles leading to pyridine compounds at room temperature.[102] The catalyst was generated *in situ* from an inorganic iron salt

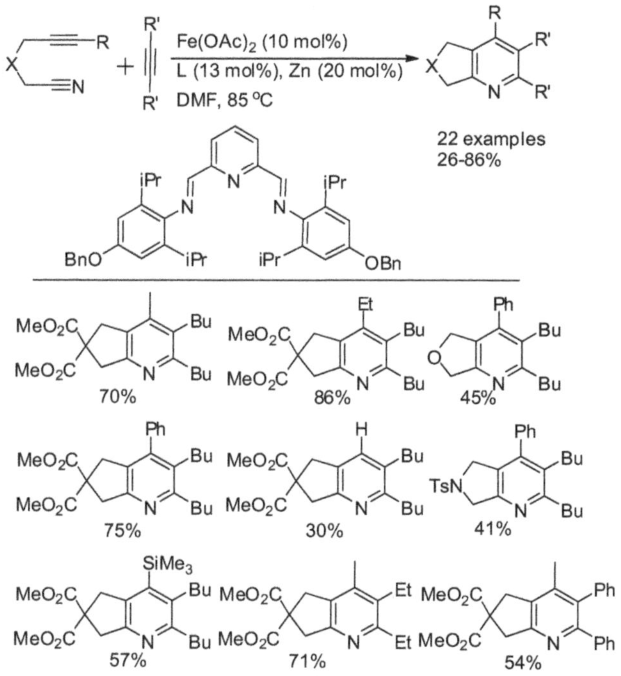

Scheme 3.65 Iron-catalyzed synthesis of pyridines.

Reaction Procedure (Scheme 3.65): In a nitrogen-filled glove-box, a solution of alkynenitrile (>1.0 M in DMF) was added to a vial containing 10 mol% Fe(OAc)$_2$ and 13 mol% L. Additional DMF was added to make the final concentration of cyanoalkyne 0.4 M (taking into account the alkyne volume). The mixture was stirred for 10 min, then 1 equiv. of alkyne and 20 mol% of zinc dust were added. The vial was capped and removed from the glove-box, then stirred at 85 °C for the indicated period of time. The crude mixture was purified *via* silica gel flash chromatography.

and a diphosphine ligand and exhibited high reactivity and regioselectivity. Various pyridine derivatives were prepared in excellent yields (Scheme 3.68). The appropriate ligand and the metal to-ligand ratio play a crucial role in the catalytic efficiency.

Wan's group succeeded in extending their methodology to the preparation of 2-aminopyridines.[103] This iron-catalyzed [2 + 2 + 2] cycloaddition reaction of diynes and cyanamides can be carried out at room temperature. Highly substituted 2-aminopyridines were obtained in good to excellent yields with high regioselectivity (Scheme 3.69). The reaction mechanism was investigated through *in situ* IR spectroscopic and control experiments. In this iron-catalyzed cycloaddition reaction, the active iron species was generated only in the presence of both alkynes and nitriles. The lower reaction temperature,

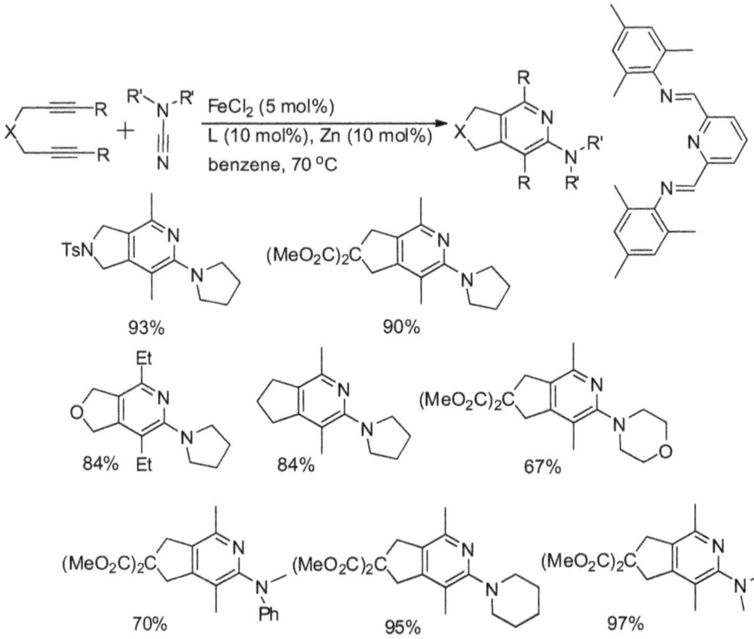

Scheme 3.66 Iron-catalyzed synthesis of 2-aminopyridines.

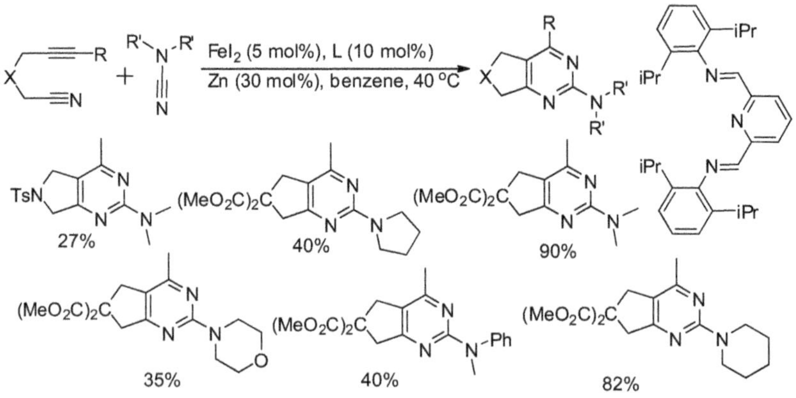

Scheme 3.67 Iron-catalyzed synthesis of 2-aminopyrimidines.

Reaction Procedure (Scheme 3.67): FeI$_2$ (15.6 mg, 0.05 mmol) and dppp (42.4 mg, 0.10 mmol) were weighed in a glove-box and placed in a dried Schlenk tube, then distilled THF (2 mL) was added. The resulting mixture was stirred at r.t. for 30 min to afford a clear orange–yellow solution, at which time Zn dust (6.5 mg, 0.10 mmol) was added. After stirring for an additional 30 min, diyne (0.5 mmol) was added followed by the nitrile

(5 mmol), and the reaction mixture was stirred for 24 h until the majority of the starting diyne had been consumed. The solvent was evaporated and the crude product was directly purified by flash column chromatography on silica gel (eluent: petroleum ether–ethyl acetate) to give the desired pyridine.

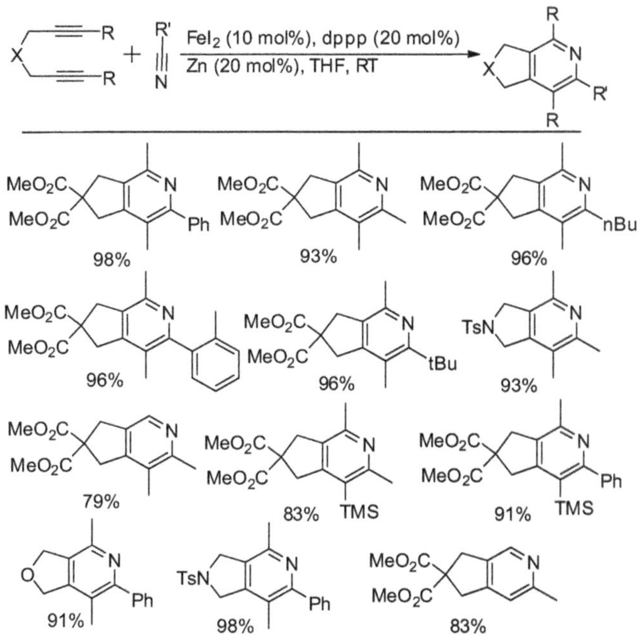

Scheme 3.68 FeI_2–dppp-catalyzed synthesis of pyridines.

broad substrates scope and inverse regioselectivity make it a complementary method. The role of Zn and ZnI_2 was also studied. When 20 mol% of ZnI_2 was added to the reaction solution it was surprisingly found that the yield of pyridine product was improved. This result indicated that ZnI_2 might act as a Lewis acid by activating the alkyne or nitrile moieties, thus allowing the generation of active low-valent iron species to be much easier.

Wang and co-workers reported an iron-promoted tandem reaction of anilines with styrene oxides to give quinolones (Scheme 3.70).[104] The reaction proceeds through a C–C bond cleavage and C–H bond activation sequence. In more detail, the first step is the reaction of aniline with styrene oxide in the presence of $FeCl_3$ as a Lewis acid to form a β-amino alcohol, which underwent dehydration to generate the corresponding enamine. Subsequently, the obtained enamine reacted with another styrene oxide assisted by $FeCl_3$, which could undergo β-H elimination from the iron–alkoxide linkage, leading to the formation of a ketone intermediate and an iron hydride species [which might eliminate HCl to reach the Fe(I) oxidation state].The ketone intermediate could then undergo Fe(I)-mediated, directed

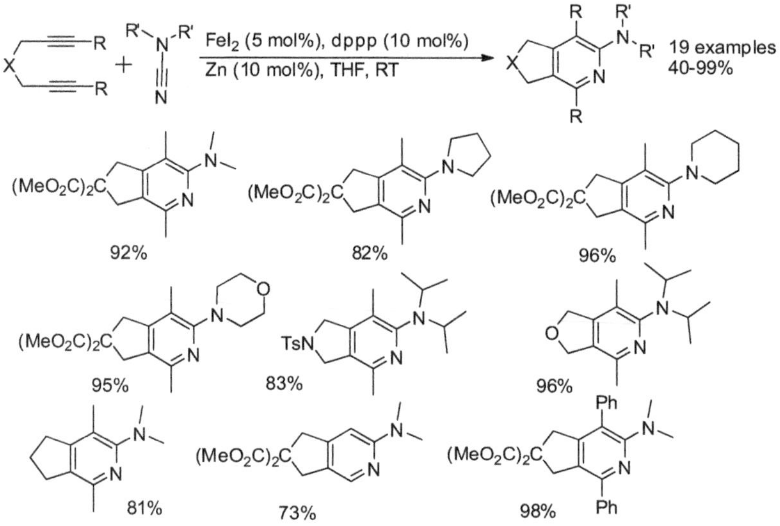

Scheme 3.69   FeI$_2$–dppp-catalyzed synthesis of 2-aminopyridines.

Reaction Procedure (Scheme 3.69): FeI$_2$ (5 mol%, 7.8 mg, 0.025 mmol) and dppp (10 mol%, 21.2 mg, 0.05 mmol) were weighed in a glove-box and placed in a dried Schlenk tube. Subsequently, 2 mL of distilled THF was added. The resulting mixture was stirred at room temperature for 30 min to afford a clear orange–yellow solution, at which time Zn dust (10 mol%, 3.3 mg, 0.05 mmol) was added. After an additional 30 min of stirring, diyne (0.5 mmol) was added followed by cyanamide (2.5 mmol, 5 equiv.), and the mixture was stirred for 24 h until most of the starting diyne had been consumed. The solvent was evaporated, and the crude product was directly purified by silica gel flash column chromatography.

C–H bond activation at the *ortho*-position of the aniline moiety to generate an iron complex, followed by chelation-assisted C–C bond activation and cleavage to form the seven-membered cyclometalate and eliminate PhCHO. After reductive elimination of FeCl, 3-phenyl-1,4-dihydroquinoline was obtained. Finally, the desired quinolone was generated through de-hydrogenation of 3-phenyl-1,4-dihydroquinoline in the presence of oxygen. It should be noted that benzaldehyde and β-amino alcohol were detected by HPLC during the reaction of the substrates by controlling the reaction conditions.

A novel one-pot synthesis of calothrixins and their analogs was achieved in 2013.[105] The reaction involving an FeCl$_3$-mediated domino reaction of the enamines as a key step. This protocol afforded calothrixin B and its de-rivatives in an overall yield of 39–50% (Scheme 3.71). Alternatively, the enamines upon interaction with CuBr$_2$ in DMF at reflux led to the formation

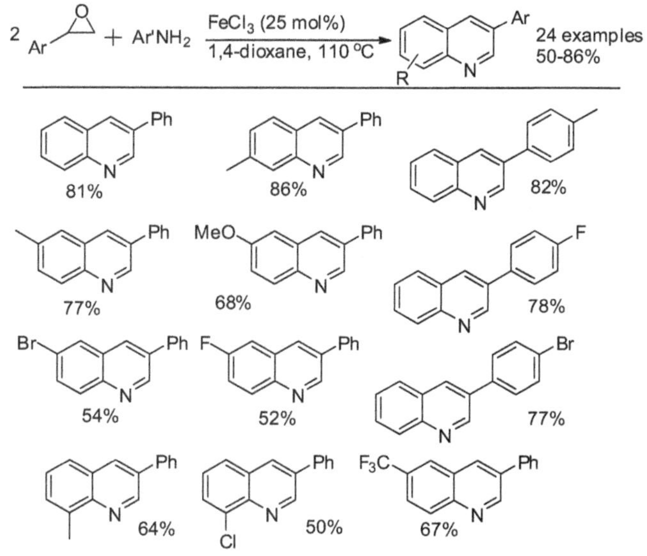

Scheme 3.70 FeCl$_3$-promoted synthesis of 3-arylquinolones.

Reaction Procedure (Scheme 3.70): Under an air atmosphere, a sealable reaction tube (25 mL) equipped with a magnetic stirrer bar was charged with styrene oxide (2.0 mmol), aniline (1.0 mmol), FeCl$_3$ (0.30 mmol) and 1,4-dioxane (2.0 mL). The rubber septum was then replaced with a Teflon-coated screw-cap and the reaction vessel placed in an oil-bath at 110 °C for 24 h. After the reaction was completed, the mixture was cooled to room temperature and diluted with ethyl acetate. The resulting solution was directly filtered through a pad of silica gel using a sintered-glass funnel and concentrated under reduced pressure. The residue was purified by flash chromatography on silica gel (eluent: hexane–ethyl acetate) to give the corresponding quinoline product.

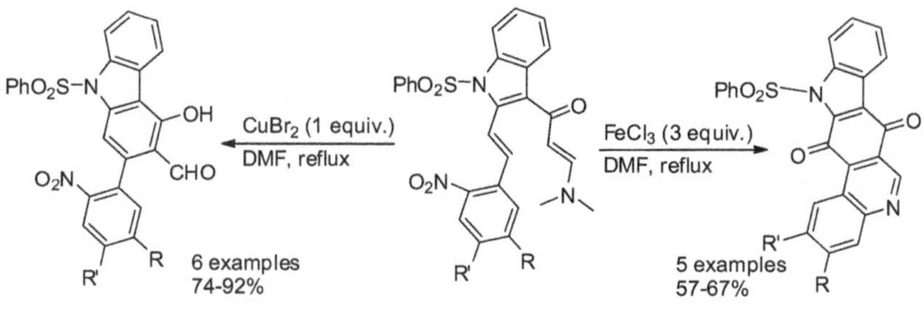

Scheme 3.71 FeCl$_3$-mediated synthesis of calothrixins.

of 1-phenylsulfonyl-2-(2'-nitroaryl)-4-hydroxycarbazole-3-carbaldehydes in excellent yields.

In 2013, a straightforward iron-catalyzed divergent oxidative tandem synthesis of dihydroquinazolines and quinolones from *N*-alkylanilines using a TEMPO oxoammonium salt as a mild and non-toxic oxidant was developed (Scheme 3.72).[106] Fe(OTf)$_2$ was the Lewis acid catalyst of choice for the formation of dihydroquinazolines, whereas FeCl$_3$ led to better results for the synthesis of quinolines. This divergent approach implies that, for both syntheses, direct oxidative functionalization of an α-C(sp^3)–H bond of the *N*-alkylanilines occurs, leading to C–N or C–C bond formation upon

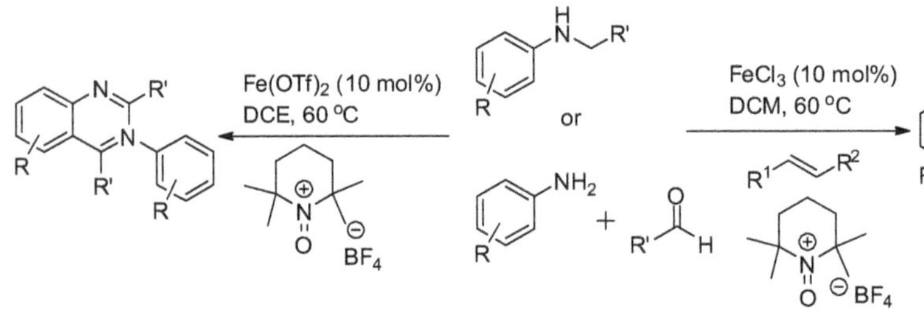

Scheme 3.72   Iron-catalyzed synthesis of dihydroquinazolines and quinolines.

General Procedure for the Synthesis of Dihydroquinazolines (Scheme 3.72): To a sealed tube equipped with a magnetic stirring bar, Fe(OTf)$_2$ (10 mol%), aniline (0.2 mmol, 1.0 equiv.), T$^+$BF$_4$ (0.40 mmol, 2.0 equiv.), and DCE (0.5 mL) were added under an argon atmosphere. The reaction mixture was then stirred at 60 °C for 24 h. The crude reaction mixture was directly purified by column chromatography on silica gel.

General Procedure A for the Synthesis of Quinolines from *N*-Alkylanilines: A mixture of substrate (0.5 mmol), olefin (1.0 mmol, 2.0 equiv.) or alkyne (4.0 equiv.), FeCl$_3$ (8.0 mg, 0.05 mmol, 10 mol%), and T$^+$BF$_4$ (243.0 mg, 1.0 mmol, 2 equiv.) in dry DCM (5.0 mL) in a Schlenk tube under an argon atmosphere was stirred at 60 °C. When the starting material had been consumed (monitored by GC–MS or TLC), the solvent was concentrated under reduced pressure and the residue was purified by column chromatography on silica gel.

General Procedure B for the Synthesis of Quinolines via a Three-Component Reaction: In a pressure Schlenk tube equipped with a magnetic stirring bar, aniline (0.2 mmol, 1.0 equiv.) and FeCl$_3$ (3.2 mg, 0.02 mmol, 10 mol%) were dissolved in dry DCM (2 mL) under an argon atmosphere. After the addition of ethyl glyoxalate (48 μL, 0.24 mmol, 1.2 equiv.; 50% in toluene), olefin or alkyne (0.4 mmol, 2.0 equiv.) and T$^+$BF$_4$ (97.2 mg, 0.4 mmol, 2.0 equiv.), the solution was stirred at 60 °C for 18 h. The crude reaction mixture was directly purified by column chromatography to obtain the corresponding quinoline derivative.

homocondensation or reaction with simple alkenes, respectively. Cyclization followed by a final oxidation generates these classes of interesting bioactive heterocycles in one synthetic transformation. Additionally, the one-pot multicomponent synthesis of quinolines from anilines, aldehydes and alkenes has also been successfully developed under these mild oxidative conditions.

Wang *et al.* reported an iron-catalyzed amination reaction for the preparation of nitrogen-containing heterocycles in 2012.[107] This is a new method for synthesizing substituted 1,2-dihydroquinolines *via* an iron-catalyzed intramolecular allylic amination of *N*-protected 2-aminophenyl-1-en-3-ols under mild conditions (Scheme 3.73). The advantages of this approach are the use of inexpensive and environmentally friendly $FeCl_3 \cdot 6H_2O$ as the catalyst and the tolerance of the presence of water and air in the reaction system. In addition, the synthetic application of this new method to the one-pot synthesis of substituted quinolines was also demonstrated. Concerning the reaction mechanism, the coordination of $FeCl_3$ catalyst to the hydroxyl group of the substrate was proposed to be the first step. Subsequent dehydration, carbocation formation and cyclization and gave a dihydroquinoline as the final product. $CuCl$, $CuCl_2$ and $PdCl_2$ were also tested as catalysts for this transformation, but only traces of the desired product were formed.

A facile and economical method for the construction of quinolines by the $FeCl_3$-catalyzed three-component coupling/hydroarylation/dehydrogenation of aldehydes, alkynes and amines was developed by Tu and co-workers in 2009.[108] A series of 2,4-disubstituted quinolines were synthesized in excellent yields from simple and readily available starting materials (Scheme 3.74). This procedure was also reported with $Fe(OTf)_3$ as catalyst.[109]

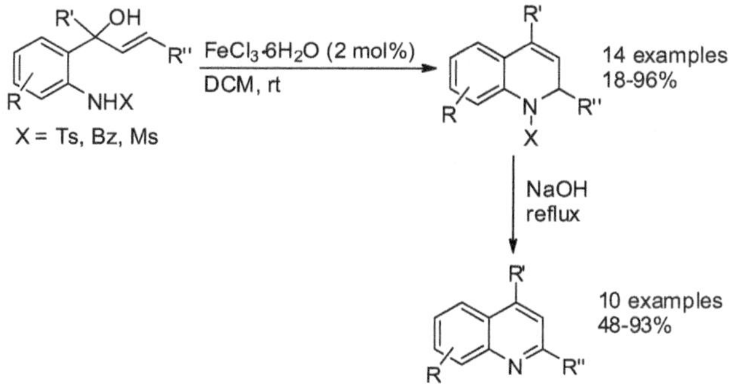

Scheme 3.73 $FeCl_3$-catalyzed intramolecular allylic amination.

Reaction Procedure (Scheme 3.73): To a solution of protected 2-aminophenyl-1-en-3-ol (0.5 mmol) in CH_2Cl_2 was added $FeCl_3 \cdot 6H_2O$ (0.01 mmol). The reaction mixture was stirred at r.t. for 1 h (monitored by TLC). The resulting mixture was purified by flash column chromatography on silica gel (eluent: petroleum ether–ethyl acetate) to give the desired pure product.

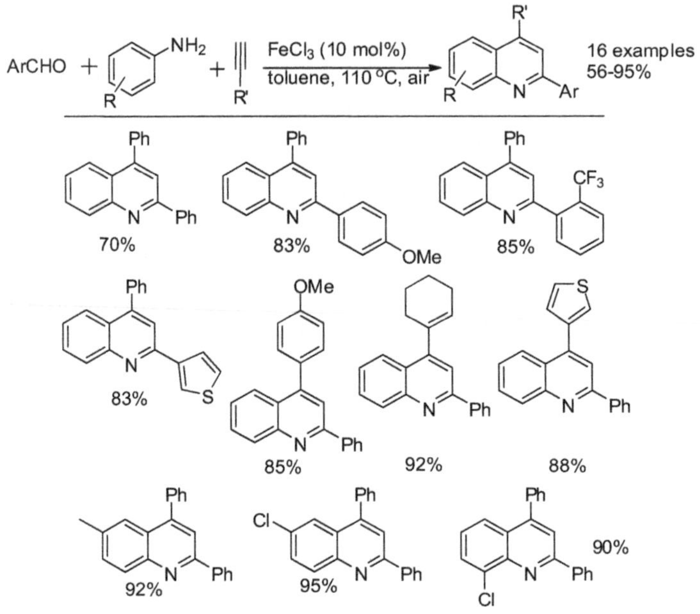

Scheme 3.74   FeCl₃-catalyzed synthesis of quinolones.

> **Reaction Procedure** (Scheme 3.74): To a 10 mL flask were added sequentially toluene (1 mL), FeCl₃ (16.2 mg, 0.1 mmol), benzaldehyde (0.104 mL, 1.0 mmol), aniline (0.096 mL, 1.05 mmol) and phenylacetylene (0.168 mL, 1.5 mmol) under an air atmosphere. The reaction mixture was stirred at 110 °C until the substrate had been completely consumed (∼24 h), then cooled to room temperature and filtered through a short silica gel column using CH₂Cl₂ as eluent. After evaporation of the solvent, the residue was purified by flash chromatography (eluent: 20:1 petroleum ether–ethyl acetate) to afford the product.

In 2009, Heravi *et al.* reported an iron(III) perchlorate [Fe(ClO₄)₃]-catalyzed synthesis of *N*-cyclohexyl-3-arylquinoxaline-2-amines.[110] The reaction proceeds through a three-component condensation of *o*-phenylenediamine, aromatic aldehydes and cyclohexyl isocyanide. All the desired products were obtained in excellent yields (Scheme 3.75). The condensation of aldehyde with *o*-phenylenediamine to give the corresponding imine was believed to be the first step, followed by the addition of cyclohexyl isocyanide to the imine, giving the final product after cyclization and rearrangement. Quinoline derivatives can also be prepared using an iron catalyst in different manners, *e.g.* a simple, inexpensive and efficient oxidation of 2-aryl-1,2,3,4-tetrahydro-4-quinolones to 4-alkoxy-2-arylquinolines employing FeCl₃ · 6H₂O (2.5 equiv.) as promoter in methanol and refluxing in a water-bath was reported in 2004.[111]

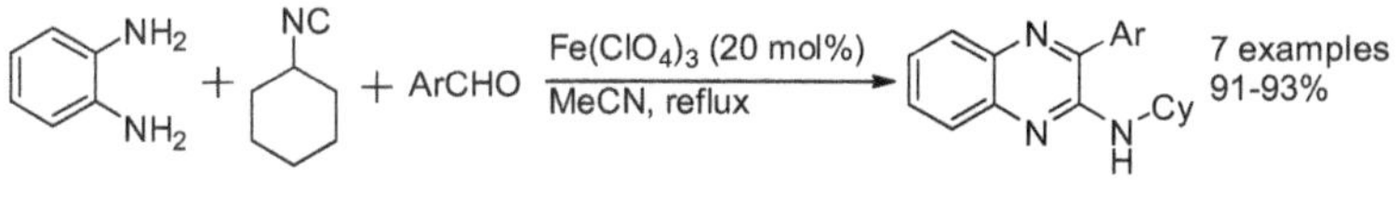

Scheme 3.75 Iron-catalyzed synthesis of quinoxalines.

Reaction Procedure (Scheme 3.75): To a mixture of *o*-phenylenediamine (1 mmol), benzaldehyde (1 mmol) and cyclohexyl isocyanide (1 mmol) in acetonitrile (5 mL), a catalytic amount of iron(III) perchlorate (0.2 mmol) was added and the mixture was refluxed for 2 h. The progress of the reaction was monitored by TLC (1 : 3 ethyl acetate–hexane). After completion of the reaction, acetonitrile was removed and the reaction mixture was diluted with water (10 mL), then CH_2Cl_2 (10 mL) was added. The organic layer was separated and dried over $MgSO_4$. The solvent was evaporated under reduced pressure and the product was obtained without any further purification.

A one-pot reaction for assembling pyrrolo[1,2-*a*]quinoxalines from 1-(2-nitrophenyl)pyrroles and various alcohols was developed in 2012.[112] The nitro reduction, alcohol oxidation, heterocycle formation and heterocycle oxidation were realized in a cascade manner. A wide range of these fused heterocycles bearing different alkyl and aryl groups at the 4-position have been elaborated from suitable substrates; thereby, 3-nitro-2-pyrrolopyridine was also compatible with this process, giving the corresponding fused tricyclic compounds (Scheme 3.76). A reaction mechanism was also proposed. In an acidic medium, iron could catalyze the reduction of the nitrophenylpyrrole to its amine counterpart giving ferric (or ferrous) salts, which in turn would be able to oxidize alcohols into aldehydes. Condensation of the latter with the amine gives the iminium salts, which spontaneously cyclize, leading to the dihydroquinoxalines. A final oxidation produces the 4-alkyl- or 4-phenylpyrroloquinoxalines in moderate to good yields.

Beifuss and co-workers developed a simple to perform and efficient method for the synthesis of 2-substituted 1,2-dihydrophthalazines employing simple starting materials.[113] The 1,2-dihydrophthalazines can be prepared in a single preparative step between 2-(bromomethyl)benzaldehydes with arylhydrazines *via* Lewis acid-catalyzed domino condensation/intramolecular substitution. The best results were achieved when 1 equiv. of 2-(bromomethyl)benzaldehyde and 1 equiv. of arylhydrazine were reacted in the presence of 2 equiv. of K_2CO_3 and 5 mol% $FeCl_3$ in CH_3CN (Scheme 3.77). Using this protocol, the 2-substituted 1,2-dihydrophthalazines were obtained selectively with yields ranging from 60 to 91%. This method is one of the few examples of the synthesis of 1,2-dihydrophthalazines that does not start from a phthalazine or a phthalazine derivative. With respect to the reaction mechanism of dihydrophthalazine formation, it was assumed that the first step is a condensation between the

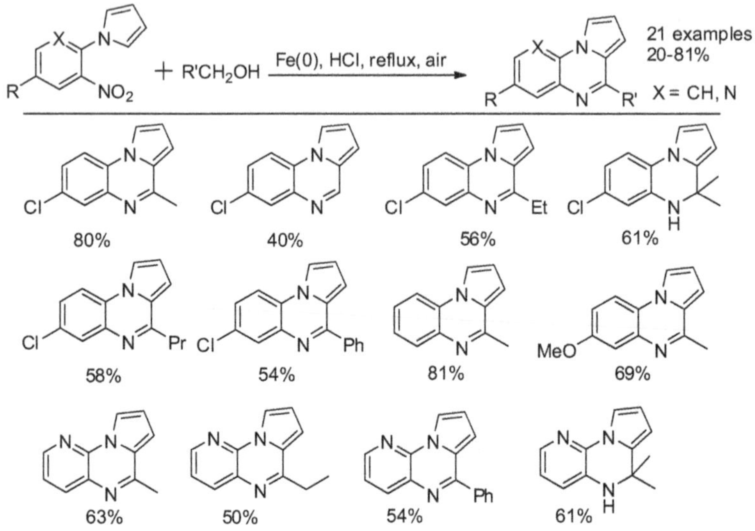

Scheme 3.76 Fe-mediated synthesis of quinoxalines.

Reaction Procedure (Scheme 3.76): To a solution of 1-(2-nitrophe-nyl)pyrrole derivative (1 equiv, 2.25 mmol) in alcohol (15 mL) was added iron powder (9 equiv., 20.25 mmol). HCl (12 M, 11 equiv., 2 mL) was added slowly *via* syringe at room temperature. After the addition was complete, the mixture was stirred at reflux for 48 h, cooled to room temperature and then quenched with a saturated aqueous solution of NaHCO$_3$. The resulting mixture was extracted with EtOAc (three times). The combined organic layers were washed with brine, dried on MgSO$_4$, filtered and evaporated. The products were purified by column chroma-tography with petroleum ether–ethyl acetate as eluent to yield the desired products.

aldehyde group of substrate and the NH$_2$ group of hydrazine to give the corresponding hydrazone. This was corroborated by the fact that, upon re-action between the two substrates in the presence of 5 mol% FeCl$_3$ for 2 h, the CHO signal of 2-(bromomethyl)benzaldehydes at $\delta = 10.26$ ppm had disappeared and was replaced by a signal at $\delta = 8.13$ ppm, which corres-ponds to the signal of a CH=N group. The hydrazone could not be isolated, but when 2 equiv. of K$_2$CO$_3$ were added to the above reaction mixture and the resulting mixture was reacted at 100 °C for 24 h, the dihydrophthalazine was formed in 78% yield.

Quinazolinones are important heterocycles with numerous reported biological activities. Various procedures have been developed for their preparation and iron salts as cheap catalysts have been applied in this area. In 2006, an iron(III) chloride hexahydrate-catalyzed synthesis of

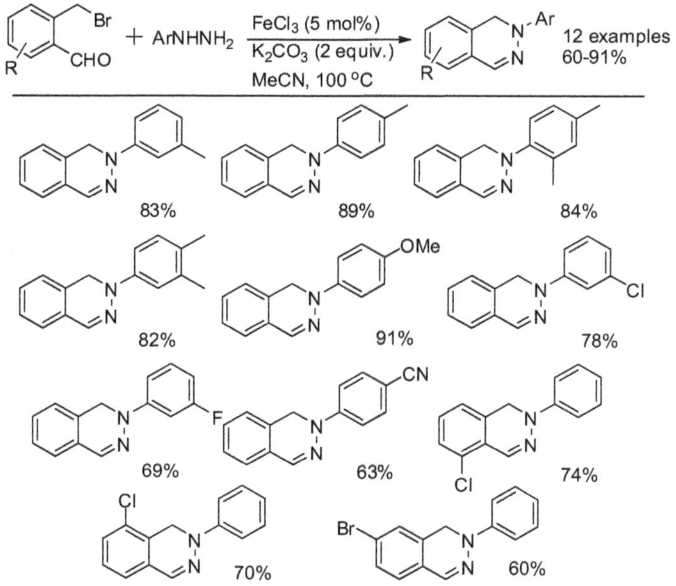

Scheme 3.77 Iron-catalyzed synthesis of 1,2-dihydrophthalazines.

Reaction Procedure (Scheme 3.77): An oven-dried vial was charged with K_2CO_3 (138 mg, 2 mmol), 2-(bromomethyl)benzaldehyde (99.5 mg, 0.5 mmol), arylhydrazine (0.5 mmol) and $FeCl_3$ (4 mg, 5 mol%) under argon. After sealing the vial, dry CH_3CN (2 mL) was added and the reaction mixture was stirred at 100 °C (oil-bath temperature) until the aldehyde was consumed. After cooling to room temperature, the vial was opened and the reaction mixture was poured into water (40 mL) and extracted with EtOAc (3×30 mL). The combined organic layers were dried over anhydrous Na_2SO_4 and concentrated *in vacuo*. The residue was purified by flash chromatography over silica gel to afford the product.

quinazolinones from anthranilamide and aldehydes was reported.[114] Condensation of aldehydes with anthranilamide in refluxing water using iron(III) chloride hexahydrate as an oxidant afforded 2-substituted 4(3*H*)-quinazolinones in good yields (77–93%) (Scheme 3.78). This method provides several advantages such as being environmentally friendly, having a simple work-up procedure and affording high yields.

Fu and co-workers developed a simple and efficient iron-catalyzed synthesis of benzothiadiazine-1,1-dioxide and quinazolinone derivatives in 2009 (Scheme 3.79).[115] These cascade coupling reactions were performed using inexpensive and environmentally friendly $FeCl_3$ as the catalyst and readily available 2-halobenzenesulfonamides, 2-bromobenzoic acids and amidine hydrochlorides as the starting materials and no additional ligand or additive was required. This is the first example of the construction of

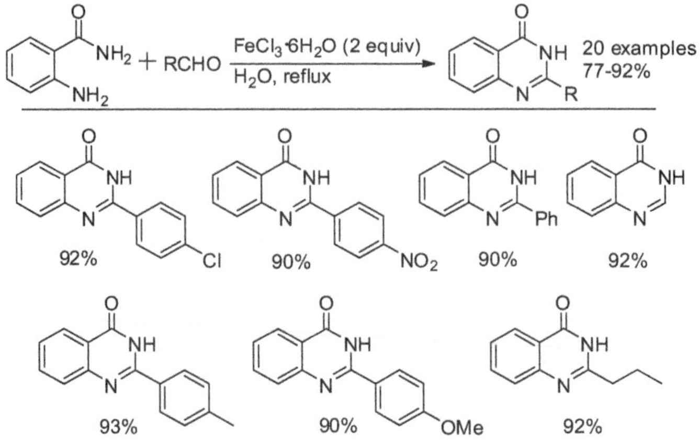

Scheme 3.78 Iron-mediated synthesis of quinazolinones.

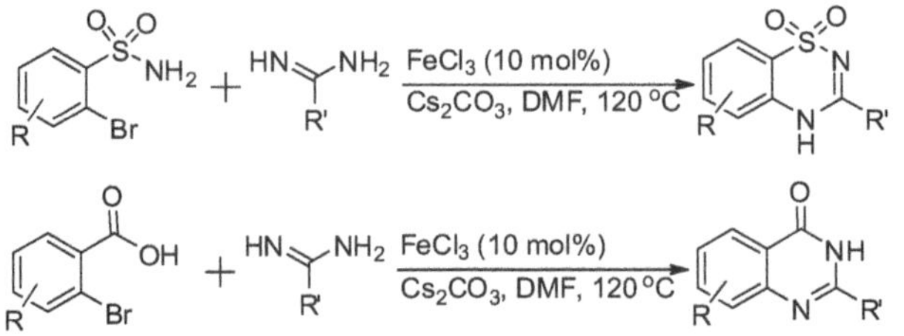

Scheme 3.79 Iron-catalyzed synthesis of benzothiadiazine-1,1-dioxides and quinazolinones.

Reaction Procedure (Scheme 3.79): A 25 mL round-bottomed flask containing a magnetic stirrer was charged with 2 mL of DMF and substituted 2-halobenzenesulfonamide or 2-bromobenzoic acid (1 mmol), Cs_2CO_3 (2 equiv.) and amidine hydrochloride (1.2 mmol) were added. After the mixture had been stirred for 10 min under a nitrogen atmosphere, $FeCl_3$ (0.1 mmol, 162 mg) was added. The mixture was stirred at 120 °C for 12 h. After completion of the reaction, the resulting solution was filtered and the solvent was removed with the aid of a rotary evaporator. The residue was purified by column chromatography on silica gel to give the desired product.

---

nitrogen-containing heterocycles *via* iron-catalyzed *N*-arylation in the absence of ligand. Therefore, this method is of practical application for the synthesis of the two different nitrogen-containing heterocycles.

2-Pyridones and their derivatives are widely naturally occurring and have a broad range of applications in the chemicals, medical and material sciences areas. In 2012, Maiti and co-workers reported a cascade cyclization of acetoacetanilide and aldehydes using iron as catalyst.[116] The reaction involves the activation of acetoacetanilides and a subsequent intermolecular C–C and

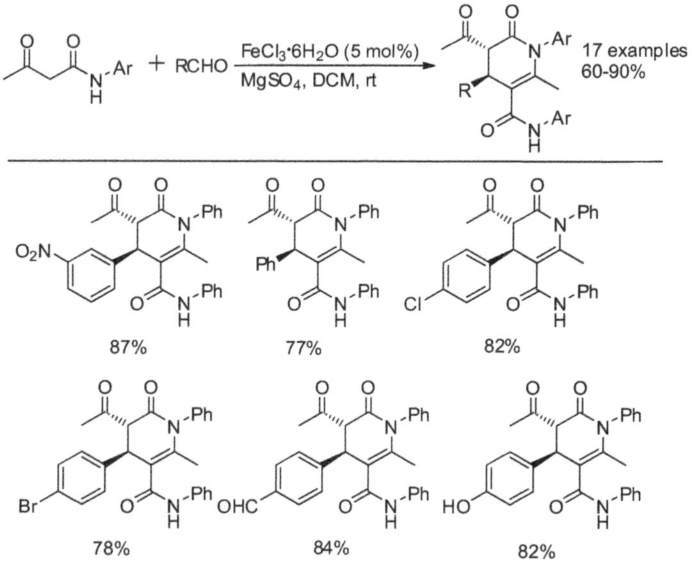

Scheme 3.80 Iron-catalyzed synthesis of 2-pyridones.

C–N bond-forming cascade cyclization process using non-toxic $FeCl_3 \cdot 6H_2O$ with its diverse Lewis acid activity as catalyst. Aromatic, aliphatic, α,β-unsaturated, sugar-based chiral and chromone aldehydes are regio- and stereoselectively cyclized with β-ketoanilide systems toward the construction of *trans*-1,2,3,4-tetrahydro-2-pyridone, *trans*-diene, sugar-based 2-oxa-5-azabicyclo[2.2.2] and 2-pyridone derivatives (Scheme 3.80).

A catalytic intramolecular aromatic C–H alkenylation of arenes with non-activated ketone carbonyls was realized in 2010.[117] With the assistance of an iron catalyst, 4-alkylenequinolin-2-ones were produced in good yields (Scheme 3.81). In this methodology, the quaternary carbon adjacent to the ketone carbonyl plays an essential role in this catalytic reaction. Based on the designed experiments and theoretical calculations, a possible reaction mechanism was proposed. The $FeCl_3$ first coordinated with carbonyl oxygen through an *n*-donor interaction, leading to the formation of an electrophilic center, which could be stabilized by the adjacent quaternary carbon. Subsequently an intramolecular electrophilic substitution proceeded to give the ring-closing product complex. Finally, the final product was formed by the fast elimination of one H_2O and the $FeCl_3$ was regenerated.

An inexpensive and environmentally benign $Fe(OTf)_3$-catalyzed alkyne-hydroarylation was reported in 2010.[118] This methodology was proven to be an effective approach for the synthesis of 1,2-dihydroquinolines and phenanthrenes and the desired products were produced in good to high yields (Scheme 3.82). Of special importance, the cationic iron catalyst permits the participation of electron-deficient aryl nucleophiles, which provides unprecedented and valuable examples of hydroarylation. The exact role of the

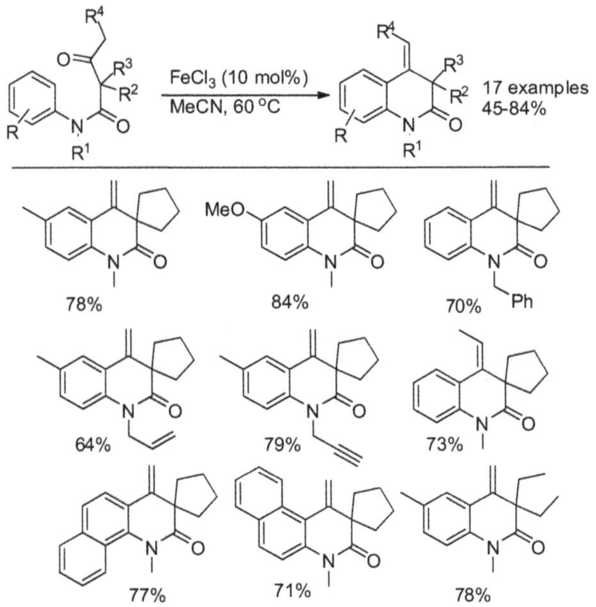

Scheme 3.81   Iron-catalyzed synthesis of 4-alkylenequinolin-2-ones.

Reaction Procedure (Scheme 3.81): Substrate (1 mmol) and FeCl₃ (0.1 mmol, 0.162 g) were added to CH₃CN (2.0 mL) with stirring. The mixture was warmed to 60 °C and stirred for 30 min. After the starting material was consumed (monitored by TLC), the reaction mixture was poured into saturated NaCl solution (20 mL). The mixture was extracted with dichloromethane (3 × 20 mL) and the combined organic phase was washed with water (3 × 20 mL), dried over MgSO₄, filtered and concentrated *in vacuo*. The crude product was purified by flash chromatography on silica gel (eluent: 10 : 1 petroleum ether–diethyl ether) to give the pure product.

---

iron catalyst in the present reaction is not clear, but the unique reaction mode would most likely be induced by the generated cationic iron–arene complexes during the reaction.

Zhou and co-workers developed an efficient method for the direct synthesis of dihydro- and tetrahydroisoquinolines from benzylamino-substituted propargylic alcohols (Scheme 3.83).[119] Advantages of the method are the easily accessible starting materials, mild conditions and a wide range of inexpensive catalysts, all of which allow it to be applied on an industrial scale. Furthermore, the results represent a good example that can illustrate the versatility of propargylic alcohols as alkylation precursors and demonstrate that their reactivity modes from propargylation through allenylation to alkenylation can be finely tuned simply by changing the substituents.

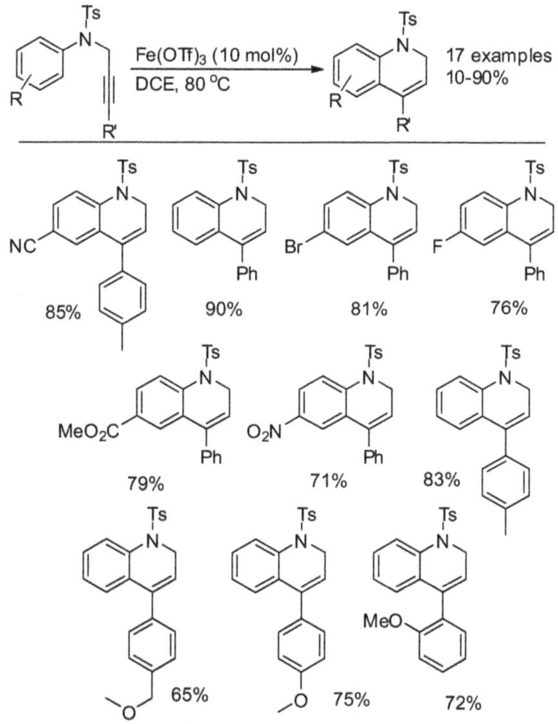

Scheme 3.82 Iron-catalyzed synthesis of 1,2-dihydroquinolines.

Reaction Procedure (Scheme 3.82): In a 20 mL Schlenk tube, a mixture of substrate (0.15 mmol), Fe(OTf)$_3$ (7.5 mg, 15 mmol) and DCE (0.5 mL) was heated at 80 °C for 3 h. After cooling to room temperature, the reaction mixture was passed through a short silica gel column with diethyl ether and then concentrated *in vacuo*. The crude product obtained was purified by column chromatography with hexane–ethyl acetate (5 : 1) as eluent to give the pure product.

The current methodology represents the first example of the intramolecular Friedel–Crafts reaction of propargylic alcohols.

Kim and co-workers developed an efficient synthetic method for the preparation of cyclic vinyl sulfides and vinylamines containing 1,2-dihydronaphthalene, 2*H*-chromene and 1,2-dihydroquinoline ring systems with a phenylsulfenyl or *N*-phenyl-*N*-tosyl group on the sp^2-hybridized benzylic carbon (Scheme 3.84).[120] The intramolecular hydroarylation of arylalkynyl phenyl sulfides and sulfonamides proceeded smoothly in a selective 6- or 7-*endo* mode in the presence of FeCl$_3$ and AgOTf. The method could be further extended to the preparation of dihydropyrano[2,3-*g*]chromene derivatives through a twofold Fe-catalyzed hydroarylation in a selective 6-*endo*

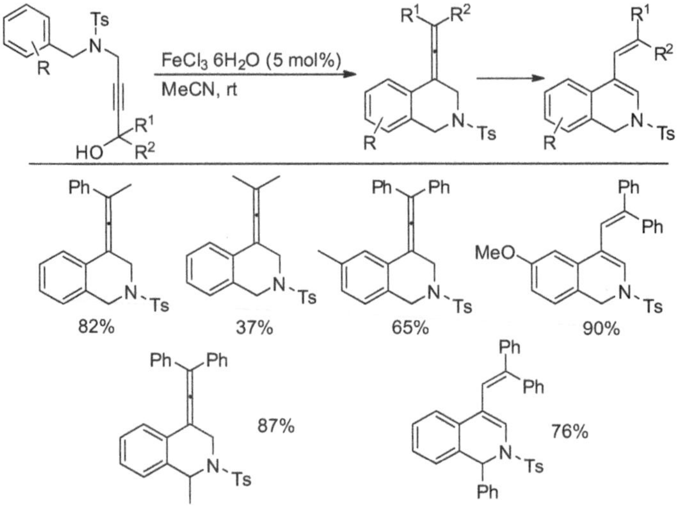

Scheme 3.83   Iron-catalyzed synthesis of isoquinolines.

Reaction Procedure (Scheme 3.83): To a solution of substrate (0.3 mmol) in CH_3CN (2.0 mL) was added $FeCl_3 \cdot 6H_2O$ (0.015 mmol). The reaction mixture was stirred at room temperature or other specified conditions (monitored by TLC), then the solvent was removed under reduced pressure and the crude material was purified by silica gel column chromatography to provide the desired product.

---

mode. Of special importance, the cationic iron catalyst permits the participation of electron-deficient aryl nucleophiles, which provide useful examples of hydroarylation.

Padrón and co-workers reported a novel iron(III) halide-mediated aza-Prins cyclization in 2006.[121] The coupling between homoallyl tosylamines and homopropargyl tosylamines provided in good yields *trans*-2-alkyl-4-halo-1-tosylpiperidines and 2-alkyl-4-halo-1-tosyl-1,2,5,6-tetrahydropyridines, respectively (Scheme 3.85). The reaction proceeded satisfactorily with aromatic and aliphatic substrates, and also with enolizable and non-enolizable aldehydes. It is noteworthy that the carbon–carbon bond formation is very rapid even at 0 °C, usually being completed within 10 min. The process is based on the consecutive generation of a γ-unsaturated iminium ion and further nucleophilic attack by the unsaturated carbon–carbon bond. Later, this reaction was realized in a catalytic manner.[122] By using a catalytic amount of Fe(acac)$_3$ (7 mol%) and a trimethylsilyl halide as halogen source, the desired products were produced in excellent yields. The method displays a broad substrate scope and is economical, environmentally friendly and experimentally simple. This catalytic method permits the construction of chloro, bromo and iodo heterocycles in high yields by the suitable combination of an iron(III) source, trimethylsilyl halide and solvent.

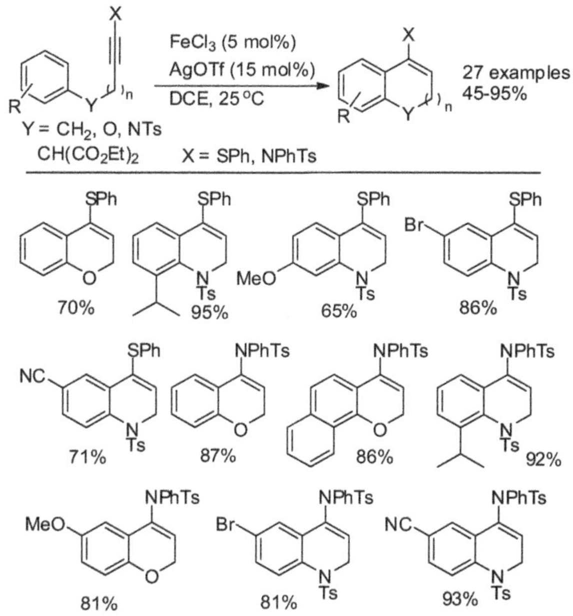

Scheme 3.84 Iron-catalyzed synthesis of 2*H*-chromenes.

Reaction Procedure (Scheme 3.84): A suspension of FeCl$_3$ (5 mol%) and AgOTf (15 mol%) in DCE (0.8 mL) was stirred at 25 °C for 5 min, then a solution of 4-methyl-*N*-phenyl-*N*-[3-(phenylthio)prop-2-ynyl]benzenesulfonamide (118.1 mg, 0.3 mmol) in DCE (0.7 mL) was added under nitrogen. After the reaction mixture had been stirred at 80 °C for 20 min, it was quenched with water. The aqueous layer was extracted with dichloromethane (2×15 mL) and the combined organic layers were washed with water and brine, filtered and dried under reduced pressure. Silica gel column chromatography (eluent: 1:15 ethyl acetate–hexane) gave the pure product.

In 2013, Ma and co-workers demonstrated a highly regioselective FeCl$_3$-catalyzed cyclization reaction of 3,4-allenylamines or alcohols with aldehydes in the presence of TMSCl.[123] This reaction produces 3-chloromethyl-1,2,5,6-tetrahydropyridine or 3-chloromethyl-5,6-dihydro-2*H*-pyran derivatives efficiently and highly selectively due to the high stability of the allyl cation intermediate (Scheme 3.86). The combination of FeCl$_3$ and TMSCl works to promote the condensation of the 3,4-allenylamines or alcohols with aldehydes, and TMSCl also serves as the halide source.

An efficient, general and one-pot procedure for the synthesis of multi-substituted xanthene derivatives through Fe(III)-catalyzed reactions of 2-aryloxybenzaldehydes with electron-rich arenes was developed by Liu and co-workers in 2009 (Scheme 3.87).[124] This method offers several advantages,

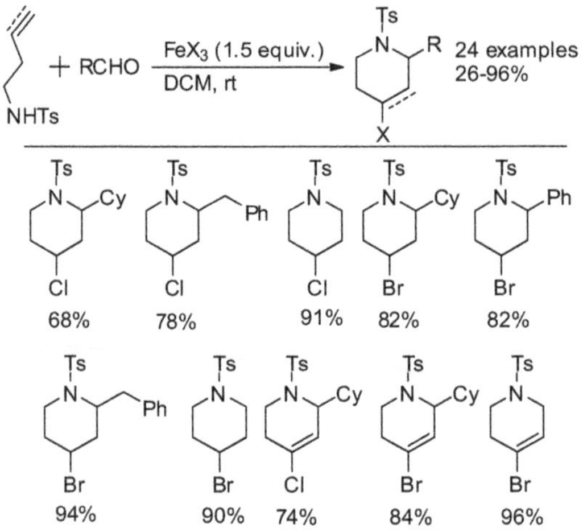

Scheme 3.85   Iron-catalyzed synthesis of piperidines.

> **Reaction Procedure** (Scheme 3.85): To a solution of homoallyl tosylamine or homopropargyl tosylamine (1 equiv.) and aldehyde (1.5 equiv.) in dry CH_2Cl_2 was added anhydrous FeX_3 (1.5 equiv.) in one portion. The reaction was completed in ~10 min, quenched by addition of water and extracted with CH_2Cl_2. The combined organic layers were dried over magnesium sulfate and the solvent was removed under reduced pressure. This crude reaction mixture was purified by flash silica gel column chromatography (eluent: *n*-hexane–ethyl acetate).

such as high selectivities, mild reaction conditions and easily accessible starting materials. A mechanistic study revealed that a C–C bond cleavage of a triarylmethane intermediate (which was detected during the optimization process) might be involved in this domino process. Additionally, a useful method to construct highly substituted tetrahydroquinolines through an iron(III) chloride (30 mol%)-mediated domino Mannich and intramolecular Friedel–Crafts alkylation followed by intermolecular Friedel-Crafts alkylation reactions of aliphatic aldehydes with aromatic amines was reported in 2011 by Liang and co-workers.[125] In general, the desired products were produced in low yields. Iron acts as a Lewis acid in this process and other traditional Lewis acids [$InCl_3$, $BF_3 \cdot Et_2O$, $Sc(OTf)_3$] also work in this transformation.

Takacs and co-workers reported the iron-catalyzed cyclization of trienes in the 1990s.[126] Various ring sizes could be produced by changing the position of the double bonds. Later, the preparation of 2,3,4-trisubstituted piperidines by intramolecular cyclization of imines and alkenes was developed.[127] In this reaction, *N*-benzyl- or *N*-tosyl-*N*-(4-methyl-3-pentenyl)aminoaldehyde benzylimines, which are obtained from alanine, leucine or phenylalanine methyl esters in five steps, can be cyclized diastereoselectively in the presence of Lewis acids to give 3-amino-2,4-dialkyl-substituted piperidines.

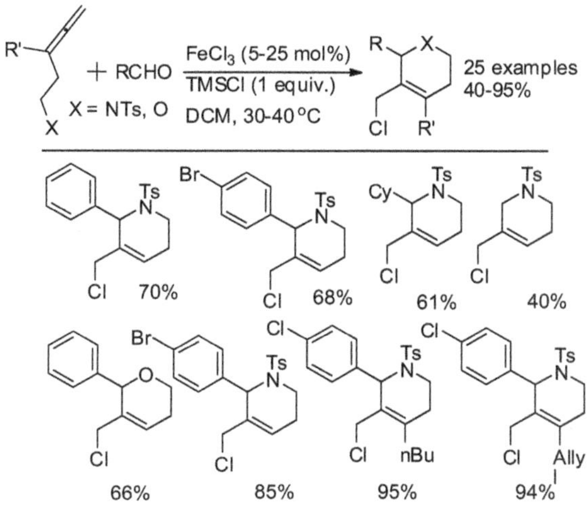

Scheme 3.86 Iron-catalyzed synthesis of pyrans.

Reaction Procedure (Scheme 3.86): To a Schlenk tube were added FeCl$_3$ (8.0 mg, 0.049 mmol) in CH$_2$Cl$_2$ (0.5 mL), N-(penta-3,4-dienyl)-4-tolylsulfonamide (44.7 mg, 0.19 mmol) in CH$_2$Cl$_2$ (0.5 mL), 4-chlorobenzaldehyde (34.1 mg, 0.24 mmol) in CH$_2$Cl$_2$ (0.5 mL) and TMSCl (22.4 mg, 0.21 mmol) in CH$_2$Cl$_2$ (0.5 mL) sequentially. The mixture was stirred at 30 °C for 8 h. When the reaction was complete as monitored by TLC (eluent: 5 : 1 petroleum ether–ethyl acetate), the resulting mixture was diluted with CH$_2$Cl$_2$ (5 mL) and ethyl acetate (10 mL). The pure product was isolated after column chromatography on silica gel.

The product distribution and diastereoselectivity depend on the type of Lewis acid and nitrogen-protecting group. Benzyl-protected imines give 2-alkyl-3-(benzylamino)-4-isopropenylpiperidines with FeCl$_3$ and 2-alkyl-3-(benzylideneamino)-4-isopropylpiperidines with TiCl$_4$. Tosyl-protected imines show a decreased level of selectivity.

An interesting iron-catalyzed hydrofunctionalization of allenes was reported by Kang's group in 2012.[128] A range of tetrahydropyrans and piperidines were produced by Fe(III)-catalyzed intramolecular hydroalkoxylation and hydroamination reactions of allenes (Scheme 3.88). Various Fe catalysts with different counterions were tested. Fe(III) salts, FeCl$_3$, Fe(TFA)$_3$, Fe(OTs)$_3$ and Fe(OTf)$_3$ allowed efficient cycloisomerization reactions under mild conditions. They are inexpensive and naturally abundant, leading to their wide industrial applicability in such reactions. Their activities toward allene and alkene activation depended sensitively on their counterion and reaction conditions. Mechanistic studies of the reaction intermediates revealed a new reaction pattern involving the Fe catalysts and diene substrates. Hydrofunctionalization involved traditional coordination, nucleophilic addition

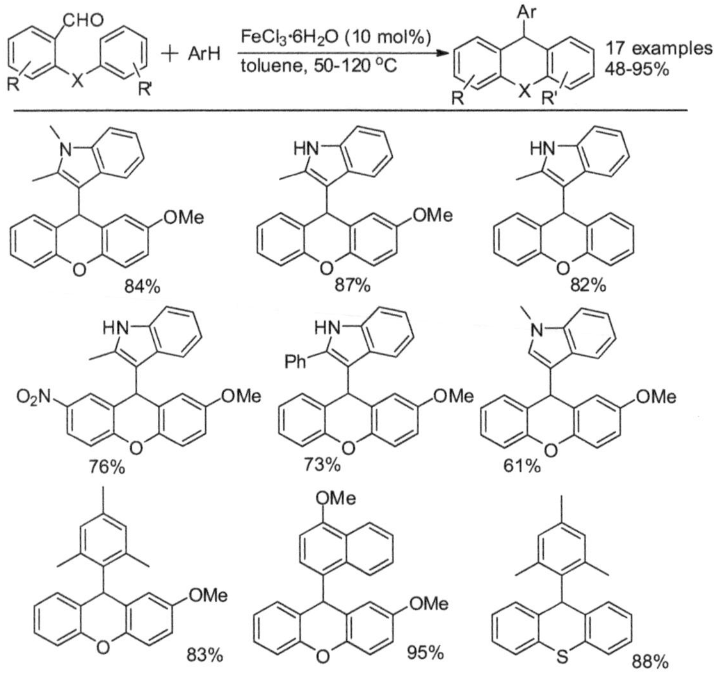

Scheme 3.87   Iron-catalyzed synthesis of xanthenes.

Typical Reaction Procedure (Scheme 3.87): A mixture of 2-(4-methoxyphenoxy)benzaldehyde (57 mg, 0.25 mmol), 1,2-dimethyl-1*H*-indole (44 mg, 0.3 mmol) and FeCl$_3$·6H$_2$O (6.7 mg, 10 mol%) in toluene (3 mL) was stirred at 50 °C for 3 h, then the reaction mixture was quenched with saturated NaHCO$_3$ solution and extracted with ethyl acetate. The organic layer was washed with brine and dried over anhydrous Na$_2$SO$_4$. After removal of the solvent, the residue was purified by column chromatography on silica gel (eluent: 10:1 petroleum ether–ethyl acetate) to afford 3-(2-methoxy-9*H*-xanthen-9-yl)-1,2-dimethyl-1*H*-indole as a white solid in 84% isolated yield.

and proto-demetalation, similar to previous observations of allenes in reactions. The cyclized product can then undergo isomerization depending on the Fe(III) catalyst and reaction conditions. For the cyclization reactions of dienes, diene–iron complexes may lead to cycloisomerization products.

An eco-friendly and highly diastereoselective synthesis of *cis*-2,6-disubstituted piperidines and *cis*-2,6-disubstituted tetrahydropyrans was developed in 2010.[129] These valuable building blocks for the synthesis of biologically active targets were produced in good yields and high selectivity (Scheme 3.89). The mild catalytic conditions developed allowed the use of substrates bearing alkyl or ester groups. The thermodynamic epimerization

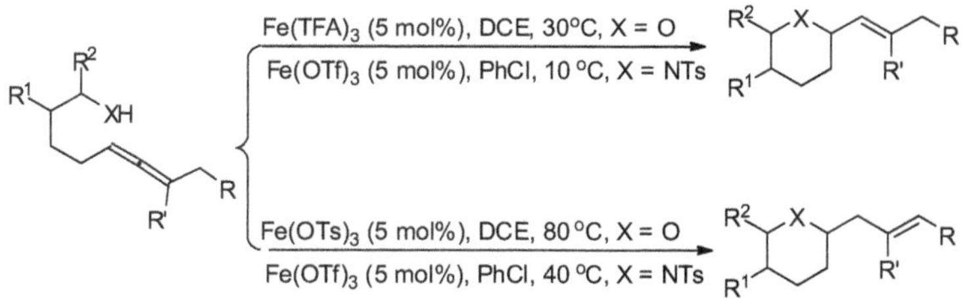

Scheme 3.88 Iron-catalyzed hydrofunctionalization of allenes.

Typical Reaction Procedure (Scheme 3.88): To a solution of allenyl alcohol (1 mmol) in DCE (0.2 M) was added 5 mol% of Fe(III) catalyst (0.05 mmol) and the resulting mixture was stirred at 30 or 80 °C. Upon completion of the reaction (monitored by TLC), the solution was concentrated and purified by flash column chromatography to afford the cyclized product.

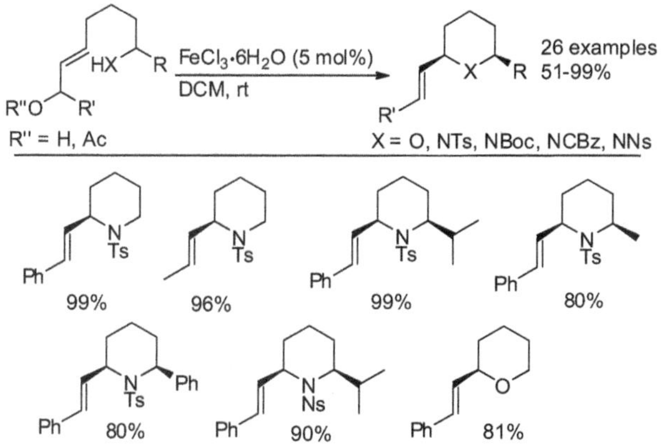

Scheme 3.89 Iron-catalyzed synthesis of tetrahydropyrans and piperidines.

Reaction Procedure (Scheme 3.89): $FeCl_3 \cdot 6H_2O$ (5 mol%) was added to a solution of substrate in CH_2Cl_2 (0.1 M) at room temperature. After the required time, the resulting mixture was directly filtered through a pad of silica gel (CH_2Cl_2) and the volatiles were removed under reduced pressure to yield the corresponding cyclized product without any further purification.

of the produced 2-alkenyl 6-substituted piperidines and tetrahydropyrans induced by $FeCl_3$ is the key to accounting for the high diastereoselectivities observed in favor of the most stable *cis*-isomers.

Additionally, pyran derivatives can be prepared by another approach. Gorman and Tomlinson prepared iron(III) 2-ethylhexanoate as a novel, mild

Lewis acid catalyst for the stereoselective Diels–Alder reaction of ethyl (*E*)-4-oxobutenoate with alkyl vinyl ethers to produce stereoselectively ethyl *cis*-2-alkoxy-3,4-dihydro-2*H*-pyran-4-carboxylates with diastereoisomeric excesses (*de*) as high as 98%.[130]

Moreover, isochromenones as biologically active molecules were also prepared with the assistance of iron catalysts. In 2011, Zeni and co-workers studied the cyclization reaction behavior of 2-alkynylaryl esters, which was promoted by FeCl₃ and diorganyl dichalcogenides, showing that the six-membered ring was selectively obtained in good yields when the reaction was carried out at room temperature under an air atmosphere (Scheme 3.90).[131] This procedure was successfully applied in the cyclization of 2-alkynylaryl esters with diorganyl dichalcogenides bearing various

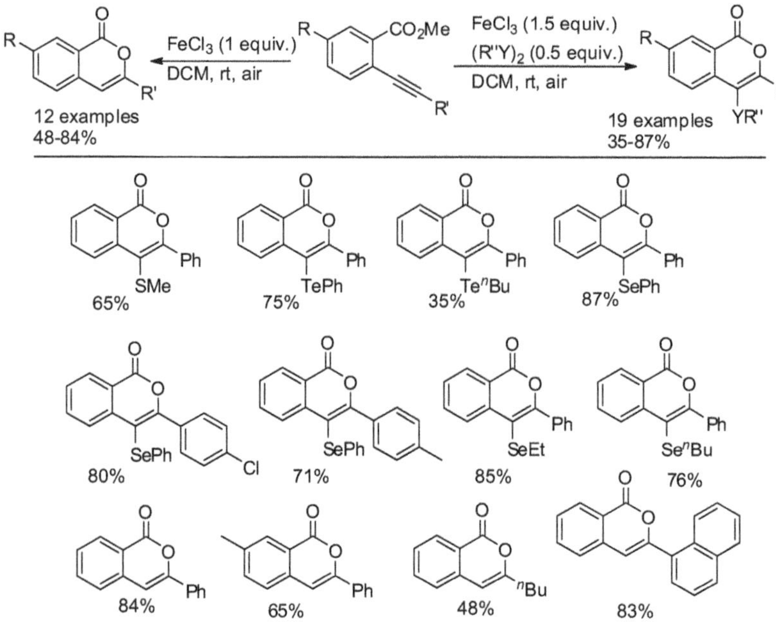

Scheme 3.90   Iron-catalyzed synthesis of isochromenones.

Reaction Procedure (Scheme 3.90): In a Schlenk tube, under air, containing CH_2Cl_2 (4 mL) were added $FeCl_3$ (0.041 g, 1 equiv., 99.99% purity from commercial suppliers) and 2-alkynylaryl esters (0.25 mmol) in CH_2Cl_2 (1 mL) and the reaction mixture was stirred overnight at room temperature. Subsequently, the mixture was diluted with CH_2Cl_2 (20 mL) and washed with a saturated solution of NH_4Cl (20 mL). The organic phase was separated, dried over $MgSO_4$ and concentrated under vacuum. The residue was purified by flash chromatography (eluent: 95 : 5 hexane–ethyl acetate).

functionalities. The reactions tolerated not only diorganyl diselenides but also diorganyl ditellurides, generally resulting in reasonable to good yields, and proceeded under relatively mild conditions. In addition, when the reaction was carried out using only $FeCl_3$ without diphenyl diselenide, isochromenones were obtained as the product, without the RY group at the 4-position. This result is significant since two classes of isochromenones can be obtained under the same reaction conditions. Furthermore, since iron salts are readily available commercially and are less expensive and relatively non-toxic, this method could be considered an economical and eco-friendly protocol. The reaction with dimethyl disulfide was also reported using this methodology. Around the same time, a facile synthetic approach to 4-sulfenylisocoumarins by the $FeCl_3$-promoted electrophilic cyclization of 2-alkynylaryl esters and disulfides was reported by Zhou and co-workers.[132] This procedure allows the simultaneous construction of the isocoumarin ring system and the installation of a sulfenyl functionality at the 4-position of the isocoumarin ring.

Zeni and co-workers developed an alternative and efficient method for the synthesis of 3-organoselenylchromenone derivatives *via* the intramolecular 6-*endo-dig* cyclization of alkynyl aryl ketones (Scheme 3.91).[133] The methodology proved to be highly regioselective, giving only the six-membered regioisomers, and was carried out using $FeCl_3$–RSeSeR at room temperature and under an ambient atmosphere, which is considered to constitute an economical and eco-friendly protocol. As many ways to transform the resulting selenium functionalities into other substituents exist, this approach to 3-organochalcogen chromenones should be useful in organic synthesis. Concerning the adverse reputation relating to the bad smell, toxicity or instability of selenium compounds, the authors stated that all compounds prepared by this methodology are solids or oils, completely odorless, very stable and can be purified and stored in the laboratory in a simple flask for more than 1 month.

In 2011, an interesting iron-catalyzed method for the preparation of flavones from terminal alkynes and 2-hydroxybenzaldehydes was reported by Maiti and co-workers.[134] Atmospheric oxygen was applied as a stoichiometric oxidant and piperidine was used as organocatalyst (Scheme 3.92), and good to excellent yields of flavones were obtained. The notable advantages of this method are operational simplicity, use of inexpensive and eco-friendly $FeCl_3$ as catalyst and ease of isolation of the products.

An iron-catalyzed cross-dehydrogenative coupling reaction occurring *via* base-promoted homolytic aromatic substitutions (BHASs) was reported more recently.[135] Fluorenones and xanthones were readily prepared *via* CDC starting with readily available *o*-formyl biphenyls and *o*-formyl biphenyl ethers, respectively (Scheme 3.93). Concerning the reaction mechanism, a radical process was proposed. Initiation of the radical chain reaction is best achieved with small amounts of $FeCp_2$ (0.1 or 1 mol%). Initiation occurs by reducing ^tBuOOH with FeX_2 to give the *tert*-butoxyl radical along with an Fe(III) complex. The *tert*-butoxyl radical then abstracts the H-atom from the

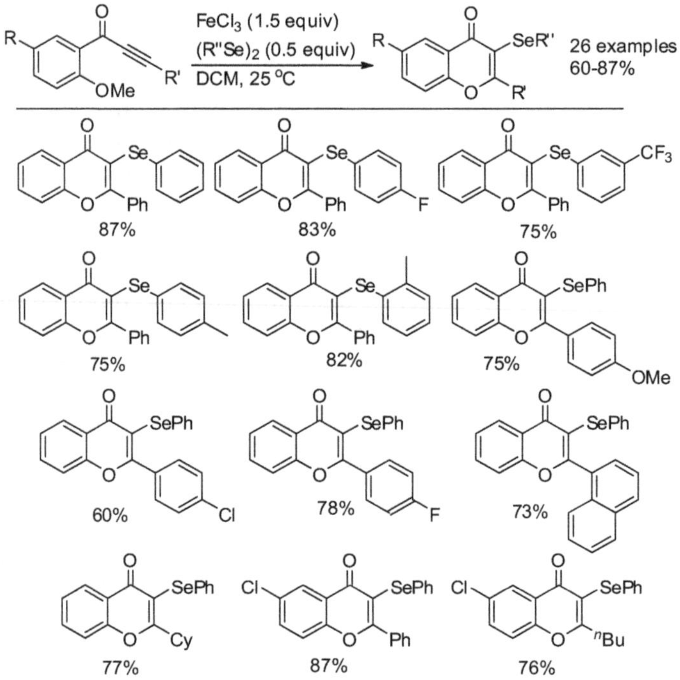

Scheme 3.91 Iron-mediated synthesis of 3-organoselenylchromenone derivatives.

Reaction Procedure (Scheme 3.91): To a Schlenck tube, under an ambient atmosphere, containing CH_2Cl_2 (4 mL) were added $FeCl_3$ (0.061 g, 1.5 equiv.) and diorganyl dichalcogenide (0.5 equiv.) and the reaction mixture was stirred for 20 min at room temperature. The appropriate alkynyl aryl ketone (0.25 mmol) in CH_2Cl_2 (1 mL) was then added and the reaction mixture stirred for a determined time. The mixture was diluted with CH_2Cl_2 (20 mL) and washed with a saturated solution of NH_4Cl (20 mL). The organic phase was separated, dried over $MgSO_4$ and concentrated under vacuum. The residue was purified by flash chromatography with hexane–acetate (95 : 5) as eluent.

---

aldehyde to give an acyl radical, which attacks the arene to generate the cyclohexadienyl radical. Deprotonation with the basic hydroxide anion leads to the biaryl radical anion. Deprotonation is facilitated by the neighboring carbonyl group. The biaryl radical anion then reduces 'BuOOH by SET to provide fluorenone and the chain-propagating *tert*-butoxyl radical along with the basic hydroxide anion. Experimentally, by running the reaction in the presence of TEMPO (1 equiv.), the TEMPO ester was isolated in 36% yield and fluorenone was not formed.

An efficient, regioselective, iron-catalyzed intramolecular hydroaryloxylation of 2-propargylphenols and -naphthols was reported by Li and co-workers

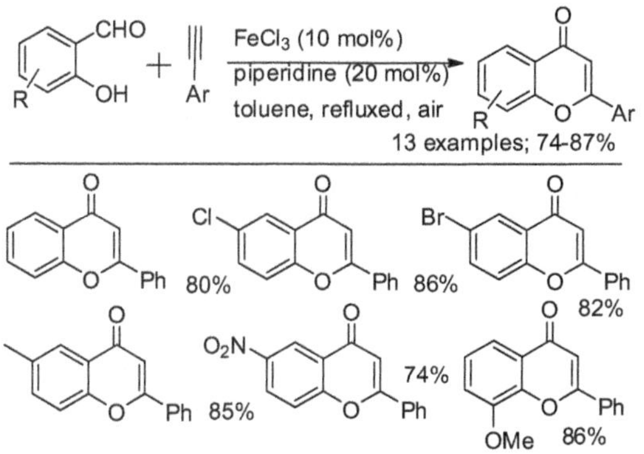

Scheme 3.92 Iron-catalyzed synthesis of flavones.

Reaction Procedure (Scheme 3.92): A mixture of 2-hydroxybenzaldehyde (122 mg, 1.0 mmol), phenylacetylene (153 mg, 1.5 mmol), anhydrous FeCl$_3$ (16 mg, 0.10 mmol) and piperidine (17 mg, 0.20 mmol) was added to dry toluene (3 mL) in a 25 mL round-bottomed flask fitted with a reflux condenser and a calcium chloride guard tube and refluxed for 6–12 h. The reaction mixture was decomposed with water and extracted with diethyl ether (3×15 mL), washed with water followed by brine solution and dried over anhydrous Na$_2$SO$_4$.

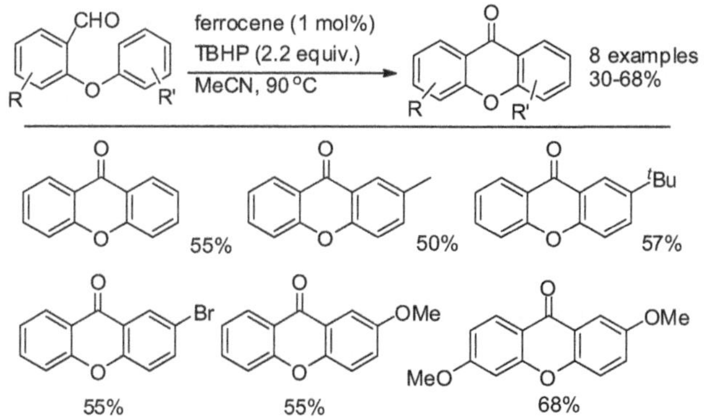

Scheme 3.93 Iron-catalyzed synthesis of xanthones.

Reaction Procedure (Scheme 3.93): A solution of ferrocene in MeCN (5.0 mM, 1 mL, 5 µmol, 1 mol%) was added to a solution of the selected 2-phenoxybenzaldehyde (0.5 mmol, 1.0 equiv.) in MeCN (1 mL) and the resulting reaction mixture was stirred for 5 min at room temperature before TBHP (aq. 70 wt%, 76 µL, 0.55 mmol, 1.1 equiv.) was added. After stirring at that temperature for an additional 5 min, the resulting reaction mixture was heated at 90 °C for 15 min and stirring was continued at that temperature for 12 h. More TBHP (1.1 equiv.) was added and stirring was continued for a further 12 h at 90 °C. After cooling to room temperature, the crude reaction mixture was filtered through a short pad of silica with CH_2Cl_2 (200 mL) as eluent and the filtrate was concentrated *in vacuo*. After crude ^1H NMR analysis using dibromomethane as an internal standard and concentration, the residue was purified by flash chromatography (petroleum ether/tert-butyl methyl ether) to afford the desired xanthones.

in 2009.[136] The reactions proceed through an *endo-dig* cyclization to afford benzopyran or naphthopyran derivatives in good to high yields using iron(III) chloride as the catalyst with the assistance of aniline in dimethylformamide (DMF). Later, a one-pot cascade reaction for the synthesis of polysubstituted benzofurans and naphthopyrans from simple phenols and propargylic alcohols catalyzed by iron(III) was developed by Yuan and Han.[137] The results demonstrate that the structural specificity for the formation of furan and pyran products is controlled by the structural nature of the propargylic alcohols. Specifically, benzofurans could be synthesized efficiently from phenols and secondary propargylic alcohols in the presence of 5 mol% of $FeCl_3 \cdot 6H_2O$ catalyst. On the other hand, pyran derivatives were obtained exclusively when tertiary propargylic alcohols were employed (Scheme 3.94). Mechanistic studies revealed that presumably due to the discriminating steric effect of secondary and tertiary propargylic alcohols, the Fe-catalyzed Friedel–Crafts reaction of phenols with the two types of alcohols proceeds *via* different routes. Most importantly, we demonstrated for the first time that fully 2,3,4-substituted naphthopyrans could be synthesized efficiently *via* an iron-catalyzed one-pot cascade reaction. Consequently, this methodology provides straightforward pathways for the versatile syntheses of valuable benzofuran and pyran derivatives from simple phenolic compounds and propargylic alcohols.

An $FeCl_3$-catalyzed intramolecular alkyne–aldehyde metathesis process for the library synthesis of functionalized 2*H*-chromenes from readily available alkynyl ethers of salicylaldehyde derivatives was developed in 2011.[138] A number of functional groups, including methoxy, phenyl, chloro, fluoro and bromo, are well tolerated under the reaction conditions (Scheme 3.95). The attractive features of this procedure are the mild reaction conditions, high atom economy and the use of inexpensive starting materials and an environmentally friendly catalyst. Moreover, this protocol can introduce a carbonyl functionality into the chromene unit. The reaction proceeds through a formal [2 + 2] cycloaddition reaction; the exact role of the iron salt is not known, but the reaction is initiated by the coordination of the carbonyl group with the iron salt through Lewis acid–base interaction to form a reactive species. Then nucleophilic attack of the alkyne unit on the activated aldehyde group generates a vinylic cation intermediate and subsequent cyclization by intramolecular nucleophilic attack of the carbonyl oxygen at

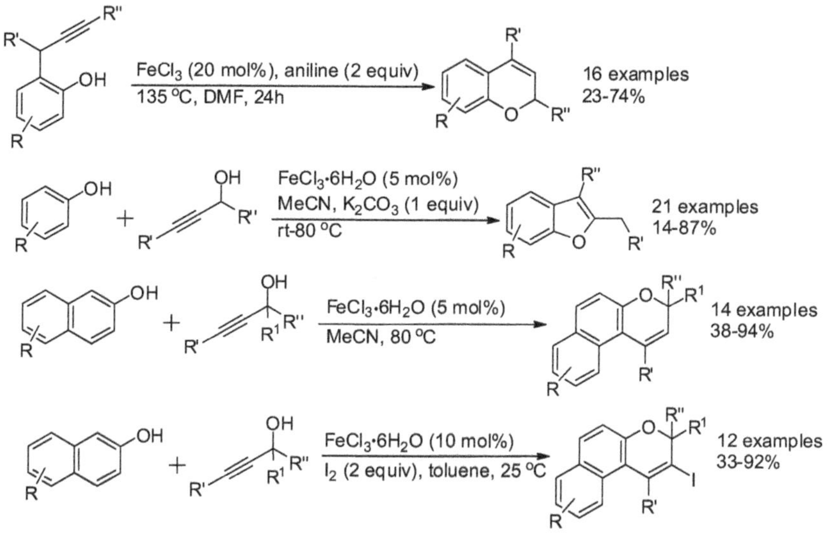

Scheme 3.94 Iron-catalyzed reactions of propargylic derivatives.

Reaction Procedure (Scheme 3.94): **General Procedure for the One-Pot Synthesis of Benzofurans:** An MeCN solution (3 mL) of phenol (0.6 mmol), propargylic alcohol (0.5 mmol) and FeCl$_3 \cdot$6H$_2$O (6.8 mg, 0.025 mmol) was stirred at 25 °C in a sealed Schlenck tube until the propargylic alcohol had disappeared (monitored by TLC). Then K$_2$CO$_3$ (69 mg, 0.5 mmol) was recharged *in situ* to the reaction vessel and the mixture was further stirred at 80 °C for several hours. After completion of the reaction (monitored by TLC), 30 mL of DCM were added and the mixture was filtered and concentrated under reduced pressure to yield the crude product, which was purified by silica gel chromatography (eluent: 10 : 1 petroleum ether–CH$_2$Cl$_2$) to give the desired pure product.

   General Procedure for the Synthesis of Naphthopyrans: An MeCN solution (3 mL) of propargylic alcohol (0.5 mmol), 2-naphthol (0.6 mmol) and FeCl$_3 \cdot$6H$_2$O (6.8 mg, 0.025 mmol) was stirred in a sealed Schlenck tube at 80 °C. After completion of the reaction (monitored by TLC), the reaction mixture was concentrated under reduced pressure and purified by silica gel chromatography (eluent: 10 : 1 petroleum ether–CH$_2$Cl$_2$) to give the desired product.

   General Procedure for the One-Pot Three-Component Synthesis of 3-Iodinated Naphthopyrans: A toluene solution (3 mL) of propargylic alcohol (0.9 mmol), 2-naphthol (0.3 mmol), I$_2$ (152 mg, 0.6 mmol) and FeCl$_3 \cdot$6H$_2$O (8.1 mg, 0.03 mmol) was stirred in a sealed Schlenck tube at 25 °C. After completion of the reaction (monitored by TLC), ethyl acetate (30 mL) was added and the mixture was washed successively three times with saturated aqueous Na$_2$S$_2$O$_3$ solution and brine. The organic layer was then dried over Na$_2$SO$_4$, filtered and concentrated. The mixture was purified by silica gel chromatography (eluent: 10 : 1 petroleum ether–CH$_2$Cl$_2$) to give the desired product.

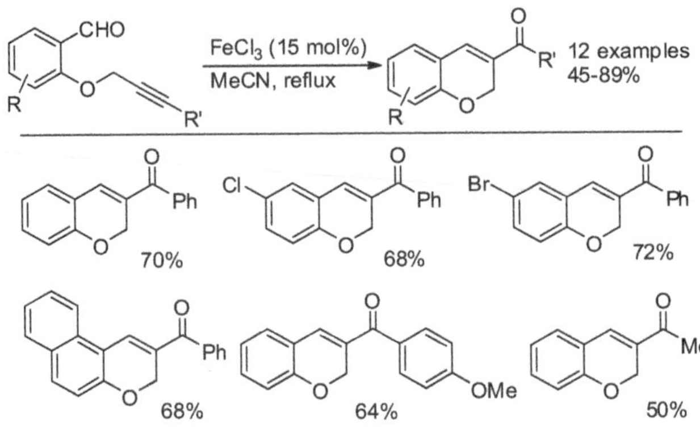

Scheme 3.95 Iron-catalyzed synthesis of 2*H*-chromenes.

Reaction Procedure: (Scheme 3.95): Substrate (0.5 mmol) was placed in a 25 mL round-bottomed flask containing 3 mL of dry acetonitrile solvent. Anhydrous FeCl$_3$ (12 mg, 0.08 mmol) was added and the reaction mixture was heated at reflux for 4 h under an argon atmosphere in the dark. After complete conversion of the starting material (monitored by TLC), acetonitrile was distilled out under reduced pressure and the residue was purified by silica gel column chromatography to afford the pure product.

the vinylic carbocation center leads to the formation of an oxetene intermediate, regenerating the iron salt for the next catalytic cycle. The oxetene intermediate then undergoes formal [2 + 2] cycloreversion, producing the desired 2*H*-chromene derivative with complete regioselectivity. The generation of a vinylic carbocation is supported by the fact that a terminal alkyne did not give the product and aryl-substituted alkynes were more efficient than alkyl-substituted alkynes in this transformation.

In 2008, Fan and Wang developed a novel annulation reaction for the construction of functionalized 4*H*-chromenes from readily available substrates under mild conditions.[139] Numerous desired products were produced in good to excellent yields (Scheme 3.96). Cu(OTf)$_2$ and Zn(OTf)$_2$ were also tested as catalysts for this transformation, but only trace amounts of the desired products were obtained.

Li and co-workers presented an iron-mediated tandem annulation strategy for the preparation of functionalized indeno[1,2-*c*]chromenes and 5*H*-naphtho[1,2-*c*]chromenes from 2-[2-(ethynyl)phenoxy]-1-arylethanones (Scheme 3.97).[140] This work was the first to disclose an iron-mediated method through

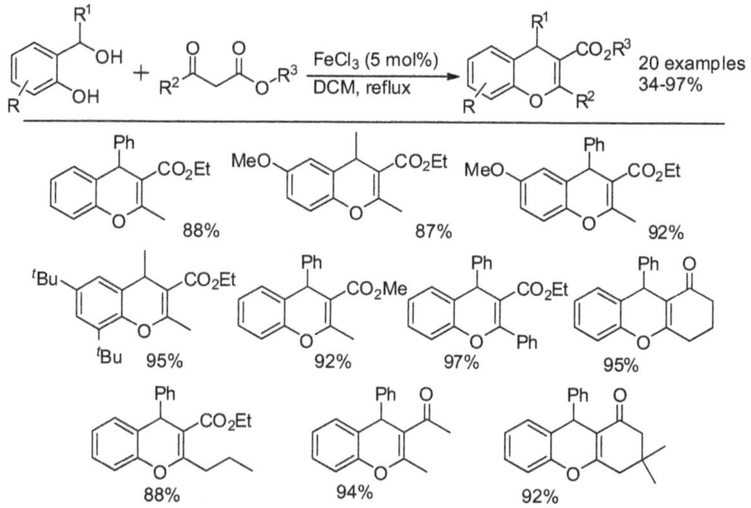

Scheme 3.96 Iron-catalyzed synthesis of 4*H*-chromenes.

Reaction Procedure (Scheme 3.96): FeCl$_3$ (8.1 mg, 0.05 mmol) and 0.50 g of 4 Å molecular sieves were added to a solution of substrate (0.5 mmol) and ethyl 3-oxobutanoate (1.0 mmol) in freshly distilled CH$_2$Cl$_2$ (1.5 mL). The resulting mixture was refluxed for 12 h, then cooled to room temperature and quenched with saturated NaHCO$_3$, and the mixture was extracted twice with CH$_2$Cl$_2$. The combined organic extracts were dried over Na$_2$SO$_4$ and filtered. Solvents were evaporated under reduced pressure. The residue was purified by column chromatography on silica gel using petroleum ether–EtOAc (10:1) as eluent.

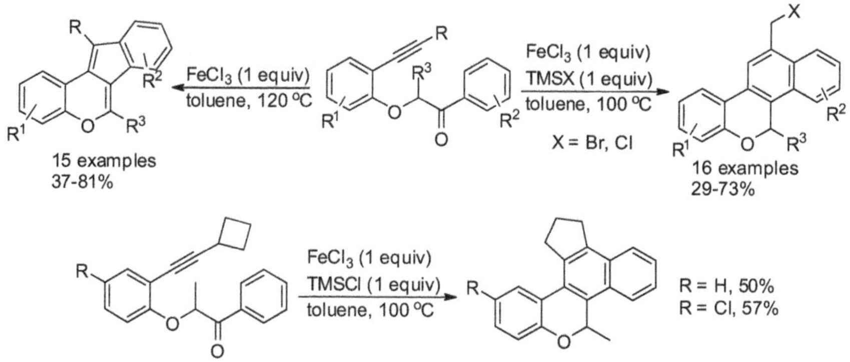

Scheme 3.97 Iron-catalyzed synthesis of 4*H*-chromenes.

sequential electrophilic addition of a ketone to an alkyne and annulation tandem reaction. Importantly, a halide is introduced into the products by a ring-opening process among the [3 + 3] annulation reactions of alkynylcyclopropanes, which makes the methodology more attractive for organic synthesis. A reaction mechanism was also proposed. Initially, coordination of FeCl$_3$ to the substrate activates the alkyne moiety, followed by sequential electrophilic addition of a ketone and FeCl$_3$ to an alkyne, affording another intermediate that subsequently undergoes electrophilic addition to an aromatic ring and ring-opening. The final product can be obtained after nucleophilic addition and rearrangement.

3.3 Other Heterocycles

Of the three-membered heterocycles, epoxides are among the most typical oxygen-containing compounds. Numerous synthetic applications have been explored for these compounds, and they have broad applications in polymer chemistry and the pharmaceutical industry.[141] Among the methodologies for epoxide preparation, epoxidation of alkenes is one of the most straightforward procedures. Iron catalysts along with other metal salts have been applied in these conversions. At the very beginning, over-stoichiometric or stoichiometric amounts of iron salts [*e.g.* Fe(acac)$_3$, FeCl$_3$] are needed to obtain certain conversions and yields with H$_2$O$_2$ as the oxidant.[142] By adding nitrogen ligands, such as 1-methylimidazole, cyclam and pyridine derivatives, the reactions can be performed in a catalytic manner.[143] Alternatively, numerous iron–nitrogen complexes have also been prepared and applied in epoxidation reactions; by using asymmetric ligands, high enantioselectivity can be achieved under mild conditions.[144]

Analogously to epoxidation, aziridination has also been achieved using the same catalysts. By using a suitable nitrenoid source, aziridines can be produced in a selective manner. The usually applied nitrenoid sources include [*N*-(*p*-tosyl)imino]phenyliodinane, bromamine-T and azides. In 2008, Che and co-workers reported a highly efficient procedure for the epoxidation and aziridination of alkenes with a well-defined iron catalyst (Scheme 3.98).[145] Alkenes were epoxidized and aziridized and sulfamate esters were intramolecularly amidated with 'iron(II) salt + 4,4′,4″-trichloro-2,2′ : 6′,2″-terpyridine' as the effective catalyst. The epoxidation of allylic-substituted cycloalkenes achieved excellent diastereoselectivities of up to 90%. ESI-MS results supported the formation of iron–oxo and –imido intermediates. Derivatization of Cl$_3$terpy to *O*-PEG-OCH$_3$-Cl$_2$terpy renders the terpyridine unit recyclable and the 'iron(II) salt + 4,4″-dichloro-4′-*O*-PEG-OCH$_3$-2,2′ : 6′,2″-terpyridine' protocol can be re-used without a significant loss of catalytic activity in alkene epoxidation.

Other iron complexes have also been synthesized and applied in the aziridination of alkenes, such as [(η^5-C$_5$H$_5$)Fe(CO)$_2$(THF)]$^+$[BF$_4$]$^-$, 1,1,4,7,7-pentamethyldiethylenetriamine (Me$_5$dien) or 1,4,7-triisopropyl-1,4,7-triazacyclononane (*i*Pr$_3$TACN) with Fe(CF$_3$SO$_3$) · 2MeCN complex and [tris(3,5-dimethylpyrazol-1-yl)methane]Fe(MeCN)$_3$(BF$_4$)$_2$ complex.[146]

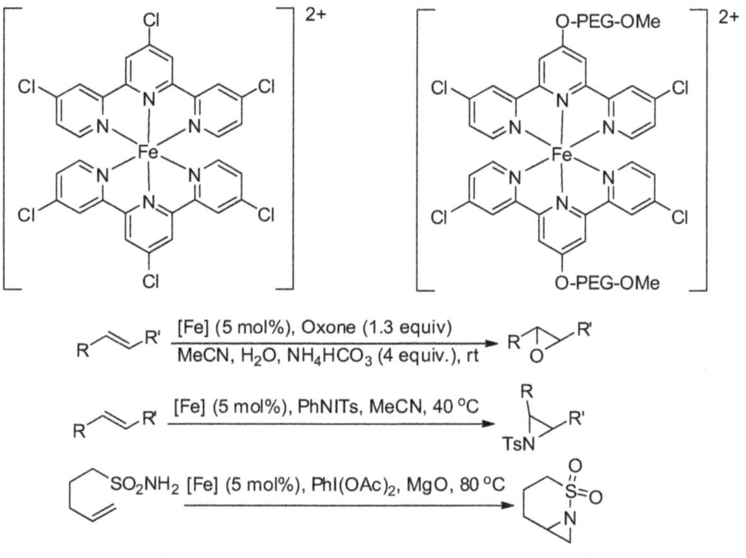

Scheme 3.98 Iron-catalyzed aziridination of alkenes.

Interestingly, a simple iron salt [Fe(OTf)$_2$]-catalyzed aziridination of alkenes was reported by Bolm and co-workers.[147] With the advantages of iron salts, this methodology is even more attractive than procedures that use an iron complex. The use of a combination of iron(II) triflate, quinaldic acid and bis[(trifluoromethyl)sulfonyl]amide (emim BTA) allows the aziridination of alkenes with equimolar amounts of iminoiodinane and provides products in good to moderate yields (Scheme 3.99).

Instead of iminoiodinane, bromamine-T has also been applied as nitrene source in the iron-catalyzed aziridination of alkenes. In 2004, iron(III)–porphyrin complexes [Fe(Por)Cl] as effective catalysts for the aziridination of alkenes using bromamine-T as the nitrene source were reported by Zhang and co-workers.[148] The catalytic system can operate under mild conditions with alkenes as limiting reagents. The aziridination reaction is general and suitable for a wide variety of alkenes, including aromatic, aliphatic, cyclic and acyclic alkenes, and also α,β-unsaturated esters (Scheme 3.100). For 1,2-disubstituted alkenes, the reactions proceeded with moderate to low stereospecificity. The reaction starts with the reaction of iron (III)–porphyrin complex with bromamine-T to generate the iron–nitrene intermediate with concomitant formation of NaBr. Nitrene transfer from the intermediate to a alkene substrate produces the aziridine product and regenerates the iron(III)–porphyrin complex to turn over the catalytic cycle. As the authors stated, the low to moderate stereoselectivity observed for 1,2-disubstituted alkenes could imply the involvement of a radical intermediate.

In 2010, [Fe^{III}(F$_{20}$-tpp)Cl] [F$_{20}$-tpp = *meso*-tetrakis(pentafluorophenyl)-porphyrinato dianion] was found to be an effective catalyst for imido/nitrene

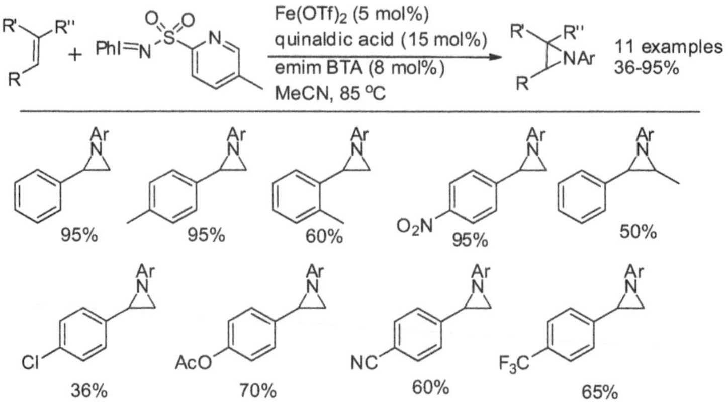

Scheme 3.99   Fe(OTf)$_2$-catalyzed aziridination of alkenes.

Reaction Procedure (Scheme 3.99): A dry Schlenk tube was charged with activated 4 Å molecular sieves (20 mg), evacuated and then filled with argon. Dry CH$_3$CN (1 mL), ionic liquid (10 µL, 38 µmol, 8 mol%) and Fe(OTf)$_2$ (8.9 mg, 25 µmol, 5 mol%) were added, resulting in a yellow suspension. Upon addition of ligand (13.0 mg, 75 µmol, 15 mol%), a color change to red was observed. The olefin (1 mmol, 1.0 equiv.) and subsequently the iminoiodinane (1 mmol, 1.0 equiv.) were added, leading to a yellow suspension. The flask was sealed under argon and placed in a preheated oil-bath at 85 °C. The reaction mixture was stirred for 1–3 h, then cooled to room temperature, diluted with distilled CH$_2$Cl$_2$ (10 mL), filtered through a short plug of Celite (eluent: CH$_2$Cl$_2$) and concentrated under reduced pressure. The residue was purified by silica gel chromatography (eluent: 95:5 to 65:35 pentane–ethyl acetate) to give the desired product.

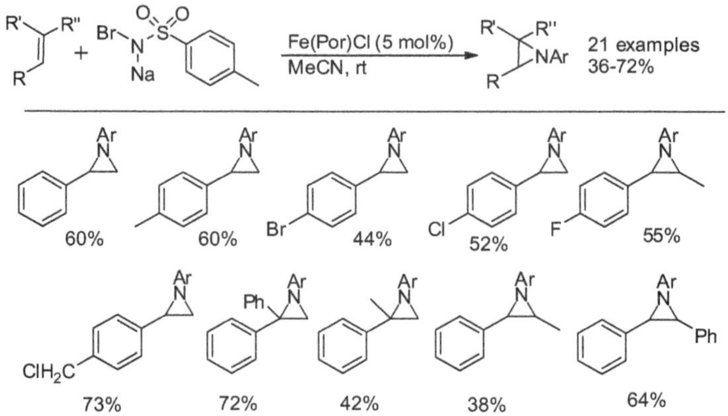

Scheme 3.100   Iron-catalyzed aziridination of alkenes with bromamine-T as nitrene source.

insertion reactions using sulfonyl and aryl azides as nitrogen source.[149] Under thermal conditions, aziridination of aryl- and alkylalkenes (16 examples, 60–95% yields), sulfimidation of sulfides (11 examples, 76–96% yields), allylic amidation/amination of α-methylstyrenes (15 examples, 68–83% yields) and amination of saturated C–H bonds including those of cycloalkanes and adamantane (eight examples, 64–80% yields) can be accomplished by using 2 mol% [Fe^{III}(F_{20}-tpp)Cl] as catalyst. Under microwave irradiation conditions, the reaction times of aziridination (four examples), allylic amination (five examples), sulfimidation (two examples) and amination of saturated C–H bonds (three examples) can be reduced by up to 16-fold (from 24–48 to 1.5–6 h) without significantly affecting the product yield and substrate conversion. One year later, a new iron aziridination catalyst supported by a macrocyclic tetracarbene ligand was synthesized by Cramer and Jenkins.[150] The catalyst, [($^{Me,Et}TC^{Ph}$)Fe(NCCH$_3$)$_2$](PF$_6$)$_2$, was synthesized from the tetraimidazolium precursor, ($^{Me,Et}TC^{Ph}$)(I)$_4$, and characterized by NMR spectroscopy, electrospray ionization mass spectrometry and single-crystal X-ray diffraction. This iron complex catalyzes the aziridination of electron-donating aryl azides and a wide variety of substituted aliphatic alkenes, including tetrasubstituted compounds, in a '$C_2 + N_1$' addition reaction (Scheme 3.101). Finally, the catalyst can be recovered and reused up to three additional times without significant reduction in yield.

In addition to the above-mentioned coupling or addition reactions, a ring expansion pathway has also been used. Baeyer–Villiger oxidation of cyclic ketones offers a promising procedure for the preparation of lactones. The methodology developed by Murahashi *et al.* applied Fe_2O_3 as the catalyst, with oxygen (1 bar) as the oxidant, in the presence of an aldehyde in benzene, and various ketones were oxidized at room temperature in good yields (Scheme 3.102).[151] By replacing the oxidant with NaN$_3$ or TMSN$_3$, lactams can be prepared instead from the same ketones in the presence of an iron catalyst.[152]

Padrón and co-workers successfully developed an efficient Prins cyclization of bis-homoallylic alcohols with aldehydes using iron(III) salts as catalyst.[153] The reaction provides a direct entry to *cis*-2,7-disubstituted

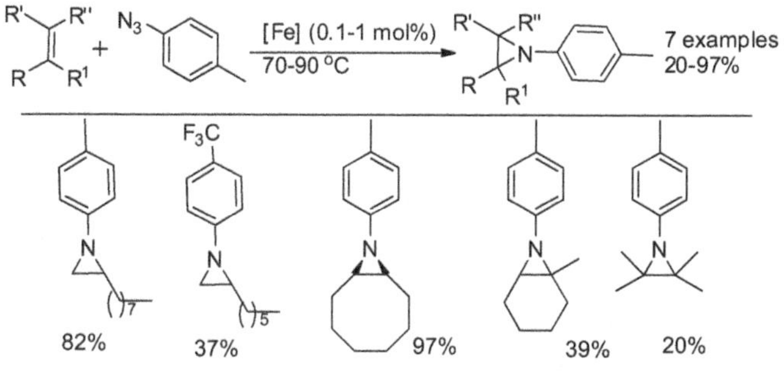

Scheme 3.101 Iron-catalyzed aziridination of alkenes with azide as nitrene source.

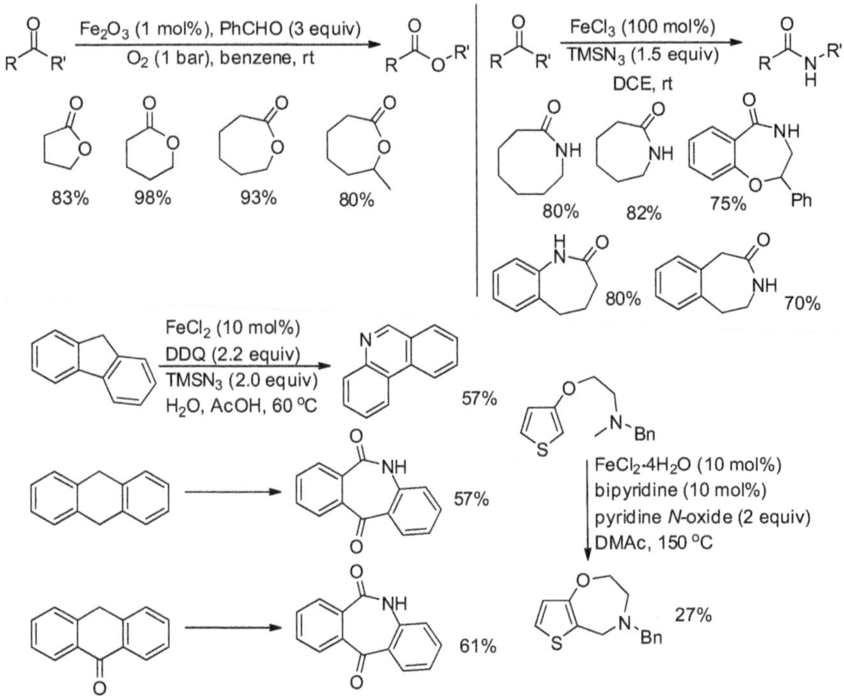

Scheme 3.102   Iron-catalyzed synthesis of lactones and lactams.

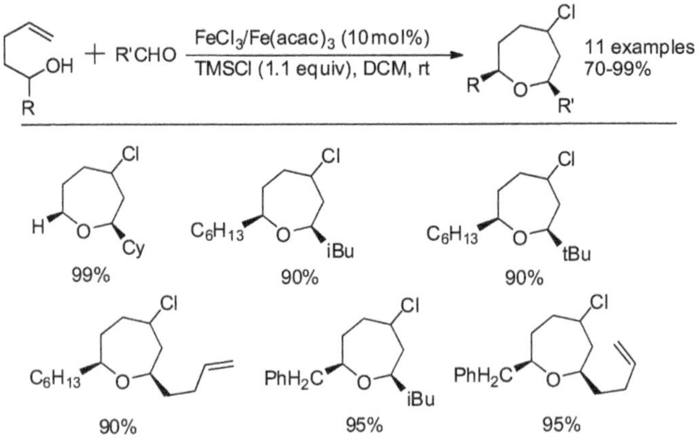

Scheme 3.103   Iron-catalyzed synthesis of oxepanes.

oxepanes with excellent yields (Scheme 3.103). Using this approach as a key step, they performed the shortest and most efficient total synthesis of (+)-isolaurepan, thus demonstrating the value of this methodology, which uses unactivated alkenes.

3.4   Summary

The main contributions to the iron-catalyzed synthesis of heterocycles have been summarized and discussed. The character of the iron catalyst in these reactions is mainly as a Lewis acid, but coupling reactions have also been explored, and even the effect of impurities cannot be excluded. Well-defined complexes are encouraged as they can help in the evaluation of reaction mechanisms; on the other hand, simple salts are also of interest, because of their easy manipulation and high stability and tolerance. In the future, the development of air-stable iron complexes and their application in organic synthesis should be investigated.

References

1. M.-C. P. Yeh, C.-W. Fang and H.-H. Lin, *Org. Lett.*, 2012, **14**, 1830–1833.
2. M.-C. P. Yeh, C.-J. Liang, C.-W. Fan, W.-H. Chiu and J.-Y. Lo, *J. Org. Chem.*, 2012, **77**, 9707–9717.
3. K. Komeyama, T. Morimoto and K. Takaki, *Angew. Chem. Int. Ed.*, 2006, **45**, 2938–2941.
4. (a) T. Taniguchi, N. Goto, A. Nishibata and H. Ishibashi, *Org. Lett.*, 2010, **12**, 112–115; (b) T. Taniguchi and H. Ishibashi, *Org. Lett.*, 2010, **12**, 124–126.
5. T. Xu, Z. Yu and L. Wang, *Org. Lett.*, 2009, **11**, 2113–2116.
6. E. T. Hennessy and T. A. Betley, *Science*, 2013, **340**, 591–595.
7. (a) M. W. Bouwkamp, A. C. Bowman, E. Lobkovsky and P. J. Chirik, *J. Am. Chem. Soc.*, 2006, **128**, 13340–13341; (b) J. M. Hoyt, K. T. Sylvester, S. P. Semproni and P. J. Chirik, *J. Am. Chem. Soc.*, 2013, **135**, 4862–4877.
8. A. Fürstner, K. Majima, R. Martin, H. Krause, E. Kattnig, R. Goddard and C. W. Lehmann, *J. Am. Chem. Soc.*, 2008, **130**, 1992–2004.
9. J. Fan, L. Gao and Z. Wang, *Chem. Commun.*, 2009, 5021–5023.
10. R. M. Carballo, M. Purino, M. A. Ramirez, V. S. Martin and J. I. Padrón, *Org. Lett.*, 2010, **12**, 5334–5337.
11. M. Nitta and T. Kobayashi, *Chem. Lett.*, 1985, 877–880.
12. S. Nakanishi, Y. Otsuji, K. Itoh and N. Hayashi, *Bull. Chem. Soc. Jpn.*, 1990, **63**, 3595–3600.
13. N. Azizi, A. Khajeh-Amiri, H. Ghafuri, M. Bolourtchian and M. R. Saidi, *Synlett*, 2009, 2245–2248.
14. B. Das, G. C. Reddy, P. Balasubramanyam and B. Veeranjaneyulu, *Synthesis*, 2010, 1625–1628.
15. S. Maiti, S. Biswas and U. Jana, *J. Org. Chem.*, 2010, **75**, 1674–1683.
16. Y. Wang, X. Bi, D. Li, P. Liao, Y. Wang, J. Yang, Q. Zhang and Q. Liu, *Chem. Commun.*, 2011, **47**, 809–811.
17. M. Zhao, H. Liang, Z. Ren and Z. Guan, *Adv. Synth. Catal.*, 2013, **355**, 221–226.

18. (a) Y. Ohshiro, K. Kinugasa, T. Minami and T. Agawa, *J. Org. Chem.*, 1970, **35**, 2136–2140; (b) A. Baba, Y. Ohshiro and T. Agawa, *J. Organomet. Chem.*, 1975, **87**, 247–256.

19. P. Mathur, R. K. Joshi, D. K. Rai, B. Jha and S. M. Mobin, *Dalton Trans.*, 2012, **41**, 5045–5054.

20. (a) W. Imhof and E. Anders, *Chem. Eur. J.*, 2004, **10**, 5717–5729; (b) W. Imhof, E. Anders, A. Göbel and H. Görls, *Chem. Eur. J.*, 2003, **9**, 1166–1181; (c) K. Kaleta, J. Fleischhauer, H. Görls, R. Beckert and W. Imhof, *J. Organomet. Chem.*, 2009, **694**, 3800–3805.

21. M. S. Sigman and B. E. Eaton, *J. Org. Chem.*, 1994, **59**, 7488–7491.

22. (a) K. M. Driller, H. Klein, R. Jackstell and M. Beller, *Angew. Chem. Int. Ed.*, 2009, **48**, 6041–6044; (b) S. Prateeptongkum, K. M. Driller, R. Jackstell and M. Beller, *Chem. Asian J.*, 2010, **5**, 2173–2176; (c) S. Prateeptongkum, K. M. Driller, R. Jackstell, A. Spannenberg and M. Beller, *Chem. Eur. J.*, 2010, **16**, 9606–9615.

23. (a) J. Falbe and F. Korte, *Angew. Chem.*, 1962, **74**, 291; *Angew. Chem. Int. Ed. Engl.*, 1962, **1**, 266–267; (b) J. Falbe and F. Korte, *Chem. Ber.*, 1962, **95**, 2680–2687.

24. W. Ji, Y. Pan, S. Zhao and Z. Zhan, *Synlett*, 2008, 3046–3052.

25. H. Jiang, W. Yao, H. Cao, H. Huang and D. Cao, *J. Org. Chem.*, 2010, **75**, 5347–5350.

26. B. M. Gai, A. L. Stein, J. A. Roehrs, F. N. Bilheri, C. W. Nogueira and G. Zeni, *Org. Biomol. Chem.*, 2012, **10**, 798–807.

27. K. Komeyama, T. Morimoto, Y. Nakayama and K. Tataki, *Tetrahedron Lett.*, 2007, **48**, 3259–3261.

28. N. Saino, D. Kogure and S. Okamoto, *Org. Lett.*, 2005, **7**, 3065–3067.

29. (a) K. I. Booker-Milburn, J. L. Jones, G. E. M. Sibley, R. Cox and J. Meadows, *Org. Lett.*, 2003, **5**, 1107–1109; (b) S. S. Velu, I. Buniyamin, L. K. Ching, F. Feroz, I. Noorbatcha, L. C. Gee, K. Awang, I. A. Wahab and J. F. Weber, *Chem. Eur. J.*, 2008, **14**, 11376–11384.

30. T. Xu, Q. Yang, D. Li, Z. Yu and Y. Li, *Chem. Eur. J.*, 2010, **16**, 9264–9272.

31. G. Hilt, P. Bolze and I. Kieltsch, *Chem. Commun.*, 2005, 1996–1998.

32. G. Hilt, P. Bolze and K. Harms, *Chem. Eur. J.*, 2007, **13**, 4312–4325.

33. C. Liu, B. Zhu, J. Zheng, X. Sun, Z. Xie and Y. Tang, *Chem. Commun.*, 2011, **47**, 1342–1344.

34. G. V. M. Sharma, K. R. Kumar, P. Sreenivas, P. R. Krishna and M. S. Chorghade, *Tetrahedron: Asymmetry*, 2002, **13**, 687–690.

35. M. Kamata, T. Kudoh, J. Kaneko, H. Kim and Y. Wataya, *Tetrahedron Lett.*, 2002, **43**, 617–620.

36. K. Komeyama, Y. Mieno, S. Yukawa, T. Morimoto and K. Takaki, *Chem. Lett.*, 2007, **36**, 752–753.

37. D. Sue, T. Kawabata, T. Sasamori, N. Tobuhiro and K. Tsubaki, *Org. Lett.*, 2010, **12**, 256–258.

38. R. Parnes, U. A. Kshirsagar, A. Werbeloff, C. Regev and D. Pappo, *Org. Lett.*, 2012, **14**, 3324–3327.

39. (a) M. S. Sigman, C. E. Kerr and B. E. Eaton, *J. Am. Chem. Soc.*, 1993, **115**, 7545–7546; (b) M. S. Sigman, B. E. Eaton, J. D. Heise and C. P. Kubiak, *Organometallics*, 1996, **15**, 2829–2832.
40. P. Mathur, R. K. Joshi, B. Jha, A. K. Singh and S. M. Mobin, *J. Organomet. Chem.*, 2010, **695**, 2687–2694.
41. S. Saha, S. K. Mandal and S. C. Roy, *Tetrahedron Lett.*, 2008, **49**, 5928–5930.
42. T. Bach, B. Schlummer and K. Barms, *Chem. Commun.*, 2000, 287–288.
43. T. Bach, B. Schlummer and K. Barms, *Chem. Eur. J.*, 2001, 7, 2581–2594.
44. H. Danielec, J. Klügge, B. Schlummer and T. Bach, *Synthesis*, 2006, 551–556.
45. D. G. Churchill and C. M. Rojas, *Tetrahedron Lett.*, 2002, **43**, 7225–7228.
46. (a) T. Yoshimitsu, T. Ino, N. Futamura, T. Kamon and T. Tanaka, *Org. Lett.*, 2009, **11**, 3402–3405; (b) D. Shigeoka, T. Kamon and T. Yoshimitsu, *Beilstein J. Org. Chem.*, 2013, **9**, 860–865.
47. T. Kamon, D. Shigeoka, T. Tanaka and T. Yoshimitsu, *Org. Biomol. Chem.*, 2012, **10**, 2363–2365.
48. G. Liu, Y. Zhang, Y. Yuan and H. Xu, *J. Am. Chem. Soc.*, 2013, **135**, 3343–3346.
49. Y. Liu, X. Guan, E. L. Wong, P. Liu, J. Huang and C. Che, *J. Am. Chem. Soc.*, 2013, **135**, 7194–7204.
50. M. Sengoden and T. Punniyamurthy, *Angew. Chem. Int. Ed.*, 2013, **52**, 572–575.
51. G. C. Senadi, W. Hu, J. Hsiao, J. K. Vandavasi, C. Chen and J. Wang, *Org. Lett.*, 2012, **14**, 4478–4481.
52. O. Debleds, E. Gayon, E. Ostaszuk, E. Vrancken and J. Campagne, *Chem. Eur. J.*, 2010, **16**, 12207–12213.
53. X. Gao, Y. Pan, M. Lin, L. Chen and Z. Zhan, *Org. Biomol. Chem.*, 2010, **8**, 3259–3266.
54. A. Speranca, B. Godoi and G. Zeni, *J. Org. Chem.*, 2013, **78**, 1630–1637.
55. F. Viton, G. Bernardinelli and E. P. Kündig, *J. Am. Chem. Soc.*, 2002, **124**, 4968–4969.
56. K. S. Williamson and T. P. Yoon, *J. Am. Chem. Soc.*, 2010, **132**, 4570–4571.
57. K. S. Williamson and T. P. Yoon, *J. Am. Chem. Soc.*, 2012, **134**, 12370–12373.
58. J. Bonnamour and C. Bolm, *Chem. Eur. J.*, 2009, **15**, 4543–4545.
59. (a) G. Qi and Y. Dai, *Chin. Chem. Lett.*, 2010, **21**, 1029–1032; (b) M. Nasrollahzadeh, Y. Bayat, D. Habibi and S. Moshaee, *Tetrahedron Lett.*, 2009, **50**, 4435–4438; (c) B. Sreedhar, A. S. Kumar and D. Yada, *Tetrahedron Lett.*, 2011, **52**, 3565–3569.
60. Y. Wang, X. Bi, W. Li, D. Li, Q. Zhang, Q. Liu and B. S. Ondon, *Org. Lett.*, 2011, **13**, 1722–1725.
61. Z. Liang, W. Hou, Y. Du, Y. Zhang, Y. Pan, D. Mao and K. Zhao, *Org. Lett.*, 2009, **11**, 4978–4981.
62. X. Guo, R. Yu, H. Li and Z. Li, *J. Am. Chem. Soc.*, 2009, **131**, 17387–17393.

63. R. M. Gay, F. Manarin, C. C. Schneider, D. A. Barancelli, M. D. Costa and G. Zeni, *J. Org. Chem.*, 2010, **75**, 5701–5706.
64. L. Tang, Y. Pang, Q. Yan, L. Shi, J. Huang, Y. Du and K. Zhao, *J. Org. Chem.*, 2011, **76**, 2744–2752.
65. H. Jasch, Y. Landais and M. R. Heinrich, *Chem. Eur. J.*, 2013, **19**, 8411–8416.
66. J. Bonnamour and C. Boim, *Org. Lett.*, 2008, **10**, 2665–2667.
67. K. Cao, Y. Tu and F. Zhang, *Sci. China Chem.*, 2010, **53**, 130–134.
68. M. Wu, X. Hu, J. Liu, Y. Liao and G. Deng, *Org. Lett.*, 2012, **14**, 2722–2725.
69. F. Li, T. Liu and G. Wang, *J. Org. Chem.*, 2008, **73**, 6417–6420.
70. H. Wang, Y. Wang, C. Peng, J. Zhang and Q. Zhu, *J. Am. Chem. Soc.*, 2010, **132**, 13217–13219.
71. S. Santra, A. K. Bagdi, A. Majee and A. Hajra, *Adv. Synth. Catal.*, 2013, **355**, 1065–1070.
72. J. Zeng, Y. J. Tan, M. L. Leow and X. Liu, *Org. Lett.*, 2012, **14**, 4386–4389.
73. J. Maes, T. R. M. Rauws and B. U. W. Maes, *Chem. Eur. J.*, 2013, **19**, 9137–9141.
74. V. Terrasson, J. Michaux, A. Gaucher, J. Wehbe, S. Marque, D. Prim and J. Campagne, *Eur. J. Org. Chem.*, 2007, 5332–5335.
75. M. T. Herrero, J. D. de Sarralde, R. SanMartin, L. Bravo and E. Domínguez, *Adv. Synth. Catal.*, 2012, **354**, 3054–3064.
76. G. Cantagrel, B. de Carné-Carnavalet, C. Meyer and J. Cossy, *Org. Lett.*, 2009, **11**, 4262–4265.
77. A. Speranca, B. Godoi, P. H. Menezes and G. Zeni, *Synlett*, 2013, **24**, 1125–1132.
78. A. L. Stein, F. N. Bilheri, A. R. Rosário and G. Zeni, *Org. Biomol. Chem.*, 2013, **11**, 2972–2978.
79. (a) A. Penoni and K. M. Nicholas, *Chem. Commun.*, 2002, 484–485; (b) A. A. Lamar and K. M. Nicholas, *Tetrahedron*, 2009, **65**, 3829–3833; (c) A. Penoni, G. Palmisano, Y. Zhao, K. N. Houk, J. Volkman and K. M. Nicholas, *J. Am. Chem. Soc.*, 2009, **131**, 653–661.
80. Y. Du, J. Chang, J. Reiner and K. Zhao, *J. Org. Chem.*, 2008, **73**, 2007–2010.
81. Z. Zheng, L. Tang, Y. Fan, X. Qi, Y. Du and D. Zhang-Negrerie, *Org. Biomol. Chem.*, 2011, **9**, 3714–3725.
82. S. Jana, M. D. Clements, B. K. Sharp and N. Zheng, *Org. Lett.*, 2010, **12**, 3736–3739.
83. Y. Liu, J. Wie and C. Che, *Chem. Commun.*, 2010, **46**, 6926–6928.
84. J. Bonnamour and C. Bolm, *Org. Lett.*, 2011, **13**, 2012–2014.
85. Q. Nguyen, T. Nguyen and T. G. Driver, *J. Am. Chem. Soc.*, 2013, **135**, 620–623.
86. Z. Guan, Z. Yan, Z. Ren, X. Liu and Y. Liang, *Chem. Commun.*, 2010, **46**, 2823–2825.
87. M. Shen and T. G. Driver, *Org. Lett.*, 2008, **10**, 3367–3370.
88. K. Bahrami, M. M. Khodaei and F. Naali, *Synlett*, 2009, 569–572.

89. E. J. Hanan, B. K. Chan, A. A. Estrada, D. G. Shore and J. P. Lyssikatos, *Synlett*, 2010, 2759–2764.
90. T. B. Nguyen, L. Ermolenko and A. Al-Mourabit, *J. Am. Chem. Soc.*, 2013, **135**, 118–121.
91. A. Correa, S. Elmore and C. Bolm, *Chem. Eur. J.*, 2008, **14**, 3527–3529.
92. J. Qiu, X. Zhang, R. Tang, P. Zhong and J. Li, *Adv. Synth. Catal.*, 2009, **351**, 2319–2323.
93. Q. Ding, B. Cao, X. Liu, Z. Zong and Y. Peng, *Green Chem.*, 2010, **12**, 1607–1610.
94. (a) Q. Ding, B. Cao, Q. Yang, X. Liu and Y. Peng, *Phosphoru, Sulfur Silicon*, 2011, **186**, 1782–1789; (b) W. Wang, W. Zhong, R. Zhou, J. Yu, J. Dai, Q. Ding and Y. Peng, *Heterocycles*, 2010, **81**, 2841–2847.
95. M. Bala, P. K. Verma, U. Sharma, N. Kumar and B. Singh, *Green Chem.*, 2013, **15**, 1687–1693.
96. H. Wang, L. Wang, J. Shang, X. Li, H. Wang, J. Gui and A. Lei, *Chem. Commun.*, 2012, **48**, 76–78.
97. (a) U. Schmidt and U. Zenneck, *J. Organomet. Chem.*, 1992, **440**, 187–190; (b) F. Knoch, F. Kremer, U. Schmidt, U. Zenneck, P. Le Floch and F. Mathey, *Organometallics*, 1996, **15**, 2713–2719.
98. K. Ferré, L. Toupet and V. Guerchais, *Organometallics*, 2002, **21**, 2578–2580.
99. B. R. D'Souza, T. K. Lane and J. Louie, *Org. Lett.*, 2011, **13**, 2936–2939.
100. T. K. Lane, B. R. D'Souza and J. Louie, *J. Org. Chem.*, 2012, **77**, 7555–7563.
101. T. K. Lane, M. H. Nguyen, B. R. D'Souza, N. A. Spahn and J. Louie, *Chem. Commun.*, 2013, **49**, 7735–7737.
102. C. Wang, X. Li, F. Wu and B. Wan, *Angew. Chem. Int. Ed.*, 2011, **50**, 7162–7166.
103. C. Wang, D. Wang, F. Xu, B. Pan and B. Wan, *J. Org. Chem.*, 2013, **78**, 3065–3072.
104. Y. Zheng, M. Wang, P. Li and L. Wang, *Org. Lett.*, 2012, **14**, 2206–2209.
105. B. M. Ramalingam, V. Saravanan and A. K. Mohanakrishnan, *Org. Lett.*, 2013, **15**, 3726–3729.
106. R. Rohlmann, T. Stopka, H. Richter and O. G. Mancheño, *J. Org. Chem.*, 2013, **78**, 6050–6064.
107. Z. Wang, S. Li, B. Yu, H. Wu, Y. Wang and X. Sun, *J. Org. Chem.*, 2012, **77**, 8615–8620.
108. K. Cao, F. Zhang, Y. Tu, X. Zhuo and C. Fan, *Chem. Eur. J.*, 2009, **15**, 6332–6334.
109. (a) H. Yao, B. Qin, H. Zhang, J. Lu, D. Wang and S. Tu, *RSC Adv.*, 2012, **2**, 3759–3764; (b) Y. Zhang, P. Li and L. Wang, *J. Heterocycl. Chem.*, 2011, **48**, 153–157.
110. M. M. Heravi, B. Baghernejad and H. A. Oskooie, *Tetrahedron Lett.*, 2009, **50**, 767–769.
111. K. H. Kumar, D. Muralidharan and P. T. Perumal, *Tetrahedron Lett.*, 2004, **45**, 7903–7906.

112. M. de F. Pereira and V. Thiéry, *Org. Lett.*, 2012, **14**, 4754–4757.
113. N. Aljaar, J. Conrad and U. Beifuss, *J. Org. Chem.*, 2013, **78**, 1045–1053.
114. G. Wang, C. Miao and H. Kang, *Bull. Chem. Soc. Jpn.*, 2006, **79**, 1426–1430.
115. D. Yang, H. Fu, L. Hu, Y. Jiang and Y. Zhao, *J. Comb. Chem.*, 2009, **11**, 653–657.
116. T. Sengupta, K. S. Gayen, P. Pandit and D. K. Maiti, *Chem. Eur. J.*, 2012, **18**, 1905–1909.
117. Y. Wang, W. Li, G. Che, X. Bi, P. Liao, Q. Zhang and Q. Liu, *Chem. Commun.*, 2010, **46**, 6843–6845.
118. K. Komeyama, R. Igawa and K. Takaki, *Chem. Commun.*, 2010, **46**, 1748–1750.
119. W. Huang, Q. Shen, J. Wang and X. Zhou, *J. Org. Chem.*, 2008, **73**, 1586–1589.
120. D. Eom, J. Mo, P. H. Lee, Z. Gao and S. Kim, *Eur. J. Org. Chem.*, 2013, 533–540.
121. (a) R. M. Carballo, M. A. Ramirez, M. L. Rodriguez, V. S. Martin and J. I. Padrón, *Org. Lett.*, 2006, **8**, 3837–3840; (b) P. O. Miranda, D. D. Diaz, J. I. Padrón, J. Bermejo and V. S. Martin, *Org. Lett.*, 2003, **5**, 1979–1982; (c) P. O. Miranda, D. D. Diaz, J. I. Padrón, M. A. Ramirez and V. S. Martin, *J. Org. Chem.*, 2005, **70**, 57–62; (d) P. O. Miranda, M. A. Ramirez, V. S. Martin and J. I. Padrón, *Chem. Eur. J.*, 2008, **14**, 6260–6268; (e) P. O. Miranda, R. M. Carballo, M. A. Ramirez, V. S. Martin and J. I. Padrón, *Arkivoc*, 2007, (iv), 331–343.
122. P. O. Miranda, R. M. Carballo, V. S. Martin and J. I. Padrón, *Org. Lett.*, 2009, **11**, 357–360.
123. J. Cheng, X. Tang and S. Ma, *ACS Catal.*, 2013, **3**, 663–666.
124. H. Li, J. Yang, Y. Liu and Y. Li, *J. Org. Chem.*, 2009, **74**, 6797–6801.
125. Y. Yang, X. Shu, H. Wie, J. Luo, S. Ali, X. Liu and Y. Liang, *Org. Biomol. Chem.*, 2011, **9**, 5028–5033.
126. (a) B. E. Takacs and J. M. Takacs, *Tetrahedron Lett.*, 1990, **31**, 2865–2868; (b) J. M. Takacs, J. J. Weidner and B. E. Takacs, *Tetrahedron Lett.*, 1993, **34**, 6219–6222; (c) J. M. Takacs, J. J. Weidner, P. W. Newsome, B. E. Takacs, R. Chidambaram and R. Shoemaker, *J. Org. Chem.*, 1995, **60**, 3473–3486; (d) J. M. Takacs, L. G. Anderson, M. W. Creswell and B. E. Takacs, *Tetrahedron Lett.*, 1987, **28**, 5627–5630.
127. (a) S. Laschat, R. Fröhlich and B. Wibbeling, *J. Org. Chem.*, 1996, **61**, 2829–2838; (b) A. Braun and J.-P. Lellouche, *Tetrahedron Lett.*, 2002, **43**, 727–730.
128. M. S. Jung, W. S. Kim, Y. H. Shin, H. J. Jin, Y- S. Kim and E. J. Kang, *Org. Lett.*, 2012, **14**, 6262–6265.
129. A. Guérinot, A. Serra-Muns, C. Gnamm, C. Bensoussan, S. Reymond and J. Cossy, *Org. Lett.*, 2010, **12**, 1808–1811.
130. D. B. Gorman and I. A. Tomlinson, *Chem. Commun.*, 1998, 25–26.
131. A. Speranca, B. Godoi, S. Pinton, D. F. Back, P. H. Menezes and G. Zeni, *J. Org. Chem.*, 2011, **76**, 6789–6797.

132. Z. Li, J. Hong, L. Wenig and X. Zhou, *Tetrahedron*, 2012, **68**, 1552–1559.
133. B. Godoi, A. Speranca, C. A. Bruning, D. F. Back, P. H. Menezes, C. W. Nogueira and G. Zeni, *Adv. Synth. Catal.*, 2011, **353**, 2042–2050.
134. G. Maiti, R. Karmakar, R. N. Bhattacharya and U. Kayal, *Tetrahedron Lett.*, 2011, **52**, 5610–5612.
135. S. Wertz, D. Leifert and A. Studer, *Org. Lett.*, 2013, **15**, 928–931.
136. X. Xu, J. Liu, L. Liang, H. Li and Y. Li, *Adv. Synth. Catal.*, 2009, **351**, 2599–2604.
137. F.-Q. Yuan and F.-S. Han, *Adv. Synth. Catal.*, 2013, **355**, 537–547.
138. K. Bera, S. Sarkar, S. Biswas, S. Maiti and U. Jana, *J. Org. Chem.*, 2011, **76**, 3539–3544.
139. J. Fan and Z. Wang, *Chem. Commun.*, 2008, 5381–5383.
140. Z. Wang, Y. Lei, M. Zhou, G. Chen, R. Song, Y. Xie and J. Li, *Org. Lett.*, 2011, **13**, 14–17.
141. (a) O. A. Wong and Y. Shi, *Chem. Rev.*, 2008, **108**, 3958–3987; (b) B. S. Lane and K. Burgess, *Chem. Rev.*, 2003, **103**, 2457–2473; (c) Q. H. Xia, H. Q. Ge, C. P. Ye, Z. M. Liu and K. X. Su, *Chem. Rev.*, 2005, **105**, 1603–1662.
142. (a) T. Yamamoto and M. Kimura, *J. Chem. Soc., Chem. Commun.*, 1977, 948–949; (b) H. Sugimoto and D. T. Sawyer, *J. Org. Chem.*, 1985, **50**, 1786–1787.
143. (a) K. Hasan, N. Brown and C. M. Kozak, *Green Chem.*, 2011, **13**, 1230–1237; (b) W. Nam, R. Ho and J. S. Valentine, *J. Am. Chem. Soc.*, 1991, **113**, 7052–7054; (c) B. F. Perandones, E. R. Nieto, C. Godard, S. Castillón, P. De Frutos and C. Claver, *ChemCatChem*, 2013, **5**, 1092–1095; (d) K. Schröder, X. Tong, B. Bitterlich, M. K. Tse, F. G. Gelalcha, A. Brückner and M. Beller, *Tetrahedron Lett.*, 2007, **48**, 6339–6342; (e) B. Bitterlich, K. Schröder, M. K. Tse and M. Beller, *Eur. J. Org. Chem.*, 2008, 4867–4870; (f) F. G. Gelalcha, B. Bitterlich, G. Anilkumar, M. K. Tse and M. Beller, *Angew. Chem. Int. Ed.*, 2007, **46**, 7293–7296; (g) B. Bitterlich, G. Anilkumar, F. G. Gelalcha, B. Spilker, A. Grotevendt, R. Jackstell, M. K. Tse and M. Beller, *Chem. Asian J.*, 2007, **2**, 521–529; (h) G. Anilkumar, B. Bitterlich, F. G. Gelalcha, M. K. Tse and M. Beller, *Chem. Commun.*, 2007, 289–291; (i) K. Schröder, S. Ethaler, B. Bitterlich, T. Schulz, A. Spannenberg, M. K. Tse, K. Junge and M. Beller, *Chem. Eur. J.*, 2009, **15**, 5471–5481.
144. (a) J.-P. Damiano, V. Munyejabo and M. Postel, *Polyhedron*, 1995, **14**, 1229–1234; (b) S. Campestrini and U. Tonellato, *J. Mol. Catal. A: Chem.*, 2001, **171**, 37–42; (c) G. Dubois, A. Murphy and T. D. P. Stack, *Org. Lett.*, 2003, **5**, 2469–2472; (d) W. Nam, S. Oh, Y. J. Sun, J. Kim, W. Kim, S. K. Woo and W. Shin, *J. Org. Chem.*, 2003, **68**, 7903–7906; (e) C. Marchi-Delapierre, A. Jorge-Robin, A. Thibon and S. Ménage, *Chem. Commun.*, 2007, 1166–1168; (f) H. L. Yeung, K. C. Sham, C. S. Tsang, T. C. Lau and H. L. Kwong, *Chem. Commun.*, 2008, 3801–3803; (g) A. Company, L. Gómez, X. Fontrodona, X. Ribas and M. Costas, *Chem. Eur. J.*, 2008, **14**, 5727–5731; (h) I. Y. Skobelev,

E. V. Kudrik, O. V. Zalomaeva, F. Albrieux, P. Afanasiev, O. A. Kholdeeva and A. B. Sorokin, *Chem. Commun.*, 2013, **49**, 5577–5579; (i) O. Y. Lyakin, R. V. Ottenbacher, K. P. Bryliakov and E. P. Talsi, *ACS Catal.*, 2012, **2**, 1196–1202; (j) M. R. dos Santos, J. R. Diniz, A. M. Arouca, A. F. Gomes, F. C. Gozzo, S. M. Tamborim, A. L. Parize, P. A. Z. Suarez and B. A. D. Neto, *ChemSusChem*, 2012, **5**, 716–726; (k) B. Wang, S. Wang, C. Xia and W. Sun, *Chem. Eur. J.*, 2012, **18**, 7332–7335; (l) T. Niwa and M. Nakada, *J. Am. Chem. Soc.*, 2012, **134**, 13538–13541; (m) Y. Nishikawa and H. Yamamoto, *J. Am. Chem. Soc.*, 2011, **133**, 8432–8435; (n) T. Akashi, J. Nakazawa and S. Hikichi, *J. Mol. Catal. A: Chem.*, 2013, **371**, 42–47; (o) U. R. Pillai, E. Sahle-Demessie, V. V. Namboodiri and R. S. Varma, *Green Chem.*, 2002, **4**, 495–497; (p) E. Rose, Q. Z. Ren and B. Andrioletti, *Chem. Eur. J.*, 2004, **10**, 224–230; (q) J. Mao, X. Hu, H. Li, Y. Sun, C. Wang and Z. Chen, *Green Chem.*, 2008, **10**, 827–831; (r) M. B. Francis and E. N. Jacobsen, *Angew. Chem. Int. Ed.*, 1999, **38**, 937–941; (s) M. C. White, A. G. Doyle and E. N. Jacobsen, *J. Am. Chem. Soc.*, 2001, **123**, 7194–7195; (t) K. Chen, M. Costas, J. Kim, A. K. Tipton and L. Que Jr, *J. Am. Chem. Soc.*, 2002, **124**, 3026–3035; (u) R. Mas-Ballesté, M. Fujita, C. Hemmila and L. Que Jr, *J. Mol. Catal. A: Chem.*, 2006, **251**, 49–53; (v) R. Mas-Ballesté and L. Que Jr, *J. Am. Chem. Soc.*, 2007, **129**, 15964–15972.

145. P. Liu, E. L.-M. Wong, A. W.-H. Yuen and C.-M. Che, *Org. Lett.*, 2008, **10**, 3275–3278.

146. (a) B. D. Heuss, M. F. Mayer, S. Dennis and M. M. Hossain, *Inorg. Chim. Acta*, 2003, **342**, 301–304; (b) K. L. Klotz, L. M. Slominski, A. V. Hull, V. M. Gottsacker, R. Mas-Ballesté, L. Que Jr and J. A. Halfen, *Chem. Commun.*, 2007, 2063–2065; (c) S. Liang and M. P. Jensen, *Organometallics*, 2012, **31**, 8055–8058.

147. (a) A. C. Mayer, A.-F. Salit and C. Bolm, *Chem. Commun.*, 2008, 5975–5977; (b) M. Nakanishi, A.-F. Salit and C. Bolm, *Adv. Synth. Catal.*, 2008, **350**, 1835–1840.

148. R. Vyas, G.-Y. Gao, J. D. Harden and X. P. Zhang, *Org. Lett.*, 2004, **6**, 1907–1910.

149. Y. Liu and C.-M. Che, *Chem. Eur. J.*, 2010, **16**, 10494–10501.

150. S. A. Cramer and D. M. Jenkins, *J. Am. Chem. Soc.*, 2011, **133**, 19342–19345.

151. S.-I. Murahashi, Y. Oda and T. Naota, *Tetrahedron Lett.*, 1992, **33**, 7557–7560.

152. (a) J. S. Yadav, B. V. S. Reddy, U. V. S. Reddy and K. Praneeth, *Tetrahedron Lett.*, 2008, **49**, 4742–4745; (b) C. Qin, W. Zhou, F. Chen, Y. Qu and N. Jiao, *Angew. Chem. Int. Ed.*, 2011, **50**, 12595–12599; (c) M. Ohta, M. P. Quick, J. Yamaguchi, B. Wünsch and K. Itami, *Chem. Asian J.*, 2009, **4**, 1416–1419.

153. M. A. Purino, M. A. Ramirez, A. H. Daranas, V. S. Martin and J. I. Padrón, *Org. Lett.*, 2012, **14**, 5904–5907.

Copper-Catalyzed Heterocycle Synthesis

The applications of copper catalysts in heterocycle synthesis have attracted the interest of synthetic chemists for many years. In this area, the ability of copper catalysts in C–X (X = C, N, O, S) bond-formation reactions have been extensively explored and many such reactions have been successfully established. The role of copper catalysts as Lewis acids has diminished and elemental steps such as oxidative addition, transmetalation and reductive elimination are usually involved in the catalysis. In this chapter, we discuss the contributions in this area.

4.1 Three-Membered Heterocycles

Aziridines are an important class of nitrogen-containing three membered heterocycles, with broad applications in pharmaceuticals and materials science. As early as 1967, Kwart and Kahn reported the first copper-catalyzed aziridination of cyclohexene by reaction with benzenesulfonyl azide.[1] In 1991, Evans and co-workers developed a copper-catalyzed aziridination of alkenes by [N-(p-toluenesulfonyl)imino]phenyliodinane.[2] Cu(I) and Cu(II) were used as the active catalyst; both electron-rich and electron-deficient alkenes were converted and gave the corresponding N-tosylaziridines in 55–95% yields (Scheme 4.1). Fe(III) was also tested as a catalyst, but gave lower yields. CuOTf was also found to be an active catalyst for this transformation. By adding bis(oxazoline) as the ligand, this reaction can performed in an enantioselective manner. In general, the yields are in the range 16–89% and with 19–97% ee.

The readily available chiral diimines were found to be active ligands for copper-catalyzed asymmetric alkene aziridination by Jacobsen's group.[3] Good yields and excellent enantiomeric excesses (ee) were achieved with

RSC Catalysis Series No. 16
Economic Synthesis of Heterocycles: Zinc, Iron, Copper, Cobalt, Manganese and Nickel Catalysts
By Xiao-Feng Wu and Matthias Beller
© Wu and Beller, 2014
Published by the Royal Society of Chemistry, www.rsc.org

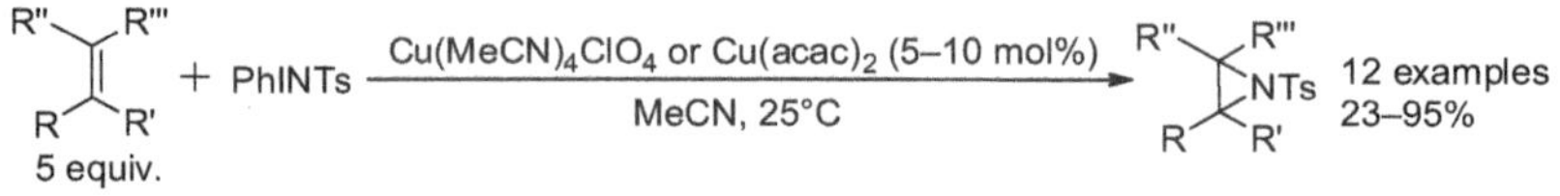

Scheme 4.1   Copper-catalyzed aziridination of alkenes.

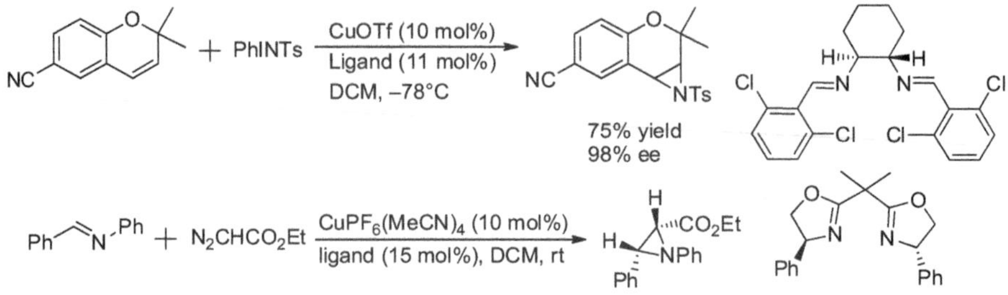

Scheme 4.2   Cu-catalyzed asymmetric aziridinations.

this system (Scheme 4.2). Based on a mechanistic study, they proposed that the reaction may proceed through copper–nitrene as the intermediate. Alternatively, they found that asymmetric azirines can also be prepared by the reaction between imines and ethyl diazoacetate in the presence of a copper catalyst.[4] At room temperature, the reaction was realized and gave the corresponding azirine in low yield and with low enantioselectivity.

A copper(I) complex, Tp′Cu(C$_2$H$_4$) [Tp′ = hydrotris(3,5-dimethyl-1-pyrazolyl)borate], was found to be an efficient catalyst for the aziridination of alkenes in 1993.[5] The aziridine from styrene was produced in 90% yield at room temperature with PhINTs as nitrene source. Interestingly, cyclopropenation of alkenes and alkynes can also be achieved with this catalyst. Andersson and co-workers reported the preparation of nitrene precursors (PhINSO$_2$Ar) and applied them in the cooper-catalyzed aziridination of alkenes.[6] N-(Arenesulfonyl)iminophenyliodinanes were produced from ArSO$_2$NH$_2$ and PhI(OAc)$_2$ in MeOH in the presence of KOH at temperatures from 0 °C to room temperature. After evaluation of their utility as nitrene precursors for the copper-catalyzed aziridination of different alkenes, the best results were obtained with p-NO$_2$-C$_6$H$_4$SO$_2$N=IPh and p-MeO-C$_6$H$_4$-SO$_2$N=IPh, both of which were found to be superior to PhI=NTs, which previously had been the reagent of choice for this type of reaction. The corresponding aziridine derivatives were obtained in good to excellent yields (60–99%) using 1.0 equiv. of alkene and 1.5 equiv. of the nitrene precursor with [Cu(MeCN)$_4$]ClO$_4$ (5 mol%) as the catalyst at room temperature in MeCN. Later, they prepared 1-phenyl-2-[(4S)-phenyl-4,5-dihydrooxazol-2-yl]propen-2-ol and {1-phenyl-2-[(4S)-phenyl-4,5-dihydrooxazol-2-yl]vinyl}-p-tolylamine in two steps from optically pure phenylglycinol or phenylalaninol as chiral ligand for the aziridination reaction. The copper–ligand complexes were isolated and tested and showed high activity in the asymmetric aziridination of styrene, giving the corresponding *N*-tosylaziridine in excellent yield and with *ee* in the range 15–34%.

Dodd and co-workers prepared the first [*N*-(alkylsulfonyl)imino]phenyliodinane, PhI=NSes [Ses = 2-(trimethylsilyl)ethanesulfonyl] and applied it in the copper-catalyzed aziridination of alkenes (Scheme 4.3).[7] In comparison with PhINTs, its isolation is much easier while their reactivities are comparable. Moreover, this new iodinane allows the formation of *N*-(Ses)aziridines that can, in turn, be opened by nucleophiles under mild conditions and/or deprotected at the nitrogen position without side reactions. In the latter case, use of the hypervalent silicon reagent TASF for deprotection of the Ses group is very convenient. These *N*-(Ses)aziridines are thus of high synthetic interest. Later, they applied this reagent in the aziridination of 11-pregnane derivatives and α-allylglycines.[8] Cu(I) salts were used the catalysts and all the products were obtained in moderate to good yields.

In addition to the exploration of aziridination reagents, many new copper complexes have also been prepared. The activities of these complexes were tested in the aziridination of alkenes and showed improved yields compared with simple copper salts. Halfen's group succeeded in preparing some novel copper–nitrogen complexes, such as (*i*Pr$_3$TACN)Cu(O$_2$CCF$_3$)$_2$ (*i*Pr$_3$TACN = 1,4,7-triisopropyl-1,4,7-triazacyclononane), LCuMeCN(PF$_6$)$_2$ [L = 1,4-bis(pyridin-2-ylmethyl)piperazine] and LCuO$_2$CCF$_3$ [L = 1-(2-pyridylmethyl)-5-methyl-1,5-diazacyclooctane], all of which gave excellent yields in the aziridination of styrene with PhINTs as nitrene source.[9]

A new catalytic enantioselective aziridination reaction of α-imino esters with diazo compounds catalyzed by chiral Lewis acid complexes [Sn(OTf)$_2$, Yb(OTf)$_3$, Zn(OTf)$_2$, AgSbF$_6$, CuClO$_4$] was studied by Jorgensen and co-workers.[10] A series of *N*-substituted α-imino esters were tested as substrates

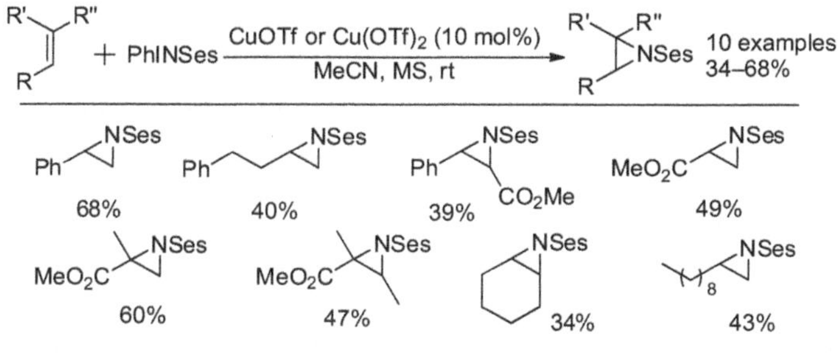

Scheme 4.3 Cu-catalyzed aziridinations using PhINSes.

Reaction Procedure (Scheme 4.3): PhINSes (1.3 equiv.) was added portionwise, over a period of 3 h, to a mixture held under argon of 4 Å molecular sieves (150 mg), Cu(OTf)$_2$ (10 mol%) and alkene (1 equiv.) in acetonitrile (1.6 mL). The green reaction mixture was stirred at room temperature for 24 h and then purified directly by flash chromatography on silica gel, affording aziridines.

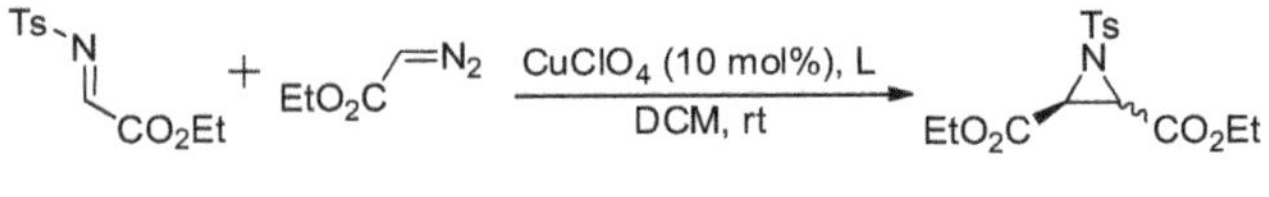

Scheme 4.4   Cu-catalyzed aziridinations of α-imino esters.

Reaction Procedure (Scheme 4.4): The Lewis acid (0.02 mmol) and ligand (0.022 mmol) were placed in a flame-dried Schlenk flask, which was evacuated twice and refilled with N_2. Dry solvent (2 mL) was added and the solution was stirred for 0.5 h at r.t. followed by the addition of 0.2 mmol of the imine, and the mixture was cooled to the desired temperature before diazo compound (1.2 equiv.) was added (EDA was dissolved in 0.5 mL of solvent and added over 9 h). The reaction mixture was stirred overnight and solvent was removed *in vacuo* and the aziridine was isolated by flash column chromatography (FC) (18 mm) on silica gel using 0.2–2% EtOAc in CH_2Cl_2 as the eluent.

---

for the aziridination reaction using trimethylsilyldiazomethane as the carbenoid fragment donor catalyzed by various chiral complexes (Scheme 4.4). Both chiral BINAP and bisoxazoline ligands, in combination with copper(I) salts in particular, can catalyze the aziridination reaction leading to the *cis*-aziridine with up to 72% *ee*, applying BINAP–copper(I) complexes as the catalyst, while the bisoxazoline–copper(I) catalysts gave an aziridination reaction with lower diastereoselectivity, but the *trans*-aziridine was formed in 69% *ee*. The influence of diazo compounds, Lewis acids, chiral ligands, solvents and counterions on the reaction course was investigated and a mechanism for the reaction was discussed in which the α-imino ester coordinates to the chiral catalyst. For trimethylsilyldiazomethane, nucleophilic attack on the coordinated α-imino ester probably takes place in the case of copper(I) as the Lewis acid, whereas for ethyl diazoacetate the reaction course is dependent on the Lewis acid; for silver(I), a nucleophilic attack is probably also operating, whereas a metal–carbene intermediate is involved when copper(I) is used.

Dauban and Dodd's group developed several interesting copper-catalyzed aziridination reactions (Scheme 4.5).[11] In 2000, they reported a copper-catalyzed synthesis of cyclic sulfonamides from unsaturated iminoiodinanes. Olefinic primary sulfonamides were treated with iodobenzene diacetate and potassium hydroxide in methanol to give intermediate iminoiodinanes. Catalytic copper(I) or copper(II) triflate then provided intramolecular nitrene delivery, leading to aziridine formation. The aziridines could be opened by various nucleophiles (methanol, thiophenol, allylmagnesium bromide, benzylamine) to give the corresponding substituted cyclic sulfonamides. Later, they reported a copper-catalyzed aziridination of olefinic sulfamates derived from primary and secondary alcohols to the corresponding bicyclic fused aziridines *via* intramolecular cyclization in the presence of iodosylbenzene. The products were opened by

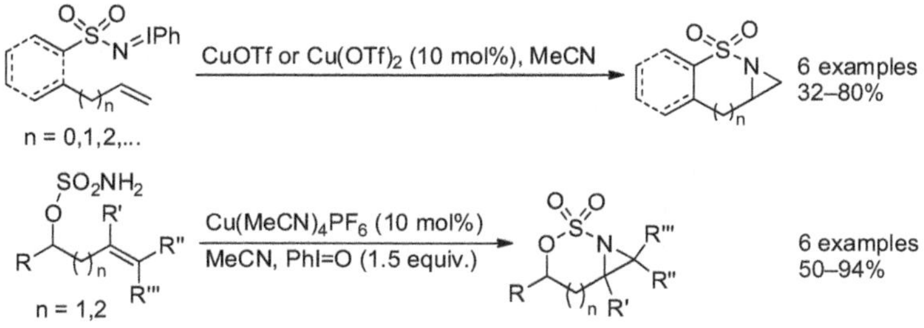

Scheme 4.5 Cu-catalyzed aziridinations of sulfonamides.

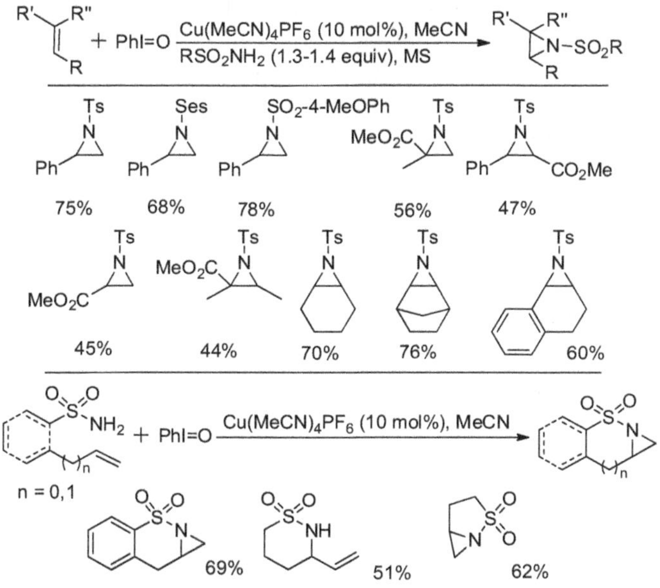

Scheme 4.6 Cu-catalyzed aziridinations of alkenes with PhIO.

various nucleophiles to give the corresponding substituted cyclic sulfamates, which in turn reacted, after nitrogen activation, with a second nucleophile at the carbon atom bearing the oxygen atom. Concomitant removal of the sulfonyloxy moiety thus gave access to polysubstituted amines.

In 2001, the same group reported a copper-catalyzed azidination of alkenes by *in situ* generation of nitrogen from iodosylbenzene (PhI=O) and sulfonamides.[12] At room temperature in MeCN, aziridines were produced in moderate yields (Scheme 4.6). By adding chiral ligand, good enantioselectivity can be achieved. This unique reaction greatly simplifies the procedure for the copper-catalyzed aziridination of alkenes and enhances its efficiency.

In 2004, the same group reported the diastereoselective copper-catalyzed aziridination of alkenes with sulfonimidamide.[13] The reaction between the iodine(III) oxidant iodosylbenzene and a sulfonimidamide in the presence of a copper catalyst leads to the formation of a highly reactive chiral metalla-nitrene. The latter adds to various types of alkenes used in stoichiometric amounts to afford the corresponding aziridines in very good yields of up to 96% (Scheme 4.7). In the case of α,β-unsaturated esters, the catalytic aziridination can also take place with good diastereoselectivities. Later, they studied in detail the effects of substituents on the sulfonimidamide in aziridination reactions.[14] They demonstrated that the addition of the nitrene derived from *p*-nitrosulfonimidamide to *tert*-butyl acrylate takes place with an optimized *de* of 94%. This methodology can therefore be applied to the synthesis of substituted amino acids, since the resulting aziridines also undergo nucleophilic ring opening under smooth conditions. Matching and mismatching effects have also been induced by combining the sulfonimidamide with a chiral copper catalyst, thereby allowing the stereoselectivity of the aziridination to be improved, although only in the case of styrene. A particularly noteworthy feature of this transformation is that the addition of the electron-deficient nitrene species occurs best in the case of electron-poor alkenes.

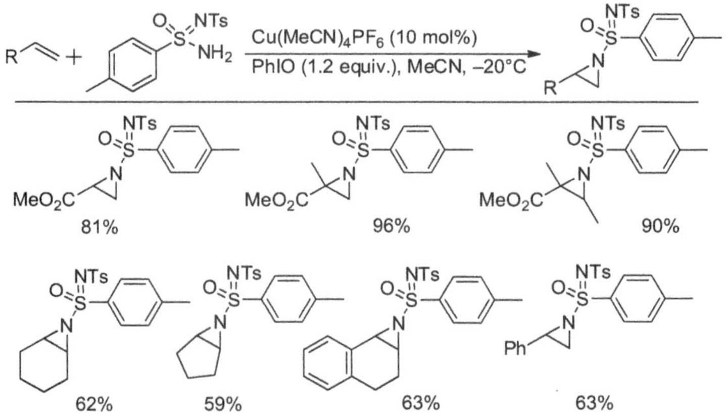

Scheme 4.7   Cu-catalyzed aziridinations of alkenes by sulfonimidamide.

Reaction Procedure (Scheme 4.7): In an oven-dried tube were introduced activated 4 Å molecular sieves (50 mg), the sulfonimidamide (0.24 mmol), and PhI=O (54 mg, 0.24 mmol). The tube was purged with argon and placed in a Sigma-Aldrich AtmosBag, which was filled with argon. Under argon, CuOTf (10 mg, 0.02 mmol) was added and the tube was tightly closed and removed from the AtmosBag. The tube was cooled to –30 °C and, under argon, acetonitrile (0.7 mL) followed by the alkene (0.2 mmol) were added. The mixture was placed in a freezer (–25 °C) for 48 h. After dilution with dichloromethane (5 mL), the molecular sieves were removed by filtration and the filtrate was evaporated to dryness under reduced pressure. The oily residue was purified by flash chromatography on silica gel, affording the aziridines as a mixture of diastereomers.

In 2000, Scott's group designed a novel bisimine ligand for the copper-catalyzed aziridination of alkenes.[15] Various aziridines were produced in a highly enantioselective manner (Scheme 4.8). Both Cu(I) and Cu(II) could efficiently drive the transformation. The performance of the catalysts is highly dependent on the substitution pattern of the ligand arylimine unit, which determines the catalyst speciation; *ortho*-disubstitution is required to furnish monometallic, active systems. The highest enantioselectivities by far are obtained for ligands that contain *o*-dichlorophenyl groups. DFT calculations were carried out and a Cu(I)–Cu(III) cycle was suggested.[16,17]

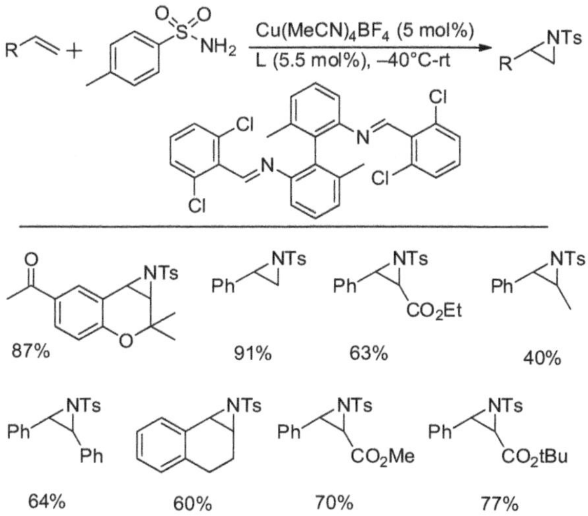

Scheme 4.8 Cu-catalyzed enantioselective aziridinations of alkenes.

Reaction Procedure (Scheme 4.8): A round-bottomed flask with a sidearm incorporating a PTFE stopcock was charged with the metal salt (5 mol%) and the diimine ligand (6 mol%) and was flushed with argon. Solvent (6 mL), usually dichloromethane, was added and the resulting mixture was stirred for 15 min. Reactions were performed at room temperature or in a bath of 2-propanol cooled to –40 °C using an immersion cooler. The alkene (1–5 equiv.) was added to the stirred solution either *via* syringe or by solid addition against a slow positive flow of argon. The solid nitrene source (1 equiv.) was added in a similar manner in a single portion to the stirred solution. The reaction mixture was quenched by dilution with hexane and filtered through a small plug of silica. The mixture was then concentrated to dryness on silica and separated by flash chromatography. The product was identified by NMR spectroscopy. *Cis/trans* ratios were calculated from ^{1}H NMR spectroscopy and enantiomeric excesses (if applicable) were determined using (+)-Eu(hfc)$_3$ as the chiral shift reagent or by chiral HPLC on a Chiracel OD column, as appropriate.

When Cu(II) was applied as pre-catalyst, it was reduced to Cu(I) *in situ* and started the catalytic cycle.

During the past 10 years, the main efforts have been focused on the development of new ligands and new copper complexes. Numerous nitrogen ligands were synthesized and applied in the aziridination of alkenes, such as pyridinophane (a three-pyridine-linked macrocycle),[18] binaphthyldiimine[19] and bisoxazolines,[20] and also many interesting copper complexes.[21] In addition to simple CuI,[22] heterogeneous copper catalysts were also developed and applied in aziridination reactions.[23] Interestingly, Lebel *et al.* developed a novel copper-catalyzed aziridination reaction based on the use of *N*-tosyloxycarbamates.[24] Meanwhile, Fleming and co-workers reported the use of a copper(I) catalyst for the intramolecular aziridination of allylic tosyloxycarbamates.[25] These reagents are advantageous in comparison with the use of hypervalent iodinane reagents, as they do not produce non-polar, high molecular weight byproducts, such as iodobenzene. The formation of troc-protected aziridines is beneficial, as the troc protecting group is easily removed under mildly basic conditions, without opening of the aziridine moiety. They showed that pyridine ligands are beneficial and clearly favor the aziridination reaction. The addition of a chiral bisoxazoline ligand can promote the reaction, which follows an enantioselective pathway.[26] The products were obtained in moderate to good yields (Scheme 4.9).

Chan and co-workers reported a method to prepare α-acyl-β-amino acid and 2,2-diacylaziridine derivatives efficiently from Cu(OTf)$_2$ + 1,10-phenanthroline (1,10-phen)-catalyzed amination and aziridination of 2-alkyl-substituted 1,3-dicarbonyl compounds with PhI=NTs (Scheme 4.10).[27] By taking advantage of

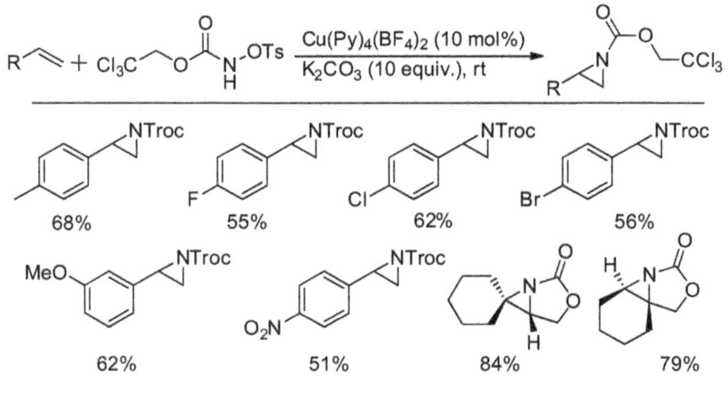

Scheme 4.9   Cu-catalyzed aziridination of alkenes with *N*-tosyloxycarbamates.

Reaction Procedure (Scheme 4.9): Cu(pyridine)$_4$(BF$_4$)$_2$ (28 mg, 0.050 mmol) was suspended in benzene (5 mL) in a 20 mL scintillation vial under air. K$_2$CO$_3$ (692 mg, 5.00 mmol) and the alkene (2.50 mmol) were successively added. The heterogeneous solution was stirred for 15 min at 25 °C and *N*-tosyloxycarbamate (0.50 mmol) was added in one portion. The solution was stirred at 25 °C overnight. Diethyl ether (15 mL) was added and the resulting mixture was filtered. The solid was washed four times with 10 mL of diethyl ether. The combined filtrate was concentrated under reduced pressure. The residue was purified by flash chromatography on silica gel using diethyl ether–hexanes as eluent. The silica gel was pretreated with 1% Et$_3$N–hexanes.

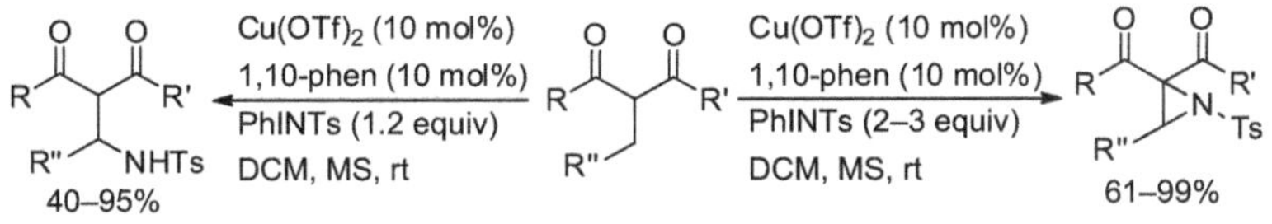

Scheme 4.10 Cu-catalyzed aziridination of 1,3-dicarbonyl compounds.

General Procedure for Cu(II)-Catalyzed Aziridination of 2-Alkyl-1,3-dicarbonyl Compounds to 2,2-Diacylaziridine Derivatives (Scheme 4.10). To a mixture of Cu(OTf)$_2$ (0.05 mmol, 18.1 mg), 1,10-phen (0.05 mmol, 9.9 mg) and powdered 4 Å molecular sieves (400 mg) were added 2 mL of CH$_2$Cl$_2$. After the mixture had been stirred for 1 h, PhI=NTs (1.0 mmol, 373 mg or 1.5 mmol, 560 mg) and substrate (0.5 mmol) were added. The reaction mixture was stirred for a further 18 h at room temperature, then filtered through Celite, washed with EtOAc (50 mL), evaporated to dryness and purified by flash column chromatography (4 : 1 *n*-hexane–EtOAc as eluent) to give the required product.

General Procedure for Cu(II)-Catalyzed Amination of 2-Alkyl-1,3-dicarbonyl Compounds to α-Acyl-β-amino Acid Derivatives. To a mixture of Cu(OTf)$_2$ (0.05 mmol, 18.1 mg), 1,10-phen (0.05 mmol, 9.9 mg) and powdered 4 Å molecular sieves (400 mg) were added 2 mL of CH$_2$Cl$_2$. After the mixture had beenstirred for 1 h, PhI=NTs (0.6 mmol, 224 mg) and substrate (0.5 mmol) were added. The reaction mixture was stirred at room temperature and monitored by TLC. Upon completion, the mixture was filtered through Celite, washed with EtOAc (50 mL), evaporated to dryness and purified by flash column chromatography (4 : 1 *n*-hexane–EtOAc as eluent) to give the required product.

the orthogonal modes of reactivity of the substrate through slight modification of the reaction conditions, a divergence in product selectivity was observed. In the presence of 1.2 equiv. of the iminoiodane, amination of the allylic C–H bond of the enolic form of the substrate, formed *in situ* through coordination to the Lewis acidic metal catalyst, was found to occur selectively and give the β-aminated adduct. On the other hand, increasing the amount of the nitrogen source from 1.2 to 2–3 equiv. resulted in preferential formal aziridination of the C–C bond of the 2-alkyl substituent of the starting material and formation of the aziridine product.

4.2   Five-membered Heterocycles

4.2.1   Copper-Catalyzed Synthesis of 1,2,3-Triazole Derivatives

Among all the five-membered heterocycles, 1,2,3-triazole compounds represent an important class of products that have numerous applications in biology and advanced materials.[28] One of the most direct pathways for the preparation of 1,2,3-triazoles is the copper-catalyzed 'click reaction,' which was reported by Sharpless and co-workers in 2001.[29] Since then, impressive progress has been made, and various copper catalysts have been developed and applied in this transformation, such as NHC-Cu,[30] Cu(OAc),[31] and other copper complexes.[32] In addition to homogeneous copper catalysts, heterogeneous coppers[33] and flow chemistry[34] were also developed. The above reactions are mainly based on the reaction of azides with alkynes (typically two components). Selected examples of three- and four-component click reactions will be discussed in detail. In 2009, Kumar *et al.* developed a simple, benign and one-pot regioselective synthesis of biologically significant 1,4-disubstituted-1*H*-1,2,3-triazoles in good yields under catalytic conditions (Scheme 4.11).[35] The protocol is applicable to a wide range of α-tosyloxy ketones and acetylenic compounds and allows the assembly of a diverse set of 1,4-disubstituted-1*H*-1,2,3-triazoles. The use of copper(I) as a reusable catalyst in aqueous poly(ethylene glycol) (PEG) makes this method facile, cost-effective and eco-friendly.

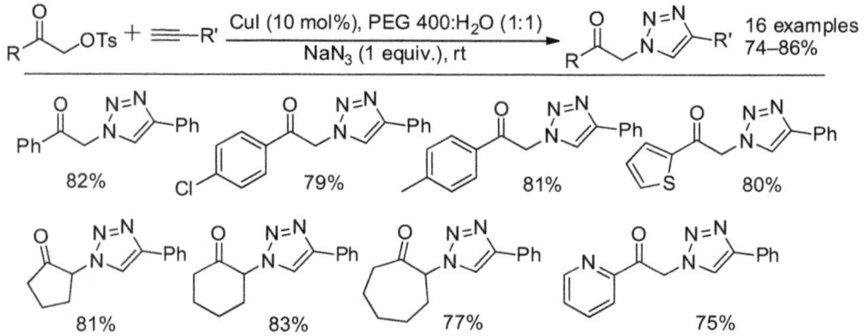

Scheme 4.11   Cu-catalyzed synthesis of 1,4-disubstituted-1*H*-1,2,3-triazoles.

General Procedure (Scheme 4.11): To a stirred solution of α-tosyloxy ketone (1 mmol), sodium azide (1 mmol) in aqueous PEG 400 (2 mL and 1:1 v/v), phenylacetylene (1 mmol) and copper iodide (10 mol%) were added and allowed stirred at room temperature for 30 min. The reaction mixture appeared turbid. On completion of the reaction (monitored by TLC), the reaction mixture was diluted with water and filtered *via* a pump to collect the product or extracted with ethyl acetate (3 × 2 mL). The organic layer was dried over anhydrous sodium sulfate and distilled using a rotary vacuum evaporator to afford pure 1,4-disubstituted-1*H*-1,2,3-triazoles.

Yadav *et al.* developed a novel approach for the preparation of α-alkoxytriazoles *via* a four-component reaction of aldehyde, alcohol, azide and alkyne.[36] The reactions proceeds through acetal formation, azidation and a subsequent 'click reaction' in the presence of copper(II) triflate and copper metal in acetonitrile. In addition to its simplicity and mild reaction conditions, this method provides a wide range of α-alkoxytriazoles in good yields in a single-step operation. Moses and co-workers reported an efficient one-pot transformation of anilines to azides (treated with *t*BuONO and TMSN₃) and applied them in an *in situ* click reaction with alkynes.[37] The *in situ* generation of azides by the reaction of organohalides with sodium azide or phenylboronic acid with sodium azide was also developed.[38,39]

Kuang and co-workers developed a simple and efficient method for the preparation of 4-aryl-1*H*-1,2,3-triazoles from *anti*-3-aryl-2,3-dibromopropanoic acids and sodium azide.[40] Using an inexpensive copper(I) iodide catalyst, the desired products were obtained in moderate to excellent yields and the reaction provides a novel access to compounds that are important in medicinal, materials and biological research (Scheme 4.12). The same group reported an alternative procedure for the preparation of 4-aryl-1*H*-1,2,3-triazoles by the reaction of 1,1-dibromoalkenes with sodium azide. In the presence of CuI (20 mol%), K₂CO₃ and sodium ascorbate (50 mol%) in DMSO at 120 °C, moderate to good yields of the desired triazoles were produced.[41]

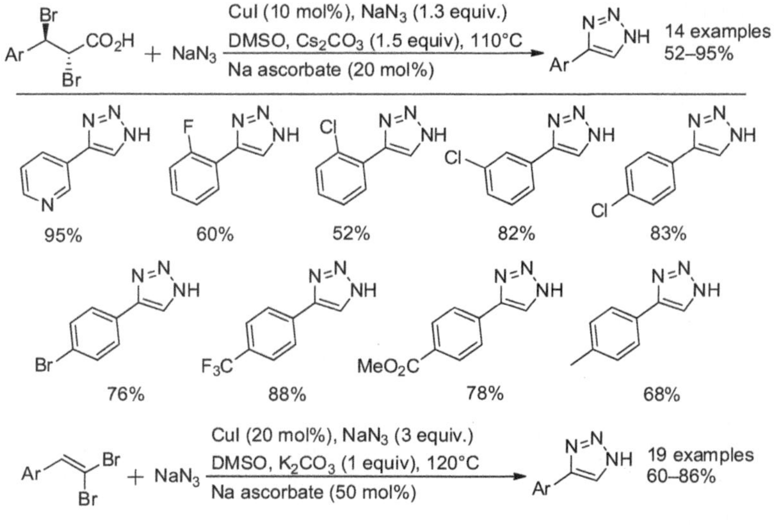

Scheme 4.12 Cu-catalyzed synthesis of 4-aryl-1*H*-1,2,3-triazoles.

General Procedure (Scheme 4.12): A solution of *anti*-3-aryl-2,3-dibromopropanic acid (0.5 mmol), NaN₃ (42 mg, 0.65 mmol), Cs₂CO₃ (408 mg, 1.25 mmol), CuI (10 mg, 0.05 mmol) and Na ascorbate (20 mg, 0.1 mmol) in DMSO (3 mL) was stirred in a sealed tube under N₂ for 5 min and then heated at 110 °C for 4 h. The mixture was allowed to cool to r.t. and then the reaction was quenched with water (25 mL). The mixture was extracted with EtOAc (3×30 mL) and the combined organic layers were washed with brine (3×10 mL), dried (Na_2SO_4) and concentrated under reduced pressure. The crude product was purified by column chromatography.

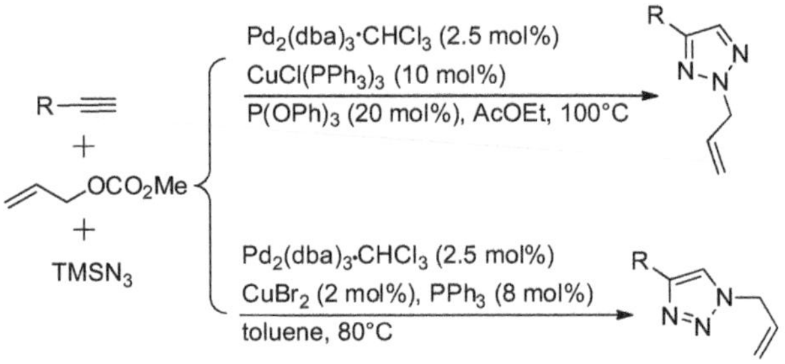

Scheme 4.13 Regiocontrolled synthesis of 2-allyl- and 1-allyl-1,2,3-triazoles.

General Procedure (Scheme 4.13): To an AcOEt solution (1 mL) of $Pd_2(dba)_3 \cdot CHCl_3$ (13.0 mg, 0.0125 mmol) and $CuCl(PPh_3)_3$ (44.3 mg, 0.05 mmol) were added $P(OPh)_3$ (0.1 mmol), phenylacetylene (0.5 mmol), allyl methyl carbonate (0.6 mmol) and $TMSN_3$ (0.6 mmol) under an argon atmosphere. The reaction mixture was stirred at 100 °C for 10 h in a tightly capped 5-mL microvial. After consumption of the starting material, the mixture was cooled to room temperature and filtered through a short Florisil pad with Et_2O (100 mL) as eluent and concentrated. The residue was purified by silica gel column chromatography.

A practical and efficient one-pot procedure for the regioselective synthesis of functionalized 1,4-disubstituted 1,2,3-triazoles from primary amines and terminal acetylenes was reported by Smith *et al.* in 2009.[42] The inexpensive, self-stable diazotransfer reagent imidazole-1-sulfonyl azide hydrochloride was utilized for the *in situ* generation of azides from primary amines (preferably aliphatic amines).[43] In 2004, a one-pot procedure for the regio-controlled synthesis of both 2-allyl- and 1-allyl-1,2,3-triazoles *via* the three-component coupling reaction between non-activated terminal alkynes, allyl carbonate and trimethylsilyl azide ($TMSN_3$) under a palladium and copper bimetallic catalyst was developed.[44] To accomplish the regioselective synthesis of the allyltriazoles, proper choice of two different catalyst systems is needed. The combination $Pd_2(dba)_3 \cdot CHCl_3$–$CuCl(PPh_3)_3$–$P(OPh)_3$ catalyzes the formation of 2-allyl-1,2,3-triazoles, while the combination $Pd(OAc)_2$–$CuBr_2$–PPh_3 promotes the formation of 1-allyl-1,2,3-triazoles (Scheme 4.13). The cooperative activity of palladium and copper catalysts plays an important role in these transformations. Most probably, the palladium catalyst works as a catalyst for generating reactive azide species, π-allylpalladium azide complex and allyl azide. The copper catalyst probably behaves as an activator of the C–C triple bond of the starting terminal alkynes by forming a copper–acetylide intermediate and thereby promotes the [3 + 2] cycloaddition reaction between the reactive azide species and the copper–acetylide to form the triazole framework.

In 2012, an efficient one-pot, three-component stepwise approach for the synthesis of N-2-aryl-substituted-1,2,3-triazoles was developed by Chen and co-workers.[45] By using this azide–chalcone oxidative cycloaddition and post-triazole arylation, a series of N-2-aryl-substituted-1,2,3-triazoles were readily prepared under mild conditions in excellent yields and with high regioselectivity (Scheme 4.14). Notably, both the catalyst and substrates are readily available. Furthermore, 4-substituted-1H-1,2,3-triazolo[4,5-c]quinolones were produced when (E)-3-(2-bromoaryl)-1-arylprop-2-en-1-ones were applied as substrates.[46] By this CuO-promoted tandem cyclization reaction, a series of triazole-fused N-heterocyclic compounds could be prepared in moderate to good yields without the addition of any additive.

Cao and co-workers reported a novel approach for the synthesis of CF_2H- or CF_3-containing 1-aryl-1,4,5-trisubstituted 1,2,3-triazoles by a one-pot, three-component reaction of arylboronic acids, sodium azide and active methylene ketones in the presence of $Cu(OAc)_2$ and piperidine using DMSO–H_2O (10 : 1) as solvent (Scheme 4.15).[47] The outstanding advantage of this method is that it avoids the isolation of the unstable and hazardous aryl

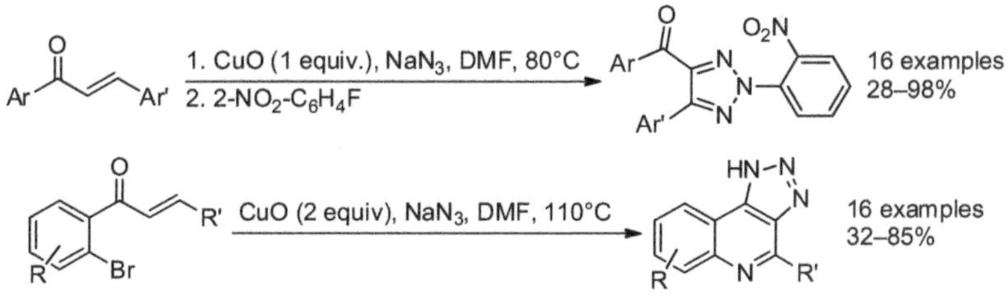

Scheme 4.14 Copper-mediated transformation of chalcones.

General Procedure (Scheme 4.14): All reactions were performed on a 0.20 mmol scale relative to chalcones. 3-(4-Chlorophenyl)-1-phenylprop-2-en-1-one (0.20 mmol), sodium azide (0.20 mmol), CuO (1.0 equiv.) and DMF (1 mL) were placed in a round-bottomed flask equipped with a stirrer. The reaction mixture was agitated at 80 °C for 24 h, then 2-F-$C_6H_4NO_2$ (0.20 mmol) was added and the reaction continued at 80 °C for 6 h. Subsequently, water (2 mL) was added to the reaction mixture and extracted with diethyl ether (3 × 10 mL). The combined organic phases were washed with brine (2 × 5 mL), dried over anhydrous $MgSO_4$ and concentrated *in vacuo*. The residue was subjected to flash column chromatography with hexanes–EtOAc (5 : 1) as eluent to obtain the desired product (93% yield).

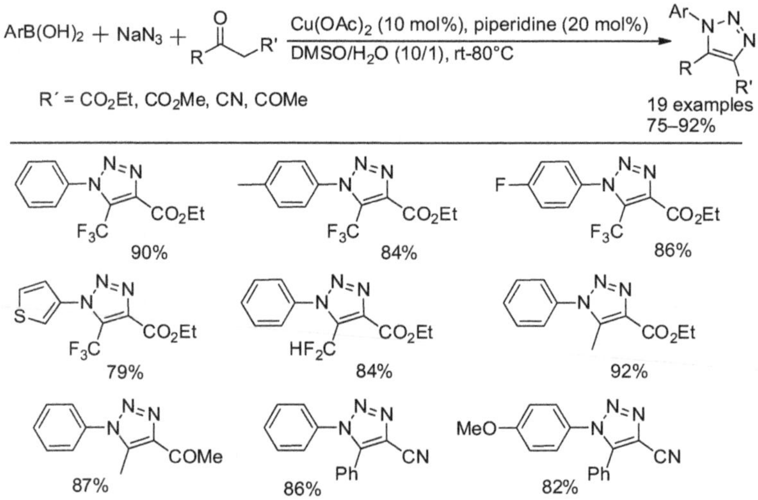

Scheme 4.15 Copper-catalyzed synthesis of 1-aryl-1,4,5-trisubstituted 1,2,3-triazoles.

General Procedure (Scheme 4.15): To a stirred solution of an arylboronic acid (2 mmol) in a mixture of DMSO (10 mL) and water (1 mL), sodium azide (0.26 g, 4 mmol) and Cu(OAc)$_2$ (0.04 g, 0.2 mmol) were added successively. The mixture was stirred for 0.5–4 h at room temperature (monitored by TLC). Then ketone (2 mmol) and piperidine (0.03 g, 0.4 mmol) were added to the solution. Reaction was continued for 2–20 h (monitored by TLC) and quenched with water (20 mL). The resulting suspension was filtered and the filtrate was diluted with CH$_2$Cl$_2$, washed successively with water and brine, dried over anhydrous MgSO$_4$ and concentrated under reduced pressure to leave the crude product. The resultant crude residue was purified by chromatography on silica gel (eluent: 5:1 petroleum ether–EtOAc) to afford the product.

---

azides. It provides a new and practical one-step route to synthesize the structural diversity of 1-aryl-1,4,5-trisubstituted 1,2,3-triazoles from readily commercially available boronic acids. First, a small organic catalyst, piperidine, reacts with an active methylene ketone, such as ethyl tri-fluoroacetoacetate, to produce the key intermediate enamine. Subsequently, dipolar cycloaddition of the enamine with the aromatic azide, which is obtained by the reaction of phenylboronic acid with sodium azide in the presence of Cu(OAc)$_2$, affords a triazoline intermediate. Finally, elimination of triazoline provides the desired product and regenerates the piperidine, closing the catalytic cycle.

In 2007, Yadav *et al.* described a direct and efficient protocol for the preparation of β-hydroxytriazoles *via* a three-component reaction of an epoxide, sodium azide and an alkyne.[48] This reaction proceeds smoothly in water at room temperature without the need for acid catalysis. In addition to its simplicity and mild reaction conditions, this method provides a wide range of β-hydroxytriazoles in excellent yields with high regioselectivity in a single-step operation. Interestingly, in 2000, a safe, efficient and improved procedure for the regioselective synthesis of 1-(2-hydroxyethyl)-1*H*-1,2,3-triazole derivatives under ambient conditions was described by Sharghi *et al.*[49] Terminal alkynes reacted with epoxides and NaN$_3$ in the presence of a copper(I) catalyst, which was prepared by *in situ* reduction of a copper(II) complex [Cu(bhppda)H$_2$O; bhppda = N^2,N^6-bis(2-hydroxyphenyl)pyridine-2,6-dicarboxamidato] with ascorbic acid in water. The regioselective reactions gave exclusively the corresponding 1,4-disubstituted 1*H*-1,2,3-triazoles in good to excellent yields (Scheme 4.16). This procedure avoids the handling of organic azides as they are generated *in situ*, making this already powerful click process even more user-friendly and safe. The remarkable features of this protocol are high yields, very short reaction times, a cleaner reaction profile in an environmentally benign solvent (water), its straightforwardness and the use of non-toxic catalysts. Furthermore, the catalyst could be recovered and recycled by simple filtration of the reaction mixture and was reused for 10 consecutive trials without significant loss of catalytic activity. No leaching of the metal complex was observed after the consecutive catalytic reactions. Later, copper nanoparticles on activated carbon were developed and applied in this conversion. The reaction used water as the only solvent and all the products were obtained in good to excellent yields.[50]

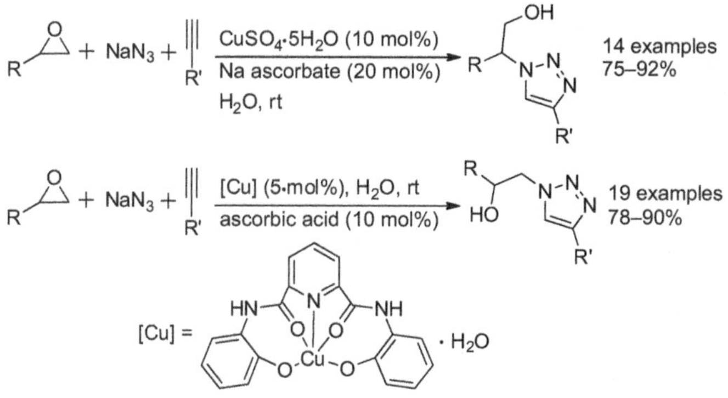

Scheme 4.16 Copper-catalyzed ring-opening of epoxides with NaN$_3$.

General Procedure (Scheme 4.16): To a suspension of styrene oxide (120 mg, 0.11 mL, 1.0 mmol) and sodium azide (78 mg, 1.2 mmol) in water (3.0 mL) were added CuSO$_4$·5H$_2$O (16.0 mg, 0.1 mmol) and sodium ascorbate (39.6 mg, 0.2 mmol). The resulting solution was stirred for 2 h at room temperature. After complete consumption of styrene oxide (monitored by TLC), phenylacetylene (102 mg, 0.11 mL, 1 mmol) was added to the reaction mixture, which was stirred for a further 2 h. The reaction mixture was extracted with ethyl acetate (3×10 mL). The combined organic layers were dried over anhydrous Na$_2$SO$_4$ and concentrated *in vacuo*. The resulting residue was purified by column chromatography (eluent: 3:1 hexane–EtOAc) to give the pure 1,2,3-triazole product.

General Procedure: For each reaction, epoxide (1 mmol), alkyne (1 mmol) and NaN$_3$ (1.2 mmol) were mixed and stirred in water (1 mL) in the presence of [Cu(bhppda)H$_2$O] (5 mol%) and ascorbic acid (10 mol%) at r.t. in an uncapped vial. After the completion of the reaction (monitored by TLC using 4:1 hexane–EtOAc), the mixture was diluted with water (5 mL), then the whole mixture was directly passed through a Celite pad and rinsed with EtOAc (3 × 10 mL). The combined organic layer was washed with saturated brine, dried (Na$_2$SO$_4$) and concentrated and the residue was recrystallized from EtOH–H$_2$O (3:1) or purified by column chromatography (eluent: 4:1 hexane–EtOAc).

Fokin and co-workers found that 2*H*-hydroxymethyl-1,2,3-triazoles can alternatively be prepared from sodium azide, formaldehyde and alkynes using a simple one-pot process (Scheme 4.17).[51] In another pathway, they can be prepared by hydroxymethylation of NH-triazoles with formaldehyde. In this three-component reaction, alkynes undergo a copper(I)-catalyzed cycloaddition with sodium azide and formaldehyde to yield 2-hydroxymethyl-2*H*-1,2,3-triazoles, which are useful intermediates that can be readily converted to polyfunctional molecules. The hydroxymethyl group can also be removed, providing convenient access to NH-1,2,3-triazoles. The reaction is experimentally simple and readily scalable.

Methods have been developed for the synthesis of 1,2,3-triazole derivatives under azide-free conditions. In 2013, a copper-catalyzed domino reaction of 2*H*-azirines with diazotetramic and diazotetronic acids to give 2*H*-1,2,3-triazoles was reported.[52] A copper(II)-catalyzed aerobic oxidative synthesis of 2,4,5-triaryl-1,2,3-triazoles and 1,3,5-triaryl-1,2,4-triazoles from bisarylhydrazones was developed in 2012 (Scheme 4.18).[53] The reaction proceeds through a cascade of C–H functionalization/C–C/N–N/C–N bond formation under mild reaction conditions. One of the interesting outcomes of this strategy is the copper(II)-catalyzed room-temperature C–H functionalization/C–N bond formation in the presence of air, which was accomplished during the synthesis of substituted 1,2,4-triazoles. This new class of compounds could give prospective luminescence as an important component in the area of pharmaceutical and biological sciences. The intermediates for both processes were isolated to elucidate the mechanism. The 1,2,4-triazoles can also be produced by oxidative intramolecular cyclization of heterocyclic hydrazones.[54] Moderate to high yields of 1,2,4-triazolo[4,3-*a*]pyridines, 1,2,4-triazolo[4,3-*a*]pyrimidines, 1,2,4-triazolo[4,3-*b*]pyridazines, 1,2,4-triazolo[4,3-*a*]-phthalazines and 1,2,4-triazolo[4,3-*a*]quinoxalines were achieved in DMF at 90–140 °C with 2 equiv. of CuCl$_2$.

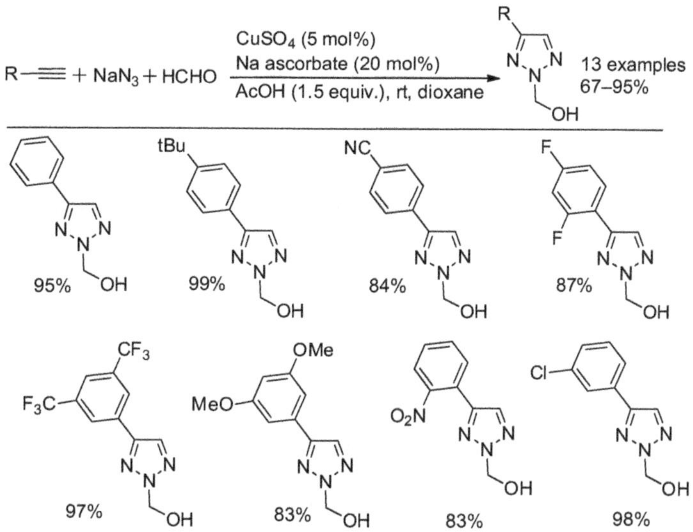

$$R\text{---}\equiv + NaN_3 + HCHO \xrightarrow[\text{AcOH (1.5 equiv.), rt, dioxane}]{\substack{\text{CuSO}_4\ (5\ \text{mol}\%)\\ \text{Na ascorbate (20 mol\%)}}}$$

13 examples
67–95%

95% ᐧOH   99% ᐧOH   84% ᐧOH   87% ᐧOH

97% ᐧOH   83% ᐧOH   83%   98%

Scheme 4.17   Copper-catalyzed synthesis of 2*H*-hydroxymethyl-1,2,3-triazoles.

General Procedure (Scheme 4.17): A mixture of 37% aqueous formaldehyde (736 mL, 9.8 mol, 10 equiv.), glacial acetic acid (84 mL, 1.47 mol, 1.5 equiv.) and 1,4-dioxane (736 mL) was stirred for 15 min, then NaN₃ (95.6 g, 1.47 mol, 1.5 equiv.) was added, followed by phenylacetylene (100 g, 0.98 mol). At this point, the pH of the reaction mixture was 6.5. After an additional 10 min of stirring, sodium ascorbate (38.8 g, 0.196 mol, 20 mol%) was added, followed by CuSO₄ solution (7.8 g, 0.049 mol, 5 mol%) in 40 mL of water. The mixture was stirred for 18 h at r.t., then diluted with water (3000 mL) and extracted using CHCl₃ (3 × 500 mL). The combined organic layers were filtered through Celite to remove solids, dried over MgSO₄, filtered and concentrated on a rotary evaporator to give a yellowish solid (157.5 g, yield 91.8%).

In the click reactions, 1*H*-tetrazoles can be produced by replacing alkynes with nitriles.[55] In 2008, a copper-catalyzed [3 + 2] cycloaddition between various nitriles and trimethylsilyl azide in the presence of a CuI catalyst in DMF–MeOH was reported. The corresponding 5-substituted 1*H*-tetrazoles were obtained in good to high yields (Scheme 4.19). The reaction probably proceeds through the *in situ* formation of a copper azide species, followed by a successive [3 + 2] cycloaddition with the nitriles. Later, several heterogeneous copper catalysts were also developed. In these cases, NaN₃ was used as a cheaper azide source.

Additionally, Jiao and co-workers demonstrated a novel Cu-promoted direct implanting of nitrogen into simple hydrocarbon molecules for triazole synthesis.[56] 1,5-Disubstituted tetrazoles were efficiently constructed by two

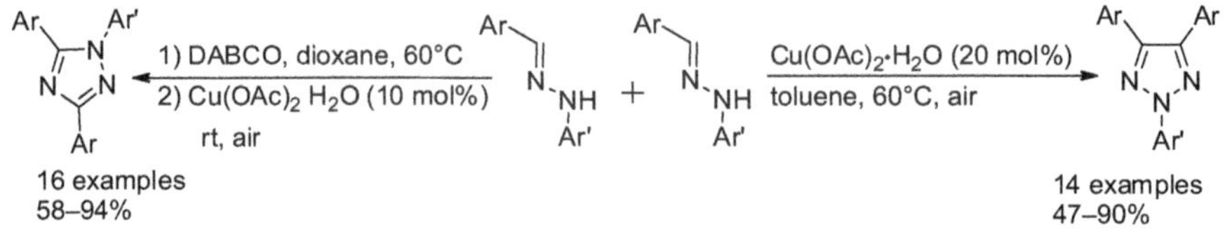

Scheme 4.18 Copper-catalyzed oxidative synthesis of 1,2,3-triazoles and 1,2,4-triazoles.

General Procedure for Copper(II)-Catalyzed Synthesis of 2,4,5-Triaryl-1,2,3-triazoles (Scheme 4.18). Bisarylhydrazones (0.5 mmol) and $Cu(OAc)_2 \cdot H_2O$ (20 mol%, 20.0 mg) were stirred at 60 °C in toluene (2 mL) under air. After stirring for the appropriate time, the reaction mixture was cooled to room temperature and passed through a short pad of silica gel using hexane followed by a mixture of ethyl acetate and hexane as eluent to give the target 2,4,5-triaryl-1,2,3-triazoles in analytically pure form.

General Procedure for Copper(II)-Catalyzed Synthesis of 1,3,5-Triaryl-1,2,4-triazoles. Bisarylhydrazones (0.5 mmol) with DABCO (1 equiv., 56.1 mg) were stirred in dioxane (2 mL) at 60 °C for 30 h under air. After cooling to room temperature, $Cu(OAc)_2 \cdot H_2O$ (10 mol%, 10.0 mg) was added in the same pot and the stirring was continued for the appropriate time at room temperature under air. After completion of the reaction, the solvent was concentrated under reduced pressure and the residue was purified by column chromatography on silica gel using ethyl acetate–hexane as eluent to afford the desired heterocycles in analytically pure form.

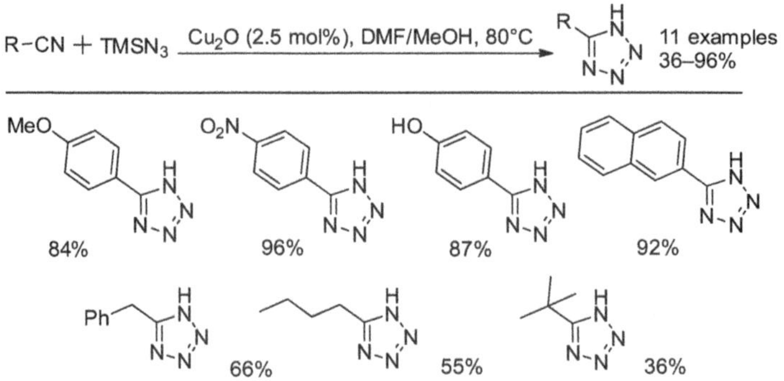

Scheme 4.19 Copper-catalyzed oxidative synthesis of 1*H*-tetrazoles.

Representative procedure (Scheme 4.19). Trimethylsilyl azide (0.1 mL, 0.75 mmol) was added to a 9 : 1 DMF–MeOH solution (1 mL, 0.5 M) of Cu_2O (1.8 mg, 0.0125 mmol) and *p*-methoxybenzonitrile (66.6 mg,

0.5 mmol) in a pressure vial. The reaction mixture was stirred at room temperature for 10 min, then heated at 80 °C for 12 h. After complete consumption of the substrate, the reaction mixture was cooled to room temperature and extracted with ethyl acetate. The organic layer was washed with 1 M HCl, dried with anhydrous Na_2SO_4 and concentrated. To the residue was added 0.25 M NaOH and the resulting mixture was stirred for 30 min at room temperature. The mixture was washed with ethyl acetate and then concentrated HCl was added until the pH of the water layer became 1. The aqueous layer was extracted (×3) with ethyl acetate and the combined organic layers were washed with 1 M HCl. The organic layer was dried over anhydrous Na_2SO_4 and concentrated.

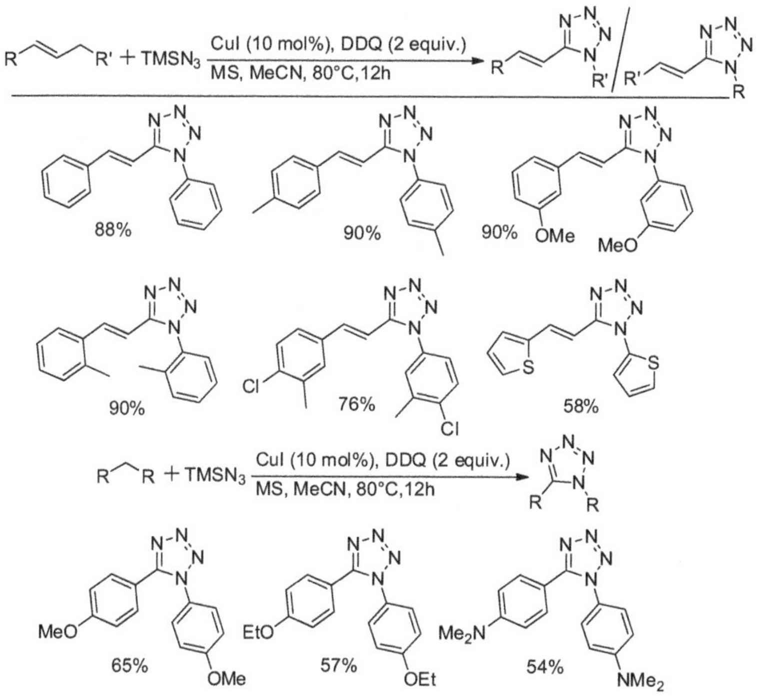

Scheme 4.20 Copper-catalyzed synthesis of 1*H*-tetrazoles *via* C–H activation.

$C(sp^3)$–H and one C–C bond cleavages under mild and neutral reaction conditions (Scheme 4.20). This protocol not only extends the application of azides in organic conversions, but also offers an alternative method to prepare 1,5-disubstituted tetrazoles, which are ubiquitous structural units in a number of biologically active compounds. This method provides a new and unique strategy to functionalize simple and readily available hydrocarbon molecules by C–H and C–C bond cleavages.

4.2.2   Copper-Catalyzed Synthesis of Imidazole Derivatives

Imidazole derivatives are known with wide utilities. Many interesting methodologies have been developed for their preparation and copper catalysts are representative catalysts. In 2012, a copper-catalyzed synthesis of the imidazo[4,5-c]pyrazole nucleus was reported.[57] Using 5-aminopyrazoles as starting material, followed by halogenation and copper-catalyzed C–N bond formation, 4-substituted imidazo[4,5-c]pyrazoles were produced in good yields (Scheme 4.21). The copper-catalyzed cyclization of the N'-(4-halopyrazol-5-yl)amidine is crucial for the success of this transformation and can be done using either microwave or thermal heating. This method is inexpensive and convenient and allows a wide range of substituents at all positions of the product. The intermediates can also be prepared by the reaction of o-bromophenyl isocyanide with amines, which gave 1-substituted benzimidazoles with the assistance of a copper catalyst.[58]

In 2006, a new, efficient synthetic procedure for 1,4-disubstituted imidazoles *via* the cross-cycloaddition between two different isocyanides was reported (Scheme 4.22).[59] Concerning the reaction mechanism, this reaction started with activation of a C–H bond of the isocyanides under the influence of a Cu_2O catalyst. The α-cuprioisocyanide or its tautomer is formed by the reaction between isocyanides and Cu_2O through the elimination of H_2O. Then, the nucleophilic addition of the intermediate to the aryl isocyanide takes place to generate another intermediate. Intramolecular attack of the nitrogen atom derived from aryl isocyanide on the carbon atom of the CN group, followed by a 1,3-hydrogen shift, would produce the cyclized

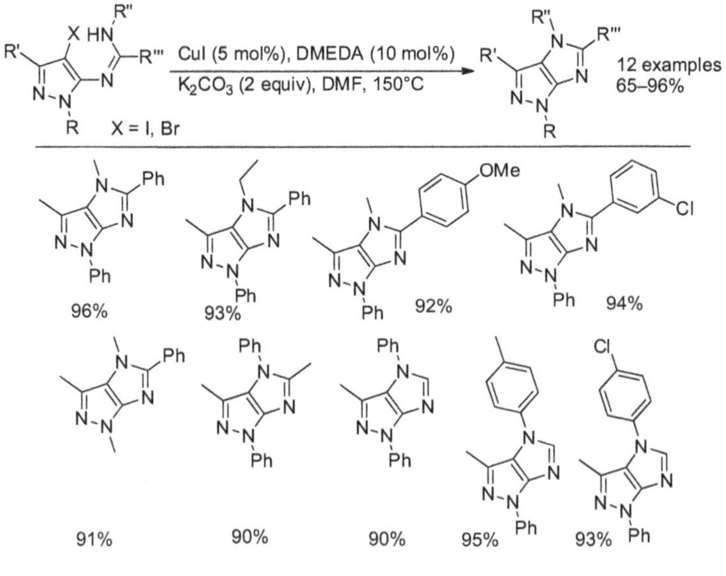

Scheme 4.21   Copper-catalyzed synthesis of imidazo[4,5-c]pyrazoles.

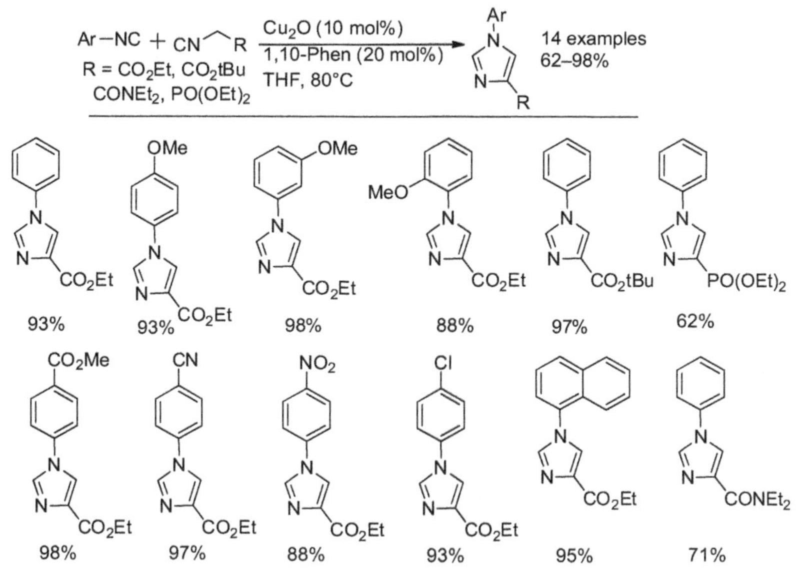

Scheme 4.22 Copper-catalyzed synthesis of imidazoles from isocyanides.

General Procedure (Scheme 4.22): To a THF solution (2.8 mL) of Cu_2O (10.0 mg, 70 μmol) and 1,10-phenanthroline (25.2 mg, 140 μmol) were added 4-methoxy-1-isocyanobenzene (66.6 mg, 0.5 mmol) and ethyl isocyanoacetate (77 μL, 0.7 mmol) under an argon atmosphere. The solution was stirred at 80 °C for 3 h. After the consumption of 4-methoxy-1-isocyanobenzene, the reaction mixture was cooled to room temperature and filtered through a short Florisil pad and concentrated. The residue was purified by column chromatography on silica gel (eluent: 10:1–1:1 *n*-hexane–AcOEt) to afford 4-ethoxycarbonyl-1-(4-methoxyphenyl)imidazole in 93% yield (114.5 mg).

intermediate; this is a formal $[3+2]$ cycloaddition process. The C–Cu bond in the intermediate is protonated by isocyanides to give the 1,4-disubstituted imidazole with regeneration of the α-cuprioisocyanide.

A novel and efficient approach for the preparation of highly substituted imidazoles in one step by using ketones and benzylamines catalyzed by commercially available CuI–$BF_3 \cdot Et_2O$ in an atmosphere of O_2 was developed.[60] $BF_3 \cdot Et_2O$ showed high reactivity as a co-catalyst combined with CuI. In addition, this reaction provides a simple, easy-to-handle and atom-economical means for the synthesis of polysubstituted imidazoles under mild conditions. This protocol involved the removal of eight hydrogen atoms, the functionalization of four $C(sp^3)$–H bonds and the formation of three new C–N bonds in one manipulation. The desired products were obtained in moderate to good yields (Scheme 4.23).

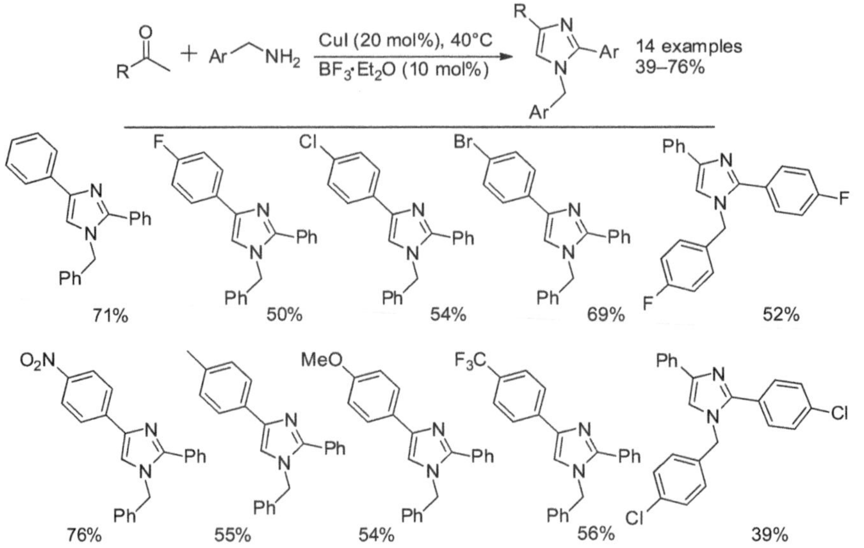

Scheme 4.23 Copper-catalyzed synthesis of imidazoles from ketones and amines.

General Procedure (Scheme 4.23): A Schlenk tube equipped with a stirrer bar was charged with acetophenone (2 mmol, 0.2401 g), benzylamine (6 mmol, 0.6424 g), CuI (0.4 mmol, 0.0759 g, 20 mol%) and BF$_3$ · Et$_2$O (0.2 mmol, 0.0284 g). The Schlenk tube was quickly evacuated, closed under vacuum and then refilled with oxygen using an oxygen balloon. The resulting mixture was stirred at 40 °C for 24 h. After completion of the reaction, the residue was directly purified by flash column chromatography using ethyl acetate–petroleum ether as eluent to afford the pure product.

   In 2012, Chiba and co-workers reported a method for the synthesis of bi- and tricyclic amidines through a copper-catalyzed aerobic [3 + 2] annulation reaction of N-alkenylamidines.[61] In the presence of CuI and bipyridine, under 1 bar of O$_2$, in DMF, the desired products were obtained in good yields. Later, they extended their system to N-allylamidines (Scheme 4.24).[62] By adding additional PhI(OAc)$_2$ and K$_3$PO$_4$, 4-acetoxymethyl-4,5-dihydroimidazoles were produced by Cu-catalyzed aminooxygenation of alkenes with amidine moieties. By reduction with aluminum hydride, diamines can be produced from the former products.

   Chen and co-workers developed a simple route for the synthesis of imidazole derivatives *via* a copper-catalyzed [3 + 2] cycloaddition reaction.[63] Nitroalkenes and amidines as abundant substrates were applied as starting materials. Additionally, high regioselectivity could be achieved by this strategy and oxygen was used as an oxidant without the addition of expensive catalysts to provide moderate to good yields (Scheme 4.25).

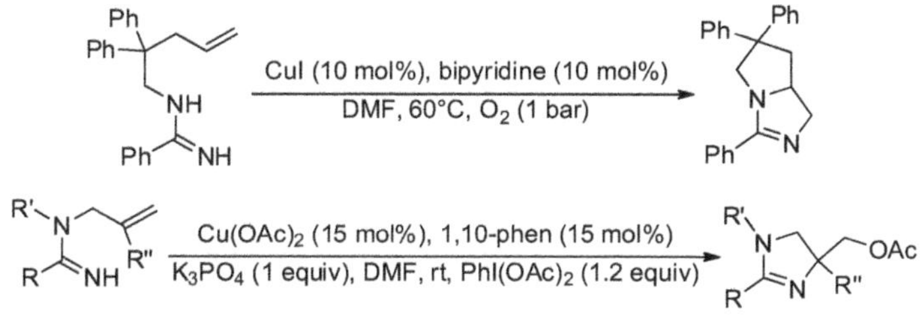

Scheme 4.24 Copper-catalyzed synthesis of amidines.

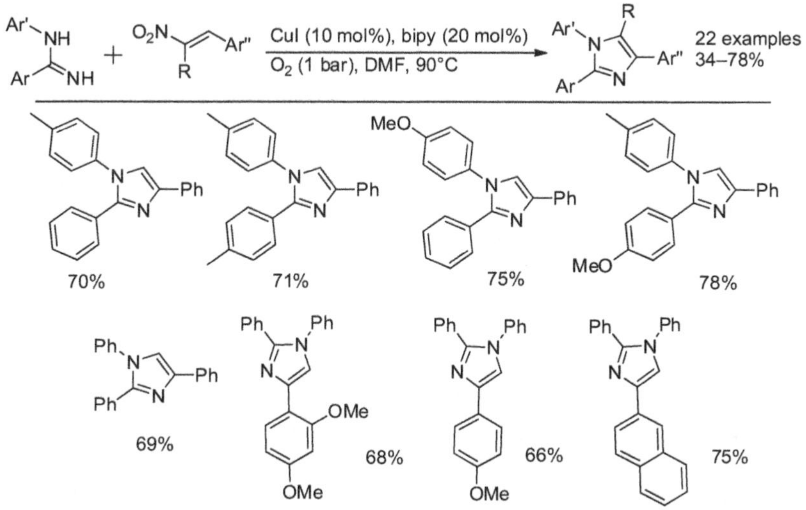

Scheme 4.25 Copper-catalyzed synthesis of imidazoles from nitroalkenes and amidines.

General Procedure (Scheme 4.25): Amidines (0.24 mmol), nitroalkenes (0.2 mmol), CuI (0.02 mmol), bipy (0.04 mmol) and DMF (2 mL) were placed in a round-bottomed side-arm flask (10 mL) with a magnetic stirrer bar under an O_2 atmosphere. The mixture was stirred at 90 °C for 5 h, cooled to room temperature, then filtered and extracted with ethyl acetate. The filtrate was concentrated under reduced pressure in order to obtain the crude product, which was further purified by silica gel chromatography (eluent: 10 : 1 petroleum ether–ethyl acetate) to give the product.

Li and Neuvilles developed an efficient and regioselective access to 1,2,4-trisubstituted imidazoles from terminal alkynes and amidines (Scheme 4.26).[64] The process, which used copper as the catalyst and oxygen as a co-oxidant, involves the selective addition of two distinct N-atoms across the triple bond, making the process particularly cheap and atom efficient.

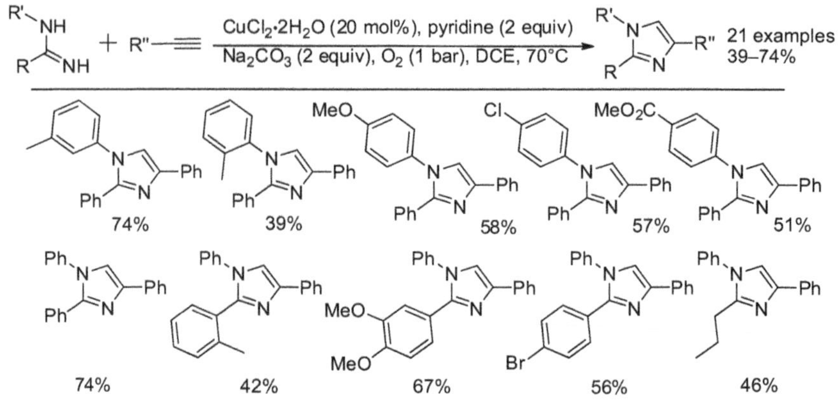

Scheme 4.26 Copper-catalyzed synthesis of imidazoles from alkynes and amidines.

> **General Procedure** (Scheme 4.26): Amidines (0.1 mmol, 1.0 equiv.), Na_2CO_3 (0.2 mmol, 2.0 equiv.) and $CuCl_2 \cdot H_2O$ (0.02 mmol, 0.2 equiv.) were introduced into a a round-bottomed flask (10 mL) sealed with rubber stopper, evacuated and backfilled with oxygen (balloon). DCE (0.5 mL) and pyridine (0.2 mmol, 2.0 equiv.) were added and the mixture was heated to 70 °C. A solution of a terminal yne (0.2 mmol, 1.0 equiv, in 1 mL of DCE) was slowly added over 10 h using a syringe pump (0.2 mL h^{-1}). The reaction mixture was further stirred for 14 h at 70 °C. The mixture was cooled to room temperature, filtered through a glass funnel and washed with EtOAc. The filtrate was concentrated under reduced pressure to give a residue that was purified by preparative TLC.

Diverse imidazoles were produced in modest to good yields in the presence of Na_2CO_3, pyridine, a catalytic amount of $CuCl_2 \cdot 2H_2O$ and oxygen (1 atm).

4.2.3 Copper-Catalyzed Synthesis of Benzimidazole Derivatives

In 2008, Brasche and Buchwald developed an interesting oxidative benzimidazole synthesis using amidines as substrates.[65] The reactions were carried out in DMSO under O_2 in the presence of AcOH and Cu(OAc)$_2$; benzimidazoles were produced in good yields *via* C–H activation (Scheme 4.27).

Zhu, Neuville and co-workers developed a mild copper-catalyzed mono-*N*-arylation of amidines using arylboronic acids under aerobic conditions.[66] The process can be extended to a one-pot synthesis of benzimidazoles involving a sequence of intermolecular C–N bond formation and intramolecular C–H functionalization/C–N bond-formation (Scheme 4.28). The

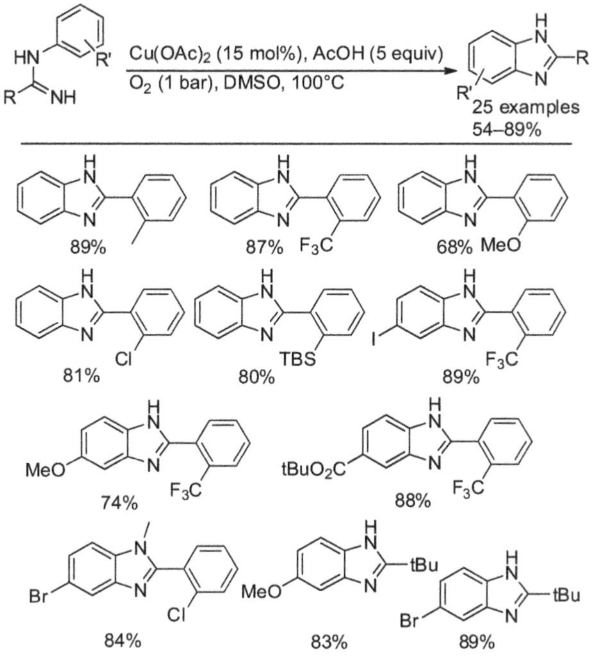

Scheme 4.27 Copper-catalyzed synthesis of benzimidazoles.

General Procedure (Scheme 4.27): An oven-dried disposable test-tube (15 mL) equipped with a stirrer bar and a rubber septum was cooled to room temperature under vacuum and backfilled with O_2. With the tube open to the air, the amidine (1.00 mmol) and Cu(OAc)$_2$ (27.2 mg, 0.15 mmol, 15 mol%) were added. The tube was evacuated and backfilled with O_2, followed by addition of DMSO (2 mL) by syringe. The reaction mixture was degassed by sonication under vacuum and backfilled with O_2 (this procedure was carried out three times). An O_2-filled balloon was attached to a needle with the aid of a small piece of rubber tubing. The needle was inserted through the rubber septum and HOAc (286 µL, 5.00 mmol, 5.00 equiv.) was added by syringe. The reaction mixture was lowered into a preheated oil-bath at 100 °C and stirred for the indicated time.

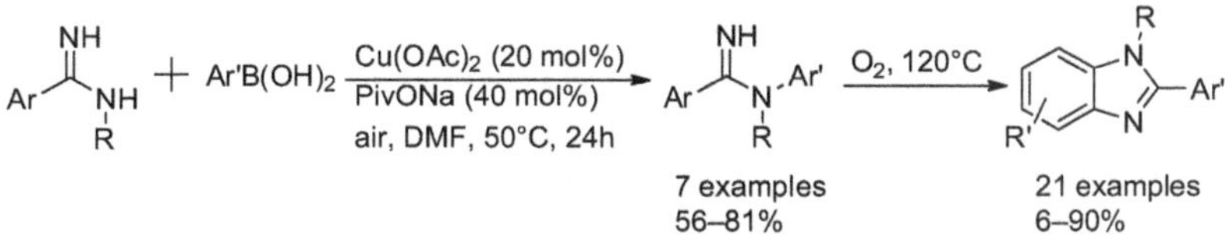

Scheme 4.28 Copper-catalyzed synthesis of benzimidazoles from amidines and arylboronic acids.

ready accessibility of the starting materials and the low cost of the catalytic system made this process a valuable alternative for the construction of these interesting heterocycles.

A new method for assembling substituted benzimidazoles was developed by Ma and co-workers that involved a one-pot coupling of 2-haloacetanilides with amines and subsequent additive cyclization.[67] A wide range of benzimidazoles, which have different functional groups such as ketone, ester, methoxy, bromo and iodo on the aromatic ring and aryl and simple and functionalized alkyl groups at the 2-position, could be elaborated from suitable substrates. Additionally, by coupling of 2-halophenylcarbamates with aqueous ammonia at room temperature and subsequent intramolecular condensation at 130 °C, several 1,3-dihydrobenzimidazol-2-ones were constructed. These two processes provide simple and reliable approaches for assembly of the pharmaceutically important heterocycles and therefore may find applications in organic synthesis (Scheme 4.29).

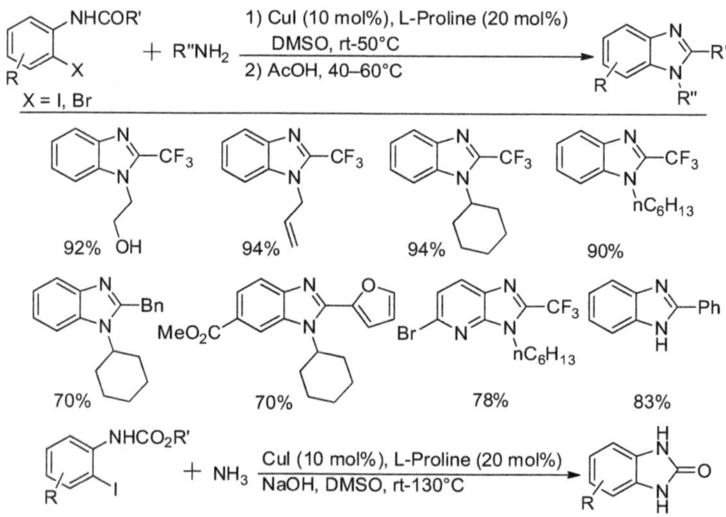

Scheme 4.29 Copper-catalyzed synthesis of 1,2-disubstituted benzimidazoles.

General Procedure (Scheme 4.29): A Schlenk tube was charged with 2-iodoacetanilides (0.25 mmol), CuI (5 mg, 0.025 mmol), L-proline (6 mg, 0.05 mmol) and NaOH (15 mg, 0.375 mmol). The tube was evacuated and backfilled with argon (three times). Aqueous ammonia (0.375 mmol) and 1 mL of DMSO were added to the tube. After the reaction mixture had been stirred at room temperature for 3–7 h, 3 mL of AcOH was added and the reaction mixture was stirred at 50–80 °C for 3–12 h. The cooled reaction mixture was partitioned between ethyl acetate and water. The organic layer was washed with aqueous NaHCO₃ and water, dried over

Na$_2$SO$_4$ and concentrated *in vacuo*. The residue was purified by column chromatography on silica gel (eluent: 3:1–1:1 petroleum ether–ethyl acetate) to provide the desired product.

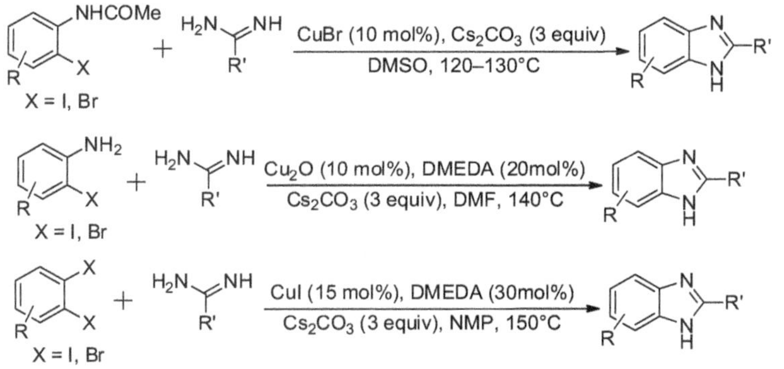

Scheme 4.30 Copper-catalyzed synthesis of benzimidazoles from amidines.

Hu, Fu and co-workers developed an efficient method for the synthesis of benzimidazoles *via* cascade reactions of *o*-haloacetoanilide derivatives with amidine hydrochlorides (Scheme 4.30).[68] The protocol uses 10 mol% CuBr as the catalyst, Cs$_2$CO$_3$ as the base and DMSO as the solvent under ligand-free conditions. The reaction proceeds *via* the sequential coupling of *o*-haloacetoanilide derivatives with amidines, hydrolysis of the intermediates (amides) and intramolecular cyclization with the loss of NH$_3$ to give 2-substituted 1*H*-benzimidazoles. Zhou and co-workers reported a methodology using *o*-haloanilines and amidines as substrates.[69] By applying Cu$_2$O as the catalyst, benzimidazole derivatives were prepared *via* a cascade C–N coupling and intramolecular transamination reaction. Deng's group studied the reaction of 1,2-dihaloarenes with *N*-substituted amidines to give benzimidazoles.[70] The first amination step is exclusively N^1-selective on the amidines and the chemoselectivity on the 1,2-dihaloarenes is predominantly controlled by the halide reactivity differences. On the basis on these findings, a highly regiospecific benzimidazole synthesis was developed. Both starting materials are readily available and the reaction scope is fairly broad. One drawback is the low yield for some substrates, which might result from the instability of the substrates and products under the high-temperature conditions.

Wang and co-workers developed a general, efficient, one-pot synthesis of functionalized benzimidazoles from terminal alkynes, *o*-aminoanilines and *p*-tolylsulfonyl azide (Scheme 4.31).[71] When using naphthalene-1,8-diamine as substrate, the reaction gave 1*H*-pyrimidines. Regarding the reaction mechanism, the authors proposed that the reaction proceeded smoothly *via* a ketenimine intermediate and afforded *N*-(2-aminophenyl)-2-phenyl-*N'*-tosylacetimid amide with excellent chemoselectivity and good yield (86%).

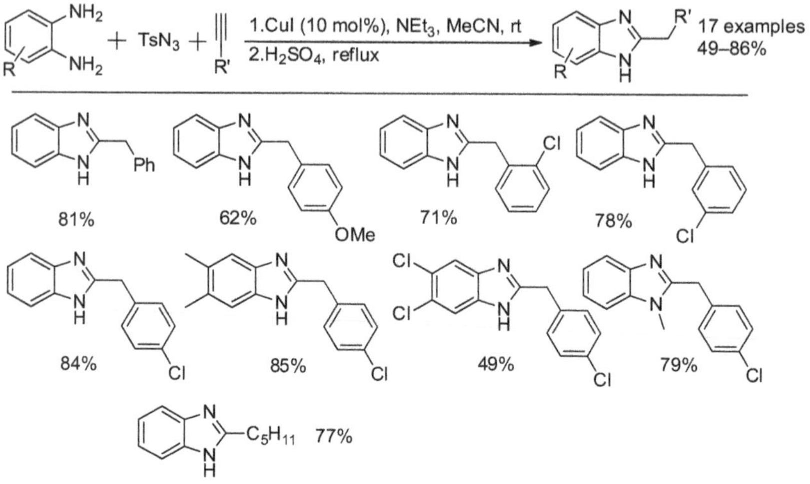

Scheme 4.31   Copper-catalyzed synthesis of benzimidazoles from *o*-aminoaniline.

General Procedure (Scheme 4.31): To a solution of *p*-tolylsulfonyl azide (2.2 mmol), terminal alkyne (2.1 mmol), diamine (*o*-aminoaniline or naphthalene-1,8-diamine; 2 mmol) and CuI (0.2 mmol) in MeCN (10 mL) was added dropwise Et$_3$N (2 mmol). The reaction mixture was stirred at r.t. under N$_2$ for 6 h, then concentrated H$_2$SO$_4$ (98%, 0.4 mL) was added and the solution was heated under reflux for 4 h. After cooling to r.t., the solution was poured into water (20 mL) and neutralized with K$_2$CO$_3$. MeCN was removed under vacuum and the resulting solution was extracted with EtOAc (3×5 mL). The organic layers were combined, dried over anhydrous Na$_2$SO$_4$ and the solvent was removed under vacuum. The residue was purified by column chromatography on silica gel (eluent: 2:1→1:2 petroleum ether–EtOAc).

When treated with 2% H$_2$SO$_4$ under reflux conditions, the latter amide compound was successfully converted into 2-benzyl-1*H*-benzimidazole in excellent yield (93%) by intramolecular nucleophilic addition and subsequent elimination.

In 2010, the same group developed another procedure for the synthesis of benzimidazoles.[72] The procedure combines the copper-catalyzed three-component cascade reaction of sulfonyl azides, alkynes and 2-bromoaniline and the copper-catalyzed intramolecular *N*-arylation of sulfonamides in one sequence, which afforded the products in moderate to good yields (Scheme 4.32). The following reaction mechanism was proposed by the authors. In the first step, in the presence of TEA and CuI, a sulfonyl azide reacts with a terminal alkyne to form the ketenimine species. The latter is quickly attacked by a nucleophile to generate an *N*-sulfonylamidine, which tautomerizes to a sulfonamide. The subsequent CuI–L-proline-catalyzed

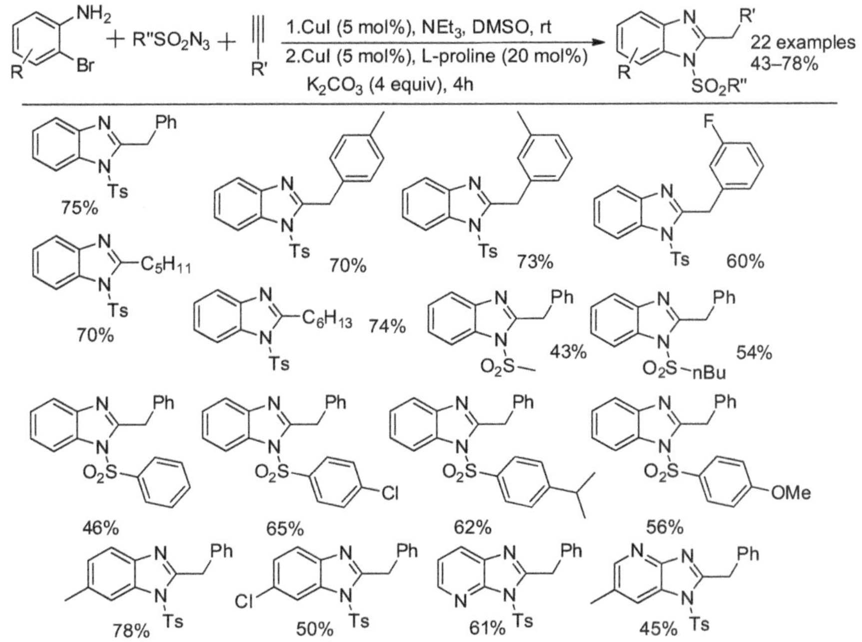

Scheme 4.32 Copper-catalyzed synthesis of benzimidazoles from 2-bromoanilines.

General Procedure (Scheme 4.32): To a solution of sulfonyl azide (0.5 mmol), alkyne (0.5 mmol), 2-bromoanline (0.55 mmol) and CuI (0.05 mmol) in DMSO (2 mL) in a Schlenk tube was slowly added TEA (0.5 mmol) *via* a syringe. The reaction solution was stirred at room temperature under N_2 for 2 h. To the solution were then added CuI (0.05 mmol), K_2CO_3 (2 mmol) and L-proline (0.2 mmol) and the mixture was stirred at room temperature under N_2 for 4 h. The reaction mixture was partitioned between ethyl acetate and saturated NH_4Cl solution and the organic layer was washed with brine, dried over Na_2SO_4 and concentrated under vacuum. The residue was purified by column chromatography on silica gel (eluent: 4:1–7:1 petroleum ether–ethyl acetate).

intramolecular C–N coupling of the sulfonamide breaks the interconvertible tautomerization and affords the desired product.

Several other copper-catalyzed intramolecular coupling methodologies for the synthesis of benzimidazoles have also been developed by different groups (Scheme 4.33). Wu and co-workers reported copper(I)-catalyzed tandem reactions for the preparation of 2-fluoroalkylbenzimidazoles using primary amines and *N*-aryltrifluoroacetimidoyl (or -bromodifluoroacetimidoyl) chlorides as substrates.[73] The reactions were carried out in DMF at 60 °C with K_2CO_3 or K_3PO_4 as base, and good to excellent yields of the desired products were obtained in the absence of ligand. Glorius and co-workers developed a

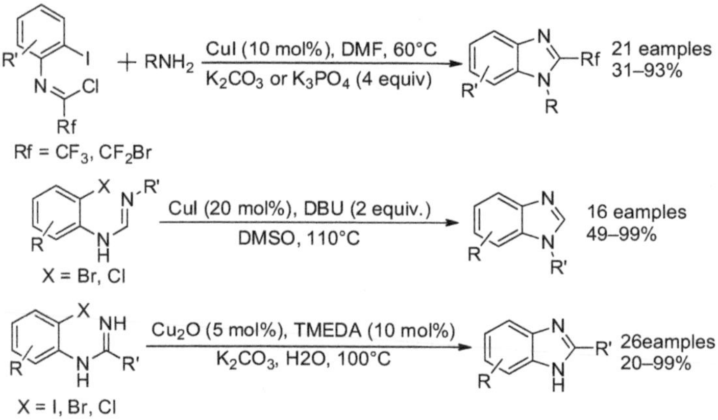

Rf = CF$_3$, CF$_2$Br

Scheme 4.33   Copper-catalyzed intramolecular synthesis of benzimidazoles.

highly efficient copper-catalyzed synthesis of *N*-substituted benzimidazoles using formamidines as substrates.[74] Sixteen examples bearing sterically demanding substituents on nitrogen such as Mes, 2,6-diisopropylphenyl and 2-*tert*-butylphenyl and various other functional groups were produced in excellent yields. More recently, a straightforward, efficient and more sustainable copper-catalyzed method was developed for intramolecular *N*-arylation to provide the benzimidazole ring system.[75] With Cu$_2$O (5 mol%) as the catalyst, DMEDA (10 mol%) as the ligand and K$_2$CO$_3$ as the base, this protocol was applied to synthesize a small library of benzimidazoles in high yields. Remarkably, the reaction was exclusively carried out in water, rendering the methodology highly valuable from both environmental and economical points of view. More recently, a copper-catalyzed intramolecular *N*-arylation of cyclic amidines was reported by Liubchak and co-workers.[76] 1,2-Disubstituted benzimidazoles were produced in excellent yields under mild conditions (CuI, DMEDA, K$_2$CO$_3$, MeCN, 80 °C). This procedure was also applied in the synthesis of alicyclic ring-fused xanthines.

An interesting procedure for the preparation of purines was developed in 2009.[77] By applying primary amides and 2-substituted heteroarene halides as the substrates and CuI and a nitrogen ligand as the catalytic system, in dioxane, using Cs$_2$CO$_3$ as base, the desired 6,7,8-trisubstituted purines were formed in good yields (Scheme 4.34). In 2013, a method for copper-catalyzed coupling between 2-iodoanilines and amides was reported.[78] 2-Substituted benzimidazoles were produced in good yields with the assistance of CuI–DMEDA. Notably, AcOH was needed to assistant the ring closure and imidazole formation, but this was not the case in the purines synthesis.

Evindar and Batey demonstrated an intramolecular aryl guanidinylation to form biologically relevant 2-aminobenzimidazoles with both palladium and copper as the catalysts (Scheme 4.35).[79] Inexpensive copper salts such as CuI were shown to be superior to their palladium counterparts, in terms of both yields and selectivities. In 2009, Bao's group developed a novel, efficient and

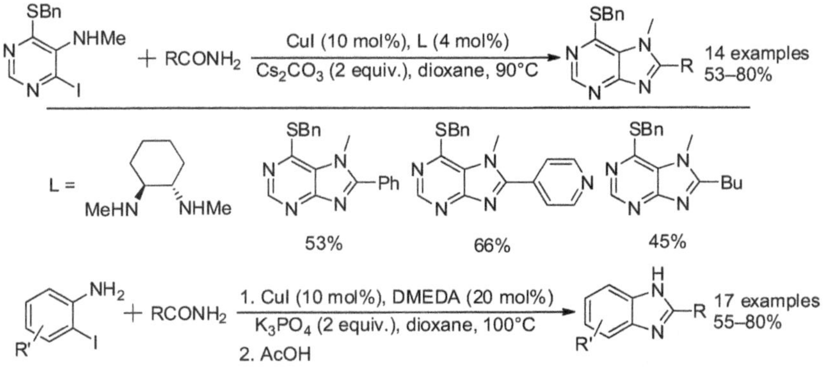

Scheme 4.34 Copper-catalyzed synthesis of purines.

General Procedure (Scheme 4.34): An oven-dried Schlenk tube was charged with 1 equiv. of substrate (100 mg), the corresponding amide (1.5 equiv.), cesium carbonate (2 equiv.), CuI (0.1 equiv, 10 mol%), *trans-N,N'*-dimethylcyclohexane-1,2-diamine (0.4 equiv, 40 mol%) and degassed dioxane (1.5 mL). The tube was capped with a rubber septum, evacuated and backfilled with argon three times, the rubber septum was replaced with a screw-cap and the mixture was stirred for 7–8 h at 90 °C.

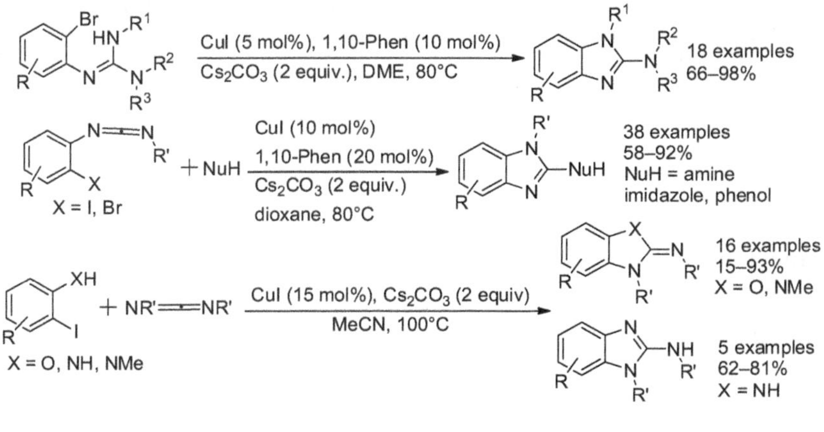

Scheme 4.35 Copper-catalyzed synthesis of 2-aminobenzimidazoles.

versatile strategy for the synthesis of various 2-heterobenzimidazoles.[80] A number of biologically and pharmaceutically active *N*-substituted 2-amino-benzimidazoles (or 2-aminopyridoimidazoles), 2-imidazylbenzimidazoles and 2-phenoxylbenzimidazoles were smoothly synthesized *via* a Cu-catalyzed one-pot addition/C–N coupling process. This protocol allows the introduction of different kinds of hetero substituents on the 2-position of the benzimidazole ring. The efficient and versatile one-pot transformation of the readily available starting materials into 2-heterobenzimidazoles would make this protocol valuable to synthetic chemists. Later, the same group reported another method for the preparation of 2-aminobenzimidazoles.[81] Various benzoxazole and benzimidazole derivatives were conveniently synthesized in moderate to excellent yields under ligand-free copper(I)-catalyzed reaction conditions.

An efficient method for the preparation of various substituted 2-mercaptobenzimidazoles from their corresponding thioureas was developed in 2009.[82] *S*-Alkylation of thioureas followed by Cu-catalyzed intramolecular *N*-arylation furnished substituted 2-mercaptobenzimidazoles in high yields with short reaction times (Scheme 4.36). Furthermore, 2-mercaptobenzimidazoles substituted with a *p*-methoxybenzyl group allowed access to benzimidazolethiones.

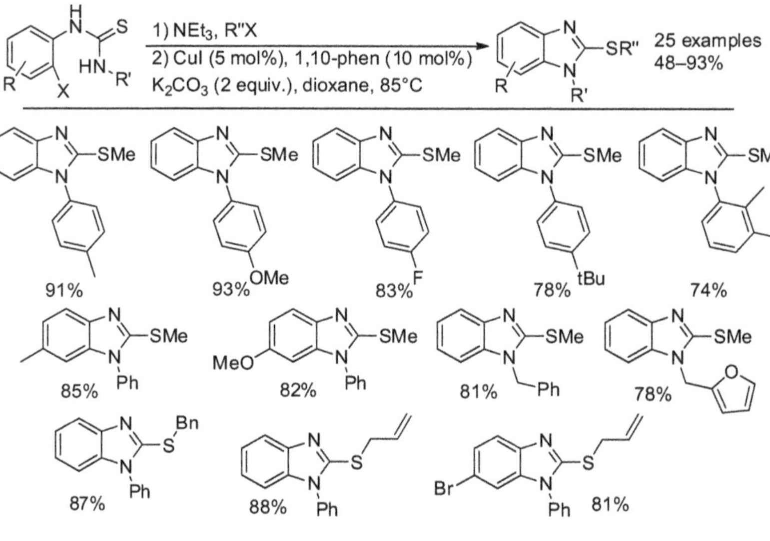

Scheme 4.36   Copper-catalyzed synthesis of 2-mercaptobenzimidazoles.

General Procedure (Scheme 4.36): NEt₃ (1.5 mmol, 210 µL) and alkyl halide (1.5 mmol) were successively added to a solution of a thiourea (1.0 mmol) in CH₃CN (5 mL) and the resulting solution was stirred for 10 min at room temperature. The solvent was evaporated, then the residue was washed with water and extracted with EtOAc (2 × 15 mL). The organic phase was dried over MgSO₄ and concentrated under vacuum to give thioether S-alkylated product in up to 98% yield, which was used for the next step without any further purification. A round-bottomed flask was charged with thioether S-alkylated product (1 mmol), CuI (0.05 mmol, 9.5 mg), 1,10-phenanthroline (0.1 mmol, 18 mg), K₂CO₃ (2 mmol, 276 mg) and 1,4-dioxane (5 mL). The resulting solution was heated at 85 °C until the disappearance of the starting material (monitored by TLC). The reaction mixture was then cooled and filtered over Celite with EtOAc. The solvent was evaporated and further purification was achieved by column chromatography.

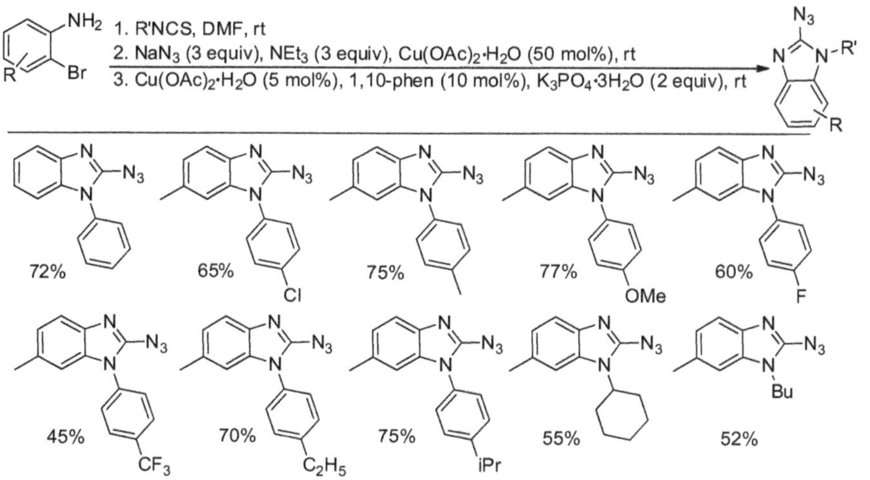

Scheme 4.37 Copper-catalyzed synthesis of 2-azidobenzimidazoles.

Additionally, a copper-catalyzed three-step synthesis of 2-azidobenzimidazoles from 2-bromoanilines, sodium azide and isothiocyanates was developed in 2012.[83] The reaction proceeds at room temperature and gives the desired products in good yields (Scheme 4.37). Under these conditions, a wide range of functional groups can be tolerated. The 2-azidobenzimidazoles formed were also converted into 2-amino- and 2-(1*H*-1,2,3-triazolyl)-1*H*-benzimidazoles in the absence of ligand or reacted further with alkynes.

Ma and co-workers developed a methodology for the preparation of *N*-substituted 1,3-dihydrobenzimidazol-2-ones.[84] The reactions started from methyl *o*-haloarylcarbamates and amines *via* a CuI–amino acid-catalyzed coupling reaction and subsequent condensation cyclization. A number of functional groups are tolerated by these reaction conditions, including vinyl, nitro, carboxylate, amide, ester, ketone and silyl ether groups (Scheme 4.38). Later, this method was applied by Zhong and Sun in the preparation of 6,9-disubstituted purin-8-ones.[85]

Fu and co-workers developed a highly efficient copper-catalyzed aerobic oxidative intramolecular sp^2 C–H amination leading to imidazobenzimidazole derivatives.[86] This protocol uses inexpensive Cu(OAc)$_2$ as the catalyst, 1,10-phenanthroline as the ligand, NaOAc as the base and economical and environmentally friendly oxygen as the oxidant and the corresponding *N*-heterocycles were obtained in excellent yields (Scheme 4.39). This method should provide a new and useful strategy for the construction of *N*-heterocycles.

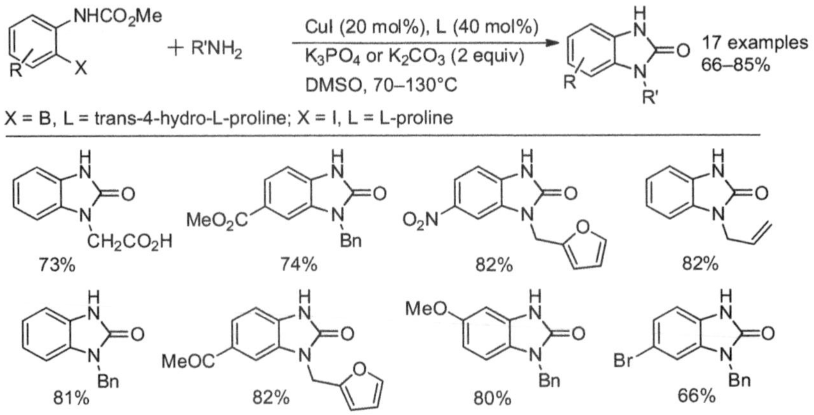

Scheme 4.38   Copper-catalyzed synthesis of 1,3-dihydrobenzimidazol-2-ones.

General Procedure (Scheme 4.38): A Schlenk tube was charged with aryl bromide (0.5 mmol), CuI (19 mg, 0.1 mmol), *trans*-4-hydro-L-proline (26 mg, 0.2 mmol) and K₃PO₄ (212 mg, 1.0 mmol), evacuated and back-filled with argon. Amine (0.5 mmol) and DMSO (1 mL) were successively added. The reaction mixture was stirred at 70 °C until the coupling was completed (monitored by TLC). The solution was then heated at 130 °C until the coupling product was consumed (monitored by TLC). The cold mixture was partitioned between ethyl acetate and saturated NH₄Cl. The organic layer was washed with brine, dried over Na₂SO₄ and concentrated *in vacuo*. The residue was purified by column chromatography on silica gel to provide the desired product.

Qiu and Wu developed a copper-catalyzed tandem $[3 + 2]$ cycloaddition/C–N coupling of carbodiimides and isocyanoacetates.[87] Benzoimidazo[1,5-*a*]imidazoles were produced in good yields by this interesting procedure (Scheme 4.40). The resulting iodo-containing benzoimidazo[1,5-*a*]imidazoles could be further decorated *via* a palladium-catalyzed cross-coupling reaction.

4.2.4   Copper-Catalyzed Synthesis of Oxazole Derivatives

A metal-catalyzed multicomponent synthesis of secondary propargylamides from trimethylsilyl-substituted imines, alkynes and acid chlorides was developed in 2005.[88] Such processes rely on the *in situ* generation of *N*-acylimines and metal acetylides, which in the presence of either BF₃ (with copper acetylides) or with the use of nucleophilic zinc(II) acetylides, couple in a catalytic fashion. By combining this process with the cycloisomerization of the secondary propargylamide product, a modular method to assemble oxazoles in a single pot can be generated (Scheme 4.41). The cyclization of propargylamides to oxazoles was applied in the preparation of steroids.[89]

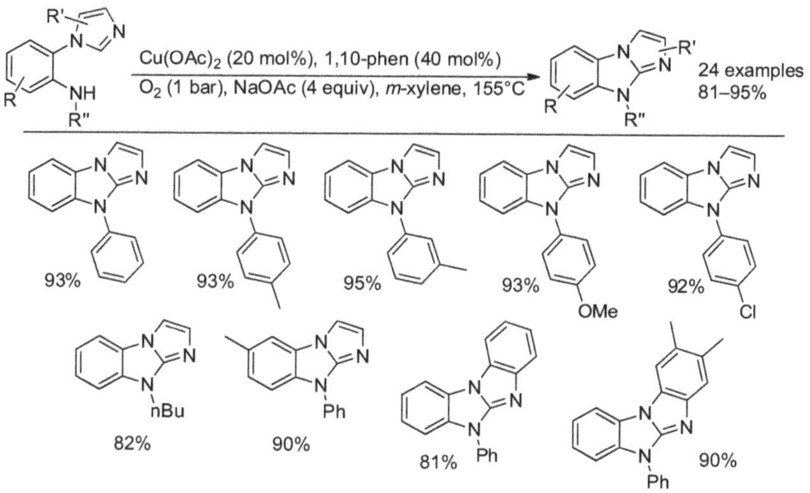

Scheme 4.39 Copper-catalyzed synthesis of imidazobenzimidazoles.

General Procedure (Scheme 4.39): Substituted 2-(1*H*-imidazol-1-yl)-*N*-alkylbenzenamine (0.3 mmol), Cu(OAc)$_2$ (0.06 mmol, 11 mg), 1,10-phenanthroline monohydrate (0.12 mmol, 23.5 mg), sodium acetate (1.2 mmol, 110 mg) and *m*-xylene (1 mL) were placed in a Schlenk tube. The reaction was performed at 155 °C for 24 or 55 h using an oxygen balloon (1 atm) and the resulting mixture was cooled to room temperature and filtered. The solid was washed with ethyl acetate (3×5 mL) and the combined filtrate was concentrated with a rotary evaporator and the residue was purified by column chromatography on silica gel using petroleum ether–ethyl acetate (5:1) as eluent to give the desired product.

An efficient two-step synthesis of 2-phenyl-4,5-substituted oxazoles involving intramolecular copper-catalyzed cyclization of highly functionalized novel β-(methylthio)enamides as the key step was reported in 2012 (Scheme 4.42).[90] These enamides are obtained by nucleophilic ring opening of newly synthesized 4-[(methylthio)hetero(aryl)methylene]-2-phenyl-5-oxazolone precursors by alkoxides, amines, amino acid esters and aryl/alkyl Grignard reagents, thus leading to the introduction of an ester, *N*-substituted carboxamide or acyl functionalities at the 4-position of the product oxazoles. Synthesis of two naturally occurring 2,5-diaryloxazoles, *i.e.* texamine and uguenenazole, *via* two-step hydrolysis–decarboxylation of the corresponding 2,5-diaryloxazole-4-carboxylates was also described. Similarly, three of the serine-derived oxazole-4-carboxamides were converted to novel trisubstituted 4,2'-bisoxazoles through a DAST/DBU-mediated cyclodehydration–dehydrohalogenation sequence.

Another efficient straightforward synthesis of 2,5,4'-trisubstituted 4,5'-bisoxazoles *via* copper(I)-catalyzed domino reactions of 2-phenyl- and

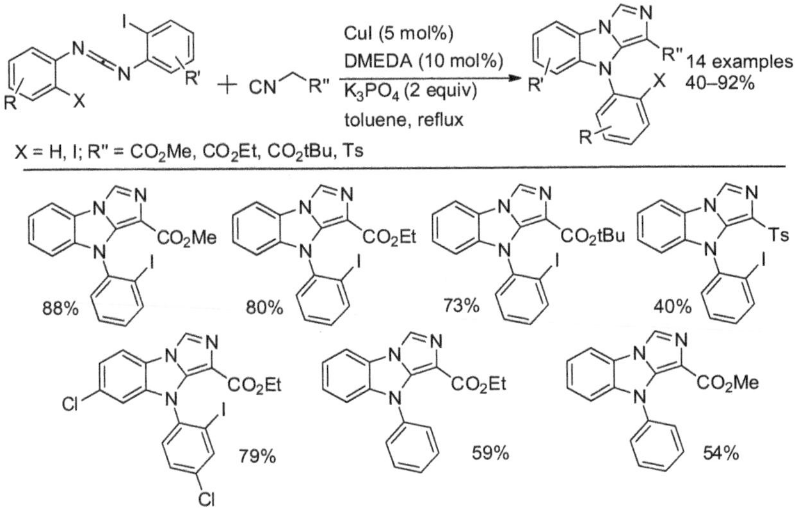

Scheme 4.40 Copper-catalyzed synthesis of benzoimidazo[1,5-*a*]imidazoles.

General Procedure (Scheme 4.40): Copper(I) iodide (10 mol%), DMEDA (20 mol%) and K₃PO₄ (2 equiv.) were added to a solution of a carbodiimide (0.5 mmol) in toluene (5 mL) under an N₂ atmosphere at room temperature. After 2 min, isocyanoacetate (1.2 equiv.) was added and the mixture was stirred at reflux. After completion of the reaction as indicated by TLC (8–12 h), the solvent was evaporated. The residue was purified on silica gel, providing the benzoimidazo[1,5-*a*]imidazole.

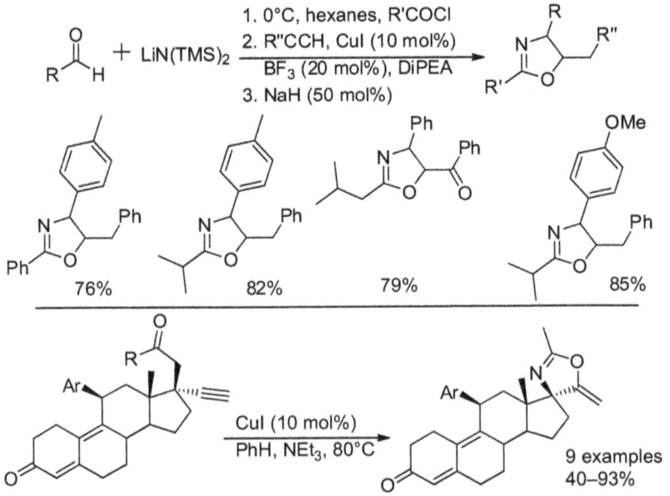

Scheme 4.41 Copper-catalyzed synthesis of oxazoles.

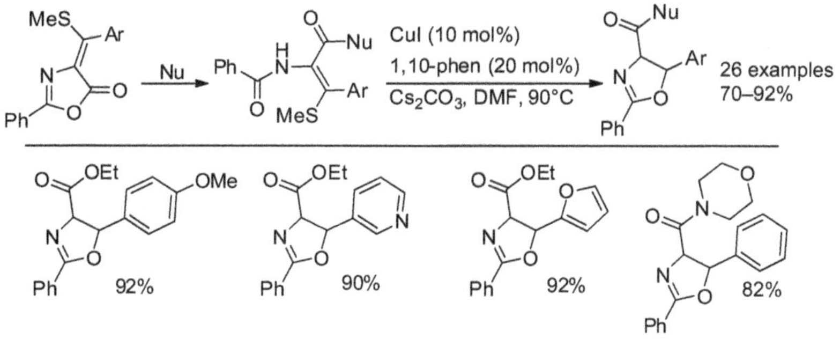

Scheme 4.42 Copper-catalyzed two-step synthesis of oxazoles.

2-(2-thienyl)-4-[(aryl/heteroaryl)-methylene]-5-oxazolones with activated methylene isocyanides was developed in 2013 by the same group.[91] The overall domino process comprised the formation of one C–C and two C–O bonds and involves initial nucleophilic ring opening of oxazolones by cupriomethylene isocyanides followed by sequential construction of two oxazole rings in the presence of a copper catalyst (Scheme 4.43). The broad substrate scope and excellent functional group compatibility of the reaction were demonstrated by employing a variety of heteroaryl- and aryl-substituted oxazolones and activated methylene isocyanides, yielding bisoxazoles with three potential points of diversity. A probable mechanism for this novel copper-catalyzed domino process was also proposed.

Wang and co-workers developed a facile and efficient oxidative synthesis of polysubstituted oxazoles from readily available starting materials.[92] This transformation from benzylamine and 1,3-diketone derivatives into oxazoles involved a tandem oxidative cyclization (Scheme 4.44). In contrast to the traditional synthetic methods for oxazoles, the reaction conditions were much milder and the reaction substrates were extended. This method is an attractive alternative for the synthesis of oxazoles. Notably, moderate yields can still be observed in the absence of CuI in MeCN and no product was obtained in toluene in the presence of catalyst.

More recently, the same group developed another facile type of one-pot, air-promoted oxidative dehydrogenation domino process for the synthesis of polyarylated oxazoles (Scheme 4.45).[93] This transformation can be realized using readily available benzylamine and benzil derivatives, involving a tandem oxidative cyclization. In contrast to the traditional synthetic methods for oxazoles, additional oxidants are not necessary and the transformation is highly efficient with the removal of aliphatic hydrogen atoms. Furthermore, the transformation can occur at room temperature and afforded moderate to good yields, which makes the method more economical.

Jiang and co-workers developed a highly efficient copper-catalyzed aerobic transformation of internal alkynes and nitriles to functionalized 1,3-oxazoles in 2012.[94] High functional-group tolerance under relatively mild conditions and high regioselectivity were exhibited. The method is amenable to the

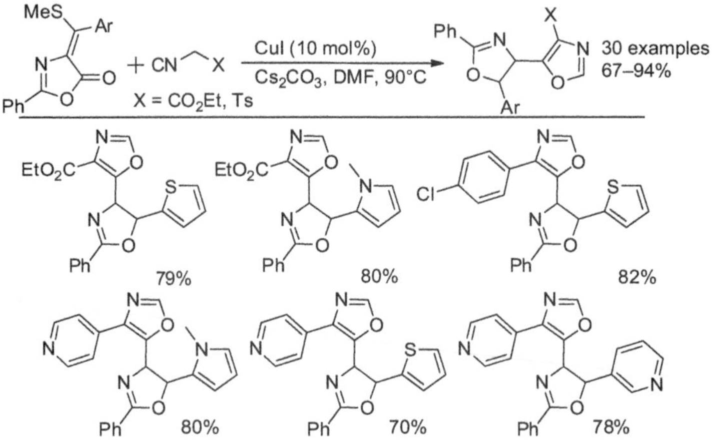

Scheme 4.43 Copper-catalyzed synthesis of bisoxazoles.

General Procedure (Scheme 4.43): To a stirred solution of the corresponding 5-oxazolone (0.3 mmol) and appropriate activated methylene isocyanides (0.3 mmol) in DMF (2 mL) was added CuI (5.7 mg, 10 mol%) under a nitrogen atmosphere, followed by addition of Cs_2CO_3 (97.7 mg, 0.3 mmol). The reaction mixture was stirred at 90 °C for 4–6 h (monitored by TLC). It was then poured into saturated NH_4Cl solution (50 mL) and extracted with EtOAc (3 × 25 mL), the extract was washed with water (2 × 30 mL) and brine (30 mL) and dried (Na_2SO_4) and the solvent was evaporated under reduced pressure to give crude bisoxazoles, which were purified by column chromatography over silica gel using EtOAc–hexane as eluent.

synthesis of a biologically active COX-2 inhibitor molecule. Isotopic labeling studies and experimental observations suggested that an enamide intermediate might be involved in this transformation. Good to excellent yields were achieved (Scheme 4.46).

Glorius's group developed a copper-catalyzed method for the regioselective, single-step preparation of 2,5-disubstituted oxazoles from readily available 1,2-dihalogenated alkenes and primary amides.[95] Many functional groups, such as halide, amino, methoxy and silyl groups, were tolerated (Scheme 4.47). In addition, use of a silylated alkene allowed the selective preparation of 2,4-disubstituted oxazoles after acid-mediated protodesilylation. The required dihalogenated alkenes were easily prepared in high yields from the corresponding alkynes by treatment with bromine or iodine.

In 2007, Buchwald and co-workers reported an oxazole synthesis using enamides as substrates.[96] The enamides were produced *in situ* from copper-catalyzed coupling between vinyl halides and primary amides. Then, in the presence of DBU and I_2, the enamides were converted into the target products in good yields. Later, they developed a copper-catalyzed oxidative

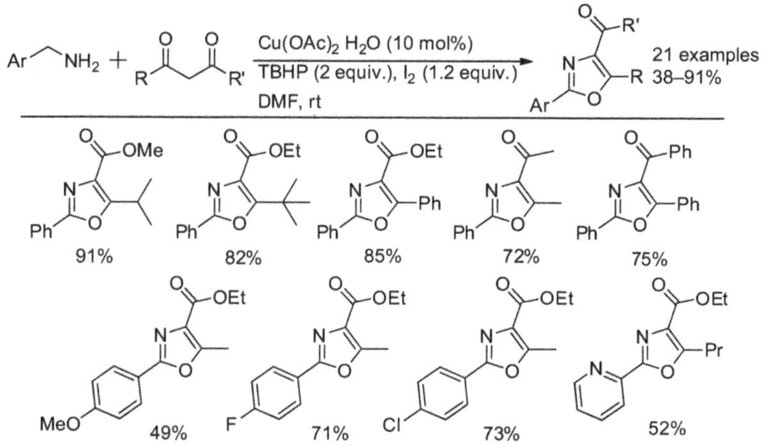

Scheme 4.44 Copper-catalyzed synthesis of polysubstituted oxazoles.

General Procedure (Scheme 4.44): To a solution of benzylamine derivatives (1.5 mmol) in DMF (3 mL) were successively added iodine (1.2 mmol), 1,3-dicarbonyl compounds (1 mmol), Cu(OAc)$_2$·H$_2$O (0.1 mmol) and TBHP (2 mmol). After the reaction mixture had been stirred for 4 h at room temperature, another portion of benzylamine derivatives (0.5 mmol) was added to the reaction system. Upon completion, the reaction mixture was extracted with EtOAc, dried over Na$_2$SO$_4$, then the organic phase was concentrated *in vacuo* and purified by silica gel column chromatography to afford the desired product.

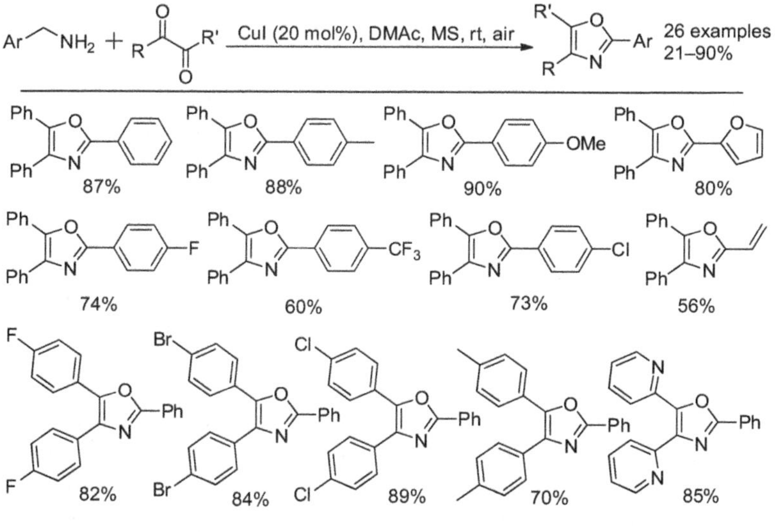

Scheme 4.45 Copper-catalyzed synthesis of oxazoles from benzylamines.

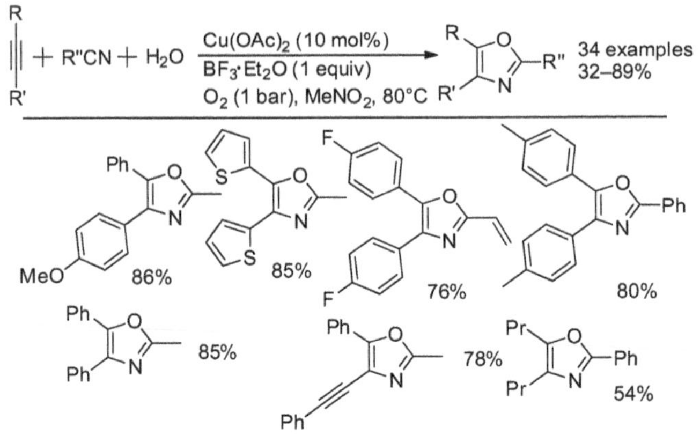

Scheme 4.46 Copper-catalyzed synthesis of oxazoles from alkynes.

General Procedure (Scheme 4.46): To a dried Schlenk tube, a mixture of alkyne (0.5 mmol), nitrile (1.5 or 0.6 mmol), H_2O (2.5 mmol), Cu(OAc)$_2$ (10 mol%), $BF_3 \cdot Et_2O$ (1 equiv.) and $MeNO_2$ (2 mL) were added successively. The mixture was stirred at 80 or 100 °C for 12 h under 1 atm O_2. After completion, the reaction mixture was purified by preparative TLC (GF254) with petroleum ether–ethyl acetate (50:1–20:1) as eluent to give the desired products.

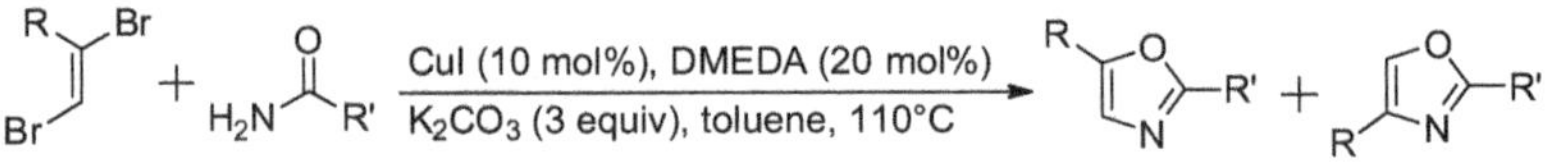

Scheme 4.47 Copper-catalyzed synthesis of oxazoles from amides.

cyclization of enamides at room temperature.[97] This efficient room-temperature catalytic oxidative cyclization of enamides generated 2,5-disubstituted oxazoles by using catalytic amounts of CuBr$_2$ in conjunction with $K_2S_2O_8$ as an oxidant. This reaction protocol can tolerate enamide substrates bearing aryl, heteroaryl, vinyl and/or alkyl substituents to afford the corresponding oxazoles in moderate to high yields. This transformation is complementary to their previously reported iodine-promoted cyclization of enamides to 2,4,5-trisubstituted oxazoles. In addition, Stahl's group reported a copper(II)-mediated oxidative cyclization of enamides to oxazoles.[98] In the presence of 2 equiv. of CuCl$_2$ and *N*-methylimidazole, under air at 140 °C, moderate to good yields of oxazoles were produced (Scheme 4.48).

Jiao and co-workers demonstrated the synthesis of oxazoles through a Cu-mediated aerobic oxidative dehydrogenative annulation of aldehydes, amines and molecular oxygen by C–H functionalization and dioxygen activation (Scheme 4.49).[99] This transformation is highly efficient with the removal of six hydrogen atoms, including the functionalization of four

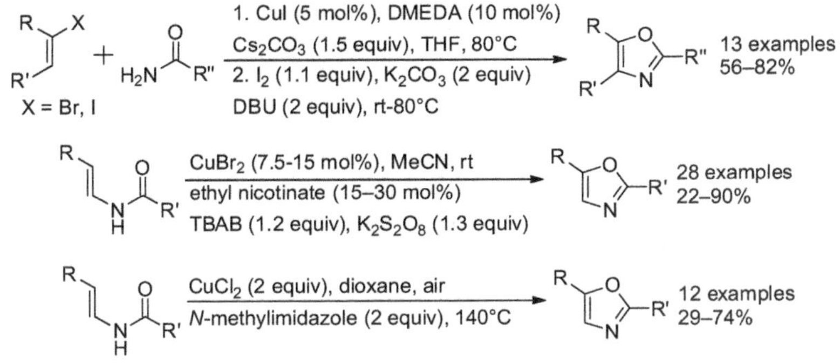

Scheme 4.48 Copper-catalyzed synthesis of oxazoles from enamides.

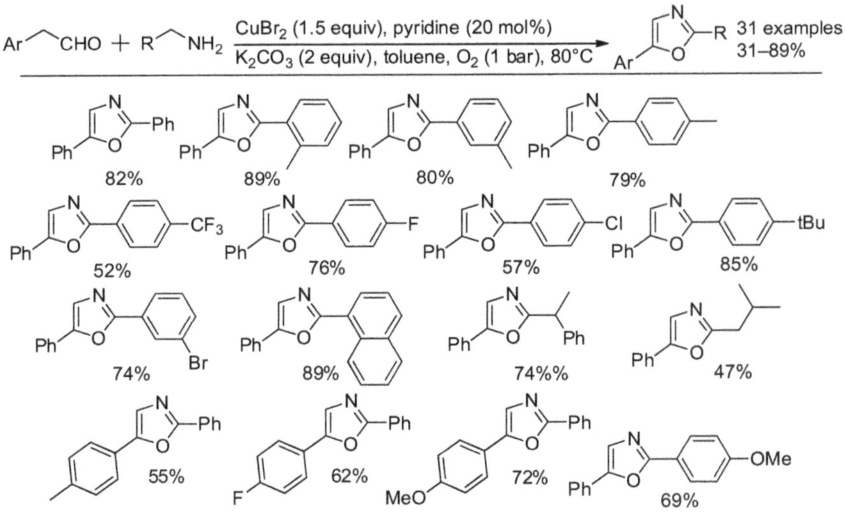

Scheme 4.49 Copper-mediated synthesis of oxazoles from aldehydes.

C(sp^3)–H bonds. Furthermore, the dehydrogenative coupling strategy and the dioxygen activation of molecular oxygen (1 atm) make this transformation very practical and atom economical. Regarding the reaction mechanism, aldehydes and anilines initially dehydrate to form imines. Under basic conditions, imines should be oxidized by molecular oxygen to form radical intermediate, a process that should be facilitated by a copper salt. Subsequently, 1,5-hydrogen atom abstraction and following intramolecular radical coupling afford the 4,5-dihydrooxazole intermediate, which can be easily oxidized to the desired oxazole product under an oxygen atmosphere.

4.2.5 Copper-Catalyzed Synthesis of Benzoxazole Derivatives

Altenhoff and Glorius developed an efficient Cu-catalyzed method for the single-step synthesis of benzoxazoles from readily available primary amides and *o*-dihalobenzene derivatives.[100] Following domino copper-catalyzed C–N and C–O bond formations, benzoxazoles were produced in good to excellent yields (Scheme 4.50). This methodology showed a broad substrate scope and various functional groups are tolerated; 2-aminobenzamide, 2-pyridylamide, 2-iodoacetamide, ethyl carbamate and 1-*tert*-butylurea did not give any desired product. This idea was adopted successfully in TBAB (tetra-butylammonium bromide) using KOH as base with CuI (10 mol%) as catalyst at 110 °C.[101]

Two domino annulation approaches for benzoxazole synthesis based on a copper catalyst using microwave and conventional thermal heating were developed by Batey and co-workers (Scheme 4.51).[102] In the first approach, copper-catalyzed intermolecular cross-coupling of 1,2-dihaloarenes with primary amides initially forms an Ar–N bond of the benzoxazole ring, followed by copper-catalyzed intramolecular cyclization to form an Ar–O bond. Benzoxazoles were formed in good yields in the reaction of 1,2-dibromobenzene, but the reaction was not regioselective in the reaction of 3,4-dibromotoluene. Furthermore, the method is limited by the availability

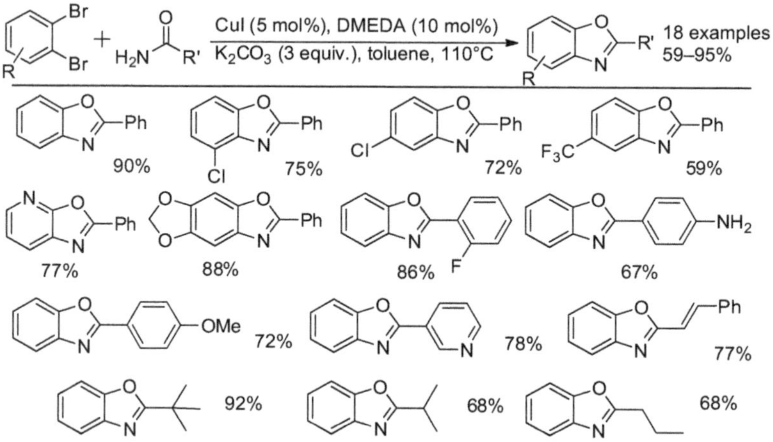

Scheme 4.50 Copper-catalyzed synthesis of benzoxazoles.

General Procedure (Scheme 4.50): 1,2-Dibromobenzene (120 mL, 1.0 mmol), benzamide (133 mg, 1.1 mmol), K$_2$CO$_3$ (414 mg, 3.0 mmol) and CuI (10 mg, 0.05 mmol) were weighed into a vial under air. The vial was evacuated and filled with argon, followed by the addition of *N,N'*-dimethylethylenediamine (11 mL, 0.1 mmol) and toluene (3 mL). The vial was closed and the reaction mixture stirred at 110 °C for 24 h. After cooling to room temperature, the reaction mixturewas poured into 25% aqueous NH$_4$OH, extracted with EtOAc, dried over Na$_2$SO$_4$, filtered and concentrated. Chromatography over silica gel (eluent: 1 : 10 EtOAc–hexane) yielded the desired product as a white powder (176 mg; yield 90%).

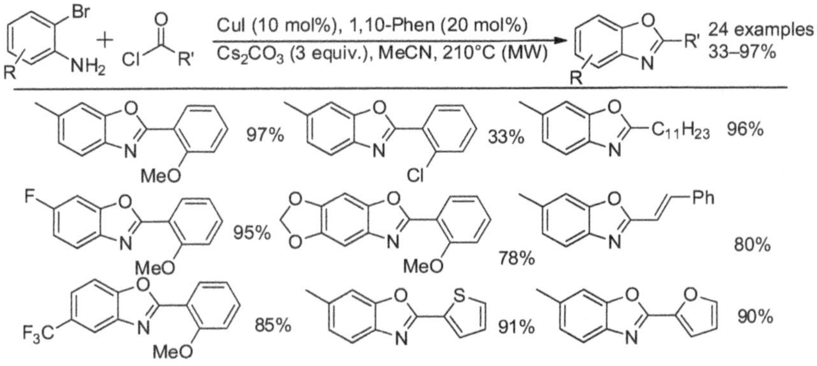

Scheme 4.51 Copper-catalyzed synthesis of benzoxazoles from acid chlorides.

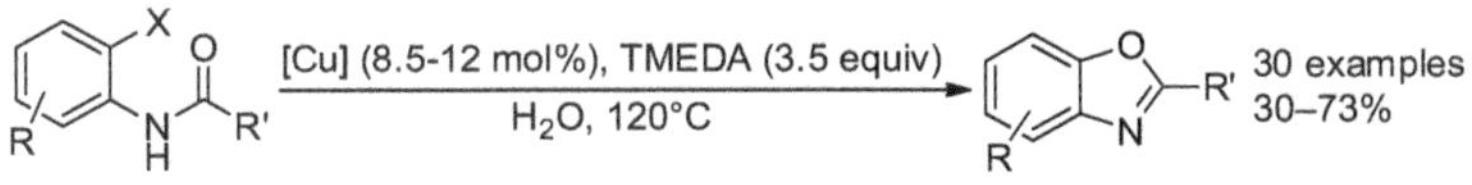

Scheme 4.52 Copper-catalyzed synthesis of benzoxazoles from 2′-haloanilides.

of 1,2-dihaloarenes. As a result of these limitations, an alternative, more versatile, one-pot domino annulation strategy was developed involving reaction of 2-bromoanilines with acyl chlorides in the presence of Cs_2CO_3, catalytic CuI and the non-acylatable ligand 1,10-phenanthroline. Under these conditions, initial acylation of the aniline is followed by copper-catalyzed intramolecular cyclization of the resultant 2-haloanilide to form the Ar–O bond of the benzoxazole ring. Optimized conditions using micro-wave irradiation achieved much shorter reaction times than conventional heating (*i.e.* 210 °C for 15 min versus 95 °C for 24 h) and were applied in the synthesis of a small library of benzoxazoles.

A general methodology for the copper-catalyzed intramolecular *o*-arylation of conveniently substituted 2′-haloanilides to deliver benzoxazoles as a valuable framework with interesting therapeutic properties was reported in 2007 (Scheme 4.52).[103] The reactions were carried out in water at 120 °C and using catalytic amounts of copper salts [both Cu(I) and Cu(II)] as the catalyst. The relatively expensive TMEDA was used as both base and ligand for this transformation. By using DMEDA as ligand, the reaction can be extended to 2′-chloroanilides.[104] In 2009, a heterogeneous copper catalyst (copper fluorapatite) was reported for this transformation.[105]

Yao and co-workers developed a general methodology for the tandem synthesis of 2-arylbenzoxazoles in 2012.[106] The reaction used 2-(2-halophe-nyl)halobenzamides and a series of nitrogen nucleophiles as starting materials and the desired products were obtained in good to excellent yields (Scheme 4.53). This procedure involves copper-catalyzed one-pot tandem C–N/C–O coupling reactions. No ligand was used in the reaction, even in the cases of primary or secondary amides, which were used as *N*-nucleophiles.

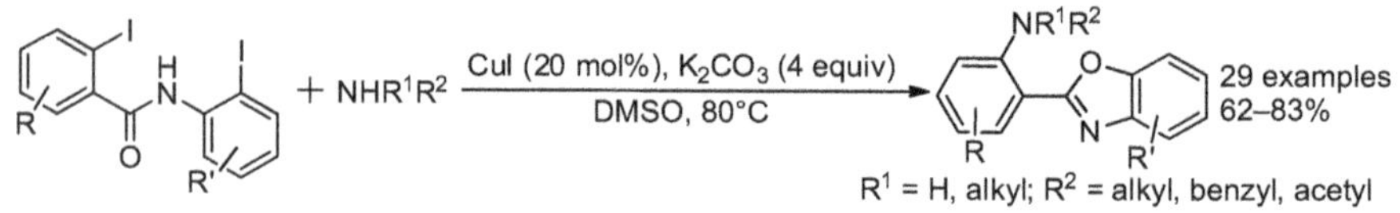

Scheme 4.53   Copper-catalyzed synthesis of 2-arylbenzoxazoles.

General Procedure (Scheme 4.53): Copper iodide (20 mol%) and potassium carbonate (2 mmol) were added to a solution of 2-iodo-*N*-(2-iodoaryl)benzamide (0.5 mmol) and a primary amine or amide (1 mmol) in DMSO and the reaction mixture was heated at 80 °C until the reaction was complete. The reaction mixture was then extracted with ethyl acetate and the ethyl acetate solution was washed with brine. The organic layer was separated, dried over anhydrous MgSO₄ and concentrated under vacuum to obtain the crude product, which was purified by column chromatography.

However, L-proline was used as ligand in the case of *N*-aromatic heterocycles such as indole, imidazole and pyrazole derivatives.

In 2008, Ueda and Nagasawa reported a copper-catalyzed intramolecular oxidative C–O coupling of benzanilides to 2-arylbenzoxazoles.[107] The advantages of this procedure include simplicity of operation, high regioselectivity and use of ready available, inexpensive and non-toxic starting materials. Moderate to good yields of the corresponding 2-arylbenzoxazoles can be isolated (Scheme 4.54).

In 2011, an efficient method for the conversion of *N*-benzyl bisarylhydrazones and bisaryloxime ethers to functionalized 2-aryl-*N*-benzylbenzimidazoles and 2-arylbenzoxazoles was described.[108] The protocol involves a copper(II)-mediated cascade C–H functionalization/C–N/C–O bond formation under neutral conditions. Substrates having either electron-donating or -withdrawing substituents undergo the cyclization to afford the target heterocycles at moderate temperature in good yields (Scheme 4.55). Using the same conditions, the reaction of bisphenyloxime thioether was also investigated. However, the substrate exhibited no desired cyclization proceeding *via* N–S cleavage to afford a mixture of benzonitrile and diphenyl disulfide in quantitative yield.

4.2.6   Copper-Catalyzed Synthesis of Indazole Derivatives

A highly efficient one-pot, two-step microwave procedure for the synthesis of 1-aryl-1*H*-indazoles was developed by Pabba *et al.* in 2005.[109] Microwave heating of 2-halobenzaldehydes or 2-haloacetophenones with phenylhydrazines at 160 °C for 10 min quantitatively yielded the arylhydrazones, which were further cyclized to give 1-aryl-1*H*-indazoles *via* CuI–diamine-catalyzed *N*-arylation under microwave heating (160 °C, 10 min). Good to

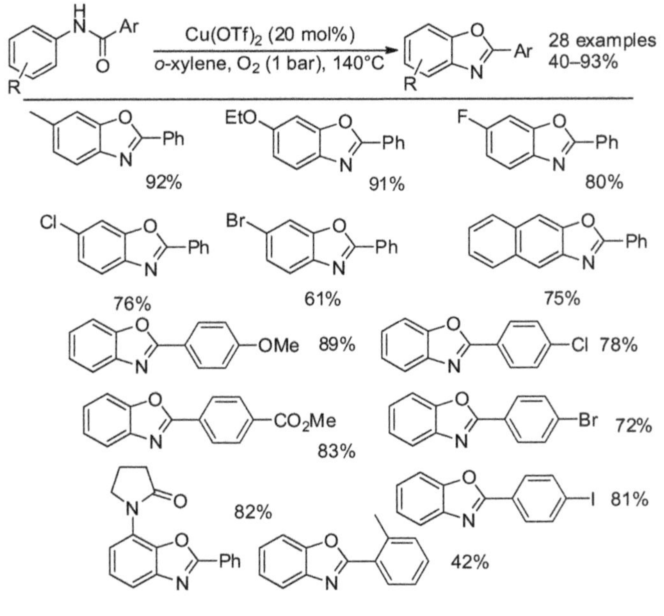

Scheme 4.54 Copper-catalyzed oxidative synthesis of 2-arylbenzoxazoles.

General Procedure (Scheme 4.54): To a dried Schlenk tube were added the benzanilide (0.25 mmol) and Cu(OTf)$_2$ (0.05 mmol). The tube and its contents were then purged under oxygen and *o*-xylene (0.5 mL) was added *via* syringe. The reaction mixture was heated with stirring at 140 °C for 28–48 h under an oxygen (balloon) atmosphere, after which a small amount of methanol was added to dissolve insoluble materials and the residue was purified *via* preparative TLC.

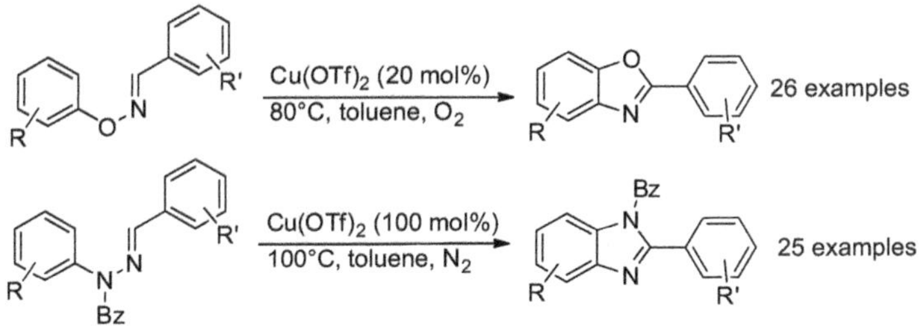

Scheme 4.55 Copper-mediated oxidative synthesis of benzoxazoles and benzimidazoles.

General Procedure (Scheme 4.55): The substrate (0.5 mmol) and Cu(OTf)$_2$ (20 mol%) were stirred at 80 °C in toluene (2 mL) under an oxygen balloon. Progress of the reaction was monitored by TLC using ethyl acetate–hexane as eluent. The reaction mixture was then cooled to room temperature and passed through a short pad of silica gel using hexane followed by ethyl acetate–hexane as eluent to afford the desired compounds in analytically pure form.

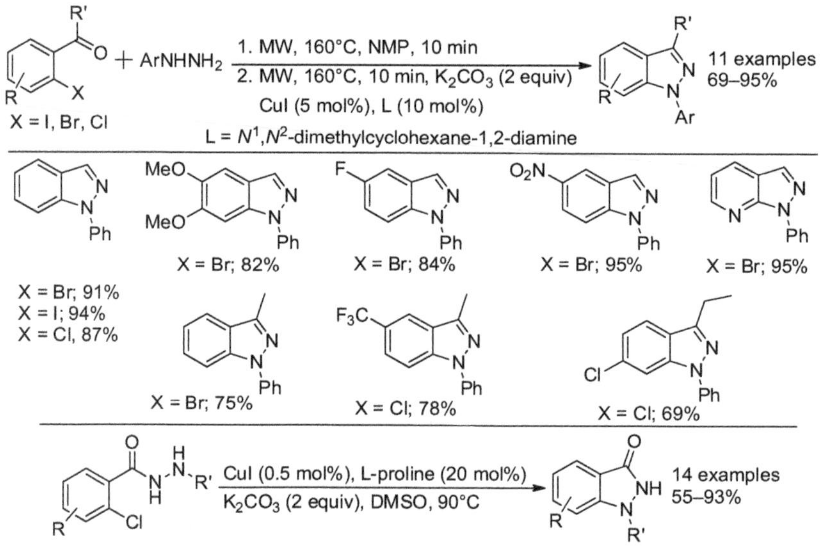

Scheme 4.56 Copper-catalyzed synthesis of indazoles.

excellent yields were observed for hydrazones of 2-iodo-, 2-bromo- and 2-chlorobenzaldehyde (Scheme 4.56). Notably, a yield of 87% was achieved for a hydrazone of unactivated 2-chlorobenzaldehyde, whereas when using the reported palladium catalysis, the yield was less than 1% for the same substrate. Furthermore, for the less reactive hydrazones of 2-bromo- and 2-chloroacetophenones, good yields were achieved using the same Cu-catalyzed microwave procedure. Later, a system using CuO (2 mol%) and K_2CO_3 as base was developed. The desired products were obtained in a regioselective manner under solvent- and ligand-free conditions at 110 °C.[110] The combination of CuI and K_3PO_4 in DMSO was also found to be an effective system.[111] Recently, a copper-catalyzed synthesis of substituted indazoles from 2-chloroarenes was reported; good to excellent yields can be achieved (Scheme 4.56).[112]

In 2013, Ma's group developed a CuBr-catalyzed coupling reaction of 2-halobenzonitriles with hydrazine carboxylic esters and a CuBr–4-hydroxy-L-proline-catalyzed coupling reaction of 2-bromobenzonitriles with N'-arylbenzohydrazides.[113] The reactions proceeded smoothly at 60–90 °C to provide substituted 3-aminoindazoles through a cascade coupling–condensation (or coupling–deacylation–condensation) process. A wide range of 3-aminoindazoles with substituents both at the 1-position and on the

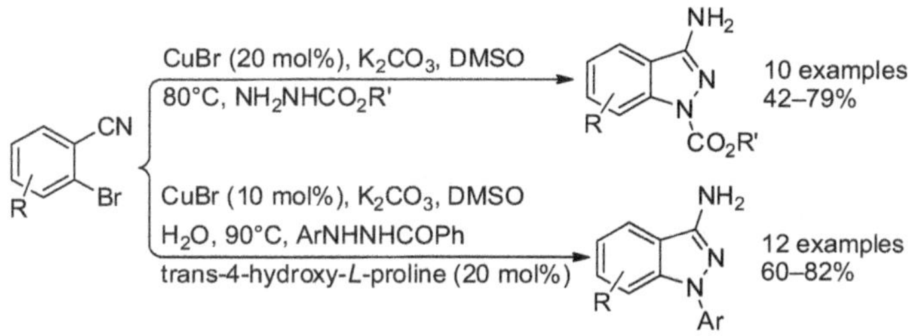

Scheme 4.57 Copper-catalyzed synthesis of 3-aminoindazoles.

General Procedure (Scheme 4.57): A Schlenk tube was charged with 2-bromobenzonitrile (0.5 mmol), hydrazines (0.6 mmol), CuBr (0.1 mmol) and K_2CO_3 (1.0 mmol). The tube was evacuated and backfilled with argon, then 1.0 mL of DMSO was added. The reaction mixture was stirred at 80 °C (60 °C for benzyl hydrazinecarboxylate) for 10–24 h. After the reaction mixture had cooled, 10 mL of saturated NH_4Cl was added. The mixture was extracted with EtOAc and then the organic layer was washed with water and brine and dried over Na_2SO_4. After concentration *in vacuo*, the residue was purified by column chromatography on silica gel (eluent: Et_3N–EtOAc–hexane) to provide the desired product.

---

phenyl ring part can be prepared from the corresponding coupling partners (Scheme 4.57).

A simple, practical and highly efficient synthesis of pyrazoles and indazoles *via* copper-catalyzed direct aerobic oxidative $C(sp^2)$–H amination was reported in 2013 by Jiang and co-workers.[114] Good functional group tolerance was exhibited in this methodology and the desired products were obtained in good to excellent yields (Scheme 4.58). Further modifications could be performed smoothly, demonstrating its potential in organic synthesis. It is noteworthy that this reaction provides a highly attractive practical synthetic strategy for the direct C–H amination that precludes the need for prefunctionalized starting materials, and the use of molecular oxygen as the oxidant makes the overall chemical transformation sustainable and practical.

Ge and co-workers developed an efficient copper-catalyzed aerobic intramolecular dehydrogenative cyclization reaction of *N,N*-disubstituted hydrazones in 2013; the reaction proceeds through by $C(sp^3)$–H oxidation, cyclization and aromatization.[115] This transformation is the first example of an intramolecular copper-catalyzed dehydrogenative coupling reaction *via* an iminium ion intermediate by a $C(sp^3)$–H bond functionalization pathway. This novel method provides a complementary, environmentally friendly and atom-efficient approach to accessing pyrazole derivatives (Scheme 4.59).

In 2013, Glorius and co-workers developed a novel methodology for the synthesis of substituted 1*H*-indazoles from readily available arylimidates

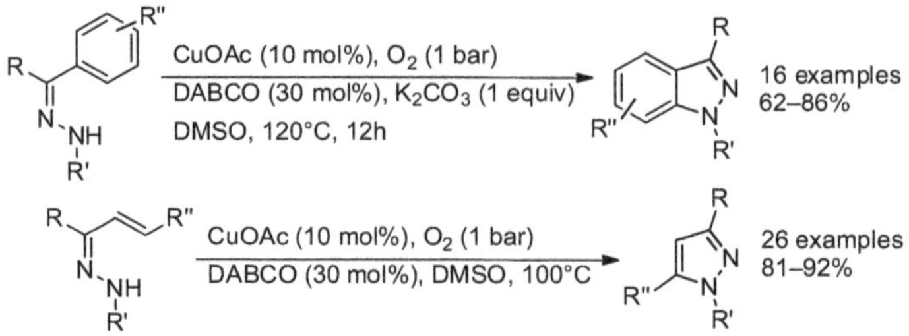

Scheme 4.58 Copper-catalyzed oxidative synthesis of indazoles.

> **General Procedure** (Scheme 4.58): To a dried Schlenk tube was added successively a mixture of hydrazone (0.2 mmol), Cu(OAc)$_2$ (10 mol%, 3 mg), DABCO (30 mol%, 6 mg) (1 equiv. of K$_2$CO$_3$ (32 mg) was also added for the synthesis of indazoles) and 2 mL of DMSO. Then the mixture was stirred at 100 °C (or 120 °C) for 12 h under 1 atm of O$_2$. Upon completion, the reaction mixture was washed by saturated NaCl aqueous solution (2 × 10 mL) and then extracted with ethyl acetate (2 × 10 mL) and the organic layers were combined, dried over anhydrous MgSO$_4$, filtered and concentrated under reduced pressure. The residue was separated by column chromatography to give the pure products.

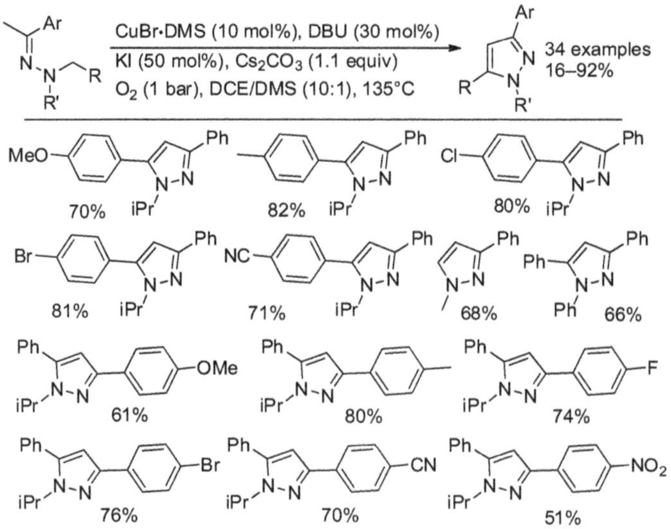

Scheme 4.59 Copper-catalyzed oxidative synthesis of pyrazoles.

> **General Procedure** (Scheme 4.59): A 50 mL Schlenk tube was charged with an *N,N*-disubstituted hydrazone (0.3 mmol), CuBr · DMS (6.1 mg, 0.03 mmol), KI (24.9 mg, 0.15 mmol), DBU (13.3 mL, 0.09 mmol), Cs$_2$CO$_3$

(107 mg, 0.33 mmol) and a solvent mixture of DCE and DMS (10:1, 3.0 mL). The vial was then evacuated and filled with 1 atm O_2 and stirred vigorously at 135 °C for 5 h. After removal of the solvent, the residue was purified by flash chromatography on silica gel (gradient eluent of 2% EtOAc in hexanes) to give the desired product.

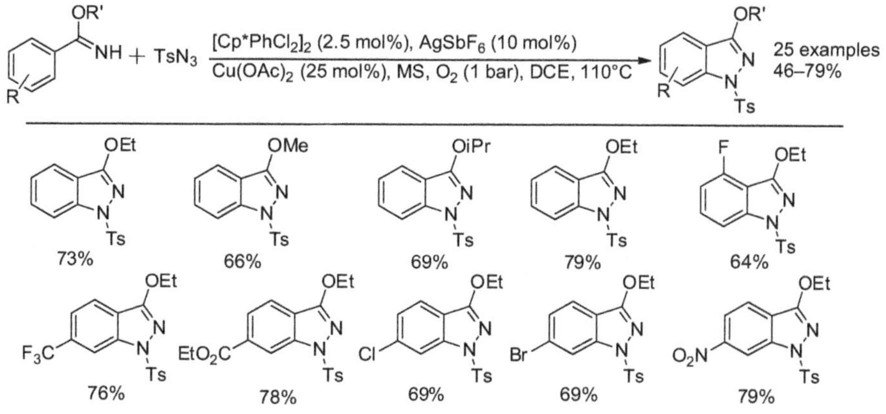

Scheme 4.60 Rh/Cu-co-catalyzed oxidative synthesis of indazoles.

General Procedure (Scheme 4.60): In a flame-dried red-cap tube, weighed molecular sieves (4 Å MS) were activated under vacuum. In a glove-box, [RhCp*Cl₂]₂ (2.5 mol%), AgSbF₆ (10 mol%) and Cu(OAc)₂ (25 mol%) were weighed into the tube. Under argon, the benzimidate substrate (0.2–0.25 mmol, 1 equiv.) was added, followed by the sequential addition of *p*-toluenesulfonyl azide (2.5 equiv.) and DCE (0.2 M). The tube was degassed and refilled with molecular oxygen (1 atm). The tube was closed and stirred at 110 °C for 24 h. The reaction mixture was then cooled to room temperature and monitored by GC–MS and TLC. The reaction mixture was diluted with EtOAc and filtered over a short pad of Celite prepacked with EtOAc. The volatiles were removed and the analytically pure product was obtained by flash chromatography on silica gel with a gradient of *n*-pentane–EtOAc as eluent.

and organo azides.[116] The reaction proceeds *via* Rh(III)-catalyzed C–H activation/C–N bond formation and Cu-catalyzed N–N bond formation. *N-H-* Imidates were applied as versatile directing groups in Rh(III)-catalyzed C–H activation, delivering substrates that undergo intramolecular N–N bond formation. The process is scalable and green, with O_2 being used as the terminal oxidant and only N_2 and H_2O formed as by-products. The corresponding 1*H*-indazoles were obtained in moderate to high yields with good functional group tolerance (Scheme 4.60). Moreover, the products could also be further converted into diverse important derivatives.

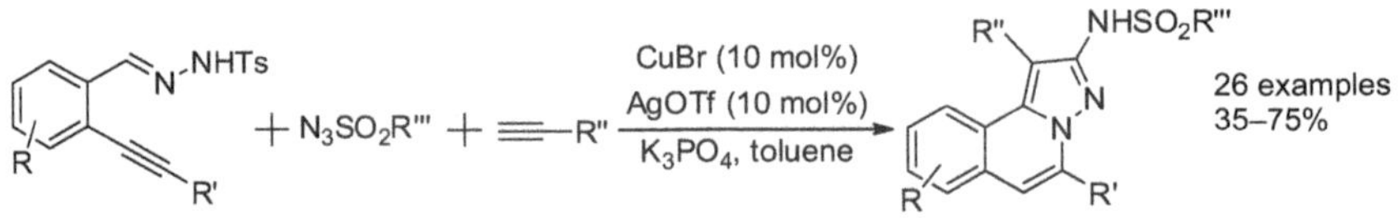

Scheme 4.61   Copper-catalyzed synthesis of isoquinolines.

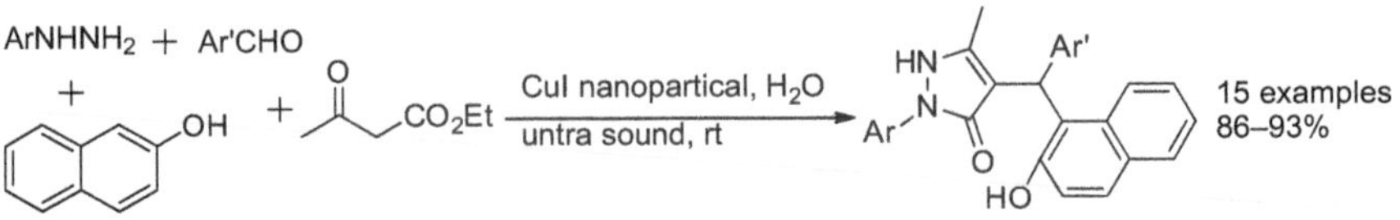

Scheme 4.62   Copper-catalyzed synthesis of pyrazolones.

Wu and co-workers described a novel and efficient route for the generation of 2-amino-*H*-pyrazolo[5,1-*a*]isoquinolines *via* a silver(I)- and copper(I)-co-catalyzed three-component reaction of *N*′-(2-alkynylbenzylidene) hydrazide, alkyne and sulfonyl azide.[117] The key intermediates were believed to be isoquinolinium-2-ylamide and ketenimine, which were generated *in situ*. This transformation proceeds with high efficiency through 6-*endo* cycliza-tion, [3 + 2] cycloaddition and subsequent aromatization. All the desired products were obtained in good yields (Scheme 4.61).

In 2013, a simple and green process to prepare copper iodide on the nano scale *via* sonication was reported. Subsequently, the nanoparticles were used as an efficient catalyst for the synthesis of 2-aryl-5-methyl-2,3-dihydro-1*H*-3-pyrazolones *via* a four-component reaction of hydrazine, ethyl acetoacetate, aldehyde and β-naphthol in water under ultrasound irradiation (Scheme 4.62).[118] This combinatorial synthesis was achieved with appli-cation of ultrasound irradiation and making use of water as a green solvent. Simple work-up, excellent yields of products and short reaction times are some of the important features of this protocol. Notably, the catalyst could be recycled and reused five times without noticeably decreasing the catalytic activity.

4.2.7   Copper-Catalyzed Synthesis of Benzothiazole Derivatives

In 2004, Batey and co-workers reported the copper-catalyzed intramolecular thiolation of *o*-halobenzothioureas.[119] 2-Aminobenzothiazoles were pro-duced in good to excellent yields (Scheme 4.63). A palladium catalyst was also checked for this transformation, but in general the copper-catalyzed protocol gave better conversions and yields. Later, they extended their pro-cedure to the preparation of 2-aromatic- and 2-alkyl-substituted benzox-azoles and benzothiazoles. This methodology was applied in the synthesis of highly substituted benzoxazoles and benzothiazoles as a further

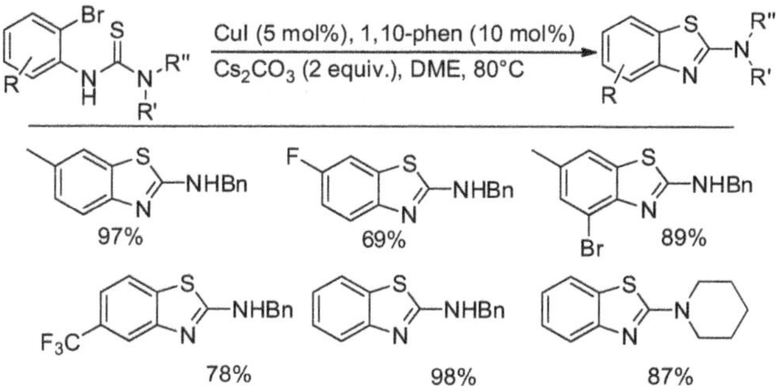

Scheme 4.63 Copper-catalyzed synthesis of 2-aminobenzothiazoles.

General Procedure (Scheme 4.63): To a mixture of thiourea (0.5 mmol), Cs_2CO_3 (1.0 mmol, 2 equiv.) and $Pd(PPh_3)_4$ (0.025 mmol, 5 mol%) or CuI (0.025 mmol, 5 mol%) and 1,10-phenanthroline (0.05 mmol, 10 mol%) was added reagent-grade dimethoxyethane (4 mL). The reaction mixture was heated at 80 °C for 16–24 h under nitrogen and subsequently diluted with EtOAc (50 mL) and washed with water (2×50 mL) and brine (1×50 mL). The organic layer was dried over $MgSO_4$, the solvent removed *in vacuo* and the crude product purified using silica gel column chromatography.

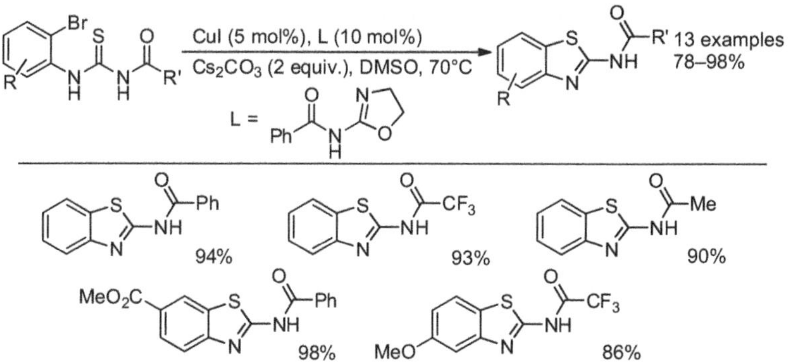

Scheme 4.64 Copper-catalyzed synthesis of 2-amidobenzothiazoles.

transformation of the Ugi reaction product.[120] Interestingly, CuO was also found to be an active catalyst for this transformation.[121]

Pan and co-workers found that *N*-(4,5-dihydrooxazol-2-yl)benzamide can be applied as an efficient *N,O*-bidentate ligand in copper-catalyzed coupling reactions.[122] They applied this ligand in the intramolecular cyclization of substituted 1-arylacyl-3-(2-bromophenyl)thioureas together with CuI. Generally, good to excellent yields of the desired products could be successfully obtained (Scheme 4.64). This method can provide more diversified

N-benzothiazol-2-ylamides under relatively mild conditions avoiding the use of the toxic bromine. Furthermore, the procedure extends the scope of carbon–heteroatom formation by using the more cost-effective Cu-catalyzed process.

Patel and co-workers developed two-ligand-assisted Cu(I)-catalyzed sequential intra- and intermolecular *S*-arylations leading to the direct synthesis of arylthiobenzothiazoles in one pot without an inert atmosphere.[123] Low catalyst loading, inexpensive metal catalyst and ligand, lower reaction temperature and shorter reaction times are the advantages of this methodology. The desired products were obtained in good to excellent yields (Scheme 4.65).

Instead of starting from the already isolated intermediate, benzothiazoles can also be prepared from 2-haloanilines and isothiocyanates. In 2009, Bao and co-workers developed a novel, efficient and concise method to synthesize *N*-substituted-2-aminobenzothiazoles under ligand-free copper(I)-catalyzed conditions from 2-haloanilines and isothiocyanates (Scheme 4.66).[124] A simple one-pot operation was conducted, readily available starting materials were employed and relatively mild conditions were applied. Various *N*-substituted-2-aminobenzothiazoles, which might be potentially applicable in the pharmaceutical and biochemical areas, were conveniently synthesized in moderate to excellent yields. The reaction is also a good example of a ligand-free copper(I)-catalyzed one-pot cascade process. Wu and co-workers found that the reaction temperature can be decreased to 50 °C by using 1,10-phenanthroline as the ligand, but more expensive DABCO is needed as base.[125] Interestingly, Li and co-workers discovered that by adding 1 equiv. of TBAB as additive, the reaction works without base at

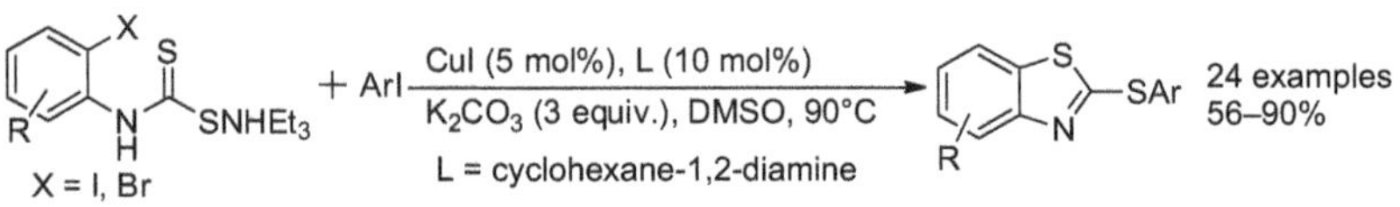

Scheme 4.65   Copper-catalyzed synthesis of benzothiazoles by *S*-arylation.

General Procedure (Scheme 4.65): An oven-dried flask was charged with CuI (5 mol%), cyclohexane-1,2-diamine (10 mol%), dithiocarbamate (1 mmol), iodobenzene (1.1 equiv.), K_2CO_3 (3 equiv.) and solvent DMSO (1 mL). The flask was kept in a preheated oil-bath at 90 °C. Heating was continued for 4 h, after which the reaction mixture was cooled and admixed with water (5 mL). The product was extracted with ethyl acetate (2×10 mL) and the organic layer was dried over anhydrous Na_2SO_4, concentrated under reduced pressure and purified over a column of silica gel (eluent: EtOAc–hexane) to give product in 86% isolated yield. The identity and purity of the products were confirmed by spectroscopic analysis.

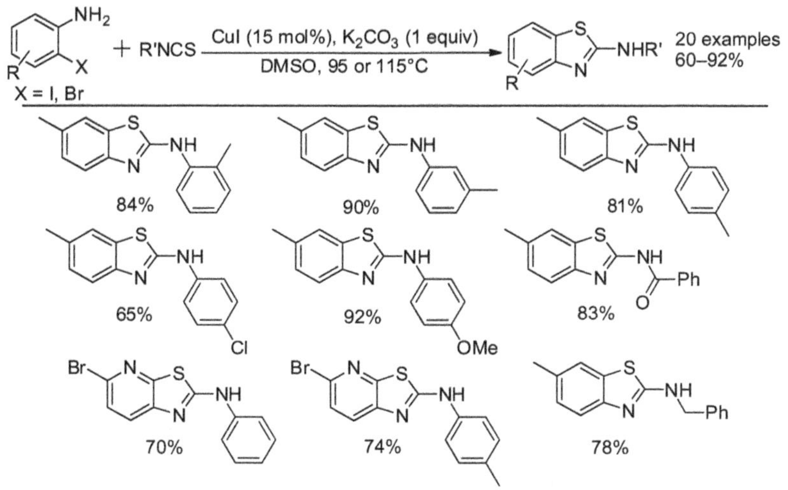

Scheme 4.66 Copper-catalyzed synthesis of benzothiazoles from isothiocaanates.

General Procedure (Scheme 4.66): An oven-dried Schlenk tube equipped with a Teflon valve was charged with a magnetic stirrer bar, K_2CO_3 (0.4 mmol, 100 mol%), CuI (0.06 mmol, 15 mol%) and 2-haloanilines (0.4 mmol). The tube was evacuated and backfilled with N_2 (this procedure was repeated three times). Under a counter-flow of N_2, DMSO (2.0 mL) was added by syringe. Then, isothiocyanates (0.7 mmol) were added by syringe under a counter-flow of N_2. The reaction mixture was stirred for 24 h at 95 (115) °C. The reaction was monitored by TLC. After the starting material had been completely consumed, the reaction was stopped and the mixture cooled to room temperature. Water (20 mL) was added to the solvent and the mixture was extracted with EtOAc (3 × 15 mL). The extracts were combined and washed with water (2 × 10 mL) and brine (15 mL) and dried ($MgSO_4$). After evaporating the solvent under reduced pressure, the residue was purified by column chromatography on silica gel to give the pure product.

40 °C in DMSO.[126] Methodologies using water as solvent[127] or solvent-free conditions have also been reported.[128] Simple CuI or $CuSO_4$ was applied as the catalyst. When the reaction was carried out in water, no base was required and the reaction proceeded at 90 °C. In comparison, NBu_3 was needed as base in the case of solvent-free conditions and with a catalytic amount of $CuSO_4$ at 80 °C. With the advantages of heterogeneous catalysts, supported copper catalysts were also developed and applied in this interesting transformation.[129,130]

Notably, an efficient protocol was developed for the preparation of 2-aminobenzothiazoles *via* a copper(I)-catalyzed tandem reaction of 2-iodoanilines with isothiocyanates at very low catalyst loadings [typically

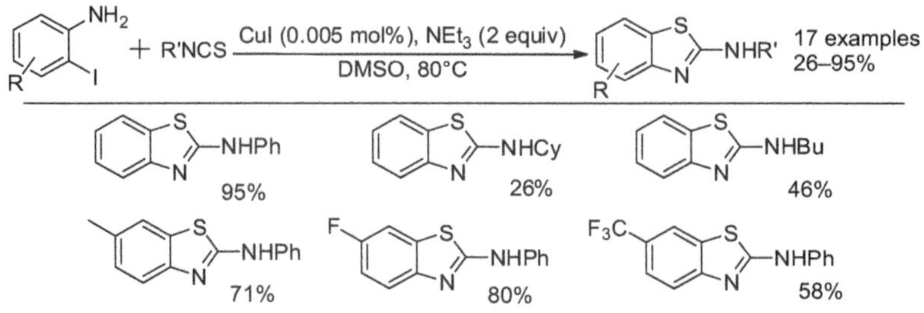

Scheme 4.67 Highly efficient copper-catalyzed synthesis of benzothiazoles.

General Procedure (Scheme 4.67): CuI solution (0.005 mol%, 2.0 mL, 5.0 mM CuI in CH$_3$CN) was placed in a 10 mL test-tube, then the CH$_3$CN was removed under vacuum. 2-Iodoaniline (0.20 mmol), isothiocyanate (0.22 mmol), Et$_3$N (58 mL, 2.0 equiv.), DMSO (2.0 mL) and a Teflon coated stirrer bar were added, the tube was placed in a preheated oil-bath at 80 °C and the mixture was stirred for 24 h under an air atmosphere and then cooled to room temperature. The solution was diluted with ethyl acetate and washed with water (10 mL) and brine (10 mL). The aqueous phase was extracted with ethyl acetate (3×10 mL). The organic layers were combined and dried over anhydrous Na$_2$SO$_4$. After evaporation under vacuum, the decired product was obtained by purification on silica gel (eluent: 15:1–4:1 petroleum ether–ethyl acetate).

50 ppm of copper(I) iodide (CuI)].[131] A variety of 2-iodoanilines could be cross-coupled with isothiocyanates, affording 2-aminobenzothiazoles in moderate to good yields (49–93%) under the given conditions (Scheme 4.67). The turnover number (TON) of this reaction reaches 67 000 and the reaction could be scaled up to at least the gram scale.

In 2009, an efficient cascade process for the synthesis of 2-substituted 1,3-benzothiazoles from 2-haloaryl isothiocyanates was developed.[132] The thiocarbamate or dithiocarbamate generated *in situ* by the reaction of 2-haloaryl isothiocyanates with *O*- or *S*-nucleophiles undergoes CuI-catalyzed intramolecular C–S bond formation, giving substituted benzothiazoles (Scheme 4.68). Both phenols and thiophenols reacted with equal ease. On the other hand, alcohols and thiols were less reactive. The rate of the reaction was faster and gave better yields when electron-withdrawing substituents were present in either of the coupling partners. 1,3-Benzothiazolones were prepared in one pot using ethanol as the solvent and nucleophile (O-source). In 2012, the same group found that this reaction also works in the absence of CuI in refluxing water.[133] In the case of 2-chloroaryl or 2-fluoroaryl isothiocyanates, CuO was still needed. K$_2$CO$_3$ was applied as base in all cases.

In 2010, a copper-catalyzed double thiolation reaction of 1,4-dihalides with sulfides was developed for selectively synthesizing 2-trifluoromethylbenzothiophenes and -benzothiazoles.[134] In the presence of CuI, a

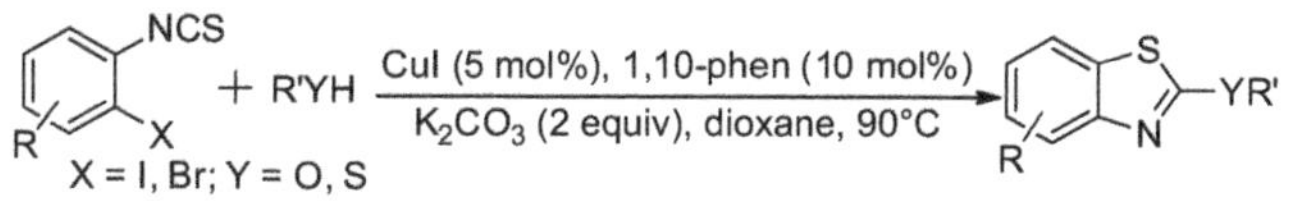

Scheme 4.68 Copper-catalyzed synthesis of benzothiazoles from 2-haloaryl isothiocyanates.

> **General Procedure** (Scheme 4.68): A round-bottomed flask with a magnetic stirrer bar, fitted with a reflux condenser, was charged with 2-bromophenyl isothiocyanate (1 mmol, 214 mg), phenol (1.1 mmol, 103 mg), CuI (0.05 mmol, 9.5 mg), 1,10-phenanthroline (0.1 mmol, 18 mg), K$_2$CO$_3$ (2 mmol, 276 mg) and dry 1,4-dioxane (3 mL). The mixture was heated at 90 °C for 12 h, protected with a guard tube. The reaction mixture was then cooled and filtered through Celite using ethyl acetate. The filtrate was evaporated to dryness and the product was purified by column chromatography.

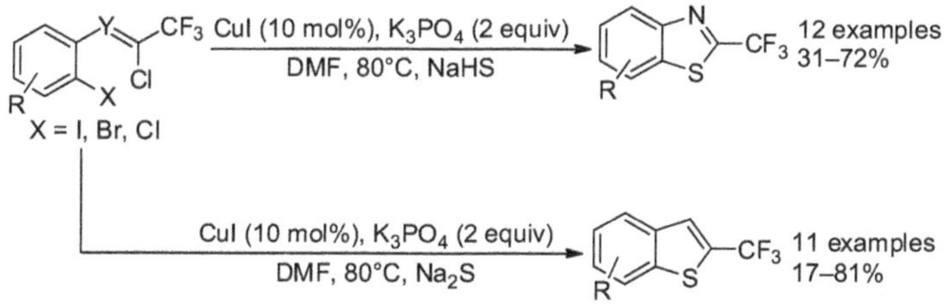

Scheme 4.69 Copper-catalyzed thiolation of 1,4-dihalides to benzothiazoles.

> **General Procedure** (Scheme 4.69): Substrate (0.2 mmol), Na$_2$S·9H$_2$O (96.0 mg, 0.4 mmol), CuI (3.8 mg, 0.02 mmol) and DMF(3 mL) were added to a two-necked flask in turn. Then the solution was stirred at 80 °C for 10 h until complete consumption of starting material as monitored by TLC and GC-MS analysis. After the reaction was finished, the mixture was washed with brine, extracted with EtOAc, dried over anhydrous Na$_2$SO$_4$ and evaporated under vacuum. The residue was purified by flash column chromatography on silica gel (EtOAc/petroleum ether) to afford the desired product.

variety of 2-halo-1-(2-haloaryl)-3,3,3-trifluoropropylenes smoothly underwent thiolation annulation with Na$_2$S to afford 2-trifluoromethylbenzothiophenes in moderate to good yields (Scheme 4.69). Moreover, the conditions are compatible with N-(2-haloaryl)trifluoroacetimidoyl chlorides in the presence of NaHS and K$_3$PO$_4$, leading to 2-trifluoromethylbenzothiazoles.

An easy and convenient one-pot method for the preparation of 2-substituted 1,3-benzothiazoles by a copper-catalyzed three-component reaction of a

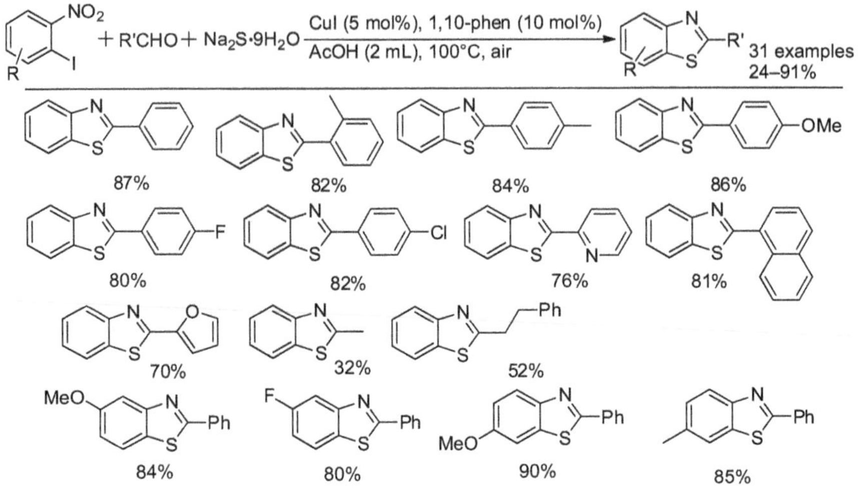

Scheme 4.70   Copper-catalyzed synthesis of benzothiazoles from 1-iodo-2-nitrobenzenes.

General Procedure (Scheme 4.70): A 25 mL tube was charged with a 1-iodo-2-nitrobenzene (1.0 mmol), an aldehyde (0.5 mmol), $Na_2S \cdot 9H_2O$ (2.5 mmol), CuI (0.05 mmol) and 1,10-phenanthroline (0.1 mmol). Glacial AcOH (2 mL) was added, the tube was capped under air and the mixture was stirred at 100 °C for 12 h. The mixture was then cooled to r.t. and diluted with EtOAc (40 mL). The resulting mixture was washed sequentially with saturated aqueous $NaHCO_3$ (3 × 10 mL) and water (2 × 10 mL) and the organic layer was dried (Na_2SO_4) and concentrated under reduced pressure. The residue was purified by column chromatography on silica gel (eluent: 12 : 1–4 : 1 petroleum ether–EtOAc).

1-iodo-2-nitrobenzene, sodium sulfide and an aldehyde was reported.[135] The reaction was best performed in acetic acid at 100 °C with copper(I) iodide and 1,10-phenanthroline as the catalyst system (Scheme 4.70). The reaction appears to tolerate a wide range of functional groups on both the 1-iodo-2-nitrobenzene and the aldehyde. The overall efficiency of the reaction favors numerous applications in organic synthesis. In this procedure, the property that sulfides can reduce aromatic nitro compounds to the corresponding anilines is the key.

Sekar's group developed a new copper-catalyzed *in situ* generation of aryl thiolates strategy for the one-pot synthesis of substituted benzothiazoles from 2-iodoanilides using xanthate as a thiol precursor.[136] A wide range of 2-iodoanilides with both electron-releasing and electron-withdrawing groups produced the corresponding benzothiazoles in good yields (Scheme 4.71). Further, this one-pot protocol was successfully utilized for the synthesis of a potent antitumor agent, 2-(3,4-dimethoxyphenyl)-5-fluorobenzo[*d*]thiazole

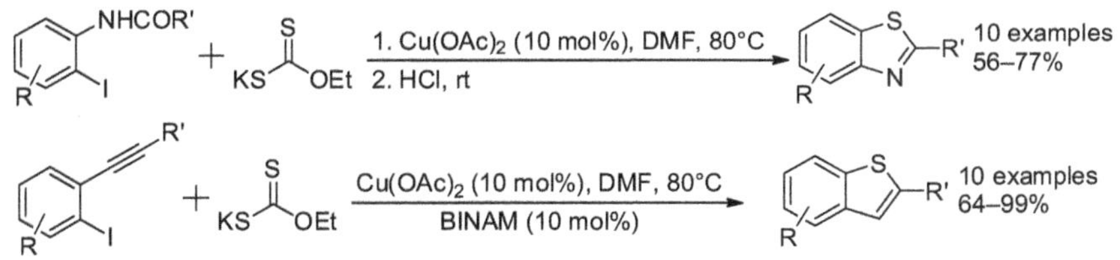

Scheme 4.71 Copper-catalyzed synthesis of benzothiazoles using xanthate.

General Procedure (Scheme 4.71): A mixture of N-(2-iodophenyl)benzamide (161.5 mg, 0.5 mmol), Cu(OAc)$_2$ (9.1 mg, 0.05 mmol) and potassium ethyl xanthate (240.4 mg, 1.50 mmol) was placed in an oven-dried reaction tube equipped with a septum. The reaction tube was evacuated and backfilled with nitrogen. N,N-Dimethylformamide (3.0 mL) was added to the reaction mixture at room temperature and the reaction tube was sealed with a glass stopper and heated for 15 h at 80 °C, then 0.5 mL of concentrated HCl was added to the cooled reaction mixture. After 8 h, 6 mL of saturated aqueous NaHCO$_3$ was added and the mixture was extracted with ethyl acetate and water. The organic layer was dried over anhydrous Na$_2$SO$_4$ and the solvent was removed under reduced pressure. The crude residue was purified by column chromatography on silica gel using ethyl acetate–hexanes as eluent to give the desired product 2-phenylbenzo[d]thiazole (65.4 mg, 62%) as a white solid.

(PMX610). Finally, the copper-catalyzed *in situ* generation of aryl thiolates strategy was successfully applied in the domino synthesis of substituted benzothiophenes from *o*-haloalkynyl benzenes using xanthate as a thiol precursor.

Ma *et al.* developed a novel method for the synthesis of substituted benzothiazoles, which relies on an unprecedented CuI-catalyzed coupling reaction of aryl halides with metal sulfides.[137] Starting from 2-haloanilides and Na$_2$S, various benzothiazoles were produced in good yields (Scheme 4.72).

In 2010, the same group reported another novel methodology for the synthesis of benzothiazoles (Scheme 4.73).[138] They identified that dithiocarbamate salts are excellent coupling partners for copper-catalyzed arylation. The products of their reaction with 2-haloanilines can undergo intramolecular condensation to afford 2-N-substituted benzothiazoles. As these salts could be prepared *in situ* from amines and nitrogen-containing heterocycles, they were able to develop a cascade three-component reaction for the synthesis of 2-N-substituted benzothiazoles from conveniently available starting materials. This method permits the assembly of 2-N-substituted benzothiazoles with great diversity and should therefore find broad application in organic synthesis.

Yu *et al.* developed an efficient copper-catalyzed approach to benzothiophene and benzothiazole derivatives using thiocarboxylic acids as a sulfur

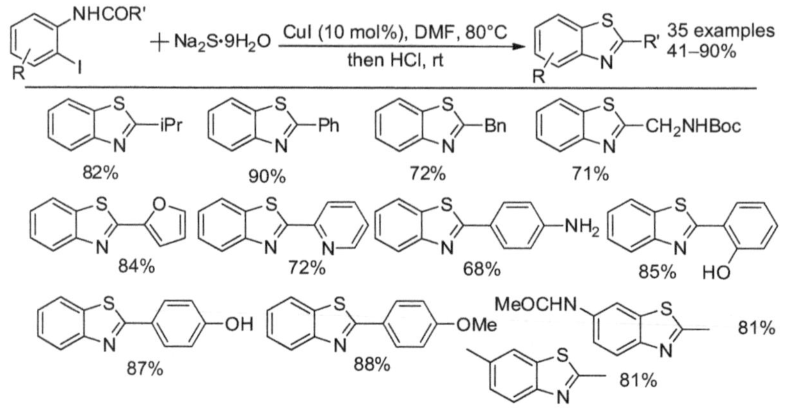

Scheme 4.72 Copper-catalyzed synthesis of benzothiazoles from 2-haloanilides.

General Procedure (Scheme 4.72): A mixture of CuI (19 mg, 0.1 mmol), *o*-iodobenzamide (1 mmol) and $Na_2S \cdot 9H_2O$ (or K_2S) (3 mmol) in DMF (2 mL) was stirred at 80 °C for 12 h. The reaction mixture was cooled to r.t., then 0.8 mL of concentrated HCl was added and the reaction mixture stirred for 5–10 h. After adding 10 mL of saturated aqueous $NaHCO_3$, it was extracted with ethyl acetate and purified by silica gel chromatography to furnish the desired product.

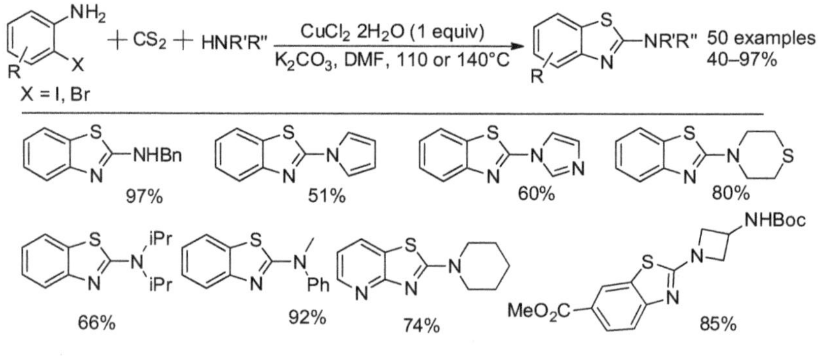

Scheme 4.73 Copper-mediated synthesis of benzothiazoles using CS_2.

General Procedure (Scheme 4.73): A mixture of 2-iodoaniline (0.5 mmol), CS_2 (0.6 mmol), an amine (0.75 mmol), $CuCl_2 \cdot 2H_2O$ (0.5 mmol) and K_2CO_3 (1.5 mmol) in DMF (1 mL) was stirred at 110 °C for 6 h. The cooled solution was partitioned between ethyl acetate and water and the organic layer was washed with water and brine and then dried over Na_2SO_4. After removal of the solvent *in vacuo*, the residue was purified by silica gel chromatography to give the desired benzothiazole.

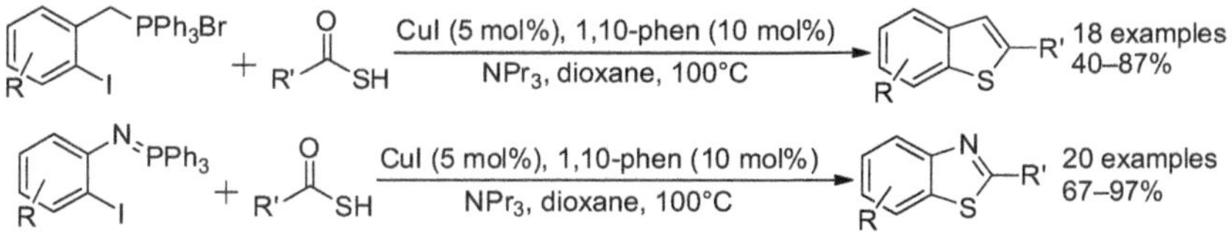

Scheme 4.74 Copper-mediated synthesis of benzothiazoles using thiocarboxylic acids.

General Procedure (Scheme 4.74): In a 15 mL two-necked flask, a mixture of (2-iodobenzyl)triphenylphosphonium bromide (168 mg, 0.3 mmol), thiobenzoic acid (53 mg, 0.33 mmol), CuI (2.8 mg, 5% mol) and 1,10-phenanthroline (5.2 mg, 10 mol%) in dioxane (3 mL) was stirred at room temperature under an N₂ atmosphere for 5 min. The solution was then stirred at 100 °C for 32 h until complete consumption of starting material (monitored by TLC). When the reaction was finished, the mixture was washed with brine, extracted with EtOAc, dried over anhydrous Na₂SO₄ and evaporated under vacuum. The residue was purified by flash column chromatography on silica gel (eluent: 8:1 petroleum ether–ethyl acetate) to give the desired product.

---

source.[139] In the presence of CuI and 1,10-phenanthroline and with n-Pr₃N as the base, (2-iodobenzyl)triphenylphosphonium bromide and (2-iodophenylimino)triphenylphosphorane reacted smoothly with thiocarboxylic acids to give benzo[b]thiophene and benzothiazole derivatives in good yields *via* sequential Ullmann-type C–S bond coupling and Wittig condensation (Scheme 4.74).

Jiang and co-workers developed an interesting and efficient method for the synthesis of 2-substituted benzothiazoles from 2-aminobenzenethiols and nitriles in the presence of Cu(OAc)₂ catalyst.[140] Both the 2-amino-benzenethiols and nitriles are cheap and commercially available. Considering the diverse substrates, ethanol as an environmentally friendly solvent, the inexpensive catalytic system, mild conditions combined with an operationally simple procedure render it a powerful component to traditional approaches for the synthesis of biologically important compounds containing a benzothiazole framework. Good to excellent yields of the target products can be isolated under standard conditions (Scheme 4.75).

Bao and co-workers developed a novel, efficient and concise method for the copper(I)-catalyzed one-pot synthesis of 2-iminobenzo-1,3-oxathioles.[141] Readily available o-iodophenols and isothiocyanates were used as substrates. This simple one-pot operation was conducted with readily available starting materials and relatively mild conditions. Various 2-iminobenzo-1,3-oxa-thioles, which might be potentially applicable in the pharmaceutical and biochemical areas, were conveniently synthesized in moderate to excellent yields (Scheme 4.76). The reaction also represents a good example of

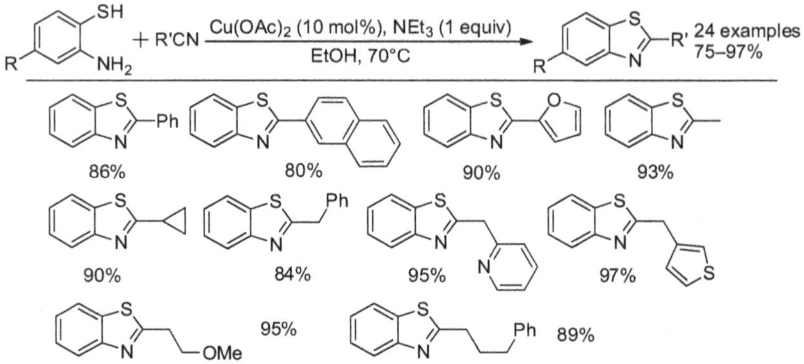

Scheme 4.75 Copper-mediated synthesis of benzothiazoles from nitriles.

General Procedure (Scheme 4.75): Substituted 2-aminobenzenethiol (0.5 mmol), nitrile (0.5 mmol), Cu(OAc)$_2$ (10 mol%), Et$_3$N (0.5 mmol) and ethanol (2.5 mL) were placed in a tube equipped with a magnetic stirrer bar. The mixture was stirred at 70 °C (oil-bath temperature) for 6 h. When the reaction was finished (monitored by TLC), the mixture was cooled to room temperature and quenched with aqueous Na$_2$CO$_3$ and the crude product was extracted with ethyl acetate. The organic extracts were concentrated *in vacuo* and the resulting residue was purified by column chromatography on silica gel with petroleum ether–ethyl acetate as eluent to afford the desired product.

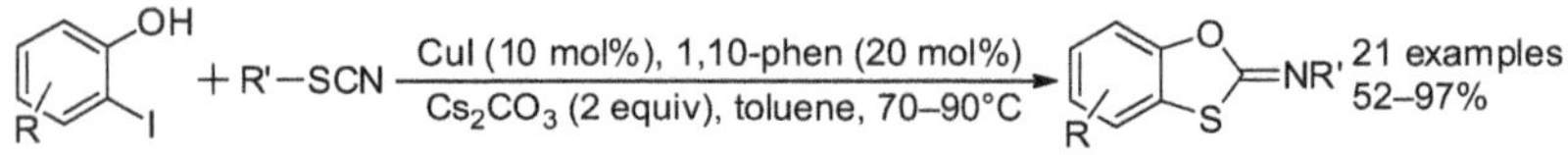

Scheme 4.76 Copper-catalyzed synthesis of 2-iminobenzo-1,3-oxathioles.

General Procedure (Scheme 4.76): An oven-dried Schlenk tube equipped with a Teflon valve was charged with a magnetic stirer bar, Cs$_2$CO$_3$ (652 mg, 2.0 mmol), CuI (19 mg, 0.10 mmol, 10 mol%), 1,10-Phen · H$_2$O (40 mg, 0.20 mmol, 20 mol%) and *o*-iodophenol (1.0 mmol). The tube was evacuated and backfilled with N$_2$ (this procedure was repeated three times). Under a counter-flow of N$_2$, toluene (1.0 mL) was added *via* syringe and the mixture was prestirred for about 0.5 h at room temperature, then a solution of isothiocyanate (1.0 ~ 1.1 mmol) in toluene (1.0 mL) was added *via* syringe under a counter-flow of N$_2$. The reaction was monitored by TLC. When the starting material was completely consumed and the bottom dot was unchanged, the reaction was stopped and cooled to room temperature. The reaction mixture was directly passed through Celite. After rinsing with 30 mL of Et$_2$O, the combined filtrate was concentrated by rotary evaporation. The residue was purified by column chromatography on silica gel to give the pure product.

intramolecular C(aryl)–S cyclization by copper(I) catalysis. Later, this procedure was modified and can be performed in water or ionic liquids.[142]

4.2.8 Copper-Catalyzed Synthesis of Furan Derivatives

A general and effective method for the synthesis of 2-monosubstituted and 2,5-disubstituted furans from readily available alkynyl ketones in the presence of catalytic amounts of Cu(I) was developed.[143] The generality of the method was demonstrated in the efficient preparation of furans possessing different functional groups, such as the sterically hindered *t*-Bu group, alkene moiety, alkoxy group directly attached to the furan ring, a remote acid-sensitive OTHP group, a base/nucleophile-sensitive ester group and an unprotected hydroxyl group. All the furans were produced in good to excellent yields (Scheme 4.77). NEt$_3$ and high temperatures are needed for the formation of allenes and for further furan synthesis. In 2012, Oh and co-workers studied the elimination of β-halovinyl ketones.[144] The corresponding allenyl and propargyl ketones were generated *in situ* and gave the corresponding furans successfully in the presence of CuCl (1 mol%).

Liang and co-workers described the first examples of copper-catalyzed [4 + 1] cycloadditions of α,β-acetylenic ketones with diazo esters.[145] This method furnished synthetically useful, highly substituted furan derivatives in a direct, one-flask process with good efficiency and regioselectivity (Scheme 4.78). Two mechanisms were proposed for the copper-catalyzed cycloaddition reaction. In one, first the diazo ester reacts with CuI to give the copper-stabilized carbene complex. Next, exposure of an α,β-acetylenic ketone to a carbenoid produces a carbonyl ylide. Then, coordination of copper to the triple bond of the carbonyl ylide would make it subject to intramolecular nucleophilic attack to produce a zwitterion. The latter undergoes a subsequent proton transfer to afford a furan with simultaneous regeneration of the CuI catalyst. An alternative mechanism is also possible, in which the copper-stabilized carbene complex directly reacts with the triple bond to produce a cyclopropenyl ketone, then a CuI-catalyzed ring-opening cycloisomerization reaction of the cyclopropenyl ketone affords a furan and regenerates the CuI catalyst. The first mechanism appears more likely since the formation of a carbonyl ylide is very easy.

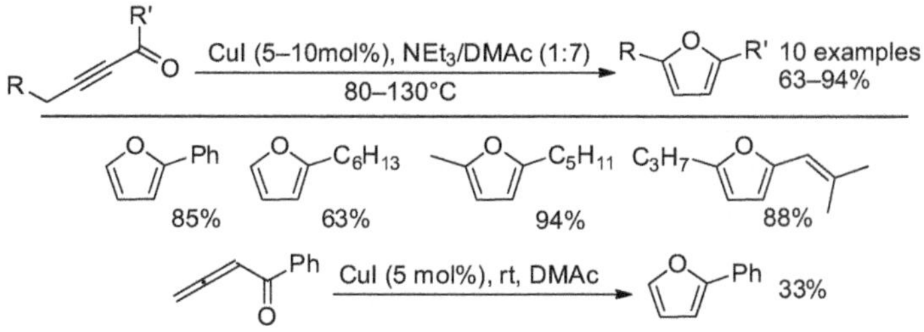

Scheme 4.77 Copper-catalyzed synthesis of furans from alkynones.

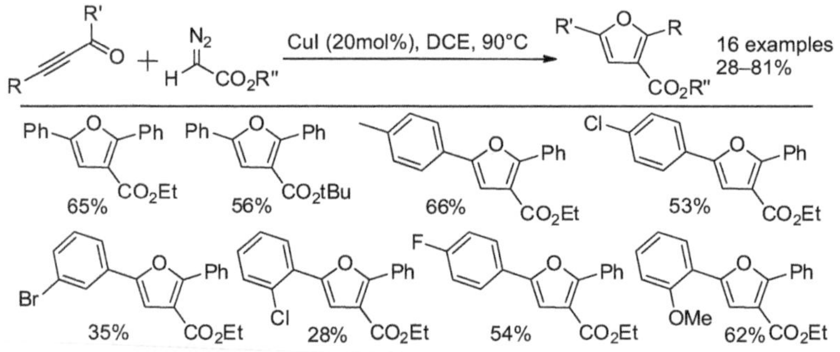

Scheme 4.78 Copper-catalyzed synthesis of furans from alkynones and diazoacetates.

General Procedure (Scheme 4.78): To a solution of α,β-acetylenic ketone (0.5 mmol) and CuI (0.1 mmol) in ClCH₂CH₂Cl (5 mL) heated at 90 °C under argon was added a solution of an EDA (1.5 mmol) in ClCH₂CH₂Cl *via* syringe pump over 5 h. Next, the reaction mixture was heated at 90 °C with stirring for 5 h. The reaction was quenched with a saturated aqueous solution of ammonium chloride and the mixture was extracted with CH₂Cl₂. The combined organic extracts were washed with water and saturated brine. The organic layer was dried (Na₂SO₄) and concentrated *in vacuo*. The residue was purified by chromatography on silica gel to afford the pure product.

Jiang and co-workers developed a new type of copper(I)-catalyzed domino process for the synthesis of furans.[146] This methodology proceeds through a sequence of rearrangement–dehydrogenation oxidation–carbene oxidation of 1,5-enynes, which are formed *in situ* from alkynols and alkynes under atmospheric pressure. This domino process provides an efficient method for the regiospecific synthesis of furan aldehydes/ketones, which are useful synthetic intermediates for bioactive compounds. Unfortunately, this domino process did take place when ethyl phenylpropionate was employed as the substrate to replace diethyl but-2-ynedioate. Subsequently, they reported a nano-Cu₂O-catalyzed synthesis of α-carbonyl furans through a cyclization-rearrangement–oxidation sequence of 1,5-enynes using phenylpropionate as substrate (Scheme 4.79).[147] This catalytic system is environmentally benign and the reaction constitutes an effective domino process for the formation of C–C and C–O bonds in modern organic chemistry. It is especially noteworthy that this domino process is applicable to a number of electron-deficient alkynes. Furthermore, these findings open up a convenient synthetic route to a variety of α-carbonyl furans, which are useful synthetic intermediates for bioactive and natural compounds. The domino reaction exhibits some unusual characteristics that are difficult to understand at present and that will be an interesting subject of mechanistic studies in the future. In 2012, the

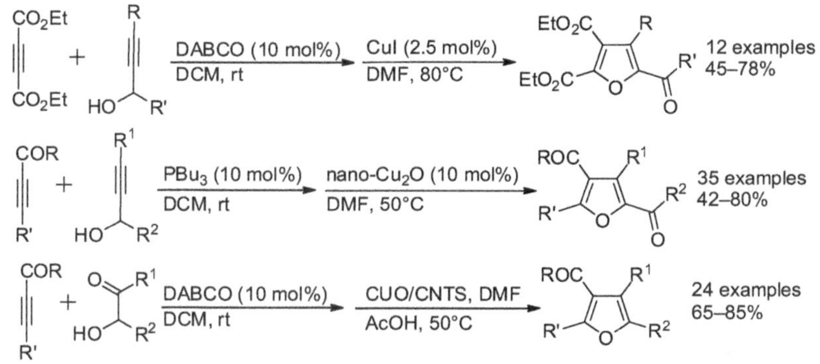

Scheme 4.79 Copper-catalyzed synthesis of furans from alkynes.

authors reported the synthesis of carbon nanotube-supported CuO and developed a highly active, easily recoverable and practical heterogeneous catalyst for cyclization reactions, which provides efficient access to the construction of highly substituted furans from electron-deficient alkynes and α-hydroxy ketones.[148] In particular, the performance of the catalyst fully promoted the transformation and shortened reaction times due to the excellent electrical conductivity of the carbon nanotubes.

In 2010, Jiang and co-workers reported another facile and efficient method to synthesize polysubstituted furans.[149] The reaction proceeds through tin(II)- and copper(I)-involved addition–oxidative cyclization of alkynoates and 1,3-dicarbonyl compounds using DDQ (2,3-dichloro-5,6-dicyano-1,4-benzoquinone) as the oxidant. This methodology not only provides a simple way to construct polysubstituted furan derivatives but also opens up a new approach to build oxygen-containing vinyl ether compounds through oxidation. The desired furans were produced in good to excellent yields (Scheme 4.80). Around the same time, the combination of CuI and O_2 in DMF was also found to be effective for this transformation.[150] Low or no yields resulted with toluene, THF, MeCN or $CHCl_3$ as solvent.

In 2010, Ma and co-workers developed a new and convenient method to assemble polysubstituted furans from vinyl iodides and terminal alkynes.[151] The reaction involves a copper–L-proline-catalyzed cross-coupling and additive cyclization process. This result provides an additional example of the synthesis of heterocycles *via* a ligand-promoted Ullmann-type coupling reaction. Compared with previous stepwise methods, this approach has the advantages of simple operation and inexpensive catalysts. The generality of this method has been demonstrated by preparing a wide range of substituted furans (Scheme 4.81). These features make this method an attractive choice for furan synthesis.

In 2013, Zhang and co-workers developed a novel copper-mediated annulation of alkyl ketones with α,β-unsaturated carboxylic acids to give the corresponding furans.[152] This reaction provides a facile and regio-defined method for the synthesis of 2,3,5-trisubstituted furans from simple chemical

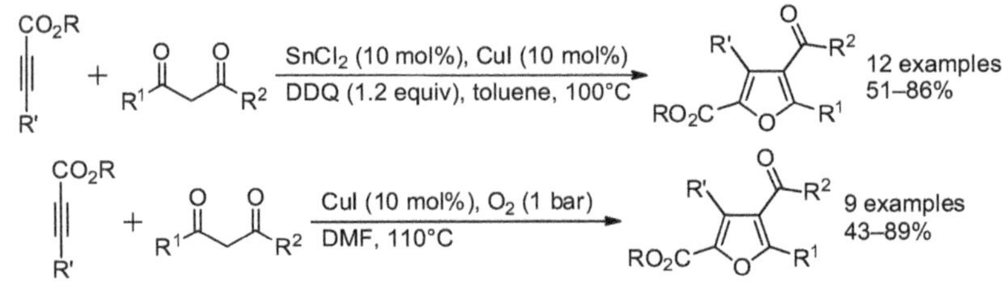

Scheme 4.80   Copper-catalyzed synthesis of furans from 1,3-dicarbonyls.

General Procedure (Scheme 4.80): A Schlenk tube was charged with substituted 3-iodoprop-2-en-1-ol (0.25 mmol), CuI (5 mg, 0.025 mmol), L-proline (9 mg, 0.075 mmol) and Cs_2CO_3 (245 mg, 0.75 mmol). The tube was evacuated and backfilled with argon and then 1-alkyne (0.3 mmol) and 1,4-dioxane (0.5 mL) were added *via* syringe. After the reaction mixture had been stirred at 80–100 °C for 36–48 h, it was partitioned between ethyl acetate and saturated NH_4Cl. The organic layer was washed with brine, dried over Na_2SO_4 and concentrated *in vacuo*. The residue was purified *via* column chromatography to provide the desired product.

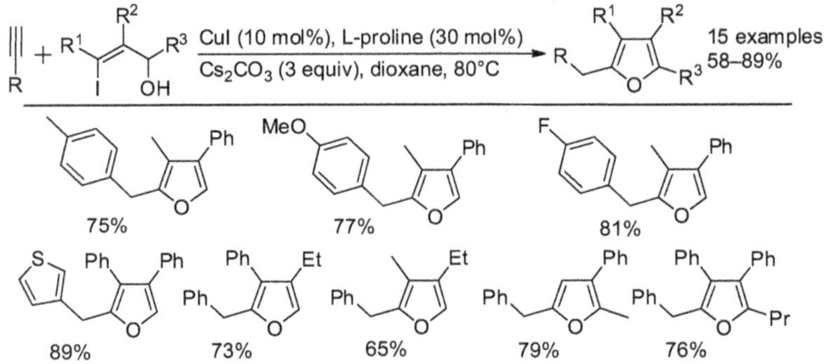

Scheme 4.81   Copper-catalyzed synthesis of furans from vinyl iodides.

reagents (Scheme 4.82). Instead of α,β-unsaturated carboxylic acids, styrenes can also be applied as substrates and good yields of furans can be isolated. Concerning the reaction mechanism, the reactions proceed through a radical pathway, which was supported by TEMPO experiments. The addition of 2 equiv. of TEMPO (2,2,6,6-tetramethylpiperidine-1-oxyl) as a radical inhibitor blocked the formation of the expected furan product.

Jiang's group reported a regioselective one-pot procedure for the synthesis of 2,5-disubstituted furans using copper(I) catalyst from haloalkynes (Scheme 4.83).[153] This chemistry proceeds through the hydration reaction of 1,3-diynes, which can be readily prepared from the coupling reaction of haloalkynes in the presence of CuI. The procedure can also be used for the facile synthesis of 2,5-disubstituted thiophenes.

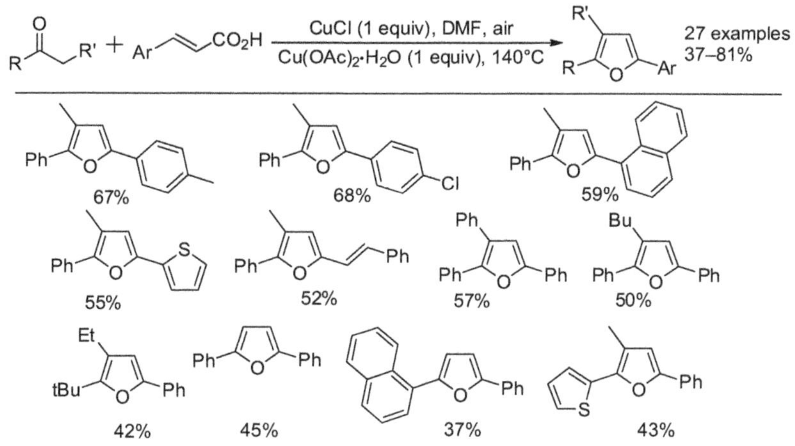

Scheme 4.82 Copper-catalyzed synthesis of furans from alkyl ketones.

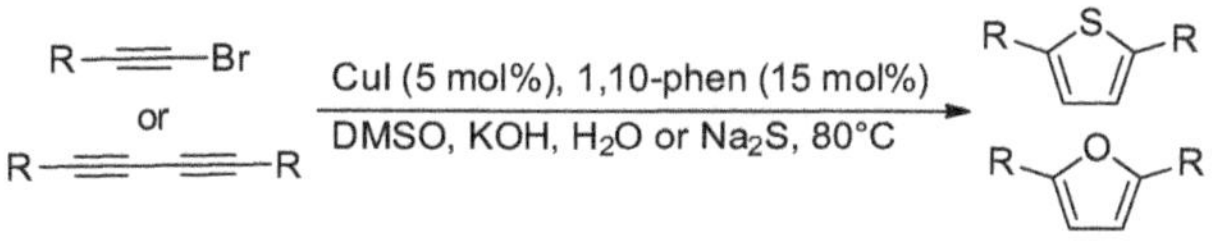

Scheme 4.83 Copper-catalyzed synthesis of furans from haloalkynes.

General Procedure (Scheme 4.83): A mixture of phenylethynyl bromide (0.5 mmol), CuI (4.8 mg, 5 mol%), 1,10-phen (13.5 mg, 15 mol%), KOH (140 mg, 5 equiv.) and H_2O (7.2 mg, 4 mmol) in DMSO (2.5 mL) was placed in a test-tube (10 mL) equipped with a magnetic stirrer bar. The mixture was stirred at 80 °C for 6 h. When the reaction was complete, the mixture was filtered through a glass filter and washed with ethyl acetate. The mixture was washed with brine and extracted with ethyl acetate. The organic layer was dried ($MgSO_4$), concentrated *in vacuo* and purified by flash silica gel chromatography (eluent: 20:1 petroleum ether–ethyl acetate) to give the desired products.

A $CuCl_2$-catalyzed heterocyclodehydration of readily available 3-yne-1,2-diol and 1-amino-3-yn-2-ol derivatives was reported in 2013 by Gabriele *et al.*[154] Substituted furans and pyrroles, respectively, were produced in good to high yields (53–99%) under mild conditions (MeOH as the solvent, 80–100 °C, 1–24 h) (Scheme 4.84). In the case of 2,2-dialkynyl-1,2-diols, bearing an additional alkynyl substituent at C2, a cascade process, corresponding to copper-catalyzed heterocyclodehydration followed by acid-catalyzed hydration of the triple bond, was realized when the reaction was carried out in the presence of both $CuCl_2$ and TsOH, leading to 3-acylfurans in one step and high yields (75–84%). Under the same conditions, *N*-Boc-2-alkynyl-1-amino-3-yn-2-ols were converted into the corresponding *N*-unsubstituted

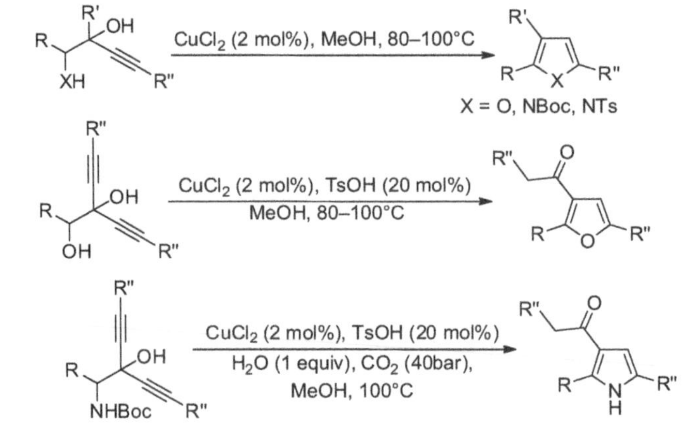

Scheme 4.84   Copper-catalyzed synthesis of furans by heterocyclodehydration.

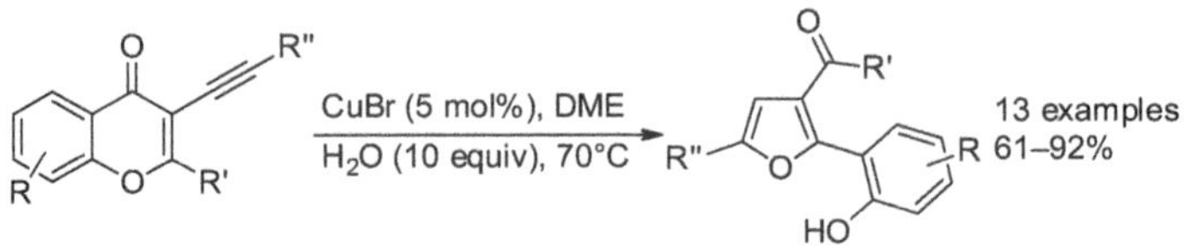

Scheme 4.85   Copper-catalyzed synthesis of 3-acylfurans.

3-acylpyrroles in low to fair yields (19–59%). However, working in the presence of added water and a large excess of CO_2 (40 atm), in addition to $CuCl_2$ and TsOH, caused a significant improvement in the yields of 3-acylpyrroles (68–87%), thus making the method of general synthetic applicability.

A highly efficient CuBr-catalyzed domino reaction of 2-substituted-3-(1-alkynyl)chromones to give functionalized 3-acylfurans was developed by Hu and co-workers.[155] The reaction is mild, environmentally friendly and easily handled without the necessity for dry solvents and an inert atmosphere. Good to excellent yields of the desired products were isolated (Scheme 4.85).

Xu and co-workers developed a general and highly efficient synthetic method for substituted 5-methylene-4,5-dihydrofuran compounds.[156] This intramolecular cyclization reaction was performed using readily available materials (2-propynyl-1,3-dicarbonyl compounds), an inexpensive catalyst (Cu^{2+} salt) and ligand (PPh_3) under very mild basic conditions with good to excellent yields (Scheme 4.86). Furthermore, the reaction exhibited excellent functional group compatibility. Additionally, 5-methylene-4,5-dihydrofurans could be easily converted into the corresponding furans in the presence of CF_3COOH in dry MeOH at room temperature within 24 h in good yields.

A series of 4-arylsulfonylimino-4,5-dihydrofurans (14 examples) were efficiently synthesized in good to excellent yields by using the copper-catalyzed three-component reaction between sulfonyl azides, phenylacetylene and β-keto esters in tetrahydrofuran (THF) at 40 °C for 8 h in the presence of triethylamine (Scheme 4.87).[157]

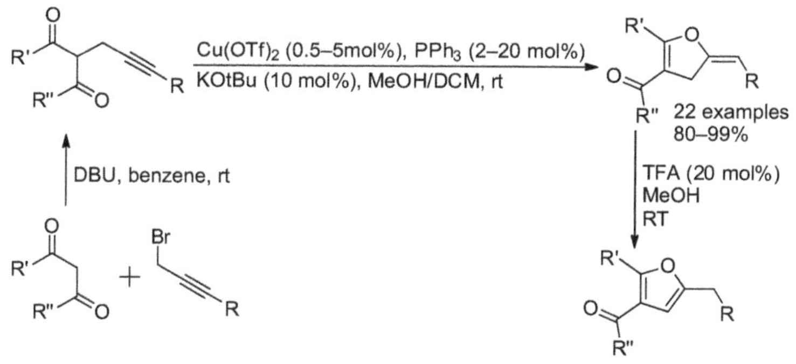

Scheme 4.86 Copper-catalyzed synthesis of dihydrofurans.

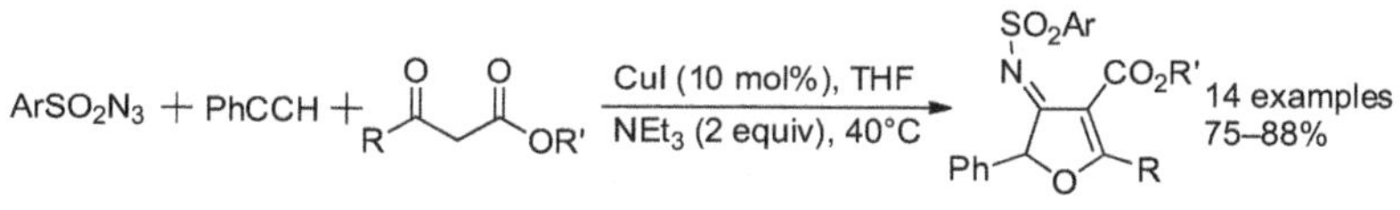

Scheme 4.87 Copper-catalyzed synthesis of dihydrofurans from azides.

> **General Procedure** (Scheme 4.87): To a stirred mixture of CuI (19.1 mg, 0.1 mmol), benzenesulfonyl azide (219 mg, 1.2 mmol), phenylacetylene (102 mg, 1 mmol) and ethyl acetoacetate (309 mg, 3.0 mmol) in anhydrous THF (5 mL) was slowly added NEt$_3$ (2 mL) *via* syringe under an N$_2$ atmosphere at 40 °C. The reaction mixture was stirred for 8 h in a sealed tube. Solvent was removed under vacuum, then the mixture was extracted with CH$_2$Cl$_2$ (3 × 10 mL). The organic layers were combined, dried over anhydrous Na$_2$SO$_4$, filtered and concentrated under vacuum. The residue was purified by flash column chromatography on silica gel (200–300 mesh) with ethyl acetate–petroleum ether (1:10–1:15) as eluent to give the desired product.

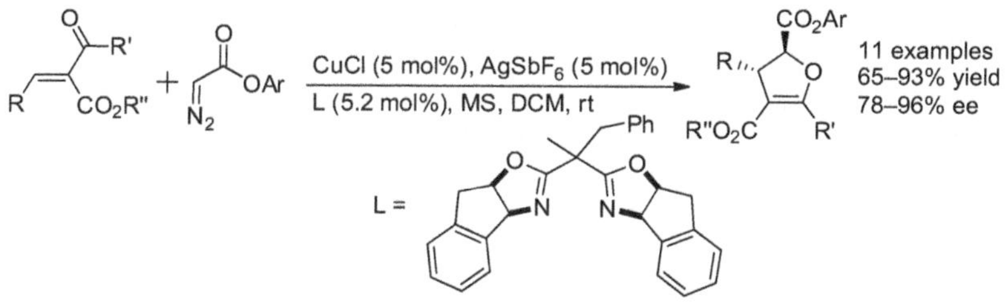

Scheme 4.88 Copper-catalyzed asymmetric synthesis of dihydrofurans.

Tang and co-workers developed an efficient and practical protocol to construct chiral multifunctionalized 2,3-dihydrofurans by combining an α-benzylidene-β-keto ester with a diazo compound.[158] Following this methodology, a series of optically active tetrasubstituted 2,3-dihydrofurans were obtained with good to excellent stereoselectivity (90:10–99:1 *dr* and 78–96% *ee*) in good to excellent yields (65–93%) (Scheme 4.88). In addition, a rational stereoinductive model was developed. This reaction should be helpful for understanding Cu(I)–side-armed BOX-catalyzed reactions and provide useful information for the rational design of new ligands to control selectivities in related asymmetric reactions.

4.2.9   Copper-Catalyzed Synthesis of Benzofuran Derivatives

In 1989, Owen and co-workers reported a copper-promoted one-step synthesis of benzofurans from *o*-iodophenols and terminal alkynes.[159] Cu$_2$O was applied as the promoter and the reactions were carried out in refluxing pyridine. Good yields of benzofurans were prepared from various alkynes and 5-*tert*-butyl-2-iodophenol. In 2002, a well-defined Cu(I) complex was prepared and applied in the synthesis of benzofurans.[160] The reactions resulted in good to excellent yields (Scheme 4.89). Later, $N^2,N^{2\prime}$-dibenzyl(1,1′-binaphthalene)-2,2′-diamine was found to be an effective ligand for this transformation under the same conditions.[161]

In 2012, a copper-catalyzed annulative amination approach to 3-aminobenzofurans and -indoles from *o*-alkynylphenols and -anilines was developed.[162] The Cu-based catalysis is based on an umpolung, electrophilic amination with *O*-benzoylhydroxylamines and permits the mild and convergent synthesis of various 3-aminobenzoheteroles of biological and pharmaceutical interest (Scheme 4.90). Some mechanistic investigations and an application of this protocol to the construction of more complex tricyclic framework were also described.

In 2010, Pyne and co-workers developed a direct and convenient method for the cyclization–halogenation and cyclization–cyanation reactions of *cis*-4-hydroxy-5-phenylethynylpyrrolidinones and *O*- and *N*-protected *o*-alkynyl-phenols and -anilines, respectively (Scheme 4.91).[163] This method allows the synthesis of 3-cyanobenzofurans and -indoles in a one-step process that otherwise would require two sequential steps from the same starting substrates, that is, iodonium ion-induced cyclization followed by a classical Rosenmund–von Braun reaction with CuCN or its more recent and milder

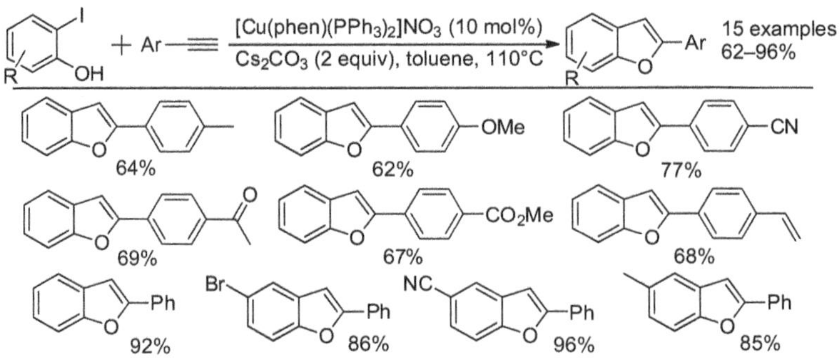

Scheme 4.89   Copper-catalyzed synthesis of benzofurans.

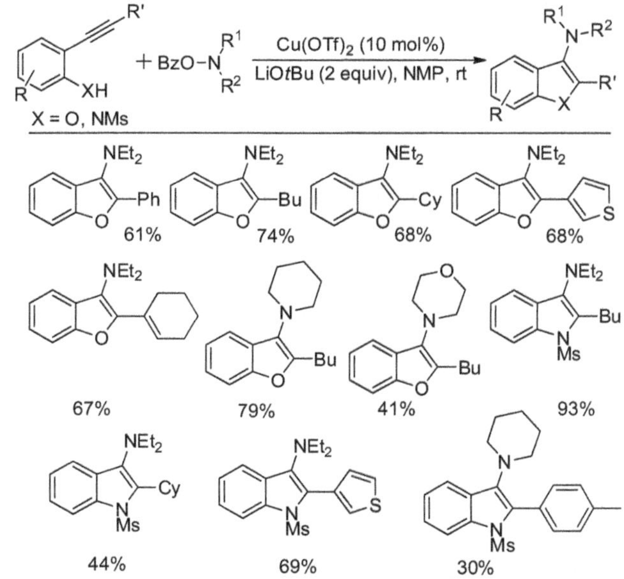

Scheme 4.90 Copper-catalyzed synthesis of 3-aminobenzofurans.

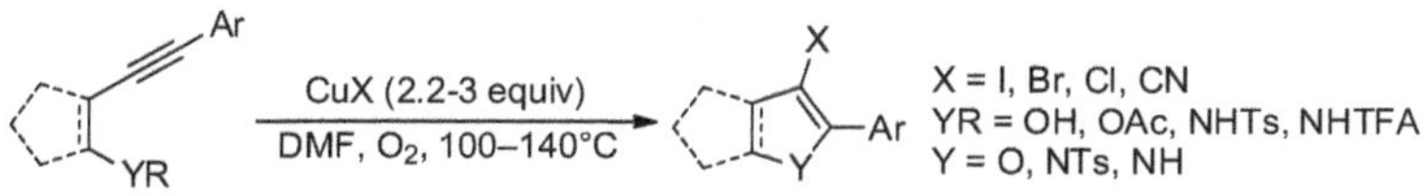

Scheme 4.91 Copper-mediated synthesis of 2,3-substituted benzofurans.

Pd-catalyzed versions, on the resulting 3-iodobenzofurans or -indoles. The majority of reactions that provided the 3-cyanoindoles, using this new one-step process, however, proceeded efficiently under much milder conditions at 100 °C. Further, this new method is also more cost-effective when one compares the relatively higher costs per mole of iodine and *N*-iodosuccinimide with the less expensive CuCN. Although the exact mechanism of these reactions is not known, the authors suggested that a Cu(II) species was not involved in the initial cyclization reaction.

A copper-mediated annulative direct coupling of *o*-alkynylphenols with 1,3,4-oxadiazoles was reported in 2011.[164] The corresponding biheteroaryls were produced in moderate yields and the reaction proceeded smoothly even under ambient conditions (Scheme 4.92). The reaction system represents a new avenue for the construction of biheteroaryl molecules of interest for their biological and physical properties. Moreover, the possibility of catalysis using CuF$_2$–MnO$_2$ was also described. With respect to the reaction mechanism, the reaction was believed to proceed through (i) direct C–H cupration of azoles, (ii) chelation-assisted second C–H cupration of arylazines and (iii) productive reductive elimination. It was envisaged that the second C–H cupration step could be replaced with the annulative cupration of *o*-alkynylphenols.

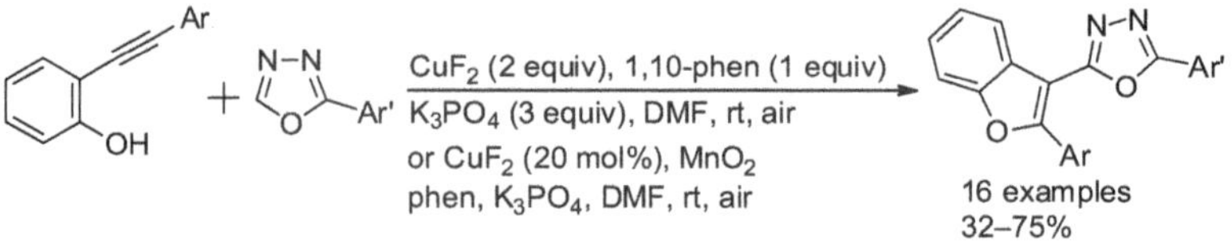

Scheme 4.92 Copper-mediated synthesis of benzofurans by annulation.

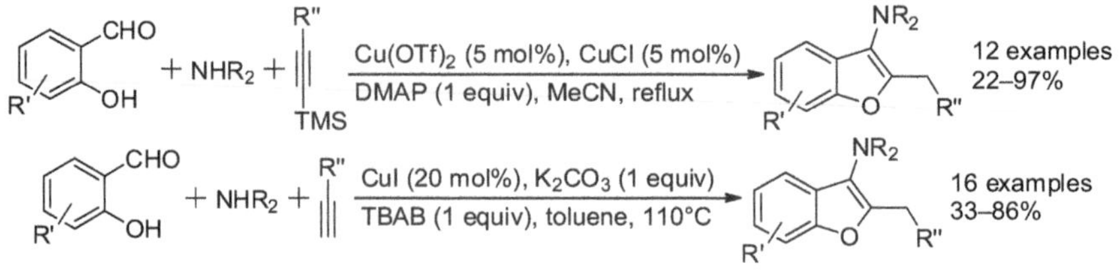

Scheme 4.93 Copper-catalyzed three-component synthesis of benzofurans.

General Procedure (Scheme 4.93): To a CH$_3$CN solution (300 μL) in a screw-capped vial under an N$_2$ atmosphere, alkynylsilane (0.45 mmol), *o*-hydroxybenzaldehyde (0.45 mmol), secondary amine (0.30 mmol), DMAP (0.3 mmol), Cu(OTf)$_2$ (5.4 mg, 0.015 mmol) and CuCl (1.5 mg, 0.015 mmol) were successively added and the vial was then sealed with a cap containing a PTFE septum. The reaction mixture was heated at 100 °C for 6 h.

General Procedure: A mixture of CuI (0.038 g, 0.2 mmol), K$_2$CO$_3$ (0.138 g, 1.0 mmol), Bu$_4$NBr (0.32 g, 1.0 mmol), salicylaldehyde (0.20 mL, 2.0 mmol), phenylacetylene (0.17 mL, 1.5 mmol) and morpholine (0.087 mL, 1.0 mmol) in toluene (5 mL) was heated at 110 °C for 3 h.

In 2008, a copper-catalyzed three-component coupling reaction to give benzofurans was reported by Sakai *et al.*[165] Alkynylsilane, *o*-hydroxybenzaldehydes and secondary amines were applied as the substrates and the combination of Cu(OTf)$_2$ and CuCl in the presence of DMAP was found to be an effective catalytic system. The desired benzofurans were produced in moderate to excellent yields (Scheme 4.93). Li's group found that normal terminal alkynes can also be used as the coupling partner.[166] The reactions were performed in toluene as 110 °C with the assistance of TBAB and K$_2$CO$_3$ using CuI as catalyst. Moderate to good yields of benzofurans were prepared by this procedure (Scheme 4.93).

In 2005, Chen and Dormer developed a highly efficient protocol for the synthesis of benzo[*b*]furans *via* a CuI-catalyzed ring closure of 2-haloaromatic ketones.[167] This process proved to be exceptionally effective with a wide variety of aromatic ketones and can be extended to aromatic aldehydes and heteroaromatic ketones. Many structurally interesting benzo[*b*]furans were readily prepared in a catalytic manner in good to excellent yields (Scheme 4.94). Later,

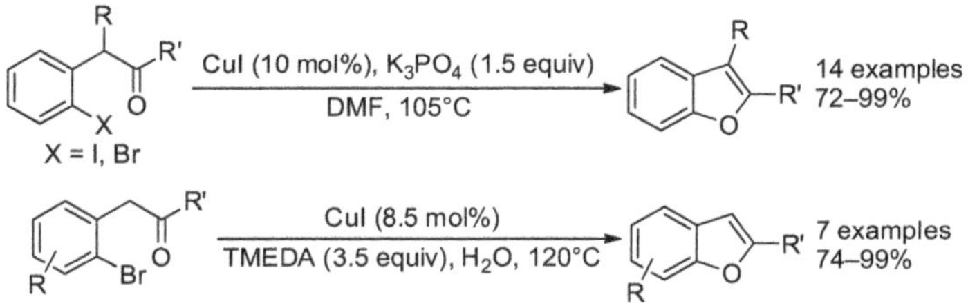

Scheme 4.94 Copper-catalyzed synthesis of benzofurans from 2-haloaromatic ketones.

this methodology was applied in a sequential reaction of TiCl$_4$-catalyzed anti-Markovnikov hydration of alkynes to 2-haloaromatic ketones, followed by copper-catalyzed synthesis of benzofurans.[168] An environmentally friendlier route to benzo[*b*]furan derivatives based on a copper-catalyzed ring-closure reaction performed in water was developed in 2006.[169] The benefits derived from the use of such a benign solvent are clear in terms of lack of toxicity, safety, cost and availability, properties that make water a desirable medium for every organic transformation. Further, the on-water methodology described here delivers a range of benzo[*b*]furans in good to excellent yields starting from readily available substrates (Scheme 4.94). In addition, the easy isolation of the target compounds, by simple extraction with dichloromethane, allows the reutilization of the aqueous solution containing the copper catalyst, thereby providing environmental and economic advantages over previously reported protocols. The main drawback of this procedure is the need for a large excess of TMEDA, an expensive base, which also acts as a ligand.

Wang and co-workers developed a general, highly efficient one-pot protocol for the synthesis of benzofuranyl- and indolylazoles in 2011.[170] The reactions proceed through a copper/palladium-catalyzed tandem reaction of an intramolecular Ullmann-type C–O(N) coupling reaction of 2-(*gem*-dibromovinyl)-phenols and 2-(*gem*-dibromovinyl)anilines followed by an intermolecular arylation of azoles through their C–H activation. The reactions of 2-(*gem*-dibromovinyl)phenols and 2-(*gem*-dibromovinyl)anilines with a variety of azoles, including oxazoles, imidazoles, thiazoles and oxadiazoles, proceeded smoothly in the presence of CuBr with Pd(PPh$_3$)$_2$Cl$_2$ used as co-catalyst and LiO^{t}Bu as a base in toluene at 100 °C for 10 h, and the corresponding products were generated in high yields. This reaction utilizes easily accessible *gem*-dibromovinyl substrates and azoles and provides a straightforward route to diverse biheteroaryl compounds. The Ullmann-type reaction of 2-(*gem*-dibromovinyl)phenols with the formation of a 2-bromobenzofuran as the intermediate was proposed in 2009[171] and was reported to be promoted by Cs$_2$CO$_3$ in EtOH at 80 °C in the absence of catalyst.[172] Wu and co-workers described a novel and efficient route for the generation of 2-(poly-fluoroaryl)benzofurans *via* a copper(I)-catalyzed tandem reaction of 2-(2,2-dibromovinyl)phenol with polyfluoroarenes in 2012 (Scheme 4.95).[173] During the reaction process, copper-catalyzed intramolecular C–O bond formation and alkenylation of the polyfluoroarene C–H bond were involved.

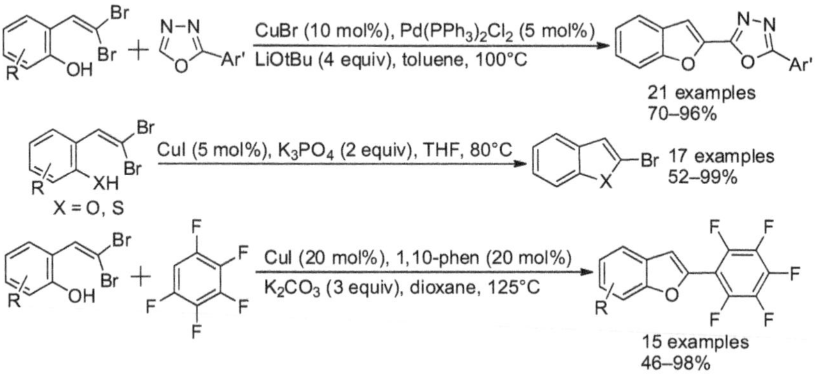

Scheme 4.95 Cu-catalyzed   synthesis   of   benzofurans   from   2-(*gem*-dibromovinyl)phenols.

General Procedure (Scheme 4.95): 2-(2,2-Dibromovinyl)phenol (0.5 mmol), CuI (0.1 mmol), 1,10-phenanthroline (0.1 mmol) and K₂CO₃ (1.5 mmol) were placed in a sealed tube under a nitrogen atmosphere. Polyfluoroarene (2.0 mmol) and 1,4-dioxane (1.0 mL) were added *via* a syringe. The reaction mixture was stirred at 125 °C for 36 h. After completion of the reaction as indicated by TLC, the residue was purified directly by flash chromatography on silica gel to afford the products.

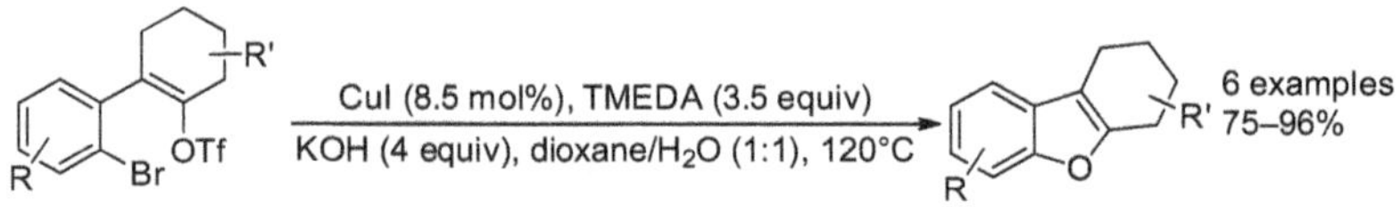

Scheme 4.96 Cu-catalyzed synthesis of benzofurans from aryl bromide–alkenyl triflates.

Willis and co-workers demonstrated a copper-catalyzed conversion of aryl bromide–alkenyl triflates to the corresponding benzofurans in 2007.[174] The combination of KOH and CuI–TMEDA was found to be an effective system for this transformation (Scheme 4.96). The reactions proceed *via* hydrolysis of the alkenyl triflates to generate enolates, which are then converted to the benzofurans under the action of Cu(I). The process tolerates variations of both the alkenyl triflate and aryl bromide portions of the substrate. The successful conversion of aryl bromide–alkenyl triflates to benzofurans is significant as it demonstrates that a single class of difunctionalized acyclic precursor can be used to access both indoles and benzofurans.

In 2007, Ma's group reported that CuI-catalyzed coupling of 1-bromo-2-iodobenzenes with β-keto esters in THF at 100 °C leads to 2,3-disubstituted benzofurans.[175] This domino transformation involves an intermolecular C–C bond formation and a subsequent intramolecular C–O bond formation

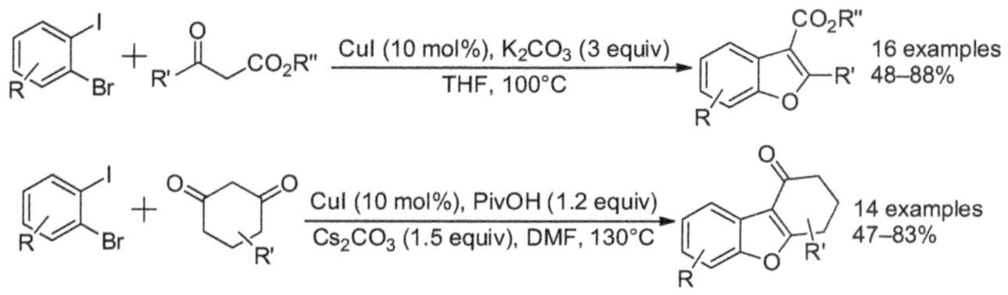

Scheme 4.97 Cu-catalyzed synthesis of benzofurans from 1,2-dihalobenzenes.

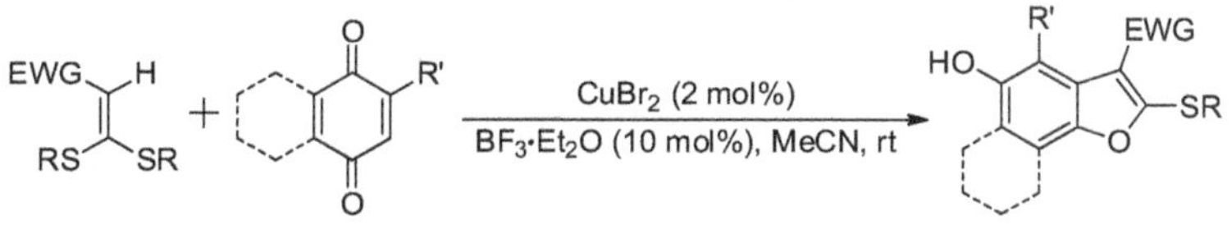

Scheme 4.98 Cu-catalyzed synthesis of benzofurans from ketene dithioacetals.

process. Benzofurans with different substituents at the 5- and 6-positions are accessible by employing the corresponding 1-bromo-2-iodobenzenes (Scheme 4.97). In 2012, the reaction of 1-bromo-2-iodobenzenes with 1,3-cyclohexanediones was developed by Beifuss and co-workers.[176] In DMF at 130 °C using Cs_2CO_3 as a base and pivalic acid as an additive, 3,4-dihydrodibenzo[*b,d*]furan-1(2*H*)-ones were selectively produced with yields ranging from 47 to 83% (Scheme 4.97). This highly regioselective domino process is based on an intermolecular Ullmann-type *C*-arylation followed by an intramolecular Ullmann-type *O*-arylation. Substituted products are accessible by employing substituted 1-bromo-2-iodobenzenes and substituted 1,3-cyclohexanediones as substrates. Reaction with an acyclic 1,3-diketone yields the corresponding benzo[*b*]furan.

A new application of copper(II) bromide-activated ketene dithioacetals as nucleophiles in organic chemistry was developed in 2010.[177] Under the co-catalysis of copper(II) bromide (2 mol%) and boron trifluoride etherate (10 mol%), the conjugate addition and sequential cyclization of α-electron-withdrawing group-substituted ketene dithioacetals with *p*-quinones in acetonitrile at room temperature gave a variety of benzofurans (Scheme 4.98). This formal [3 + 2] cycloaddition provides a general method for the catalytic synthesis of polyfunctionalized benzofurans with the advantages of readily available starting materials, cheap catalysts, mild reaction conditions, good yields and wide range of synthetic potential for the benzofuran products. Further transformations of the resulting benzofurans to 2-aminobenzofurans and benzofuro[2,3-*d*]pyrimidine derivatives were also investigated.

An efficient synthesis of 2- and 4-iododibenzofurans through CuI-mediated sequential iodination–cycloetherification of two aromatic C–H bonds in *o*-arylphenols was developed by Zhu and co-workers.[178] The reaction was mediated by 1.5 equiv. of CuI, which serves as an iodinating reagent first and

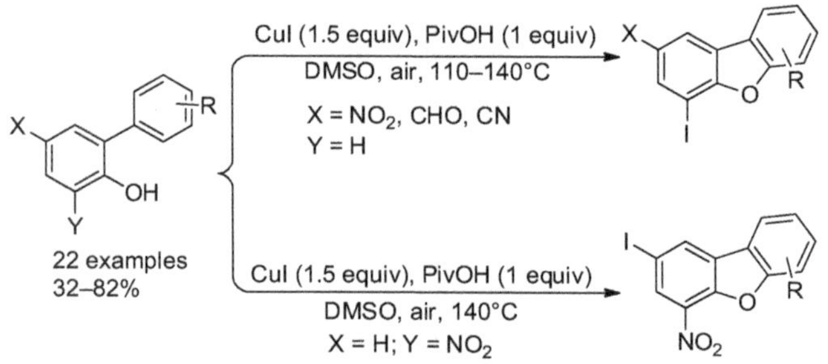

Scheme 4.99 Cu-mediated oxidative synthesis of benzofurans.

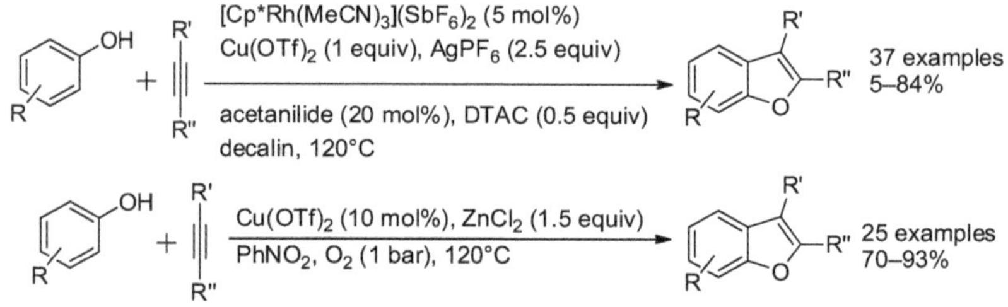

Scheme 4.100 Cu-catalyzed oxidative synthesis of benzofurans from phenols.

then as a promoter for C–H cycloetherification (Scheme 4.99). Both the pre-existing electron-withdrawing groups (NO_2, CN and CHO) and the newly introduced iodide are readily transferred to diversify the dibenzofuran skeleton. Deuterium labeling experiments suggested that an irreversible rate-limiting C–H cleavage step is involved. Results of DFT calculations preferred Cu(III) to Cu(II) species in the pivalate-assisted CMD process.

In 2013, Shi and co-workers reported a copper-mediated oxidative annulation of phenols and internal alkynes to afford benzofuran derivatives.[179] In the presence of a rhodium catalyst and DTAC (dodecyltrimethylammonium chloride) as additive, various phenols and unactivated internal alkynes were successfully employed, and the corresponding benzofurans were obtained in good yields (Scheme 4.100). Jiang's group successfully developed a copper-catalyzed version of this interesting reaction.[180] Polysubstituted benzofurans were produced efficiently *via* copper-catalyzed nucleophilic addition and aerobic oxidative cyclization of phenols and alkynes. It is noteworthy that this method exhibits good functional group tolerance and provides an attractive synthetic strategy for the synthesis of benzofuran derivatives. Furthermore, the use of readily available starting materials and molecular oxygen as the oxidant makes the overall chemical transformation sustainable and practical.

In 2000, Zhu *et al.* described a copper-catalyzed intramolecular cyclization of 2-(2′-chlorophenyl)ethanol.[181] 2,3-Dihydrobenzofuran can be prepared by this procedure in good yield (Scheme 4.101). NaH was needed as a base and

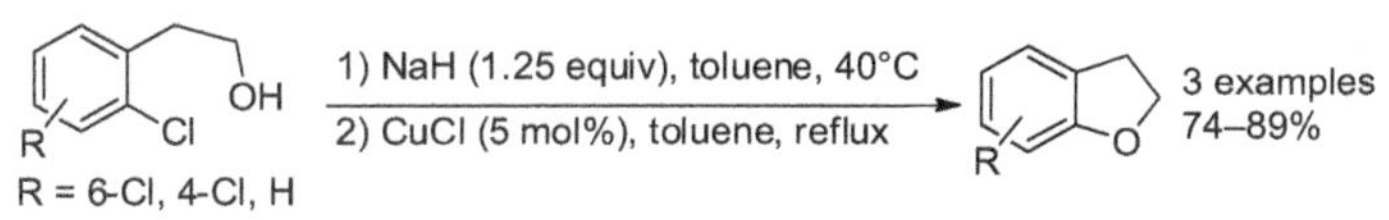

Scheme 4.101 Cu-catalyzed synthesis of 2,3-dihydrobenzofurans.

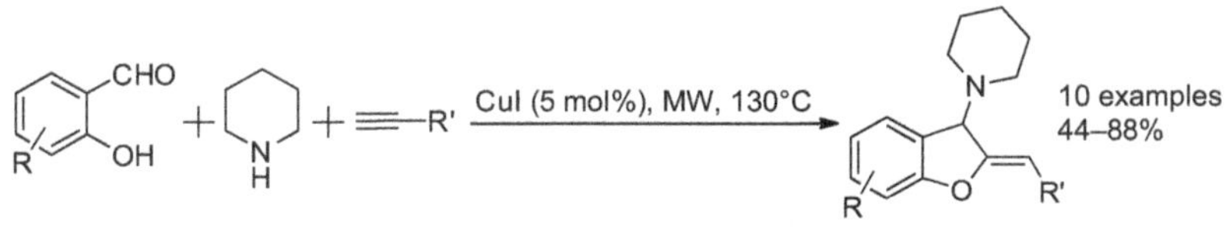

Scheme 4.102 Cu-catalyzed three-component synthesis of dihydrobenzofurans.

should be treated with substrates in advance. This method was later applied in the asymmetric synthesis of corsifuran A.[182]

Li's group developed a simple and efficient procedure for the synthesis of dihydrobenzofurans in 2008.[183] Salicylaldehydes, amines and aliphatic alkynes were applied as substrates and the desired products were obtained in good yields (Scheme 4.102). The use of aliphatic alkynes containing a heteroatom is critical to the success of the reaction. AgCl (5 mol%) can also catalyze this transformation effectively.

4.2.10 Copper-Catalyzed Synthesis of Pyrrole Derivatives

Li and co-workers developed an efficient copper-catalyzed double alkenylation of amides with (1Z,3Z)-1,4-diiodo-1,3-dienes.[184] The reactions proceed to afford di- or trisubstituted N-acylpyrroles in good to excellent yields using CuI as the catalyst, Cs$_2$CO$_3$ as the base and *rac-trans-N,N'*-dimethylcyclohexane-1,2-diamine as the ligand (Scheme 4.103). Buchwald's group reported an efficient Cu-catalyzed method for the conversion of 1,3-dihalo-1,3-dienes into valuable heterocycles, such as N-Boc-substituted pyrroles and heteroarylpyrroles.[185] The transformation tolerated of a wide variety of functional groups in a range of substitution patterns. In 2010, Xi and co-workers described a simple and efficient approach to various substituted N-arylpyrroles.[186] The method is based on the copper-catalyzed sequential inter- and intramolecular N-vinylation of aromatic amines. The reactions proceed to afford substituted N-arylpyrroles in good to excellent yields using CuI as the precatalyst, *t*BuONa as the base and N^1,N^2-dimethylethane-1,2-diamine (DMEDA) as the ligand. Anilines with electron-donating and -withdrawing substituents and also a heteroaromatic amine performed very well under the conditions used. Tri- and tetrasubstituted dienyl diiodides also performed well under the reaction conditions and afforded the corresponding substituted N-arylpyrroles in good yields. Products were also obtained in high yields with CuBr or CuCl as precatalyst. Interestingly, when 1,4-diiodobut-1-enes and amides were applied as substrates, only 2,3-dihydropyrroles were isolated.[187]

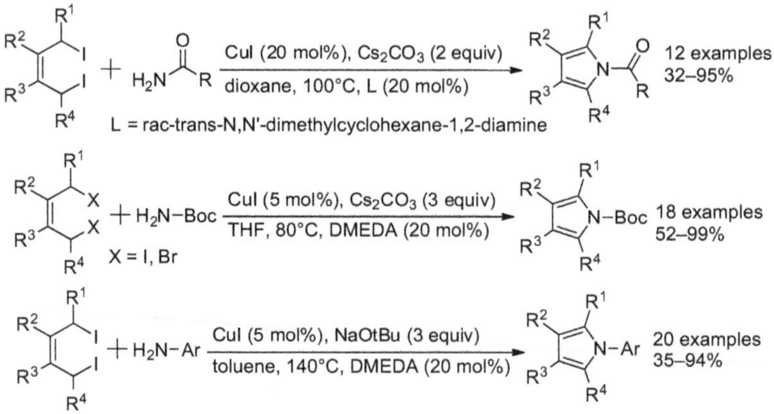

Scheme 4.103 Cu-catalyzed synthesis of pyrroles from 1,3-dihalo-1,3-dienes.

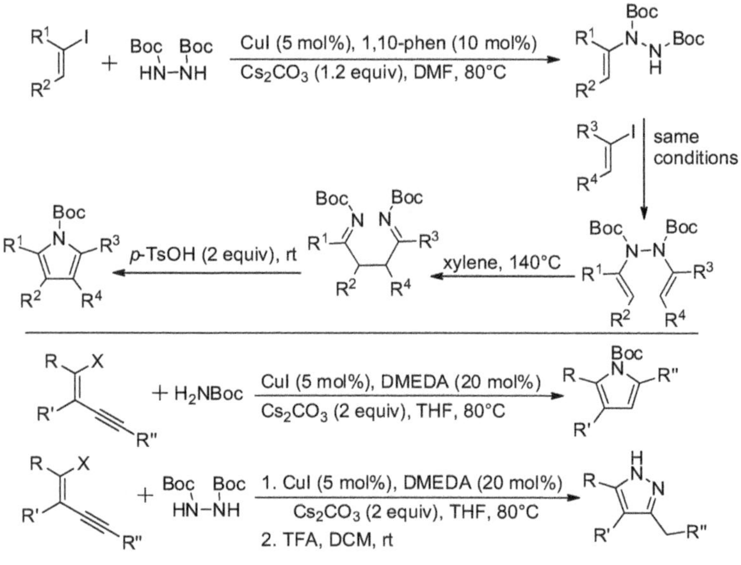

Scheme 4.104 Cu-catalyzed sequential synthesis of pyrroles.

In 2007, Rivero and Buchwald reported another reaction procedure for the preparation of pyrroles.[188] This procedure offers a new route to highly substituted pyrroles. This transformation consists of two sequential copper-catalyzed vinylations of bis-Boc-hydrazine followed by thermal rearrangement/cyclization. A wide variety of functionalized pyrroles can be prepared in a selective manner from simple and easily accessible precursors in good yields (Scheme 4.104). The reactions of haloenynes with amines to give pyrroles and with hydrazine to give pyrazoles were also reported by the same groups (Scheme 4.104).[189]

Li's group reported an additional procedure for the synthesis of pyrroles.[190] This CuI–*N*,*N*-dimethylglycine hydrochloride-catalyzed reactions of amines with γ-bromo-substituted γ,δ-unsaturated ketones in the presence of K₃PO₄ and NH₄OAc led to the formation of the corresponding polysubstituted pyrroles in good to excellent yields (Scheme 4.105). The presence of NH₄OAc might function as the NH₃ source, which served as the co-ligand to improve the efficiency of the copper catalyst. A reaction mechanism was also proposed. The amines undergo condensation with the carbonyls first to give the corresponding imines, which rearrange to the enamine structures. The enamines then undergo intramolecular *N*-vinylation under the catalysis of Cu(I) to give 2-methylene-2,3-dihydropyrroles, which then isomerize to the pyrroles as the final products.

An interesting procedure for the rapid intramolecular capture of a highly active organocopper intermediate produced in the formal [3 + 2] cycloaddition of isocyanides to ynones, subsequently leading to the formation of an intramolecular aryl C–C bond, which gave indenone fused pyrroles as the final products, was developed in 2011.[191] This reaction represents a new type of copper-catalyzed tandem reaction, which offers a simple and efficient method for the synthesis of 4-oxoindeno[1,2-*b*]pyrroles in good to excellent yields (Scheme 4.106). The process takes place efficiently when a variety of 1-(2-iodoaryl)-2-yn-1-ones are used and it displays wide functional group compatibility.

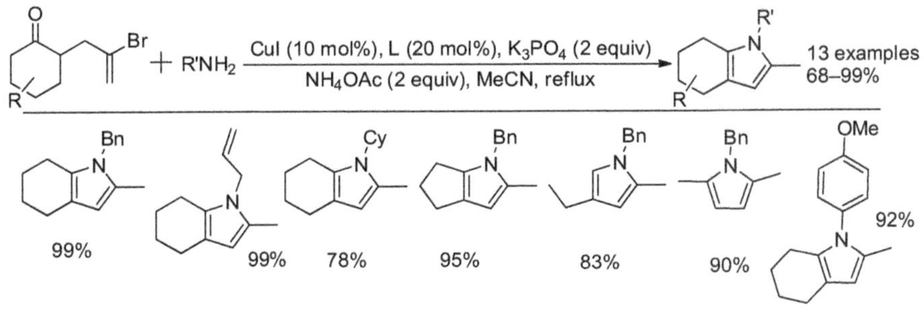

Scheme 4.105 Cu-catalyzed synthesis of polysubstituted pyrroles.

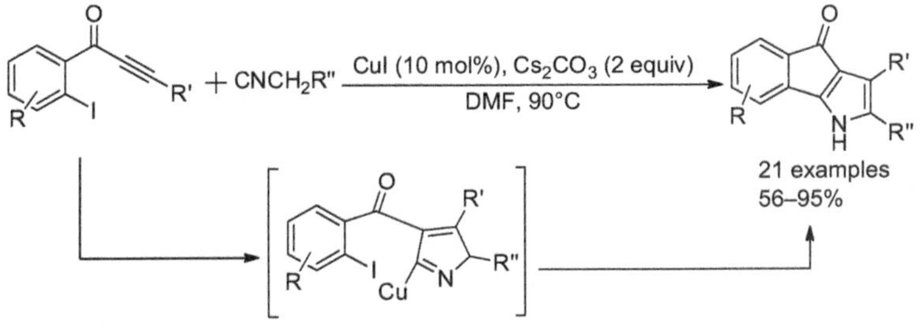

Scheme 4.106 Cu-catalyzed synthesis of 4-oxoindeno[1,2-*b*]pyrroles.

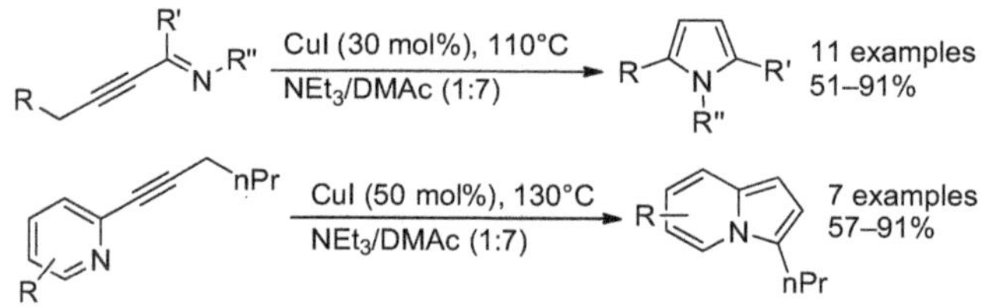

Scheme 4.107   Cu-catalyzed synthesis of pyrroles from alkynylimines.

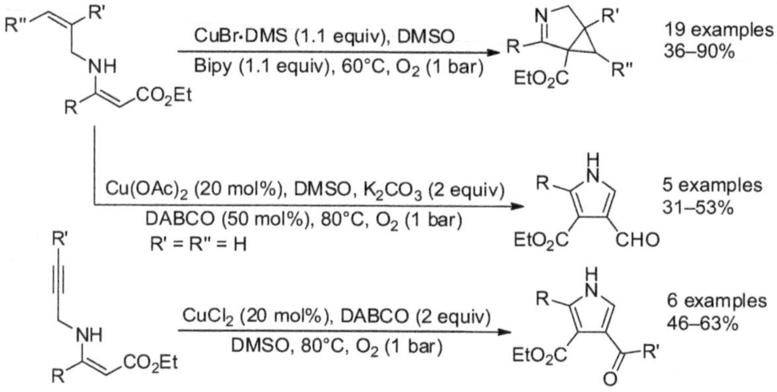

Scheme 4.108   Cu-mediated synthesis of 4-carbonylpyrroles.

A novel, general and efficient method, the Cu-assisted cycloisomerization of alkynylimines to the pyrrole ring, was developed in 2001 (Scheme 4.107).[192] The generality and synthetic usefulness of this methodology were demonstrated in the efficient synthesis of pyrroles and various types of fused heteroaromatic compounds and the expeditious synthesis of (±)-monomorine. Soon afterwards, this procedure was applied in the total synthesis of (±)-tetraponerine T6.[193]

In 2011, Chiba and co-workers developed synthetic methods for the preparation of 3-azabicyclo[3.1.0]hex-2-enes and 4-carbonylpyrroles (Scheme 4.108).[194] They used copper-mediated/catalyzed reactions of *N*-allyl/propargyl enamine carboxylates under an O_2 atmosphere that involved intramolecular cyclopropanation and carbooxygenation, respectively. These methodologies take advantage of orthogonal modes of chemical reactivity of readily available *N*-allyl/propargyl enamine carboxylates; complementary pathways can be accessed by slight modification of the reaction conditions.

A straightforward method for the synthesis of polysubstituted pyrroles was achieved easily *via* oxidative cyclization of β-enamino ketones or esters and alkynoates catalyzed by CuI in the presence of O_2.[195] Michael additional was proposed as the first step (Scheme 4.109).

The formal cycloaddition of α-metallated methyl isocyanides to the triple bond of electron-deficient acetylenes represents a direct and convenient approach to oligosubstituted pyrroles (Scheme 4.110).[196] The scope and

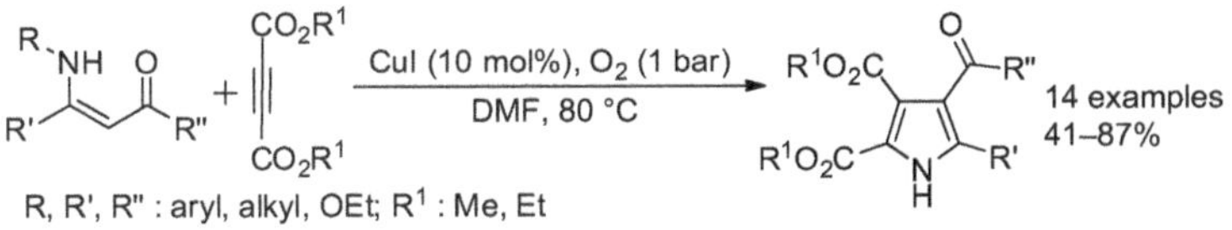

Scheme 4.109 Cu-catalyzed synthesis of pyrroles from β-enamino ketones.

General Procedure (Scheme 4.109): An oven-dried Schlenk tube was charged with CuI (5.7 mg, 0.03 mmol), (Z)-1,3-diphenyl-3-(phenylamino)prop-2-en-1-one (90 mg, 0.30 mmol) and dimethyl but-2-ynedioate (43 mg, 0.30 mmol). The tube was sealed and then evacuated and backfilled with oxygen (three cycles). Then 2 mL of DMF was added to the reaction system, which was stirred at 80 °C under O_2 (1 atm) for 4 h. After cooling to room temperature, the solution was diluted with 10 mL of ethyl acetate, washed with 5 mL of brine and dried over anhydrous Na_2SO_4. After the solvent had been evaporated *in vacuo*, the residues were purified by column chromatography with petroleum ether–EtOAc (5 : 1) as eluent to afford the product.

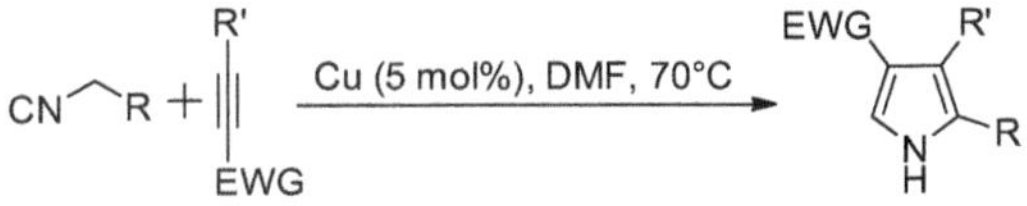

Scheme 4.110 Cu-catalyzed synthesis of pyrroles from α-metallated methyl isocyanides.

limitations of this reaction (24 examples, 25–97% yield) were reported along with a rationalization of the mechanism. In addition, a related newly developed CuI-mediated synthesis of 2,3-disubstituted pyrroles by the reaction of copper acetylides derived from unactivated terminal alkynes with substituted methyl isocyanides was described (11 examples, 5–88% yield). Interestingly, dppp can also catalyze this reaction and afforded the 2,3-di-EWG-substituted pyrroles.

An efficient copper-catalyzed one-step, three-component strategy for preparing a library of aminoindolizino[8,7-*b*]indoles from *N*-substituted 1-formyl-9*H*-β-carbolines, secondary amines and substituted alkynes with high atom economy was developed by Batra and co-workers in 2012 (Scheme 4.111).[197] Regarding the mechanism, it was assumed that initially morpholine reacts with the aldehyde, leading to the formation of an iminium ion with loss of a water molecule. Subsequently, a Cu-coordinated alkyne formed *in situ* reacts with the iminium ion, whereby nucleophilic attack of the pyridyl nitrogen on the Cu-coordinated allenyl double bond occurs, resulting in the formation of a cationic intermediate. The morpholine captures a proton from the cationic intermediate to furnish the final product after protonolysis.

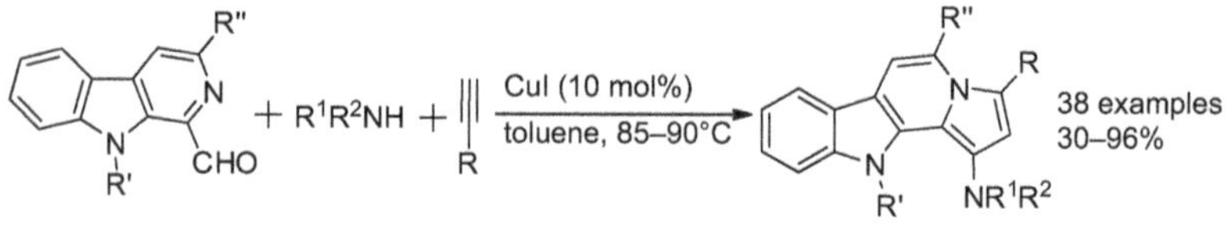

Scheme 4.111   Cu-catalyzed three-component synthesis of pyrroles.

> **General Procedure** (Scheme 4.111): To a reaction vessel containing 10 mL of toluene were added substrate (150 mg, 0.51 mmol), morpholine (67.0 μL, 0.76 mmol), phenylacetylene (61.0 μL, 0.56 mmol) and CuI (10 mg, 0.05 mmol), then nitrogen was bubbled for 10 min to deoxygenate the reaction mixture. Thereafter, the resulting solution was stirred at 85 °C for 7 h. On completion, the reaction mixture was cooled to room temperature and purified by silica gel column chromatography (eluent: 97 : 3 hexanes–EtOAc) to obtain pure product.

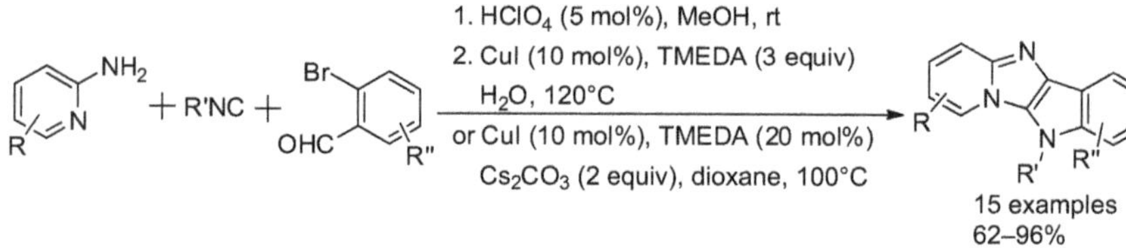

Scheme 4.112   Cu-catalyzed three-component synthesis of fused indoles.

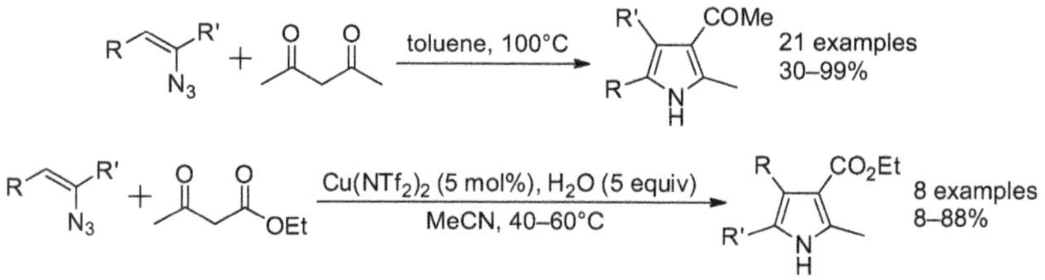

Scheme 4.113   Cu-catalyzed synthesis of pyrroles from vinyl azides.

A two-step sequence was reported in 2012 which opening access to diverse 5H-pyrido[2′,1′ : 2,3]imidazo[4,5-b]indoles and analogs.[198] The reaction involves a Bienaymé multicomponent reaction followed by N-arylation. The desired polycyclic scaffolds were obtained in high yield using copper iodide in association with TMEDA in aqueous or dioxane media (Scheme 4.112).

Two synthetic methods for tetra- and trisubstituted N-H-pyrroles were presented in 2008 (Scheme 4.113):[199] (i) thermal pyrrole formation by the reaction of vinyl azides with 1,3-dicarbonyl compounds *via* the 1,2-addition of 1,3-dicarbonyl compounds to a 2H-azirine intermediates generated *in situ* from vinyl azides; (ii) Cu(II)-catalyzed synthesis of pyrroles from α-ethoxycarbonyl

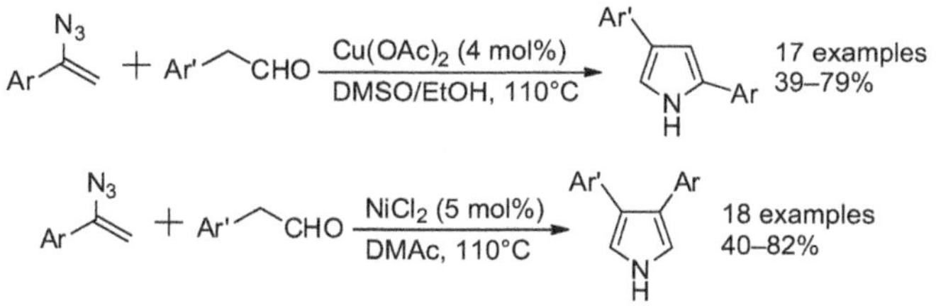

Scheme 4.114 Cu-catalyzed synthesis of pyrroles from vinyl azides and aldehydes.

vinyl azides and ethyl acetoacetate through 1,4-addition of the acetoacetate to the vinyl azides. By applying these two methods, regioisomeric pyrroles can be prepared selectively starting from the same vinyl azides.

In 2012, Jiao and co-workers reported a novel and efficient copper- or nickel-catalyzed highly selective denitrogenative annulation of vinyl azides with arylacetaldehydes;[200] 2,4- and 3,4-diaryl-substituted pyrroles can be highly regioselectively prepared by this protocol, simply switched by the selection of the transition metal catalyst. Compared with the reported acidic or basic conditions for polysubstituted pyrrole synthesis, the present reaction conditions are mild, neutral and very simple, without any additives (Scheme 4.114).

4.2.11 Copper-Catalyzed Synthesis of Indole Derivatives

As early as 1966, the reaction of copper(I) acetylides with *o*-halophenols, *o*-haloanilines and *o*-halocarboxylic acids was reported.[201] 2-Substituted indoles and benzofurans and 3-alkylidenephthalides were produced in moderate to good yields. However, the main drawback is the use of pyridine as solvent at 120 °C in most cases. Cacchi and co-worker developed a general method for the synthesis of 2-aryl- and 2-heteroarylindoles from aryl iodides and 1-alkynes through a domino copper-catalyzed process (Scheme 4.115).[202] The best results were obtained with [Cu(phen)(PPh$_3$)$_2$]NO$_3$ in the presence of K$_3$PO$_4$ in toluene or dioxane at 110 °C. 2-Aryl- and 2-heteroarylindoles can also be isolated in good yields by using catalysts derived from CuI and PPh$_3$ in dioxane at 110 °C. If an *N*-acetamido-protected substrate was applied as starting material, alkylidenebenzoxazine formation as a competing *O*-cyclization process occurred. Later, they reported another procedure based on copper-catalyzed cyclization of *N*-(2-iodoaryl)enaminones for the synthesis 3-aroylindoles.[203] They applied readily available 2-iodoanilines and α,β-ynones as substrates. The reaction tolerates a variety of useful functionalities including ether, keto, cyano, bromo and chloro substituents. This indole synthesis can also be carried out from 2-iodoanilines and α,β-ynones through a sequential process that omits the isolation of enaminone intermediates. However, when the *N*-(2-bromoaryl)enaminone was subjected to their standard conditions, the formation of the expected indole product was accompanied by the formation of the benzoxazepine derivative.

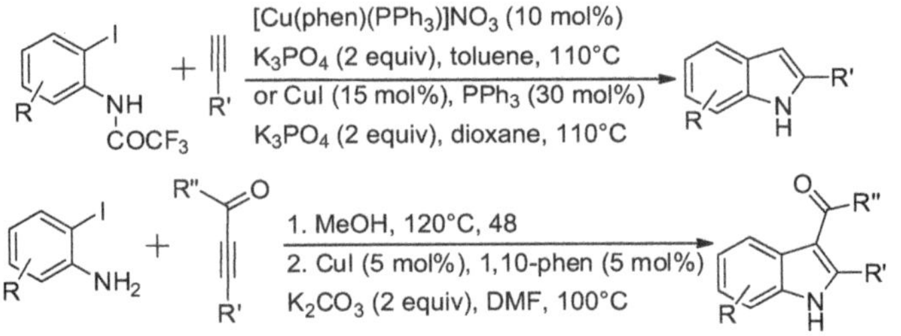

Scheme 4.115 Cu-catalyzed synthesis of indoles from 2-iodoanilines.

General Procedure (Scheme 4.115): In a Carusel Tube Reactor (Radely Discovery Technology), to a solution of *o*-iodotrifluoroacetanilide (0.157 g, 0.50 mmol) and 1-chloro-4-ethynylbenzene (0.068 g, 0.50 mmol) in 2.0 mL of toluene, [Cu(phen)(PPh$_3$)]NO$_3$ (0.050 mmol, 0.041 g) and K$_3$PO$_4$ (0.212 g, 1.00 mmol) were added. The mixture was stirred for 6 h at 110 °C under an argon atmosphere. After cooling, the reaction mixture was diluted with ethyl acetate, washed with water, dried over Na$_2$SO$_4$ and concentrated under reduced pressure. The residue was purified by chromatography (silica gel, 35 g; 85 : 15 *n*-hexane–EtOAc as eluent) to give 0.0900 g of the 2-(*p*-chlorophenyl)indole product (80 % yield).

General Procedure: To a stirred solution of 2-iodoaniline (109.5 mg, 0.5 mmol) in MeOH (1.0 mL), 1,3-diphenylprop-2-yn-1-one (154.5 mg, 0.75 mmol) was added at r.t. The reaction mixture was warmed to 120 °C and stirred for 48 h, then the volatile materials were evaporated at reduced pressure and CuI (4.8 mg, 0.025 mmol), 1,10-phenanthroline (4.5 mg, 0.025 mmol), K$_2$CO$_3$ (138.0 mg, 1.0 mmol) and DMF (4 mL) were added. The reaction mixture was warmed to 100 °C and stirred for 2.5 h. After cooling, the reaction mixture was diluted with Et$_2$O, washed with 1 M HCl and brine, dried over Na$_2$SO$_4$ and concentrated under reduced pressure. The residue was purified by chromatography on silica gel (eluent: 75 : 25 *n*-hexane–EtOAc) to afford the product.

In 2007, Liu and Ma reported a copper-catalyzed coupling of terminal alkynes with 2-bromotrifluoroacetanilides.[204] By using L-proline as ligand, indoles were produced in good yields (Scheme 4.116). Cross-coupling of 1-alkynes with vinyl iodides using CuI–*N*,*N*-dimethylglycine as the catalytic system at 80 °C in dioxane to afford conjugated enynes was also described. This conversion involves a CuI–L-proline-catalyzed coupling between aryl bromide and the 1-alkyne followed by a CuI-mediated cyclization process. An *ortho*-substituent effect directed by NHCOCF$_3$ enables the reaction to proceed under these mild conditions. Both arylacetylenes and *O*-protected propargyl alcohol can be applied, leading to 5-, 6- or 7-substituted 2-aryl- and protected 2-hydroxymethylindoles in good yields. With simple aliphatic alkynes, however, lower yields were observed.

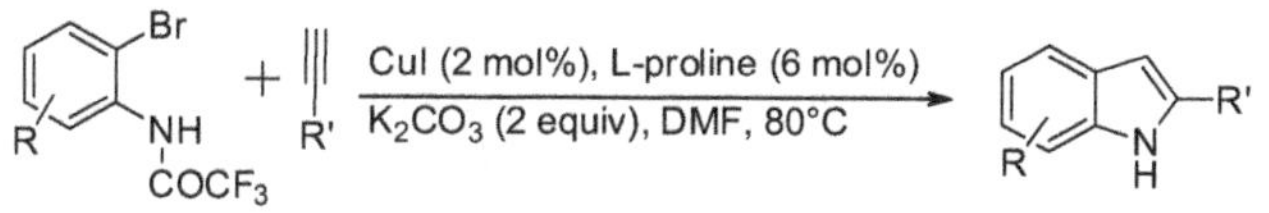

Scheme 4.116 Cu-catalyzed synthesis of indoles from 2-bromoanilines.

> **General Procedure** (Scheme 4.116): A mixture of bromide (1.1 mmol), phenylacetylene (1.0 mmol), CuI, additive and K_2CO_3 (2.0 mmol) in 2 mL of DMF was heated in a sealed tube at 80 °C under argon. When the reaction was completed (monitored by TLC), the cooled mixture was partitioned between ethyl acetate and water. The organic layer was separated and the aqueous layer was extracted twice with ethyl acetate. The combined organic layers were washed with brine, dried over Na_2SO_4 and concentrated *in vacuo*. The residual oil was loaded on a silica gel column and eluted with ethyl acetate–petroleum ether (1:10–1:8) to afford the pure product.

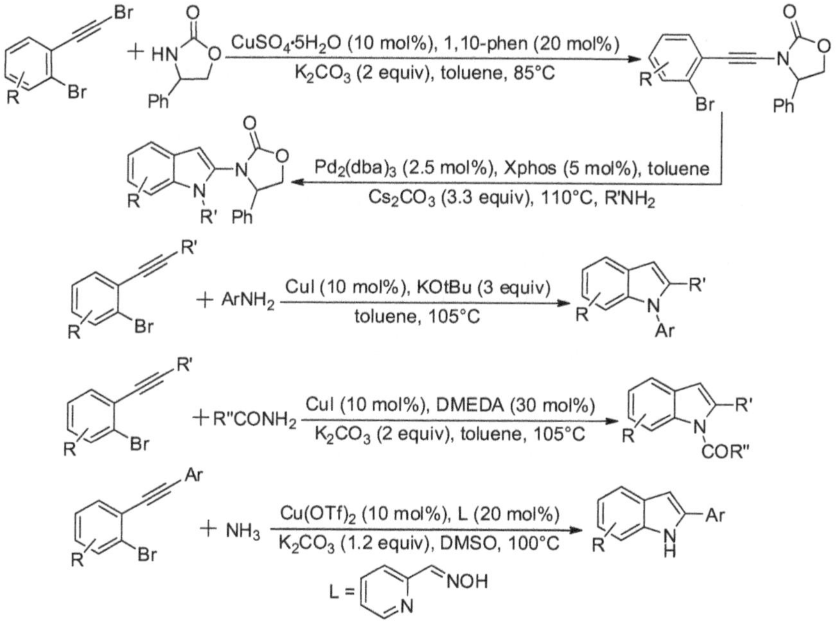

Scheme 4.117 Cu-catalyzed synthesis of indoles from *o*-alkynylhaloarenes.

A sequential metal-catalyzed C–N bond formation employing *o*-haloaryl acetylenic bromides was described in 2008.[205] The initial amidation is highly selective for C(sp)–N bond formation, leading to *o*-haloaryl-substituted ynamides that can be useful building blocks, while the overall sequence provides a facile construction of 2-amidoindoles (Scheme 4.117). Ackermann reported a copper- or palladium-catalyzed amination of *o*-alkynylhaloarenes to indoles in 2005.[206] Various indoles were produced in good yields (Scheme 4.117). Furthermore, a multicatalytic one-pot indole synthesis starting from *o*-chloroiodobenzene was also studied using a single catalyst consisting of an *N*-heterocyclic carbene–palladium complex and CuI. In 2009, they successfully extended their procedure to primary amides.[207] *N*-Aryl-, *N*-acyl- and *N-H*-(aza)indoles were produced in moderate to good yields. In the system, DMEDA showed better activity than the other ligands, such as L-proline, 1,10-phenanthroline, phosphine and NHCs. More recently, Li and co-workers realized this transformation with aqueous ammonia.[208] a wide range of electron-rich and electron-deficient *N-H* free 2-arylindoles were produced in excellent yields from the corresponding 2-bromoarylacetylenes *via* a sequential amination and cyclization.

The above-mentioned procedures were all based on the generation of 2-ethynylanilines *in situ* followed by copper-catalyzed cyclization. A more straightforward approach would be to start directly from 2-ethynylanilines. In 2002, Hiroya and co-workers reported a copper-catalyzed construction of indoles from 2-ethynylanilines and later they found that this conversion can be carried out in an aqueous medium at room temperature (Scheme 4.118).[209] In 2003, a microwave-assisted solid-phase organic synthesis version was adopted and resin-bonded 1-acyl-2-alkyl-5-arenesulfamoylindoles were produced in good yields.[210]

In 2007, Fujii, Ohno and co-workers developed a novel synthesis of 2-(aminomethyl)indoles through a copper-catalyzed domino three-component coupling-cyclization (Scheme 4.119).[211] This domino reaction forming two C–N bonds and one C–C bond was the first catalytic multicomponent indole construction producing water as the only theoretical waste. The use of the chiral ligand PINAP in the reaction with alkylaldehydes produced the corresponding indole bearing a branched substituent with moderate *ee* values. This reaction is synthetically useful for the diversity-oriented synthesis of not only 2-(aminomethyl)indoles but also tetrahydropyridine- and benzazepine-fused indoles, using readily available reaction components. The benzo[*e*][1,2]thiazine and indene motif can also be constructed by using a similar domino three-component coupling and cyclization strategy.

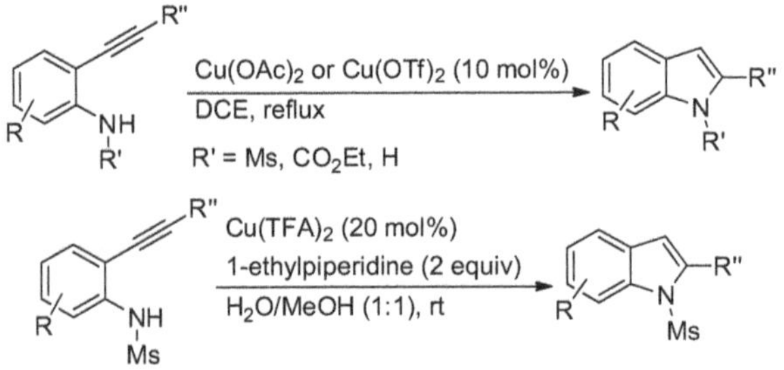

Scheme 4.118 Cu-catalyzed synthesis of indoles from 2-ethynylanilines.

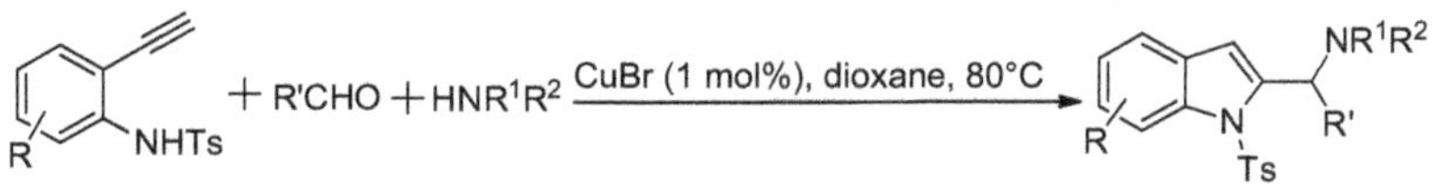

Scheme 4.119 Cu-catalyzed three-component synthesis of indoles.

General Procedure (Scheme 4.119): To a stirred mixture of 2-ethynylaniline (50.0 mg, 0.18 mmol), $(HCHO)_n$ (11.1 mg, 0.37 mmol) and CuBr (0.3 mg, 0.0018 mmol) in dioxane (3.0 mL) was added diisopropylamine (28.6 μL, 0.20 mmol) at room temperature under argon and the reaction mixture was stirred at 80 °C for 15 min. Concentration under reduced pressure followed by column chromatographic purification over silica gel with hexane–EtOAc (10:1) as eluent afforded the indole.

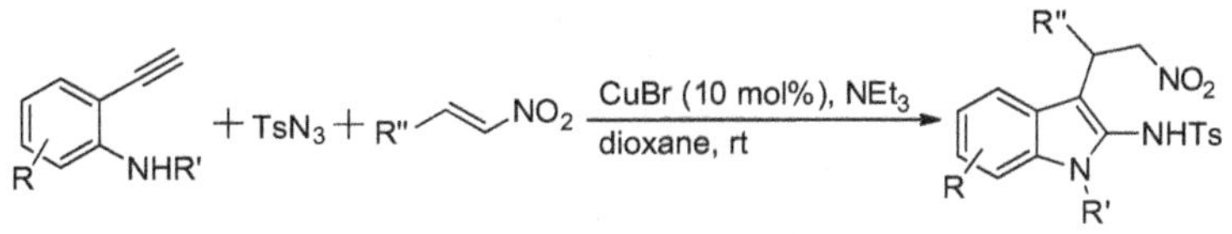

Scheme 4.120 Cu-catalyzed three-component synthesis of indoles from 2-ethynylanilines.

Wu and co-workers developed an efficient copper(I)-catalyzed three-component reaction of 2-ethynylanilines, sulfonyl azide and nitroalkenes in 2011.[212] A variety of polysubstituted indole derivatives were obtained in moderate to good yields (Scheme 4.120). In this transformation, a broad substrate scope was demonstrated. Additionally, the reaction conditions are extremely mild. The biological activity of these products was also tested. Some hits as an HCT-116 inhibitor were found. Regarding the reaction mechanism, the reactive ketenimine would be generated by the ring-opening rearrangement of a triazole intermediate, which was formed from an alkyne and sulfonyl azide in the presence of CuBr and triethylamine. Subsequently intramolecular nucleophilic addition occurs, leading to the intermediate, which then undergoes intermolecular Michael addition and tautomerization, forming the desired products.

A convenient and efficient method for the synthesis of 3-haloindoles was developed by Shen and Lu in 2009.[213] Both 3-chloro- and 3-bromoindole derivatives can be obtained in high yields by the reaction of *N*-electron-withdrawing group-substituted 2-alkynylanilines with copper(II) halide in dimethyl sulfoxide (DMSO) within a short period (Scheme 4.121). Investigation of the reaction mechanism revealed that at least 2 equiv. of copper(II) halide are necessary.

A new and efficient catalytic approach to the synthesis of 3-acylindoles under Pd–Cu co-catalyzed oxidative conditions was demonstrated by Liang and co-workers in 2013.[214] *tert*-Butyl hydroperoxide (TBHP) acts not only as

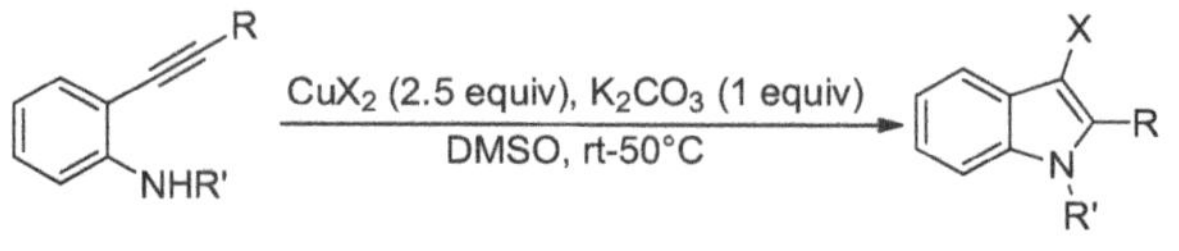

Scheme 4.121   Cu-catalyzed synthesis of 3-haloindoles.

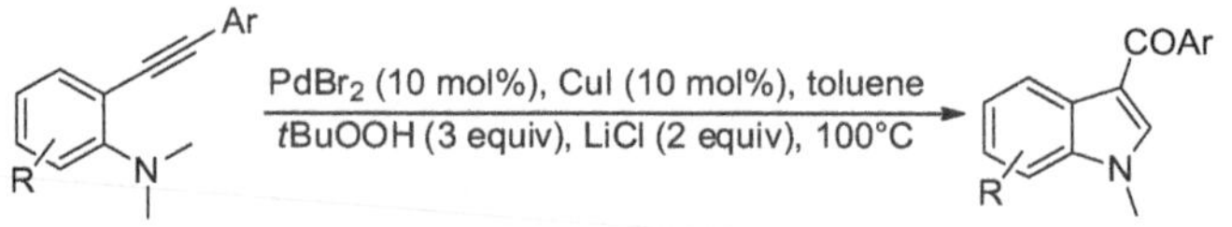

Scheme 4.122   Cu-catalyzed synthesis of 3-acylindoles.

General Procedure (Scheme 4.122): To a solution of *N,N*-dimethyl-2-(phenylethynyl)aniline (44.2 mg, 0.200 mmol) in toluene (2 mL) was added PdBr$_2$ (5.3 mg, 10 mol%), CuI (3.8 mg, 10 mol%), LiCl (17 mg, 2.0 equiv.) and 120 μL of TBHP (5 M in decane). The reaction mixture was stirred in air at 100 °C for 8 h and the resulting mixture was quenched with water and extracted twice with EtOAc. The combined organic extracts were washed with brine, dried over MgSO$_4$ and concentrated. Purification of the crude product by flash column chromatography afforded the pure product.

the oxidant, but also as an oxygen source in this approach. The process allows the rapid and atom-economical assembly of 3-acylindoles from readily available starting materials and tolerates a broad range of functional groups. The mechanism was discussed based on a labeling experiment and ESI-MS analysis. This method provided an alternative approach to the current CDC reaction *via* an intramolecular oxidative coupling and allowed the rapid construction of diverse indole derivatives in one step from readily available starting materials and in a direct, efficient and atom-economical process (Scheme 4.122).

A new strategy for accessing *N*-indolyl- and *N*-benzofuranylindoles and -benzimidazoles through a series of tandem reactions was reported in 2012.[215] The potential of this method is shown by its adaptability to a wide variety of reagents, mild reaction conditions and simple manipulation. Starting from 2-alkynylcyclohexadienimines or cyclohexadienones and 2-alkynylanilines or *N*^{1}-benzylbenzene-1,2-diamine, the desired products were produced in good yields (Scheme 4.123).

Yamamoto's group developed a novel procedure for the synthesis of 3-allyl-*N*-(alkoxycarbonyl)indoles.[216] This methodology proceeds *via* the reaction of 2-alkynylphenyl isocyanates and allyl carbonates in the presence of Pd(PPh$_3$)$_4$ (1 mol%) and CuCl (4 mol%) bimetallic catalyst. Good yields of the products were obtained (Scheme 4.124). Concerning the reaction

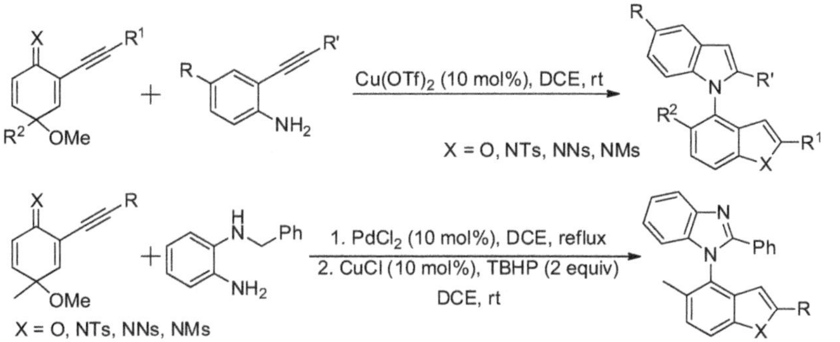

Scheme 4.123 Cu-catalyzed synthesis of *N*-indolyl- and *N*-benzofuranylindoles.

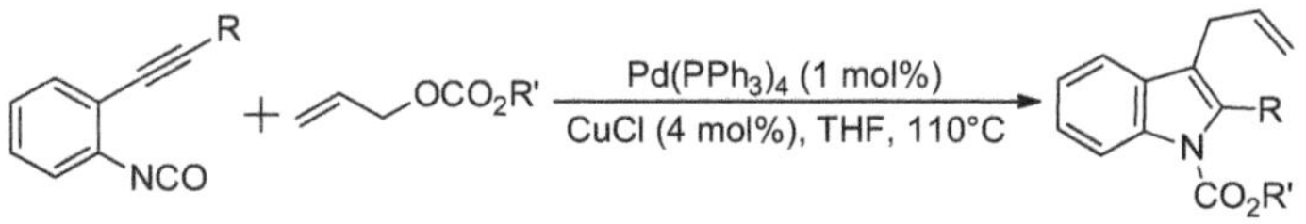

Scheme 4.124 Cu–Pd-catalyzed synthesis of 3-allylindoles.

General Procedure (Scheme 4.124): To a THF solution (0.5 mL) of 2-(1-pentynyl)phenyl-1-isocyanate (92.7 mg, 0.5 mmol), Pd(PPh$_3$)$_4$ (5.8 mg, 0.005 mmol) and CuCl (2.0 mg, 0.02 mmol) was added allyl methyl carbonate (69 µL, 0.6 mmol) under an argon atmosphere. The solution was stirred at 100 °C for 1 h in a tightly capped 5 mL microvial. The reaction mixture was cooled to room temperature, filtered through a short Florisil pad and concentrated. The residue was purified by column chromatography on silica gel (eluent: 100 : 1–20 : 1 hexane–AcOEt), yielding the pure product.

mechanism, the authors proposed a likely pathway. It is most probable that Pd(0) acts as a catalyst for the formation of a π-allylpalladium alkoxide intermediate and CuI behaves as a Lewis acid to activate the isocyanate, and the cyclization step proceeds with the cooperative catalytic activity of Pd and Cu. On the other hand, *N*-(alkoxycarbonyl)indoles are produced *via* the reaction of 2-alkynylphenyl isocyanates and alcohols under a catalytic amount of Na$_2$PdCl$_4$ (5 mol%) or PtCl$_2$ (5 mol%). Pd(II) and Pt(II) catalysts exhibit dual roles: they act as a Lewis acid to accelerate the addition of alcohols to isocyanates and as a typical transition metal catalyst to activate the alkyne for the subsequent cyclization.

May and co-workers reported a procedure for the synthesis of *N*-substituted indoles from primary amines through a tandem reaction sequence.[217] Initial condensation of the amine with an α-(*o*-haloaryl) ketone or aldehyde was followed by intramolecular aryl amination using CuI as the catalyst. A variety of anilines and alkylamines, including those with significant steric

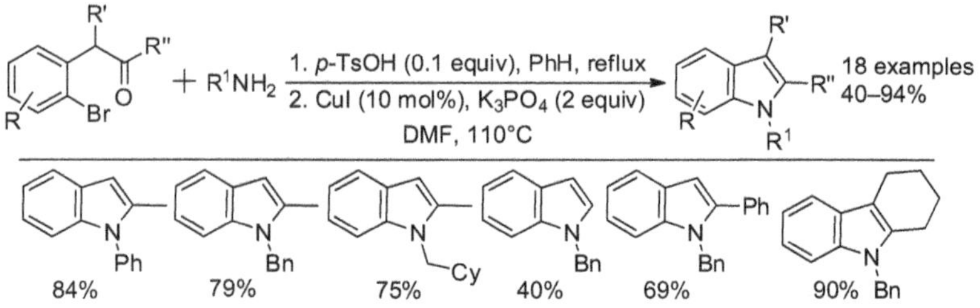

Scheme 4.125   Cu-catalyzed synthesis of indoles from α-(*o*-haloaryl) ketones.

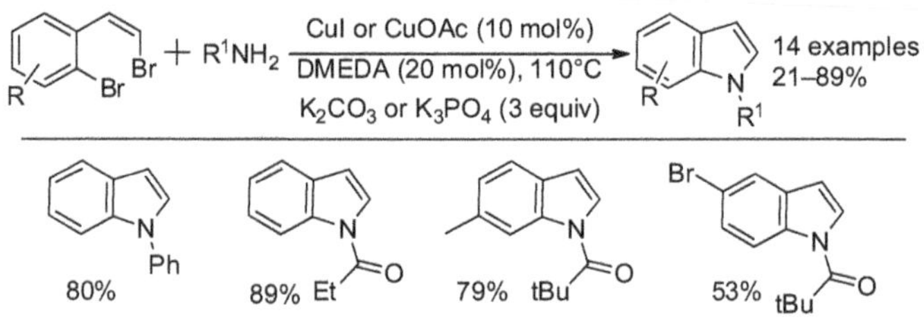

Scheme 4.126   Cu-catalyzed synthesis of indoles from 2-(2-haloalkenyl)aryl halides.

demands, were converted to indoles in high yields and with varying indole substitution (Scheme 4.125).

Willis and co-workers demonstrated tandem Cu-catalyzed amination reactions for the preparation of *N*-functionalized indoles.[218] Employing 2-(2-haloalkenyl)aryl halides and amines as substrates, the products were obtained in a straightforward manner in good yields (Scheme 4.126). This pathway had been reported previously using a palladium catalyst, and this was compared with the present system. The range of *N*-coupling partners that can be used complements that achievable using Pd catalysis, the major advantage when employing the Cu system being the successful preparation of *N*-acylindoles. Conversely, couplings employing simple amines were less efficient when using the Cu chemistry. Ultimately, the choice of Pd or Cu catalysis will be determined on a case-by-case basis, balancing economic considerations regarding the catalyst and other reaction components against the reactivity and efficiency possible with the two systems.

A method for the preparation of *N*-substituted 1*H*-indole-3-carboxylates by CuI-mediated intramolecular *N*-arylation was reported in 2008.[219] 1-Aminoindole-3-carboxylic acid derivatives can be synthesized, starting from suitable substituted hydrazines and methyl 2-(2-bromophenyl)-3-formylacetates, in two steps. Only *N,N,N'*-trisubstituted enehydrazines can be employed as cyclization precursors for 1-aminoindoles. However, use of Boc-protected substrates in the cyclization step and subsequent deprotection led to *N*-monosubstituted and *N*-unsubstituted 1-aminoindole-3-carboxylates.

Tanimori's group reported the copper-catalyzed one-step synthesis of 2,3-disubstituted indoles.[220] Readily available starting materials, 2-iodoaniline

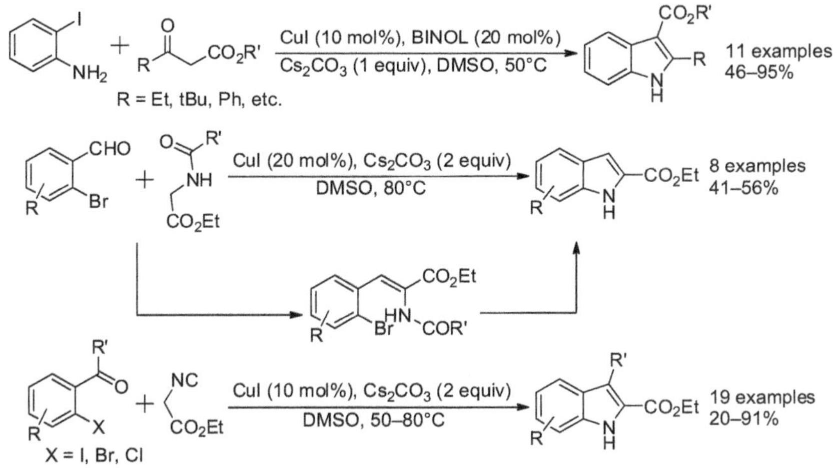

Scheme 4.127 Cu-catalyzed synthesis of 2,3-disubstituted indoles.

and various β-keto esters, were applied as substrates. The advantage of this method is the use of cheap catalysts and simple experimental procedures under mild reaction conditions. Substituted indole derivatives were synthesized in good yields by this procedure (Scheme 4.127). In 2010, Koenig *et al.* reported a ligand-free, copper-catalyzed cascade sequence to give indole-2-carboxylic esters.[221] A variety of indole-2-carboxylic esters are accessible in yields of up to 61% through a copper-catalyzed reaction of a series of commercially available 2-haloarylaldehydes with benign glycine amido esters, including the common reagent ethyl acetamidoacetate. This one-pot, three-reaction format allows ready entry to the desired heterocycles from starting substrates in the reactivity order iodo > bromo > chloro substituents. An assortment of functional groups are tolerated, adding to the generality of this methodology. Methodology started from the intermediates was also developed.[222] Another efficient ligand-free copper-catalyzed procedure for the synthesis of indole-2-carboxylic acid esters was reported in 2008.[223] The reaction followed a condensation–coupling–deformylation cascade process using 2-haloarylaldehydes or ketones with ethyl isocyanoacetates. The reactions performed well at room temperature or 50 °C for iodo- and bromo-substituted substrates. Chloride-substituted substrates also successfully yielded the desired indole-2-carboxylic acid esters with acceptable yields when the temperature was elevated to 80 °C. Furthermore, the reaction displayed excellent functional group compatibility and high chemical selectivity in the presence of a broad range of functional groups.

Fu and co-workers developed a simple and efficient method for the synthesis of 2-amino-1*H*-indole-3-carboxylate derivatives.[224] These cascade reactions of substituted *N*-(2-halophenyl)-2,2,2-trifluoroacetamides with an alkyl 2-cyanoacetate or malononitrile performed well under mild conditions and the corresponding indole derivatives containing amino and carboxylate groups were obtained in good yields (Scheme 4.128). The present method

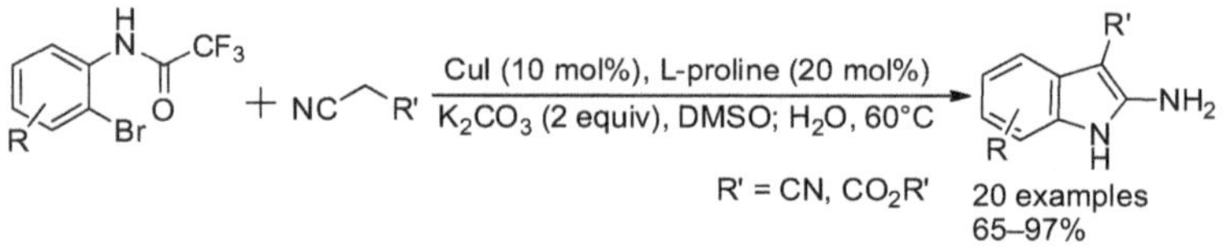

Scheme 4.128   Cu-catalyzed synthesis of 2-amino-1*H*-indole-3-carboxylates.

> **General Procedure** (Scheme 4.128): A 10 mL round-bottomed flask was charged with a magnetic stirrer bar, DMSO (0.5 mL) and water (0.5 mL) [water was not added until the Ullmann-type coupling was completed (monitored by TLC) when methyl 2-cyanoacetate was used as the substrate], substituted *N*-(2-halophenyl)-2,2,2-trifluoroacetamide (0.5 mmol), alkyl 2-cyanoacetates or malononitrile (0.6 mmol), L-proline (0.1 mmol, 12 mg) and K$_2$CO$_3$ (1 mmol, 138 mg). After stirring the mixture for 15 min under a nitrogen atmosphere, CuI (0.05 mmol, 10 mg) was added to the flask and the mixture was stirred at 60 °C for 9 or 12 h under a nitrogen atmosphere. The resulting mixture was cooled to room temperature and filtered. The solid was washed with methanol (2×3 mL), the combined filtrate was concentrated on a rotary evaporator and the residue was purified by column chromatography on silica gel using petroleum ether–ethyl acetate (1 : 1–1 : 2) as eluent to give the desired product.

shows practical advantages over the previous methods and its starting materials are readily available, so it will provide an opportunity for the construction of diverse and useful molecules in organic and medicinal chemistry. Interestingly, when the –NHCOCF$_3$ group in the *N*-(2-halophenyl)-2,2,2-trifluoroacetamide was replaced with –NHCOMe, only a trace amount of the target product was observed.

   Ma and co-workers developed an efficient and high-yielding cascade process for the assembly of 2,3-substituted indoles in 2007 (Scheme 4.129).[225] The system is based on the CuI–L-proline-catalyzed coupling of 2-halotrifluoroacetanilides with β-keto esters and amides followed by *in situ* acidic hydrolysis. It is noteworthy that this procedure can also be carried out by *in situ* basic hydrolysis of the trifluoroacetanilide moiety. However, this procedure requires a strong electron-withdrawing group in the 4-position of the 2-halotrifluoroacetanilides. Various functional groups survived the reaction conditions, including ketones, esters and nitro, iodo, olefinic, chloro and benzoxy moieties. It was found that under anhydrous conditions, coupling of 4-nitro-2-iodotrifluoroacetanilide with methyl acetoacetate produced 2-(trifluoromethyl)-5-nitroindole in 78% yield. This discovery stimulated the authors to develop another methodology for the preparation of 2-(trifluoromethyl)indoles.[226] By varying the substituents of 2-halotrifluoroacetanilides and β-keto esters, a wide range of substituted 2-(trifluoromethyl)indoles were prepared, including two indoles that are

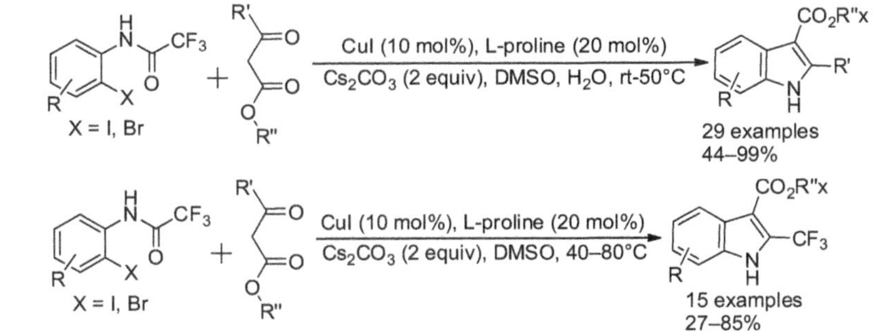

Scheme 4.129 Cu-catalyzed synthesis of indoles from β-keto esters.

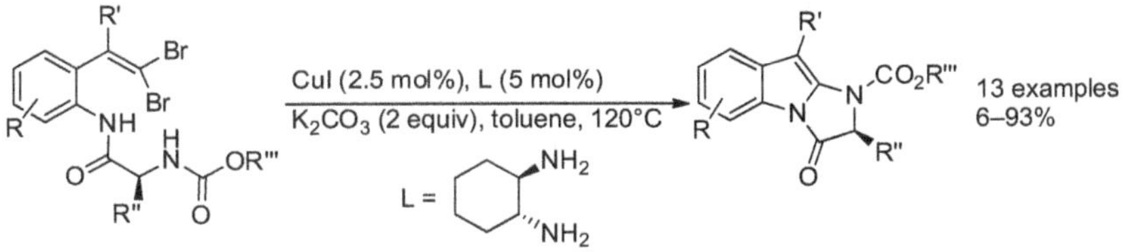

Scheme 4.130 Cu-catalyzed synthesis of imidazoindolones.

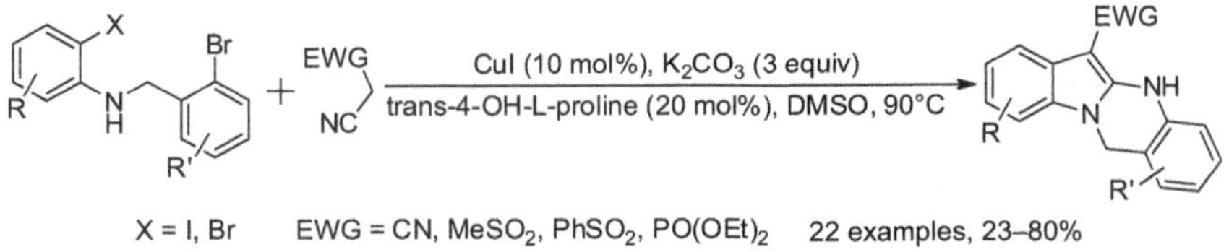

X = I, Br   EWG = CN, MeSO₂, PhSO₂, PO(OEt)₂   22 examples, 23–80%

Scheme 4.131 Cu-catalyzed synthesis of 5,12-dihydroindolo[2,1-*b*]quinazolines.

useful precursors for known biologically important compounds. This reaction occurs *via* a novel mechanism and many functional groups, such as nitro, ketone, ester, hydroxy, iodo, chloro and olefinic, are tolerated under these conditions.

Lautens and co-workers developed an efficient one-pot procedure giving rapid access to a wide range of substituted imidazoindolones from readily prepared *gem*-dibromovinyl substrates *via* a novel tandem intramolecular C–N bond formation (Scheme 4.130).[227] This process is based on copper iodide as the catalyst, which is a cheap, air-stable compound.

A domino synthesis of 5,12-dihydroindolo[2,1-*b*]quinazoline derivatives *via* copper-catalyzed Ullmann-type intermolecular C–C and intramolecular C–N couplings was reported in 2012.[228] Good yields of various 5,12-dihydroindolo[2,1-*b*]quinazoline derivatives were obtained (Scheme 4.131). Reaction scopes and limitations and the reaction mechanism were discussed.

An interesting method was devised for the synthesis of 2,4,5-trisubstituted pyrrole derivatives by Reddy *et al.* in 2013.[229] The reaction proceeds through

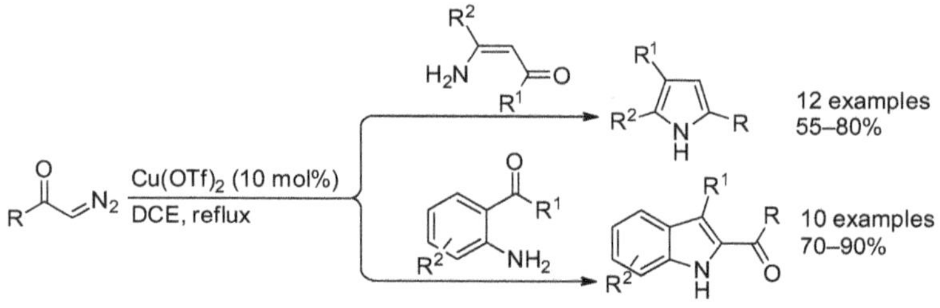

Scheme 4.132 Cu-catalyzed synthesis of indoles and pyrroles.

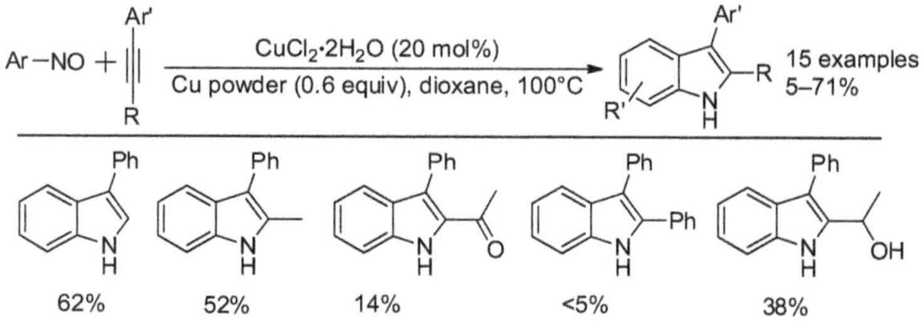

Scheme 4.133 Cu-catalyzed synthesis of 3-arylindoles.

> **General Procedure** (Scheme 4.133): A Schlenk flask was charged with
> $CuCl_2 \cdot 2H_2O$ (0.2 equiv.), Cu powder (0.6 equiv.), dioxane (5 mL) and
> phenylacetylene (5 equiv.). The flask was placed in a preheated oil-bath at
> 100 °C and then a solution of nitrosobenzene (1 mmol) in dioxane (5 mL)
> was added slowly with the help of a syringe pump over a period of 4 h
> under a positive pressure of nitrogen. The reaction mixture was cooled
> and filtered through Celite using diethyl ether. The solvent was removed
> under vacuum and further purification of the crude product was achieved
> by column chromatography (eluent: hexane–ethyl acetate).

the coupling of α-diazo ketones with β-enamino ketones and esters using
10 mol% of $Cu(OTf)_2$. A wide range of 2,3-disubstituted indole derivatives
were also prepared from α-diazo ketones and 2-aminoaryl or alkyl ketones
(Scheme 4.132). The synthetic versatility of this approach was exemplified in
the formal synthesis of homofascaplysin C. Regarding the reaction mechanism, $Cu(OTf)_2$ act as a Lewis acid here. First, the Lewis acid activates the
carbonyl group of the diazo compound, which acts as an electrophile and
reacts with the amino group of the enamine, leading to the pyrrole.

A direct method for the preparation of various 3-arylindoles from their
corresponding nitrosoarenes and alkynes was developed in 2011.[230] Various
substituted nitrosoarenes and alkynes were employed to obtain substituted
indoles by using a Cu(II)–Cu(0) catalytic system (Scheme 4.133). The proposed mechanism involves the formation of an *N*-hydroxyindole in this

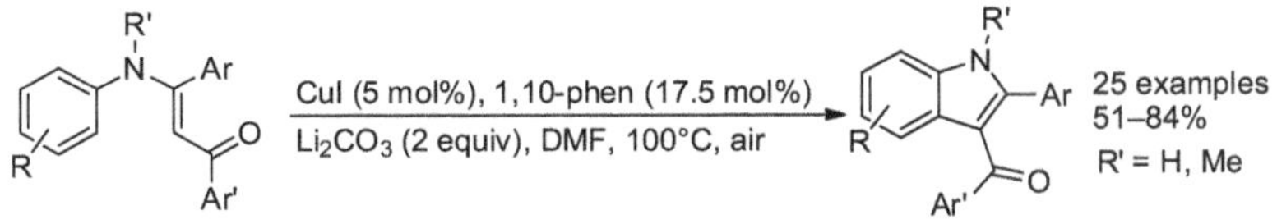

Scheme 4.134 Cu-catalyzed oxidative synthesis of indoles.

> **General Procedure** (Scheme 4.134): An oven-dried Schlenk tube was charged with α,β-ynones (1 mmol), anilines (1 mmol) and anhydrous MeOH (1 mL). The tube was sealed and stirred at 80 °C for the indicated period. The reaction mixture was cooled to room temperature, the solvent was evaporated and the residue was purified by silica gel chromatography with *n*-hexane–ethyl acetate mixtures as eluent.

catalytic process, which is deoxygenated further to give a 3-arylindole, and this was confirmed by a controlled experiment starting from *N*-hydroxyindole. The Cu(I) species generated *in situ* from Cu(II) and Cu(0) is responsible for deoxygenation of *N*-hydroxyindole, and in turn is converted into Cu(II) and the cyclic process continues. This is a two-step method that involves annulation of the nitrosoarene and alkyne followed by deoxygenation to give 3-arylindoles.

Cacchi and co-workers developed an efficient copper-catalyzed approach to the construction of ££ a multisubstituted indole skeleton from readily available *N*-arylenaminones (Scheme 4.134).[231] This novel method tolerates a variety of useful functionalities, including the whole range of halogen substituents. With *N*-(2-bromophenyl)enaminone, remarkable selectivity was observed that favors C–H functionalization in comparison with C–Br functionalization and affords 7-bromoindoles, key intermediates in the synthesis of biologically active compounds. Indole products can also be prepared from α,β-ynones and primary amines by a sequential process that omits the isolation of the enaminone intermediates. Since multisubstituted indoles are formed by assembling 2-haloaroyl chlorides, terminal alkynes and primary amines, a wide variety of indole derivatives can be synthesized by using this protocol.

A copper-catalyzed dimerization of 2-arylindoles was developed in 2012 by Zou, Zhang and co-workers.[232] This conversion provides a novel route for accessing fused nitrogen-containing heterocycles (Scheme 4.135). The use of an inexpensive copper catalyst and O_2 or air as the oxidant is a practical advantage. The incorporation of an oxygen atom into the organic frameworks from atmospheric molecular oxygen offers the most ideal oxidation process. In this methodology, Cu(I) salts, such as CuBr, CuI and CuCl, gave low yields of the products, whereas Cu(II) gave better yields in general.

A copper-catalyzed cascade reaction for alkyl 2*H*-isoindole-1-carboxylates synthesis was described in 2011 by Batra and co-workers.[233] The reaction

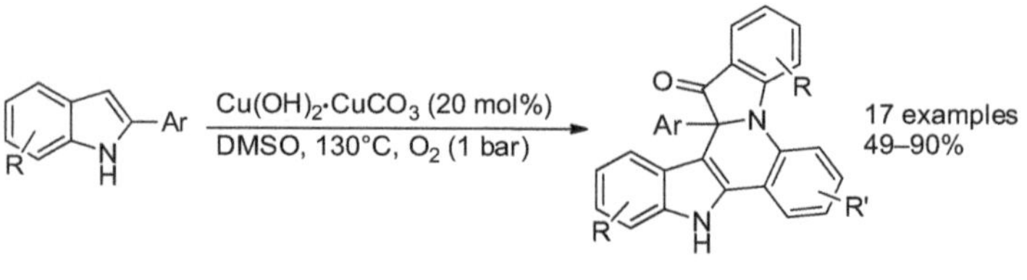

Scheme 4.135 Cu-catalyzed oxidative dimerization of indoles.

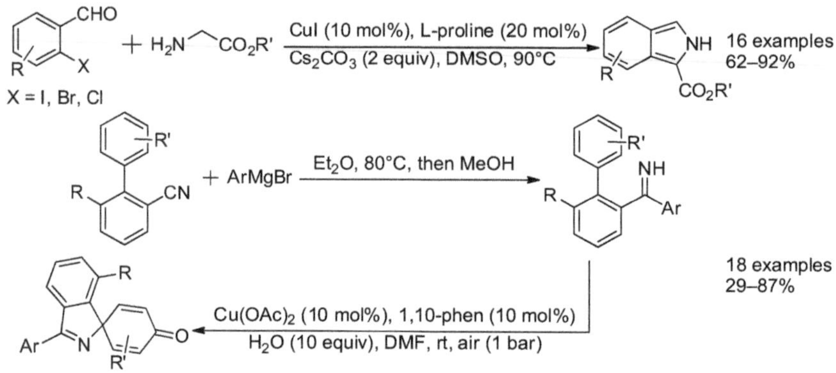

Scheme 4.136 Cu-catalyzed synthesis of isoindole derivatives.

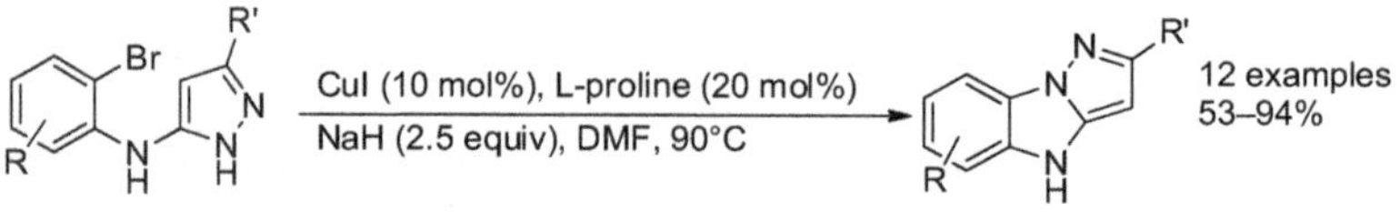

Scheme 4.137 Cu-catalyzed synthesis of indoles and benzimidazoles.

proceeds *via* condensation/α-arylation between 2-halobenzaldehydes and α-amino acid esters. In 2012, Chiba's group developed a method for the Cu-catalyzed aerobic spirocyclization of biaryl *N-H*-imines which could be prepared concisely from readily available biaryl-2-carbonitriles and Grignard reagents.[234] Molecular oxygen is a prerequisite for achieving the catalytic spirocyclization, where one of the oxygen atoms of O_2 is regioselectively incorporated into the benzene ring with dearomatization through 1,4-aminooxygenation. All the products were obtained in good yields (Scheme 4.136).

Ila and co-workers reported two efficient routes to pyrazolo[3,4-*b*]indoles and pyrazolo[1,5-*a*]benzimidazoles based on palladium and copper catalysts.[235] Starting from 5-(2-bromoanilino)pyrazoles, indole derivatives can be prepared by using a palladium catalyst when the starting material is 1-aryl substituted. On the other hand, benzimidazoles were produced by copper-catalyzed cyclization of 1-unsubstituted 5-(2-bromoanilino)pyrazole precursors *via* intramolecular C–N bond formation (Scheme 4.137).

4.2.12 Copper-Catalyzed Synthesis of Pyrrolidine Derivatives

The intramolecular hydroamination of alkenes offers a straightforward route for the synthesis of saturated heterocylces. Copper as an efficient catalyst was also explored in this area. In 2002, Noack and Göttlich reported a copper(I)-catalyzed cyclization of unsaturated *N*-benzoyloxyamines.[236] The reaction was proposed to proceed *via* a radical pathway and substituted pyrrolidine was produced in promising yield. In 2009, a Cu–xantphos system [Cu(O-*t*Bu)–xantphos, 10–15 mol%] was found to be efficient for the intramolecular hydroamination of unactivated terminal alkenes bearing an unprotected aminoalkyl substituent.[237] The reactions were carried out in alcoholic solvents and gave the desired pyrrolidine and piperidine derivatives in excellent yields (Scheme 4.138). This system is applicable to both primary and secondary amines and tolerates a variety of functional groups.

Chemler's group found that a copper(II) carboxylate can promote the deamination of terminal alkenes.[238] The reaction uses copper(II) neodecanoate [Cu(ND)$_2$] as the promoter, which allowed the less polar solvent dichloroethane (DCE) to be used and, as a consequence, decomposition of less reactive substrates could be avoided. High diastereoselectivity was observed in the synthesis of 2,5-disubstituted pyrrolidines. Ureas, bis(anilines) and α-amidopyrroles derived from 2-allylaniline could also participate in the diamination reaction. Later, the same group developed an intramolecular catalytic diastereoselective aminooxygenation protocol for unactivated alkenes (Scheme 4.139).[239] The reaction of α-substituted 4-pentenylsulfonamides with catalytic Cu(EH)$_2$ (EH = 2-ethylhexanoate) under O$_2$ affords 2,5-*cis*- and 2,5-*trans*-pyrrolidines in good to excellent yields and >20:1 selectivity. In contrast, β- and γ-substituted 4-pentenylsulfonamides gave higher diastereoselectivities with the use of a Cu(OTf)$_2$-bis(oxazoline) catalyst. The reaction of substituted *N*-allylureas uniformly provided high 4,5-*trans* selectivity irrespective of the ligands on the copper atom. The size of

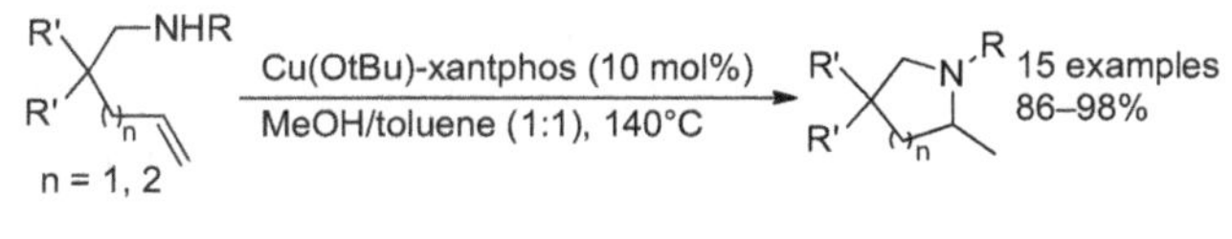

Scheme 4.138 Cu-catalyzed synthesis of pyrrolidines.

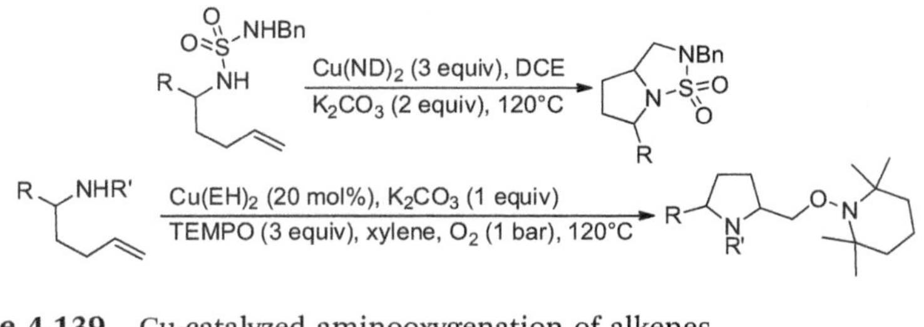

Scheme 4.139 Cu-catalyzed aminooxygenation of alkenes.

the N-substituent also influenced the level of diastereoselectivity with some substrates. Catalytic enantioselective desymmetrization was investigated with *meso*-α- and -β-substituted 4-pentenylsulfonamides. Whereas the α-substituted sulfonamides gave no enantioselectivity, enantioselective desymmetrization was achieved (up to 98% *ee*) with β-allyl-4-pentenylsulfonamide. The origin of the enantioselectivity is best rationalized by invoking chair-like transition states. It is possible that the participation of a boat-like transition state is responsible for the poor enantioselectivity observed in reactions with the *meso* substrate.

Li and co-workers reported a new method for the intramolecular chloroamination of unfunctionalized alkenes in 2013.[240] In the presence of 1 equiv. of CuCl$_2$·2H$_2$O, N-substituted pent-4-en-1-amines or N-substituted hex-5-en-1-amines were converted to 2-chloromethylpyrrolidines or 2-chloromethylpiperidines in good isolated yields (Scheme 4.140). MgCl$_2$ can be used as a chlorine source and reactions can be carried out in the presence of a catalytic amount of Cu(II). Cu(I) itself is ineffective in the absence of air, but can promote the reaction in open-air systems. The reaction is easy to carry out and the reagents are inexpensive and readily available. The reaction can also be carried out in a catalytic manner when a suitable chlorine source such as MgCl$_2$ is used. Using pure oxygen instead of air leads to an accelerated conversion of the substrates. As 2-chloromethylpyrrolidine compounds can be easily converted to 3-chloropiperidines, this approach essentially provides access to both 2-chloromethylpyrrolidine and 3-chloropiperidine compounds.

In addition to the mentioned alkenes, reactions of allenes have also been reported. In 2008, Okamoto's group demonstrated an intramolecular hydroamination of allenylamines to 3-pyrolines or 2-alkenylpyrrolidines with various copper salts as the catalysts.[241] Copper salts, such as CuCl, CuI, CuCl$_2$ and Cu(OTf)$_2$, which exhibited good catalytic reactivity, are inexpensive and are relatively less toxic, both of which are characteristics that should be synthetically usefu,l especially for application to a large-scale process.

In addition to the amination reactions, 1,3-dipolar cycloaddition of N-alkylideneglycine esters with activated alkenes offers another interesting pathway for five-membered heterocycle synthesis (Scheme 4.141). The presence of Cu(I) or Cu(II) salts can catalyze this transformation efficiently; by

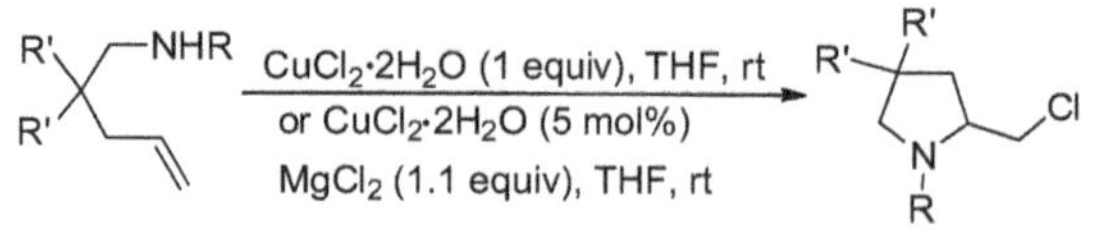

Scheme 4.140   Cu-catalyzed chloroamination of alkenes.

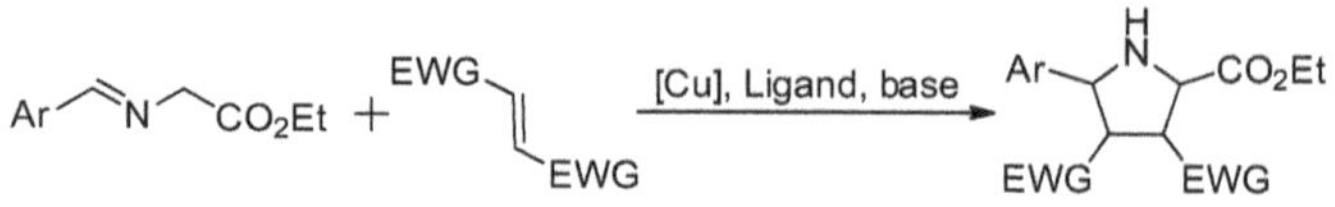

Scheme 4.141   Cu-catalyzed 1,3-dipolar cycloaddition.

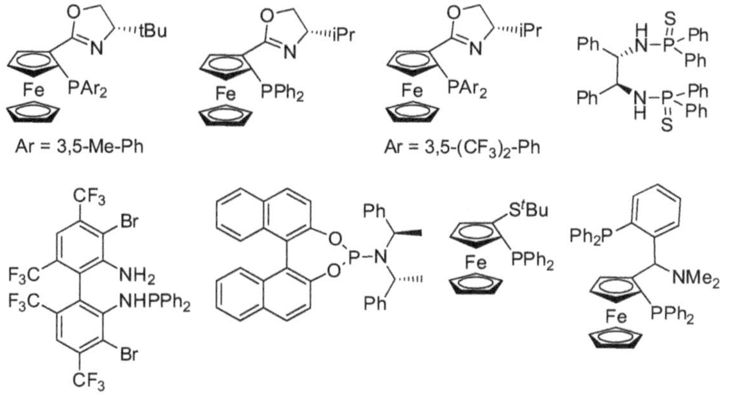

Scheme 4.142 Ligands for Cu-catalyzed asymmetric 1,3-dipolar cycloaddition.

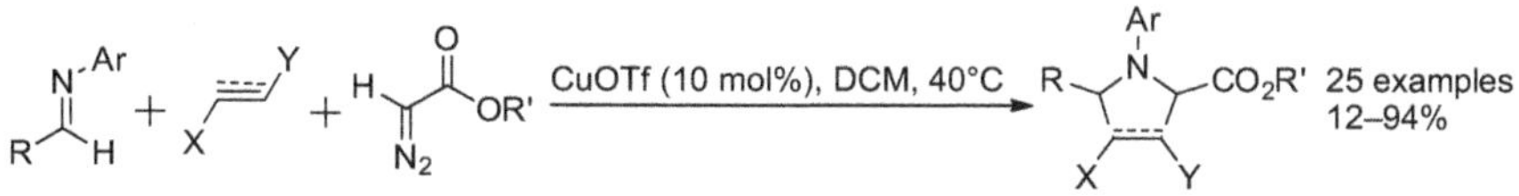

Scheme 4.143 Cu-catalyzed three-membered cycloaddition reaction.

adding chiral ligands, this reaction can be performed in an enantioselective manner. The reported ligands include BINAP,[242] DIOP, ferrocenes,[243,244] thiophosphoramide and selenophosphoramide,[245] BiphamPhos,[246] phosphoramidite,[247] fesulfos, taniaphos,[248] and so on (Scheme 4.142).

A three-membered reaction based on the combination of an α-diazo ester and an imine in the presence of a copper(I) catalyst generates a transient azomethine ylid and gives pyrrolidines after reaction with alkenes or alkynes (Scheme 4.143).[249] This 1,3-dipole, generated *in situ*, undergoes diastereoselective cycloadditions with activated dipolarophiles to afford highly substituted pyrrolidines in a convergent, three-component assembly reaction. Notably, this process is capable of generating four contiguous stereogenic centers in one operation by employing a readily available catalyst.

4.2.13 Copper-Catalyzed Synthesis of Indoline Derivatives

As discussed in previous section, intramolecular amination of alkenes can provide saturated heterocycles, whereas the cyclization of *o*-allylanilines affords a promising pathway for benzo-fused heterocycles, such as indolines. Chemler's group has made significant contributions in this area.[250] Numerous indolines have been produced in a selective manner in good yields (Scheme 4.144). By the addition of a chiral ligand, enantioselectivity can also be observed. In the early stages, the procedure needed an equal amount of copper salt and was developed into a real catalytic version soon afterwards.

Additionally, copper-catalyzed coupling reactions have also been explored. In 2002, a combination of copper iodide and cesium acetate was found to

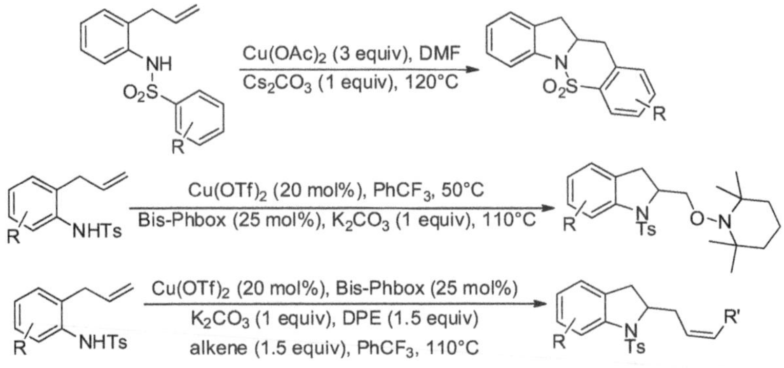

Scheme 4.144   Cu-catalyzed amination reactions for the synthesis of indolines.

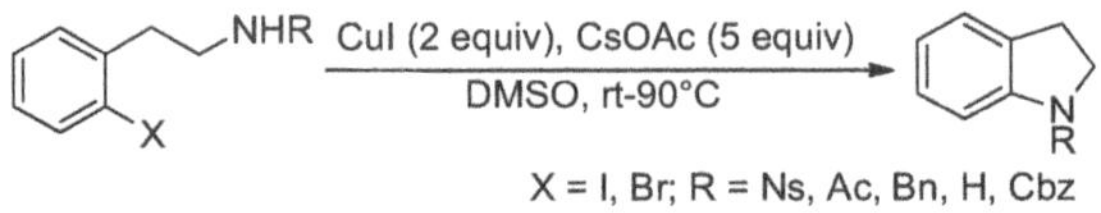

Scheme 4.145   Cu-mediated intramolecular amination of aryl halides.

mediate the intramolecular amination of aryl halides under mild conditions.[251] The reaction proceeds at room temperature with primary amines or N-benzylamines and at moderately elevated temperatures with other amine derivatives (Scheme 4.145). The reaction has been applied to the formation of five-, six- and seven-membered rings. Remarkably, halogens at the *meta*-position were retained, providing a definitive advantage over palladium-catalyzed systems.

In 2007, Zhu *et al.* developed an interesting method for the synthesis of pyrazolo[1,5-*a*]indoles from pyrazole derivatives using copper(I)-catalyzed intramolecular cyclization.[252] This method provides a general, simplified and easily operated route to pharmacologically attractive compounds and can afford products in good yields (Scheme 4.146).

Minatti and Buchwald developed a highly efficient one-pot procedure for the synthesis of indolines and their homologs based on a domino Cu-catalyzed amidation–nucleophilic substitution reaction in 2008.[253] Substituted 2-iodophenethyl mesylates and related compounds afforded the corresponding products in excellent yields (Scheme 4.147). No erosion of optical purity was observed when transforming enantiomerically pure mesylates under the reaction conditions. The mild reaction conditions and the broad substrate scope render this method attractive and complementary to existing methods for the synthesis of indolines.

The first highly enantioselective copper-catalyzed intramolecular Ullmann C–N coupling reaction was developed in 2012.[254] Yu, Cai and co-workers found that the asymmetric desymmetrization of 1,3-bis(2-iodoaryl)propan-2-amines catalyzed by CuI–(*R*)-BINOL-derived ligands led to the enantioselective formation of indolines in high yields and excellent *ee* (Scheme 4.148).

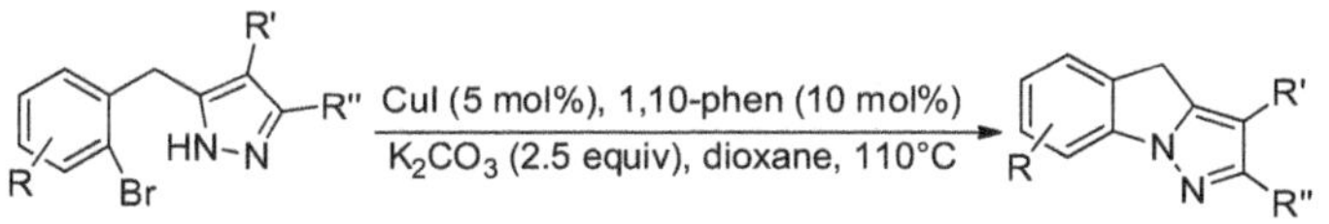

Scheme 4.146 Cu-catalyzed intramolecular amination of aryl halides.

General Procedure (Scheme 4.146): CuI (10 mg, 0.05 mmol), 1,10-phe-nanthroline (18 mg, 0.1 mmol) and K$_2$CO$_3$ (345 mg, 2.5 mmol) were added to a solution of substrate (1.0 mmol) dissolved in anhydrous 1,4-dioxane (3 mL) in a pressure tube under the protection of argon. The reaction mixture was stirred at 110 °C for 20 h and filtered through a pad of Celite. After removal of the solvent under reduced pressure, the residue was purified by chromatography (Silica Gel H, 10 g; petroleum ether–ethyl acetate as eluent) to give the product.

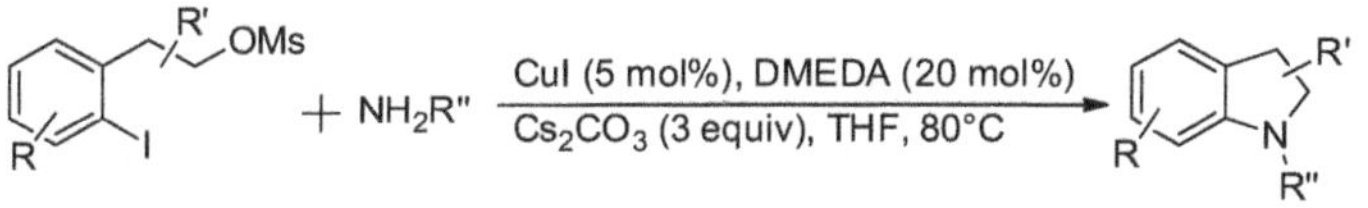

Scheme 4.147 Cu-catalyzed synthesis of indolines *via* a domino process.

General Procedure (Scheme 4.147): An oven-dried Schlenk tube equipped with a Teflon screw-cap was charged with a magnetic stirrer bar, CuI (4.8 mg, 5 mol%), Cs$_2$CO$_3$ (489 mg, 3 equiv.), the corresponding mesylate (0.5 mmol, if solid) and the corresponding carbamate or amide (1.2 equiv.). The tube was evacuated and backfilled with argon; this procedure was performed three times. *N,N'*-Dimethylethylenediamine (11 µl, 20 mol%) was added, followed by THF (1 mL) or a solution of the corresponding mesylate (0.5 mmol, if liquid) in THF (1 mL). The tube was sealed and the reaction mixture was stirred in a preheated oil-bath at 80 °C for 16 h. The reaction mixture was cooled to room temperature and diluted with hexanes–ethyl acetate (7 : 1) (2 mL). This crude reaction mixture was directly subjected to purification by column chromatography on alumina.

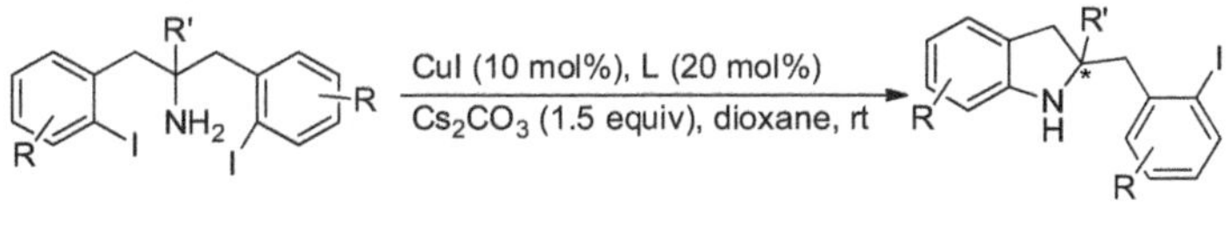

Scheme 4.148 Cu-catalyzed nantioselective synthesis of indolines.

This method was also applied to the formation of 1,2,3,4-tetra-hydroquinolines in high yields and with excellent enantioselectivity. In 2011, a facile CuI-mediated *N*-arylation of diketopiperazine using the Fukuyama modification of the Ullmann–Goldberg reaction was exploited in the synthesis of enantiopure polycyclic diketopiperazines from easily assembled dipeptides or functionalized Schöllkopf reagents.[255]

Wu and co-workers discovered a novel cascade reaction of 2-ethynylarylmethylenecyclopropanes with sulfonyl azide to give fused indolines.[256] The conversion was catalyzed by copper(I) iodide under mild conditions and the target products were produced in good yields (Scheme 4.149). Regarding the reaction mechanism, a ketenimine species and 6π-electrocyclization were believed to be involved. In detail, a 2-ethynylarylmethylenecyclo-propane would react with sulfonyl azide catalyzed by a copper salt affording the triazole intermediate, which was then converted to the reactive ketenimine *via* a ring-opening rearrangement. A consecutive 6π-electrocyclization occurred to form an intermediate, which subsequently underwent a rearrangement to produce the fused indolines.

In 2010, Li and co-workers developed a novel copper-catalyzed intramolecular C–H oxidation/acylation protocol for the preparation of indoline-2,3-diones.[257] This protocol used two C–H bonds as the reaction partners and O_2 as the terminal oxidant. Substituted indoline-2,3-diones with high tolerance of functional groups were produced in good yields. The mechanism was also discussed on the basis of kinetic isotope effect experiments and *in situ* FTIR and HRMS (ESI) analysis. In 2013, Fu's group developed another interesting copper-catalyzed process for the synthesis of indoline-2,3-dione derivatives.[258] The protocol uses inexpensive $Cu(OAc)_2$ as the catalyst, picolinic acid as the ligand and *tert*-butyl hydroperoxide as the oxidant, readily available secondary anilines and ethyl glyoxalate as the starting materials. The corresponding indoline-2,3-dione derivatives were prepared in moderate to good yields (Scheme 4.150). The procedure involves

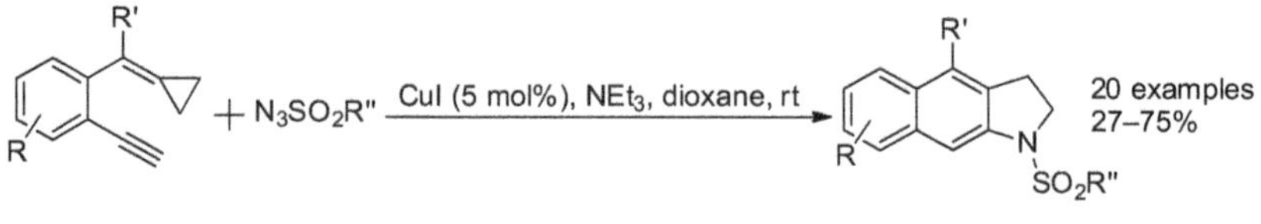

Scheme 4.149   Cu-catalyzed synthesis of fused indolines.

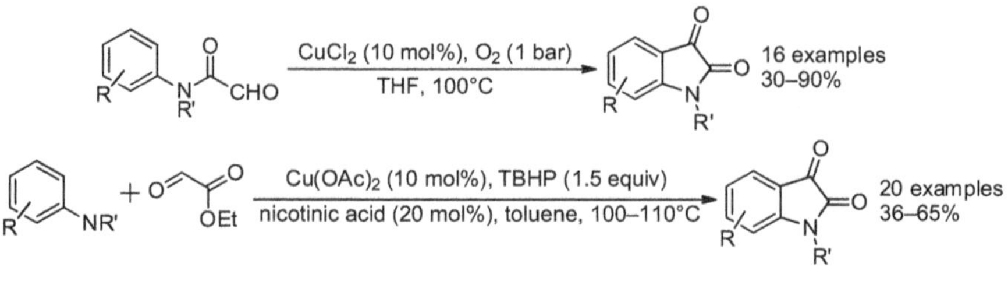

Scheme 4.150   Cu-catalyzed synthesis of indoline-2,3-diones.

intermolecular copper-catalyzed selective oxidative *ortho*-site aromatic acylation of the NH group in secondary anilines and intramolecular nucleophilic attack of the NH group on the ester. This method represents the first copper-catalyzed example of the synthesis of dicarbonyl compounds using ethyl glyoxalate as the dicarbonylating agent.

In 2010, Yoon and co-workers demonstrated that the copper(II)-catalyzed oxyamination of indoles affords aminals in high yields and with excellent regioselectivity (Scheme 4.151).[259] The regiochemical outcome of this process was consistent with a radical intermediate. The aminal products can be converted readily into 3-aminoindoles, α-carbolines, pyrroloindolines and other synthetically valuable heterocyclic structures. Additionally, a chiral auxiliary-controlled asymmetric oxyamination of tryptamines was also carried out. This transformation permits the enantioselective construction of 3-aminopyrroloindolines, which may be useful in the synthesis of oxidized indole natural products. In 2012, You and co-workers developed an efficient method to provide functionalized furoindolines *via* Cu(I)-catalyzed arylation or vinylation of 2-substituted tryptophols and the subsequent cyclization reaction (Scheme 4.151).[260] This cascade dearomatization sequence of tryptophols provided versatile furoindoline derivatives with two quaternary carbon centers in excellent yields under mild conditions.

More recently, Huang, Wen and co-workers developed an efficient method for the formation of a broad range of functionalized carbazoles under mild conditions.[261] In this method, inexpensive copper acetate was employed as the catalyst and, regardless of the electron properties attached on the diphenyleniodoniums or amines, afforded the expected carbazoles in moderate to good yields (Scheme 4.152). Halogens, including fluorine, chlorine and bromine, were well tolerated in the reactions, providing an opportunity to functionalize the obtained carbazoles further. Various amines, including

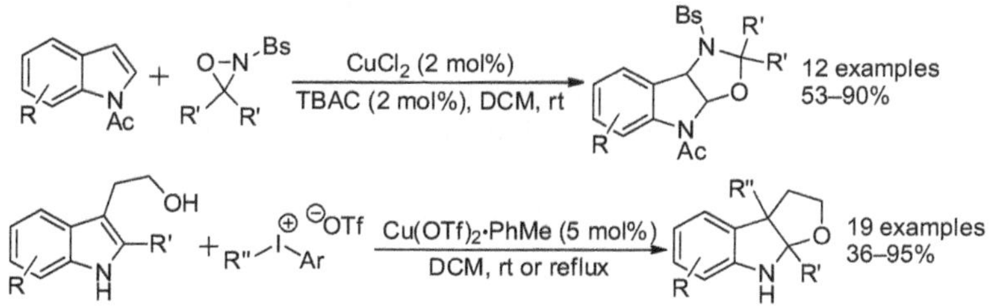

Scheme 4.151 Cu-catalyzed synthesis of indolines from indoles.

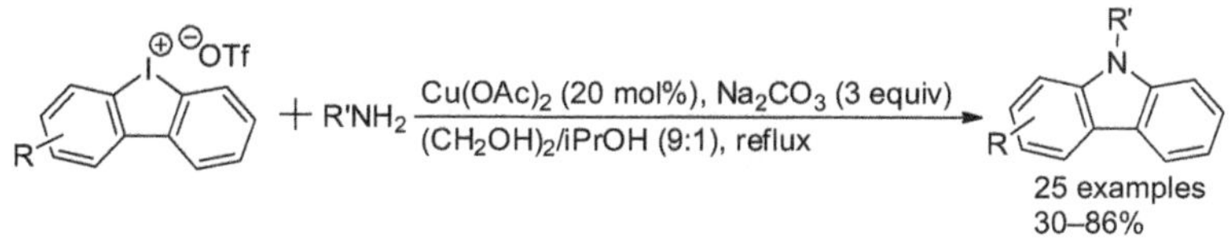

Scheme 4.152 Cu-catalyzed synthesis of carbazoles.

aniline, aliphatic amines and sulfonamides, gave carbazoles in modest to good yields. The N-substituted carbazoles obtained from sulfonamides can be easily unblocked from the sulfonyl group, resulting in free carbazoles. From a single diphenyleneiodonium or amine, a diverse range of N-substituted carbazoles were easily obtained. This method makes it possible to generate effectively drug-like carbazoles and offers a potential scaffold for drug screening. A compound that has an outstanding ability to protect HT-22 neuronal cells from damage induced by glumatate and homocysteic acid was also prepared.

4.2.14   Copper-Catalyzed Synthesis of Indolinone Derivatives

In 2013, a simple protocol for the synthesis of 2,2-disubstituted-3-indolinones was reported (Scheme 4.153).[262] This method uses a catalytic Cu(I)–ascorbate redox system for an S_NAr reaction with azide, followed by Smalley cyclization of α-bromophenyl sec-alkyl/sec-alkenyl ketones. A comparison of the optical properties of all the synthesized compounds revealed interesting results.

Jiang, Ma and co-workers developed a CuI–L-proline-catalyzed coupling of 2-bromobenzamides and terminal alkynes in 2009.[263] The reactions were performed in iPrOH (or DMF and DMSO) at 85–110 °C and subsequent additive cyclization produced substituted 3-methyleneisoindolin-1-ones (Scheme 4.154). Variations of the N-substituents, aromatic ring and methylene part are possible using suitable starting materials. This process gives straightforward access to substituted 3-methyleneisoindolin-1-ones.

In 2007, an efficient and general synthetic method for 3-unsubstituted 4-bromo-N-substituted oxindoles was reported (Scheme 4.155).[264] The key step is an intramolecular copper-catalyzed amidation reaction. Steric factors have a negative effect on both the reaction rate and yield. Amination of the 4-bromo substituent using copper catalysis failed whereas the corresponding palladium-catalyzed aminations were more promising. Later, a Cu_2O–benzene-1,2-diamine system was developed for the intramolecular N-arylation to prepare oxindoles.[265] This method efficiently provided polysubstituted oxindoles in moderate to excellent yields with good tolerance of various substrates and required only a very small amount of catalyst under

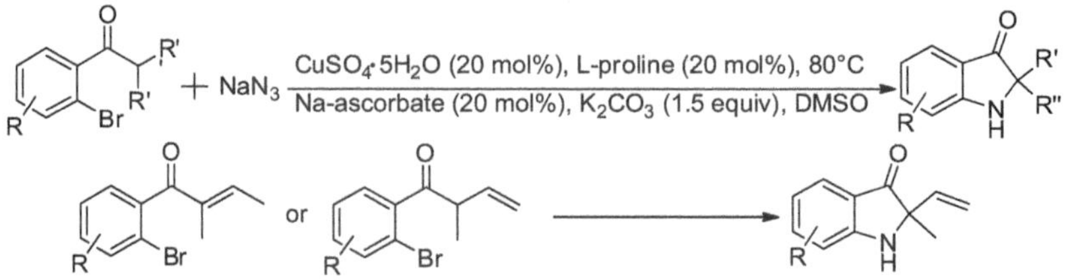

Scheme 4.153   Cu-catalyzed synthesis of 3-indolinones.

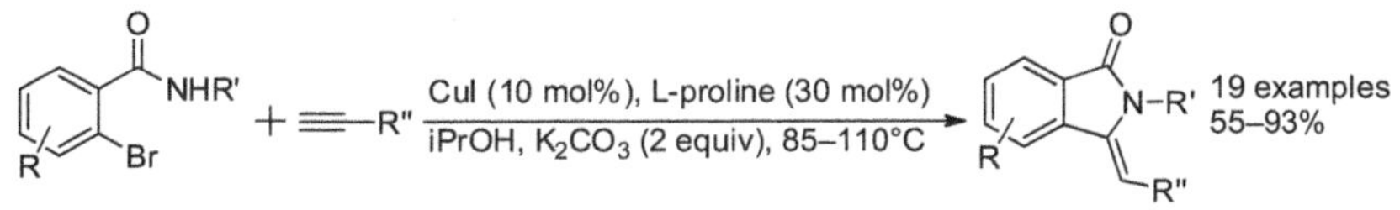

Scheme 4.154 Cu-catalyzed synthesis of 1-indolinones.

> **General Procedure** (Scheme 4.154): An oven-dried Schlenk tube was charged with CuI (10 mg, 0.05 mmol), L-proline (18 mg, 0.15 mmol), potassium carbonate (138 mg, 1 mmol) and 2-bromobenzylamide (0.5 mmol). The tube was evacuated and backfilled with argon (three times), then phenylacetylene (83 μl, 0.75 mmol) and *i*PrOH (1.0 mL) were added. The reaction mixture was stirred at 85 °C for 24–28 h. After cooling, 30 mL of ethyl acetate was added. The organic layer was separated and washed with 1 M HCl (15 mL) and brine (20 mL) and dried over Na₂SO₄. After removal of the solvent *in vacuo*, the residue was purified by silica gel chromatography to give the desired product.

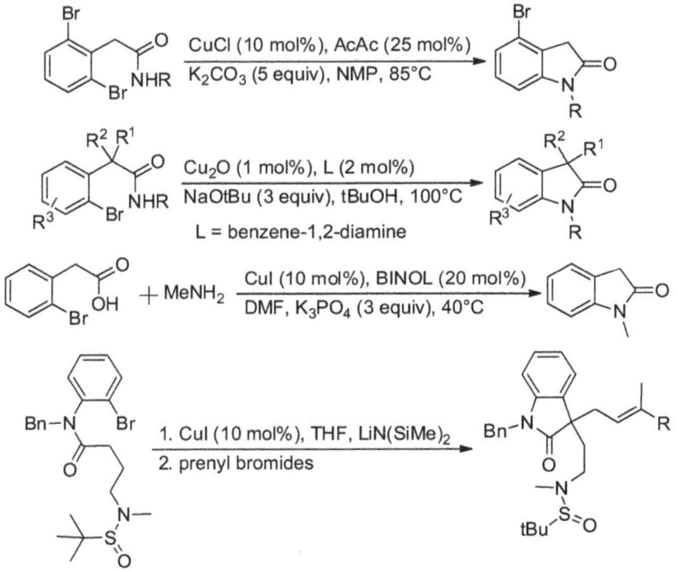

Scheme 4.155 Cu-catalyzed synthesis of oxindoles.

aerobic conditions. This type of idea was also applied in the preparation of 2-alkylaminobenzo[*b*]furans.[266] Fu and co-workers found that oxindoles can be synthesized by the copper-catalyzed reaction of methylamine hydrochloride with 2-(2-bromophenyl)acetic acid.[267] This approach was explored further by Viswanathan's group.[268] Zhang and co-workers developed a highly useful copper-catalyzed intramolecular arylation–alkylation of *o*-bromoanilides.[269] On the basis of this method, a general synthetic strategy was established for the synthesis of (−)-debromoflustramines B and E and pseudophrynamine alkaloids. They also developed a ligand-free copper-catalyzed process for the synthesis of spirocyclic oxindoles and revealed a remote aza-assisted effect. These sequential reactions led to medicinally interesting oxindoles bearing a C3a full quaternary carbon center in a flexible and practical way.

Hsieh *et al.* developed a novel methodology for a copper-catalyzed domino coupling reaction in the synthesis of oxindoles.[270] This is a very efficient synthetic route to provide various oxindole derivatives in moderate to excellent yields with tolerance of a wide variety of substrates (Scheme 4.156). Moreover, they also applied this method to synthesize (±)-coerulescine and (±)-horsfiline in a very efficient way with few steps and high total yields. The reaction was believed to be initiated by the coordination of the nitrile to copper complex.

Yin *et al.* developed a procedure for the preparation of three kinds of novel spirofurooxindoles from readly available furan derivatives in a concise and effective way (Scheme 4.157),[271] involving a copper sulfate pentahydrate-catalyzed intramolecular Friedel–Crafts reaction using an oxocarbenium

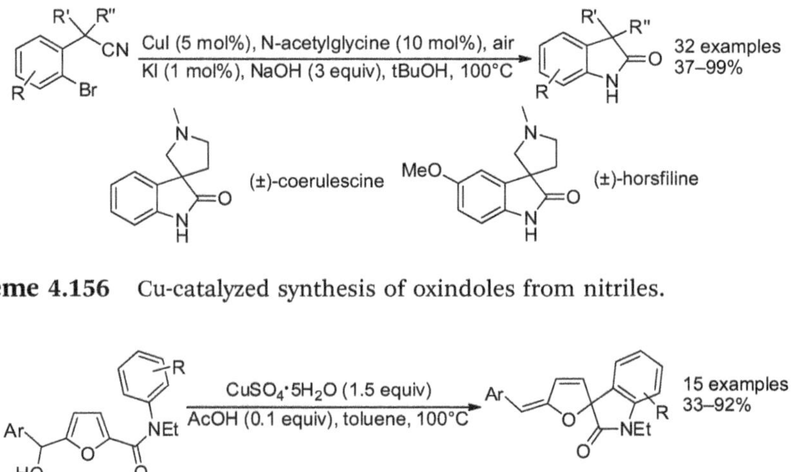

Scheme 4.156 Cu-catalyzed synthesis of oxindoles from nitriles.

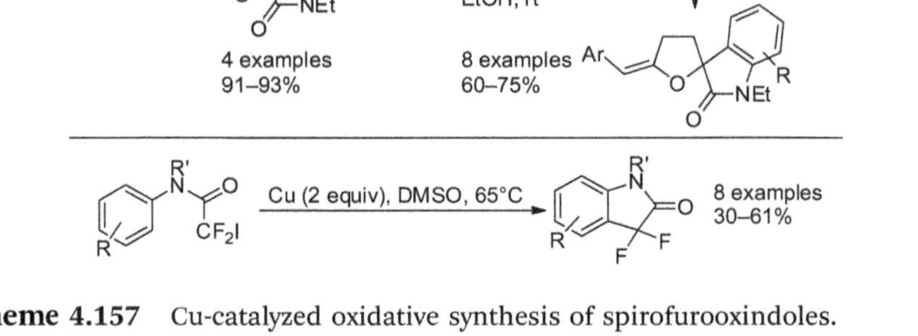

Scheme 4.157 Cu-catalyzed oxidative synthesis of spirofurooxindoles.

species derived from a furan ring as the alkylating agent. Using this protocol, spirofurooxindoles with multi-reactive sites were synthesized simply and concisely. In addition, selective hydrogenation of the endocyclic double bond and full hydrogenation of the endo- and exocyclic double bonds of spirofurooxindoles provided spirofurooindoles. In 2010, Hu and co-workers reported a copper-mediated intramolecular cyclization of iododifluoro-acetamides.[272] The corresponding fluorinated oxindiles were prepared in moderate yields.

An interesting method for the synthesis of pyrrolopyrimidine derivatives *via* C–H activation was reported in 2013.[273] By copper-catalyzed dehydrogenative C–H activation of uracils in an open atmosphere, the corresponding pyrrolopyrimidines were produced in good yields (Scheme 4.158). Kündig and co-workers developed a CuCl$_2$-mediated direct intramolecular oxidative coupling of C(sp^2)–H and C(sp^3)–H centers to 3,3-disubstituted oxindoles.[274] Aromatic, heteroaromatic and alkyl substituents and also a heteroatom at the quaternary center are well tolerated and the reaction gave the desired products in good to excellent yields. The reaction was carried out in the presence of NaO*t*Bu and CuCl$_2$ in DMF at 110 °C. The key step of this reaction is the formation of an amidyl radical by one-electron oxidation of an amide enolate followed by an intramolecular radical cyclization reaction (homolytic aromatic substitution reaction). A detailed DFT study showed that the cyclization of the amidyl radical is the rate-limiting step in the oxindole synthesis, whereas the second single-electron transfer (SET) becomes the rate-determining step in the aza-oxindole formation. Computational data are in agreement with the experimentally observed relative reactivity and regioselectivity.

In 2006, Lu and Ma reported a copper-catalyzed intramolecular coupling of β-keto-2-iodoanilides.[275] CuI–L-proline was applied as the catalytic system and numerous substituted 3-acyloxindoles were produced in good yield in DMSO at room temperature (Scheme 4.159). Electronic effects on the aromatic ring have little influence on this reaction. Variations at the 1-, 3-, 4-, 5- and 6-positions of the oxindoles were achieved by employing the corresponding amides. Later, Li's group developed a novel one-pot multicatalytic route for the synthesis of 3-methyleneindolin-2-ones,[276] involving sequential copper-catalyzed amination and palladium-catalyzed C–H functionalization

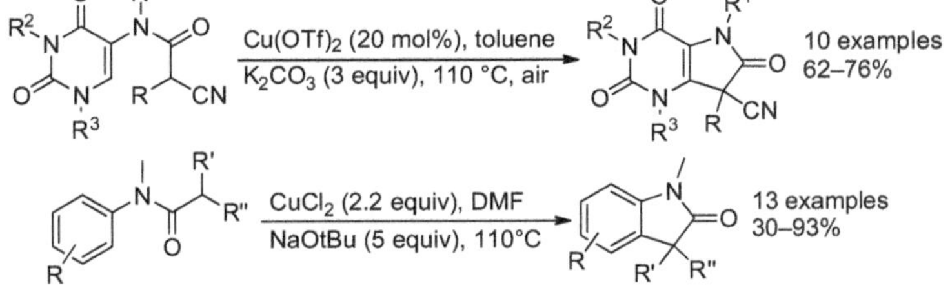

Scheme 4.158 Cu-catalyzed oxidative synthesis of oxindoles *via* C–H activation.

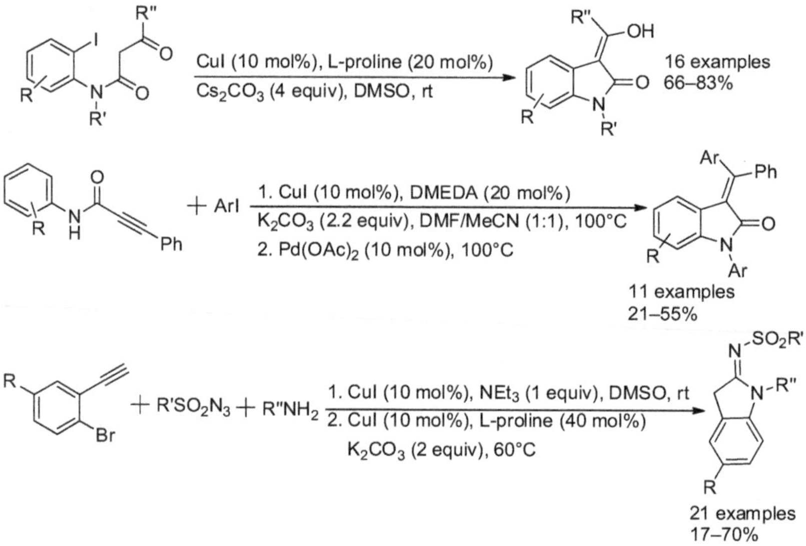

Scheme 4.159 Cu-catalyzed synthesis of 3-methyleneindolinones.

processes. In the presence of CuI and Pd(OAc)$_2$, a variety of propiolamides underwent the reaction with iodides to afford the corresponding 3-methyleneindolin-2-ones in moderate yields. Wang and co-worker described a one-pot synthesis of 2-(N-sulfonylimino)indolines.[277] The procedure combines the copper-catalyzed three-component reaction of sulfonyl azides, o-bromophenylacetylenes and amines and copper-catalyzed intramolecular C–N coupling in one sequence, which afforded the products in moderate to good yields. The resulting 2-(N-sulfonylimino)indolines could be easily converted to pharmaceutically valuable oxindoles (indolin-2-ones).

4.2.15 Copper-Catalyzed Synthesis of Imidazopyridine Derivatives

In 2008, a novel organocopper-mediated two-component coupling–cycloisomerization cascade transformation was reported.[278] This mild and efficient method allows the facile synthesis of C1 alkyl- and aryl-substituted N-fused heterocycles, such as indolizines, pyrroloquinolines and pyrroloisoquinolines, from readily available starting materials (Scheme 4.160). It deserves mention that these important heterocyclic scaffolds with an alkyl or aryl substituent at C1 are not available *via* the existing cycloisomerization methods.

A copper(I)-catalyzed regioselective [3 + 2] cyclization of pyridines toward alkenyldiazoacetates leading to functionalized indolizine derivatives was reported in 2010 (Scheme 4.161).[279] All of the heterocyclic substrates used were commercially available and did not require additional purification.

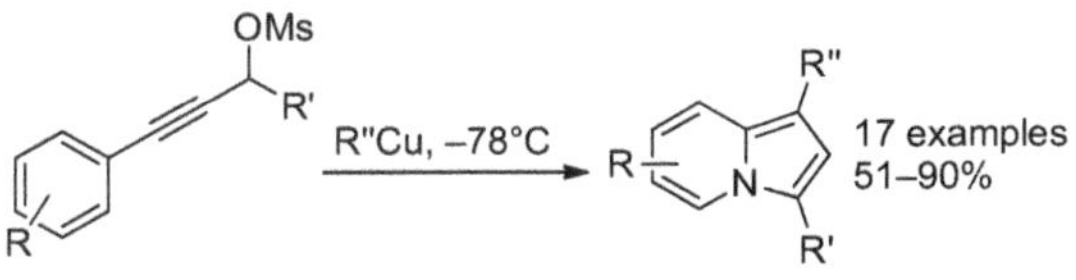

Scheme 4.160 Cu-mediated annulation of alkynes.

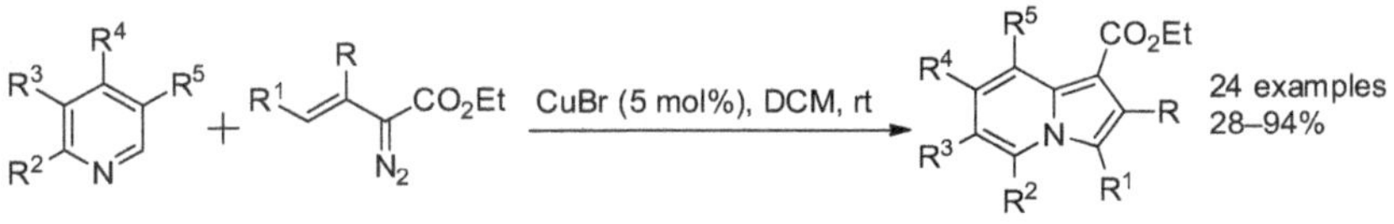

Scheme 4.161 Cu-catalyzed synthesis of indolizine derivatives.

General Procedure (Scheme 4.161): CuBr (3.6 mg, 0.025 mmol) was added to a solution of pyridine (0.5 mmol) and diazo compound (0.5 mmol) in CH_2Cl_2 (5 mL). The mixture was stirred at room temperature with protection from light until dissaperance of the starting diazo compound (monitored by TLC; 4–14 h). The solvent was removed under reduced pressure and the residue was purified by flash chromatography over a short, light-shielded column of deoxygenated SiO_2 using hexane–ethyl acetate (5 : 1) as eluent to give the corresponding indolizine derivatives.

A broad range of pyridine substrates are compatible with this operationally simple and mild copper-catalyzed cyclization, allowing heterocyclic nuclei as important as indolizidines and benzoindolines (pyrroloquinolines and -isoquinolines) to be readily obtained with a broad array of substitution and functionalization patterns. It is noteworthy that this is the first successful example of metal-catalyzed cyclization of a π-deficient heterocyclic system with alkenyldiazo compounds, in contrast to the extensive chemistry performed on π-excessive heterocycles. A tentative rationale for this cyclization was proposed by the authors. Initially, decomposition of diazo would form the copper(I) alkenylcarbene species, which then would evolve to the cuprate by Michael-type addition of the pyridine nitrogen, probably directed by pyridine-copper coordination. The cyclization of the latter would generate the copper(III) metallacycle, which would undergo reductive elimination assisted by the proximal ring C–C double bond, giving rise to the Cu(I) complex. Further metal decoordination would afford the dihydroindolizine, thus concluding the catalytic cycle. Finally, oxidative aromatization would result in the formation of the corresponding indolizidine.

Guo, Fossey and co-workers developed an intramolecular C–H activation–amination reaction of purine nucleosides to synthesize a series of multi-heterocyclic compounds, which are of great importance in medicinal chemistry (Scheme 4.162).[280] This represents the first example of the use of an intramolecular C–H activation–amination reaction for the synthesis of purine

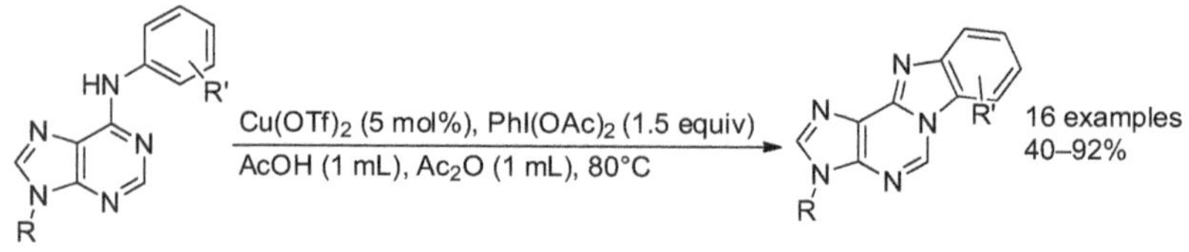

Scheme 4.162   Cu-catalyzed synthesis of purine fused-ring polycyclics.

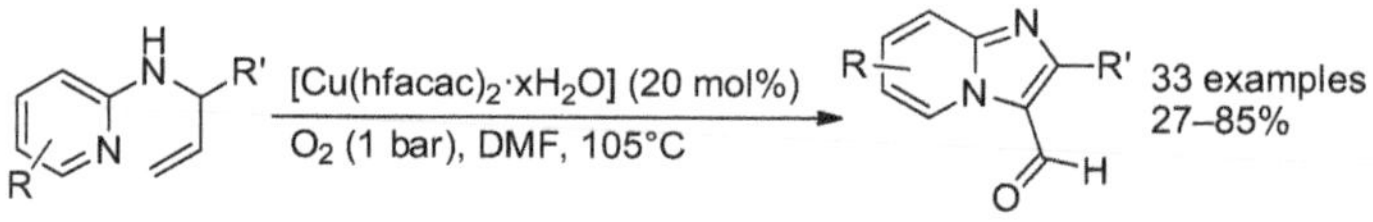

Scheme 4.163   Cu-catalyzed synthesis of formyl-substituted heterocycles.

General Procedure (Scheme 4.163): A mixture of substrate (0.5 mmol), [Cu(hfacac)$_2 \cdot x$H$_2$O] (47.8 mg, 0.1 mmol, 20 mol%) in DMF (1.5 mL) was stirred at 105 °C under O$_2$ (balloon pressure). The reaction was cooled to room temperature after complete consumption of the starting material (monitored by TLC). Saturated aqueous NaHCO$_3$ (10 mL) and EtOAc (10 mL) were added successively to the reaction mixture. The organic phase was separated and the aqueous phase was further extracted with EtOAc (2×10 mL). The combined organic layers were dried over anhydrous Na$_2$SO$_4$ and concentrated. The residue was purified by flash chromatography on silica gel (eluent: 3 : 1 petroleum ether–ethyl acetate) to provide the desired product.

nucleosides, which offers facile alternative access to some useful multifused-ring purine heterocyclic compounds. The fluorescence of these purine nucleoside compounds will be helpful for the study of their medicinal relevance.

Zhang, Zhu and co-workers reported their achievement in 2011 of the development of an unprecedented IDA (intramolecular dehydrogenative aminooxygenation) process that produces imidazo[1,2-*a*]pyridine-3-carbaldehydes and 1,2-disubstituted imidazole-4-carbaldehydes from readily available *N*-allyl-2-aminopyridines and substituted *N*-allylamidines, respectively.[281] The reaction, carried out using 20 mol% of Cu(II) catalyst in DMF or DMAc under dioxygen, is efficient and environmentally benign because it does not require additional organic or inorganic oxidants. Substituted imidazo[1,2-*a*]pyridine-3-carbaldehydes and imidazole-4-carbaldehydes, which can serve as versatile synthetic intermediates, are obtained in moderate to good yields by a reaction that possesses a broad substrate scope and good functionality tolerance (Scheme 4.163). The process opens up a new path towards the direct formation of aromatic *N*-heterocycles substituted with a formyl group from acyclic substrates. Mechanistic studies directly showed that the carbonyl oxygen in the aldehyde products is derived from dioxygen by a pathway that takes place *via* a peroxy–copper(III) intermediate.

In 2010, the same group described a novel and efficient approach for the preparation of pyrido[1,2-*a*]benzimidazoles by using direct intramolecular aromatic C–H amination of *N*-aryl-2-aminopyridines, a process that is co-catalyzed by inexpensive Cu(OAc)$_2$ and Fe(NO$_3$)$_3 \cdot$ 9H$_2$O.[282] In this process, the pyridinyl nitrogen in the substrates acts as both a directing group and a nucleophile. Diverse pyrido[1,2-*a*]benzimidazoles with substituents at different positions were generated in moderate to excellent yields (Scheme 4.164). However, electron-withdrawing substituents on the *meta*-position of the aniline ring and any position of the pyridine ring are unfavorable. In addition, mechanistic studies, aimed at uncovering the unique role played by iron(III), were performed. The results suggested that iron(III) aids the reaction by enabling the formation of the more electrophilic Cu(III) species, which facilitates the subsequent S_EAr process. Meanwhile, a copper-only system was reported by Maes and co-workers.[282b] O$_2$ was used as oxidant and acid as additive.

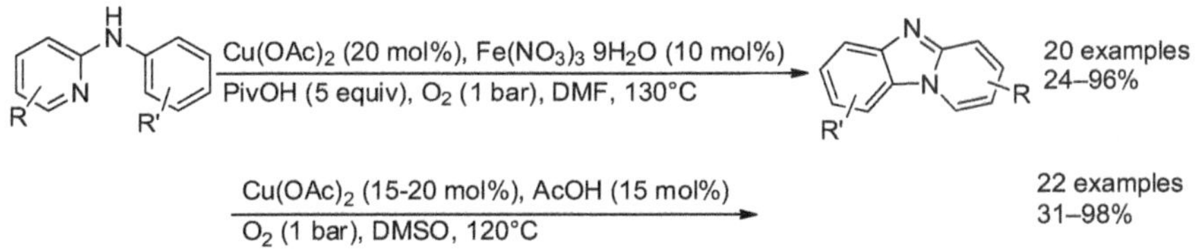

Scheme 4.164 Cu-catalyzed synthesis of pyrido[1,2-*a*]benzimidazoles.

General Procedure (Scheme 4.164): A mixture of *N*-aryl-2-aminopyridine (0.5 mmol), Cu(OAc)$_2$ (20 or 100 mol%), Fe(NO$_3$)$_3 \cdot$ 9H$_2$O (10 mol%) and PivOH (2.5 mmol) in DMF (1.0 mL) was stirred at 130 °C under balloon pressure of O$_2$. The reaction was cooled to room temperature after complete consumption of starting material (monitored by TLC). Water (10 mL), triethylamine (1.0 mL) and EtOAc (10 mL) were added successively to the reaction mixture. The organic phase was separated and the aqueous phase was further extracted with EtOAc (3×10 mL). The combined organic layers were dried over anhydrous Na$_2$SO$_4$ and concentrated. The residue was purified by flash chromatography to provide the desired product.

General Procedure: Substrate (0.50 mmol), Cu(OAc)$_2 \cdot$ H$_2$O (see reference: conditions a and b, 0.075 mmol, 15 mol%; conditions c, 0.1 mmol, 20 mol%), additive (conditions a, acetic acid, 0.075 mmol; conditions b, 3,4,5-trifluorobenzoic acid (3,4,5-TFBA), 0.075 mmol; conditions c, 3,4,5-TFBA, 0.1 mmol) and solvent (DMSO, 1.00 mL) were added to a 10 mL microwave vial. The resulting reaction mixture was stirred under a flow of oxygen for 5 min prior to sealing under oxygen with a pressure cap, then supplied with additional oxygen by a balloon and heated by a temperature-calibrated aluminum hotplate at 120 °C.

In 2011, a simple and efficient method for the preparation of pyrido[1,2-
a]benzimidazoles by a copper-catalyzed inter- and intramolecular C–N
coupling cascade process was designed and carried out by Zhou and co-
workers.[283] A wide range of functional groups were well tolerated under the
reaction conditions and a series of pyrido[1,2-*a*]benzimidazoles with substituents at different positions were generated in moderate to excellent yields
(Scheme 4.165). Considering the low cost of the catalytic system and the
starting materials, this strategy will be highly useful in organic and medicinal chemistry. Notably, the inter- and intramolecular transition metal-
catalyzed amination of 2-chloro-3-iodopyridine and 2,3-dibromopyridine,
respectively, with benzodiazinamines yielded six tetracyclic azaheteroaro-
matic cores.[284] C–N bond formation was achieved *via* auto-tandem (Pd
catalyst) and also one-pot (sequential use of a Pd and a Cu catalyst) catalysis.

A general and highly efficient method for the synthesis of imidazopyridine
derivatives by the copper-catalyzed three-component coupling reaction of
aryl-, heteroaryl- and alkylaldehydes with 2-aminopyridines and terminal
alkynes was developed in 2010.[285] The employment of 2-aminoquinoline and
2-aminoisoquinoline as coupling partners in this transformation led to
imidazoquinoline and imidazoisoquinoline frameworks in good yields. The
synthetic utility of this novel three-component reaction was illustrated in a
highly efficient one-pot synthesis of alpidem and zolpidem (Scheme 4.166).
The reaction was proposed to proceed through propargylamine as an
intermediate.

Liu and co-workers reported a Cu(II) and Fe(III) co-catalyzed diamination
of 2-aminopyridines and 2-aminoisoquinolines with readily available alkynes (Scheme 4.167).[286] The strategy allows for the direct synthesis of
imidazo[1,2-*a*]pyridines and imidazo[1,2-*a*]isoquinolines in yields of up to

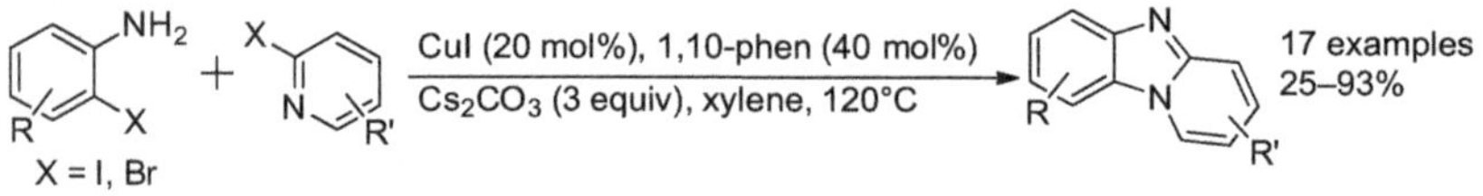

Scheme 4.165   Cu-catalyzed synthesis of pyrido[1,2-*a*]benzimidazoles from
2-haloanilines.

General Procedure (Scheme 4.165): An oven-dried Schlenk tube was
charged with 2-iodoaniline (0.25 mmol, 54.8 mg), Cs_2CO_3 (0.75 mmol,
244.4 mg), CuI (0.05 mmol, 9.5 mg) and 1,10-phenanthroline (0.1 mmol,
18 mg). The tube was evacuated and backfilled with N_2 and then
2-iodopyridine (0.3 mmol, 61.5 mg) and xylene (0.5 mL) were added. The
reaction mixture was stirred at 120 °C for 12 h and then allowed to cool to
room temperature. The mixture was diluted with water and extracted with
ethyl acetate. The extracts were combined and dried with anhydrous
Na_2SO_4. The residue was purified by silica gel column chromatography
and the solvent was removed to afford the corresponding product.

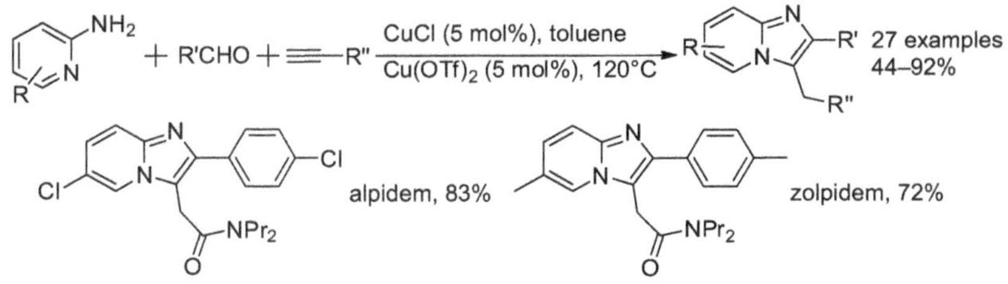

Scheme 4.166 Cu-catalyzed synthesis of imidazopyridines.

> **General Procedure** (Scheme 4.166): 2-Aminopyridine (0.5 mmol), CuCl
> (2.5 mg, 0.025 mmol, 5 mol%), Cu(OTf)$_2$ (9.04 mg, 0.025 mmol, 5 mol%)
> and aldehyde were added to a Weaton (1 mL) microreactor in a glove-box
> under an inert atmosphere. Subsequently, anhydrous toluene (500 mL)
> and alkyne (0.75 mmol, 1.5 equiv.) were added to the mixture. The
> microreactor was capped with a Teflon pressure cap and placed in a
> preheated (120 °C) aluminum heating block. The reaction mixture was
> heated at 120 °C. Upon completion of the reaction (12–16 h), the mixture
> was filtered through a plug of neutral alumina (eluent: EtOAc). The fil-
> trate was concentrated under reduced pressure and then purified by
> column chromatography on silica gel (eluent: 4:17:84 Et$_3$N–EtOAc–
> hexanes) to afford the imidazopyridine.

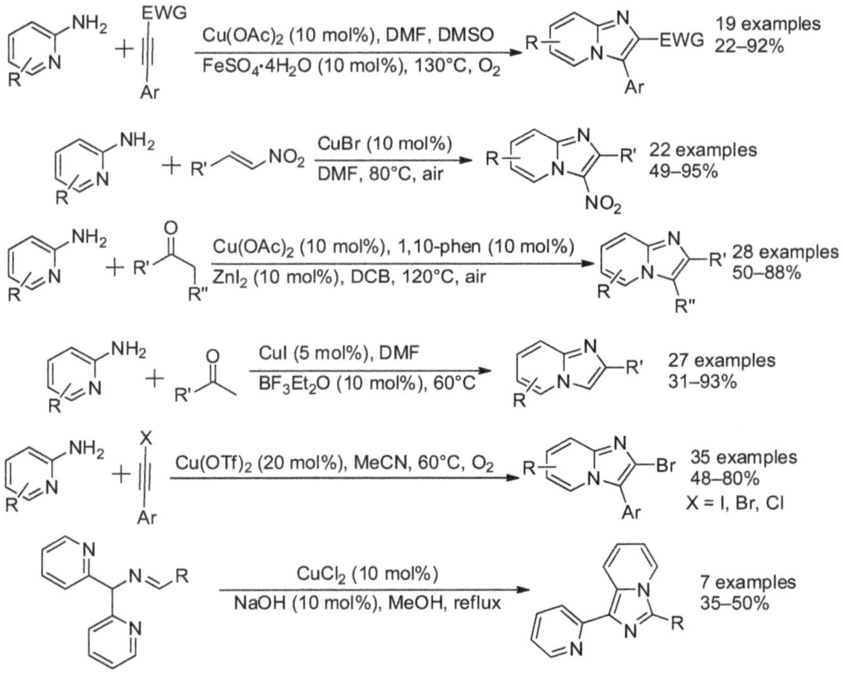

Scheme 4.167 Cu-catalyzed synthesis of imidazo[1,2-*a*]pyridines.

92%. These structures are ubiquitously found in a large variety of compounds which possess important pharmacological properties that are essential to the biopharmaceutical and chemical sectors. This strategy provides simple, facile and straightforward access to an extensive array of compounds and synergizes the well-explored intramolecular diamination of alkenes to alkynes in an intermolecular manner. For aminopyridines and aminoquinolines, the high levels of chemoselectivity and regioselectivity of the two nitrogens were demonstrated for a range of compounds, illustrating the flexibility and good reactivity of this strategy. Yan *et al.* developed another copper-catalyzed method for the synthesis of imidazo[1,2-*a*]pyridines with aminopyridines and nitroalkenes using air as oxidant in a one-pot procedure.[287] In this process, the reaction appears to be very general and suitable for construction of a variety of imidazo[1,2-*a*]pyridines. Hajra's group developed a new copper-catalyzed oxidative cyclization *via* C–H amination between 2-aminopyridines and methyl aryl/heteroaryl ketones under ambient air.[288] Imidazo[1,2-*a*]pyridines containing a wide range of functional groups were synthesized from basic and readily available starting materials. This simple, one-pot reaction protocol is applicable for the direct preparation of zolimidine (a marketed anti-ulcer drug) on a large scale. The same procedure was also reported with CuI and O_2 as the catalytic system.[289] Around the same time, an efficient copper-catalyzed method for the synthesis of 2-haloimidazopyridines with aminopyridines and haloalkynes using molecular oxygen as oxidant in a one-pot manner was developed.[290] In this process, the reaction appears to be very general and suitable for the construction of a variety of 2-halo-substituted imidazopyridines, imidazopyrazines and imidazopyrimidines. The intermolecular oxidative diamination of haloalkynes was achieved for the first time. Importantly, the mild reaction conditions and the efficient conversion of the alkyl-substituted haloalkynes are great improvements over the previous methods. Moreover, the resultant 2-haloimidazo[1,2-*a*]pyridines could be efficiently converted to other functionalized imidazopyridine products *via* substitution, coupling reactions and other transformations, which further indicates potential applications of this method in synthetic and pharmaceutical chemistry. A copper-catalyzed oxidative transformation of imines was also reported by Döring and co-workers.[291]

Ueda and Nagasawa developed an efficient copper-catalyzed oxidative synthesis of 1,2,4-triazole derivatives from 2-aminopyridines and nitriles in 2009.[292] The catalytic cycle was achieved by use of molecular oxygen (air at 1 atm) as the oxidant, which produces water as the sole theoretical by-product. The reaction allows the facile generation of diversity in the construction of 1,2,4-triazoles, which are important structures. Under the standeard conditions, all the desired triazoles were produced in good yields (Scheme 4.168). Regarding the reaction mechanism, copper first promotes nucleophilic attack of 2-aminopyridine on the nitrile, probably by forming a coordinated intermediate, to provide an amidine. Subsequent copper-induced oxidative cyclization of the amidine provides a triazolopyridine and reduced copper species.

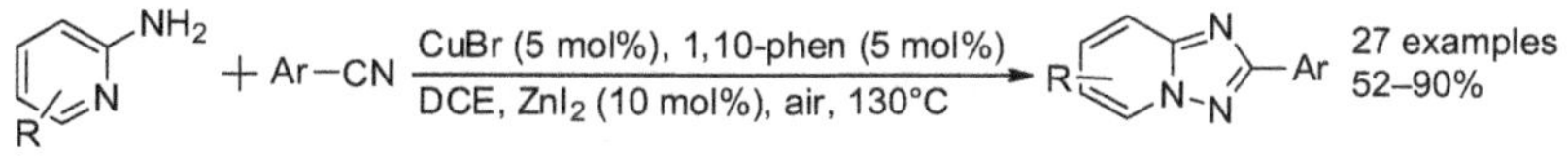

Scheme 4.168 Cu-catalyzed synthesis of triazoles.

General Procedure (Scheme 4.168): In a dried screw-capped vial were placed nitrile (0.30 mmol), 2-aminopyridine (0.36 mmol), CuBr (2.2 mg, 0.015 mmol), 1,10-phenanthroline (2.7 mg, 0.015 mmol, 5 mol%) and ZnI$_2$ (9.6 mg, 0.03 mmol, 10 mol%). 1,2-Dichlorobenzene (0.6 mL) was then added and the vial was sealed under atmospheric air. The reaction mixture was stirred at 130 °C for 24 h in a preheated oil-bath. After cooling to room temperature, the reaction mixture was diluted with EtOAc and filtered over a glass filter. The filtrate was concentrated and purified by column chromatography on silica gel.

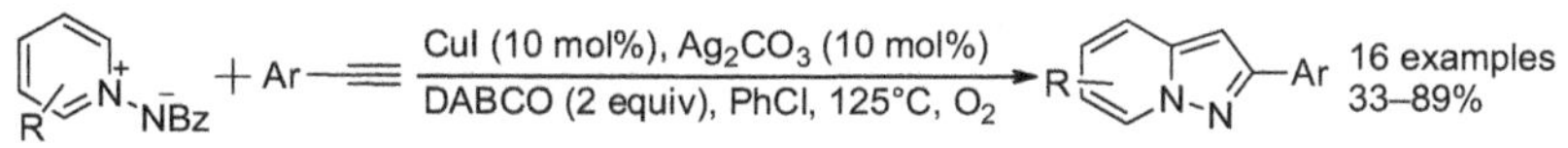

Scheme 4.169 Cu-catalyzed synthesis of pyrazolo[1,5-*a*]pyridines.

The second step (oxidative N–N bond formation) was confirmed by conversion of the amidine the desired product under the reaction conditions. The presence of oxygen reoxidizes copper to the active species.

In 2013, Jiao and co-workers demonstrated an aerobic direct dehydrogenative annulation of terminal alkynes to pyrazolo[1,5-*a*]pyridine derivatives.[293] Good yields of the desired products were obtained under the standard conditons (Scheme 4.169). Concerning the reaction mechansiem, the initial copper acetylide is generated from an alkyne with Cu(I) catalyst assisted by Ag$_2$CO$_3$ under basic conditions. Subsequently, the oxidative insertion of copper acetylide into the pyridinium ylide forms a Cu(III) intermediate *via* C–H bond functionalization. The intermediate then undergoes reductive elimination to afford the direct alkynylation intermediate, in which the alkyne can be activated by copper catalyst. Subsequent 5-*endo* cyclization of the formed alkyne *via* the anti-aminocupration of the alkyne attacked by the amido nitrogen occurs with conversion into the final product after protonation and rearomatization. DABCO assists the expulsion of the benzoyl moiety in this process, but the role of the ligand on Cu to facilitate the reaction cannot be excluded.

4.2.16 Copper-Catalyzed Synthesis of Sulfur Containing Heterocycles

Xi and co-workers described an efficient copper-catalyzed method for the synthesis of thiophenes *via* double vinylation of potassium sulfide with dienyl diiodides.[294] This methodology provided a facile route for the synthesis of di-, tri- and tetrasubstituted thiophenes in excellent yields (Scheme 4.170). The transformation is distinguished by simplicity, low cost and tolerance of a wide variety of substituted groups in a range of

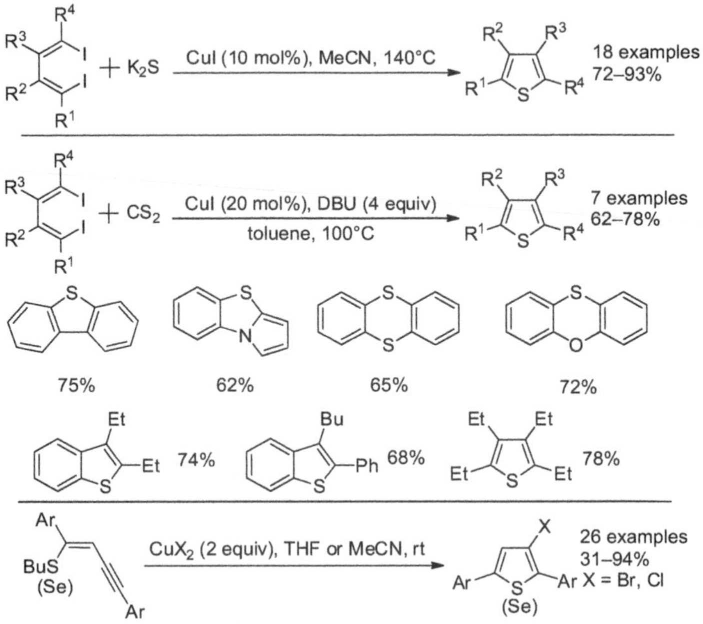

Scheme 4.170 Cu-catalyzed synthesis of thiophenes.

General Procedure (Scheme 4.170): CuI (19 mg, 0.1 mmol), 1,4-diiodo-1,3-diene (1.0 mmol), K_2S (330 mg, 3.0 mmol) and acetonitrile (5.0 mL) were placed at room temperature in a screw-capped test-tube with a Teflon-lined septum. The tube was then sealed and the reaction mixture was stirred at 140 °C (external temperature) for 24 h. The reaction was quenched with water (5 mL) and the resulting mixture was extracted with diethyl ether (3 × 5 mL). The combined organic layer was dried with Na_2SO_4 and then concentrated under vacuum. The residue was purified by column chromatography on silica gel (eluent: petroleum ether) to obtain the product.

General Procedure: A sealed tube was charged with a mixture of diiodoaryl compound (0.5 mmol), carbon disulfide (1 mmol), CuI (0.1 mmol) and DBU (2 mmol) and then stirred in toluene (1.5 mL) at 100 °C under a nitrogen atmosphere for the indicated time. After completion, water (5 mL) was added and the mixture was extracted with EtOAc (3×5 mL) and dried with anhydrous Na_2SO_4. Evaporation of the solvent followed by purification on silica gel (eluent: petroleum ether) provided the corresponding product.

substitution patterns. Soon afterwards, the same group discovered the possibility of using CS_2 as a sulfur source in combination with DBU.[295] Using diiodo compounds as substrates, various heterocycles can be prepared effectively (Scheme 4.170). Zeni and co-workers developed an alternative and efficient method for the synthesize of 3-halochalcogenophene and thiophene derivatives through intramolecular 5-*endo-dig* cyclization of (Z)-chalcogenoenynes.[296] The reactions were carried out under an ambient atmosphere using $CuCl_2$ at 50 °C or $CuBr_2$ at room temperature, which is a relatively economical and eco-friendly protocol. 3-Bromoselenophene derivatives were applied as substrates in palladium-catalyzed cross-coupling reactions with boronic acids to give Suzuki-type products in good yields.

Zhang and co-workers reported a copper-catalyzed thiolation annulation reaction of 2-bromoalkynylbenzenes with sodium sulfide in 2011.[297] In the presence of CuI and TMEDA, a variety of 2-substituted benzo[*b*]thiophenes were readily prepared in moderate to good yields by the reaction of 2-bromoalkynylbenzenes and $Na_2S \cdot 9H_2O$ (Scheme 4.171). In contrast to the harsh reaction conditions (−78 or 180 °C) of the traditional methods, the relatively mild reaction conditions of the present procedure provide a new optional route for constructing the benzo[*b*]thiophene ring. However, aliphatic alkynes did not give any of the desried product. More recently, Kesharwani and co-workers developed a green method for the synthesis of benzo[*b*]thiophenes by employing the environmentally friendly solvent ethanol and inexpensive inorganic salts.[298] A variety of functional groups are tolerated under these conditions. This method can be used not only to produce well-established iodo- and bromocyclized products, but also to produce relatively scarce chlorocyclized products in high yields. Alkynes bearing aryl, alkyl, vinyl and TMS groups were also successfully cyclized.

In 2012, a novel method to prepare benzisothiazol-3(2*H*)-ones was presented by Xi and co-workers based on the copper-catalyzed reaction of *o*-halobenzamides with potassium thiocyanate (KSCN) in water.[299] The reaction proceeds *via* a tandem reaction with S–C and S–N bond formation. Moderate yields of the desired products were isolated (Scheme 4.172). A mthodo for the preparation of the seleno analog was reported in 2010 by Kumar's group.[300] As an efficient copper-catalyzed method for the synthesis of biologically important ebselen and related analogs containing an Se–N bond, good to excellent yields can be achieved (Scheme 4.172). This is the

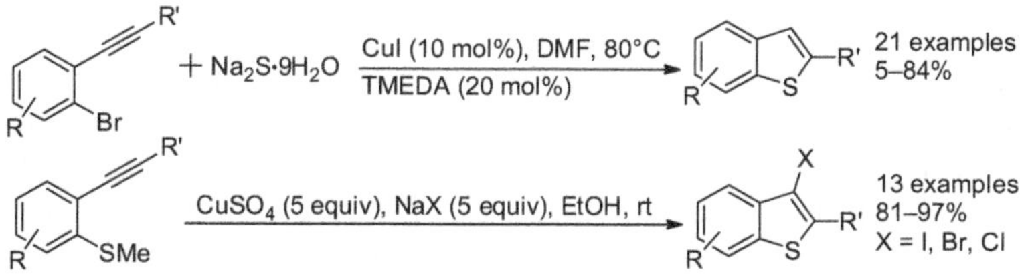

Scheme 4.171 Cu-promoted synthesis of benzo[*b*]thiophenes.

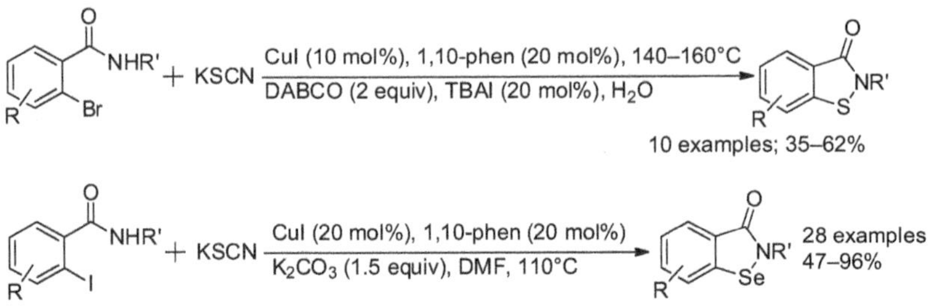

Scheme 4.172   Cu-catalyzed synthesis of benzisothiazol-3(2*H*)-ones.

first report of a catalytic reaction of selenation and Se–N bond formation. The copper-catalyzed reaction tolerates functional groups such as amide, hydroxyl, ether, nitro, fluoride and chloride groups. The best results are obtained by using a combination of potassium carbonate as a base, iodo- or bromoarylamide substrates, selenium powder and copper iodide catalyst.

4.2.17   Copper-Catalyzed Synthesis of Other Five-Membered Heterocycles

Lactams are an important class of heterocyclic compounds found in many biologically active natral products and medicinal agents. In 2010, Li and co-workers developed a copper-catalyzed intramolecular *C*-vinylation of activated methylene compounds to give their lactams.[301] Using CuI as the catalyst and 3,4,7,8-tetramethyl-1,10-phenanthroline as the ligand, 61–99% yields can be isolated in THF at 50 °C (Scheme 4.173).

Ma and Xie reported a methodology for the synthesis of 4-halo-5-hydro-xypyrrol-2(5*H*)-ones *via* an efficient sequential halolactamization–hydroxylation reaction of 4-monosubstituted 2,3-allenamides with CuX_2 (X = Br, Cl) as the promoter. The products were obtained in high yields (Scheme 4.174).[302] Halolactamization of fully substituted 2,3-dienamides afforded 4-halopyrrol-2(5*H*)-ones. In 2008, a copper-catalyzed reaction of carbonyl-ene-nitriles with carbon nucleophiles, such as aromatics and ketones, affording pyrrolin-2-ones (γ-lactam) was reported.[303] The desired products were obtained in excellent yield. The reaction mechanism involves addition reactions with a ketamine moiety of the 2-aza-2,4-cyclopentadie-none intermediate, which is formed *via* hydration of a nitrile moiety followed by dehydrative cyclization. This method might provide an efficient tool for the synthesis of various natural products containing a γ-lactam skeleton, such as salinosporamide A–C, omuralide, lactacystin, PI-091 and lucilactaene. More recently, a four-component procedure was reported for the synthesis of lactams.[304] Copper was applied as the catalyst and aldehydes, amines and alkynes were used as substrates. The desired products were obtained in good to excellent yields.

In 2013, a novel Cu(II)-catalyzed cyclization of α-diazo-β-oxoamides with amines was developed,[305] constituting a straightforward method to

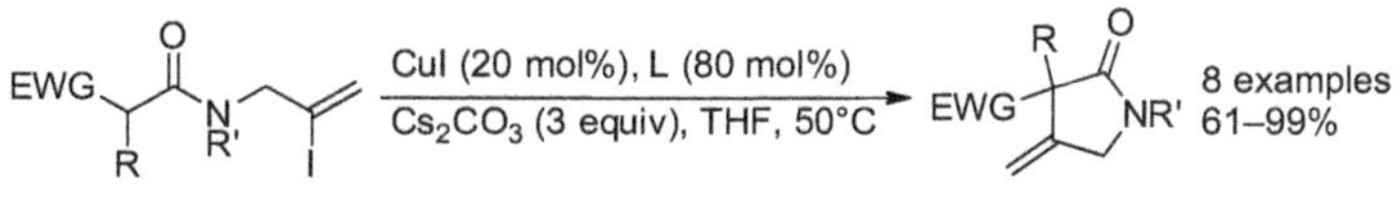

Scheme 4.173 Cu-catalyzed synthesis of lactams.

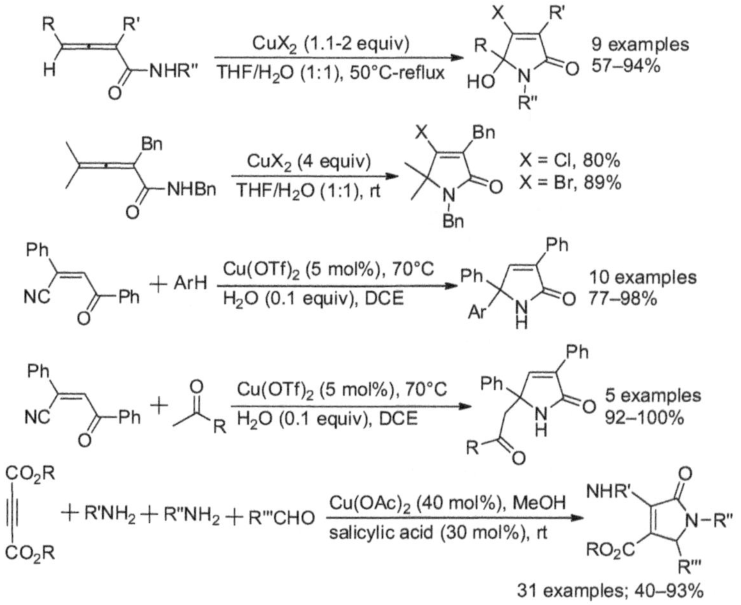

Scheme 4.174 Cu-promoted cyclization of allenamides.

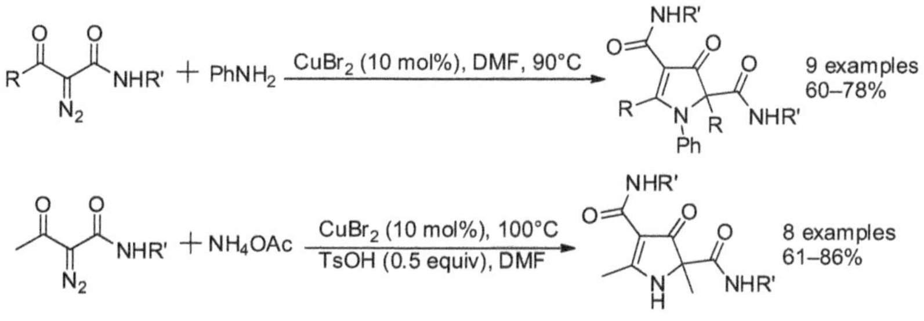

Scheme 4.175 Cu-catalyzed synthesis of pyrrol-3(2H)-ones.

construct pyrrol-3(2H)-one rings in good yields (Scheme 4.175). The intramolecular hydrogen-bonding effect in α-diazo-β-oxoamides plays an essential role in this reaction. A plausible reaction mechanism involving divergent generation and subsequent [2 + 3] cyclization of ketene and α-diazoimine intermediates was proposed.

Fu and co-workers developed a new copper-catalyzed 1,3-dipolar cycloaddition of terminal alkynes.[306] This transformation may rely on the transient formation of a copper acetylide to enhance the reactivity of the dipolar-ophile. By employing a phosphaferroceneoxazoline as a chiral bidentate ligand, they efficiently coupled a wide range of azomethine imines and alkynes to generate useful heterocycles in very good enantiomeric excess (Scheme 4.176). In 2011, Mizuno and co-workers found that the hetero-geneous catalyst $Cu(OH)_x$–Al_2O_3 can catalyze this transformation efficiently. Various desired products were produced in excellent yields in toluene at 40 °C.[307]

In 2011, Zhou's group developed an intermolecular cycloaddition of conjugated dienes and nitrene precursors to 3-pyrrolines, which usually give aziridines as their product.[308] This interesting transformation was realized by using copper(II) 1,1,1,5,5,5-hexafluoroacetylacetonate [$Cu(hfacac)_2$] as the active catalyst. This method is applicable to a wide array of dienes with good yields (Scheme 4.177). When 1,4-disubsituted dienes are used as substrates, good-to-excellent *cis* or *trans* selectivity can be obtained. Notably, the *cis* or *trans* preference depends on the nature of the substituents, rather than the diene geometry. Mechanistic studies revealed that the $[4+1]$ cycloaddition proceeds through diene aziridination and subsequent ring expansion. Among common copper catalysts, only [$Cu(hfacac)_2$] can efficiently catalyze both steps, which explains the unique efficiency of this catalyst.

Chemler and co-workers reported a novel copper-catalyzed intramolecular carboetherification of 4-pentenols in 2012.[309] This reaction efficiently provides fused- and bridged-ring tetrahydrofurans from common and unactivated alkenols (Scheme 4.178). Evidence for oxycupration as the al-kene addition mechanism was presented, notably in the form of ligand-based asymmetric catalysis. The C–C bond-forming step in the carboether-ification reaction is thought to occur *via* carbon radical arene addition and this step constitutes an efficient C–H functionalization. A novel, efficient, diastereoselective and enantioselective hydroetherification reaction, the

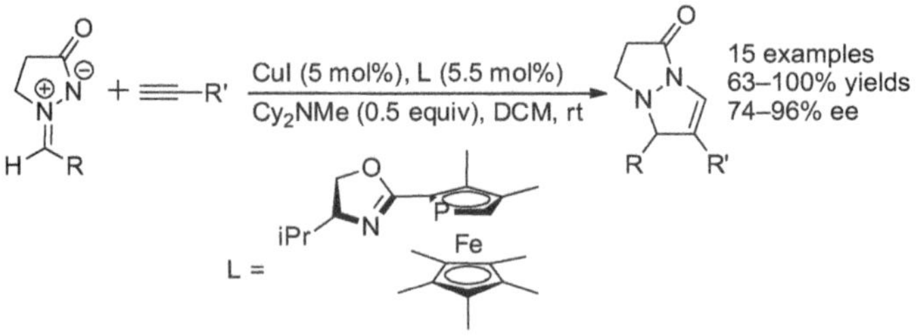

Scheme 4.176　Cu-catalyzed reaction of alkynes and azomethine imines.

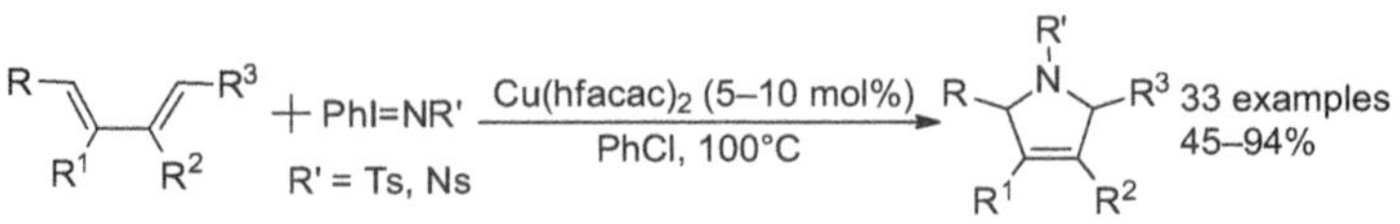

Scheme 4.177　Cu-catalyzed synthesis of 3-pyrrolines.

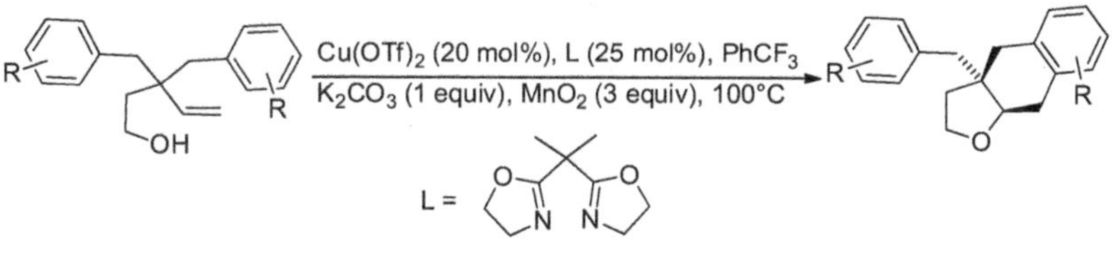

Scheme 4.178 Cu-catalyzed synthesis of tetrahydrofurans.

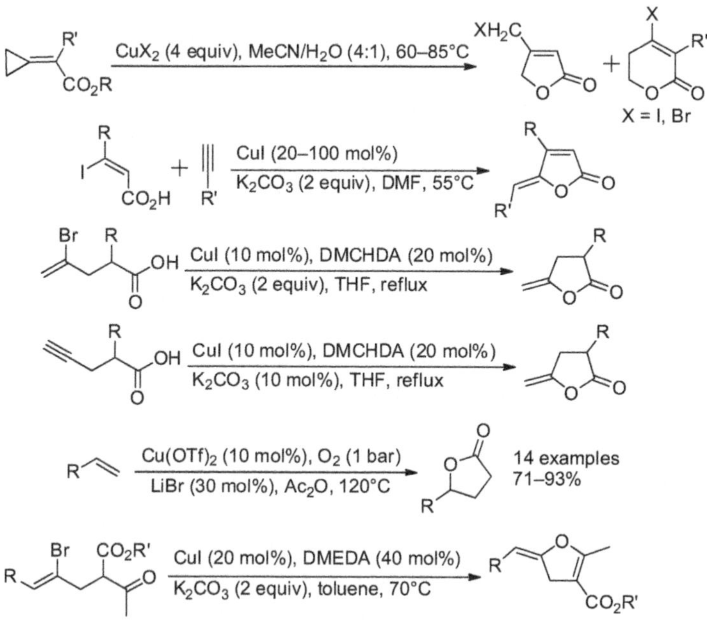

Scheme 4.179 Cu-catalyzed synthesis of lactones.

result of trapping the carbon radical intermediate with 1,4-cyclohexadiene in an intermolecular H-atom abstraction step, was also presented. This hydroetherification reaction could prove useful in generating 5-methyl-2,2-disubstituted tetrahydrofurans currently of interest to the perfume industry.

In 2002, a procedure for converting cyclopropylideneacetic acids (or esters) to 4-substituted 2(5H)-furanones or 3,4-substituted 5,6-dihydro-2H-pyran-2-ones was reported.[310] The reactions were performed with CuBr$_2$ (or CuI/I$_2$) as the catalyst system and moderate to good yields could be achieved in aqueous acetonitrile (Scheme 4.179). The selectivity of the reaction depended greatly on the reaction temperature. In 2009, an efficient and cheaper method for the synthesis of a (Z)-5-ylidene-5H-furan-2-one library under mild and more environmentally friendly conditions was developed.[311] These issues were addressed by replacing the expensive Pd catalysts with less expensive 20 mol% CuI. In general, a practical and general catalytic system [Cu(I) salts in DMF] for the efficient tandem coupling–heterocyclization reaction of (Z)-3-iodopropenoic acid derivatives with terminal alkynes was developed, and an important range of (*Z*)-5-alkylidene- or arylidene-5*H*-furan-2-ones and also 3-substituted isocoumarins can be easily formed. Compared with other copper-catalyzed Sonogashira-type reactions, it was found that the reaction temperature is significantly lower (90–140 *versus* 50 °C), indicating a strong effect of the close carboxylate function. As a possible explanation, the authors believe that the oxidative addition of copper into the C–I bond would be favored through an intramolecular process from a copper carboxylate. Also in 2009, a CuI–trans-*N*,*N'*-dimethylcyclohexane-1,2-diamine catalyst system was reported for the *o*-vinylation of carboxylic acids.[312] A number of carboxylic acids underwent efficient intramolecular *o*-vinylation with vinyl bromides, leading to the synthesis of the corresponding five- and six-membered enol lactones. The same catalytic system also led to the efficient cycloisomerization of alkynoic acids. Fang and Li found that the reaction of bromoenones can lead to the formation of dihydrofurans with copper as the catalyst.[313] By this copper-catalyzed intramolecular *o*-vinylation of carbonyl compounds with vinyl bromides, five-, six- and even seven-membered cyclic alkenyl ethers were obtained efficiently with β-keto esters as nucleophiles. In 2010, Jiang and co-workers developed a new copper-catalyzed oxidative [3 + 2] cycloaddition of alkenes with anhydrides using oxygen as the sole oxidant to afford γ-lactones.[314] This catalyzed cyclization process has a broad substrate scope and affords γ-lactones in good to excellent yields.

Ma and co-workers developed a convenient and efficient method for the synthesis of 4-bromo- or 4-chloro-2,5-dihydro[1,2]oxaphosphole 2-oxides by the reaction of monoesters of 1,2-allenylphosphonic acids with CuX_2.[315] The desired products were produced in high yields under standeard conditions (Scheme 4.180). This should be a very useful synthetic strategy for the construction of relatively more complex phosphorus-containing compounds and provides the possibility of finding potentially bioactive hits in medicinal chemistry.

A novel cascade reaction of nitrones with sulfonyl azides and alkynes mediated by copper(I) for the one-step synthesis of a diverse range of imidazolidin-4-ones was developed in 2011.[316] This reaction proceeds *via* sulfonyl azide–alkyne and ketenimine–nitrone double cycloaddition sequences, which are normally rare. The reported protocol involves nitrone as the cycloaddition partner, which was significant in itself, as the addition proceeds across C=N bonds of *N*-sulfonylketenimines, where additions to C=C bonds are most common. The use of a copper(I)–Y zeolite as a heterogeneous Cu(I) source allows for the rapid and simple isolation of the reaction products by simple filtration, in addition to other advantages such as recyclability

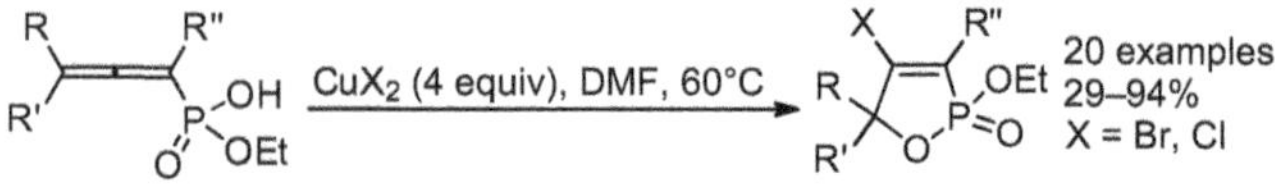

Scheme 4.180   Cu-catalyzed cyclization of 1,2-allenylphosphonic acids.

and minimization of metallic waste. In 2009, Shi and co-workers reported a methodology for the synthesis of 4,4-disubstituted 2-imidazolidinones.[317] A variety of disubstituted terminal alkenes were effectively diaminated using CuCl as catalyst and di-*tert*-butyldiaziridinone as a nitrogen source, and various 4,4-disubstituted 2-imidazolidinones were produced in good to excellent yields (Scheme 4.181). In addition, the synthesis of the 4,4-disubstituted 2-imidazolidinone Sch 425078 (a potent NK1 antagonist) was achieved in three steps using this diamination. The ability to remove selectively one or two protecting groups would provide opportunities to introduce different substituents on the nitrogens if desired. Further, the same group described a process for the synthesis of imidazolinones *via* an α-amination reaction of aryl ketones using CuCl as catalyst and di-*tert*-butyldiaziridinone as the nitrogen source.[318] A variety of imidazolinone derivatives were prepared in moderate yields under mild conditions.

Lee *et al.* describled a copper-catalyzed synthesis of 1,3-oxazolidines in 2002.[319] The desired products were obtained in high yields by copper-catalyzed addition of ethyl diazoacetate to imines in the presence of acetone (Scheme 4.182). Hydrolysis of the oxazolidines with 6 M HCl to give 1,2-amino

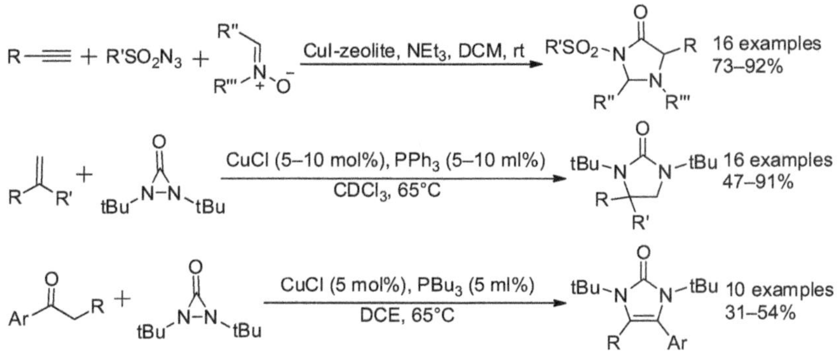

Scheme 4.181 Cu-catalyzed synthesis of imidazolidinones.

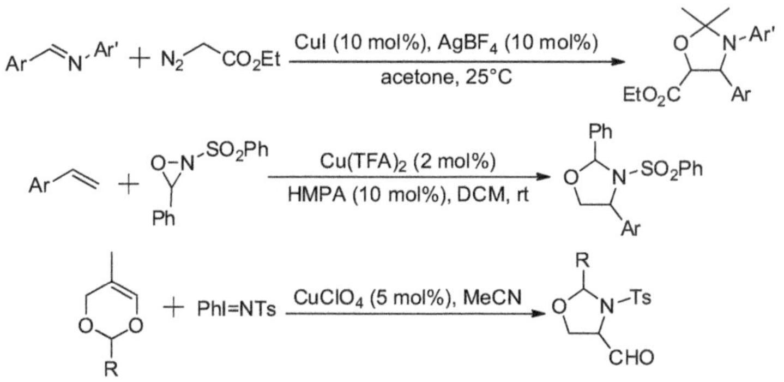

Scheme 4.182 Cu-catalyzed synthesis of 1,3-oxazolidines.

alcohols was also carried out. In 2007, Yoon and co-workers also found that oxazolidines can be prepared from alkenes.[320] In the presence of a copper(II) catalyst, oxazolidines were produced by the aminohydroxylation of styrenes and electron-rich alkenes with oxaziridines. Later, they succeeded in extending their methodology to 1,3-dienes. On adding a chiral ligand, high enantios- electivity was observed. Additionally, the reaction of dioxins with nitrenes generated from PhI=NTs could also lead to oxazolidine derivatives.[321]

A simple and efficient synthesis of oxazolidinones through a copper- catalyzed coupling of aldehydes, amines and terminal alkynes under at- mospheric pressure of carbon dioxide was reported in 2008 by Yoo and Li.[322] Propargylamine was found to be the reaction intermediate. The electron rich amine can react with CO_2 to form carbamic acid and essentially increase the effective concentration of CO_2 in solution and aid the carboxylative cycliza- tion step. In 2012, a facile approach to polysubstituted oxazolidin-2-ones *via* a copper(I)-catalyzed tandem decarboxylative–carboxylative cyclization of a propiolic acid, a primary amine and an aldehyde was developed.[323] This new multicomponent coupling constitutes an efficient methodology to provide the corresponding oxazolidin-2-ones in good yields (Scheme 4.183).

Isoxazolidines are useful in organic synthesis, drug discovery and chemical biology endeavors and many novel procedures have been de- veloped for the preparation of these compounds (Scheme 4.184). In 1999, a copper(II)–bisoxazoline-catalyzed diastereo- and enantioselective 1,3-dipolar

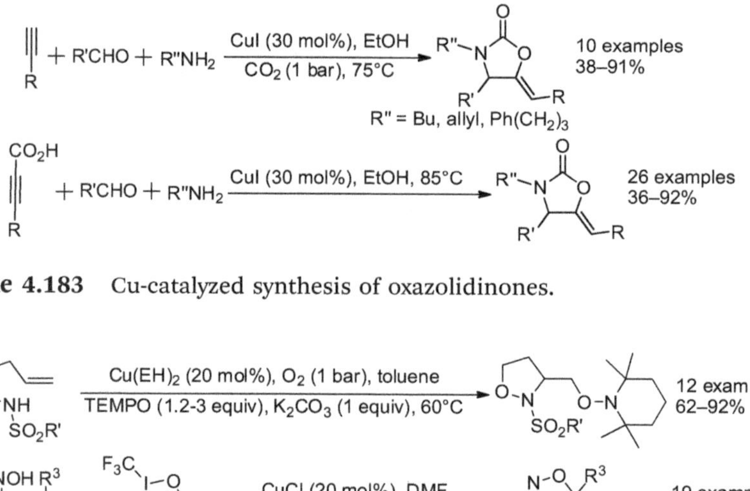

Scheme 4.183 Cu-catalyzed synthesis of oxazolidinones.

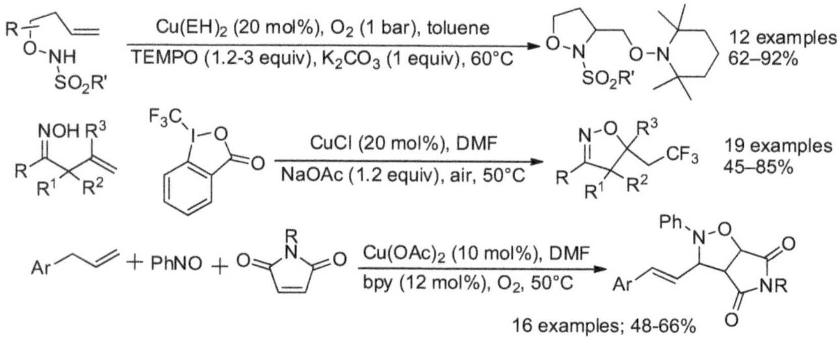

Scheme 4.184 Cu-catalyzed synthesis of isoxazolidines.

cycloaddition reaction between electron-poor nitrones that can coordinate to the chiral Lewis acid in a bidentate fashion and electron-rich alkenes was developed. In the presence of $Cu(OTf)_2$–bisoxazoline as the catalyst, the nitrones react smoothly with vinyl ethers at room temperature, giving iso-xazolidines in good yields and diastereoselectivity and with enantioselec-tivities of up to 94% *ee*, especially for ethyl vinyl ether and 2-methoxypropene.[324] The reaction involves the approach of the alkene fragment of the vinyl ether to the *re*-face of the nitrone. On the basis of the absolute stereochemistry of the isoxazolidine formed, a pentacoordinated or a tetrahedral intermediate is proposed to account for the absolute stereo-chemistry. In the former intermediate, the chiral bisoxazoline ligand and the nitrone occupy four of the five available coordination sites at the copper(II) center and the vinyl ether is coordinated to the fifth coordination site. This strategy has been studied by different groups and extended to α'-hydroxyenones and -alkenones.[325,326] Alternatively, a new stereoselective synthesis of methyleneoxy-substituted isoxazolidines was reported by Chemler and co-workers in 2012.[327] The method involves copper-catalyzed aminooxygenation–cyclization of *N*-sulfonyl-*O*-butenylhydroxylamines in the presence of (2,2,6,6-tetramethylpiperidin-1-yl)oxyl radical (TEMPO) and O_2 and provides substituted isoxazolidines in excellent yields and diastereos-electivities. They also demonstrated selective mono N–O reduction followed by oxidation of the remaining N–O bond to give a 2-amino-γ-lactone. Re-duction of the γ-lactone provides the corresponding aminodiol. In 2011, a catalytic method employing economical copper–diimine catalysts for the oxidative cyclization of carbamates and sulfamates to corresponding five- and six-membered *N,O*-heterocycles was developed.[328] Both benzylic and less reactive saturated substrates are effectively cycloaminated, in contrast to corresponding (diimine)Cu-catalyzed intermolecular reactions that are typically ineffective for unactivated hydrocarbons. The hydrolytic convert-ibility of these products to 1,2- and 1,3-amino alcohol derivatives further amplifies their synthetic value and portends their valuable application in the stereoselective synthesis of complex molecules. Additonally, a mild and ef-ficient copper-catalyzed trifluoromethylation reaction that involves the cyclization of oximes was developed by Liang and co-workers in 2013.[329] This method provides convenient access to a variety of useful CF_3-containing 4,5-dihydroisoxazoles by constructing a C–CF_3 bond and a C–O bond in one step. Moreover, a triple cascade was developed by Huang's group using a simple copper catalyst to *trans*-selectively access bicyclic isoxazolidines in a one-pot manner.[330] This strategy features the *in situ* generation of nitrones and subsequent trapping by [3 + 2] cycloaddition. In this method, copper serves three catalytic functions: as a Lewis acid for the ene reaction, as an organometallic for aerobic oxidation and as a Lewis acid for an *endo*-selective [3 + 2] cycloaddition. The successful merging of aerobic oxidation and Lewis acid catalysis demonstrated efficient cascade synergy.

A copper(II) triflate catalyzed the ring-opening of aryl- and vinyl-substi-tuted epoxides with various carbonyl compounds to give 1,3-dioxolanes was reported in 2005, and earlier with $CuSO_4$ in 1978.[331] The reaction conditions are mild and the desired products are obtained in good yields (Scheme 4.185). Alkyl- and alkoxycarbonyl-substituted epoxides remain unchanged under the reaction conditions. This allows the selective opening of aryl-substituted epoxides in the presence of alkyl-substituted epoxides. In 2006, copper(II) triflate was shown to be an efficient catalyst for the isomerization–cyclization of α-methallyloxycarboxylic acids, and aluminum(III) triflate is also active for this transformation.[332] This tandem reaction was carried out efficiently and allowed the formation of 2,5-disubstituted 1,3-dioxolan-4-ones in yields of 60–70% (Scheme 4.185). The same isomerization–cyclization reaction was extended to α-thiomethallyloxycarboxylic acids for the synthesis of oxathiolanones and oxathianones. The tandem process also occurred with α-hydroxymethallyl esters. Other allylic ethers underwent allylic deprotection. The simplicity of the catalytic system, the use of commercially available Cu(II) or Al(III) triflates without the need for a ligand and the relatively mild reaction conditions are to be emphasized.

In 2013, Wang and co-workers developed the first catalytic asymmetric synthesis of fluorinated imidazolidines via Cu(I)/(S,R_p)-PPFOMe-catalyzed 1,3-dipolar cycloaddition of azomethine ylides with various fluorinated imines.[333] The optimal catalytic system exhibited extremely high reactivity, excellent diastereoselectivity, good enantioselectivity and broad substrate scope (Scheme 4.186). The ready availability of the starting materials and the great importance of the enantioenriched fluorinated compounds make this methodology particularly interesting in synthetic chemistry.

An interesting copper-catalyzed intra- and intermolecular alkoxylation of azoles was reported in 2013 (Scheme 4.187).[334] This reaction is a rare

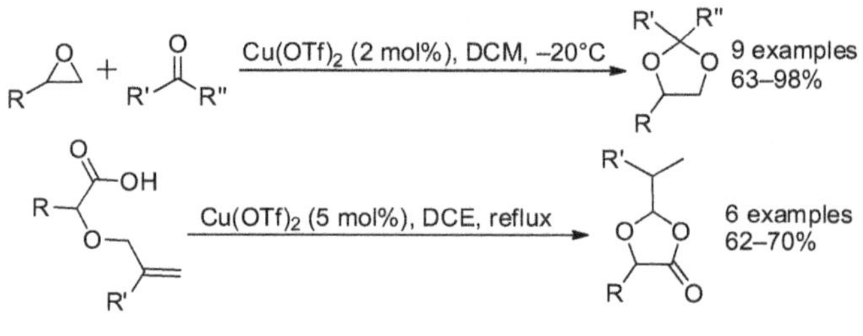

Scheme 4.185   Cu-catalyzed synthesis of 1,3-dioxolanes.

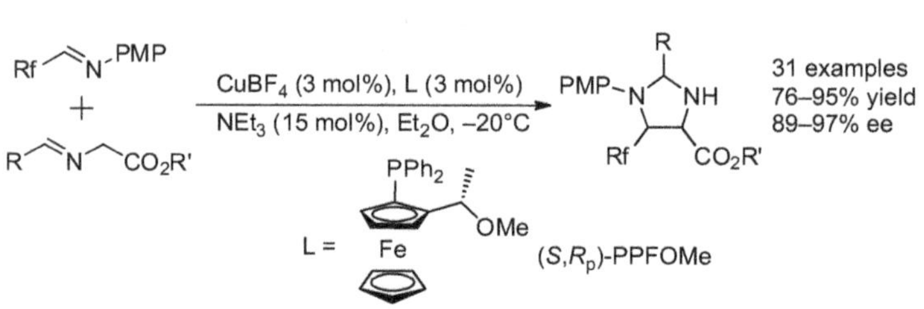

Scheme 4.186   Cu-catalyzed synthesis of imidazolidines.

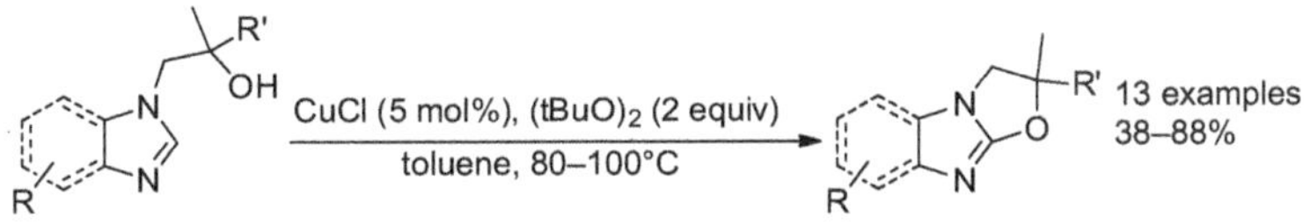

Scheme 4.187   Cu-catalyzed alkoxylation of azoles.

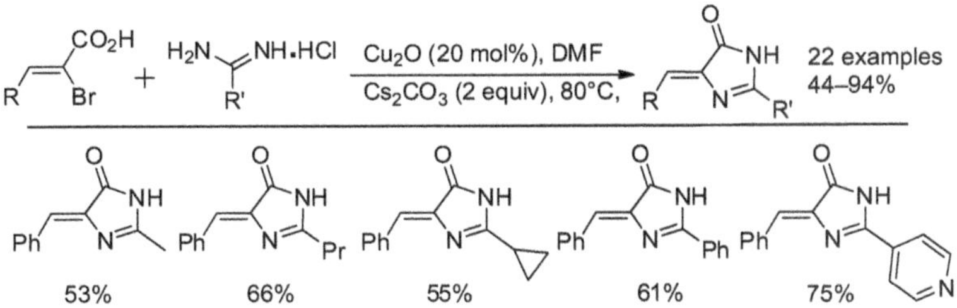

Scheme 4.188   Cu-catalyzed synthesis of imidazolone derivatives.

example of transition metal-catalyzed C–H alkoxylation of heteroaromatic compounds. In addition, the alkoxylation reaction proceeded well even on the gram scale. In most intermolecular alkoxylations, the use of an excess amount of alcohol (in some cases alcohols are used as a solvent) is indispensable to promote the alkoxylation reaction efficiently, but this alkoxylation reaction proceeded using only 1 equiv. of alcohol. To gain some insight into the reaction mechanism to establish if any radical intermediates are generated, a modol reaction was carried out in the presence of a radical scavenger, (2,2,6,6-tetramethylpiperidin-1-yl)oxyl (TEMPO) (5.0 mol%). It was found that the yield of the product was not affected (87%), hence the authors suggested that this alkoxylation does not proceed *via* a radical pathway. However, in our opinion, in order to exclude a radical pathway, at least an equimolar amount of TEMPO should be added.

In 2010, Fu and co-workers reported the synthesis of imidazolone derivatives whihc are biologically and pharmaceutically active molecules and the chromophores of the fluorescent proteins.[335] They found a simple and efficient approach to 4-arylidene-2-alkyl-4,5-dihydro-1*H*-imidazol-5-ones (2,4-disubstituted imidazolones) and the protocol uses readily available 2-bromo-3-alkylacrylic acids and amidines as the starting materials without the addition of any ligand or additive. The reactions were performed under mild conditions. All the products were obtained in good to excellent yields (Scheme 4.188).

In 2011, a direct access to symmetrical and unsymmetrical 2,5-disubstituted [1,3,4]-oxadiazoles was developed by Patel and co-workers (Scheme 4.189).[336] The reaction proceeds through an imine C–H functionalization of *N*-arylidenearoylhydrazides using a catalytic quantity of Cu(OTf)$_2$. This is the first example of amidic oxygen functioning as a nucleophile in a Cu-catalyzed oxidative coupling of an imine C–H bond.

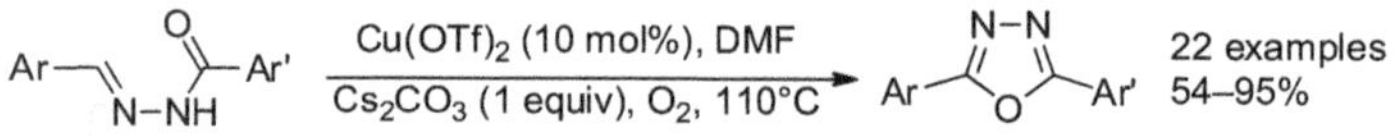

Scheme 4.189 Cu-catalyzed synthesis of oxadiazoles.

> **General Procedure** (Scheme 4.189): An oven-dried flask was charged with Cu(OTf)$_2$ (0.0362 g, 0.10 mmol), Cs$_2$CO$_3$ (0.326 g, 1 mmol), *N*-benzylide-nebenzohydrazide (0.224 g, 1 mmol) and solvent DMF (2 mL). The flask was heated in a preheated oil-bath at 110 °C for 16 h until complete consumption of the starting material (monitored by TLC). The reaction mixture was then cooled and admixed with water (5 mL). The mixture was extracted with ethyl acetate (2×15 mL) and the combined organic layer was dried over anhydrous Na$_2$SO$_4$. Concentration *in vacuo* followed by silica gel column purification (eluent: EtOAc–hexane) gave the desired product.

These reactions can be performed in an air atmosphere and in the presence of moisture, making it exceptionally practical for application in organic synthesis.

4.3   Six-Membered Heterocycles

In 2009, a cascade coupling–condensation process to produce poly-substituted 1,2-dihydroisoquinolines was reported.[337] The starting material could be easily assembled *via* a CuI-catalyzed three-component reaction of 2-bromobenzaldehydes, 1-alkynes and primary amines. This advantage allows the elaboration of functionalized 1,2-dihydroisoquinolines in a convenient manner. All the desired products were isolated in moderate to good yields (Scheme 4.190). In case of ethyl 3-oxo-3-phenylpropanoate as a substrate, no desired product was isolated. These results could be rationalized by the steric effect of the β-keto esters.

Chan and co-workers developed an efficient synthetic route to *trans*-2,5-dihydro-1*H*-pyrroles and 1,2-dihydroquinolines in 2010.[338] The procedure relies on copper(II) triflate-catalyzed intramolecular hydroamination of homoallylic amino alcohols under mild and operationally straightforward conditions; all the products were prepared in good to excellent yields (Scheme 4.191). For reactions leading to the *trans*-2,5-dihydro-1*H*-pyrrole product, yields of 52–83% along with *trans* selectivities of up to >99 : 1 *dr* and *ee* values up to 97% were accomplished from enantioenriched 1-(tosyl-amino)pent-4-en-2-ols ranging from 91 to 99% *ee*. Without the need for inert and moisture-free conditions, reactions involving 1-[2-(tosylamino)phenyl]-but-3-en-1-ols afforded the corresponding 1,2-dihydroquinoline products in excellent yields of up to 99% and with complete chemoselectivity. The mechanism was suggested to involve copper(II)-mediated dehydration of the homoallylic amino alcohol. Protonation of the resultant copper(II)-activated aminodiene is then thought to trigger subsequent intramolecular hydro-amination to give the partially hydrogenated nitrogen heterocycle.

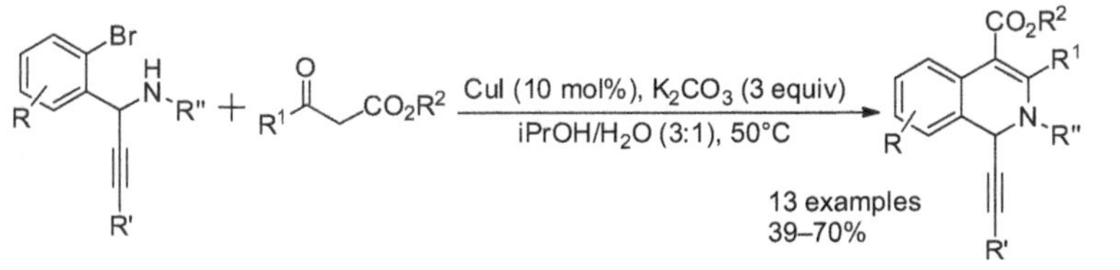

Scheme 4.190 Cu-catalyzed synthesis of 1,2-dihydroisoquinolines.

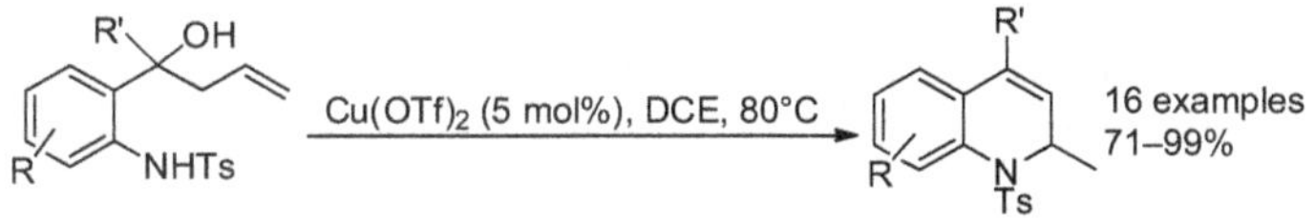

Scheme 4.191 Cu-catalyzed synthesis of 1,2-dihydroisoquinolines.

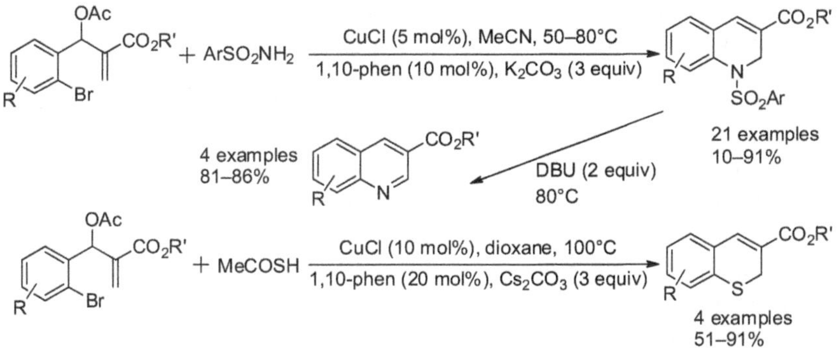

Scheme 4.192 Cu-catalyzed cyclization of Baylis–Hillman acetates.

More recently, a variety of substituted quinoline/pyridine, thiochromene and naphthalene derivatives, which might be of biological and medicinal value, were synthesized by copper-catalyzed domino S_N2'–coupling, S_N2'–deacylation–coupling and S_N2'–coupling–elimination reactions.[339] The method provides a general and convenient approach to the synthesis of various substituted cyclic compounds from the corresponding Baylis–Hillman acetates and *N-/S-/C-*nucleophiles. The starting materials are readily available, the procedures are simple and the cyclized products could be synthesized in moderate to excellent yields (Scheme 4.192). The quinolines and dihydroquinolines were selectively assembled in one pot. Furthermore, this Cu-catalyzed domino method can be successfully applied for constructing three different types of cyclic moieties by varying the nucleophiles. In 2012, a copper-catalyzed intramolecular amination followed by *in situ* oxidation of primary allylamines generated from the Morita–Baylis–Hillman adducts of 4-iodopyrazolecarbaldehydes for preparing substituted pyrazolo[4,3-*b*]pyridines was described.[340]

In 2012, a novel conversion of mono- or dialkyl-substituted furans into 1,2-dihydropyridines by reaction with PhI = NTs at room temperature was reported.[341] The reaction is catalyzed by complexes of general formula TpxM (M = Cu, Ag) and consists of a one-pot procedure with four consecutive catalytic cycles. Furan aziridination is followed by aziridine ring opening, transimination reaction, inverse electronic demand aza-Diels–Alder reaction and a final hydrogen elimination reaction. A mechanism of the overall transformation is proposed in which the metal complex plays a crucial role along the reaction pathway. Latern, they found that the presence of Tp^{Br3} Cu(NCMe) [Tp^{Br3} = hydrotris(3,4,5-tribromopyrazolyl)borate] as the catalyst can promote this reaction under very mild conditions.[342] The use of a furan and ethyl vinyl ether with PhI=NTs led to the corresponding 1,2,3,4-tetrahydropyridine.

A Lewis acid catalyst-tuned reaction of *N'*-(2-alkynylbenzylidene)hydrazides with diethyl phosphite was described by Wu and co-workers in 2009.[343] Isoquinolin-1-yl phosphonate was generated when copper triflate was utilized as the catalyst, whereas 2-amino-1,2-dihydroisoquinolin-1-yl phosphonate was obtained in the presence of palladium acetate. Good yields can be achieved in general (Scheme 4.193).

In 2009, a facile and direct synthesis of diversely substituted, medicinally -useful indolo- and pyrrolo[2,1-*a*]isoquinolines in good yields and with excellent regioselectivity was reported.[344] The chemistry appears to involve the preferential nucleophilic addition of indoles and pyrroles to the *o*-haloarylalkynes over *N*-arylation of the aryl halide. From a synthetic point of view, the net transformation involves a one-step conversion of simple, readily available starting materials into an interesting class of heterocyclic derivatives in good yields (Scheme 4.194). An inexpensive compound, hydroxymethylbenzotriazole, is used as a ligand along with inexpensive CuI, thereby increasing the overall utility of this reaction.

Liang and co-workers reported a copper-catalyzed transformation of 2-azido-3-(2-iodophenyl)acrylates and terminal alkynes in 2010.[345] In the presence of a copper(II) catalyst without ligands, a wide range of [1,2,3]triazolo[5,1-*a*]isoquinolines were produced in good yields (Scheme 4.195). These compounds favor expulsion of N$_2$ in refluxing acetic

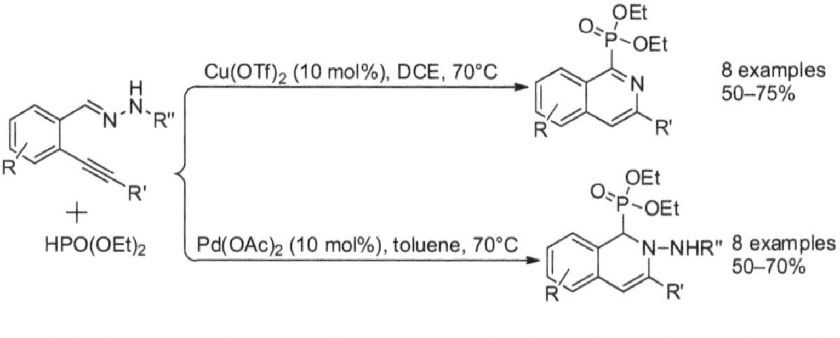

Scheme 4.193   Cu-catalyzed cyclization of *N'*-(2-alkynylbenzylidene)hydrazides.

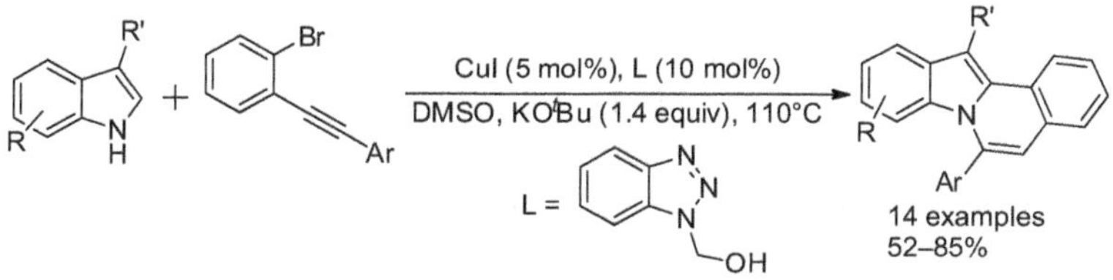

Scheme 4.194 Cu-catalyzed synthesis of pyrrolo[2,1-*a*]isoquinolines.

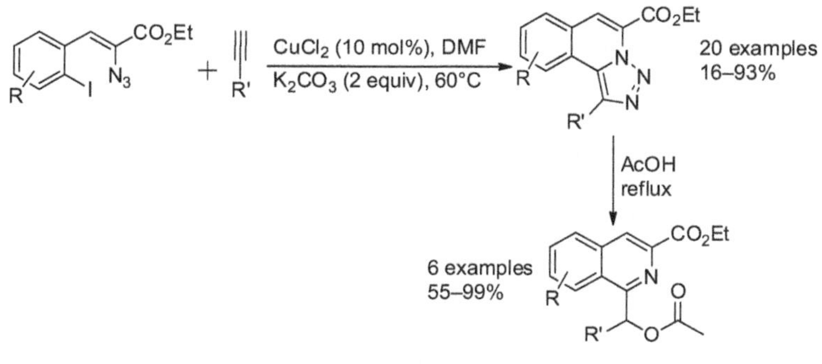

Scheme 4.195 Cu-catalyzed synthesis of [1,2,3]triazolo[5,1-*a*]isoquinolines.

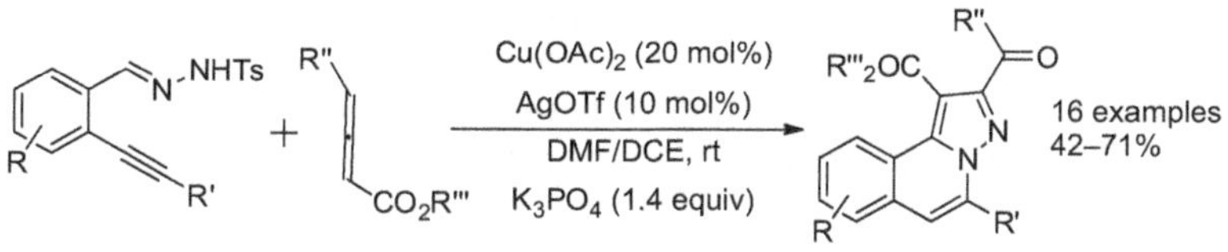

Scheme 4.196 Cu-catalyzed synthesis of *H*-pyrazolo[5,1-*a*]isoquinolines.

acid to form 1,3-disubstituted isoquinolines. The procedure is simple, economical and efficient. The tandem reaction proceeds under mild conditions and tolerates various functional groups.

In 2012, a cascade reaction of *N'*-(2-alkynylbenzylidene)hydrazides with allenoate in the presence of dioxygen co-catalyzed by silver triflate and copper(II) acetate under mild conditions was described by Wu and co-workers.[346] This methodology provides an efficient approach to 2-carbonyl-*H*-pyrazolo[5,1-*a*]isoquinolines (Scheme 4.196). The silver triflate–copper(II) acetate cooperative catalysis is essential for successful conversion. A possible mechanism is illustrated in which the reaction proceeds through a peroxy-copper(III) intermediate.

More recently, a novel and efficient copper-catalyzed method for the synthesis of tetrazolo[5,1-*a*]isoquinolines was developed by Fu and co-workers.[347] The protocol uses readily available substituted 5-(2-halophenyl)-1*H*-tetrazoles and alkynes as the starting materials and inexpensive copper(I) iodide as the catalyst. The domino reaction underwent sequential copper-catalyzed Sonogashira cross-coupling and intramolecular addition of the NH from a tetrazole group to an internal alkyne. Further, a Cu-catalyzed Sonogashira-type coupling of 3-(2-bromophenyl)pyrazoles with terminal alkynes combined with a subsequent electrophilic cyclization in the same pot was developed to afford pyrazolo[5,1-*a*]isoquinolines.[348] The transformation proceeds without theneed to use Pd or ligands, while an Ag additive is beneficial. A copper-catalyzed three-component reaction of 1-(2-bromophenyl)-3-phenylprop-2-yn-1-one, hydrazine hydrochloride and ethyl 2-cyanoacetate to give *H*-pyrazolo[5,1-*a*]isoquinolines was also developed.[349] This new, concise and efficient one-pot approach offers a straightforward method for the synthesis of *H*-pyrazolo[5,1-*a*]isoquinolines, and products containing various functional groups were obtained in moderate to good yields. The reaction proceeds through 3-(2-bromophenyl)pyrazoles as intermediates. Lu and Fu reported a convenient and efficient copper-catalyzed cascade method for the synthesis of benzimidazoisoquinoline derivatives in 2011.[350] The protocol uses inexpensive CuI as the catalyst and readily available substituted 2-(2-halophenyl)benzoimidazoles (from reactions of substituted benzene-1,2-diamines with 2-halobenzoic acids in acidic medium) and alkyl cyanoacetates as the starting materials. The copper-catalyzed reactions of substituted 2-(2-bromophenyl)benzoimidazoles with alkyl cyanoacetates proceeded very well under mild conditions and the corresponding benzimidazoisoquinolines with amino and carboxylate groups were obtained in good to excellent yields. These methods provide diverse and useful pathways for the synthesis of isoquinolines (Scheme 4.197).

In 2010, a new and efficient method for the synthesis of aza-fused polycyclic quinolines was described.[351] The protocol includes an intermolecular

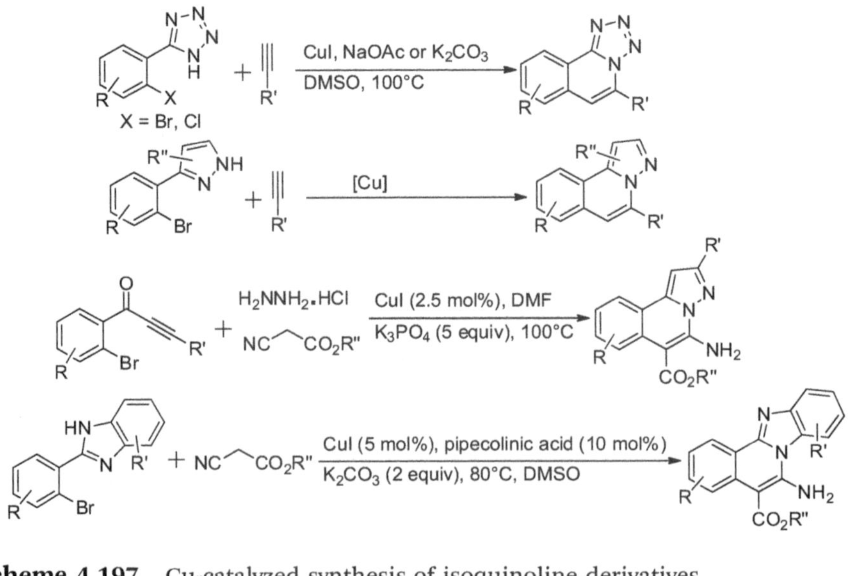

Scheme 4.197   Cu-catalyzed synthesis of isoquinoline derivatives.

condensation followed by a copper-catalyzed intramolecular C–N coupling reaction. The method was applied to a wide range of 2-iodo-, 2-bromo- and 2-chloroarylaldehyde substrates to yield the aza-fused polycyclic quinolines in good yields (Scheme 4.198). Copper-catalyzed tandem coupling of 1,4-dihalo-1,3-dienes with azoles *via* an N–H bond and its adjacent C–H bond activation was described by Xi and co-workers.[352] The reaction exhibits good regioselectivity when an unsymmetrical 1,4-dihalo-1,3-diene is employed. Notably, this approach constitutes an unprecedented direct alkenylation-based domino process that is applicable to the synthesis of nitrogen-bridgehead azolopyridine derivatives.

N-Propargylic β-enaminones were used as intermediates for the synthesis of polysubstituted pyrroles and pyridines in 2008.[353] The best results were obtained using DMSO as solvent. In the presence of Cs$_2$CO$_3$, *N*-propargylic β-enaminones were cyclized to pyrroles in good to high yields, whereas omitting bases and using CuBr led to the selective formation of pyridines. More recently, Jiang and co-workers reported a copper-catalyzed oxidative synthesis of pyridines (Scheme 4.199).[354] With neat conditions and facile operation, the fragment-assembling strategy affords a broad range of 2,4,6-trisubstituted pyridines in up to 95% yield from simple and readily available

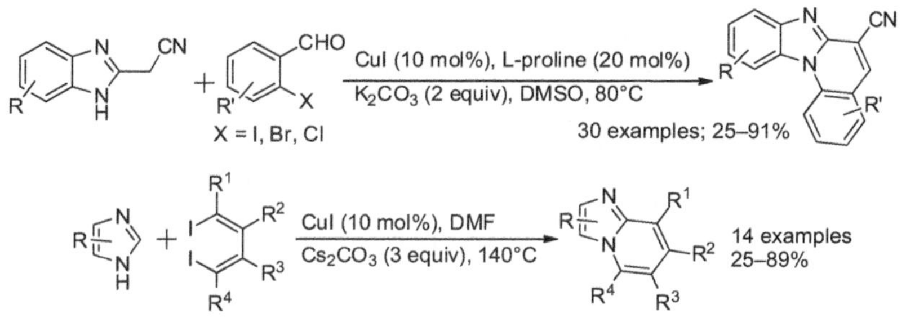

Scheme 4.198 Cu-catalyzed synthesis of azolopyridines.

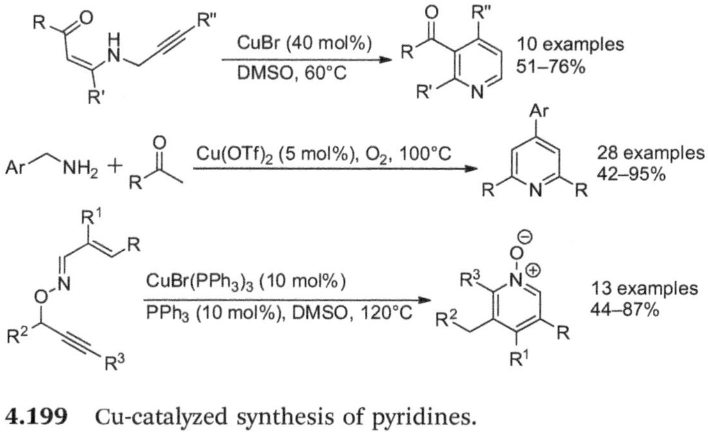

Scheme 4.199 Cu-catalyzed synthesis of pyridines.

starting materials. Interestingly, when pyridin-2-yl methylamine was employed as the substrate, an α-alkylation reaction of ketones readily occurred to give β-(pyridin-2-yl) ketones instead of the 2,4,6-trisubstituted pyridines. Pyridine *N*-oxides have recently attracted much attention in catalytic C–H functionalization studies; a new approach to multi-substituted pyridine derivatives was reported in 2010.[355] The reaction works in a catalytic, efficient and regioselective manner and the products were obtained in good yields (Scheme 4.199).

A copper(I)-catalyzed domino four-component coupling–cyclization of 2-ethynylbenzaldehydes, paraformaldehyde, secondary amine and *t*-BuNH₂ was reported in 2008.[356] The reactions works in DMF and leads to the direct and efficient formation of 3-(aminomethyl)isoquinolines in good to high yields (Scheme 4.200). Ma and co-workers developed a CuI-catalyzed coupling of 2-halobenzylamines with β-keto esters or 1,3-diketones in 2-propanol using K₂CO₃ as base.[357] 1,2-Dihydroisoquinolines were produced *via* coupling–condensative cyclization, and underwent smooth dehydrogenation under an air atmosphere to afford substituted isoquinolines. More recently, Fu's group developed a novel and efficient copper-catalyzed approach to 1,3-diaminoisoquinoline derivatives.[358] The protocol uses readily available substituted 5-(2-halophenyl)-1*H*-tetrazoles, alkyl 2-cyanoacetates and malononitrile as the starting materials and inexpensive CuI as the catalyst, and the corresponding 1,3-diaminoisoquinoline derivatives were obtained in moderate to good yields. The domino reaction proceeded through sequential copper-catalyzed *C*-arylation, intramolecular nucleophilic addition and denitrogenation of tetrazoles. The synthesized 1,3-diaminoisoquinoline derivatives containing various functional groups are important biological and pharmaceutical molecules.

In 2010, Wu and co-workers developed a method for the synthesis of 2-fluoromethylated quinolones.[359] The mechanism proceeds through the reaction of *N*-aryl-fluorinated imidoyl iodides with terminal alkynes and gave

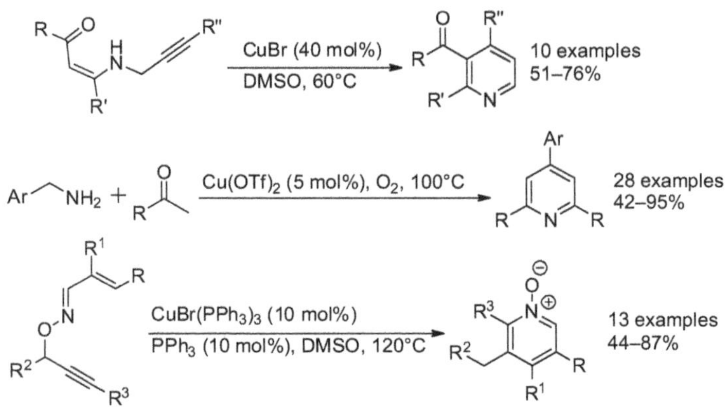

Scheme 4.200   Cu-catalyzed synthesis of isoquinolines.

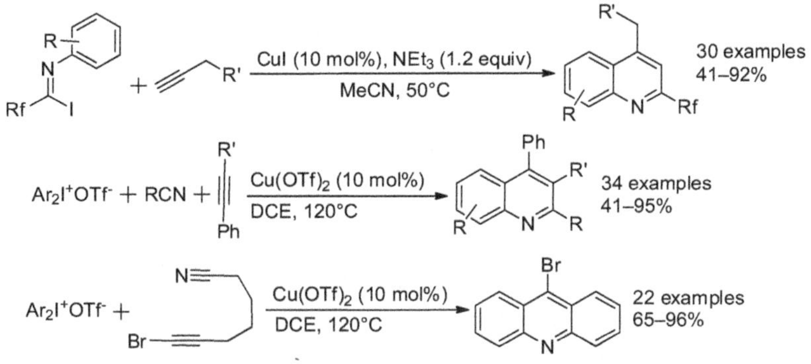

Scheme 4.201 Cu-catalyzed synthesis of quinolones.

the desired products in good yields with catalysis by copper(I) iodide alone (Scheme 4.201). In 2013, Chen and co-workers reported an efficient and regioselective procedure for the synthesis of multiply substituted quinolines from diaryliodoniums, alkynes and nitriles.[360] This $[2 + 2 + 2]$ cyclization is catalyzed by Cu(OTf)$_2$ and the aryl group of the diaryliodoniums serves as a C$_2$ building block. This strategy marks a significant departure from known approaches based on condensation chemistry (*e.g.* Combes synthesis, Conrad–Limpach–Knorr synthesis and Friedländer synthesis) and permits variations in the substitution patterns on the quinolines. The cascade annulation is believed to involve a series of cationic intermediates, thus ensuring an efficient process and high regioselectivity. Soon afterwards, they reported another concise construction of polycyclic quinolines *via* intramolecular $[2 + 2 + 2]$ annulation of ω-cyano-1-alkynes with diaryliodonium salts.[361] The process produced polycyclic quinolines in high yields with readily available starting materials and was tolerant to halogen substituents.

In 2006, a methodology based on 2-aminobenzyl alcohol and ketones in dioxane at 100 °C and using a catalytic amount of CuCl$_2$ along with KOH under an O$_2$ atmosphere to afford the corresponding quinolines in good yields was developed.[362] 2-Aminobenzyl alcohol was oxidatively coupled and cyclized with various aldehydes in a step-by-step procedure. Initial treatment of 2-aminobenzyl alcohol in the presence of CuCl$_2$ and KOH in dioxane under an O$_2$ atmosphere and subsequent addition of aldehyde to the mixture followed by stirring under an argon atmosphere gave 3-substituted quinolines in moderate to good yields (Scheme 4.202). Liu and co-workers developed a simple and practical copper-catalyzed tandem Grignard-type imine addition–Friedel–Crafts alkenylation of arenes with alkynes for the efficient synthesis of quinoline-2-carboxylates *via* activation of C–H bonds.[363] The target products were obtained in high yields from a variety of readily available alkynes and imines. This new method only requires one copper catalyst, which is both simpler and less costly than those methods with two metals for both steps, and tolerates a variety of useful functional groups. The system's efficiency allowed the reactions to be carried out at

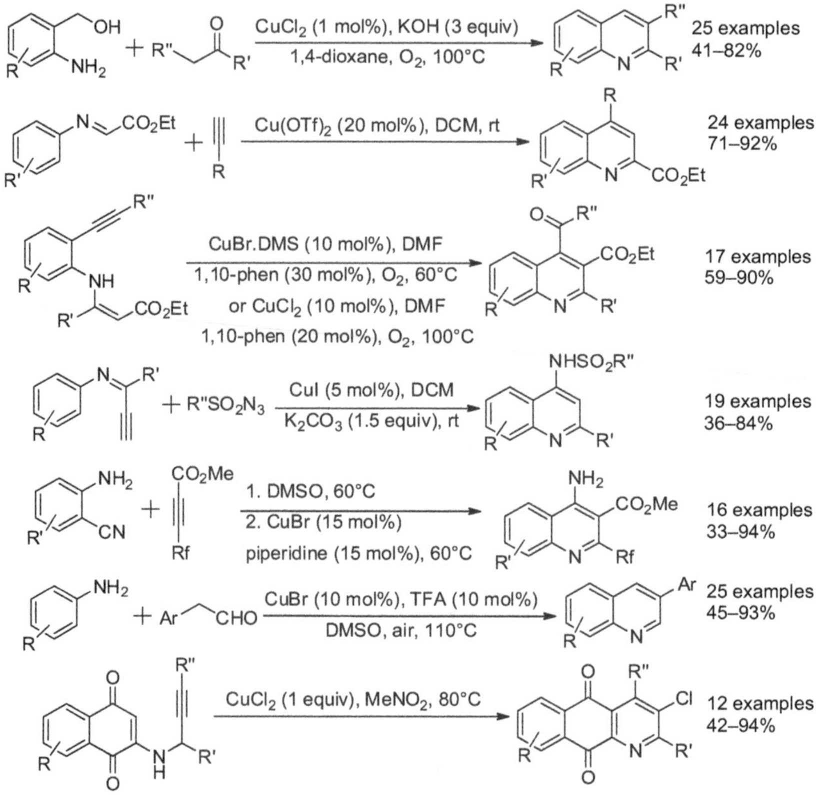

Scheme 4.202   Cu-catalyzed synthesis of quinolones *via* cyclization.

room temperature. In 2012, Chiba and co-workers developed a synthetic method for the preparation of highly substituted quinolones from *N*-(2-alkynylaryl)enamine carboxylates under Cu-catalyzed aerobic conditions *via* intramolecular carbo-oxygenation of alkynes.[364] This strategy was further applied to *N*-alkynylamidines for amino-oxygenation of alkynes, leading to imidazole and quinazoline derivatives. A novel copper-catalyzed oxidative cyclization of enynes and *in situ*-formed enynes leading to 4-carbonylquinolines using dioxygen as an oxygen source was developed by Liang and co-workers.[365] The proposed method proceeded using inexpensive copper as the catalyst and molecular oxygen as the oxygen source, making this transformation sustainable and practical. The mechanism was also discussed based on labeling experiments, EPR studies and ESI-MS analysis. First, Cu(III) and a superoxide radical ($O_2^{\bullet-}$) are formed through the reaction of $CuCl_2$ and O_2 in the presence of an organic base (DABCO) and 1,10-phenanthroline, which could be detected by EPR. Then, Cu(III) can combine with substrate to form a Cu(III) complex. Successive carbocupration to the alkyne moiety gives a vinyl copper peroxy intermediate, which undergoes deprotonative O–O bond cleavage to give a 4-carbonylquinoline along with the generation of Cu(II). More recently, a novel and efficient approach to 4-sulfonamidoquinolines *via* copper-catalyzed cascade reaction of sulfonyl azides with alkynylimines was developed.[366] A 1,3-dipole cycloaddition–ketenimine formation–6π-electrocyclization–[1,3]-H shift cascade reaction was involved. Various 4-sulfonamidoquinolines were afforded in up to 84% yield with 19 examples. This synthetic strategy features atom economy, concise steps, easy operation and mild reaction conditions. Cao and co-workers described an efficient route to 2-perfluoroalkylated quinoline derivatives through the copper(I)-mediated coupling-cyclization of 2-aminobenzonitriles with methyl perfluoroalk-2-ynoates.[367] Moderate to excellent yields were achieved under mild conditions. The reaction mechanism was also discussed. Huang and co-workers developed a simple, novel and efficient method for the synthesis of polysubstituted quinolones in a one-pot manner.[368] The reactions involve a copper-catalyzed aerobic oxidative dehydrogenative annulation of aniline and aldehydes by C–H functionalization, C–C formation and cleavage. Air was used as the oxidant, which makes the transformation very practical and economical. Moreover, an efficient synthetic methodology was developed to assemble 1-azaanthraquinones from *N*-propargylaminoquinones by copper(II)-promoted sequential 6-*endo-dig* chlorocyclization and oxidative aromatization.[369] The approach can be extended to prepare chlorinated alkaloids such as cleistophine and sampangine. A possible mechanism involving C–C bond formation triggered by regioselective electrophilic activation and C–Cl bond formation *via* reductive elimination was proposed.

In 2012, Zhou *et al.* demonstrated a copper-catalyzed approach for the synthesis of *N*-arylacridones.[370] The reactions proceed *via* sp^2 C–H bond amination using air as oxidant under neutral conditions and gave the desired products in good yields (Scheme 4.203). Soon afterwards, a procedure for the synthesis of *N*-methyl-substituted acridones was reported.[371] This is an efficient synthesis of a diverse set of 10-methylacridin-9(10*H*)-ones from 2-(methylamino)benzophenones, which proceeded though Cu-catalyzed intramolecular aromatic C–H amination using O$_2$ as the sole oxidant to provide the desired products in moderate to good yields. In addition, 2-allylamino- and 2-(benzylamino)benzophenones and unprotected substrates can also undergo the C–H amination reaction to deliver the corresponding cyclization products smoothly. Preliminary mechanistic studies suggest that C–H activation is involved in a rate-limiting step. Alternativelly, Cheng and co-workers successfully developed a CuI–bpy-catalyzed synthesis of acridone derivatives through C–H functionalization and C–N bond formation of 2-aminobenzophenone in one pot.[372] The protocol generally requires a very low catalyst loading (1.0 mol%) and was successfully applied to the total synthesis of arborinine with excellent yield. Furthermore, this catalytic reaction involves an unusual pseudo-1,2-migration of the substituent on the arene ring, leading to the formation of two regioisomers. Notably, Zhou *et al.* reported another copper-catalyzed approach for the synthesis of acridones.[373] The reactions proceeded *via* C–C bond cleavage

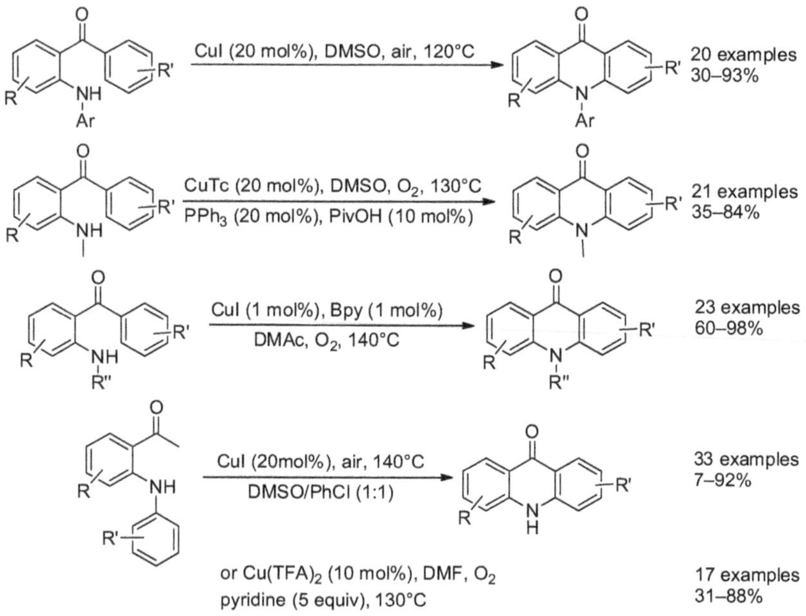

Scheme 4.203 Cu-catalyzed synthesis of acridones.

and intramolecular cyclization using air as the oxidant under neutral conditions. This transformation offers an alternative method to prepare medicinally important acridones and a new strategy for C–C bond cleavage. Meanwhile, Fu's group carried out a systematic study on this topic.[374] Their protocol used inexpensive $Cu(O_2CCF_3)_2$ as catalyst, pyridine as additive and economical and environmentally friendly oxygen as the oxidant, and the corresponding acridones with various functional groups were obtained in moderate to good yields.

Hu and co-workers reported a novel procedure for the preparation of 3-aryl-β-carbolin-1-ones.[375] Ethyl acetamidocyanoacetate was employed as a nucleophilic donor in a Michael addition reaction for the efficient introduction of two nitrogen-containing functional groups to the adduct, then a very mild intramolecular ketone–nitrile annulation of the adduct gave the desired pyridine intermediate conveniently. Finally, the indole ring was assembled efficiently by an intramolecular N-arylation of amide catalyzed by CuI to yield the target compound. Benzo-fused β-carbolin-1-ones are also known as isoquinolinones. Fu and co-workers developed a simple and highly efficient method for the synthesis of 3,4-disubstituted isoquinolin-1(2H)-one derivatives in 2009 (Scheme 4.204).[376] The couplings of 2-bromo- and 2-iodobenzamides or 2-chloronicotinamide with β-keto esters were performed efficiently at 80 °C without addition of any ligand or additive. The method shows economical and practical advantages over the previous methods and uses a readily available starting material. Later, they developed another simple and efficient copper-catalyzed one-pot tandem method for

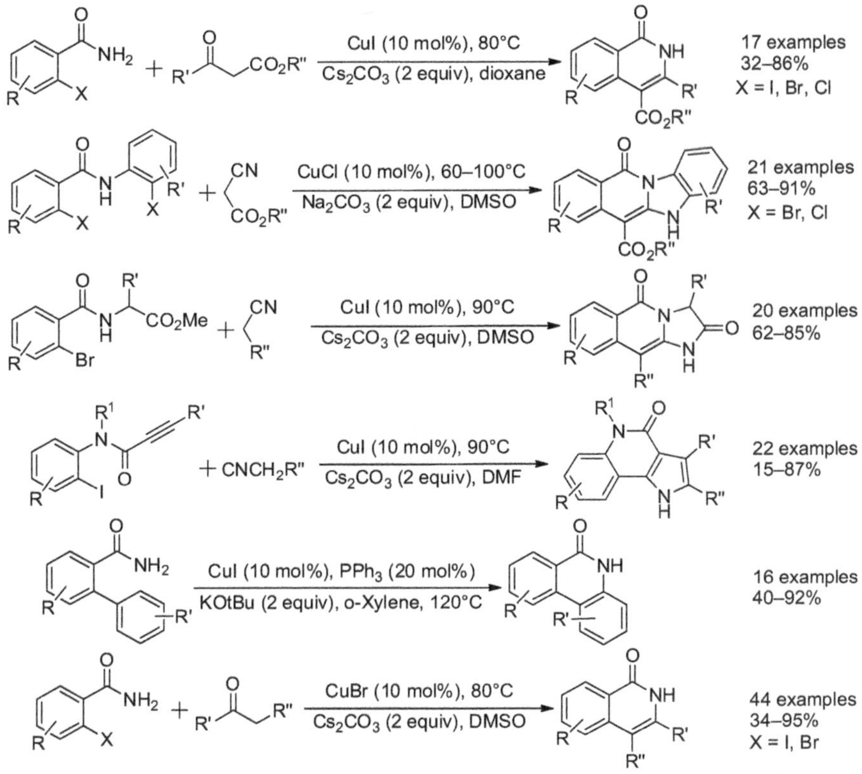

Scheme 4.204 Cu-catalyzed synthesis of isoquinolinones.

the synthesis of benzimidazo[1,2-*b*]isoquinolin-11-one derivatives.[377] The couplings of substituted 2-halo-*N*-(2-halophenyl)benzamides with alkyl 2-cyanoacetates or malononitrile were performed effi under mild conditions without addition of any ligand or additive. Advantages of this method are an economical procedure, practicality and readily available starting materials. Further, they applied readily available 2-halobenzamides containing amino acids and their methyl esters, substituted phenylacetonitriles and malononitrile as the starting materials for the preparation of heterocycles with isoquinolinones as the key structure.[378] Cai and co-workers found that the copper-catalyzed tandem reaction of isocyanides with *N*-(2-haloaryl)propiolamides is very efficient approach for the synthesis of pyrrolo[3,2-*c*]quinolin-4-ones.[379] Highly reactive cyclic organocopper intermediates were proposed to be generated in the copper-catalyzed formal [3 + 2] cycloaddition reaction of isocyanides with triple bonds. Intramolecular trapping of the organocopper intermediates can lead to aryl C–C bond formation, which offered an efficient method for constructing fused pyrrole structures. More recently, a copper-catalyzed cyclization of 2-phenylbenzamides to phenanthridin-6(5*H*)-ones using air as the oxidant and KO*t*Bu as the base was developed.[380] It was discovered that, in addition to PPh₃, other ligands such as

1,10-phenanthroline, TMEDA and L-proline could also be used as the ligand to effect the transformation. In 2013, Zhu and co-workers developed a novel and efficient method to afford isoquinolin-1(2*H*)-ones, which are ubiquitous structural units in a number of biologically active compounds.[381] The combination of simple starting materials without addition of any ligand was developed through sequential α-arylation and dehydrative cyclization processes. This synthetic route tolerates significant substrate variations to deliver a broad range of substituted products.

In 2011, an efficient copper-catalyzed method for the synthesis of quinolinones, pyridinones and heteroannulated pyridinones *via* cascade reactions of substituted 2-iodo-, 2-bromo- and 2-chlorobenzocarbonyls with 2-arylacetamides was reported.[382] Good to excellent yields of the desried products were isolated (Scheme 4.205). In 2013, Li's group described a new method for the carbocyclization of α-C(sp³)–H bonds of amides with alkynes followed by ketonization of the alkynes that utilizes inexpensive copper(II) acetoacetonate [Cu(acac)₂] as catalyst under an oxygen atmosphere to synthesize quinolinones.[383] It is noteworthy that the oxygen atom of the newly formed carbonyl group is from atmospheric molecular oxygen. Qian *et al.* demonstrated that the potential of copper catalysis in the construction of complex heterocycles can be greatly expanded by merging Ullmann-type coupling reactions with the highly compatible azide–acetylene cycloaddition reaction along with a 'click-and-activate' approach.[384] Facile triazole formation not only introduced additional diversity, but also facilitated the Camps cyclization by activating the adjacent methylene group as a nucleophile. After cyclization, the triazole subunit seemed to play a subsequent role in

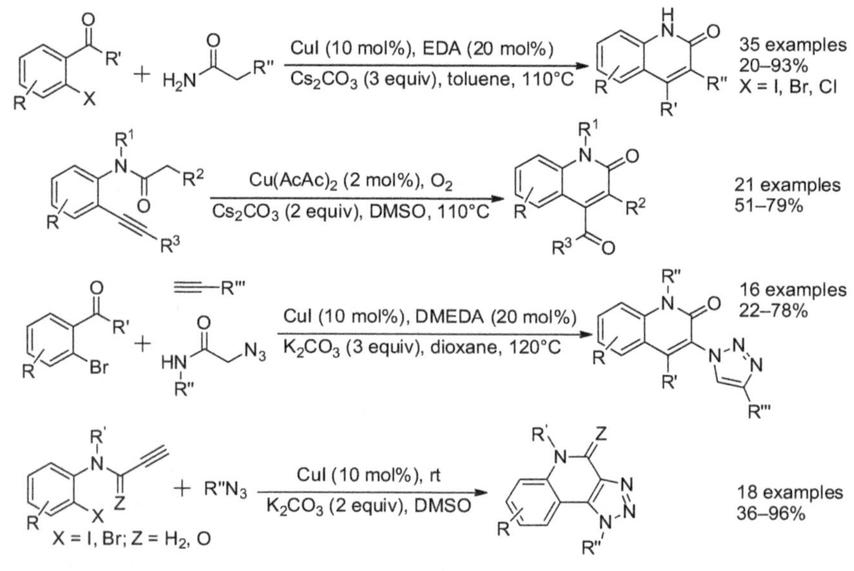

Scheme 4.205 Cu-catalyzed synthesis of quinolinones.

activating the 2′-bromoaryl group as an electrophile in the intermediate to permit an intramolecular direct C–H functionalization of itself under mild conditions. With just a single copper catalyst, up to five new bonds (three C–N and two C–C) and three new rings can be created through a cascade of four types of reactions involving three different copper catalytic cycles. Perumal and co-workers developed a three-component, one-pot sequential reaction for the synthesis of a series of 1,2-dihydrobenzo[*g*]quinoline-5,10-diones in good yields,[385] starting from 2-hydroxynaphthalene-1,4-dione, substituted anilines and propargyl and 3-ethylpropargyl bromides, furnished with *N*-propargylaminonaphthoquinones and followed by copper(II) triflate-catalyzed intramolecular 6-*endo-dig* cyclization. Cai's group developed a novel tandem reaction for the synthesis of heterocycles by utilizing two well-known copper-catalyzed reactions, CuAAC and Ullmann coupling.[386] The intramolecular trapping of the C–Cu intermediate produced in CuAAC led to further formation of an aryl C–C bond through Ullmann C–C coupling. The process took place efficiently when a variety of *N*-(2-iodoaryl)propiolamides or 2-iodo-*N*-(prop-2-ynyl)benzenamines were used and it displayed a wide range of functional group compatibility.

Urabe and co-worker developed a copper-catalyzed 1,2-double amination of 1-halo-1-alkynes in 2008.[387] This procedure provides a concise pathway for the synthesis of protected tetrahydropyrazines and related heterocyclic compounds in good yields (Scheme 4.206). In 2012, Zhou and co-workers reported that phenazines can be synthesized by Cu-catalyzed homocoupling of 2-iodoanilines or 2-bromoanilines in water, and moderate to excellent yields were achieved.[388] This is a simple, efficient, economical and environmentally friendly protocol for the synthesis of phenazines *via* Cu-catalyzed homocoupling of 2-haloanilines in water. This methodology could also be successfully applied to coupling reactions between two different 2-iodoanilines with moderate yields. The reaction of 2-formylpyrroles and *o*-aminoiodoarenes using CuI and sparteine as the catalyst system was reported in 2010.[389] Substituted pyrrolo[1,2-*a*]quinoxalines and related heterocycles were obtained in good yields. The reaction also works for the annulation of 2-formylindoles, 2-formylimidazole, 2-formylbenzimidazole and a 3-formylpyrazole. Fu and co-workers developed an efficient copper-catalyzed aerobic oxidative intramolecular alkene C–H amination leading to *N*-heterocycles.[390] The protocol uses cheap and readily available $Cu(O_2CCF_3)_2$ as the catalyst, substituted 3-methyleneisoindolin-1-ones as the starting material and economical and environmentally friendly air as the oxidant, and the corresponding *N*-heterocycles were obtained in good to excellent yields. More recently, another straightforward procedure was described to prepare substituted isoindolo[2,1-*a*]quinoxalin-6(5*H*)-ones.[391] This is an operationally simple and efficient cascade strategy for access to substituted isoindolo[2,1-*a*]quinoxalines through a one-step copper-catalyzed C–N coupling reaction between substituted 2*H*-isoindole-1-carbaldehydes and substituted 2-halophenylamines. Liu and co-workers developed a novel and efficient methodology involving the Ugi four-component

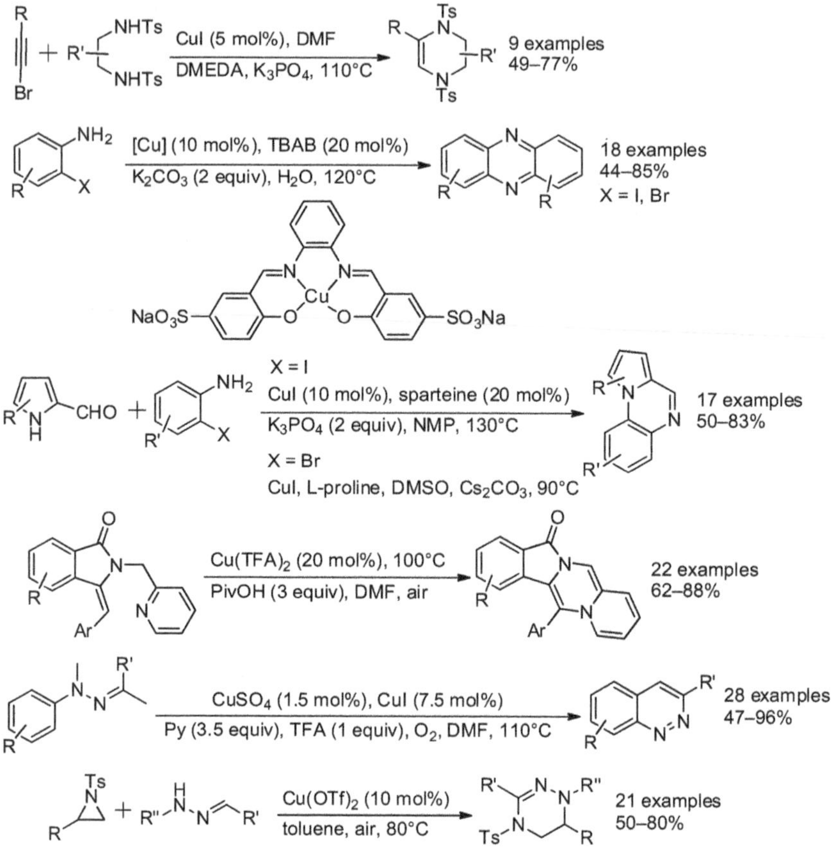

Scheme 4.206 Cu-catalyzed synthesis of (benzo-fused)pyrazines and other heterocycles.

reaction to synthesize selectively two distinct sets of indole-based heterocyclic compounds.[392] From the same set of Ugi adducts, 5,6-dihydroindolo[1,2-*a*]quinoxalines were rapidly generated in excellent yields by a copper-catalyzed N–H arylation pathway, whereas 6,7-dihydroindolo[2,3-*c*]quinolones were obtained in good yields under palladium-catalyzed conditions without protection of the indole N1 moiety. Microwave heating was used to accelerate these intramolecular C–N and C–C bond-forming reactions. The two-step protocols have several advantages, such as high selectivity, simple starting materials, broad substrate scope, operational simplicity, step economy, high yields and short reaction times. Additionally, in 2012 Ge and co-workers described an efficient procedure based on copper-catalyzed aerobic intramolecular dehydrogenation for the cyclization of *N*-methyl-*N*-phenylhydrazones.[393] The reactions proceed through sequential C(sp^3)–H oxidation, cyclization and aromatization processes. This novel method provides efficient access to cinnoline derivatives. Moreover, Wang's group developed a single-step approach to tetrahydrotriazines *via* the Cu(II)-catalyzed tandem reaction of *N*-tosylaziridines and hydrazones under aerobic conditions.[394] The process was proposed to involve a copper-catalyzed nucleophilic ring-opening reaction and a copper-catalyzed intramolecular C–H amidation. The use of inexpensive Cu catalyst and air as the ideal oxidant is a significant practical advantage.

In 2011, Fu and co-workers developed a convenient and efficient copper-catalyzed cascade method for the synthesis of benzimidazoquinazoline derivatives (Scheme 4.207).[395] This reaction usesg readily available substituted 2-(2-halophenyl)benzoimidazoles with amidines or guanidines as substrates under mild conditions (even at room temperature). Benzimidazoquinazoline and benzimidazo[1,2-*c*]quinazolin-5-amine derivatives were obtained in good to excellent yields. Soon afterwards, they developed another simple and efficient copper-catalyzed cascade method for the synthesis of

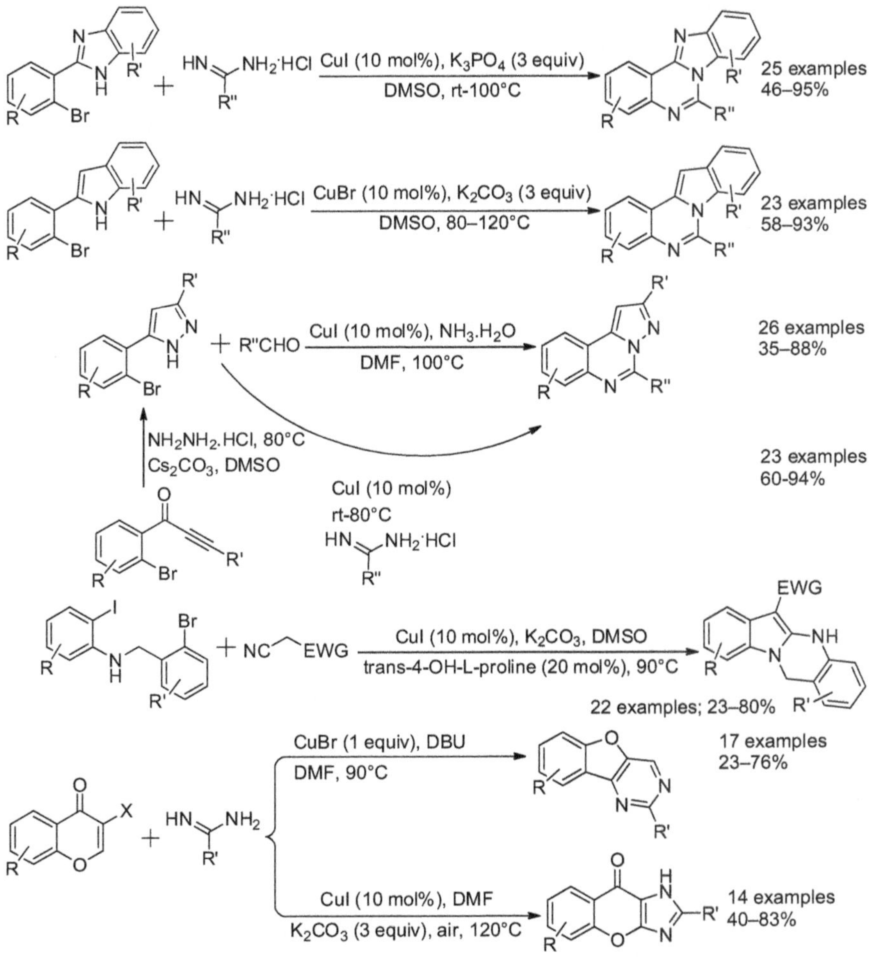

Scheme 4.207 Cu-catalyzed synthesis of benzimidazoquinazolines.

1*H*-indolo[1,2-*c*]quinazoline derivatives.[396] This method uses inexpensive CuBr as the catalyst and readily-available substituted 2-(2-halophenyl)-indoles and amidines as starting materials. Remarkably, no ligand or additive was required and the reaction conditions were mild in general. More recently, Guo *et al.* developed a facile methodology for the synthesis of pyrazolo[1,5-*c*]quinazolines or 5,6-dihydropyrazolo[1,5-*c*]quinazolines *via* a copper-catalyzed tandem reaction of 5-(2-bromoaryl)-1*H*-pyrazoles with carbonyl compounds and aqueous ammonia under an air atmosphere.[397] Interestingly, with cyclic ketones, the corresponding spiro-5,6-dihydro-pyrazolo[1,5-*c*]quinazolines were successfully synthesized in modest yields. Moreover, they found that the tandem reaction could also be carried out in a one-pot, four-component manner by starting with 1-(2-bromophenyl)-1,3-diones, hydrazine hydrate, carbonyl compounds and aqueous ammonia. In 2012, Fu and co-workers reported an easy and efficient method for the synthesis of pyrazolo[1,5-*c*]quinazolines.[398] This is a one-pot, two-step reaction using readily available substituted 1-(2-halophenyl)-3-akylprop-2-yn-1-ones, hydrazine hydrochloride and amidine hydrochlorides as substrates under mild conditions. The corresponding pyrazolo[1,5-*c*]quinazolines were obtained in good to excellent yields. Zhao, Dong and co-workers developed a domino procedure for the synthesis of 5,12-dihydroindolo[2,1-*b*]quinazoline derivatives *via* copper-catalyzed Ullmann-type intermolecular C–C and intramolecular C–N couplings.[399] Good yields of various 5,12-dihy-droindolo[2,1-*b*]quinazoline derivatives were obtained. The protocol displays a wide functional group compatibility and provides economical and practical advantages over the previous methods. Hu and co-workers described the development of a novel cascade reaction for the conversion of 3-chloro-chromones into diverse benzofuro[3,2-*d*]-pyrimidines and related polycyclic heterocylic scaffolds.[400] The process was promoted by CuBr and takes place *via* a chemoselective Michael addition–elimination and double intra-molecular cyclization sequence. Significantly, the conditions utilized for the tandem process are mild and economical. In 2013, the same group de-veloped a simple and efficient one-pot copper-catalyzed direct C–O coupling method to form a novel imidazochromone scaffold.[401] Only a catalytic amount of CuI and 3.0 equiv. of K_2CO_3 were employed to promote the whole reaction efficiently. A novel six-membered heterocyclic skeleton of imida-zochromone was prepared *via* an efficient one-pot reaction including a key step of copper-catalyzed aerobic C–H intramolecular cycloetherification. Notably, this process does not require the presence of strong *para* electron-withdrawing groups on the phenol component. Also, the results of this effort show that acylphenols containing electron-rich heterocycles participate in an efficient C–H activation/C–O formation process.

Fu's group developed a general and highly efficient copper-catalyzed method for the synthesis of quinazoline and quinazolinone derivatives.[402] The target products were obtained in good to excellent yields *via* cascade reactions of amidine hydrochlorides with substituted 2-halobenzaldehydes, 2-halophenyl ketones or methyl 2-halobenzoates (Scheme 4.208). The

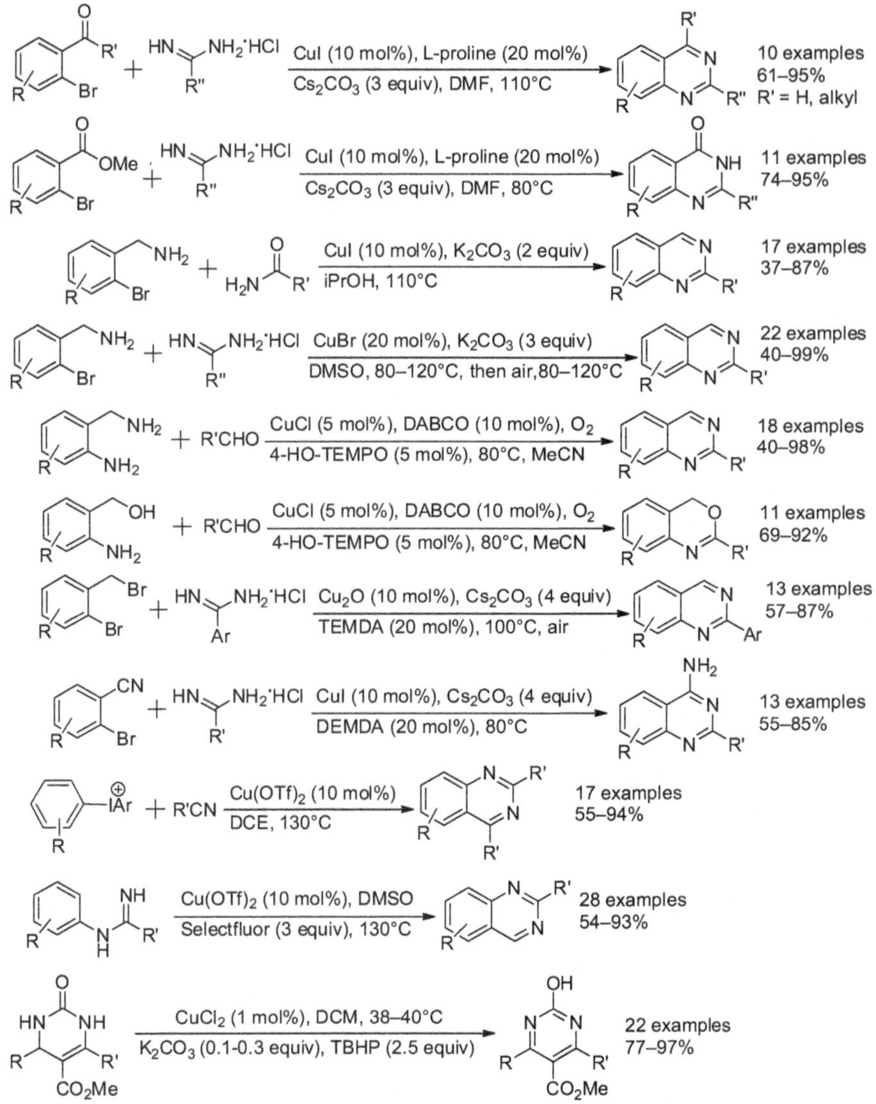

Scheme 4.208 Cu-catalyzed synthesis of quinazolines.

reaction of *o*-iodobenzaldehydes under ligand-free conditions was reported later.[403] In 2010, Fu and co-workers reported a simple and efficient copper-catalyzed approach to quinazoline derivatives.[404] This interesting protocol used readily available substituted (2-bromophenyl)methylamines and amides as the starting materials and the cascade reactions were performed under air *via* sequential Ullmann-type coupling and aerobic oxidation without addition of any ligand or additive. They succeeded in extending their methodology to amidine hydrochlorides as the starting materials.[405]

Inexpensive CuBr was used as catalyst and air as oxidant and the corresponding quinazoline derivatives were obtained in moderate to good yields. The procedure underwent sequential intermolecular *N*-arylation, intramolecular nucleophilic substitution and aerobic oxidation. In 2012, the Cu–N-ligand–TEMPO catalytic system was applied to the aerobic oxidative synthesis of heterocycles.[406] 2-Substituted quinazolines and 4*H*-3,1-benzoxazines were synthesized efficiently from the one-pot reaction of aldehydes with 2-aminobenzylamines and 2-aminobenzyl alcohols, respectively, by employing CuCl–DABCO–4-HO-TEMPO as the catalyst and oxygen as the terminal oxidant. Alternatively, Beifuss and co-workers found a simple and efficient one-pot method for the synthesis of substituted quinazolines from easily accessible *o*-bromobenzyl bromides and benzamidines.[407] This new Cu$_2$O-catalyzed reaction delivered the products selectively and with yields ranging from 57 to 85% by using water as the solvent under mild reaction conditions. Fu and co-workers developed a simple and efficient method for the synthesis of 2-aminoquinazoline and 2,4-diaminoquinazoline derivatives.[408] The couplings of substituted 2-bromobenzonitriles with amidines or guanidines performed well under mild conditions and the target products were obtained in good yields when the reaction temperature was raised to 80 °C. More recently, an efficient Cu-catalyzed synthesis of quinazolines was reported.[409] The reaction proceeds through C–N bond-formation reactions between N–H bonds of amidines and C(sp^3)–H bonds adjacent to sulfur or nitrogen atoms in commonly used solvents, such as DMSO, DMF, DMA, NMP or TMEDA, and followed by intramolecular C–C bond formation reactions. The generality and high selectivity of the annulation reaction towards various C(sp^3)–H bonds together with employing readily available amidines as the substrates made this method very attractive. Additionlly, Chen and co-workers found that quinazolines can even be prepared from diaryliodonium salts and two nitriles.[410] A mild, practical procedure for the dehydrogenation of dihydropyrimidines and dihydropyrimidinones has also been reported.[411] The proposed mechanism involves coordination of the metal catalyst to an NH moiety, followed by oxidative elimination of the resulting metal complex.

Fu's group developed a simple and highly efficient method for the synthesis of quinazolinone derivatives in 2009 (Scheme 4.209).[412] The coupling reactions of 2-bromo- and 2-iodobenzoic acid derivatives with amidines performed well at room temperature without the addition of a ligand or an additive. The target products were also obtained in higher yields from the non-active substrates, such as 2-chlorobenzoic acid or guanidines, when the reaction temperature was raised to 80 °C. This method is economical and practical and the starting materials are readily available. In 2011, the same group reported a copper-catalyzed domino method for the synthesis of quinazolinones using readily available α-amino acids as the nitrogen-containing motifs.[413] This domino process underwent Ullmann-type *N*-arylation, decarboxylation, aerobic oxidation and intramolecular addition. They also described another efficient approach to quinazolinone derivatives by

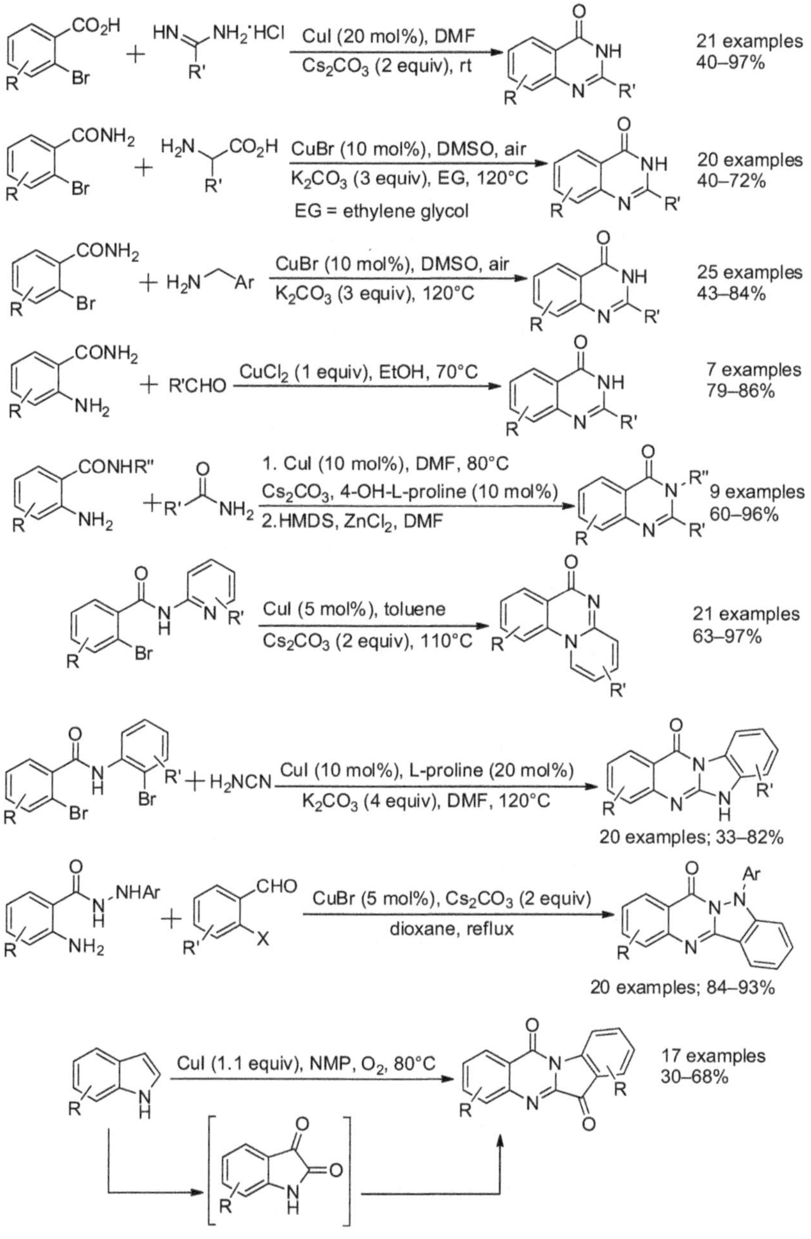

Scheme 4.209 Cu-catalyzed synthesis of quinazolinones.

using substituted 2-halobenzamides and (aryl)methanamines as the starting materials and air as oxidant.[414] This is a representative example of constructing *N*-heterocycles *via* sequential Ullmann-type coupling under air and with aerobic oxidative C–H amidation. Later, this prodcure was applied in the synthesis of pyrimido[4,5-*b*]carbazolone derivatives.[415] Alternatively, the condensation of anthranilamide with aryl-, alkyl- or heteroarylaldehydes in refluxing ethanol and in the presence of $CuCl_2$ was found to give the corresponding 2-substituted quinazolinones in excellent yields.[416] In 2011, a simple and efficient copper-catalyzed method for the synthesis of fused pyridoquinazolones was developed by Fu and co-workers.[417] No addition of any ligand or additive was needed and it can tolerate various functional groups. By using inexpensive CuI as the catalyst and readily available substituted 2-halo-*N*-(pyridin-2-yl)benzamides as the starting materials, the corresponding target products were obtained in good to excellent yields. Soon afterwards, the same group reported a convenient copper-catalyzed domino method for the synthesis of benzimidazo[2,1-*b*]quinazolin-12(6*H*)-ones.[418] The reactions use readily available substituted 2-bromo-*N*-(2-halophenyl)benzamides with cyanamide as substrates, in which cyanamide acts as a useful building block, and gave the desired products in good yields. Ma and co-workers reported a facile and efficient approach for assembling substituted quinazolinones, which relies on a copper-catalyzed aryl amidation reaction of *N*-substituted *o*-bromobenzamides with amides and subsequent spontaneous or $HMDS/ZnCl_2$-mediated condensative cyclization.[419] The protocol could tolerate a variety of functional groups, providing a wide range of 3-substituted and 2,3-disubstituted quinazolinones in good to excellent yields. Its application has been demonstrated by the formal synthesis of dictyoquinazol A and the facile preparation of methaqualone. Wang's group developed a recoverable and reusable Fe_3O_4 magnetic nanoparticle-supported Cu(I) catalyst for the synthesis of quinazolinones and bicyclic pyrimidinones.[420] In the presence of a supported Cu(I) catalyst (10 mol%), amidines reacted with substituted 2-halobenzoic acids and 2-bromocycloalk-1-enecarboxylic acids to generate the corresponding *N*-heterocyclic products in good to excellent yields at roomtemperature in DMF. In addition, the supported Cu(I) catalyst could be recovered ad reused at least 10 times without a significant decrease in its catalytic activity. More recently, Wang and co-workers found that the treatment of 2-amino-*N'*-arylbenzohydrazides and *o*-halogenated benzaldehyde in the presence of CuBr and Cs_2CO_3 can give the corresponding 5-arylindazolo[3,2-*b*]quinazolin-7(5*H*)-one in high yields.[421] This procedure involves an Ullmann-type reaction and provides an efficient method to construct fused tetracyclic heterocycles. Fresneda, Molina and co-workers reported the preparation of quinazolinones *via* iminophosphoranes, which were derived from *N*-substituted *o*-azidobenzamides by a combination of the aza-Wittig methodology and CuI-catalyzed heteroarylation.[422] was It was reported that tryptanthrins, as analogs of quinazolinones, can be prepared from indoles using copper as catalyst under oxidative conditions.[423] This cascade process was believed to involve copper-catalyzed aerobic oxidation of indole, hydrolysis of amide, copper-catalyzed decarboxylative coupling, intramolecular nucleophilic addition and oxidative aromatization. The reactions could be carried out under mild conditions with varying functional group tolerance. Moreover, the substituent diversity could be quickly expanded by the condensation between the substituted indoles and substituted isatins *via* this strategy.

In 2013, a novel synthetic protocol for the one-pot chemo- and stereo-selective construction of diversely functionalized pyrido[1,2-*a*]pyrimidin-4-imines was reported (Scheme 4.210).[424] The reaction proceeds through a copper(I)-catalyzed [3 + 2] cycloaddition–ring-opening rearrangement–[4 + 2] cycloaddition–aromatization cascade with sulfonyl azides, alkynes and *N*-arylidenepyridin-2-amines. In addition, the catalytic activity of copper(I)-modified zeolite, a recyclable, heterogeneous catalyst, was also investigated, and gave improved yields compared with its homogeneous equivalents.

Xu and Fu developed a simple, practical and highly efficient copper-catalyzed one-pot synthesis of imidazo- and benzoimidazoquinazolinones in 2012 (Scheme 4.211).[425] The procedure was based on a sequential copper-catalyzed Ullmann-type *N*-arylation and aerobic oxidative intramolecular C–H amidation. The protocol uses readily available substituted 2-halo-*N*-alkylbenzamides, 2-chloro-*N*-propylpyridine-3-carboxamide and imidazole and benzimidazole derivatives as the starting materials, inexpensive

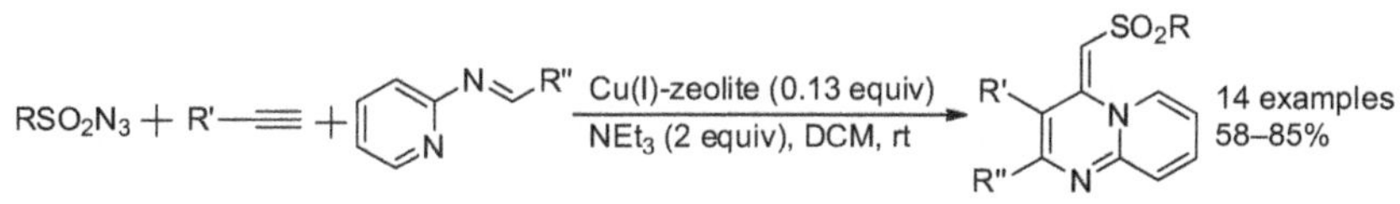

Scheme 4.210   Cu-catalyzed synthesis of pyrido[1,2-*a*]pyrimidin-4-imines.

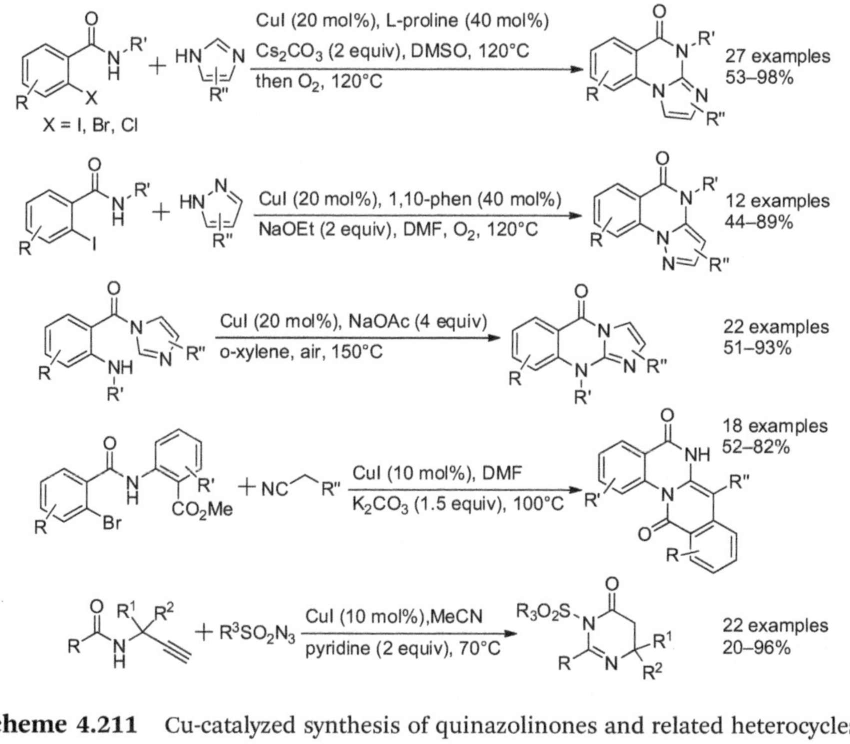

Scheme 4.211   Cu-catalyzed synthesis of quinazolinones and related heterocycles.

CuI–L-proline as the catalyst system and economical and environmentally friendly oxygen gas as the oxidant. The imidazo- and benzoimidazoquinazolinones were obtained in good to excellent yields. Instead of imidazoles, the reaction with pyrazoles was reported later.[426] Qiao and co-workers developed another copper-catalyzed aerobic oxidative method for the synthesis of imidazo- and benzimidazoquinazolinones.[427] The protocol uses inexpensive CuCl as the catalyst, the safe, economical and environmentally friendly air as the oxidant in the absence of ligand and is tolerant towards functional groups in the substrates. The reaction proceeds through direct intramolecular C–H amination of (1*H*-imidazol-1-yl)[2-(alkylamino)phenyl]methanones and the corresponding target products were obtained in moderate to good yields. Safari and Gandomi-Ravandi prepared Cu carbon nanotubes (CNTs) and applied them in organic synthesis.[428] They efficiently catalyzed the condensation reaction of isatoic anhydride, aldehydes and primary amines or ammonium acetate to afford the corresponding 2,3-dihydroquinazolin-4(1*H*)-one derivatives in high yield. The protocol proved to be efficient and environmentally benign in terms of very easy work-up, high yields and ease of recovery of the catalyst. Fu and co-workers developed a copper-catalyzed domino method for synthesis of isoquinolino[2,3-*a*]quinazolinones.[429] They used readily available, substituted methyl 2-(2-haloobenzamido)benzoates and nitriles as the starting materials. This domino process comprises an Ullmann-type *C*-arylation, intramolecular addition of NH to CN and nucleophilic attack of amino on ester groups. More recently, an efficient method for the preparation of dihydropyrimidin-4-ones from *N*-(prop-2-yn-1-yl)amides and sulfonyl azides was reported.[430] The cascade reaction underwent copper-catalyzed alkyne–azide cycloaddition, formation and intramolecular nucleophilic addition of ketenimine and subsequent rearrangement in a single step. Moreover, the starting materials, propargylamides, could be easily prepared from propargylic alcohols *via* the Ritter reaction or from propargylamines *via* aminolysis of acid chlorides, and sulfonyl azides could be obtained from sulfonyl chlorides and sodium azide.

Bao and co-workers reported a novel and convenient one-pot synthesis of pyrimido[1,6-*a*]indol-1(2*H*)-one derivatives in 2010.[431] The reaction proceeds through a nucleophilic addition–Cu-catalyzed *N*-arylation–Pd-catalyzed C–H activation sequential process. The reaction of easily prepared *o-gem*-dibromovinyl isocyanates with *N*-alkylanilines gave the desired indole derivatives in moderate to good yields (Scheme 4.212). Lee's group developed a novel divergent tandem one-pot method for the synthesis of 3,5,6-trisubstituted 1*H*-pyrimidin-2,4-dione derivatives.[432] In the presence of 10 mol% of Cu(OAc)$_2$, the α-substituted Blaise reaction intermediates (R$_2 \neq$ H) reacted with isocyanates chemoselectively to afford pyrimidin-2,4-diones, whereas the α-unsubstituted Blaise reaction intermediate (R$_2$ = H) showed a propensity to be a *C*-nucleophile towards electrophiles, permitting the installation of different functionalities at the 5-position through sequential tandem reactions. An efficient copper(II)-catalyzed two-step, one-pot domino process for the rapid synthesis of novel *N,N*-multiheterocycles was reported

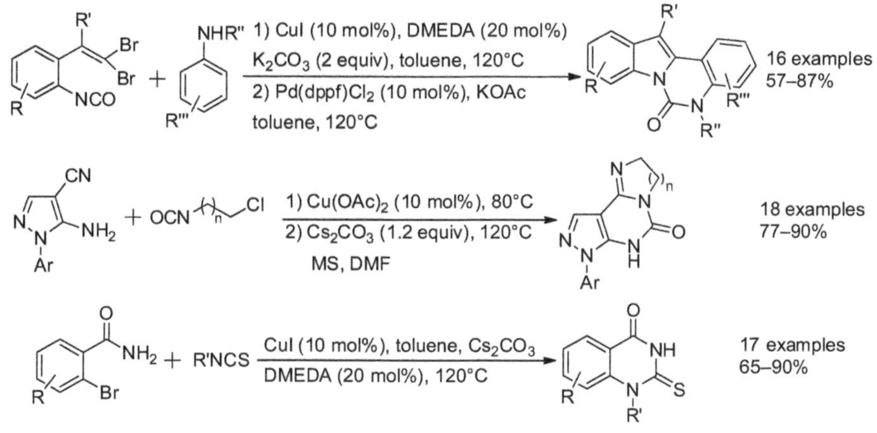

Scheme 4.212 Cu-catalyzed synthesis of 1*H*-pyrimidin-2,4-diones.

in 2013.[433] The copper(II) acetate-promoted amine–isocyanate coupling and the Cs$_2$CO$_3$-induced one-pot domino ring-closure process afforded novel pyrazole-fused imidazo- and pyrimido[1,2-*c*]pyrimidinone scaffolds, with isolated yields of up to 90%. Copper(II) acetate was also introduced as a catalyst in the reactions of β-aminocarbaldehydes and chloroalkyl isocyanates, furnishing pyrazolo[3,4-*d*]pyrimidin-6(5*H*)-ones *via* a cyclocondensation procedure. Additionally, copper-catalyzed tandem reaction of *o*-bromobenzamides and isothiocyanates was also described.[434] This method provides an efficient and practical route for the synthesis of 2-thioxo-2,3-dihydroquinazolin-4(1*H*)-ones. The optimal conditions involved the following parameters: CuI as precatalyst, Cs$_2$CO$_3$ as base, *N*,*N*′-dimethylethane-1,2-diamine as ligand and toluene as solvent, with a reaction temperature of 120 °C. The Biginelli reaction was also explored with copper catalysts.[435]

In 2008, Yuan and Ma developed a CuI–L-proline-catalyzed synthesis of pyrrolo[1,2-*a*]quinoxalines.[436] The reaction proceeds *via* the coupling of 2-halotrifluoroacetanilides with pyrrole-2-carboxylate esters in DMSO at 80–90 °C followed by *in situ* hydrolysis at 60 °C. Indole-2-carboxylate esters underwent the same process smoothly to provide the corresponding tetracyclic products. A wide range of these fused heterocycles bearing different functional groups, such as ketone, ester, methoxy, bromo and chloro, were prepared in good yields (Scheme 4.213). In 2010, Chen and Bao reported an efficient copper-catalyzed method for the synthesis of quinoxalin-2(1*H*)-one derivatives.[437] The reactions proceed through domino S$_N$Ar–coupling–demesylation reactions of *N*-(2-halophenyl)methylsulfonamides with 2-haloamides. Various quinoxalinones with diversity at three positions on their scaffold were obained. In the same year, a general and practical route to chiral quinoxalinones was developed by Tanimori *et al*.[438] The method is based on the copper(I) chloride–dimethylethylenediamine (DMEDA) catalyst system. With the use of 1 mol% copper(I) chloride, structurally diverse

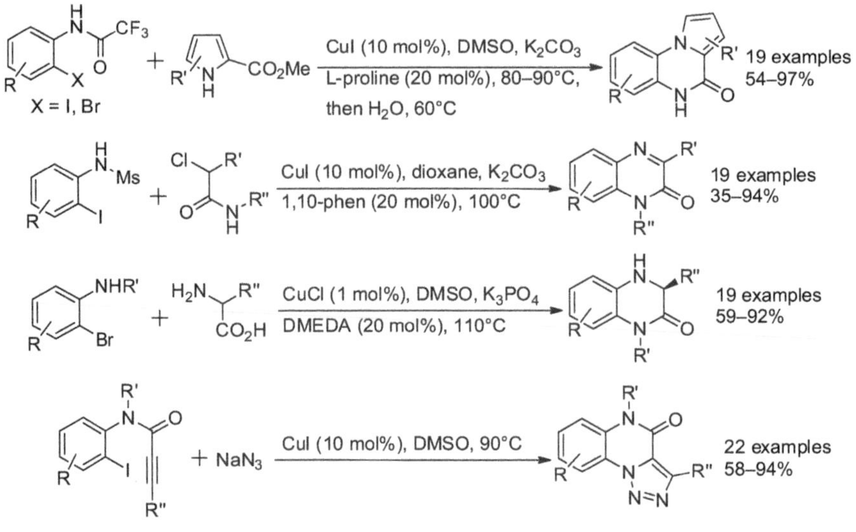

Scheme 4.213 Cu-catalyzed synthesis of quinoxalinones.

quinoxalin-2-ones were generated with high optical purity from readily available starting materials, 2-haloanilines and α-amino acids, in a one-pot manner. This method tolerates a variety of functional groups on both coupling partners. Ligand-free conditions also provide excellent efficiency. The conditions employing an organic base instead of potassium phosphate also gave good yields of products. This method was successively applied to a concise synthesis of the chiral drug GW420867X. Another interesting methodology for quinoxalinone synthesis was developed in 2012 by Ding and Cai's groups.[439] They described a simple and efficient approach for the synthesis of [1,2,3]triazolo[1,5-*a*]quinoxalin-4(5*H*)-ones. The methodology is based on a tandem reaction of 1-(2-haloaryl)propiolamides with sodium azide through a [3 + 2] azide–alkyne cycloaddition and intramolecular Ullmann-type C–N coupling process. A wide of functional groups are compatible under the conditions used.

   Fang and Li reported a copper-catalyzed intramolecular coupling of aryl bromides with 1,3-dicarbonyls *via* a six-membered ring closure in 2006 (Scheme 4.214).[440] With CuI (10 mol%) as the catalyst, *N,N'*-dimethylethylenediamine as the ligand and Cs₂CO₃ as the base, the reactions of α-(2-bromobenzyl)-β-keto esters in THF at reflux temperature afforded the corresponding substituted 4*H*-1-benzopyrans in high yields *via* O-arylation. On the other hand, the reactions of δ-(2-bromophenyl)-β-keto esters in refluxing dioxane led to the formation of 3,4-dihydronaphthalen-2(1H)-one derivatives *via* C-arylation. Thus, with the appropriate choice of substrates, chemoselective O-arylation or C-arylation could be well implemented. In 2011, Beifuss and co-workers reported the intermolecular verison of this reaction.[441] Depending on the ratio of the substrates and the reaction conditions, the Cu(I)-catalyzed domino reaction between bromobenzyl bromides and β-keto

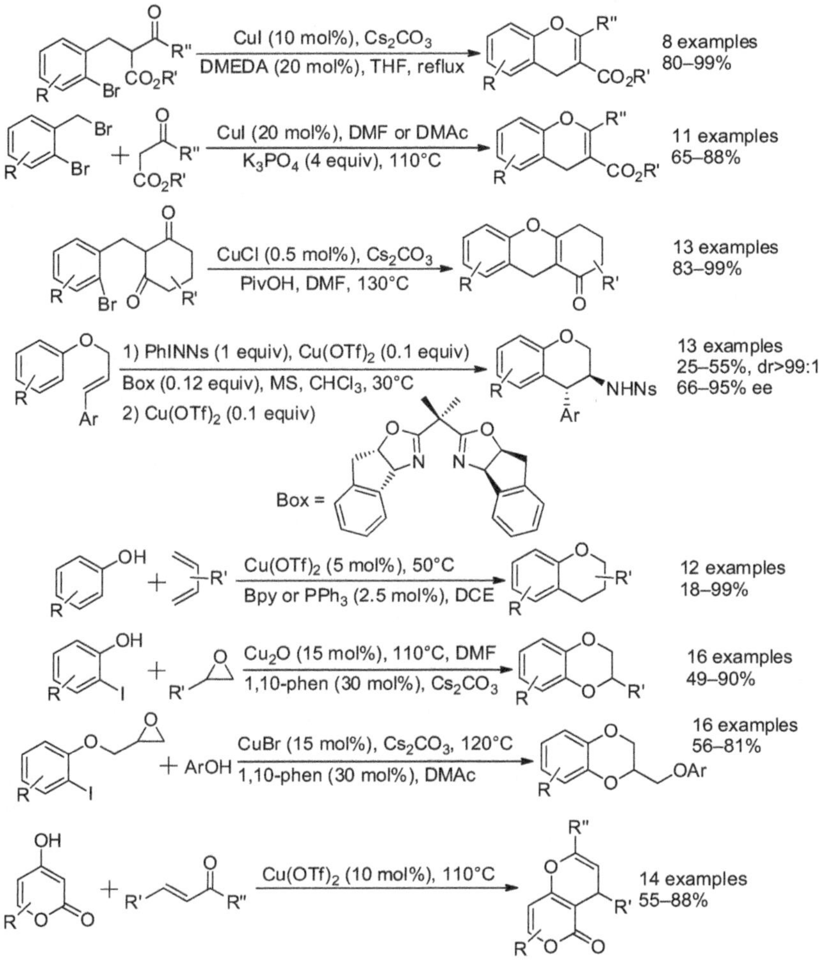

Scheme 4.214 Cu-catalyzed synthesis of chromenes and dioxins.

esters exclusively yields either 4*H*-chromenes or naphthalenes. Soon afterwards, they published a detailed study on the *O*-arylation of easily accessible 2-(2-bromobenzyl)cyclohexane-1,3-diones.[442] The desired products can be synthesized in a one-pot reaction between 2-bromobenzyl bromides and 1,3-cyclohexanediones *via* Cu(I)-catalyzed domino intermolecular *C*-benzylation/intramolecular *O*-arylation, but the competing *O*-benzylation in the initial step could not be suppressed under the conditions and gave rise to the formation of benzyl ethers as side products. The synthesis of the 2,3,4,9-tetrahydro-1*H*-xanthen-1-ones in two steps has proven to be a valuable alternative to the domino process, as no side product formation occurred. The required *C*-benzylated 1,3-diones could be obtained selectively by reacting 2-bromobenzyl bromides with 1,3-diones under basic conditions with yields ranging from 45 to 83%. Subsequently, the

2-(2-bromobenzyl)cyclohexane-1,3-diones were cyclized to the corresponding 2,3,4,9-tetrahydro-1*H*-xanthen-1-ones by Cu(I)-catalyzed intramolecular *O*-arylation in 83–99% yield. The best results were obtained when the cyclizations were performed with 0.5 mol% CuCl as the catalyst, 1.2 equiv. of pivalic acid as an additive and 1 equiv. of Cs_2CO_3 as the base. Hajra and Sinha developed a catalytic enantioselective one-pot aziridoarylation reaction of aryl cinnamyl ethers.[443] The combination of a suitable copper catalyst and a chiral bisoxazoline ligand was found to be very efficient for asymmetric aziridination followed by an intramolecular arylation (Friedel–Crafts) reaction to provide a general and direct method for the synthesis of *trans*-3-amino-4-arylchromans with high regio-, diastereo- (*dr* > 99:1) and enantioselectivity (up to 95% *ee*) with moderate yields. The final product, *trans*-3-amino-4-arylchroman, is an advanced intermediate for the synthesis of chromenoisoquinoline compounds such as doxanthrine, a potent and selective full agonist for the dopamine D_1 receptor. Adrio and Hii found that $Cu(OTf)_2$–bipy is an efficient catalyst system for the addition of phenols to 1,3-dienes under aerobic conditions.[444] The reaction involves a tandem hydroalkoxylation–rearrangement–hydroalkylation sequence, furnishing *O*-heterocycles in moderate to good yields, and the catalyst can be recycled without any decrease in catalytic activity, which is comparable to that of Ag and Au catalysts in intermolecular annulation reactions of ArOH with 1,3-dienes. The application of a copper catalyst in the aminoacetoxylation of alkenes to piperidines has also been reported,[445] and it was reported that chromenes can be prepared from aryloxy propargylated aldehydes, various azides, active methylene compounds and 1,3-cyclohexanediones using catalytic amounts of $Cu(OAc)_2$–sodium ascorbate and diammonium hydrogenphosphate in aqueous ethanol or using ionic liquids such as [bmim][NO_3] in the presence of 30 mol% CuI as efficient media for the domino Knoevenagel–hetero-Diels–Alder reaction of *o*-propargyloxy-benzaldehydes as unactivated terminal alkynes with some active methylene compounds.[446] In 2008, Bao and co-workers found that 2,3-dihydro-1,4-benzodioxins can be prepared in a tandem one-pot procedure by reaction of *o*-iodophenols with epoxides catalyzed by a Cu_2O–1,10-phenanthroline–Cs_2CO_3 system.[447] The reaction was suggested to occur *via* a novel ring-opening–coupling mechanism, giving moderate to good yields. Moreover, both aryl and aliphatic epoxides are tolerated under these conditions. Later, the same group reported another efficient method for the preparation of various 2-substituted-1,4-benzodioxanes by CuBr-catalyzed tandem reactions of 2-[(2-iodophenoxy)methyl]oxiranes with phenols.[448] The reaction involves the ring opening of the 2-[(2-iodophenoxy)methyl]oxirane followed by an intramolecular C–O cross-coupling cyclization. All reactions were performed using simple and cheap materials and the products were obtained with considerably increased molecular complexity and the typical structural features of known biofunctional scaffolds. More importantly, through modification of the reactant structure, the problem of producing isomers *via* nuleophilic attack on different sites of epoxides can be successfully avoided in this new procedure. More recently, an efficient procedure for the synthesis of pyrano[3,2-*c*]coumarins was achieved *via* a copper(II) triflate-catalyzed tandem reaction of 4-hydroxycoumarin with α,β-unsaturated carbonyl compounds.[449] Copper triflate was found to be a novel and highly efficient Lewis acid catalyst for the synthesis of pyranocoumarins *via* Michael addition followed by intramolecular cyclization. The reaction is highly regioselective and water is produced as the sole by-product. Operational simplicity, environmentally friendly reaction conditions avoiding toxic reagents and volatile solvents and compatibility with various functional groups are the advantages of this procedure.

In 2007, a simple and highly effective copper-catalyzed C–O coupling reaction was developed and applied to the synthesis of benzopyranones (Scheme 4.215).[450] The reaction of various 2-halobiarylcarboxylic acids was examined using microwave irradiation. A new class of pyrroloisoquinoline alkaloids, isolamellarin, was synthesized based on the annulation of dihydroisoquinoline with aryl pyruvates under basic condition and Cu-mediated/microwave-assisted C–O$_{carboxylic}$ lactonization. Yamamoto and Kirai developed a procedure for the preparation of 4-arylcoumarins in 2008.[451] In the presence of 2–4 mol% of CuOAc, methyl phenylpropiolates having a MOM-protected hydroxy group at the *ortho*-position underwent hydroarylation with various arylboronic acids in MeOH at ambient temperature, resulting in the formation of 4-arylcoumarins in high yields after acidic workup. This

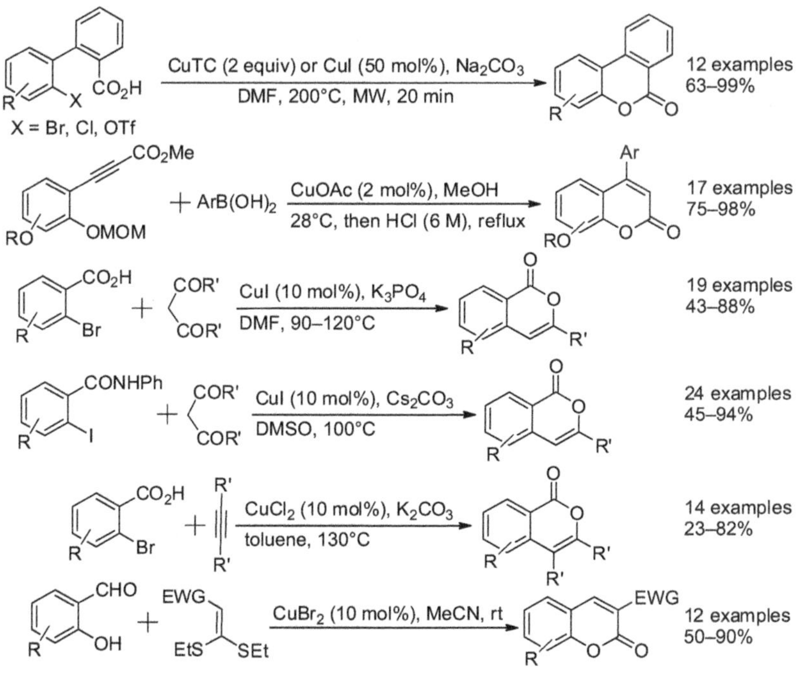

Scheme 4.215 Cu-catalyzed synthesis of (iso)coumarins.

protocol is compatible with phenylboronic acids having various functional groups including C–H bonds and an aldehyde. In 2012, Xi and co-workers developed an efficient strategy for the synthesis of a variety of 3-substituted isocoumarins.[452] The reaction proceeded from *o*-halobenzoic acids and 1,3-diketones *via* a copper(I)-catalyzed domino reaction in DMF under the action of K$_3$PO$_4$ at 90–120 °C without a ligand, and the corresponding 3-substituted isocoumarin derivatives were produced in good to excellent yields. The *o*-Halobenzoic acids could be *o*-iodobenzoic acid, *o*-bromobenzoic acid and *o*-chlorobenzoic acid derivatives. 1,3-Diketones could be alkyl- and aryl-substituted 1,3-diketones. Yao's group report a facile and rapid synthesis of isocoumarin derivatives using a copper-catalyzed tandem C–C and C–O coupling strategy from readily available substrates.[453] The reactions of a wide range of 2-iodo-*N*-phenylbenzamides and acyclic diketones as starting materials were investigated. Structurally diverse isocoumarin derivatives were produced in good to excellent yields. In addition, this methodology can also be utilized for the synthesis of various pyranoquinolinone derivatives, which are important constituents in many pharmaceutically active substances. Interestingly, the reaction involving 1,3-cyclohexanedione gave an isoquinolinone derivative. More recently, Guo reported the addition of *o*-halobenzoic acids to active internal alkynes for producting isocoumarin derivatives.[454] CuCl$_2$ was used as the catalyst and gave the corresponding isocoumarins in moderate to good yields. This method showed good functional group tolerance. Alternatively, the reactions of α-electron-withdrawing group (EWG)-substituted ketene *S,S*-acetals with aldehydes or ketones also gave coumarins.[455] The reactions were catalyzed by cheap and commercially available copper(II) bromide (10 mol%) in acetonitrile at room temperature, and a variety of densely functionalized products were produced in excellent yields (80–98%).

In 2009, Ohno and co-workers developed a copper-mediated oxidative *ortho* C–H functionalization using tetrahydropyrimidine as a directing group (Scheme 4.216).[456] This reaction applies to 2-phenyl-1,4,5,6-tetrahydropyrimidines having an electron-donating or a weak electron-withdrawing group and affords the corresponding phenol derivatives within 1 h. Use of

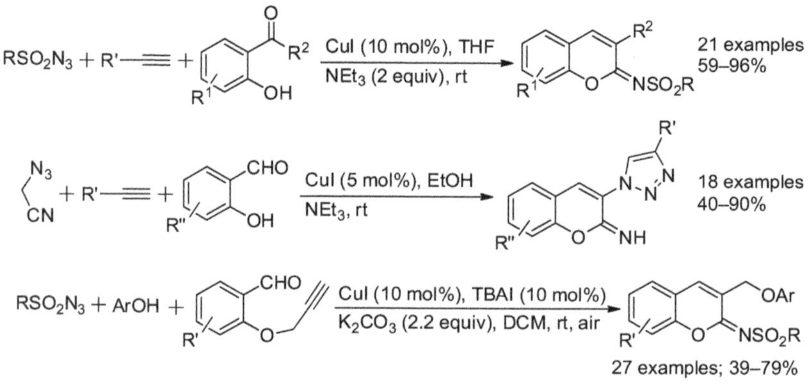

Scheme 4.216   Cu-catalyzed synthesis of iminochromenes.

tert-butyl carbamate or tosylamide instead of water promotes the introduction of a nitrogen functionality to give aniline derivatives. Wang and co-workers reported a procedure for the preparation of substituted iminocoumarins in 2006.[457] The desired products were prepared in good to excellent yields *via* a copper-catalyzed multicomponenet reaction of sulfonyl azides, terminal alkynes and salicylaldehydes or *o*-hydroxylacetophenones. More recently, Qian *et al.* reported a one-pot, three-component catalytic condensation reaction for the synthesis of a variety of 3-triazolyl-2-iminochromenes.[458] A Cu(I)-catalyzed cycloaddition between 2-azidoacetonitrile and an acetylene formed a triazole and activated the neighboring methylene group, inducing an aldol–cyclization–dehydration sequence in the presence of a salicylaldehyde. Additionally, a copper(I)-catalyzed one-pot synthesis of iminocoumarin aryl methyl ethers from an ynal, phenol and sulfonyl azide *via* a cascade [3 + 2]-cycloaddition, 1,3-pseudopericyclic ketenimine rearrangement, 1,4-conjugate addition and aldol-type condensation was reported.[459] This protocol provides a potential route for the construction of a library of iminocoumarin aryl methyl ethers in good yields.

In 2009, Sekar and co-workers reported an interesting protocol for the synthesis of the *trans*-1,4-benzoxazine moiety.[460] This domino reaction started with the ring opening of aziridine, followed by a Goldberg coupling cyclization using the readily available ethylenediamine–CuI complex as catalyst and K_2CO_3 as base. A variety of *trans*-1,4-benzoxazines were synthesized from the corresponding aziridines and *o*-iodophenols in good to excellent isolated yields under relatively mild conditions (Scheme 4.217).

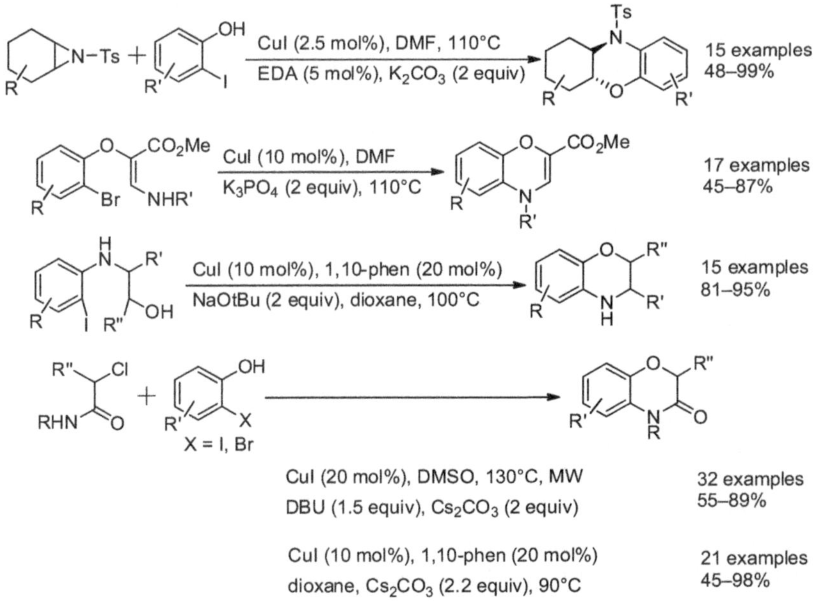

Scheme 4.217 Cu-catalyzed synthesis of 1,4-benzoxazin-3-(4*H*)-ones.

Alumina-supported copper(II) was also explored in this methodlgy.[461] A method starting from *o*-bromo-substituted phenoxyacetates or (phenythio)acetates and primary amines to give *N*-substituted 4*H*-1,4-benzoxazine- and 4*H*-1,4-benzothiazine-2-carboxylates was reported in 2011.[462] This approach involved a Cu(I)-catalyzed Ullmann-type cyclization as a key step and gave the corresponding products in moderate to high yields. A number of groups, including alkoxy, fluoro, bromo, cyclopropyl and alkoxycarbonyl, are welltolerated under the reaction conditions. A recently published procedure started from the adducts formed by the treatment of 1,2-cyclic sulfamidates with 2-halophenols.[463] The method is based on intramolecular copper-catalyzed *O*-arylation of β-amino alcohols as reported by Liu and Chen.[464] The reaction works well without *N*-protection. In addition, the process tolerates variations of both the aryl iodide and amino alcohol portions of the substrate. Liu and co-workers developed an efficient and convenient method for preparing *N*-substituted 2*H*-1,4-benzoxazin-3-(4*H*)-ones in 2009.[465] The reaction started from 2-halophenols *via* a nucleophilic substitution with 2-chloroacetamides followed by a CuI-catalyzed coupling cyclization. A broad range of substrates can be effectively employed to afford the desired products in good yields. Bao's group developed a similar procedure based on thermal heating.[466] Various 2*H*-1,4-benzoxazin-3-(4*H*)-ones with diversity at three substituents on their scaffold were synthesized conveniently in good to excellent yields.

Sequeira and Chemler developed a new copper(II) 2-ethylhexanoate-promoted addition of an alcohol and an amine across an alkene (oxyamination).[467] Several 2-aminomethylmorpholines were synthesized in good to excellent yields and diastereoselectivities. Maycock and co-worekrs developed a copper-mediated intramolecular α-functionalization of tertiary amines (Scheme 4.218).[468] The reaction proceeds *via* C–H bond oxidative activation (benzylic and non-benzylic) to produce diverse dihydro-1,3-oxazines through C–O bond formation. The method is very simple and uses inexpensive $Cu(OAc)_2 \cdot H_2O$ as the catalyst. This conversion can be efficiently performed in an open vessel without the addition of external co-oxidants or additives. Neither dry solvent nor an inert atmosphere are required. Most importantly, naphthoxazines can be produced with 100% diastereoselectivity. In 2011, a novel $Cu(OTf)_2$-catalyzed annulation for the construction of benzoxazine derivatives from readily available *N-p*-tolylamides in the presence of Selectfluor and water was reported.[469] The reaction proceeds through the first intermolecular C–H activated dehydrogenative cross-coupling reaction of benzylic methyl $C(sp^3)$–H and aromatic $C(sp^2)$–H bonds and subsequent intramolecular C–O bond formation. Soon afterwards, they developed another novel highly selective Selectfluor-mediated copper-catalyzed benzylic C–O cyclization for the synthesis of 4*H*-3,1-benzoxazines.[470] The predominant selectivity for a benzylic $C(sp^3)$–H over an aromatic $C(sp^2)$–H bond in *N-o*-tolylbenzamides was achieved. Lv, Wang and co-workers reported a mild and efficient Cu_2O-catalyzed domino intramolecular C–N coupling–C–Y (Y = O, S, N) bond formation reaction.[471] Oxazino[3,2-*a*]indole, thiazino[3,2-*a*]indole

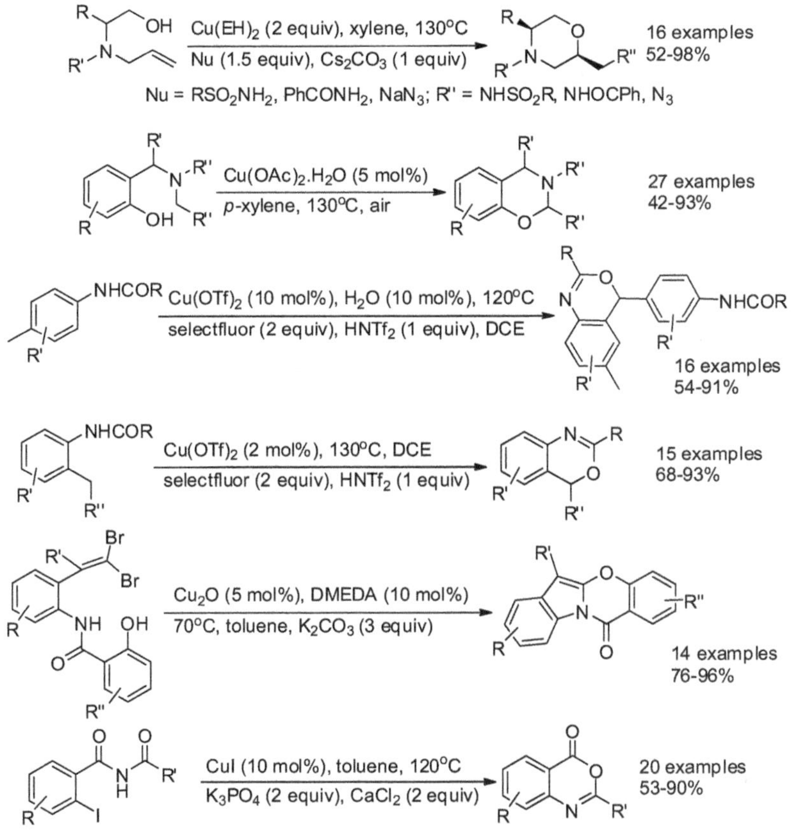

Scheme 4.218 Cu-catalyzed synthesis of morpholines and other heterocycles.

and indolo[2,1-*b*]quinazoline derivatives were facilely assembled from readily accessible *gem*-dibromovinyl systems in good to excellent yields even under an air atmosphere. Moreover, Zhu and co-workers developed a facile and efficient copper-catalyzed method for the synthesis of 4*H*-3,1-benzoxazin-4-one derivatives.[472] This procedure was based on a tandem intramolecular C–N coupling–rearrangement process. *N*-Acyl-2-halobenzamides were applied as substrates and the C–N bond formation–rearrangement sequence is the key step in this transformation.

In 2009, Prasad and Sekar demonstrated a novel and efficient protocol for the synthesis of phenothiazine moieties by a domino ring opening of the aziridine ring with *o*-halothiophenols followed by Goldberg coupling-cyclization using the readily available ethylenediamine–CuI complex as the catalyst.[473] It is important to mention that less reactive *o*-chlorothiophenols were also successfully used for the domino reaction to produce phenothiazine moieties. Hexahydro-1*H*-phenothiazine moieties were synthesized in good to excellent yields (Scheme 4.219). In 2010, Ma *et al.* developed a new approach to construct functionalized phenothiazines.[474] This methodology

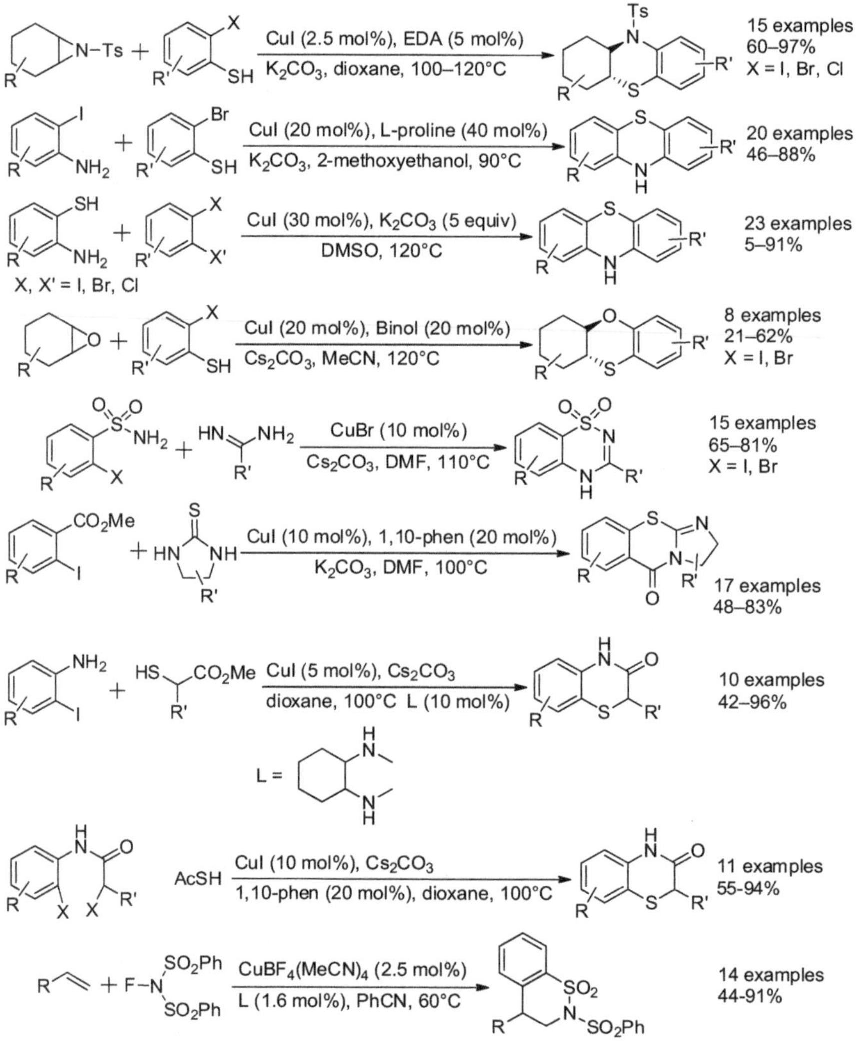

Scheme 4.219 Cu-catalyzed synthesis of sulfur-containing heterocycles.

started from substituted 2-iodoanilines and 2-bromobenzenethiols and was based on a sequentially controlled CuI–L-proline-catalyzed cascade process. The efficiency and substituent tolerance of this procedure were fully demonstrated by synthesizing a number of functionalized phenothiazines. Some of these products are known psychotropic drugs or intermediates for preparing bioactive compounds. Recently, a ligand-free CuI-catalyzed cascade C–S and C–N cross-coupling of (hetero)aryl o-dihalides and o-aminobenzenethiols was developed by Zeng and co-workers.[475] Various phenothiazines were synthesized with excellent regioselectivity. The ring opening of epoxides with 2-bromobenzenethiols was also reported with a copper catalyst.[476] 1,4-Benzoxathiine moieties were synthesized by domino S_N2 ring opening of epoxide with *o*-halothiophenols followed by Cu(I)–BINOL-catalyzed Ullmann-type coupling cyclization [intramolecular C(aryl)–O bond formation] with moderate to good yields. In 2009, Fu, Hu and co-workers developed a simple and practical copper-catalyzed method for the synthesis of 1,2,4-benzothiadiazine 1,1-dioxide derivatives.[477] The reaction proceeded *via* cascade reactions of substituted 2-halobenzenesulfonamides with amidines. Bao's group reported an efficient cascade method for the synthesis of aza[2,1-*b*][1,3]benzothiazinones from cyclic thiourea and methyl 2-iodobenzoate *via* a Cu(I)-catalyzed C–S coupling–amidation process.[478] A number of aza[2,1-*b*][1,3]benzothiazinones containing five-, six- and seven-membered rings and different substituents were obtained in good yield. Huang *et al.* developed a novel synthesis of 1,4-benzothiazin-3-ones *via* Cu-catalyzed coupling of readily available substituted 2-iodoanilines with 2-mercaptoacetate.[479] The method has advantages such as one-step simple operation, wide reaction scope and moderate to excellent isolated yields. Bao and co-workers reported a copper-catalyzed cascade method for the synthesis of 2*H*-benzo[*b*][1,4]thiazin-3(4*H*)-ones from 2-halo-*N*-(2-halophenyl)acetamides and AcSH *via* an S_N2–deacetylation–coupling process and of quinoxalin-2(1*H*)-ones from 2-halo-*N*-(2-halophenyl)acetamides and TsNH$_2$ *via* an S_N2–coupling–desulfonation process.[480] A Cu-catalyzed intermolecular carboamination of alkenes was described in 2013.[481] The reaction of terminal alkenes and an internal alkene with *N*-fluorobenzenesulfonimide was promoted by 2.5 mol% of a Cu(I) salt at 60 °C and six-membered ring sultams were obtained in 44–91% yields. In 2002, a method using olefinic sulfamates from primary and secondary alcohols *via* copper-catalyzed intramolecular aziridination in the presence of iodosylbenzene to afford novel bicyclic fused aziridines was reported.[482] The latter were opened by various nucleophiles to give the corresponding substituted cyclic sulfamates, which in turn reacted, after nitrogen activation, with a second nucleophile at the carbon atom bearing the oxygen atom. Concomitant removal of the sulfonyloxy moiety gave access to polysubstituted amines.

4.4 Other Heterocycles

The Kinugasa reaction is a copper-catalyzed addition reaction of alkynes with nitrones to give β-lactams.[483] This reaction was discovered in 1972 by the reaction of copper(I) phenylacetylide with nitrones at room temperature. Since then, impresive progress has been achieved, with the *in situ* formation of copper(I) phenylacetylide from a copper salt and terminal alkynes in the presence of base. In the case of nitrones, in addition to normal liner nitrones, cyclic nitrones and even sugar-based nitrones were also applied. By the addition of a chiral ligand, enantioselectivity can be achieved.[484] In 2006, Zhao and Li developed a CuCl–2,2′-bipyridine-catalyzed three-component reaction of *N*-substituted hydroxylamines, aldehydes and phenylacetylene.[485] In the presence of NaOAc under neat conditions, the corresponding β-lactams were produced in good to excellent yields. Aromatic, heteroaromatic and aliphatic aldehydes are tolerated in this reaction. The electronic effects of the aldehydes were studied for the reaction with *N*-methylhydroxylamine and *N*-benzylhydroxylamine. For *N*-methylhydroxylamine, electron-rich aldehydes provided higher yields than electron-deficient aldehydes, whereas for *N*-benzylhydroxylamine, no significant electronic effect was observed for the aldehydes. In 2009, Zhang and co-workers developed an alternative reaction for the synthesis of highly functional spiro-β-lactams.[486] This is a representive examples of an oxidative C–C bond-forming procedure that led to spiro-β-lactams and spiropyrrolidonea (Scheme 4.220).

In 2007, an efficient CuI-catalyzed MCR of sulfonyl azides, terminal alkynes and carbodiimides was successfully established by Xu *et al.*[487] No base was needed. The reaction described is mild, general and efficient, and a variety of 2-(sulfonylimino)-4-(alkylimino)azetidine derivatives were produced in good yields (Scheme 4.221). The mechanism may involve a [2 + 2] cycloaddition. Shipman's group developed an efficient, two-step synthesis of 3-methylene-1,2-diazetidines through a Cu(I)-catalyzed 4-*exo* ring closure reaction.[488] This methodology is high yielding and has broad substrate scope. This new class of *N*-heterocycles will serve as useful templates for the synthesis of functionalized 1,2-diazetidines, vicinal diamines and other valued *N*-containing systems. The double bond of this new class of strained heterocycle can be functionalized in a stereocontrolled manner by using palladium-catalyzed Heck reactions. Moreover, chemoselective reduction of 3-alkylidene-1,2-diazetidines gives access to saturated 1,2-diazetidines and vicinal diamines. More recently, an efficient and diastereoselective synthesis of highly functionalized azetidin-2-imines was achieved by Lu, Wang and co-workers.[489] The catalysis strategies include a copper-catalyzed azide–alkyne cycloaddition, a copper-catalyzed C(sp)–C(sp^2) cross-coupling reaction and an intermolecular [2 + 2] cycloaddition. The products could be conveniently

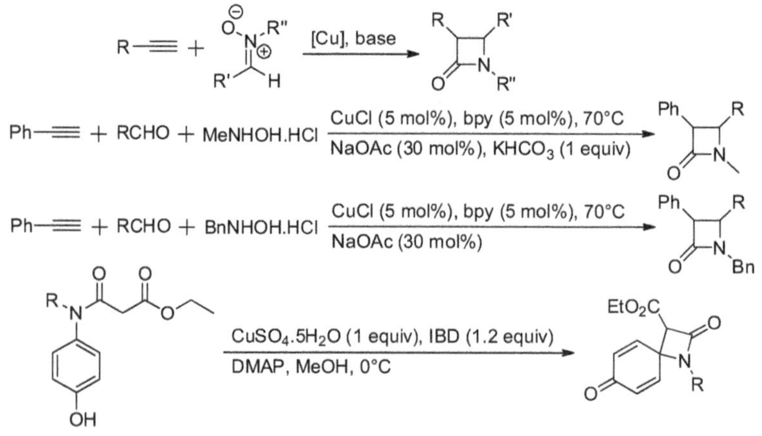

Scheme 4.220   Cu-catalyzed synthesis of β-lactam derivatives.

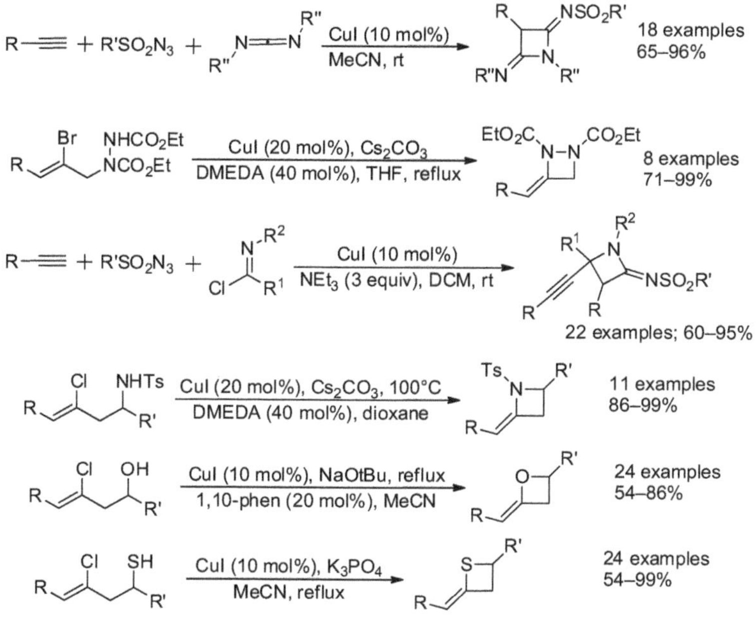

R——≡ + R'SO$_2$N$_3$ + N≡N—R''(R'') → [CuI (10 mol%), MeCN, rt] → product 18 examples 65–96%

R—(Br, NHCO$_2$Et, NCO$_2$Et) → [CuI (20 mol%), Cs$_2$CO$_3$, DMEDA (40 mol%), THF, reflux] → product 8 examples 71–99%

R——≡ + R'SO$_2$N$_3$ + (Cl, R^1, N–R^2) → [CuI (10 mol%), NEt$_3$ (3 equiv), DCM, rt] → product 22 examples; 60–95%

R—(Cl, NHTs, R') → [CuI (20 mol%), Cs$_2$CO$_3$, 100°C, DMEDA (40 mol%), dioxane] → product 11 examples 86–99%

R—(Cl, OH, R') → [CuI (10 mol%), NaOtBu, reflux, 1,10-phen (20 mol%), MeCN] → product 24 examples 54–86%

R—(Cl, SH, R') → [CuI (10 mol%), K$_3$PO$_4$, MeCN, reflux] → product 24 examples 54–99%

Scheme 4.221 Cu-catalyzed synthesis of β-lactams and related compounds.

converted into structurally interesting dihydroazeto[1,2-*a*]benzo[*e*]azepin-2(4*H*)-imines. Hu and Li found that 2-alkylideneazetidines can be prepared efficiently from *N*-tosyl-3-halo-3-butenylamines using Ullmann-type coupling with CuI–*N,N'*-dimethylethylenediamine catalysis.[490] The 2-alkylideneazetidines could be readily converted to the corresponding β-lactams by oxidation with O$_3$. Soon afterwards, they reported a copper-catalyzed intramolecular coupling of vinyl bromides with alcohols (4-*exo* ring closure) and later of thiols.[491,492] In the latter case, K$_3$PO$_4$·3H$_2$O was used as base and vinyl chloride or bromide was successfully implemented without the help of an additional ligand.

In 2004, Buchwald's group reported a simple method for the preparation of medium-ring heterocycles (seven-, eight-, nine- and ten-membered) (Scheme 4.222).[493] The process employs a Cu-catalyzed coupling of a β-lactam with an aryl bromide or iodide followed by intramolecular attack of a pendant amino group. In some instances, the intermediate β-lactam was observable and can be converted to the aza-heterocycle by catalysis. Acetic acid was found to be superior to transition metal complexes as a catalyst for this ring-expansion process. Gan and Ma reported a novel protocol to assemble 1,5-benzothiazepines based on applying two copper-catalyzed coupling reactions.[494] Various *N*-substituted 1,5-benzothiazepines were prepared by changing the amino acid coupling partners. Later, they found that *N*-substituted aminocarbonyl groups possess a strong *ortho*-substituent effect, allowing the ligandless CuCl-catalyzed coupling of 2-bromo-*N*-phenylbenzamides and primary amines at room temperature with a low catalyst

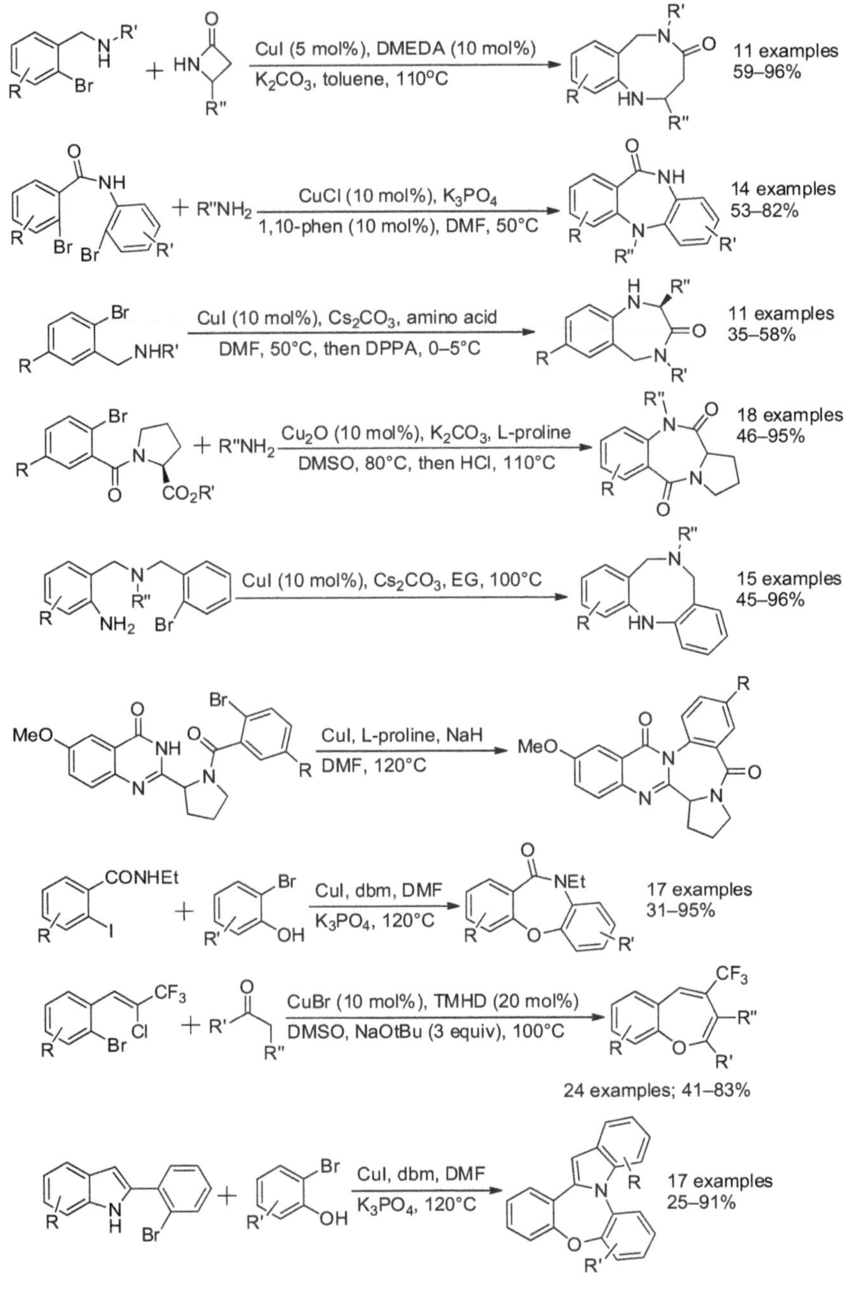

Scheme 4.222 Cu-catalyzed synthesis of dibenzodiazepines.

loading.[495] On the basis of this observation, they developed a facile method for assembling 5-substituted 11-oxodibenzodiazepines *via* double aryl amination. In 2009, Ma, Gao and co-workers developed a CuI-catalyzed coupling of 2-bromobenzylamines and α-amino acids and subsequent condensative cyclization reactions (directly or mediated with DPPA), and 1,4-benzodiazepin-3-ones were produced in moderate yields.[496] Using L-proline and L-valine as the starting materials, enantiopure products were obtained, although partial racemization occurred for other amino acids. Subsequently the same group reported another procedure for the synthesis of N-substituted pyrrolo[2,1-c][1,4]benzodiazepine-5,11-diones.[497] By varying the primary amines and substituents on the aromatic ring of aryl iodides, a wide range of these heterocycles were assembled. In 2011, Spring and co-worekrs reported a new copper-catalyzed strategy for the synthesis of nitrogen-linked seven-, eight- and nine-membered biaryl ring systems.[498] The process is technically simple, proceeds under relatively mild conditions, displays a broad substrate scope and forms biologically valuable products that are difficult to synthesize by other methods. A copper-catalyzed intramolecular N-arylation of a quinazolinone nucleus that furnished the central benzodiazepine core unit was demonstrated in 2010.[499] This methodology was also applied in the total synthesis of (–)-circumdatin H and J. A regioselective copper-initiated domino reaction to give dibenzoxazepinones from 2-iodobenzamides and 2-bromophenols was reported in 2012.[500] As supported by detailed mechanistic studies, the reactions proceeds first by a copper-catalyzed Ullmann coupling, followed by a base-mediated Smiles rearrangement and final ring-closing process. Zhang and co-workers described a copper-catalyzed cyclization reaction of trifluoromethyl-containing o-halo-β-chlorostyrenes with ketones in 2013.[501] Using a combination of copper(I) bromide, 2,2,6,6-tetramethylheptane-3,5-dione and sodium *tert*-butoxide, a variety of 4-trifluoromethylbenzoxepines were prepared in moderate to good yields by the tandem α-alkenylation of ketones with subsequent O-arylation. Zhang's group demonstrated that fused dibenzo[b,f][1,4]oxazepines can be prepared from 2-halophenols and 2-(2-halophenyl)-1H-indoles with copper as catalyst.[502] The Smiles rearrangement was included as a key step.

In 2008, a novel method for the preparation of fused indoles *via* copper-catalyzed domino three-component coupling–indole formation–N-arylation was reported.[503] Starting from simple 2-ethynylanilines and o-bromobenzylamines, complex indole-fused tetracyclic compounds were easily and directly synthesized in a single reaction vessel. More recently, a general synthesis of dibenzoxepine lactams was developed using a one-pot Cu-catalyzed etherification–aldol condensation cascade reaction.[504] The reaction of 4-hydroxyisoindolin-1-one with a wide range of 2-bromobenzaldehydes in the presence of a copper catalyst provided various aristoyagonine derivatives in good yields. The total synthesis of aristoyagonine was successfully achieved in a highly efficient manner using this protocol, which also afforded a variety of aristoyagonine derivatives. In 2013, an efficient method for the synthesis of multisubstituted oxazepine derivatives in a single operation was reported (Scheme 4.223).[505] Oxazepine derivatives were efficiently prepared from O-propargylic oximes and dipolarophiles through copper-catalyzed cascade reactions which proceed through a

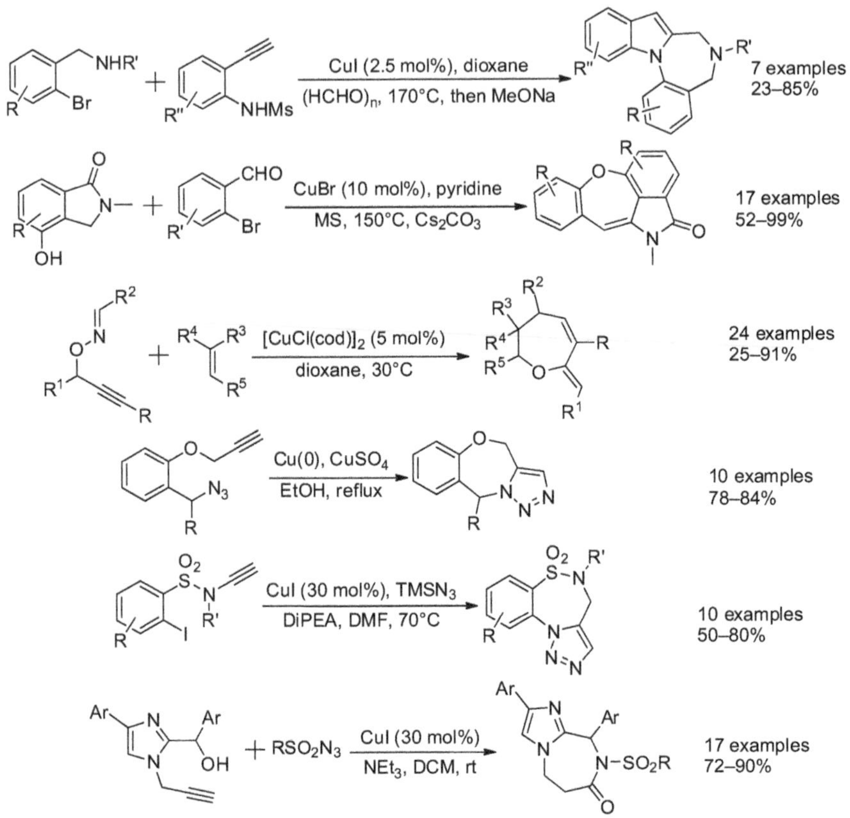

Scheme 4.223 Cu-catalyzed synthesis of oxazepine derivatives and other compounds.

2,3-rearrangement, [3 + 2] cycloaddition and a subsequent 1,3-oxygen rearrangement. The process involves the cleavage of C–O and N–O bonds. A series of [1,2,3]triazolo [5,1-*c*][1,4]benzoxazepine derivatives was synthesized in 2011 by the intramolecular Cu(I)-catalyzed cycloaddition of azidoalkynes derived from salicylaldehyde.[506] The biological profile of these heterocyclic structural scaffolds toward antibacterial and antifungal activity was also illustrated. A practical and efficient synthesis of triazolothiadiazepine-1,1-dioxide derivatives *via* copper-catalyzed [3 + 2] cycloaddition followed by *N*-arylation was described.[507] This method was also applied in the synthesis of indoline- and thiophene-fused triazolothiadiazepine 1,1-dioxide derivatives. In 2012, a strategy based on copper(I)-catalyzed cascade intramolecular nucleophilic attack on *N*-sulfonylketenimine followed by rearrangement of sulfonimidates to sulfonamides and giving substituted 8,9-dihydro-5*H*-imidazo[1,2-*a*][1,4]diazepin-7(6*H*)-ones was developed.[508] Wu and co-workers discovered that triazolo-fused 3′,5′-cyclic nucleoside analogs can be synthesized by an intramolecular 1,3-dipolar cycloaddition of nucleoside-derived azidoalkynes in a regio- and stereospecific manner.[509]

The thymine nucleoside base in these target compounds was converted successfully into the corresponding 5-methylcytosine component. The synthesized compounds were examined in a MAGI assay for exploring anti-HIV activity and in an H9 T lymphocytes assay for measuring cell toxicity.

Copper-catalyzed coupling reactions have also been applied in the preparation of macromolecules. In 1991, Boger's group reported a copper-catalyzed Ullmann reaction to prepare macrocyclic diaryl ethers, and applications in the total synthesis of vancomycin and ristocetin have also been realized.[510–512] In 2001, a copper acetate-mediated intramolecular *O*-arylation of phenols with phenylboronic acid pseudopeptides was reported,[513] and applied as a key step in the preparation of macrocyclic biphenyl ether hydroxamic acid inhibitors of collagenase 1 and gelatinase A and B. The intramolecular macrocyclization was found to be mild and tolerant of common chemical functionality. Fu's group developed an efficient method for the preparation of medium- and large-sized nitrogen heterocycles *via* copper-catalyzed intramolecular *N*-arylation of phosphoramidates and carbamates.[514,515] Introduction of a phosphoryl or *tert*-butoxycarbonyl (Boc) group at N-termini can improve intramolecular cyclization under copper catalysis and the phosphoryl and Boc can easily be removed under mild conditions; hence this convenient and efficient method is suitable for the preparation of medium- and large-sized nitrogen heterocycles. In 2012, an asymmetric total synthesis and determination of the absolute configuration of the neuroactive marine macrolide palmyrolide A was described by Maio and co-workers.[516] A copper(I) iodide and cesium carbonate-based catalyst system is the key step for macrocyclization *via trans*-enamide formation. A two-step macrocyclization strategy for the synthesis of 12- and 14-membered cyclic peptidotriazoles by combining a one-pot, four-component reaction and an intramolecular [3 + 2] azide–alkyne click cycloaddition reaction was described.[517] Macrocycles are obtained in good to excellent yield from the aqueous work-up of the reaction mixture and it is possible to expand or contract the ring size by adjusting the length of the nitrile moiety used in the MCR stage. This idea was successfully applied in the total synthesis of a triazole–epothilone analog. The key step to generate the macrocyclic ring and the triazole ring was to apply Cu_2O nanoparticles to catalyze the 1,3-dipolar cycloaddition. In 2006, a general synthetic route to β-lactam-fused enediynes by intramolecular Kinugasa reaction was successfully developed.[518]

4.5 Summary

The main contributions of copper catalysts in the synthesis of heterocycles have been discussed. The power of copper catalysts has been proven by the impresive achievements in this area. The main ligands applied in copper-catalyzed reactions are nitrogen-containing ligands (TMEDA, DMEDA, 1,10-phenanthroline and so on) and amino acid ligands (L-proline). CuI is the most commonly used salt in coupling chemistry. If the Lewis acid properties are applied, copper salts such as $CuSO_4$ or $Cu(OTf)_2$ have more advantanges.

References

1. H. Kwart and A. A. Kahn, *J. Am. Chem. Soc.*, 1967, **89**, 1951–1958.
2. (a) D. A. Evans, K. A. Woerpel, M. M. Hinman and M. M. Faul, *J. Am. Chem. Soc.*, 1991, **113**, 726–728; (b) D. A. Evans, M. M. Faul and M. T. Bilodeau, *J. Org. Chem.*, 1991, **56**, 6744–6746; (c) D. A. Evans, M. M. Faul and M. T. Bilodeau, *J. Am. Chem. Soc.*, 1994, **116**, 2742–2753; (d) D. A. Evans, M. M. Faul, M. T. Bilodeau, B. A. Anderson and D. M. Barnes, *J. Am. Chem. Soc.*, 1993, **115**, 5328–5329.
3. (a) Z. Li, K. R. Conser and E. N. Jacobsen, *J. Am. Chem. Soc.*, 1993, **115**, 5326–5327; (b) Z. Li, R. W. Quan and E. N. Jacobsen, *J. Am. Chem. Soc.*, 1995, **117**, 5889–5890.
4. K. B. Hansen, N. S. Finney and E. N. Jacobsen, *Angew. Chem. Int. Ed. Engl.*, 1995, **34**, 676–678.
5. (a) P. J. Pérez, M. Brookhart and J. L. Templeton, *Organometallics*, 1993, **12**, 261–262; (b) M. M. Díaz-Requejo, P. J. Pérez, M. Brookhart and J. L. Templeton, *Organometallics*, 1997, **16**, 4399–4402.
6. (a) M. J. Södergren, D. A. Alonso, A. V. Bedekar and P. G. Andersson, *Tetrahedron Lett.*, 1997, **38**, 6897–6900; (b) S. K. Bertilsson, L. Tedenborg, D. A. Alonso and P. G. Andersson, *Organometallics*, 1999, **18**, 1281–1286.
7. (a) P. Dauban and R. H. Dodd, *J. Org. Chem.*, 1999, **64**, 5304–5307; (b) P. Dauban and R. H. Dodd, *Tetrahedron Lett.*, 1998, **39**, 5739–5742.
8. (a) P. H. Di Chenna, P. Dauban, A. Ghini, R. Baggio, M. T. Garland, G. Burton and R. H. Dodd, *Tetrahedron*, 2003, **59**, 1009–1014; (b) L. Sanière, L. Leman, J.-J. Bourguignon, P. Dauban and R. H. Dodd, *Tetrahedron*, 2004, **60**, 5889–2897.
9. (a) J. A. Halfen, J. K. Hallman, J. A. Schultz and J. P. Emerson, *Organometallics*, 1999, **18**, 5435–5437; (b) J. A. Halfen, J. M. Uhan, D. C. Fox, M. P. Mehn and L. Que Jr, *Inorg. Chem.*, 2000, **39**, 4913–4920; (c) J. A. Halfen, D. C. Fox, M. P. Mehn and L. Que Jr, *Inorg. Chem.*, 2001, **40**, 5060–5061.
10. K. Juhl, R. G. Hazell and K. A. Jorgensen, *J. Chem. Soc., Perkin Trans.*, 1999, **1**, 2293–2297.
11. (a) P. Dauban and R. H. Dodd, *Org. Lett.*, 2000, **2**, 2327–2329; (b) F. Duran, L. Leman, A. Ghini, G. Burton, P. Dauban and R. H. Dodd, *Org. Lett.*, 2002, **4**, 2481–2483; (c) A. Estéoule, F. Durán, P. Retailleau, R. H. Dodd and P. Dauban, *Synthesis*, 2007, 1251–1260.
12. (a) P. Dauban, L. Sanière, A. Tarrade and R. H. Dodd, *J. Am. Chem. Soc.*, 2001, **123**, 7707–7708; (b) L. Leman, L. Sanière, P. Dauban and R. H. Dodd, *Arkivoc*, 2003, (vi), 126–134.
13. P. H. Di Chenna, F. Robert-Peilippe, P. Dauban and R. H. Dodd, *Org. Lett.*, 2004, **6**, 4503–4505.
14. F. Robert-Peilippe, P. H. Di Chenna, C. Liang, C. Lescot, F. Collet, R. H. Dodd and P. Dauban, *Tetrahedron: Asymmetry*, 2010, **21**, 1447–1457.

15. (a) C. J. Sanders, K. M. Gillespie, D. Bell and P. Scott, *J. Am. Chem. Soc.*, 2000, **122**, 7132–7133; (b) K. M. Gillespie, C. J. Sanders, P. O'Shaughnessy, I. Westmoreland, C. P. Thickitt and P. Scott, *J. Org. Chem.*, 2002, **67**, 3450–3458.

16. K. M. Gillespie, E. J. Crust, R. J. Deeth and P. Scott, *Chem. Commun.*, 2001, 785–786.

17. P. Brandt, M. J. Södergren, P. G. Andersson and P.-O. Norrby, *J. Am. Chem. Soc.*, 2000, **122**, 8013–8020.

18. A. N. Vedernikov and K. G. Caulton, *Org. Lett.*, 2003, **5**, 2591–2594.

19. (a) M. Shi, C. Wang and A. S. C. Chan, *Tetrahedron: Asymmetry*, 2001, **12**, 3105–3111; (b) M. Shi and C. Wang, *Chirality*, 2002, **14**, 412–416.

20. (a) L. Ma, D. Du and J. Xu, *J. Org. Chem.*, 2005, **70**, 10155–10158; (b) L. Ma, P. Jiao, Q. Zhang and J. Xu, *Tetrahedron: Asymmetry*, 2005, **16**, 3718–3734; (c) L. Ma, D. Du and J. Xu, *Chirality*, 2006, **18**, 575–580; (d) L. Ma, P. Jiao, Q. Zhang, D. Du and J. Xu, *Chirality*, 2007, **19**, 878–884; (f) J. Xu, L. Ma and P. Jiao, *Chem. Commun.*, 2004, 1616–1617.

21. (a) W. Sun, E. Herdtweck and F. E. Kühn, *New J. Chem.*, 2005, **29**, 1577–1580; (b) Y. Li, B. Diebl, A. Raith and F. E. Kühn, *Tetrahedron Lett.*, 2008, **49**, 5954–5956; (c) Y. Li, J. He, V. Khankhoje, E. Herdtweck, K. Köhler, O. Storcheva, M. Cokoja and F. E. Kühn, *Dalton Trans.*, 2011, **40**, 5746–5754; (d) M. A. Mairena, M. Diaz-Requejo, T. R. Belderrain, M. C. Nicasio, S. Trofimenko and P. J. Pérez, *Organometallics*, 2004, **23**, 253–256; (e) S. T. Handy and M. Czopp, *Org. Lett.*, 2001, **3**, 1423–1425; (f) S. T. Handy, A. Ivanow and M. Czopp, *Tetrahedron Lett.*, 2006, **47**, 1821–1823; (g) P. Comba, M. Merz and H. Pritzkow, *Eur. J. Org. Chem.*, 2003, 1711–1718; (h) P. Comba, C. Lang, C. L. de Laorden, A. Muruganantham, G. Rajaraman, H. Wadepohl and M. Zajaczkowski, *Chem. Eur. J.*, 2008, **14**, 5313–5328; (i) P. Comba, C. Haaf, A. Lienke, A. Muruganantham and H. Wadepohl, *Chem. Eur. J.*, 2009, **15**, 10880–10887; (j) H. V. R. Dias, H. Lu, H. Kim, S. A. Polach, T. K. H. H. Goh, R. G. Browning and C. J. Lovely, *Organometallics*, 2002, **21**, 1466–1473; (k) J. A. Flores, V. Badarinarayana, S. Singh, C. J. Lovely and H. V. R. Dias, *Dalton Trans.*, 2009, 7648–7652; (l) H. Han, I. Bae, E. J. Yoo, J. Lee, Y. Do and S. Chang, *Org. Lett.*, 2004, **6**, 4109–4112; (m) H. Han, S. B. Park, S. K. Kim and S. Chang, *J. Org. Chem.*, 2008, **73**, 2862–2870; (n) D. B. Llewellyn, D. Adamson and B. A. Arndtsen, *Org. Lett.*, 2000, **2**, 4165–4168; (o) L. D. Amisial, X. Dai, R. A. Kinney, A. Krishnaswamy and T. H. Warren, *Inorg. Chem.*, 2004, **43**, 6537–6539; (p) H. Kwong, D. Liu, K. Chan, C. Lee, K. Huang and C. Che, *Tetrahedron Lett.*, 2004, **45**, 3965–3968; (q) R. R. Conry, A. A. Tipton, W. S. Striejewske, E. Erkizia, M. A. Malwitz, A. Caffaratti and J. A. Natkin, *Organometallics*, 2004, **23**, 5210–5218; (r) T. C. H. Lam, W. Mak, W. Wong, H. Kwong, H. H. Y. Sung, S. M. F. Lo, I. D. Williams and W. Leung, *Organometallics*, 2004, **23**, 1247–1252; (s) F. Mohr, S. A. Binfield, J. C. Fettinger and A. N. Vedernikov, *J. Org. Chem.*, 2005, **70**, 4833–4839; (t) X. Wang and K. Ding, *Chem. Eur. J.*, 2006, **12**, 4568–4575; (u) T. Dhanalakshmi,

E. Suresh, H. Stoeckli-Evans and M. Palaniandavar, *Eur. J. Inorg. Chem.*, 2006, 4687–4695; (v) J. Pérez, D. Morales, L. A. García-Escudero, H. Martinez-Garcia, D. Miguel and P. Bernad, *Dalton Trans.*, 2009, 375–382; (w) H. Martinez-Garcia, D. Morales, J. Pérez, M. Puerto and D. Miguel, *Inorg. Chem.*, 2010, **49**, 6974–6985; (x) N. Nebra, C. Lescot, P. Dauban, S. Mallet-Ladeira, B. Martin-Vaca and D. Bourissou, *Eur. J. Org. Chem.*, 2013, 984–990; (y) Q. Xu and D. H. Appella, *Org. Lett.*, 2008, **10**, 1497–1500.

22. J. W. W. Chang, T. M. U. Ton, Z. Zhang, Y. Xu and P. W. H. Chan, *Tetrahedron Lett.*, 2009, **50**, 161–164.

23. (a) S. Taylor, J. Gullick, P. McMorn, D. Bethell, P. C. B. Page, F. E. Hancock, F. King and G. J. Hutchings, *Top. Catal.*, 2003, **24**, 43–50; (b) C. Langham, S. Taylor, D. Bethell, P. McMorn, P. C. B. Page, D. J. Willock, C. Sly, F. E. Hancock, F. King and G. J. Hutchings, *J. Chem. Soc., Perkin Trans.*, 1999, **2**, 1043–1049; (c) C. Langham, P. Piaggio, D. Bethell, D. F. Lee, P. McMorn, P. C. B. Page, D. J. Willock, C. Sly, F. E. Hancock, F. King and G. J. Hutchings, *Chem. Commun.*, 1998, 1601–1602; (d) J. Gullick, D. Ryan, P. McMorn, D. Bethell, F. King, F. Hancock and G. Hutchings, *New J. Chem.*, 2004, **28**, 1470–1478; (e) J. Gullick, S. Taylor, D. Ryan, P. McMorn, M. Coogan, D. Bethell, P. C. B. Page, F. E. Hancock, F. King and G. J. Hutchings, *Chem. Commun.*, 2003, 2808–2809; (f) L. Jeffs, D. Arquier, B. Kariuki, D. Bethell, P. C. B. Page and G. J. Hutchings, *Org. Biomol. Chem.*, 2011, **9**, 1079–1084; (g) S. Taylor, J. Gullick, P. McMorn, D. Bethell, P. C. B. Page, F. E. Hancock, F. King and G. J. Hutchings, *J. Chem. Soc., Perkin Trans.*, 2001, **2**, 1724–1728; (h) S. Taylor, J. Gullick, P. McMorn, D. Bethell, P. C. B. Page, F. E. Hancock, F. King and G. J. Hutchings, *J. Chem. Soc., Perkin Trans.*, 2001, **2**, 1714–1723.

24. H. Lebel, S. Lectard and M. Parmentier, *Org. Lett.*, 2007, **9**, 4797–4800.

25. R. Liu, S. R. Herron and S. A. Fleming, *J. Org. Chem.*, 2007, **72**, 5587–5591.

26. (a) H. Lebel, S. Lectard and M. Parmentier, *Pure Appl. Chem.*, 2010, **82**, 1827–1833; (b) H. Lebel, M. Parmentier, O. Leogane, K. Ross and C. Spitz, *Tetrahedron*, 2012, **68**, 3396–3409.

27. T. M. U. Ton, C. Tejo, D. L. Y. Tiong and P. W. H. Chan, *J. Am. Chem. Soc.*, 2012, **134**, 7344–7350.

28. (a) J.-F. Lutz, *Angew. Chem. Int. Ed.*, 2007, **46**, 1018–1025; (b) S. Bräse, C. Gil, K. Knepper and V. Zimmermann, *Angew. Chem. Int. Ed.*, 2005, **44**, 5188–5240; (c) J. Hu, J. R. Lu and Y. Ju, *Chem. Asian J.*, 2011, **6**, 2636–2647; (d) S. G. Agalave, S. R. Maujan and V. S. Pore, *Chem. Asian J.*, 2011, **6**, 2696–2718; (e) J. M. Holub and K. Kirshenbaum, *Chem. Soc. Rev.*, 2010, **39**, 1325–1337; (f) J. E. Moses and A. D. Moorhouse, *Chem. Soc. Rev.*, 2007, **36**, 1249–1262; (g) A. Qin, J. W. Y. Lam and B. Z. Tang, *Chem. Soc. Rev.*, 2010, **39**, 2522–2544.

29. (a) H. C. Kolb, M. G. Finn and K. B. Sharpless, *Angew. Chem. Int. Ed.*, 2001, **40**, 2004–2021; (b) L. Liang and D. Astruc, *Coord. Chem. Rev.*,

2011, **255**, 2933–2945; (c) V. D. Bock, H. Hiemstra and J. H. van Maarseveen, *Eur. J. Org. Chem.*, 2006, 51–68; (d) M. Meldal and C. W. Tornoe, *Chem. Rev.*, 2008, **108**, 2952–3015; (e) J. E. Hein and V. V. Fokin, *Chem. Soc. Rev.*, 2010, **39**, 1302–1315; (f) E. Lallana, R. Riguera and E. Fernandez-Megia, *Angew. Chem. Int. Ed.*, 2011, **50**, 8794–8804.

30. (a) S. Díez-González and S. P. Nolan, *Angew. Chem. Int. Ed.*, 2008, **47**, 8881–8884; (b) S. Díez-González, A. Correa, L. Cavallo and S. P. Nolan, *Chem. Eur. J.*, 2006, **12**, 7558–7564; (c) M.-L. Teyssot, A. Chevry, M. Traikia, M. El-Ghozzi, D. Avignant and A. Gautier, *Chem. Eur. J.*, 2009, **15**, 6322–6326.

31. (a) C. Shao, G. Cheng, D. Su, J. Xu, X. Wang and Y. Hu, *Adv. Synth. Catal.*, 2010, **352**, 1587–1592; (b) Q. Zhang, X. Wang, C. Cheng, R. Zhu, N. Liu and Y. Hu, *Org. Biomol. Chem.*, 2012, **10**, 2847–2854; (d) R. Berg, J. Straub, E. Schreiner, S. Mader, F. Rominger and B. E. Straub, *Adv. Synth. Catal.*, 2012, **354**, 3445–3450.

32. (a) Z. Gonda and Z. Novák, *Dalton Trans.*, 2010, **39**, 726–729; (b) L. S. Campbell-Verduyn, L. Mirfeizi, R. A. Dierckx, P. H. Elsinga and B. L. Feringa, *Chem. Commun.*, 2009, 2139–2141; (c) P. Ji, J. H. Atherton and M. I. Page, *Org. Biomol. Chem.*, 2012, **10**, 7965–7969; (d) F. Wang, H. Fu, Y. Jiang and Y. Zhao, *Green Chem.*, 2008, **10**, 452–456; (e) N. Candelon, D. Lastécoueres, A. K. Diallo, J. R. Aranzaes, D. Astruc and J.-M. Vincent, *Chem. Commun.*, 2008, 741–743; (f) J. Zhang, J. Wu, L. Shen, G. Jin and S. Cao, *Adv. Synth. Catal.*, 2011, **353**, 580–584; (g) B. R. Buckley, S. E. Dann and H. Heaney, *Chem. Eur. J.*, 2010, **16**, 6278–6284; (h) D. Lubriks, I. Sokolovs and E. Suna, *J. Am. Chem. Soc.*, 2012, **134**, 15436–15442; (i) F. Wang, H. Fu, Y. Jiang and Y. Zhao, *Adv. Synth. Catal.*, 2008, **350**, 1830–1834; (j) Y. Zhang, X. Han, C. Ren and Y. Zhao, *Synth. Commun.*, 2013, **43**, 2119–2126; (k) A. Z. Ahmady, F. Heidarizadeh and M. Keshavarz, *Synth. Commun.*, 2013, **43**, 2100–2109; (l) B. R. Buckley, S. E. Dann, H. Heaney and E. C. Stubbs, *Eur. J. Org. Chem.*, 2011, 770–776; (m) T. Jin, S. Kamijo and Y. Yamamoto, *Eur. J. Org. Chem.*, 2004, 3789–3791; (n) Z. Li, T. S. Seo and J. Ju, *Tetrahedron Lett.*, 2004, **45**, 3143–3146; (o) K. A. Dururgkar, R. G. Gonnade and C. V. Ramana, *Tetrahedron*, 2009, **65**, 3974–3979; (p) J. Garcia-Alvarez, J. Diez, J. Gimeno, F. J. Suarez and C. Vincent, *Eur. J. Inorg. Chem.*, 2012, 5854–5863; (q) O. A. Attanasi, G. Favi, P. Filippone, F. Mantellini, G. Moscatelli and F. R. Perrulli, *Org. Lett.*, 2010, **12**, 468–471; (r) B. H. M. Kuijmers, G. C. T. Dijkmans, S. Groothuys, P. J. L. M. Quaedflieg, R. H. Blaauw, F. L. van Delft and F. P. J. T. Rutjes, *Synlett*, 2005, 3059–3062; (s) L. Wu, Y. Xie, Z. Chen, Y. Niu and Y. Liang, *Synlett*, 2009, 1453–1456.

33. (a) L. Liang, J. Ruiz and D. Astruc, *Adv. Synth. Catal.*, 2011, **353**, 3434–3450; (b) T. Katayama, K. Kamata, K. Yamaguchi and N. Mizuno, *ChemSusChem*, 2009, **2**, 59–62; (c) F. Alonso, Y. Moglie, G. Radivoy and M. Yus, *Eur. J. Org. Chem.*, 2010, 1875–1884; (d) H. Sharghi, R. Khalifeh and M. M. Doroodmand, *Adv. Synth. Catal.*, 2009, **351**, 207–218;

(e) A. Coelho, P. Diz, O. Caamano and E. Sotelo, *Adv. Synth. Catal.*, 2010, **352**, 1179–1192; (f) J. Albadi, M. Keshavarz, M. Abedini and M. Vafaie-Nezhad, *Chin. Chem. Lett.*, 2012, **23**, 797–800; (g) C. Zhang, B. Huang, Y. Chen and D.-M. Cui, *New J. Chem.*, 2013, **37**, 2606–2609; (h) S. Chassaing, A. S. S. Sido, A. Alix, M. Kumarraja, P. Pale and J. Sommer, *Chem. Eur. J.*, 2008, **14**, 6713–6721; (i) Z. Zhang, F. Ye, L. Zheng, K. Yang, G. Lai and L. Xu, *Chem. Eur. J.*, 2012, **18**, 14094–14099; (j) F. Alonso, Y. Moglie, G. Radivoy and M. Yus, *Adv. Synth. Catal.*, 2010, **352**, 3208–3214; (k) J. Albadi, M. Keshavarz, F. Shirini and M. Vafaie-Nezhad, *Catal. Commun.*, 2012, **27**, 17–20; (l) N. Mukherjee, S. Ahammed, S. Bhadra and B. C. Ranu, *Green Chem.*, 2013, **15**, 389–397; (m) A. N. Prasad, B. Thirupathi, G. Raju, R. Srinivas and B. M. Reddy, *Catal. Sci. Technol.*, 2012, **2**, 1264–1268; (n) R. Hudson, C. Li and A. Moores, *Green Chem.*, 2012, **14**, 622–624; (o) F. Alonso, Y. Moglie, G. Radivoy and M. Yus, *Org. Biomol. Chem.*, 2011, **9**, 6385–6395; (p) C. W. Tornoe, C. Christensen and M. Meldal, *J. Org. Chem.*, 2002, **67**, 3057–3064; (q) M. A. Ermeydan, F. Dumoulin, T. V. Basova, D. Bouchu, A. G. Gürek, V. Ahsen and D. Lafont, *New J. Chem.*, 2010, **34**, 1153–1162.

34. (a) A. C. Varas, T. Noel, Q. Wang and V. Hessel, *ChemSusChem*, 2012, **5**, 1703–1707; (b) M. Fuchs, W. Goessler, C. Oilger and C. O. Kappe, *Adv. Synth. Catal.*, 2010, **352**, 323–328; (c) S. B. Ötvös, I. M. Mándity, L. Kiss and F. Fülöp, *Chem. Asian J.*, 2013, **8**, 800–808.

35. D. Kumar, V. B. Reddy and R. S. Varma, *Tetrahedron Lett.*, 2009, **50**, 2065–2068.

36. J. S. Yadav, B. V. S. Reddy, G. M. Reddy and S. R. Anjum, *Tetrahedron Lett.*, 2009, **50**, 6029–6031.

37. K. Barral, A. D. Moorhouse and J. E. Moses, *Org. Lett.*, 2007, **9**, 1809–1811.

38. (a) A. K. Feldman, B. Colasson and V. V. Fokin, *Org. Lett.*, 2004, **6**, 3897–3899; (b) A. Kolarovic, M. Schnürch and M. D. Mihovilovic, *J. Org. Chem.*, 2011, **76**, 2613–2618.

39. (a) B. Kaboudin, Y. Abedi and T. Yokomatsu, *Org. Biomol. Chem.*, 2012, **10**, 4543–4548; (b) D. Yang, N. Fu, Z. Liu, Y. Li and B. Chen, *Synlett*, 2007, 278–282.

40. Y. Jiang, C. Kuang and Q. Yang, *Synthesis*, 2010, 4256–4260.

41. X. Wang, C. Kuang and Q. Yang, *Eur. J. Org. Chem.*, 2012, 424–428.

42. N. M. Smith, M. J. Greaves, R. Jewell, M. W. D. Perry, M. J. Stocks and J. P. Stonehouse, *Synlett*, 2009, 1391–1394.

43. E. D. Goddard-Borger and R. V. Stick, *Org. Lett.*, 2007, **9**, 3797–3800.

44. S. Kamijo, T. Jin, Z. Huo and Y. Yamamoto, *J. Org. Chem.*, 2004, **69**, 2386–2393.

45. Y. Zhang, X. Li, J. Li, J. Chen, X. Meng, M. Zhao and B. Chen, *Org. Lett.*, 2012, **14**, 26–29.

46. K. Li, J. Chen, J. Li, Y. Chen, J. Qu, X. Guo, C. Chen and B. Chen, *Eur. J. Org. Chem.*, 2013, 6246–6248.
47. J. Zhang, G. Jin, S. Xiao, J. Wu and S. Cao, *Tetrahedron*, 2013, **69**, 2352–2356.
48. J. S. Yadav, B. V. S. Reddy, G. M. Reddy and D. N. Chary, *Tetrahedron Lett.*, 2007, **48**, 8773–8776.
49. H. Sharghi, M. Hosseini-Sarvari, F. Moeini, R. Khalifeh and A. S. Beni, *Helv. Chim. Acta*, 2010, **93**, 435–449.
50. F. Alonso, Y. Moglie, G. Radivoy and M. Yus, *J. Org. Chem.*, 2011, **76**, 8394–8405.
51. J. Kalisiak, K. B. Sharpless and V. V. Fokin, *Org. Lett.*, 2008, **10**, 3171–3174.
52. N. V. Rostovskii, M. S. Novikov, A. F. Khlebnikov, S. M. Korneev and D. S. Yufit, *Org. Biomol. Chem.*, 2013, **11**, 5535–5545.
53. M. M. Guru and T. Punniyamurthy, *J. Org. Chem.*, 2012, 77, 5063–5073.
54. M. Ciesielski, D. Pufky and M. Döring, *Tetrahedron*, 2005, **61**, 5942–5947.
55. (a) T. Jin, F. Kitahara, S. Kamijo and Y. Yamamoto, *Tetrahedron Lett.*, 2008, **49**, 2824–2827; (b) T. Jin, F. Kitahara, S. Kamijo and Y. Yamamoto, *Chem. Asian J.*, 2008, **3**, 1575–1580; (c) T. Abe, G. Tao, Y. Joo, R. W. Winter, G. L. Gard and J. M. Shreeve, *Chem. Eur. J.*, 2009, **15**, 9897–9904; (d) B. Sreedhar, A. S. Kumar and D. Yada, *Tetrahedron Lett.*, 2011, **52**, 3565–3569; (e) U. Yapuri, S. Palle, O. Gudaparthi, S. R. Narahari, D. K. Rawat, K. Mukkanti and J. Vantikommu, *Tetrahedron Lett.*, 2013, **54**, 4732–4734.
56. F. Chen, C. Qin, Y. Cui and N. Jiao, *Angew. Chem. Int. Ed.*, 2011, **50**, 11487–11491.
57. K. Liubchak, A. Tolmachev and K. Nazarenko, *J. Org. Chem.*, 2012, 77, 3365–3372.
58. A. V. Lygin and A. de Meijere, *Eur. J. Org. Chem.*, 2009, 5138–5141.
59. C. Kanazawa, S. Kamijo and Y. Yamamoto, *J. Am. Chem. Soc.*, 2006, **128**, 10662–10663.
60. Z.-J. Cai, S.-Y. Wang and S.-J. Ji, *Org. Lett.*, 2012, **14**, 6068–6071.
61. Y.-F. Wang, X. Zhu and S. Chiba, *J. Am. Chem. Soc.*, 2012, **134**, 3679–3682.
62. S. Sanjaya and S. Chiba, *Org. Lett.*, 2012, **14**, 5342–5345.
63. D. Tang, P. Wu, X. Liu, Y.-X. Chen, S.-B. Guo, W.-L. Chen, J.-G. Li and B.-H. Chen, *J. Org. Chem.*, 2013, **78**, 2746–2750.
64. J. Li and L. Neuville, *Org. Lett.*, 2013, **15**, 1752–1755.
65. G. Brasche and S. L. Buchwald, *Angew. Chem. Int. Ed.*, 2008, **47**, 1932–1934.
66. J. Li, S. Bénard, L. Neuville and J. Zhu, *Org. Lett.*, 2012, **14**, 5980–5983.
67. (a) B. Zou, Q. Yuan and D. Ma, *Angew. Chem. Int. Ed.*, 2007, **46**, 2598–2601; (b) X. Diao, Y. Wang, Y. Jiang and D. Ma, *J. Org. Chem.*, 2009, **74**, 7974–7977.

68. D. Yang, H. Fu, L. Hu, Y. Jiang and Y. Zhao, *J. Org. Chem.*, 2008, **73**, 7841–7844.

69. Y. Qu, L. Pan, Z. Wu and X. Zhou, *Tetrahedron*, 2013, **69**, 1717–1719.

70. (a) X. Deng, H. McAllister and N. S. Mani, *J. Org. Chem.*, 2009, **74**, 5742–5745; (b) X. Deng and N. S. Mani, *Eur. J. Org. Chem.*, 2010, 680–686.

71. J. She, Z. Jiang and Y. Wang, *Synlett*, 2009, 2023–2027.

72. H. Jin, X. Xu, J. Gao, J. Zhong and Y. Wang, *Adv. Synth. Catal.*, 2010, **352**, 347–350.

73. J. Zhu, H. Xie, Z. Chen, S. Li and Y. Wu, *Chem. Commun.*, 2009, 2338–2340.

74. K. Hirano, A. T. Biju and F. Glorius, *J. Org. Chem.*, 2009, **74**, 9570–9572.

75. J. Peng, M. Ye, C. Zong, F. Hu, L. Feng, X. Wang, Y. Wang and C. Chen, *J. Org. Chem.*, 2011, **76**, 716–719.

76. (a) K. Liubchak, A. Tolmachev, O. O. Grygorenko and K. Nazarenko, *Tetrahedron*, 2012, **68**, 8564–8571; (b) K. Liubchak, K. Nazarenko and A. Tolmachev, *Tetrahedron*, 2012, **68**, 2993–3000.

77. N. Ibrahim and M. Legraverend, *J. Org. Chem.*, 2009, **74**, 463–465.

78. H. Yuan, Y. Chen, J. Song, C. Chen and B. Chen, *Chin. J. Chem.*, 2013, **31**, 1247–1249.

79. G. Evindar and R. A. Batey, *Org. Lett.*, 2003, **5**, 133–136.

80. X. Lv and W. Bao, *J. Org. Chem.*, 2009, **74**, 5618–5621.

81. G. Shen and W. Bao, *Adv. Synth. Catal.*, 2010, **352**, 981–986.

82. S. Murru, B. K. Patel, J. Le Bras and J. Muzart, *J. Org. Chem.*, 2009, **74**, 2217–2220.

83. T. Ramana and T. Punniyamurthy, *Chem. Eur. J.*, 2012, **18**, 13279–13283.

84. B. Zou, Q. Yuan and D. Ma, *Org. Lett.*, 2007, **9**, 4291–4294.

85. Q.-F. Zhong and L.-P. Sun, *Tetrahedron*, 2010, **66**, 5107–5111.

86. X. Wang, Y. Jin, Y. Zhao, L. Zhu and H. Fu, *Org. Lett.*, 2012, **14**, 452–455.

87. G. Qiu and J. Wu, *Chem. Commun.*, 2012, **48**, 6046–6048.

88. D. A. Black and B. A. Arndtsen, *Tetrahedron*, 2005, **61**, 11317–11321.

89. C. Jin, J. P. Burgess, J. A. Kepler and C. E. Cook, *Org. Lett.*, 2007, **9**, 1887–1890.

90. S. V. Kumar, B. Saraiah, N. C. Misra and H. Ila, *J. Org. Chem.*, 2012, **77**, 10752–10763.

91. S. Yugandar, A. Acharya and H. Ila, *J. Org. Chem.*, 2013, **78**, 3948–3960.

92. C. Wan, J. Zhang, S. Wang, J. Fan and Z. Wang, *Org. Lett.*, 2010, **12**, 2338–2341.

93. P. Hu, Q. Wang, Y. Yan, S. Zhang, B. Zhang and Z. Wang, *Org. Biomol. Chem.*, 2013, **11**, 4304–4307.

94. X. Li, L. Huang, H. Chen, W. Wu, H. Huang and H. Jiang, *Chem. Sci.*, 2012, **3**, 3463–3467.

95. K. Schuh (née Müller) and F. Glorius, *Synthesis*, 2007, 2297–2306.

96. R. Martin, A. Cuenca and S. L. Buchwald, *Org. Lett.*, 2007, **9**, 5521–5524.

97. C. W. Cheung and S. L. Buchwald, *J. Org. Chem.*, 2012, **77**, 7526–7537.

98. A. E. Wendlandt and S. S. Stahl, *Org. Biomol. Chem.*, 2012, **10**, 3866–3870.

99. Z. Xu, C. Zhang and N. Jiao, *Angew. Chem. Int. Ed.*, 2012, **51**, 11367–11370.

100. G. Altenhoff and F. Glorius, *Adv. Synth. Catal.*, 2004, **346**, 1661–1664.

101. M. A. Ali, M. Suri and T. Punniyamurthy, *Synthesis*, 2013, 501–506.

102. R. D. Wiirre, G. Evindar and R. A. Batey, *J. Org. Chem.*, 2008, **73**, 3452–3459.

103. N. Barbero, M. Carril, R. SanMartin and E. Domínguez, *Tetrahedron*, 2007, **63**, 10425–10432.

104. D. Xu, L.-P. Sun and Q.-D. You, *Tetrahedron*, 2012, **68**, 4248–4251.

105. M. L. Kantam, G. T. Venkanna, K. B. S. Kumar, V. Balasubrahmanyam and S. Bhargava, *Synlett*, 2009, 1753–1756.

106. V. Kavala, D. Janreddy, M. J. Raihan, C.-W. Kuo, C. Ramesh and C.-F. Yao, *Adv. Synth. Catal.*, 2012, **354**, 2229–2240.

107. (a) S. Ueda and H. Nagasawa, *Angew. Chem. Int. Ed.*, 2008, **47**, 6411–6413; (b) S. Ueda and H. Nagasawa, *J. Org. Chem*, 2009, **74**, 4272–4277.

108. (a) M. M. Guru, M. A. Ali and T. Punniyamurthy, *Org. Lett.*, 2011, **13**, 1194–1197; (b) M. M. Guru, M. A. Ali and T. Punniyamurthy, *J. Org. Chem.*, 2011, **76**, 5295–5308.

109. C. Pabba, H.-J. Wang, S. R. Mulligan, Z.-J. Chen, T. M. Stark and B. T. Gregg, *Tetrahedron Lett.*, 2005, **46**, 7553–7557.

110. D. Vina, E. del Olmo, J. L. López-Pérez and A. S. Feliciano, *Org. Lett.*, 2007, **9**, 525–528.

111. M. Gao, X. Liu, X. Wang, Q. Cai and K. Ding, *Chin. J. Chem.*, 2011, **29**, 1199–1204.

112. S. Tanimori, Y. Kobayashi, Y. Iesaki, Y. Ozaki and M. Kirihata, *Org. Biomol. Chem.*, 2012, **10**, 1381–1387.

113. L. Xu, Y. Peng, Q. Pan, Y. Jiang and D. Ma, *J. Org. Chem.*, 2013, **78**, 3400–3404.

114. X. Li, L. He, H. Chen, W. Wu and H. Jiang, *J. Org. Chem.*, 2013, **78**, 3636–3646.

115. G. Zhang, Y. Zhao and H. Ge, *Angew. Chem. Int. Ed.*, 2013, **52**, 2559–2563.

116. D.-G. Yu, M. Suri and F. Glorius, *J. Am. Chem. Soc.*, 2013, **135**, 8802–8805.

117. S. Li, Y. Luo and J. Wu, *Org. Lett.*, 2011, **13**, 4312–4315.

118. A. Ziarati, J. Safaei-Ghomi and S. Rohani, *Ultrason. Sonochem.*, 2013, **20**, 1069–1075.

119. (a) L. L. Joyce, G. Evindar and R. A. Batey, *Chem. Commun.*, 2004, 446–447; (b) G. Evindar and R. A. Batey, *J. Org. Chem.*, 2006, **71**, 1802–1808.

120. J. H. Spatz, T. Bach, M. Umkehrer, J. Bardin, G. Ross, C. Burdack and J. Kolb, *Tetrahedron Lett.*, 2007, **48**, 9030–9034.

121. P. Saha, T. Ramana, N. Purkait, M. A. Ali, R. Paul and T. Punniyamurthy, *J. Org. Chem.*, 2009, **74**, 8719–8725.

122. J. Wang, F. Peng, J.-L. Jiang, Z.-J. Lu, L.-Y. Wang, J. Bai and Y. Pan, *Tetrahedron Lett.*, 2008, **49**, 467–470.

123. S. Murru, H. Ghosh, S. K. Sahoo and B. K. Patel, *Org. Lett.*, 2009, **11**, 4254–4257.
124. G. Shen, X. Lv and W. Bao, *Eur. J. Org. Chem.*, 2009, 5897–5901.
125. Q. Ding, X. He and J. Wu, *J. Comb. Chem.*, 2009, **11**, 587–591.
126. Y.-J. Guo, R.-Y. Tang, P. Zhang and J.-H. Li, *Tetrahedron Lett.*, 2010, **51**, 649–652.
127. N. Khatun, L. Jamir, M. Ganesh and B. K. Patel, *RSC Adv.*, 2012, **2**, 11557–11565.
128. R. Yao, H. Liu, Y. Wu and M. Cai, *Appl. Organomet. Chem.*, 2013, **27**, 109–113.
129. J. Yang, P. Li and L. Wang, *Tetrahedron*, 2011, **67**, 5543–5549.
130. R. Xiao, W. Hao, J. Ai and M.-Z. Cai, *J. Organomet. Chem.*, 2012, **705**, 44–50.
131. Y.-L. Sun, Y. Zhang, X.-H. Cui and W. Wang, *Adv. Synth. Catal.*, 2011, **353**, 1174–1178.
132. (a) S. Murru, P. Mondal, R. Yella and B. K. Patel, *Eur. J. Org. Chem.*, 2009, 5406–5413; (b) S. K. Sahoo, N. Khatun, A. Gogoi, A. Deb and B. K. Patel, *RSC Adv.*, 2013, **3**, 438–446.
133. (a) S. K. Rout, S. Guin, J. Nath and B. K. Patel, *Green Chem.*, 2012, **14**, 2491–2498; (b) S. K. Sahoo, A. Banerjee, S. Chakraborty and B. K. Patel, *ACS Catal.*, 2012, **2**, 544–551.
134. C.-L. Li, X.-G. Zhang, R.-Y. Tang, P. Zhong and J.-H. Li, *J. Org. Chem.*, 2010, **75**, 7037–7040.
135. J. Liu, Q. Gui, Z. Yang, Z. Tan, R. Guo and J.-C. Shi, *Synthsis*, 2013, 943–951.
136. D. J. C. Prasad and G. Sekar, *Org. Biomol. Chem.*, 2013, **11**, 1659–1665.
137. D. Ma, S. Xie, P. Xue, X. Zhang, J. Dong and Y. Jiang, *Angew. Chem. Int. Ed.*, 2009, **48**, 4222–4225.
138. D. Ma, X. Lu, L. Shi, H. Zhang, Y. Jiang and X. Liu, *Angew. Chem. Int. Ed.*, 2011, **50**, 1118–1121.
139. H. Yu, M. Zhang and Y. Li, *J. Org. Chem.*, 2013, **78**, 8898–8903.
140. Y. Sun, H. Jiang, W. Wu, W. Zeng and X. Wu, *Org. Lett.*, 2013, **15**, 1598–1601.
141. X. Lv, Y. Liu, W. Qian and W. Bao, *Adv. Synth. Catal.*, 2008, **350**, 2507–2512.
142. (a) Q. Ding, X. Liu, B. Cao, Z. Zong and Y. Peng, *Tetrahedron Lett.*, 2011, **52**, 1964–1967; (b) F. Yao, W. Hao and M.-Z. Cai, *J. Organomet. Chem.*, 2013, **723**, 137–142.
143. A. V. Kel'in and V. Gevorgyan, *J. Org. Chem.*, 2002, **67**, 95–98.
144. H. Y. Kim, J.-Y. Li and K. Oh, *J. Org. Chem.*, 2012, **77**, 11132–11145.
145. L.-B. Zhao, Z.-H. Guan, Y. Han, Y.-X. Xie, S. He and Y.-M. Liang, *J. Org. Chem.*, 2007, **72**, 10276–10278.
146. H. Cao, H. Jiang, W. Yao and X. Liu, *Org. Lett.*, 2009, **11**, 1931–1933.
147. H. Cao, H. Jiang, G. Yuan, Z. Chen, C. Qi and H. Huang, *Chem. Eur. J.*, 2010, **16**, 10553–10559.

148. H. Cao, H. Jiang, X. Zhou, C. Qi, Y. Lin, J. Wu and Q. Liang, *Green Chem.*, 2012, **14**, 2710–2714.

149. W. Liu, H. Jiang, M. Zhang and C. Qi, *J. Org. Chem.*, 2010, **75**, 966–968.

150. R. Yan, J. Huang, J. Luo, P. Wen, G. Huang and Y. Liang, *Synlett*, 2010, 1071–1074.

151. Y. Wang, L. Xu and D. Ma, *Chem. Asian J.*, 2010, **5**, 74–76.

152. Y. Yang, J. Yao and Y. Zhang, *Org. Lett.*, 2013, **15**, 3206–3209.

153. H. Jiang, W. Zeng, Y. Li, W. Wu, L. Huang and W. Fu, *J. Org. Chem.*, 2012, **77**, 5179–5183.

154. B. Gabriele, L. Veltri, P. Plastina, R. Mancuso, M. V. Vetere and V. Maltese, *J. Org. Chem.*, 2013, **78**, 4919–4928.

155. L. Huang, F. Hu, Q. Ma and Y. Hu, *Tetrahedron Lett.*, 2013, **54**, 3410–3414.

156. Y.-F. Chen, H.-F. Wang, Y. Wang, Y.-C. Luo, H.-L. Zhu and P.-F. Xu, *Adv. Synth. Catal.*, 2010, **352**, 1163–1168.

157. Y. Shang, K. Ju, X. He, J. Su, S. Yu, M. Zhang, K. Liao, L. Wang and P. Zhang, *J. Org. Chem.*, 2010, **75**, 5743–5745.

158. J.-L. Zhou, L.-J. Wang, H. Xu, X.-L. Sun and Y. Tang, *ACS Catal.*, 2013, **3**, 685–688.

159. G. J. S. Doad, J. A. Barltrop, C. M. Petty and T. C. Owen, *Tetrahedron Lett.*, 1989, **30**, 1597–1598.

160. C. G. Bates, P. Saejueng, J. M. Murphy and D. Venkataraman, *Org. Lett.*, 2002, **4**, 4727–4729.

161. E. A. Jaseer, D. J. C. Prasad and G. Sekar, *Tetrahedron*, 2010, **66**, 2077–2082.

162. (a) K. Hirano, T. Satoh and M. Miura, *Org. Lett.*, 2011, **13**, 2395–2397; (b) N. Matsuda, K. Hirano, T. Satoh and M. Miura, *J. Org. Chem.*, 2012, **77**, 617–625.

163. N. K. Swamy, A. Yazici and S. G. Pyne, *J. Org. Chem.*, 2010, **75**, 3412–3419.

164. H. Hachiya, K. Hirano, T. Satoh and M. Miura, *Org. Lett.*, 2011, **13**, 3076–3079.

165. N. Sakai, N. Uchida and T. Konakahara, *Tetrahedron Lett.*, 2008, **49**, 3437–3440.

166. H. Li, J. Liu, B. Yan and Y. Li, *Tetrahedron Lett.*, 2009, **50**, 2353–2357.

167. C.-Y. Chen and P. G. Dormer, *J. Org. Chem.*, 2005, **70**, 6964–6967.

168. L. Ackermann and L. T. Kaspar, *J. Org. Chem.*, 2007, **72**, 6149–6153.

169. (a) M. Carril, R. SanMartin, I. Tellitu and E. Domínguez, *Org. Lett.*, 2006, **8**, 1467–1470; (b) M. Carril, R. SanMartin, E. Domínguez and I. Tellitu, *Green Chem.*, 2007, **9**, 219–220.

170. W. Chen, M. Wang, P. Li and L. Wang, *Tetrahedron*, 2011, **67**, 5913–5919.

171. S. G. Newman, V. Aureggi, C. S. Bryan and M. Lautens, *Chem. Commun.*, 2009, 5236–5238.

172. Y. Ji, P. Li, X. Zhang and L. Wang, *Org. Biomol. Chem.*, 2013, **11**, 4095–4101.

173. S. Ye, G. Liu, S. Pu and J. Wu, *Org. Lett.*, 2012, **14**, 70–73.
174. A. C. Tadd, M. R. Fielding and M. C. Willis, *Tetrahedron Lett.*, 2007, **48**, 7578–7581.
175. B. Lu, B. Wang, Y. Zhang and D. Ma, *J. Org. Chem.*, 2007, **72**, 5337–5341.
176. N. Aljaar, C. C. Malakar, J. Conrad, S. Strobel, T. Schleid and U. Beifuss, *J. Org. Chem.*, 2012, **77**, 7793–7803.
177. Y. Liu, M. Wang, H. Yuan and Q. Liu, *Adv. Synth. Catal.*, 2010, **352**, 884–892.
178. J. Zhao, Q. Zhang, L. Liu, Y. He, J. Li, J. Li and Q. Zhu, *Org. Lett.*, 2012, **14**, 5362–5365.
179. R. Zhu, J. Wie and Z. Shi, *Chem. Sci.*, 2013, **4**, 3706–3711.
180. W. Zeng, W. Wu, H. Jiang, L. Huang, Y. Sun, Z. Chen and X. Li, *Chem. Commun.*, 2013, **49**, 6611–6613.
181. J. Zhu, B. A. Price, S. X. Zhao and P. M. Skonezny, *Tetrahedron Lett.*, 2000, **41**, 4011–4014.
182. H. Adams, N. J. Gilmore, S. Jones, M. P. Muldowney, S. H. von Reuss and R. Vemula, *Org. Lett.*, 2008, **10**, 1457–1460.
183. R.-V. Nguyen and C.-J. Li, *Synlett*, 2008, 1897–1901.
184. (a) X. Yuan, X. Xu, X. Zhou, J. Yuan, L. Mai and Y. Li, *J. Org. Chem.*, 2007, **72**, 1510–1513; (b) E. Li, X. Xu, H. Li, H. Zhang, X. Xu, X. Yuan and Y. Li, *Tetrahedron*, 2009, **65**, 8961–8968.
185. R. Martín, C. H. Larsen, A. Cuenca and S. L. Buchwald, *Org. Lett.*, 2007, **9**, 3379–3382.
186. Q. Liao, L. Zhang, F. Wang, S. Li and C. Xi, *Eur. J. Org. Chem.*, 2010, 5426–5431.
187. X. Zhou, H. Zhang, J. Yuan, L. Mai and Y. Li, *Tetrahedron Lett.*, 2007, **48**, 7236–7239.
188. M. R. Rivero and S. L. Buchwald, *Org. Lett.*, 2007, **9**, 973–976.
189. R. Martín, M. R. Rivero and S. L. Buchwald, *Angew. Chem. Int. Ed.*, 2006, **45**, 7079–7082.
190. Y. Pan, H. Lu, Y. Fang, X. Fang, L. Chen, J. Qian, J. Wang and C. Li, *Synthesis*, 2007, 1242–1246.
191. Q. Cai, F. Zhou, T. Xu, L. Fu and K. Ding, *Org. Lett.*, 2011, **13**, 340–343.
192. A. V. Kel'in, A. W. Sromek and V. Gevorgyan, *J. Am. Chem. Soc.*, 2001, **123**, 2074–2075.
193. J. T. Kim and V. Gevorgyan, *Org. Lett.*, 2002, **4**, 4697–4699.
194. K. K. Toh, Y. F. Wang, E. P. J. Ng and S. Chiba, *J. Am. Chem. Soc.*, 2011, **133**, 13942–13945.
195. R.-L. Yan, J. Luo, C.-X. Wang, C.-W. Ma, G.-S. Huang and Y.-M. Liang, *J. Org. Chem.*, 2010, **75**, 5395–5397.
196. (a) O. V. Larionov and A. de Meijere, *Angew. Chem. Int. Ed.*, 2005, **44**, 5664–5667; (b) A. V. Lygin, O. V. Larionov, V. S. Korotkov and A. de Meijere, *Chem. Eur. J.*, 2008, **15**, 227–236; (c) S. Kamijo, C. Kanazawa and Y. Yamamoto, *J. Am. Chem. Soc.*, 2005, **127**, 9260–9266.
197. S. U. Dighe, S. Hutait and S. Batra, *ACS Comb. Sci.*, 2012, **14**, 665–672.

198. A. E. Akkaoui, M. Hiebel, A. Mouaddib, S. Berteina-Raboin and G. Guillaumet, *Tetrahedron*, 2012, **68**, 9131–9138.

199. S. Chiba, Y.-F. Wang, G. Lapointe and K. Narasaka, *Org. Lett.*, 2008, **10**, 313–316.

200. F. Chen, T. Shen, Y. Cui and N. Jiao, *Org. Lett.*, 2012, **14**, 4926–4929.

201. C. E. Castro, E. J. Gaughan and D. C. Owsley, *J. Org. Chem.*, 1966, **31**, 4071–4078.

202. (a) S. Cacchi, G. Fabrizi and L. M. Parisi, *Org. Lett.*, 2003, **5**, 3843–3846; (b) S. Cacchi, G. Fabrizi, L. M. Parisi and R. Bernini, *Synlett*, 2004, 287–290.

203. R. Bernini, S. Cacchi, G. Fabrizi, E. Filisti and A. Sferrazza, *Synlett*, 2009, 1480–1484.

204. F. Liu and D. Ma, *J. Org. Chem.*, 2007, **72**, 4844–4850.

205. P.-Y. Yao, Y. Zhang, R. P. Hsung and K. Zhao, *Org. Lett.*, 2008, **10**, 4275–4278.

206. L. Ackermann, *Org. Lett.*, 2005, **7**, 439–442.

207. L. Ackermann, S. Barfüsser and H. K. Potukuchi, *Adv. Synth. Catal.*, 2009, **351**, 1064–1072.

208. H. Wang, Y. Li, L. Jiang, R. Zhang, K. Jin, D. Zhao and C. Duan, *Org. Biomol. Chem.*, 2011, **9**, 4983–4986.

209. (a) K. Hiroya, S. Itoh, M. Ozawa, Y. Kanamori and T. Sakamoto, *Tetrahedron Lett.*, 2002, **43**, 1277–1280; (b) K. Hiroya, S. Itoh and T. Sakamoto, *Tetrahedron*, 2005, **61**, 10958–10964.

210. W. M. Dai, D. S. Guo, L. P. Sun and X. H. Huang, *Org. Lett.*, 2003, **5**, 2919–2922.

211. (a) H. Ohno, Y. Ohta, S. Oishi and N. Fujii, *Angew. Chem. Int. Ed.*, 2007, **46**, 2295–2298; (b) Y. Ohta, S. Oishi, N. Fujii and H. Ohno, *Org. Lett.*, 2009, **11**, 1979–1982; (c) Y. Ohta, H. Chiba, S. Oishi, N. Fujii and H. Ohno, *J. Org. Chem.*, 2009, **74**, 7052–7058; (d) Y. Suzuki, Y. Ohta, S. Oishi, N. Fujii and H. Ohno, *J. Org. Chem.*, 2009, **74**, 4246–4251.

212. Z. Chen, D. Zheng and J. Wu, *Org. Lett.*, 2011, **13**, 848–851.

213. Z. Shen and X. Lu, *Adv. Synth. Catal.*, 2009, **351**, 3107–3112.

214. X. F. Xia, L. L. Zhang, X. R. Song, Y. N. Niu, X. Y. Liu and Y. M. Liang, *Chem. Commun.*, 2013, **49**, 1410–1412.

215. M. Yang, J. Tang and R. Fan, *Chem. Commun.*, 2012, **48**, 11775–11777.

216. (a) S. Kamijo and Y. Yamamoto, *Angew. Chem. Int. Ed.*, 2002, **41**, 3230–3233; (b) S. Kamijo and Y. Yamamoto, *J. Org. Chem.*, 2003, **68**, 4764–4771.

217. R. Besandre, M. Jaimes and J. A. May, *Org. Lett.*, 2013, **15**, 1666–1669.

218. R. C. Hodgkinson, J. Schulz and M. C. Willis, *Org. Biomol. Chem.*, 2009, **7**, 432–434.

219. (a) F. Melkonyan, A. Topolyan, M. Yurovskaya and A. Karchava, *Eur. J. Org. Chem.*, 2008, 5952–5956; (b) F. S. Melkonyan, A. V. Karchava and M. A. Yurovskaya, , *J. Org. Chem.*, 2008, **73**, 4275–4278.

220. S. Tanimori, H. Ura and M. Kirihata, *Eur. J. Org. Chem.*, 2007, 3977–3980.
221. S. G. Koenig, J. W. Dankwardt, Y. Liu, H. Zhao and S. P. Singh, *Tetrahedron Lett.*, 2010, **51**, 6549–6551.
222. C. Barberis, T. D. Gordon, C. Thomas, X. Zhang and K. P. Cusack, *Tetrahedron Lett.*, 2005, **46**, 8877–8880.
223. Q. Cai, Z. Li, J. Wie, C. Ha, D. Pei and K. Ding, *Chem. Commun.*, 2009, 7581–7583.
224. X. Yang, H. Fu, R. Qiao, Y. Jiang and Y. Zhao, *Adv. Synth. Catal.*, 2010, **352**, 1033–1038.
225. Y. Chen, X. Xie and D. Ma, *J. Org. Chem.*, 2007, **72**, 9329–9334.
226. Y. Chen, Y. Wang, Z. Sun and D. Ma, *Org. Lett.*, 2008, **10**, 625–628.
227. J. Yuen, Y. Q. Fang and M. Lautens, *Org. Lett.*, 2006, **8**, 653–656.
228. M. Jiang, J. Li, F. Wang, Y. Zhao, F. Zhao, X. Dong and W. Zhao, *Org. Lett.*, 2012, **14**, 1420–1423.
229. B. V. S. Reddy, M. R. Reddy, Y. G. Rao, J. S. Yadav and B. Sridhar, *Org. Lett.*, 2013, **15**, 464–467.
230. S. Murru, A. A. Gallo and R. S. Srivastava, *Eur. J. Org. Chem.*, 2011, 2035–2038.
231. R. Bernini, G. Fabrizi, A. Sferrazza and S. Cacchi, *Angew. Chem. Int. Ed.*, 2009, **48**, 8078–8081.
232. P. Sang, Y. Xie, J. Zou and Y. Zhang, *Adv. Synth. Catal.*, 2012, **354**, 1873–1878.
233. S. Biswas, M. Nayak, S. Kanojiya and S. Batra, *Adv. Synth. Catal.*, 2011, **353**, 3330–3334.
234. Y. L. Tnay, C. Chen, Y. Y. Chua, L. Zhang and S. Chiba, *Org. Lett.*, 2012, **14**, 3550–3553.
235. S. Kumar, H. Ila and H. Junjappa, *J. Org. Chem.*, 2009, **74**, 7046–7051.
236. M. Noack and R. Göttlich, *Chem. Commun.*, 2002, 536–537.
237. H. Ohmiya, T. Moriya and M. Sawamura, *Org. Lett.*, 2009, **11**, 2145–2147.
238. T. P. Zabawa and S. R. Chemler, *Org. Lett.*, 2007, **9**, 2035–2038.
239. (a) M. C. Paderes and S. R. Chemler, *Org. Lett.*, 2009, **11**, 1915–1918; (b) M. C. Paderes and S. R. Chemler, *Eur. J. Org. Chem.*, 2011, 3679–3684.
240. G.-Q. Liu, W. Li and Y.-M. Li, *Adv. Synth. Catal.*, 2013, **355**, 395–402.
241. A. Tsuhako, D. Oikawa, K. Sakai and S. Okamoto, *Tetrahedron Lett.*, 2008, **49**, 6529–6532.
242. Y. Oderaotoshi, W. Cheng, S. Fujitomi, Y. Kasano, S. Minakata and M. Komatsu, *Org. Lett.*, 2003, **5**, 5043–5046.
243. W. Gao, X. Zhang and M. Raghunath, *Org. Lett.*, 2005, **7**, 4241–4244.
244. X.-X. Yan, Q. Peng, Y. Zhang, K. Zhang, W. Hong, X.-L. Hou and Y.-D. Wu, *Angew. Chem. Int. Ed.*, 2006, **45**, 1979–1983.
245. M. Shi and J.-W. Shi, *Tetrahedron: Asymmetry*, 2007, **18**, 645–650.
246. C.-J. Wang, G. Liang, Z.-Y. Xue and F. Gao, *J. Am. Chem. Soc.*, 2008, **130**, 17250–17251.

247. L. M. Castelló, C. Nájera, J. M. Sansano, O. Larranaga, A. de Cózar and F. P. Cossío, *Org. Lett.*, 2013, **15**, 2902–2905.

248. (a) S. Cabrera, R. G. Arrayás and J. C. Carretero, *J. Am. Chem. Soc.*, 2005, **127**, 16394–16395; (b) T. Llamas, R. G. Arrayás and J. C. Carretero, *Org. Lett.*, 2006, **8**, 1795–1798; (c) T. Llamas, R. G. Arrayás and J. C. Carretero, *Synthesis*, 2007, 950–956; (d) S. Cabrera, R. G. Arrayás, B. Martín-Matute, F. P. Cossío and J. C. Carretero, *Tetrahedron*, 2007, **63**, 6587–6602; (e) B. Martín-Matute, S. I. Pereira, E. Pena-Cabrera, J. Adrio, A. M. S. Silva and J. C. Carretero, *Adv. Synth. Catal.*, 2007, **349**, 1714–1724; (f) A. López-Pérez, J. Adrio and J. C. Carretero, *Angew. Chem. Int. Ed.*, 2009, **48**, 340–343; (g) R. Robles-Machín, M. González-Esguevillas, J. Adrio and J. C. Carretero, *J. Org. Chem.*, 2010, **75**, 233–236.

249. C. V. Galliford, M. A. Beenen, S. T. Nguyen and K. A. Scheidt, *Org. Lett.*, 2003, **5**, 3487–3490.

250. (a) E. S. Sherman, S. R. Chemler, T. B. Tan and O. Gerlits, *Org. Lett.*, 2004, **6**, 1573–1575; (b) T. P. Zabawa, D. Kasi and S. R. Chemler, *J. Am. Chem. Soc.*, 2005, **127**, 11250–11251; (c) W. Zeng and S. R. Chemler, *J. Am. Chem. Soc.*, 2007, **129**, 12948–12949; (d) P. H. Fuller and S. R. Chemler, *Org. Lett.*, 2007, **9**, 5477–5480; (e) E. S. Sherman, P. H. Fuller, D. Kasi and S. R. Chemler, *J. Org. Chem.*, 2007, **72**, 3896–3905; (f) P. H. Fuller, J.-W. Kim and S. R. Chemler, *J. Am. Chem. Soc.*, 2008, **130**, 17638–17639; (g) E. S. Sherman and S. R. Chemler, *Adv. Synth. Catal.*, 2009, **351**, 467–471; (h) F. C. Sequeira, B. W. Turnpenny and S. R. Chemler, *Angew. Chem. Int. Ed.*, 2010, **49**, 6365–6368; (i) M. T. Bovino and S. R. Chemler, *Angew. Chem. Int. Ed.*, 2012, **51**, 3923–3927; (j) T. W. Liwosz and S. R. Chemler, *J. Am. Chem. Soc.*, 2012, **134**, 2020–2023; (k) B. W. Turnpenny, K. L. Hyman and S. R. Chemler, *Organometallics*, 2012, **31**, 7819–7822; (l) M. C. Paderes, J. B. Keister and S. R. Chemler, *J. Org. Chem.*, 2013, **78**, 506–515.

251. (a) K. Yamada, T. Kubo, H. Tokuyama and T. Fukuyama, *Synlett*, 2002, 231–234; (b) T. Kubo, C. Katoh, K. Yamada, K. Okano, H. Tokuyama and T. Fukuyama, *Tetrahedron*, 2008, **64**, 11230–11236.

252. Y.-M. Zhu, L.-N. Qin, R. Liu, S.-J. Ji and H. Katayama, *Tetrahedron Lett.*, 2007, **48**, 6262–6266.

253. A. Minatti and S. L. Buchwald, *Org. Lett.*, 2008, **10**, 2721–2724.

254. F. Zhou, J. Guo, J. Liu, K. Ding, S. Yu and Q. Cai, *J. Am. Chem. Soc.*, 2012, **134**, 14326–14329.

255. H. J. Lim, J. C. Gallucci and T. V. RajanBabu, *Org. Lett.*, 2010, **12**, 2162–2165.

256. S. Li, Y. Luo and J. Wu, *Org. Lett.*, 2011, **13**, 3190–3193.

257. B. X. Tang, R. J. Song, C. Y. Wu, Y. Liu, M. B. Zhou, W. T. Wie, G. B. Deng, D. L. Yin and J. H. Li, *J. Am. Chem. Soc.*, 2010, **132**, 8900–8902.

258. T. Liu, H. Yang, Y. Jiang and H. Fu, *Adv. Synth. Catal.*, 2013, **355**, 1169–1176.

259. T. Benkovics, I. A. Guzei and T. P. Yoon, *Angew. Chem. Int. Ed.*, 2010, **49**, 9153–9157.
260. C. Liu, W. Zhang, L. X. Dai and S. L. You, *Org. Lett.*, 2012, **14**, 4525–4527.
261. D. Zhu, Q. Liu, B. Luo, M. Chen, R. Pi, P. Huang and S. Wen, *Adv. Synth. Catal.*, 2013, **355**, 2172–2178.
262. Y. Goriya and C. V. Ramana, *Chem. Commun.*, 2013, **49**, 6376–6378.
263. L. Li, M. Wang, X. Zhang, Y. Jiang and D. Ma, *Org. Lett.*, 2009, **11**, 1309–1312.
264. A. van den Hoogenband, J. H. M. Lange, J. A. J. den Hartog, R. Henzen and J. W. Terpstra, *Tetrahedron Lett.*, 2007, **48**, 4461–4465.
265. Y. H. Jhan, T. W. Kang and J. C. Hsieh, *Tetrahedron Lett.*, 2013, **54**, 1155–1159.
266. G. Feng, J. Wu and W. M. Dai, *Tetrahedron Lett.*, 2007, **48**, 401–404.
267. L. Zeng, H. Fu, R. Qiao, Y. Jiang and Y. Zhao, *Adv. Synth. Catal.*, 2009, **351**, 1671–1676.
268. V. A. Ignatenko, N. Deligonul and R. Viswanathan, *Org. Lett.*, 2010, **12**, 3594–3597.
269. Y. Zhou, Y. Xi, J. Zhao, X. Sheng, S. Zhang and H. Zhang, *Org. Lett.*, 2012, **14**, 3116–3119.
270. J. C. Hsieh, A. Y. Cheng, J. H. Fu and T. W. Kang, *Org. Biomol. Chem.*, 2012, **10**, 6404–6409.
271. B. L. Yin, J. Q. Lai, Z. R. Zhang and H. F. Jiang, *Adv. Synth. Catal.*, 2011, **353**, 1961–1965.
272. J. Zhu, W. Zhang, L. Zhang, J. Liu, J. Zheng and J. Hu, *J. Org. Chem.*, 2010, **75**, 5505–5512.
273. B. Roy, S. Hazra, B. Mondal and K. C. Majumdar, *Eur. J. Org. Chem.*, 2013, 4570–4577.
274. C. Dey, E. Larionov and E. P. Kündig, *Org. Biomol. Chem.*, 2013, **11**, 6734–6743.
275. B. Lu and D. Ma, *Org. Lett.*, 2006, **8**, 6115–6118.
276. R. J. Song, Y. Liu, R. J. Li and J. H. Li, *Tetrahedron Lett.*, 2009, **50**, 3912–3916.
277. H. Jin, B. Zhou, Z. Wu, Y. Shen and Y. Wang, *Tetrahedron Lett.*, 2011, **67**, 1178–1182.
278. D. Chernyak, S. B. Gadamsetty and V. Gevorgyan, *Org. Lett.*, 2008, **10**, 2307–2310.
279. J. Barluenga, G. Lonzi, L. Riesgo, L. A. López and M. Tomás, *J. Am. Chem. Soc.*, 2010, **132**, 13200–13202.
280. G. R. Qu, L. Liang, H. Y. Niu, W. H. Rao, H. M. Guo and J. S. Fossey, *Org. Lett.*, 2012, **14**, 4494–4497.
281. H. Wang, Y. Wang, D. Liang, L. Liu, J. Zhang and Q. Zhu, *Angew. Chem. Int. Ed.*, 2011, **50**, 5678–5681.
282. (a) H. Wang, Y. Wang, C. Peng, J. Zhang and Q. Zhu, *J. Am. Chem. Soc.*, 2010, **132**, 13217–13219; (b) K. R. Masters, T. R. M. Rauws, A. K. Yadav, W. A. Herrebout, B. V. der Veken and B. U. W. Maes, *Chem. Eur. J.*, 2011, **17**, 6315–6320.

283. Z. Wu, Q. Huang, X. Zhou, L. Yu, Z. Li and D. Wu, *Eur. J. Org. Chem.*, 2011, 5242–5245.
284. T. R. M. Rauws, C. Biancalani, J. W. De Schutter and B. U. W. Maes, *Tetrahedron*, 2010, **66**, 6958–6964.
285. N. Chernyak and V. Gevorgyan, *Angew. Chem. Int. Ed.*, 2010, **49**, 2743–2746.
286. J. Zeng, Y. J. Tan, M. L. Leow and X. W. Liu, *Org. Lett.*, 2012, **14**, 4386–4389.
287. R. L. Yan, H. Yan, C. Ma, Z. Y. Ren, X. A. Gao, G. S. Huang and Y. M. Liang, *J. Org. Chem.*, 2012, **77**, 2024–2028.
288. A. K. Bagdi, M. Rahman, S. Santra, A. Majee and A. Hajra, *Adv. Synth. Catal.*, 2013, **355**, 1741–1747.
289. (a) D. C. Mohan, R. R. Donthiri, S. N. Rao and S. Adimurthy, *Adv. Synth. Catal.*, 2013, **355**, 2217–2221; (b) Z.-J. Cai, S.-Y. Wang and S.-J. Ji, *Adv. Synth. Catal.*, 2013, **355**, 2686–2692.
290. Y. Gao, M. Yin, W. Wu, H. Huang and H. Jiang, *Adv. Synth. Catal.*, 2013, **355**, 2263–2273.
291. M. E. Bluhm, M. Ciesielski, H. Görls and M. Döring, *Angew. Chem. Int. Ed.*, 2002, **41**, 2962–2965.
292. S. Ueda and H. Nagasawa, *J. Am. Chem. Soc.*, 2009, **131**, 15080–15081.
293. S. Ding, Y. Yan and N. Jiao, *Chem. Commun.*, 2013, **49**, 4250–4252.
294. W. You, X. Yan, Q. Liao and C. Xi, *Org. Lett.*, 2010, **12**, 3930–3933.
295. P. Zhao, H. Yin, H. Gao and C. Xi, *J. Org. Chem.*, 2013, **78**, 5001–5006.
296. D. A. Barancelli, R. E. Schumacher, M. R. Leite and G. Zeni, *Eur. J. Org. Chem.*, 2011, 6713–6718.
297. L.-L. Sun, C.-L. Deng, R.-Y. Tang and X.-G. Zhang, *J. Org. Chem.*, 2011, **76**, 7546–7550.
298. S. Kim, N. Dahal and T. Kesharwani, *Tetrahedron Lett.*, 2013, **54**, 4373–4376.
299. F. Wang, C. Chen, G. Deng and C. Xi, *J. Org. Chem.*, 2012, **77**, 4148–4151.
300. S. J. Balkrishna, B. S. Bhakuni, D. Chopra and S. Kumar, *Org. Lett.*, 2010, **12**, 5394–5397.
301. L. Chen, M. Shi and C. Li, *Chin. J. Chem*, 2010, **28**, 1660–1664.
302. S. Ma and H. Xie, *Org. Lett.*, 2000, **2**, 3801–3803.
303. M. Murai, K. Miki and K. Ohe, *J. Org. Chem.*, 2008, **73**, 9174–9176.
304. L. Lv, S. Zheng, X. Cai, Z. Chen, Q. Zhu and S. Liu, *ACS Comb. Sci.*, 2013, **15**, 183–192.
305. Z. Wang, X. Bi, P. Liao, X. Liu and D. Dong, *Chem. Commun.*, 2013, **49**, 1309–1311.
306. (a) R. Shintani and G. C. Fu, *J. Am. Chem. Soc.*, 2003, **125**, 10778–10779; (b) A. Suárez, C. W. Downey and G. C. Fu, *J. Am. Chem. Soc.*, 2005, **127**, 11244–11245.
307. K. Yoshimura, T. Oishi, K. Yamaguchi and N. Mizuno, *Chem. Eur. J*, 2011, **17**, 3827–3831.
308. Q. Wu, J. Hu, X. Ren and J. Zhou, *Chem. Eur. J.*, 2011, **17**, 11553–11558.

309. Y. Miller, L. Miao, A. S. Hosseini and S. R. Chemler, *J. Am. Chem. Soc.*, 2012, **134**, 12149–12156.
310. X. Huang and H. Zhou, *Org. Lett.*, 2002, **4**, 4419–4422.
311. S. Inack-Ngi, R. Rahmani, L. Commeiras, G. Chouraqui, J. Thibonnet, A. Duchene, M. Abarbri and J.-L. Parrain, *Adv. Synth. Catal.*, 2009, **351**, 779–788.
312. C. Sun, Y. Fang, S. Li, Y. Zhang, Q. Zhao, S. Zhu and C. Li, *Org. Lett.*, 2009, **11**, 4084–4087.
313. Y. Fang and C. Li, *Chem. Commun.*, 2005, 3574–3576.
314. L. Huang, H. Jiang, C. Qi and X. Liu, *J. Am. Chem. Soc.*, 2010, **132**, 17652–17654.
315. F. Yu, X. Lian, J. Zhao, Y. Yu and S. Ma, *J. Org. Chem.*, 2009, **74**, 1130–1134.
316. K. Namitharan and K. Pitchumani, *Org. Lett.*, 2011, **13**, 5728–5731.
317. Y. Wen, B. Zhao and Y. Shi, *Org. Lett.*, 2009, **11**, 2365–2368.
318. B. Zhao, H. Du and Y. Shi, *J. Org. Chem.*, 2009, **74**, 4411–4413.
319. S.-H. Lee, J. Yang and T.-D. Han, *Tetrahedron Lett.*, 2001, **42**, 3487–3490.
320. (a) D. J. Michaelis, C. J. Shaffer and T. P. Yoon, *J. Am. Chem. Soc.*, 2007, **129**, 1866–1867; (b) D. J. Michaelis, M. A. Ischay and T. P. Yoon, *J. Am. Chem. Soc.*, 2008, **130**, 6610–6615; (c) T. Benkovics, J. Du, I. A. Guzei and T. P. Yoon, *J. Org. Chem.*, 2009, **74**, 5545–5552; (d) D. J. Michaelis, K. S. Williamson and T. P. Yoon, *Tetrahedron*, 2009, **65**, 5118–5124; (e) S. M. DePorter, A. C. Jacobsen, K. M. Partridge, K. S. Williamson and T. P. Yoon, *Tetrahedron Lett.*, 2010, **51**, 5223–5225.
321. S. Flock and H. Frauenrath, *Arkivoc*, 2007, (x), 245–259.
322. W.-J. Yoo and C.-J. Li, *Adv. Synth. Catal.*, 2008, **350**, 1503–1506.
323. H. Feng, D. S. Ermolatév, G. Song and E. V. Van der Eycken, *Adv. Synth. Catal.*, 2012, **354**, 505–509.
324. K. B. Jensen, R. G. Hazell and K. A. Jorgensen, *J. Org. Chem.*, 1999, **64**, 2353–2360.
325. (a) T. Saito, T. Yamada, S. Miyazaki and T. Otani, *Tetrahedron Lett.*, 2004, **45**, 9581–9584; (b) T. Saito, T. Yamada, S. Miyazaki and T. Otani, *Tetrahedron Lett.*, 2004, **45**, 9585–9587.
326. C. Palomo, M. Oiarbide, E. Arceo, J. M. García, R. López, A. González and A. Linden, *Angew. Chem. Int. Ed.*, 2005, **44**, 6187–6190.
327. S. D. Karyakarte, T. P. Smith and S. R. Chemler, *J. Org. Chem.*, 2012, **77**, 7755–7760.
328. D. N. Barman and K. M. Nicholas, *Eur. J. Org. Chem.*, 2011, 908–911.
329. Y. T. He, L. L. Li, Y. F. Yang, Y. Q. Wang, J. Y. Luo, X. Y. Liu and Y. M. Liang, *Chem. Commun.*, 2013, **49**, 5687–5689.
330. H. Chen, Z. Wang, Y. Zhang and Y. Huang, *J. Org. Chem.*, 2013, **78**, 3503–3509.
331. (a) P. Krasik, M. Bohemier-Bernard and Q. Yu, *Synlett*, 2005, 854–856; (b) R. P. Hanzlik and M. Leinwetter, *J. Org. Chem.*, 1978, **43**, 438–440.
332. C. Chaminade, L. Coulombel, S. Olivero and E. Dunach, *Eur. J. Org. Chem.*, 2006, 3554–3557.

333. Q. H. Li, L. Wie, X. Chen and C. J. Wang, *Chem. Commun.*, 2013, **49**, 6277–6279.
334. N. Takemura, Y. Kuninobu and M. Kanai, *Org. Lett.*, 2013, **15**, 844–847.
335. X. Gong, H. Yang, H. Liu, Y. Jiang, Y. Zhao and H. Fu, *Org. Lett.*, 2010, **12**, 3128–3131.
336. S. Guin, T. Ghosh, S. K. Rout, A. Banerjee and B. K. Patel, *Org. Lett.*, 2011, **13**, 5976–5979.
337. Y. Wu, Y. Zhang, Y. Jiang and D. Ma, *Tetrahedron Lett.*, 2009, **50**, 3683–3685.
338. W. Rao, P. Kothandaraman, C. B. Koh and P. W. H. Chan, *Adv. Synth. Catal.*, 2010, **352**, 2521–2530.
339. Q. Niu, H. Mao, G. Yuan, J. Gao, H. Liu, Y. Tu, X. Wang and X. Lv, *Adv. Synth. Catal.*, 2013, **355**, 1185–1192.
340. M. Nayak and S. Batra, *RSC Adv.*, 2012, **2**, 3367–3373.
341. M. R. Fructos, E. Álvarez, M. M. Díaz-Requejo and P. J. Pérez, *J. Am. Chem. Soc.*, 2010, **132**, 4600–4607.
342. L. Maestre, M. R. Frzctos, M. M. Díaz-Requejo and P. J. Pérez, *Organometallics*, 2012, **31**, 7839–7843.
343. X. Yu, Q. Ding, Z. Chen and J. Wu, *Tetrahedron Lett.*, 2009, **50**, 4279–4282.
344. A. K. Verma, T. Kesharwani, J. Singh, V. Tandon and R. C. Larock, *Angew. Chem. Int. Ed.*, 2009, **48**, 1138–1143.
345. Y. Y. Hu, J. Hu, X. C. Wang, L. N. Guo, X. Z. Shu, Y. N. Niu and Y. M. Liang, *Tetrahedron*, 2010, **66**, 80–86.
346. Z. Chen, L. Gao, S. Ye, Q. Ding and J. Wu, *Chem. Commun.*, 2012, **48**, 3975–3977.
347. L. Shi, R. Wang, H. Yang, Y. Jiang and H. Fu, *RSC Adv.*, 2013, **3**, 6278–6281.
348. L. Yang, H. Ren, D. Wang, F. Shi and C. Wu, *RSC Adv.*, 2013, **3**, 10434–10441.
349. X. Yang, Y. Luo, Y. Jin, H. Liu, Y. Jiang and H. Fu, *RSC Adv.*, 2012, **2**, 8258–8261.
350. J. Lu and H. Fu, *J. Org. Chem.*, 2011, **76**, 4600–4605.
351. Q. Cai, Z. Li, J. Wie, L. Fu, C. Ha, D. Pei and K. Ding, *Org. Lett.*, 2010, **12**, 1500–1503.
352. Q. Liao, L. Zhang, S. Li and C. Xi, *Org. Lett.*, 2011, **13**, 228–231.
353. S. Cacchi, G. Fabrizi and E. Filisti, *Org. Lett.*, 2008, **10**, 2629–2632.
354. H. Huang, X. Ji, W. Wu, L. Huang and H. Jiang, *J. Org. Chem.*, 2013, **78**, 3774–3782.
355. I. Nakamura, D. Zhang and M. Terada, *J. Am. Chem. Soc.*, 2010, **132**, 7884–7886.
356. Y. Ohta, S. Oishi, N. Fujii and H. Ohno, *Chem. Commun.*, 2008, 835–837.
357. B. Wang, B. Lu, Y. Jiang, Y. Zhang and D. Ma, *Org. Lett.*, 2008, **10**, 2761–2763.
358. L. Shi, H. Yang, Y. Jiang and H. Fu, *Adv. Synth. Catal.*, 2013, **355**, 1177–1184.

359. S. Li, Y. Yuan, J. Zhu, H. Xie, Z. Chen and Y. Wu, *Adv. Synth. Catal.*, 2010, **352**, 1582–1586.

360. Y. Wang, C. Chen, J. Peng and M. Li, *Angew. Chem. Int. Ed.*, 2013, **52**, 5323–5327.

361. Y. Wang, C. Chen, S. Zhang, Z. Lou, X. Su, L. Wen and M. Li, *Org. Lett.*, 2013, **15**, 4794–4797.

362. C. S. Cho, W. X. Ren and S. C. Shim, *Tetrahedron Lett.*, 2006, **47**, 6781–6785.

363. H. Huang, H. Jiang, K. Chen and H. Liu, *J. Org. Chem.*, 2009, **74**, 5476–5480.

364. K. K. Toh, S. Sanjaya, S. Sahnoun, S. Y. Chong and S. Chiba, *Org. Lett.*, 2012, **14**, 2290–2292.

365. X. F. Xia, L. L. Zhang, X. R. Song, X. Y. Liu and Y. M. Liang, *Org. Lett.*, 2012, **14**, 2480–2483.

366. G. Cheng and X. Cui, *Org. Lett.*, 2013, **15**, 1480–1483.

367. J. Han, L. Cao, L. Bian, J. Chen, H. Deng, M. Shao, Z. Jin, H. Zhang and W. Cao, *Adv. Synth. Catal.*, 2013, **355**, 1345–1350.

368. R. Yan, X. Liu, C. Pan, X. Zhou, X. Li, X. Kang and G. Huang, *Org. Lett.*, 2013, **15**, 4876–4879.

369. N. Fei, H. Yin, S. Wang, H. Wang and Z. J. Yao, *Org. Lett.*, 2011, **13**, 4208–4211.

370. W. Zhou, Y. Liu, Y. Yang and G. J. Deng, *Chem. Commun.*, 2012, **48**, 10678–10680.

371. J. Huang, C. Wan, M. F. Xu and Q. Zhu, *Eur. J. Org. Chem.*, 2013, 1876–1880.

372. P. C. Huang, K. Parthasarathy and C. H. Cheng, *Chem. Eur. J.*, 2013, **19**, 460–464.

373. W. Zhou, Y. Yang, Y. Liu and G. J. Deng, *Green Chem.*, 2013, **15**, 76–80.

374. J. Yu, H. Yang, Y. Jiang and H. Fu, *Chem. Eur. J.*, 2013, **19**, 4271–4277.

375. S. Wang, J. Sun, G. Yu, X. Hu, J. O. Liu and Y. Hu, *Org. Biomol. Chem.*, 2004, **2**, 1573–1574.

376. F. Wang, H. Liu, H. Fu, Y. Jiang and Y. Zhao, *Org. Lett.*, 2009, **11**, 2469–6472.

377. J. Lu, X. Gong, H. Yang and H. Fu, *Chem. Commun.*, 2010, **46**, 4172–4174.

378. T. Liu, R. Wang, H. Yang and H. Fu, *Chem. Eur. J.*, 2011, **17**, 6765–6771.

379. F. Zhou, J. Liu, K. Ding, J. Liu and Q. Cai, *J. Org. Chem.*, 2011, **76**, 5346–5353.

380. Q. Gui, Z. Yang, X. Chen, J. Liu, Z. Tan, R. Guo and W. Yu, *Synlett*, 2013, 1016–1020.

381. Y. Shi, X. Zhu, H. Mao, H. Hu, C. Zhu and Y. Cheng, *Chem. Eur. J.*, 2013, **19**, 11553–11557.

382. L. Fu, X. Huang, D. Wang, P. Zhao and K. Ding, *Synthesis*, 2011, 1547–1554.

383. P. Xie, Z. Q. Wang, G. B. Deng, R. J. Song, J. D. Xia, M. Hu and J. H. Li, *Adv. Synth. Catal.*, 2013, **355**, 2257–2262.

384. W. Qian, H. Wang and J. Allen, *Angew. Chem. Int. Ed.*, 2013, **52**, 10992–10996.
385. B. D. Bala, S. Muthusaravanan and S. Perumal, *Tetrahedron Lett.*, 2013, **54**, 3735–3739.
386. Q. Cai, J. Yan and K. Ding, *Org. Lett.*, 2012, **14**, 3332–3335.
387. Y. Fukudome, H. Naito, T. Hata and H. Urabe, *J. Am. Chem. Soc.*, 2008, **130**, 1820–1821.
388. L. Yu, X. Zhou, D. Wu and H. Xiang, *J. Organomet. Chem.*, 2012, **705**, 75–78.
389. J. T. Reeves, D. R. Fandrick, Z. Tan, J. J. Song, H. Lee, N. K. Yee and C. H. Senanayake, *J. Org. Chem.*, 2010, **75**, 992–994.
390. J. Lu, Y. Jin, H. Liu, Y. Jiang and H. Fu, *Org. Lett.*, 2011, **13**, 3694–3697.
391. S. Biswas and S. Batra, *Eur. J. Org. Chem.*, 2013, 4895–4902.
392. L. Zhang, F. Zhao, M. Zheng, Y. Zhai, J. Wang and H. Liu, *Eur. J. Org. Chem.*, 2013, 5710–5715.
393. G. Zhang, J. Miao, Y. Zhao and H. Ge, *Angew. Chem. Int. Ed.*, 2012, **51**, 8318–8321.
394. D. Hong, X. Lin, Y. Zhu, M. Lei and Y. Wang, *Org. Lett.*, 2009, **11**, 5678–5681.
395. S. Xu, J. Lu and H. Fu, *Chem. Commun.*, 2011, **47**, 5596–5598.
396. H. Zhang, Y. Jin, H. Liu, Y. Jiang and H. Fu, *Eur. J. Org. Chem.*, 2012, 6798–6803.
397. S. Guo, J. Wang, X. Fan, X. Zhang and D. Guo, *J. Org. Chem.*, 2013, **78**, 3262–3270.
398. X. Yang, Y. Jin, H. Liu, Y. Jiang and H. Fu, *RSC Adv.*, 2012, **2**, 11061–11066.
399. M. Jiang, J. Li, F. Wang, Y. Zhao, F. Zhao, X. Dong and W. Zhao, *Org. Lett.*, 2012, **14**, 1420–1423.
400. B. Chao, S. Lin, Q. Ma, D. Lu and Y. Hu, *Org. Lett.*, 2012, **14**, 2398–2401.
401. J. Sheng, B. Chao, H. Chen and Y. Hu, *Org. Lett.*, 2013, **15**, 4508–4511.
402. (a) C. Huang, Y. Fu, H. Fu, Y. Jiang and Y. Zhao, *Chem. Commun.*, 2008, 6333–6335; (b) X. Huang, H. Yang, H. Fu, R. Qiao and Y. Zhao, *Synthesis*, 2009, 2679–2688.
403. (a) V. L. Truong and M. Morrow, *Tetrahedron Lett.*, 2010, **51**, 758–760; (b) A. V. Vypolzov, D. V. Dar'in, S. G. Ryazanov and P. S. Lobanov, *Chem. Heterocycl. Compd.*, 2011, **46**, 1481–1485.
404. C. Wang, S. Li, H. Liu, Y. Jiang and H. Fu, *J. Org. Chem.*, 2010, **75**, 7936–7938.
405. Q. Liu, Y. Zhao, H. Fu and C. Cheng, *Synlett*, 2013, **24**, 2089–2094.
406. B. Han, X. L. Yang, C. Wang, Y. W. Bai, T. C. Pan, X. Chen and W. Yu, *J. Org. Chem.*, 2012, **77**, 1136–1142.
407. C. C. Malakar, A. Baskakova, J. Conrad and U. Beifuss, *Chem. Eur. J.*, 2012, **18**, 8882–8885.
408. X. Yang, H. Liu, H. Fu, R. Qiao, Y. Jiang and Y. Zhao, *Synlett*, 2010, 101–106.
409. Y. Lv, Y. Li, T. Xiong, W. Pu, H. Zhang, K. Sun, Q. Liu and Q. Zhang, *Chem. Commun.*, 2013, **49**, 6439–6441.

410. X. Su, C. Chen, Y. Wang, J. Chen, Z. Lou and M. Li, *Chem. Commun.*, 2013, **49**, 6752–6754.
411. K. Yamamoto, Y. G. Chen and F. G. Buono, *Org. Lett.*, 2005, **7**, 4673–4676.
412. X. Liu, H. Fu, Y. Jiang and Y. Zhao, *Angew. Chem. Int. Ed.*, 2009, **48**, 348–351.
413. W. Xu and H. Fu, *J. Org. Chem.*, 2011, **76**, 3846–3852.
414. W. Xu, Y. Jin, H. Liu, Y. Jiang and H. Fu, *Org. Lett.*, 2011, **13**, 1274–1277.
415. D. K. Sreenivas, N. Ramkumar and R. Nagarajan, *Org. Biomol. Chem.*, 2012, **10**, 3417–3423.
416. (a) R. J. Abdel-Jalil, W. Voelter and M. Saeed, *Tetrahedron Lett.*, 2004, **45**, 3475–3476; (b) G. A. N. K. Durgareddy, R. Ravikumar, S. Ravi and S. R. Adapa, *J. Chem. Sci.*, 2013, **125**, 175–182.
417. L. Chen, H. Fu and R. Qiao, *Synlett*, 2011, 1930–1936.
418. D. Yang, Y. Wang, H. Yang, T. Liu and H. Fu, *Adv. Synth. Catal.*, 2012, **354**, 477–482.
419. L. Xu, Y. Jiang and D. Ma, *Org. Lett.*, 2012, **14**, 1150–1153.
420. L. Yu, M. Wang, P. Li and L. Wang, *Appl. Organomet. Chem.*, 2012, **26**, 576–582.
421. D. S. Chen, G. L. Dou, Y. L. Li, Y. Liu and X. S. Wang, *J. Org. Chem.*, 2013, **78**, 5700–5704.
422. J. A. Bleda, P. M. Fresneda, R. Orenes and P. Molina, *Eur. J. Org. Chem.*, 2009, 2490–2504.
423. C. Wang, L. Zhang, A. Ren, P. Lu and Y. Wang, *Org. Lett.*, 2013, **15**, 2982–2985.
424. K. Namitharan and K. Pitchumani, *Adv. Synth. Catal.*, 2013, **355**, 93–98.
425. H. Xu and H. Fu, *Chem. Eur. J.*, 2012, **18**, 1180–1186.
426. D. Chen, Q. Chen, M. Liu, S. Dai, L. Huang, J. Yang and W. Bao, *Tetrahedron*, 2013, **69**, 6461–6467.
427. L. Chen, C. Li, X. Bi, H. Liu and R. Qiao, *Adv. Synth. Catal.*, 2012, **354**, 1773–1779.
428. J. Safari and S. Gandomi-Ravandi, *J. Mol. Catal. A: Chem.*, 2013, **371**, 135–140.
429. T. Liu, C. Zhu, H. Yang and H. Fu, *Adv. Synth. Catal.*, 2012, **354**, 1579–1584.
430. J. Wang, J. Wang, P. Lu and Y. Wang, *J. Org. Chem.*, 2013, **78**, 8816–8820.
431. Z. J. Wang, J. G. Yang, F. Yang and W. Bao, *Org. Lett.*, 2010, **12**, 3034–3037.
432. Y. S. Chun, Z. Xuan, J. H. Kim and S. G. Lee, *Org. Lett.*, 2013, **15**, 3162–3165.
433. M. Gyuris, L. G. Puskás, G. K. Tóth and I. Kanizsai, *Org. Biomol. Chem.*, 2013, **11**, 6320–6327.
434. F. Wang, P. Zhao and C. Xi, *Tetrahedron Lett.*, 2011, **52**, 231–235.
435. (a) L. M. Ramos, A. Y. P. de L. Tobio, M. R. dos Santos, H. C. B. de Oliveira, A. F. Gomes, F. C. Gozzo, A. L. de Oliveria and B. A. D. Neto, *J. Org. Chem.*, 2012, **77**, 10184–10193; (b) H. G. O. Alvim, T. B. de Lima,

H. C. B. de Oliveira, F. C. Gozzo, J. L. de Macedo, P. V. Abdelnur, W. A. Silva and B. A. D. Neto, *ACS Catal*, 2013, **3**, 1420–1430.

436. Q. Yuan and D. Ma, *J. Org. Chem.*, 2008, **73**, 5159–5162.

437. D. Chen and W. Bao, *Adv. Synth. Catal.*, 2010, **352**, 955–960.

438. S. Tanimori, H. Kashiwagi, T. Nishimura and M. Kirihata, *Adv. Synth. Catal.*, 2010, **352**, 2531–2537.

439. J. Yan, F. Zhou, D. Qin, T. Cai, K. Ding and Q. Cai, *Org. Lett.*, 2012, **14**, 1262–1265.

440. Y. Fang and C. Li, *J. Org. Chem.*, 2006, **71**, 6427–6431.

441. C. C. Malakar, D. Schmidt, J. Conrad and U. Beifuss, *Org. Lett.*, 2011, **13**, 1972–1975.

442. K. Sudheendran, C. C. Malakar, J. Conrad and U. Beifuss, *J. Org. Chem.*, 2012, **77**, 10194–10210.

443. S. Hajra and D. Sinha, *J. Org. Chem.*, 2011, **76**, 7334–7340.

444. L. A. Adrio and K. K. Hii, *Chem. Commun.*, 2008, 2325–2327.

445. D. E. Mancheno, A. R. Thornton, A. H. Stoll, A. Kong and S. B. Blakey, *Org. Lett.*, 2010, **12**, 4110–4113.

446. (a) D. I. MaGee, M. Salehi, M. Dabiri and M. Bahramnejad, *Synth. Commun.*, 2013, **43**, 486–497; (b) S. Balalaie, J. Azizian, A. Shameli and H. R. Bijanzadeh, *Synth. Commun.*, 2013, **43**, 1787–1795; (c) M. Mollazadeh, M. J. Khoshkholgh, S. Balalaie, F. Rominger and H. R. Bijanzadeh, *J. Heterocycl. Chem.*, 2010, **47**, 1200–1208; (d) M. J. Khoshkholgh, M. Lotfi, S. Balalaie and F. Rominger, *Tetrahedron*, 2009, **65**, 4228–4234.

447. W. Bao, Y. Liu, X. Lv and W. Qian, *Org. Lett.*, 2008, **10**, 3899–3902.

448. Y. Liu and W. Bao, *Org. Biomol. Chem.*, 2010, **8**, 2700–2703.

449. A. K. Bagdi, A. Majee and A. Hajra, *Tetrahedron Lett.*, 2013, **54**, 3892–3895.

450. N. Thasanan, R. Worayuthakarn, P. Kradanrat, E. Hohn, L. Young and S. Ruchirawat, *J. Org. Chem.*, 2007, **72**, 9379–9382.

451. Y. Yamamoto and N. Kirai, *Org. Lett.*, 2008, **10**, 5513–5516.

452. S. Cai, F. Wang and C. Xi, *J. Org. Chem.*, 2012, **77**, 2331–2336.

453. V. Kavala, C. C. Wang, D. K. Barange, C. W. Kuo, P. M. Lei and C. F. Yao, *J. Org. Chem.*, 2012, **77**, 5022–5029.

454. X. X. Guo, *J. Org. Chem.*, 2013, **78**, 1660–1664.

455. H. J. Yuan, M. Wang, Y. J. Liu and Q. Liu, *Adv. Synth. Catal.*, 2009, **351**, 112–116.

456. T. Mizuhara, S. Inuki, S. Oishi, N. Fujii and H. Ohno, *Chem. Commun.*, 2009, 3413–3415.

457. S. L. Cui, X. F. Lin and Y. G. Wang, *Org. Lett.*, 2006, **8**, 4517–4520.

458. W. Qian, A. Amegadzie, D. Winternheimer and J. Allen, *Org. Lett.*, 2013, **15**, 2986–2989.

459. G. Murugavel and T. Punniyamurthy, *Org. Lett.*, 2013, **15**, 3828–3831.

460. R. K. Rao, A. B. Naidu and G. Sekar, *Org. Lett.*, 2009, **11**, 1923–1926.

461. S. Bhadra, L. Adak, S. Samanta, A. K. M. M. Islam, M. Mukherjee and B. C. Ranu, *J. Org. Chem.*, 2010, **75**, 8533–8541.

462. F. Melkonyan, A. Topolyan, A. Karchava and M. Yurovskaya, *Tetrahedron*, 2011, **67**, 6826–6832.
463. P. Jangili, J. Kashanna and B. Das, *Tetrahedron Lett.*, 2013, **54**, 3453–3456.
464. Z. Liu and Y. Chen, *Tetrahedron Lett.*, 2009, **50**, 3790–3793.
465. E. Feng, H. Huang, Y. Zhou, D. Ye, H. Jiang and H. Liu, *J. Org. Chem.*, 2009, **74**, 2846–2849.
466. D. Chen, G. Shen and W. Bao, *Org. Biomol. Chem.*, 2009, **7**, 4067–4073.
467. F. C. Sequeira and S. R. Chemler, *Org. Lett.*, 2012, **14**, 4482–4485.
468. M. L. Deb, S. S. Dey, I. Bento, M. T. Barros and C. D. Maycock, *Angew. Chem. Int. Ed.*, 2013, **52**, 9791–9795.
469. T. Xiong, Y. Li, X. Bi, Y. Lv and Q. Zhang, *Angew. Chem. Int. Ed.*, 2011, **50**, 7140–7143.
470. Y. Li, Z. Li, T. Xiong, Q. Zhang and X. Zhang, *Org. Lett.*, 2012, **14**, 3522–3525.
471. Z. Xia, K. Wang, J. Zheng, Z. Ma, Z. Jiang, X. Wang and X. Lv, *Org. Biomol. Chem.*, 2012, **10**, 1602–1611.
472. Z. Y. Ge, Q. M. Xu, X. D. Fei, T. Tang, Y. M. Zhu and S. J. Ji, *J. Org. Chem.*, 2013, **78**, 4524–4529.
473. D. J. C. Prasad and G. Sekar, *Org. Biomol. Chem.*, 2009, **7**, 5091–5097.
474. D. Ma, Q. Geng, H. Zhang and Y. Jiang, *Angew. Chem. Int. Ed.*, 2010, **49**, 1291–1294.
475. C. Dai, X. Sun, X. Tu, L. Wu, D. Zhan and Q. Zeng, *Chem. Commun.*, 2012, **48**, 5367–5369.
476. C. Korupalli, A. Dandapat, D. J. C. Prasad and G. Sekar, *Org. Chem. Int.*, 2011, 980765.
477. D. Yang, H. Liu, H. Yang, H. Fu, L. Hu, Y. Jiang and Y. Zhao, *Adv. Synth. Catal.*, 2009, **351**, 1999–2004.
478. D. Chen, J. Wu, J. Yang, L. Huang, Y. Xiang and W. Bao, *Tetrahedron Lett.*, 2012, **53**, 7104–7107.
479. W. S. Huang, R. Xu, R. Dodd and W. C. Shakespeare, *Tetrahedron Lett.*, 2013, **54**, 5214–5216.
480. D. Chen, Z. J. Wang and W. Bao, *J. Org. Chem.*, 2010, **75**, 5768–5771.
481. K. Kaneko, T. Yoshino, S. Matsunaga and M. Kanai, *Org. Lett.*, 2013, **15**, 2502–2505.
482. F. Duran, L. Leman, A. Ghini, G. Burton, P. Dauban and R. H. Dodd, *Org. Lett.*, 2002, **4**, 2481–2483.
483. (a) L. K. Ding and W. J. Irwin, *J. Chem. Soc., Perkin Trans.*, 1976, **1**, 2382–2386; (b) M. Kinugasa and S. Hashimoto, *J. Chem. Soc., Chem. Commun.*, 1972, 466–467; (c) K. Okuro, M. Enna, M. Miura and M. Nomura, *J. Chem. Soc., Chem. Commun.*, 1993, 1107–1108; (d) M. Miura, M. Enna, K. Okuro and M. Nomura, *J. Org. Chem.*, 1995, **60**, 4999–5004; (e) A. Basak, G. Bhattacharya and H. M. M. Bdour, *Tetrahedron*, 1998, **54**, 6529–6538; (f) A. Basak, S. C. Ghosh, T. Bhowmick, A. K. Das and V. Bertolasi, *Tetrahedron Lett.*, 2002, **43**, 5499–5501; (g) R. K. Khangarot and K. P. Kaliappan, *Eur. J. Org. Chem.*, 2011, 6117–6127;

(h) B. Grzeszczyk, K. Polawska, Y. M. Shaker, S. Stecko, A. Mames, M. Woznica, M. Chmielewski and B. Furman, *Tetrahedron*, 2012, **68**, 10633–10639; (i) A. G. Coyne, H. Müller-Bunz and P. J. Guiry, *Tetrahedron: Asymmetry*, 2007, **18**, 199–207; (j) T. Saito, T. Kikuchi, H. Tanabe, J. Yahiro and T. Otani, *Tetrahedron Lett.*, 2009, **50**, 4969–4972; (k) A. Mames, S. Stecko, P. Mikolajczyk, M. Soluch, B. Furman and M. Chmielewski, *J. Org. Chem.*, 2010, **75**, 7580–7587; (l) M. C. Ye, J. Zhou and Y. Tang, *J. Org. Chem.*, 2006, **71**, 3576–3582; (m) S. Stecko, A. Mames, B. Furman and M. Chmielewski, *J. Org. Chem.*, 2009, **74**, 3094–3100; (n) X. Zhang, R. P. Hsung, H. Li, Y. Zhang, W. L. Johnson and R. Figueroa, *Org. Lett.*, 2008, **10**, 3477–3479.

484. (a) M. M. C. Lo and G. C. Fu, *J. Am. Chem. Soc.*, 2002, **124**, 4572–4573; (b) R. Shintani and G. C. Fu, *Angew. Chem. Int. Ed.*, 2003, **42**, 4082–4085; (c) M. C. Ye, J. Zhou, Z. Z. Huang and Y. Tang, *Chem. Commun.*, 2003, 2554–2555.

485. L. Zhao and C. J. Li, *Chem. Asian J.*, 2006, **1–2**, 203–209.

486. J. Liang, J. Chen, F. Du, X. Zeng, L. Li and H. Zhang, *Org. Lett.*, 2009, **11**, 2820–2823.

487. X. Xu, D. Cheng, J. Li, H. Guo and J. Yan, *Org. Lett.*, 2007, **9**, 1585–1587.

488. M. J. Brown, G. J. Clarkson, G. G. Inglis and M. Shipman, *Org. Lett.*, 2011, **13**, 1686–1689.

489. Y. Xing, H. Zhao, Q. Shang, J. Wang, P. Lu and Y. Wang, *Org. Lett.*, 2013, **15**, 2668–2671.

490. (a) H. Lu and C. Li, *Org. Lett.*, 2006, **8**, 5365–5367; (b) T. Hu and C. Li, *Org. Lett.*, 2005, **7**, 2035–2038.

491. Y. Fang and C. Li, *J. Am. Chem. Soc.*, 2007, **129**, 8092–8093.

492. Q. Zhao, L. Li, Y. Fang, D. Sun and C. Li, *J. Org. Chem.*, 2009, **74**, 459–462.

493. A. Klapars, S. Parris, K. W. Anderson and S. L. Buchwald, *J. Am. Chem. Soc.*, 2004, **126**, 3529–3533.

494. J. Gan and D. Ma, *Org. Lett.*, 2009, **11**, 2788–2790.

495. X. Diao, L. Xu, W. Zhu, Y. Jiang, H. Wang, Y. Guo and D. Ma, *Org. Lett.*, 2011, **13**, 6422–6425.

496. H. Wang, Y. Jiang, K. Gao and D. Ma, *Tetrahedron*, 2009, **65**, 8956–8960.

497. X. Lu, L. Shi, H. Zhang, Y. Jiang and D. Ma, *Tetrahedron*, 2010, **66**, 5714–5718.

498. J. L. Kenwright, W. R. J. D. Galloway, D. T. Blackwell, A. Isidro-Llobet, J. Hodgkinson, L. Wortmann, S. D. Bowden, M. Welch and D. R. Spring, *Chem. Eur. J.*, 2011, **17**, 2981–2986.

499. U. A. Kshirsagar and N. P. Argade, *Org. Lett.*, 2010, **12**, 3716–3719.

500. M. O. kitching, T. E. Hurst and V. Snieckus, *Angew. Chem. Int. Ed.*, 2012, **51**, 2925–2929.

501. L. L. Sun, B. L. Hu, R. Y. Tang, C. L. Deng and X. G. Zhang, *Adv. Synth. Catal.*, 2013, **355**, 377–382.

502. P. Sang, M. Yu, H. Tu, J. Zou and Y. Zhang, *Chem. Commun.*, 2013, **49**, 701–703.

503. Y. Ohta, H. Chiba, S. Oishi, N. Fujii and H. Ohno, *Org. Lett.*, 2008, **10**, 3535–3538.
504. H. S. Lim, Y. L. Choi and J. N. Heo, *Org. Lett.*, 2013, **15**, 4718–4721.
505. I. Nakamura, Y. Kudo and M. Terada, *Angew. Chem. Int. Ed.*, 2013, **52**, 7536–7539.
506. S. Chandrasekhar, M. Seenaiah, A. Kumar, C. R. Reddy, S. K. Mamidyala, C. G. Kumar and S. Balasubramanian, *Tetrahedron Lett.*, 2011, **52**, 806–808.
507. D. K. Barange, Y. C. Tu, V. Kavala, C. W. Kuo and C. F. Yao, *Adv. Synth. Catal.*, 2011, **353**, 41–48.
508. M. Nagaraj, M. Boominathan, D. Perumal, S. Muthusubramanian and N. Bhuvanesh, *J. Org. Chem.*, 2012, **77**, 6319–6326.
509. J. Sun, X. Liu, H. Li, R. Duan and J. Wu, *Helv. Chim. Acta*, 2012, **95**, 772–779.
510. (a) D. L. Boger and D. Yohannes, *J. Org. Chem.*, 1991, **56**, 1763–1767; (b) D. L. Boger, Y. Nomoto and B. R. Teegarden, *J. Org. Chem.*, 1993, **58**, 1425–1433.
511. (a) K. C. Nicolaou, C. N. C. Boddy, S. Natarajan, T. Y. Yue, H. Li, S. Bräse and J. M. Ramanjulu, *J. Am. Chem. Soc.*, 1997, **119**, 3421–3422; (b) K. C. Nicolaou and C. N. C. Boddy, *J. Am. Chem. Soc.*, 2002, **124**, 10451–10455.
512. J. C. Collins, K. A. Farley, C. Limberakis, S. Liras, D. Price and K. James, *J. Org. Chem.*, 2012, **77**, 11079–11090.
513. C. P. Decicco, Y. Song and D. A. Evans, *Org. Lett.*, 2001, **3**, 1029–1032.
514. T. Yang, C. Lin, H. Fu, Y. Jiang and Y. Zhao, *Org. Lett.*, 2005, 7, 4781–4784.
515. Y. Hitotsuyanagi, H. Ishikawa, S. Naito and K. Takeya, *Tetrahedron Lett.*, 2003, **44**, 5901–5903.
516. R. Tello-Aburto, E. M. Johnson, C. K. Valdez and W. A. Maio, *Org. Lett.*, 2012, **14**, 2150–2153.
517. (a) D. Bahulayan and S. Arun, *Tetrahedron Lett.*, 2012, **53**, 2850–2853; (b) X. Duan, Y. Zhang, Y. Ding, J. Lin, X. Kong, Q. Zhang, C. Dong, G. Luo and Y. Chen, *Eur. J. Org. Chem.*, 2012, 500–508.
518. R. Pal and A. Basak, *Chem. Commun.*, 2006, 2992–2994.

Cobalt-Catalyzed Heterocycle Synthesis

Cobalt has the chemical configuration $[Ar]4s^2 3d^7$ and has oxidation states Co(II) and Co(III). Cobalt has many applications in a wide range of areas. In materials, cobalt is primarily used as the metal, for example, in the preparation of magnetic, wear-resistant and high-strength alloys. In biology, cobalt is the active center of coenzymes called cobalamins, the most common example of which is vitamin B_{12}. As such, it is an essential trace dietary mineral for all animals. Cobalt in inorganic form is also an active nutrient for bacteria, algae and fungi. In chemistry, various cobalt compounds are used as oxidation catalysts in chemical reactions. Cobalt acetate is used for the conversion of xylene to terephthalic acid, the precursor to the bulk polymer poly(ethylene terephthalate). Typical catalysts are the cobalt carboxylates (known as cobalt soaps). They are also used in paints, varnishes and inks as 'drying agents' through the oxidation of drying oils. The same carboxylates are also used to improve the adhesion of steel to rubber in steel-belted radial tires. Cobalt-based catalysts are also important in reactions involving carbon monoxide. In this chapter, the applications of cobalt catalysts in the synthesis of heterocycles is discussed.

5.1 Five-Membered Heterocycles

In 1986, Pattenden and co-workers reported a cobalt-promoted oxidative free-radical cyclization of alkyl bromides,[1] by which functionalized butyrolactones were selectively formed in good yields. In 1990, a convenient method for the stereoselective preparation of trans-2-hydroxymethyltetrahydrofurans by the oxidative cyclization of 5-hydroxy-1-alkenes was developed by Inoki and Mukaiyama.[2] They used bis(1-morpholinocarbamoyl-4,4-dimethyl-1,3-penta-nedionato)cobalt(II) [Co(modp)$_2$] as the catalyst with molecular oxygen as the

RSC Catalysis Series No. 16
Economic Synthesis of Heterocycles: Zinc, Iron, Copper, Cobalt, Manganese and Nickel Catalysts
By Xiao-Feng Wu and Matthias Beller
© Wu and Beller, 2014
Published by the Royal Society of Chemistry, www.rsc.org

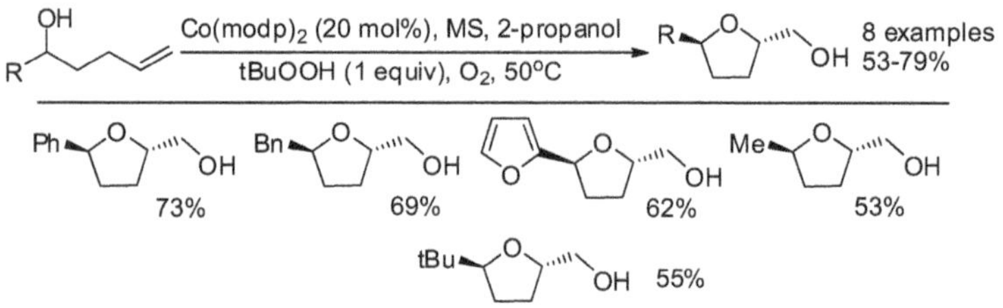

Scheme 5.1   Cobalt-catalyzed synthesis of furans.

oxidant, and good yields of the desired products were obtained (Scheme 5.1). The reaction proceeded with extremely high *trans* selectivity and the yield of the final product could be improved by the addition of *tert*-butyl hydroperoxide.

Hartung and co-workers found that bishomoallylic alcohols (pent-4-en-1-ols) can undergo efficient oxidative cyclizations with O_2 as oxidant.[3] Bis{2,2,2-trifluoromethyl-1-[(1R,4S)-1,7,7-trimethyl-2-(oxo-$_\kappa O$)bicyclo[2.2.1]hept-3-yliden]ethanolato-$_\kappa O$}cobalt(II) was used as catalyst in solutions of 2-propanol at 60 °C. Ring closures occurred diastereoselectively and afforded 2,3-*trans*- (96% *de*), 2,4-*cis*- (~60% *de*) and 2,5-*trans*-substituted (>99% *de*) (phenyl)tetrahydrofur-2-ylmethanols as the major components. Formation of bicyclic compounds and a 2,3,4,5-substituted oxolane was feasible as exemplified by syntheses of oxabicyclo[4.3.0]nonylmethanols and a derivative of the natural product magnosalicin in 61–72% yield (90–99% *de*). The effectiveness of tetrahydrofuran synthesis was critically dependent on (i) solvent, (ii) reaction temperature, (iii) initial cobalt concentration, (iv) chain length between hydroxyl and vinyl groups and (v) substitution at reacting entities. Solvent variation from *i*PrOH to CHD (cyclohexa-1,4-diene)–C_6H_6 was associated with a change in product selectivity from oxidative cyclization (formation of substituted tetrahydrofur-2-ylmethanols) toward alkenol cyclization (synthesis of substituted 2-alkyltetrahydrofurans). This finding suggests that the reaction is likely to be terminated *via* carbon radical trapping.[4] This implies that groups other than the H-atom should be transferable in the final step of the synthesis. Soon afterwards, they reported the reductive and brominative termination of alkenol cyclization under aerobic conditions (Scheme 5.2).[5] Reductive ring closures were achieved in cyclohexa-1,4-diene, and a change in selectivity from reductive termination to bromocyclization was attainable upon addition of $BrCCl_3$ or diethyl dibromomalonate (DBM) to standard reaction mixtures. This CoL$_2$ ({4-[3,5-bis(trifluoromethyl)phenyl]-4-oxybut-3-en-2-one}cobalt(II))–O_2 catalyst system was assumed to serve as a one-electron oxidant for transformation of the alkene into a radical cation and subsequently into an intermediate that is proposed to be a free carbon radical. Notably, this type of cyclization was applied in the total synthesis of gigantetrocin A and asimilobin by Shi and co-workers.[6–8]

More recently, the oxidative cyclization of alkenols and alkenes/alkynes was reported (Scheme 5.3).[9] In detail, the aerobic oxidation of alkyl- and

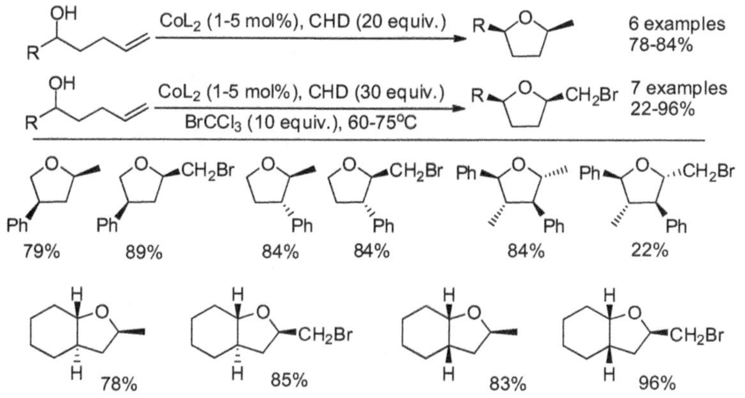

Scheme 5.2 Cobalt-catalyzed reductive and brominative cyclization of alkenols.

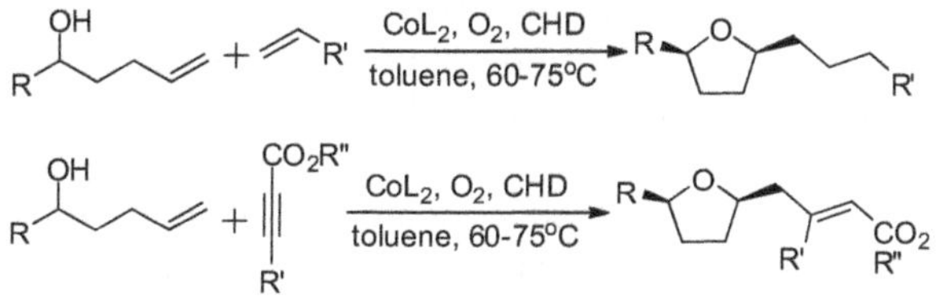

Scheme 5.3 Cobalt-catalyzed cyclization of alkenols and alkenes/alkynes.

phenyl-substituted 4-pentenols (bishomoallyl alcohols), catalyzed by cobalt(II) complexes in solutions of γ-terpinene or cyclohexa-1,4-diene, stereoselectively gave tetrahydrofurylmethyl radicals. Cyclized radicals were trapped with monosubstituted alkenes (*e.g.* acrylonitrile, methyl acrylate), (*E*)- and (*Z*)-1,2-diacceptor-substituted alkenes (*e.g.* dimethyl fumarate, fumarodinitrile, *N*-phenylmaleimide) and ester-substituted alkynes (*e.g.* ethyl propynoate). Oxidative addition cascades thus furnished side chain-substituted (CN, CO$_2$R, COR or SO$_2$R) di- and trisubstituted tetrahydrofurans in stereoselective reactions (2,3-*trans*, 2,4-*cis* and 2,5-*trans*). A diastereomerically pure bistetrahydrofuran was prepared in a cascade consisting of two aerobic oxidations, one alkyne addition and one final H-atom transfer.

In 2009, Pagenkopf and co-workers reported a second-generation catalyst for the oxidative cyclization of alkenols.[10] The catalyst preparation was straightforward at the gram scale with centrifugation as the only means of purification. In all cases examined, cyclization reactions using Co(nmp)$_2$ [nmp = 5,5-dimethyl-1-(4-methylpiperazin-1-yl)hexane-1,2,4-trione] showed improved yields and fewer side products. Most importantly, Co(nmp)$_2$ greatly simplified post-reaction purification by replacing difficult column chromatography with an aqueous workup.

In 2011, a density functional theory (DFT) calculation was performed by Fink and co-workers in order to understand how molecular oxygen binds to cobalt catalysts.[11] Bis[trifluoroacetylacetonato(−1)]cobalt(II) [Co(tfa)$_2$] was chosen as an example of cobalt(II) diketonate complexes to catalyze the aerobic oxidation of alkenols into functionalized tetrahydrofurans. To gain insight into the activation of triplet dioxygen by $Co(tfa)_2$ in a protic solvent, as used in oxidation catalysis, the electronic structure of aquabis[trifluoroacetylacetonato(-1)]cobalt(II) [$Co(tfa)_2(H_2O)$] and the derived dioxygen adduct were characterized using *ab initio* (CASSCF, NEVPT2) and DFT (BP86, TPSSh, B3LYP) methods. The ground state of $Co(tfa)_2(H_2O)$ was a high-spin quartet state. As dioxygen approaches the cobalt atom, the quartet state couples with a triplet dioxygen molecule and forms a sextet, a quartet and a doublet spin state with the high-spin state being the lowest in energy. At the equilibrium $Co-O_2$ distance of 1.9 Å, $Co(tfa)_2(H_2O)(O_2)$ has a doublet superoxo Co(III) ground state with the unpaired electron residing on the oxygen moiety, in a nearly unchanged O_2 p* orbital.

In addition to the cyclization of alkenols, a cobalt catalyst was also applied in the tandem radical cyclization of 6-halo-1-hexene derivatives and 6-halo-1-hexynes.[12,13] Oshima and co-workers found that benzyl-substituted heterocycles can be produced in moderate to good yields with $CoCl_2(dppe)$ as the catalyst (Scheme 5.4).[14] Concerning the reaction mechanism, this transformation started with the reaction of $CoCl_2(dppe)$ with 4 equiv. of PhMgBr gives [$Co(0)Ph_2(dppe)$]($MgBr$)$_2$ with concomitant production of 1 equiv. of biphenyl. The zerovalent -ate complex undergoes a single-electron transfer to a substrate to yield an anion radical of the substrate and cobalt(I) complex. The immediate loss of bromide from the anion radical affords a 5-hexenyl radical intermediate, which is transformed into a cyclopentylmethyl radical. Then, the cobalt(I) complex would recombine with the carbon-centered radical to form a divalent cobalt species. The subsequent reductive elimination provides the product and the Co(0) complex, which can be reconverted into [$Co(0)Ph_2(dppe)$]($MgBr$)$_2$ by the action of the remaining PhMgBr.

Interestingly, the same group found that when an aromatic Grignard reagent is replaced with a trimethylsilylmethyl Grignard reagent, the selectivity of this reaction was changed.[15] Methylenecyclopentanes as Heck-type products were isolated in good yields (Scheme 5.5). During the optimization of this reaction, the choice of the ligand was found to be crucial. DPPB was found to be best, whereas DPPE gave mainly the reduced product. Later, they extend their methodology to the cyclization of

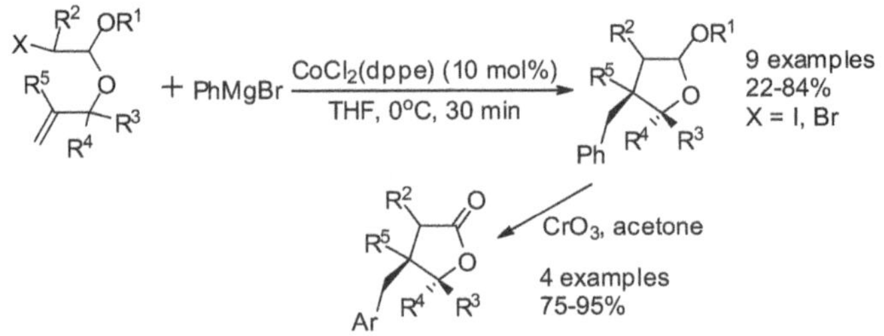

Scheme 5.4   Cobalt-catalyzed synthesis of benzyl-substituted heterocycles.

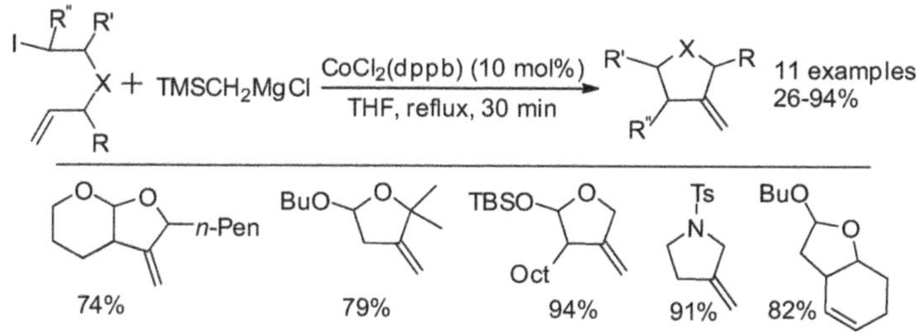

Scheme 5.5 Cobalt-catalyzed synthesis of methylenecyclopentanes.

Reaction Procedure (Scheme 5.5): Anhydrous cobalt(II) chloride (6.5 mg, 0.050 mmol) was placed in a 20 mL flask and heated with a hair dryer *in vacuo* for 2 min. After the color of the cobalt salt became blue, dppb (25 mg, 0.060 mmol) and THF (1 mL) were added sequentially under argon at 25 °C. The mixture was stirred for about 5 min and a blue mixture was obtained. Trimethylsilylmethylmagnesium chloride (1.0 M THF solution, 1.5 mL, 1.5 mmol) was added to the mixture at 0 °C. After 5 min, iodo acetal (0.18 g in 1.0 mL of THF, 0.50 mmol) was introduced *via* a syringe. The whole mixture was heated at reflux (bath temperature 80 °C) for 5 min, then was poured into saturated ammonium chloride solution. The products were extracted with ethyl acetate (2 × 20 mL) and the combined organic layer was dried over Na₂SO₄ and concentrated. Careful silica gel column purification of the crude product provided the pure product.

silicon-tethered 6-iodo-1-hexene derivatives, giving the corresponding benzyl-substituted oxasilacyclopentanes in good yield, which were further converted to 4-aryl-1,3-diols after Tamao–Fleming oxidation.[16]

The above silicon-containing heterocycles could also be prepared by [2 + 2 + 2] cycloaddition,[17] and in 2007 Tacke's group reported a simple catalytic system for [2 + 2 + 2] alkyne cycloadditions consisting of cobalt(II) iodide and zinc powder in acetonitrile. This system was applied to the direct synthesis of hydroxyalkyl-substituted 1,3-disilaindanes, 1,4-disilatetralines and 1,3-disila-1,3-dihydroisobenzofurans starting from silicon-containing diynes and unprotected propargyl alcohols (Scheme 5.6). The reactions were performed under mild conditions and a relatively low catalyst load (2.5%) gave almost complete consumption of the starting materials in a very short time

Polyfunctionalized furans constitute an important class of five-membered *O*-heterocycles with wide applications. In 2012, Zhang and co-workers reported a regioselective synthesis of multisubstituted furans via metallor-adical cyclization of alkynes with α-diazocarbonyls.[18] In their procedure, Co(III)–carbene radicals generated from activation of α-diazocarbonyls by Co(II)–porphyrin complexes undergo a tandem radical addition reaction with alkynes and affords five-membered furan structures (Scheme 5.7). The Co(II)

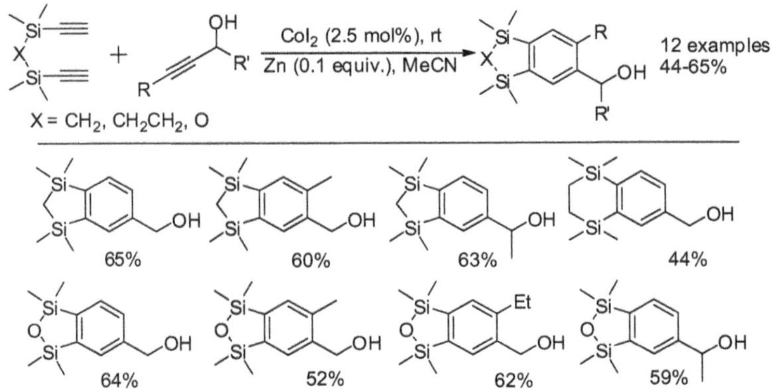

Scheme 5.6 Cobalt-catalyzed synthesis of silicon-containing heterocycles.

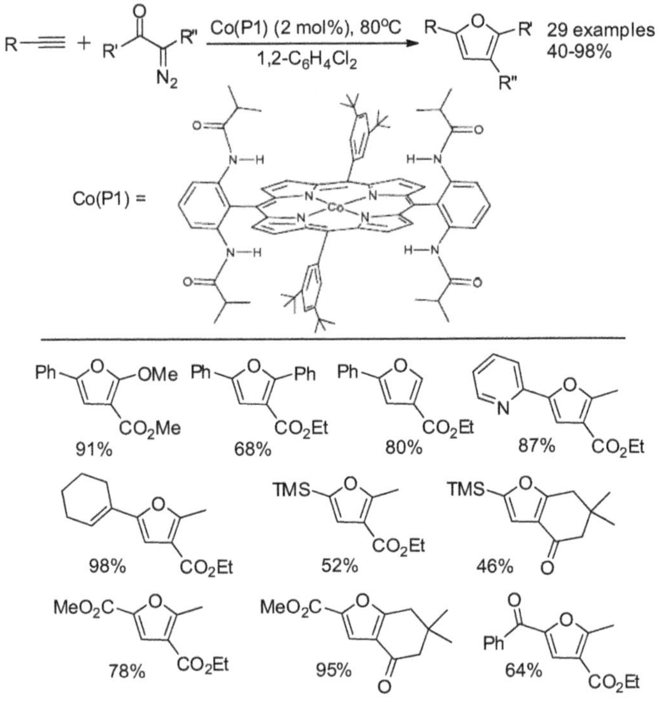

Scheme 5.7 Cobalt-catalyzed synthesis of furans from alkynes.

Reaction Procedure (Scheme 5.7): Catalyst [Co(P1)] (2 mol%) was placed in an oven-dried, resealable Schlenk tube, which was capped with a Teflon screw-cap, evacuated and backfilled with nitrogen. The screw-cap was replaced with a rubber septum and 2 equiv. of acetylene (0.2 mmol) and 1.0 equiv. of diazo compound (0.1 mmol) in 0.5 mL 1,2-dichlorobenzene (anhydrous) were added *via* a syringe. The Schlenk tube was capped with a Teflon screw-cap in place of the rubber septum and stirred at 80 °C for 24 h. When the reaction was completed, the resulting mixture was concentrated and the residue was purified by flash silica gel chromatography to give the products.

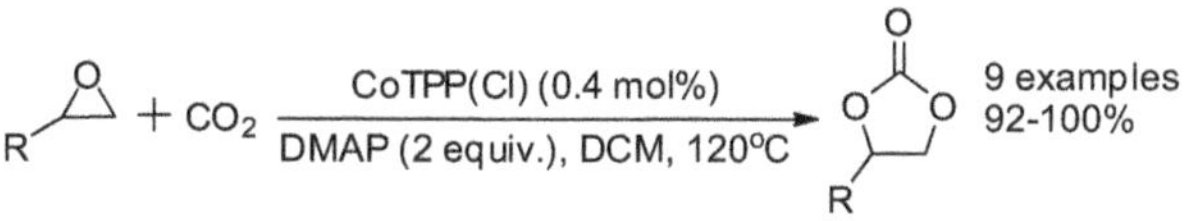

$$R \overset{O}{\triangle} + CO_2 \xrightarrow[\text{DMAP (2 equiv.), DCM, 120°C}]{\text{CoTPP(Cl) (0.4 mol\%)}} \quad \begin{array}{c} 9 \text{ examples} \\ 92\text{-}100\% \end{array}$$

Scheme 5.8 Cobalt-catalyzed synthesis of cyclic carbonates.

complex of 3,5-Di-*t*Bu-IbuPhyrin, [Co(P1)] was found to be effective in catalyzing the metalloradical cyclization reaction under neutral and mild conditions. The [Co(P1)]-catalyzed process tolerates a wide range of α-diazocarbonyls and terminal alkynes with varied steric and electronic properties, producing polyfunctionalized furans with complete regioselectivity. The catalytic synthesis features a high degree of functional group tolerance and can be applied iteratively to construct functionalized α-oligofurans.

Nguyen and co-workers demonstrated that Co(TPP)Cl–DMAP is an excellent catalyst system for the coupling of CO_2 with a variety of epoxides to yield cyclic carbonates selectively and stereospecifically.[19] The corresponding cyclic carbonate products were produced in high yield and selectivity for a variety of terminal mono- and disubstituted epoxides (Scheme 5.8). 1,2-Disubstituted internal epoxides were also investigated as substrates and found to react with very high stereospecificity.

In 2005, Cheng and co-workers developed a cobalt-catalyzed [2 + 2 + 2] cyclotrimerization of alkynyl alcohols with propiolates.[20] The reaction is highly regio- and chemoselective, affording benzolactone derivatives in good to excellent yields (Scheme 5.9a). Later, they found that cobalt-catalyzed co-cyclization of 2-iodobenzoates with aldehydes could afford substituted benzolactone derivatives.[21] The reaction proceeded in a one-pot manner under mild conditions and gave the desired products in good to excellent yields (Scheme 5.9b). In addition, high enantioselectivity of the cyclization was obtained by employing cobalt complexes with a suitable bidentate chiral ligand. These results revealed that this cobalt-catalyzed cyclization is a good complement to nickel-catalyzed cyclization reactions. Concerning the reaction mechanism, the reduction of cobalt(II) to cobalt(I) by Zn metal likely initiates the catalytic reaction. Oxidative addition of methyl 2-iodobenzoate with the cobalt(I) species yields an *ortho*-metalated methybenzoate complex with both the *ortho*-carbon atom and the ester oxygen atom bonded to the cobalt(III) center. Coordination of the aldehyde molecule to the cobalt center adjacent to the *ortho*-metalated methyl benzoate group followed by insertion of a cobalt–carbon bond in the aldehyde affords a cobalt alkoxide intermediate. Intramolecular nucleophilic addition of the coordinated alkoxy

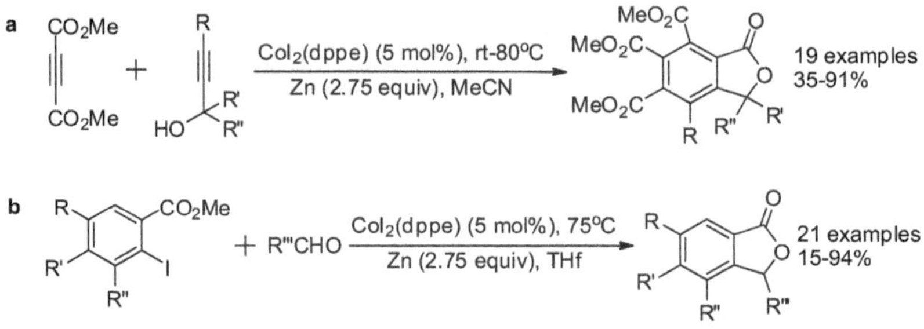

Scheme 5.9   Cobalt-catalyzed synthesis of benzolactone derivatives.

group to the ester group gives the final product and a cobalt(III) species. The latter is reduced by zinc metal to regenerate the active cobalt(I) species.

The γ-butyrolactone subunit is a characteristic feature of many natural products that display significant biological activities. With this background, Le Gall and co-workers developed a one-pot, three-component procedure for the synthesis of butyrolactones.[22] They used aryl bromides, dimethyl itaconate and either aldehydes or ketones as substrates and obtained the desired products in good yields (Scheme 5.10). This cobalt-catalyzed domino process formally involves the *in situ* metalation of an aromatic bromide, conjugate addition onto dimethyl itaconate, an aldolization reaction with a carbonyl compound and a final cyclization into a five-membered lactone. This procedure was applied to the concise synthesis of a range of functionalized γ-butyrolactones with a methyl paraconate subunit.

In 2008, Chung and co-workers demonstrated that Co_2Rh_2 can be used as an effective catalyst in the carbonylative cycloaddition of 2-alkynylanilines in the presence of carbon monoxide.[23] Various oxindoles were formed in moderate to excellent yields (Scheme 5.11). The catalytic system can be used at least three times without loss of catalytic activity.

Tetrahydrofurans are an important class of heterocycle that exist in numerous natural products and biologically active molecules. Cobalt-catalyzed cyclization of alkynes offers an interesting pathway for the preparation of these compounds. In 1999, Chung and co-workers reported a $Co_2(CO)_8$-catalyzed tandem $[2+2+1]/[2+2+2]$ cycloaddition reaction of terminal diynes under CO pressure.[24] This procedure gave novel tetracyclic compounds as the terminal products (Scheme 5.12a). Later, they reported that novel tetracyclic compounds containing cyclopentenones can be constructed by the cobalt octacarbonyl-catalyzed double $[2+2+1]$ cyclo-addition reaction of triynes under CO pressure (Scheme 5.12b).[25] They demonstrated that through judicious design of the substrate, either a double $[2+2+1]$ or a $[2+2+2]$ cycloaddition reaction can be carried out selectively, providing a viable route to tetracyclic or tricyclic compounds. These reactions provide rapid access to tetracyclic skeletons of 5–5–5–5 and/ or tricyclic skeletons of 5–6–5 ring systems. Important features of this catalytic reaction are the experimental simplicity and a high conversion rate.

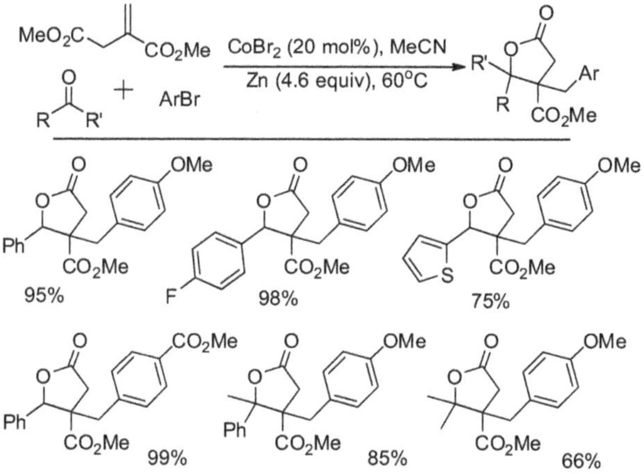

Scheme 5.10 Cobalt-catalyzed synthesis of lactones.

Reaction Procedure (Scheme 5.10): A dried 100 mL round-bottomed flask was flushed with argon and charged with acetonitrile (20 mL). Dodecane (0.2 mL), zinc dust (3 g, 46 mmol), dimethyl itaconate (7.9 g, 50 mmol), aromatic aldehyde (10 mmol) and aryl bromide (15 mmol) were added while stirring. Cobalt bromide (0.44 g, 2 mmol), trifluoroacetic acid (0.1 mL) and 1,2-dibromoethane (0.2 mL) were added successively to the mixture, which was heated at 60 °C until consumption of the aryl bromide was complete (45 min to 3 h, monitored by gas chromatography). The reaction mixture was then filtered through Celite. The Celite was washed several times with diethyl ether and the combined organic fractions were concentrated *in vacuo*. The crude reaction product was purified by flash column chromatography over silica gel (eluent: 1 : 0–0 : 1 pentane–diethyl ether) to afford the lactone.

Green's group reported that cycloheptyne–dicobalt hexacarbonyl complexes, substituted by propargylic ether functions, undergo $[2+2+2]$ cycloaddition reactions with alkynes to give tricyclic benzocycloheptanes.[26] Good yields of the desired products were isolated. Mechanistically, they proposed a $[2+2+2]$ cycloaddition mechanism analogous to that invoked for CoCp complexes, involving either a cobaltacycloheptatriene intermediate and subsequent reductive elimination or a $[4+2]$ cycloaddition of the added alkyne with a tethered cobaltacyclopentadiene.

In 2005, Okamoto and co-workers reported a cobalt-catalyzed cyclotrimerization of triynes.[27] Annulated benzenes were formed in good yields by treatment with a catalytic amount of zinc powder, *N*-heterocyclic carbene and $CoCl_2$ or $FeCl_3$ (Scheme 5.13a). The reagent combinations of imidazolium carbene–$CoCl_2$ and –$FeCl_3$ were good catalyst precursors and could transform a variety of triynes to annulated benzenes in good to excellent

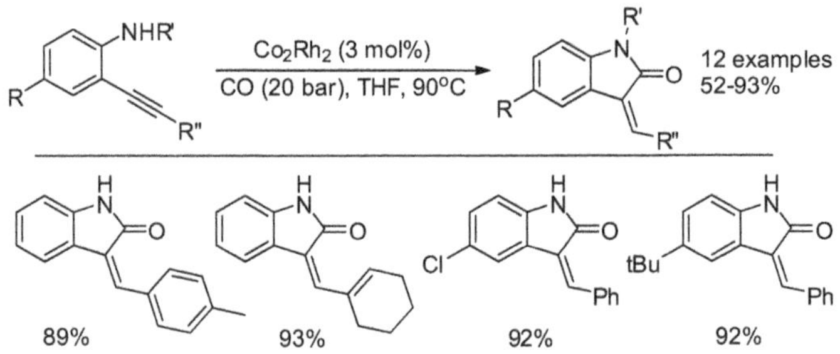

Scheme 5.11   Cobalt-catalyzed synthesis of oxindoles.

Reaction Procedure (Scheme 5.11): Immobilization of Metal Nano-particles on Charcoal: In a two-necked flask were placed *o*-dichloro-benzene (24 mL), oleic acid (0.2 mL) and trioctylphosphine oxide (0.4 g). While the solution was heated at 180 °C, a solution of the metal carbonyl $Co_2Rh_2(CO)_{12}$ (0.8 g) in 20 mL of *o*-dichlorobenzene was injected into the flask. The resulting solution was heated at 180 °C for 2 h, concentrated to 5 mL and cooled to room temperature, then 30 mL of THF were added. After the solution had been well stirred for 10 min, flame-dried charcoal (2.0 g) was added. After the resulting solution had been refluxed for 12 h, the precipitates were filtered and washed with diethyl ether (20 mL), dichloromethane (20 mL), acetone (20 mL) and methanol (20 mL). Vacuum drying gave a black solid.

General procedure for Co_2Rh_2-Catalyzed Carbocyclization of 2-Ethyny-lanilines to Oxindoles: Reactions were performed in a 100 mL stainless-steel autoclave equipped with a stirrer bar and 2-ethynylaniline (0.3 mmol), THF (5 mL) and 3 mol% Co_2Rh_2 (15 mg of the immobilized Co_2Rh_2) were placed in the autoclave in that order,. The reactor was charged with 20 atm of CO and heated at 90 °C for 5 h. After the reactor had cooled to room temperature, the solution was filtered and concentrated and the product was isolated by chromatography on a silica gel column, eluting with hexane–ethyl acetate (4 : 1).

---

yields. In 2007, they developed a cobalt-catalyzed cyclization of diynes and alkynes.[28] They found a unique additive effect of silver salts in a ligand–$CoCl_2 \cdot 6H_2O$–Zn catalyst system for the [2 + 2 + 2] cycloaddition reaction of alkynes. The effect was specific to reactions with the 2-iminomethylpyridine ligand but no effect was observed for the phosphine-based system. The catalysis thus developed introduces a practical and quick route to highly substituted benzenes from various alkynes (Scheme 5.13b). The specificity of the additive effect to the dipimp ligand can be explained as follows: it can be assumed that the Co(I) complex such as dipimp–Co(I)Cl (A), derived from the reaction of dipimp with $CoCl_2$ and followed by reduction with Zn, may be converted to the corresponding cationic complex by reaction with AgX.

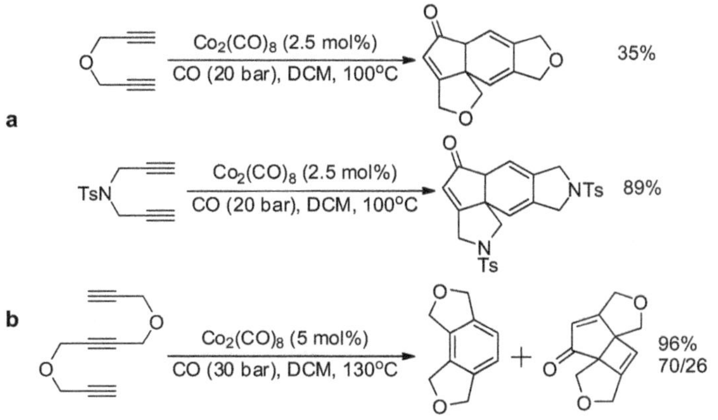

Scheme 5.12 Cobalt-catalyzed cyclization of diynes.

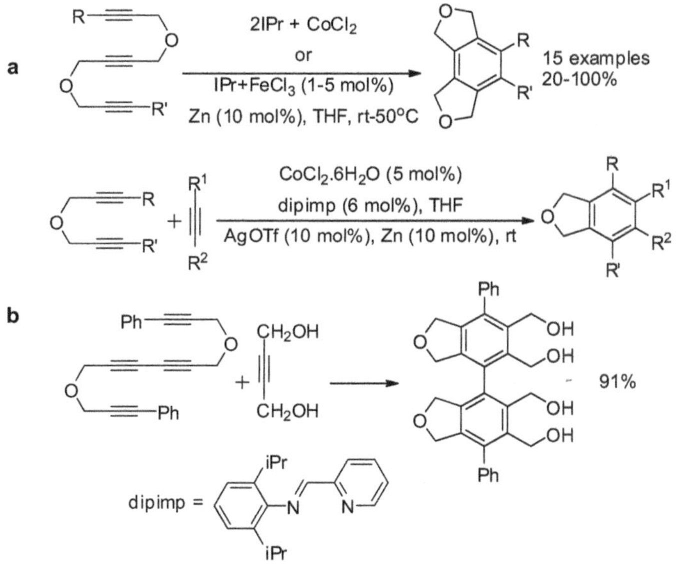

Scheme 5.13 CoCl₂-catalyzed cyclization of diynes and triynes.

The positive charge present in the complex can be delocalized to the pyridine moiety of the ligand. The specific additive effect to the dipimp system can be addressed by the contribution of such a resonance stabilization, whereas such a stabilization is not possible in the case of the dppe–Co complex. The electron-deficient nature of the cationic complex with a weaker s-donative dipimp ligand than dppe in the intermediate complex may facilitate the reaction with unactivated monoalkynes. This type of transition metal complex-catalyzed [2 + 2 + 2] cyclotrimerization of α,ω-diynes was applied in the preparation of anomerically pure *C*-aryldeoxyribosides with *C*-ethynyldeoxyriboside under simple and mild reaction conditions.[29] Rh, Ru and Ni complexes proved to be effective for catalyzing the reaction; the best results were obtained with the stable and easy to handle Wilkinson's catalyst [Rh(PPh$_3$)$_3$Cl]. From a synthetic point of view, the advantages of this strategy are the modular approach with regard to structural variety of the α,ω-diynes, the functional group compatibility of the transition metal catalysis with a wide variety of functional groups in the reactants and an easily removable *p*-toluoyl protective group in the deoxyriboside.

Sugihara *et al.* prepared a cobalt carbonyl cluster, methylidynetricobalt nonacarbonyl, by reaction of dicobalt octacarbonyl with trihaloalkanes and applied it in the inter- and intramolecular cyclotrimerization of alkynes.[30] Substituted benzene derivatives were isolated in good to excellent yields. The advantage of these catalysts is that they are more stable against auto-oxidation than the parent dicobalt octacarbonyl. Later, Gandon, Aubert and co-workers prepared air-stable {(C$_5$H$_5$)Co} complexes and found that they are versatile catalysts for various [2 + 2 + 2] cycloadditions and related reactions, for forming benzenes, pyridines and 1,3-cyclohexadienes (Scheme 5.14a).[31] Although, in some cases, they exhibit similar activity, they present major advantages compared with common catalysts such as [CpCo(CO)$_2$] and [CpCo(C$_2$H$_4$)$_2$]. They are air stable and retain their activity after several months of storage in simple vials. With the exception of pyridines, the cycloaddition products were synthesized without visible-light irradiation or under microwave conditions. The solvents could be used as found in the laboratory without further purification. More recently, Hapke and co-workers described a novel and facile synthetic strategy to obtain [CpCo(alkene)(phosphite)] complexes with a variety of phosphites and al-kene ligands from commercially available CpCo(CO)$_2$ (Scheme 5.14b).[32] Precatalysts of the structure [CpCo(dimethyl fumarate)(phosphite)] were air-stable and recyclable complexes, which were active in inter- and intra-molecular [2 + 2 + 2] cycloaddition reactions for the synthesis of pyridines and benzenes. The reaction times were short, non-dried toluene was used as the solvent and no inert gas atmosphere was required, demonstrating the robustness of the precatalyst.

In addition to oxygen-containing compounds, nitrogen-containing five-membered heterocycles have also been prepared using cobalt catalysts. In 2007, a cobalt-catalyzed sequential cyclization/cross-coupling reaction of

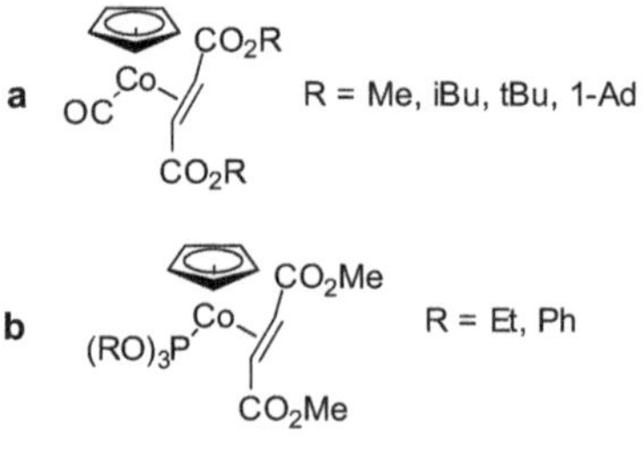

Scheme 5.14   Cobalt complexes for cyclization of diynes.

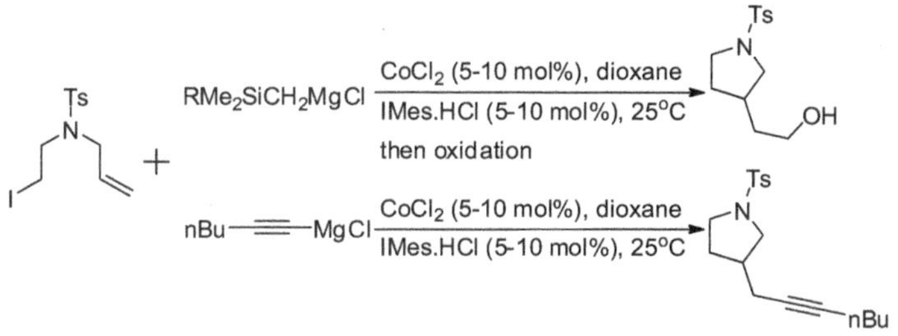

Scheme 5.15   Cobalt-catalyzed synthesis of pyrrolidine derivatives.

6-halo-1-hexene derivatives with Grignard reagents was developed.[33] Pyrrolidine derivatives were isolated in good yields (Scheme 5.15). *N*-Heterocyclic carbene was found to be an effective ligand for this transformation, whereas phosphine and amine ligands were not active.

In 2002, Yamada and co-workers prepared a cationic cobalt(III) complex with an optically active β-ketoiminato ligand and applied it in the enantioselective 1,3-dipolar cycloaddition reaction of α,β-unsaturated aldehydes and nitrones derived from 2-halobenzaldehydes.[34] The corresponding isoxazolidines were produced in high yields with excellent *endo* selectivities and high enantioselectivities (Scheme 5.16).

Alternatively, Yu and co-workers recently developed a cobalt-catalyzed aerobic oxidative cyclization of β,γ-unsaturated oximes.[35] The cobalt complex Co(nmp)₂ {bis[5,5-dimethyl-1-(4- methylpiperazin-1-yl)hexane-1,2, 4-trione]cobalt (II)} was found to be an effective catalyst for this transformation and gave the corresponding isoxazolines in good yields (Scheme 5.17). The key cyclization step involves the generation of carbon-centered radicals. The products are largely dependent on the reaction conditions. The oxidative termination products were produced predominantly when the reaction was carried out in 2-propanol, whereas the reductive termination products were selectively obtained in toluene in the presence of cyclohexa-1,4-diene (CHD). This work provides an alternative protocol for the synthesis of 4,5-dihydroisoxazoles.

Roberto and Alper reported a cobalt-catalyzed carbonylation of azetidines in 1989.[36] Pyrrolidinones were produced in good yields using cobalt carbonyl as the catalyst (Scheme 5.18). Interestingly, tetrahydroazepinones were formed on applying 2-vinylazetidines as substrates. Regarding the mechanism for transforming 2-alkylazetidines to 5-substituted pyrrolidin-2-ones, the coordination of substrates to the cobalt complex was believed to be the first step. Then insertion of cobalt in the least-substituted ring carbon–nitrogen bond gave a metallacycle. After insertion of one molecule of CO and reductive elimination, the final product was formed.

In 2009, Cheng and co-workers reported a cobalt-catalyzed regioselective synthesis of pyrrolidinone derivatives by reductive coupling of nitriles and acrylamides.[37] This novel procedure for pyrrolidinone synthesis proceeds *via* cobalt-catalyzed reductive coupling of nitrile and acrylamide followed by

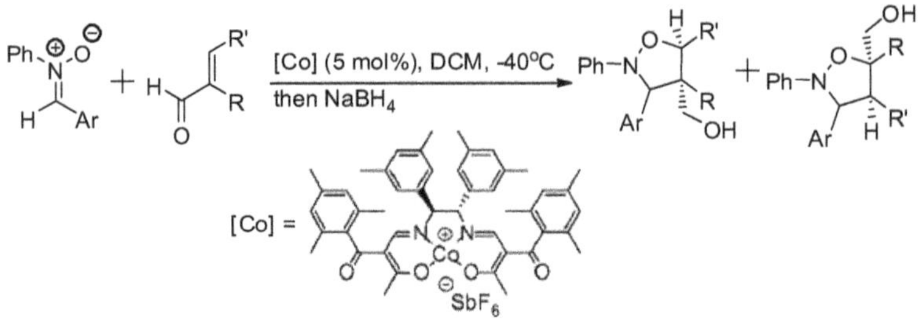

Scheme 5.16   Cobalt-catalyzed synthesis of isoxazolidines.

Reaction Procedure (Scheme 5.16): To a stirred solution of Co(III)–SbF$_6$ (14.8 mg, 0.015 mmol) in DCM (0.5 mL) at −60 °C were added α-benzylpropenal (222.2 mg, 1.5 mmol) in DCM (0.5 mL) and nitrone (79.8 mg, 0.3 mmol) in DCM (1.0 mL). After stirring the reaction mixture at −60 °C for 84 h, the reaction was quenched by the addition of NaBH$_4$ (85.1 mg, 2.25 mmol) in EtOH. The product was extracted with EtOAc. The organic extract was washed with brine and dried over anhydrous Na$_2$SO$_4$. The solvent was removed under reduced pressure and the residue was purified by silica-gel column chromatography (eluent: 20 : 1 hexane–EtOAc) to afford the product.

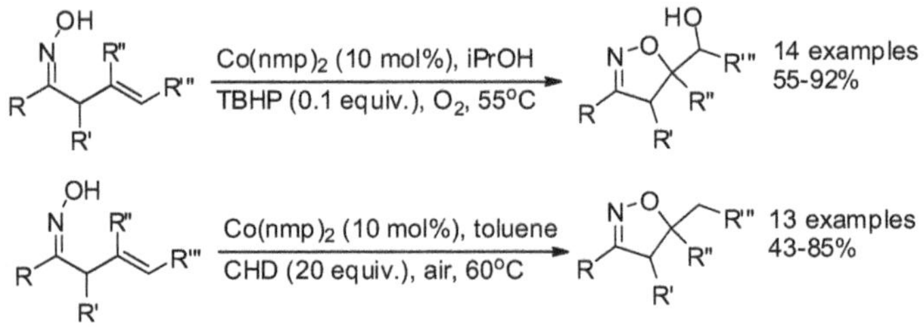

Scheme 5.17   Cobalt-catalyzed oxidative synthesis of isoxazolidines.

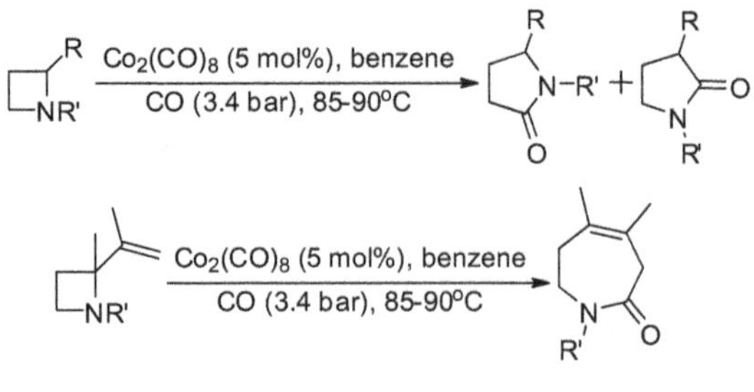

Scheme 5.18   Cobalt-catalyzed carbonylative synthesis of pyrrolidinones.

keto-amide cyclization and dehydration in one pot, the desired products being isolated in good yields (Scheme 5.19). In respect of the reaction mechanism, the catalytic cycle was likely initiated by the reduction of Co(II) to Co(I) by zinc dust. This was followed by the chemoselective cyclometala-tion of Co(I) with nitrile and acrylamide to form cobaltaazacyclopentene intermediate. Protonation of the intermediate followed by hydrolysis gives amide intermediate and a Co(III) species. The Co(III) species is reduced by zinc to regenerate the active Co(I) species for the next cycle. The amide intermediate further undergoes keto–amide cyclization and elimination of water to give the final pyrrolidinone derivative. This mechanism explains the regioselectivity of the reductive coupling product. In the catalytic reaction, ZnI$_2$ probably acts as a Lewis acid to remove a halide from the Co(I) center, assisting the coordination of the nitrile and acrylamide to the metal center. In addition, it can also activate the keto group in the intermediate, assisting in the cyclization of amide to give the final product.

Additionally, cobalt catalysts have also been applied in the radical cycli-zation and carbonylative C–H activation. Using a combination of cobalt(II) chloride and Grignard reagent, tin hydride could be replaced and gave the corresponding indol-2(3*H*)-ones from their parent *N*-methyl-*N*-acryloyl-2-haloanilines.[38] In the case of carbonylative C–H activation, 2-phenyl-1*H*-indazolones were produced from azobenzenes.[39] Sasaki and co-workers

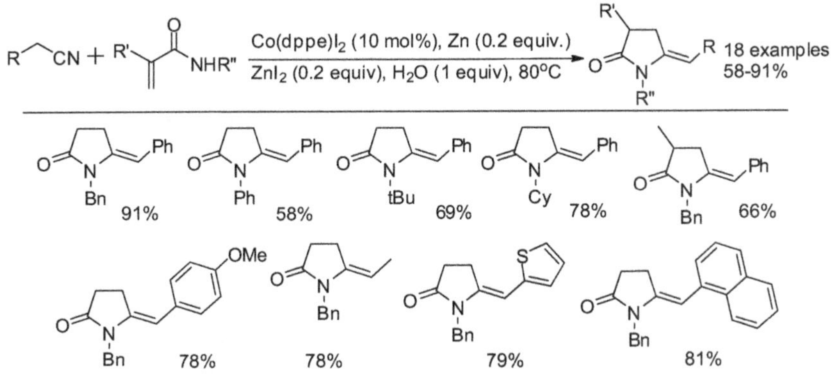

Scheme 5.19 Cobalt-catalyzed reductive synthesis of pyrrolidinones.

Reaction Procedure (Scheme 5.19): A sealed tube containing CoI$_2$(dppe) (0.1 mmol), zinc powder (1.5 mmol) and zinc iodide (0.2 mmol) was evacuated and purged three times with nitrogen gas. Then, nitrile (2.5 mmol), acrylamide (1.0 mmol) and water (1.0 mmol) were added sequentially and the reaction mixture was stirred at 80 °C for 12 h. The mixture was filtered through a short Celite pad, which was washed several times with dichloromethane. The filtrate was concentrated and the resi-due was purified on a silica gel column using hexanes–ethyl acetate as eluent to afford the pyrrolidinone derivative.

demonstrated that benzimidazole derivatives can be synthesized by the coupling of 1,2-phenylenediamine with aldehydes using commercially available Co(OH)$_2$ or CoO as an efficient recyclable catalyst under an open oxygen atmosphere at room temperature using ethanol as solvent.[40] Benzimidazoles could be synthesized successfully using these catalysts with acid-sensitive, sterically hindered and substituted aromatic and aliphatic aldehydes. The reactions were performed in ethanol and the catalyst could be reused for several cycles without much decrease in activity. The salient features of the method include mild conditions, short reaction times, high yields, recyclable catalyst, large-scale synthesis and simple procedure. Sharghi *et al.* found that a cobalt–salen complex supported on activated carbon [Co$_x$O–Co(salen)] could catalyze the condensation between phenylenediamine and aldehydes at room temperature.[41] The heterogeneous catalyst was characterized by powder X-ray diffraction (XRD), scanning electron microscopy (SEM), atomic force microscopy (AFM), thermogravimetric (TG) methods for analysis of nitrogen adsorption and Fourier transform infrared (FT-IR) spectroscopy. Leaching experiments showed that the catalyst is most strongly anchored to the activated support after 10 cycles of reuse.

More recently, a novel and practical Co(OAc)$_2 \cdot$4H$_2$O-catalyzed direct insertion reaction of isocyanides into active N–H and O(S, NR)–H bonds was developed by Ji *et al.*[42] This method not only expands the scope of the cobalt-catalyzed cross-coupling reaction but also provides a general, reliable and diverse approach leading to valuable substituted 2-aminobenzimidazole, 2-aminobenzothiazole and 2-aminobenzoxazole frameworks (Scheme 5.20).

Alternatively, a cobalt(II) complex-catalyzed intramolecular C–N and C–O cross-couplings of (Z)-N'-(2-halophenyl)-N-phenylamidines and N-(2-bromophenyl)benzamides to afford the corresponding substituted benzimidazoles and benzoxazoles in the presence of K$_2$CO$_3$ at moderate temperature was reported in 2010.[43] The protocol is general, air stable and affords the products selectively in moderate to high yields (Scheme 5.21). Regarding the reaction mechanism, the coordination of cobalt to the substrate was proposed as the first step. The final products were eliminated after oxidative addition and reductive elimination.

In 2013, a cobalt-catalyzed oxidative [3 + 2] cycloaddition cascade of dihydroisoquinoline esters with nitroalkenes or N-sulfuryl aldimines was developed by Wang and co-workers.[44] As a multi-component reaction for the synthesis of 5,6-dihydroimidazo[2,1-a]isoquinolines, the reaction works under almost identical conditions at room temperature and gave the desired products in good yields (Scheme 5.22). This method is particularly suitable for the synthesis of tricyclic nitrogen heterocycles owing to its simple manipulation, wide scope of reaction substrates and excellent regioselectivity. On the basis of mechanistic studies, a reaction pathway was proposed. First, dihydroisoquinoline ester was oxidized to the cation radical through a cobalt-catalyzed single-electron transfer process, which was further oxidized to give the iminium ion. Under alkaline conditions, the iminium ion

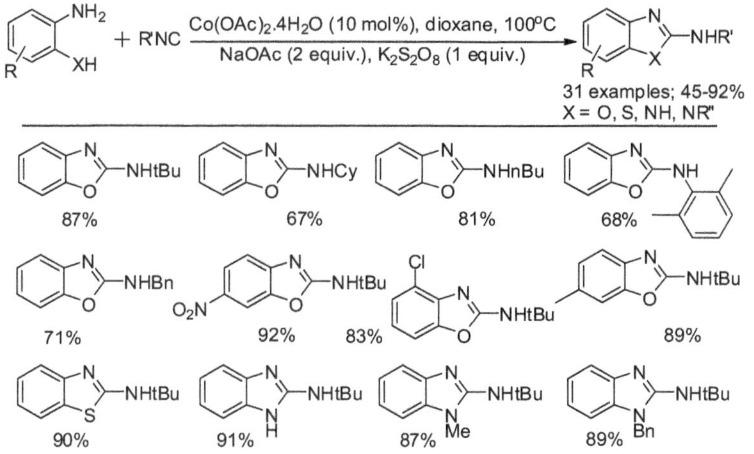

Scheme 5.20 Cobalt-catalyzed oxidative insertion of isocyanides.

Reaction Procedure (Scheme 5.20): 1,4-Dioxane or DMF (3 mL) was added to a mixture of amino compound (0.5 mmol), isocyanide (0.6 mmol), Co(OAc)$_2 \cdot$4H$_2$O (10 mol%), NaOAc (2 equiv.) and K$_2$S$_2$O$_8$ (1 equiv.). The mixture was stirred by magnetic stirrer for an appropriate time at 100 °C until the amino compound was completely consumed (monitored by TLC analysis), then the solvent was evaporated under reduced pressure. The residue was purified by flash column chromatography with ethyl acetate–petroleum ether as the eluent to afford the pure product.

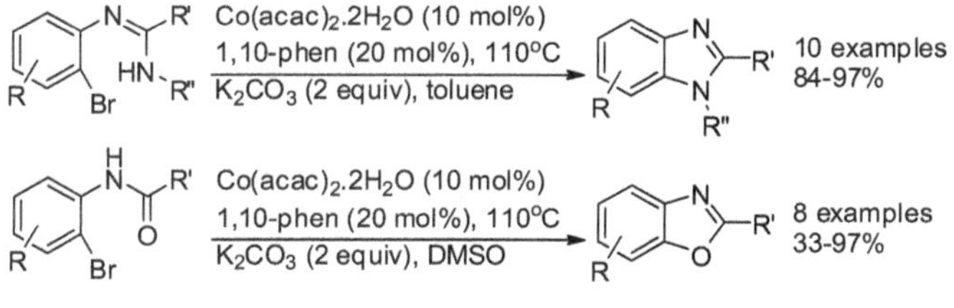

Scheme 5.21 Cobalt-catalyzed synthesis of benzimidazoles and benzoxazoles *via* coupling reactions.

Reaction Procedure (Scheme 5.21): **General Procedure for Benzimidazole Synthesis:** 2-Bromoarylamidine (0.5 mmol), Co(acac)$_2 \cdot$2H$_2$O (10 mol%, 14.7 mg), 1,10-phenanthroline (20 mol%, 19.8 mg) and K$_2$CO$_3$ (1.0 mmol, 138.0 mg) were stirred in toluene (1.0 mL) at 110 °C under air. The progress of the reaction was monitored by TLC using ethyl acetate–hexane as eluent. The reaction mixture was cooled to room temperature and passed through a short pad of silica gel using ethyl acetate–hexane as eluent to provide benzimidazoles.

General Procedure for Benzoxazole Synthesis: 2-Bromoarylbenzamide (0.5 mmol), Co(acac)$_2 \cdot$2H$_2$O (10 mol%, 14.7 mg), 1,10-phenanthroline (20 mol%, 19.8 mg) and K$_2$CO$_3$ (1.0 mmol, 138.0 mg) were stirred in DMSO (1.0 mL) at 110 °C under air. The progress of the reaction was monitored by TLC using ethyl acetate–hexane as eluent. The reaction mixture was cooled to room temperature and diluted with ethyl acetate (20 mL). The organic layer was washed successively with brine (1 × 5 mL) and water (2 × 5 mL). Drying and evaporation of the solvent provided a residue that was purified by silica gel chromatography using ethyl acetate–hexane as eluent.

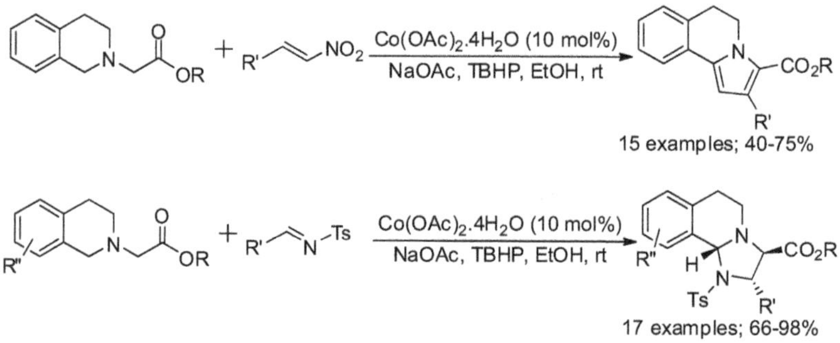

Scheme 5.22   Cobalt-catalyzed oxidative cyclization reactions.

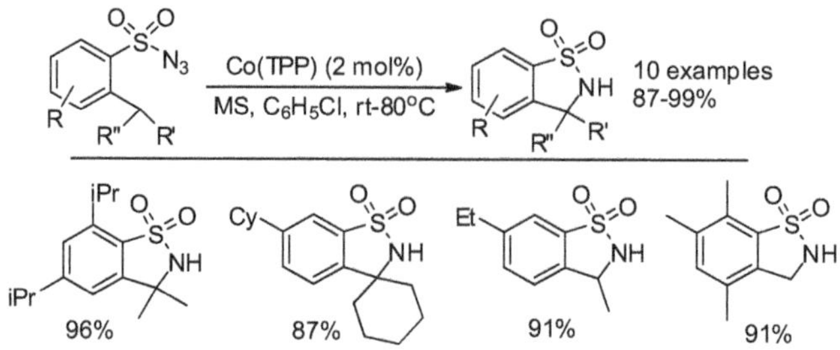

Scheme 5.23   Cobalt-catalyzed cyclization of arylsulfonyl azides.

can be deprotonated to yield an isoquinolinium ylide. Subsequently, 1,3-dipolar cycloaddition between the isoquinolinium ylide and dipolarophiles affords the hexahydropyrrolo[2,1-*a*]isoquinolines or hexahydroimidazo[2,1-*a*] isoquinolines.

In addition, in 2007, Zhang and co-workers developed the first Co-based catalytic system for intramolecular C–H amination with azides.[45] As they demonstrated, the commercially available Co(TPP) is an effective and general catalyst for catalyzing intramolecular C–H amination with arylsulfonyl azides, leading to valuable benzosultam derivatives in excellent yields (Scheme 5.23). In addition to primary, secondary and tertiary benzylic C–H bonds, non-benzylic C–H bonds can also be intramolecularly aminated. Later, they reported a [Co(II)(Por)]-based catalytic system for the highly effective intramolecular C–H amination of phosphoryl azides, which produced a wide range of cyclophosphoramidates in high yields with nitrogen gas as the only by-product.[46] In addition to the neutral and non-oxidative conditions, this new catalytic system is highlighted by features such as amination of primary C–H bonds and formation of seven-membered ring structures.

5.2 Six-Membered and Other Heterocycles

Pyridines have been found widely distributed in natural and biologically active compounds and also act as important reagents in organic and organometallic reactions. With this background, many useful procedures have been developed for their preparation. Among them, the cycloaddition of alkynes with nitriles offers a straightforward procedure for the synthesis of substituted pyridines and reactions have been reported with Co, Fe, Ni, Rh and Ru complexes. Cobalt was applied in this context as early as 1973.[47] In the presence of acetylene and acetonitrile, using π-cyclopentadienyl-(triphenylphosphine)cobalt-tetraphenylcyclopentadiene as the catalyst in benzene, the corresponding 2-picolines can be produced in good yields. The reaction temperature can be decreased by using a cobalt–diphenylacetylene complex as the catalyst. Later, a stoichiometric study was performed in order to evaluate the reaction mechanism. Pyridines were successfully prepared from nitriles and acetylenes *via* caboltacyclopentadienes.[48]

In 1977, Vollhardt's group developed the cobalt-catalyzed synthesis of annelated pyridines, and using diynes and nitriles as substrates, the desired products were formed in refluxing *o*-xylene in low to good yields (Scheme 5.24a).[49] Later, they showed that annelated pyridines can be prepared by cyclization of α,ω-cyanoalkynes with alkynes as well (Scheme 5.24b).[50] This methodology was applied in the total synthesis of vitamin B$_6$.[51,52] In 1983, they succeeded in applying η^5-cyclopentadienyl-dicarbonylcobalt in the cyclization of 1,2-bis(propargyl)-1,2,3,4-tetra-hydroisoquinolines with alkynes.[53] This procedure provides a novel synthetic entry into various tetrahydroprotoberberine nuclei. By judicious

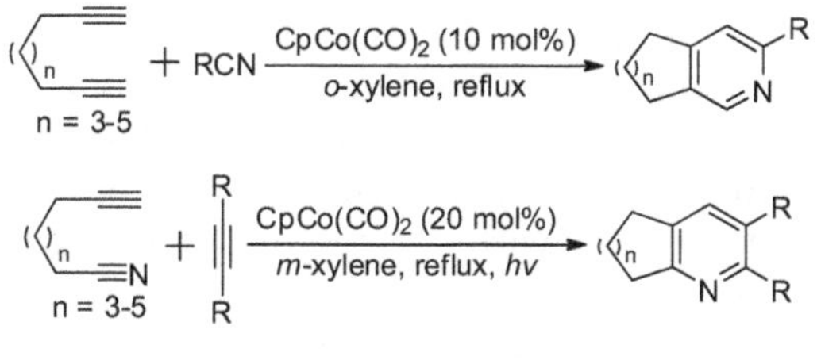

Scheme 5.24 Cobalt-catalyzed cyclization of alkynes with nitriles.

choice of trimethylsilyl substituents, regiocontrol in the D-ring can be achieved.

The correlation between the ^{13}C and ^{59}Co NMR spectra of substituted cyclopentadienylcobalt complexes and their catalytic properties in the synthesis of pyridine derivatives was examined in 1984.[54] Since the correlations can be expressed as linear relationships, it was proved that the direct screening of potential catalysts by NMR is possible. In that study, the effect of the ligand was found have a strong influence on the selectivity. Following this study, it was found that the cyclization of acetylene with nitriles could be realized at room temperature in the presence of CpCo(COD) as catalyst and light irradiation.[55–57]

Heller and co-workers found that the selectivity of the photochemical synthesis of pyridines based on alkynes (*e.g.* ethyne, propyne, butyne and dimethylbutyne) and nitriles (*e.g.* benzo-, aceto- and trimethoxybenzonitrile, *n*-butyl cyanoacetate and *tert*-butyl cyanide) with cobalt(I) complexes [YCo(COD)] (Y = η^5-cyclopentadienyl, η^5-indenyl, η^5-tetraphenylcyclopentadienyl, η^5-acetylcyclopentadienyl, η^3-cyclooctenyl, η^6-1-phenylborinato; COD = cycloocta-1,5-diene) can be controlled.[58] Thus, very high chemoselectivities were obtained at low alkyne concentrations. For gaseous reactants, the appropriate concentrations can be adjusted by choosing the correct partial pressures. Advantage can also be taken of the temperature-dependent solubility of gases in different solvents. In water, for example, the product distribution can be shifted in favor of the formation of pyridines. The regioselectivity can be advantageously influenced by the properties of the ligands in the precatalyst and by the application of bulky nitriles and alkynes. Instead of artificial light, sunlight can also be employed as a radiation source without loss of selectivity. Vitamin B_6 synthesis was also reported.[59] The key step involves the light-promoted [2 + 2 + 2] cyclization of 3,3-bissilyl di-2-propynyl ethers and acetonitrile in the presence of [CpCo(COD)] (1 mol%) as catalyst. Oxidation and iodination procedures were elaborated to convert 3-silylpyridines into corresponding 3-hydroxy and 3-iodo derivatives.

Interestingly, an asymmetric version was developed by Gutnov *et al.* in 2004.[60] This is the first asymmetric approach to biaryl atropoisomers by the [2 + 2 + 2] cycloaddition of alkynes and nitriles using chiral cobalt(I) complexes as the catalyst. Homochiral 2-arylpyridines were successfully prepared (Scheme 5.25). Notably, a diastereomeric cobaltacyclopentadiene complex was also isolated; the corresponding pyridine can be formed by mixing this intermediate with nitrile.

Later, Hapke *et al.* performed a systematic study on this topic.[61] A range of axially chiral 2-arylpyridines were produced by cobalt-catalyzed [2 + 2 + 2] cycloaddition reactions. The use of a planar chiral (1-neomenthylindenyl)cobalt(COD) complex under photochemical conditions was the key for reacting the 1-naphthyldiynes with a range of differently functionalized nitriles, giving the enantiomeric atropoisomers with high chemical yields and up to 94% *ee*. In 2013, the same group reported their achievements in the preparation of cobalt complexes.[62,63] They developed new and efficient

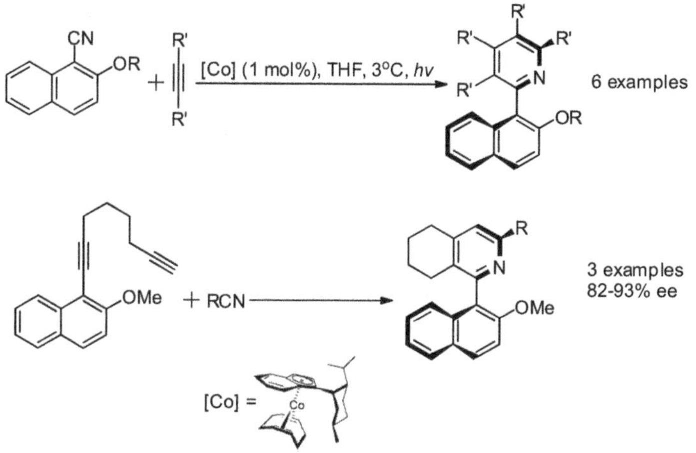

Scheme 5.25 Cobalt-catalyzed asymmetric cyclization of alkynes with nitriles.

Reaction Procedure (Scheme 5.25): A thermostated (3 °C) reaction vessel was loaded with diyne (524 mg, 2 mmol), catalyst (8.4 mg, 0.02 mmol), THF (20 mL) and benzonitrile (412 mL, 4 mmol) under an argon atmosphere. The mixture was stirred and irradiated with two 460 W lamps (420 nm) for 24 h. The reaction was quenched by switching off the lamps and simultaneously letting in air. The extent of reaction, *i.e.* conversion of the starting diyne, was determined by GC. The reaction mixture was filtered through a thin pad of silica gel with THF as eluent. The solvent was removed *in vacuo* to give an oily residue, which was further dissolved in Et$_2$O (10 mL). Colorless crystals of the product (413 mg, 57% yield) were filtered off and washed with Et$_2$O. The optical purity was determined as >98% *ee* by HPLC.

synthetic routes to a large number of novel CpCo^I complexes containing either alkene–phosphite ligand combinations ([CpCo(H$_2$C=CHSiMe$_3$)-{P(OR)$_3$}]) or two phosphite ligands ([CpCo{P(OR)$_3$}$_2$]) by selective substitution from [CpCo(H$_2$C=CHSiMe$_3$)$_2$] or by photochemical exchange reactions from [CpCo(COD)], respectively. The screening of the catalytic activity in [2 + 2 + 2] cycloadditions showed that [CpCo{(OEt)$_3$}$_2$] and [CpCo{(O*i*Pr)$_3$}$_2$] are the most reactive bisphosphite complexes at a reaction temperature as mild as 50 °C. On the other hand, the mixed CpCo^I–alkene-phosphite complexes exhibited much higher catalytic activities but were also stable compounds at room temperature. This highly beneficial combination of stability and high catalytic activity was proven by comparison of the reaction temperatures for [CpCo(H$_2$C=CHSiMe$_3$)$_2$], [CpCo(H$_2$C=CHSiMe$_3$){(OPh)$_3$}] and [CpCo{(OPh)$_3$}$_2$]. Computational results showed that in the mixed complexes [CpCo(H$_2$C=CHSiMe$_3$){P(OR)$_3$}], the dissociation of the phosphate ligand is preferred to that of the alkene ligand. Also, the reactivity and stability differences between [CpCo(H₂C=CHSiMe₃)₂] and [CpCo{P(OR)₃}₂] can be demonstrated. The synthetic scope of pyridine synthesis from diynes and nitriles was investigated for [CpCo(H₂C=CHSiMe₃){(OPh)₃}] in both intramolecular and intermolecular reactions, demonstrating the high utility of this new precatalyst. The concept of using two different ligands for complexes of the type [CpCo(L¹)(L²)] to combine the properties of the respective complexes [CpCo(L¹)₂] and [CpCo(L²)₂] proved to be very useful for the development of new, reactive and stable CpCoᴵ precatalysts.

Saa's group developed a short alternative for the synthesis of annelated 3-substituted bipyridines and symmetric 3,3″-substituted terpyridines in 1997 (Scheme 5.26a).[64] These compounds are interesting building blocks for the assembly of supramolecular coordination compounds. Later, they described a new, one-step method for synthesizing annelated symmetric 3,3′-substituted 2,2′-bipyridines from acyclic precursors (Scheme 5.26b).[65] This method relies on the regioselective Co(I)-catalyzed co-cyclization between 5-hexynenitrile and 1,3-diynes and reverses the usual strategy in bipyridine synthesis, the future biaryl bond being present prior to the construction of either of the two aryl rings. Spiropyridines, a novel series of C₂-symmetric ligands, were also prepared (Scheme 5.26c).[66] By Co(I)-catalyzed double co-cyclization between bisalkynenitriles and alkynes, the corresponding products were formed in moderate yields. This one-step method allowed the first synthesis of the interesting 7,7′- and 8,8′-spyropyridines from acyclic precursors.

Subsequently, many new cobalt complexes were prepared and applied in alkyne–nitrile cyclotrimerization reactions. Fatland and Eaton prepared a water-soluble cobalt(I) catalyst and used it in the preparation of highly functionalized pyridines in aqueous solution (Scheme 5.27).[67] Ward's group

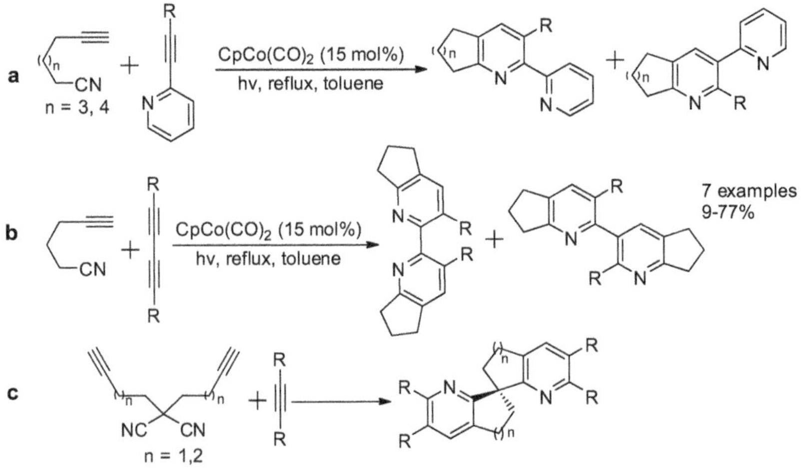

Scheme 5.26 Cobalt-catalyzed synthesis of bipyridines.

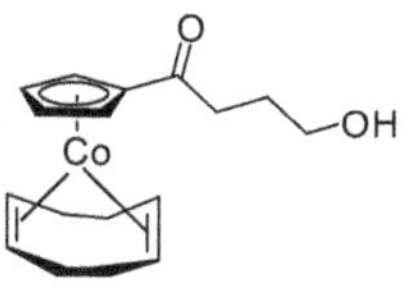

Scheme 5.27   Water soluble cobalt catalyst.

reported the combinatorial synthesis of substituted pyridines in the solution phase.[68] They used $[CpCo(CO)_2]$ as a commercially available catalyst; amine-*N*-oxide was needed as a CO scavenger to activate the catalyst. This idea was later applied by Deiters and co-workers, who reported the first solid-support $[2 + 2 + 2]$ cyclotrimerizative synthesis of pyridines.[69] Tetra-methylammonium oxide was used as the catalyst-activating additive. The desired pyridines were formed in good yields and excellent purities. Maryanoff and co-workers investigated the formation of macrocycles from α,π-diynes in cobalt-mediated co-cyclotrimerization reactions.[70,71] Long-chain α,π-diynes underwent metal-mediated $[2 + 2 + 2]$ cycloadditions with nitriles, cyanamides or isocyanates in the presence of $CpCo(CO)_2$ to yield pyridine-containing macrocycles, *i.e. m-* and *p*-pyridinophanes. The regio-selectivity of these reactions was affected by the length and type of linker unit between the alkyne groups, and also by certain stereoelectronic factors. An analogous α,π-cyanoalkyne combined with an alkyne to yield two isomeric *m*-pyridinophanes. By using this improved protocol, they were able to co-cyclotrimerize long-chain α,π-diynes with alkynes in certain cases to demonstrate a successful macrocyclic variant of the Vollhardt reaction. They succeeded in extending their methodology to diynes and cyanamides, and multi-substituted 2-aminopyridines were formed in low to good yields.[72] $CpCo(CO)_2$ was also applied as a catalyst in the cyclization of various substrates, such as nitrilediynes[73,74] and diynes with benzonitriles,[75–77] which was used as a key step in the total synthesis of (+)-complanadine A and (±)-strychnine,[78] silane-containing compounds,[79,80] and so on. Re-placing thermal heating with microwave heating was found to increase the reaction efficiency and shorten the reaction time.[81,82]

In 2009, the first $[2 + 2 + 2]$ co-cyclizations between ynamides, nitriles and alkynes were reported by Aubert, Malacria and co-workers.[83] They opened up a new access route to unprecedented nitrogen-containing heterocycles of the type 2-trimethylsilyl-3-aminopyridines. Using $[CpCo(C_2H_4)_2]$ (Cp = cyclo-pentadienyl) as the catalyst, intramolecular cyclizations could be achieved in up to 100% yield. The presence of the trimethylsilyl group allowed a rare type of Hiyama cross-coupling: one of the silylated pyridines could be coupled with *p*-iodoanisole to give a new type of biaryl system. Later, they together with Gandon's group performed a comprehensive study of the cobalt-catalyzed $[2 + 2 + 2]$ cycloaddition between yne-ynamides and nitriles to afford aminopyridines.[84,85] About 30 nitriles exhibiting a broad range of steric demand and electronic properties were evaluated, some of which opened up new perspectives in metal-catalyzed arene formation

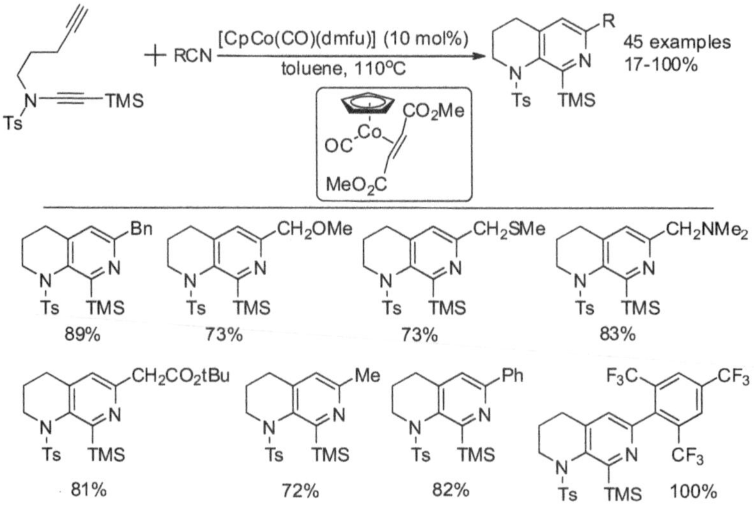

Scheme 5.28 [CpCo(CO)(dmfu)]-catalyzed synthesis of pyridines.

(Scheme 5.28). In particular, the use of [CpCo(CO)(dmfu)] (dmfu = dimethyl fumarate) as a precatalyst made possible the incorporation of electron-deficient nitriles into the pyridine core. Modification of the substitution pattern at the yne-ynamide allows the regioselectivity to be switched toward 3- or 4-aminopyridines. Application of this synthetic methodology to the construction of the aminopyridone framework using an yne-ynamide and an isocyanate was also briefly examined. DFT computations suggested that 3-aminopyridines are formed by formal [4 + 2] cycloaddition between the nitrile and the intermediate cobaltacyclopentadiene, whereas 4-amino-pyridines arise from an insertion pathway.

In addition to the above-mentioned Co(I) complexes, more air-stable Co(II) complexes were also applied in co-cyclotrimerization reactions. Cheng's group demonstrated the first intramolecular [2 + 2 + 2] co-cyclo-trimerization of nitrilediynes.[86] Co(dppe)I$_2$–Zn was used as the catalyst system and polycyclic pyridine derivatives were produced at relatively low temperature in a highly atom-economical manner (Scheme 5.29a). The presence of a bulkier substituent at the terminal alkyne carbon of the nitrilediynes was essential for this catalytic intramolecular [2 + 2 + 2] co-cyclotrimerization to proceed smoothly. The cobalt system appears to be an alternative to the widely used CoCp(CO)$_2$ as the catalyst for [2 + 2 + 2] cycloaddition. Around the same time, Okamoto and co-workers reported another CoCl$_2$-based system.[87–89] A variety of substituted 2,2'-bipyridines were synthesized by a 1,2-bis(diphenylphosphino)ethane (dppe)–cobalt chloride hexahydrate (CoCl$_2 \cdot$ 6H$_2$O)–zinc-catalyzed [2 + 2 + 2] cycloaddition reaction of diynes and nitriles, with all reactions exhibiting exclusive regioselectivity (Scheme 5.29b). Thus, symmetrical and unsymmetrical 1,6-diynes and 2-cyanopyridine reacted in the presence of 5 mol% of dppe,

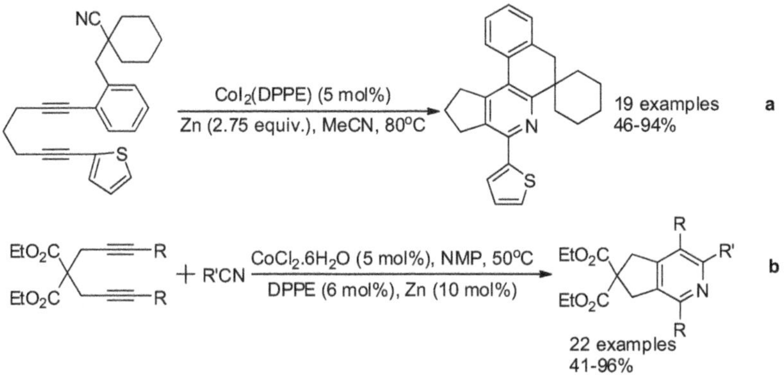

Scheme 5.29 Co(II)-catalyzed synthesis of pyridines.

Reaction Procedure (Scheme 5.29): A 25 mL round-bottomed side-arm flask containing CoI_2(dppe) (0.005 mmol) and Zn (2.75 mmol) was evacuated and purged three times with nitrogen gas. To the flask were then added nitrilediyne (1.00 mmol) and CH_3CN (3.0 mL) *via* syringes. The reaction mixture was stirred at 80 °C for 12 h. At the end of the reaction, the reaction mixture was diluted with CH_2Cl_2, filtered through Celite and the filtrate was concentrated. The crude residue was purified through a silica gel column using hexanes–ethyl acetate as eluent to give pure product.

Reaction Procedure: To a stirred mixture of zinc powder (3.5 mg, 0.05 mmol), diyne or tetrayne (0.5 mmol) and nitrile (1.5–80 equiv.) in NMP (1 mL) was added a solution of $CoCl_2 \cdot 6H_2O$ (6 mg, 0.025 mmol) and dppe (12 mg, 0.03 mmol) in NMP (1 mL) at room temperature. The mixture was then stirred at room temperature or at 50 °C. The reaction progress was monitored by TLC. After completion of the reaction, a small portion of EtOAc or Et_2O was added and the mixture was passed through a pad of Celite with EtOAc or Et_2O as eluent. The filtrate was concentrated to dryness and the residue was chromatographed on silica gel using hexane– EtOAc as eluent to give the corresponding bipyridine derivative.

---

5 mol% of $CoCl_2 \cdot 6H_2O$ and 10 mol% of zinc powder to provide the corresponding 2,2'-bipyridines. Under identical reaction conditions, 1-(2-pyridyl)-1,6-diynes and nitriles reacted smoothly with exclusive regioselectivity to produce 2,2'-bipyridines in good yield. 2,2'-Bipyridines were also obtained by the double $[2 + 2 + 2]$ cycloaddition reaction of 1,6,8,13-tetraynes with nitriles. Similarly, 2,2':6',2''-terpyridines were synthesized from a 1-(2-pyridyl)-1,6-diyne and 2-cyanopyridine. The regiochemistry observed can be explained by considering the electronic nature of cobaltacyclopentadiene intermediates and nitriles. A survey of the exclusive regiochemical trend gives reasonable credence to the synthetic potential of the method.

In order to evaluate the reaction mechanism, computational studies were carried out by different groups.[90–93] Koga and co-workers used B3LYP-level calculations to study the reaction of acetonitrile with a cobaltacyclopentadiene complex in singlet and triplet electronic states. Whereas the most favorable path for the singlet reaction passes through a [4 + 2] cycloaddition transition state, the most favorable triplet reaction follows the reaction mechanism through azacobaltacycloheptatriene. Since the transition state for the singlet reaction path is lower in energy than those for the triplet reaction path, the singlet reaction seems to be more favorable. However, the reactant in the triplet state is more stable than that in the singlet state, suggesting that the two-state reactivity (TSR) mechanism with spin changes is followed. Determination of energy minimum crossing points between singlet and triplet energy surfaces led to the conclusion that the TSR mechanism is more favorable than both the singlet- and triplet-state (single-state reactivity) (SSR) mechanisms. Comparison of a reaction profile between trimerization of acetylene and co-cyclotrimerization of acetonitrile with two acetylene molecules in the presence of the CpCo catalyst was also made. Additionally, the beneficial role of microwave activation was studied. It was found that microwave irradiation can decrease the catalytic induction period through thermal effects and can also increase the triplet lifetime and promote the reaction, thus improving the final yield.

In 2002, Jones and co-workers reported a simple procedure for the synthesis of quinolines from diallylanilines.[94] The raw materials are cheap and the catalyst is readily available. By using dicobaltoctacarbonyl as the catalyst, diallylanilines were converted into quinolines in moderate yields (Scheme 5.30). Arylimines are also found to undergo heteroannulation in the presence of diallylaniline as an allyl fragment donor to give quinolines. Imines can also be allylated to give quinoline derivatives.

As early as 1989, Gleiter and Schehlmann reported the synthesis of CpCo-complexed α-pyrans.[95] Treatment of 5-cyclononynone, 5-cyclodecynone, dodecanal-1,7-diyne and tridecanal-1,7-diyne with $CpCo(CO)_2$ or CpCo(COD) produced the corresponding CpCo-complexed α-pyrans in moderate yields. Mechanistically, the reactions proceed *via* an *in situ*-formed diene as the intermediate, followed by reaction with aldehyde. Nowadays, these types of transformation are called hetero-Diels–Alder reactions.

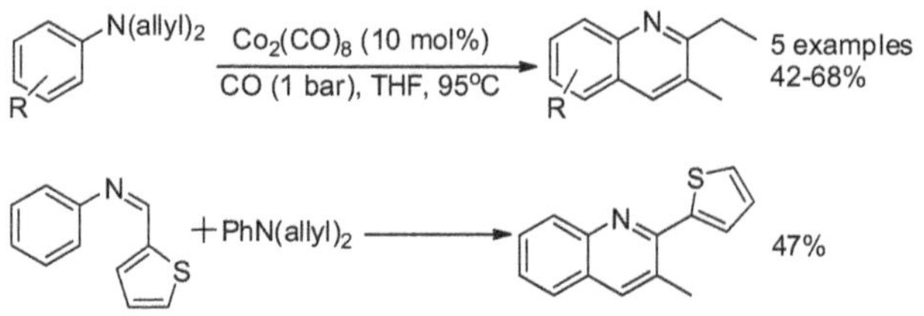

Scheme 5.30 $Co_2(CO)_8$-catalyzed synthesis of quinolines.

In 1998, Wu and co-workers performed a formal total synthesis of 3-deoxy-D-manno-2-octulosonic acid (KDO).[96] They used a (salen)Co(II) complex as the catalyst, and an electron-rich diene and ethyl glyoxylate underwent a hetero Diels–Alder reaction in a highly double-stereoselective manner. A facial-specific hydroboration followed by oxidative workup led to a diol system with a transdiequatorial arrangement of hydroxyl groups at the C4 and C5 positions. Inversion of the configuration of the C5-hydroxyl group in intermediate and then ketal formation afforded the desired target diisopropylidene-2-deoxy-KDO methyl ester, which could be converted to KDO. They examined some achiral dienes and other salen-derived catalysts. Jurczak and co-workers reported the discovery of a high-pressure (10–11 kbar) reaction of 1-methoxybuta-1,3-diene with *tert*-butyldimethylsilyl-oxyacetaldehyde, using chiral (salen)Co(II) or (salen)Cr(III)Cl complexes as catalyst. Good yields (up to 90%) with very good diastereoselectivity (up to 92%) and enantioselectivity (up to 94% *ee*) could be achieved.[97–99]

Feng and co-workers found that the chiral compound ethyl 2-amino-5-oxo-4-aryl-5,6,7,8-tetrahydro-4*H*-chromene-3-carboxylate can be synthesized through a tandem Michael addition–cyclization process catalyzed by a salen–cobalt(II) complex (Scheme 5.31).[100] The corresponding products, which have extensive biological and pharmacological activities, can be obtained in moderate to good yields (up to 81%) with high enantioselectivities (up to 89% *ee*). This pathway is air tolerant and the catalyst is prepared easily with readily available reagents. A possible catalytic cycle was proposed to explain the formation of the products. The key step for the process involves

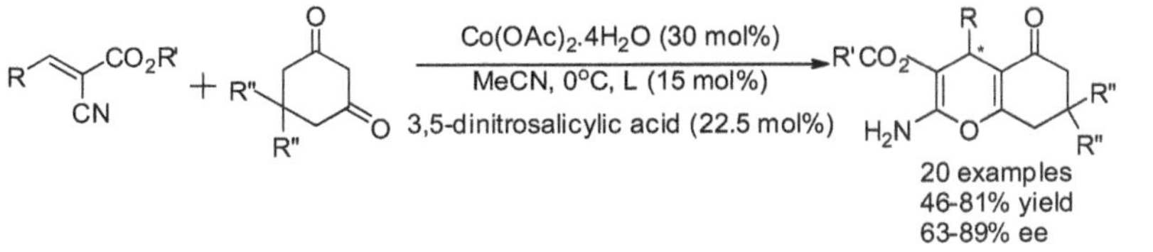

Scheme 5.31 Cobalt-catalyzed synthesis of chromenes.

Reaction Procedure (Scheme 5.31): Chiral salen ligand (9.7 mg, 0.015 mmol), Co(OAc)$_2$ · 4H$_2$O (7.5 mg, 0.03 mmol) and 3,5-dinitrosalicylic acid (5.1 mg, 0.0225 mmol) were placed in a test-tube, then MeCN (0.5 mL) was added. After stirring for 1 h at room temperature, the mixture was cooled to 0 °C. A solution of substrate 1 (22.1 mg, 0.11 mmol) and substrate 2 (11.2 mg, 0.10 mmol) in MeCN (0.5 mL) was added and the reaction mixture was stirred at 0 °C for 48 h, then diluted with ethyl acetate (40 mL). After extraction with 1 M NaOH (4 × 30 mL), the organic layers were combined and dried with sodium sulfate and evaporated under reduced pressure. The crude product was purified by column chromatography (eluent: 1:2 ethyl acetate–petroleum ether) to afford the product.

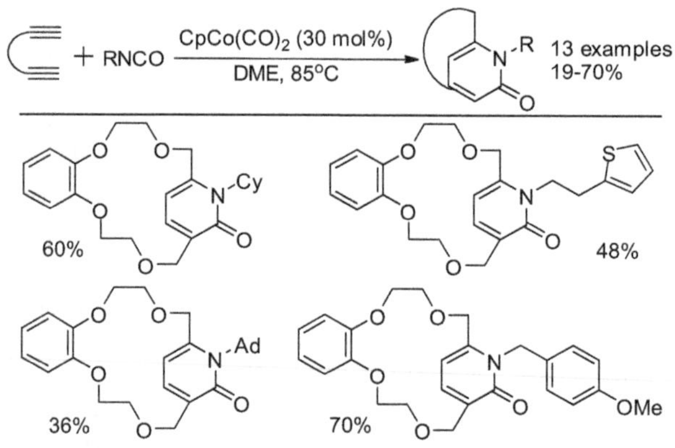

Scheme 5.32 Cobalt-catalyzed synthesis of pyridones.

deprotonation–reprotonation. The chiral salen–cobalt complex first abstracts a proton from the enol to form an intermediate, which then reacts with ethyl 2-cyano-3-phenylacrylate through a Michael addition process. Cyclization then occurs quickly so that no Michael reaction intermediate is detected. The final products were eliminated after reprotonation and tautomerization.

Maryanoff and co-workers reported the cobalt-catalyzed macrocyclization of α,ω-diynes and isocyanates in 2003.[101] This procedure offers a straight-forward approach to 2-oxopyridinophanes (Scheme 5.32). This reaction occurs more efficiently with aliphatic than with aromatic isocyanates. Contrary to standard protocols for CpCo(CO)$_2$ diyne cycloadditions, these macrocyclizations are conveniently carried out at reduced temperature (85 °C *versus* 140 °C), without irradiation or syringe-pump addition. In the [2 + 2 + 2] cycloaddition of 1,15-diynes, there was a strong predominance of the 3,6-disubstituted pyridine macrocycles (>20 : 1) relative to 4,6-dis-ubstitution. In the case of 1,17-diynes, although the 3,6-disubstituted pyri-dine macrocycles were the major isomer, the 4,6-disubstituted pyridones had a significant presence. It is noteworthy that these positive results were achieved for a very challenging reaction involving incorporation of an external reactive species in a bimolecular macrocyclization. More recently, Lv's group performed a theoretical study on this topic.[102] The two-state reaction (TSR) mechanism of CpCo(C$_4$H$_4$) with isocyanate on the triplet and singlet potential energy surfaces was investigated at the B3LYP level. The minimal energy crossing point (CPs) in the crossing seam between the potential energy surfaces were located using the method of Harvey *et al.* and possible spin inversion processes were considered by means of spin–orbit coupling (SOC) calculations. As a result, two distinct reaction pathways for the formation of a pyridin-2-one–cobalt complex were found. For the first pathway, there are two key crossing points along the reaction pathway. The first crossing point, CP2, exists near ^{1}B. The reacting system will change its spin multiplicity from the triplet state to the singlet state near this crossing region because the magnitude of the spin-multiplicity mixing ($SOC1_{\text{-e},1-\text{center}} = 393.37$ cm^{-1}) increases in a small energy gap between high- and low-spin states will greatly enhance the probability of intersystem crossing. The single (P_1^{ISC}) and double (P_2^{ISC}) passes estimated at CP2 are approximately 0.28 and 0.48, respectively. The second crossing point, CP3, will again change its spin multiplicity from the singlet state to the triplet state in the Co–C_γ bond activation pathway, leading to a decrease in the barrier height of ^{1}TS(CF) from 19.5 to 9.5 kcal mol^{-1}. Hence the reaction system will access a lower energy pathway and then move on the triplet potential energy surface as the reaction proceeds. As for the second pathway, the formation of the initial transition is a very unfavorable process kinetically. However, from the beginning of ^{1}H, TSR is a very favorable process kinetically and thermodynamically. After passing point CP5, the triplet potential energy surface can provide a low-cost reaction pathway toward the product complex.

Cobalt has also been applied in hydroformylation reactions. In 2011, Mulzer and Coates reported a catalytic domino reaction that efficiently provides access to ampakines.[103] This approach was based on the cobalt-catalyzed hydroformylation of dihydrooxazines and allows for the facile synthesis of the pharmaceutically interesting compound CX-614 and related substances. All the desired products were produced in good to excellent yields (Scheme 5.33).

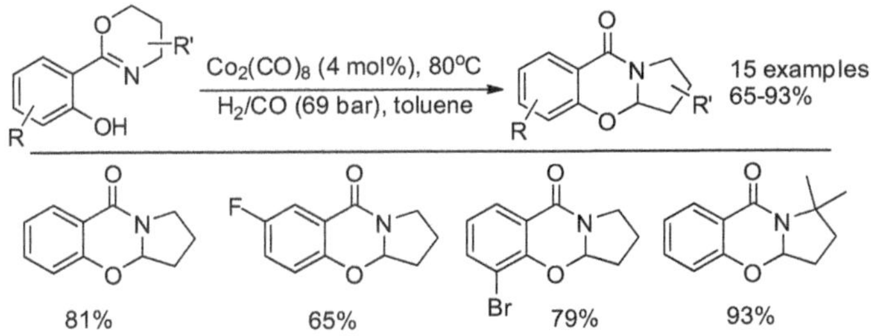

Scheme 5.33 Cobalt-catalyzed hydroformylations.

Reaction Procedure (Scheme 5.33): In a glove-box, vials equipped with Teflon-coated magnetic stirrer bars were charged with Co$_2$(CO)$_8$, the appropriate dihydrooxazine and toluene. The vials were then placed in a custom-made six-well high-pressure reactor which was subsequently closed, removed from the glove-box and pressurized with carbon monoxide (500 psi partial pressure) and hydrogen (500 psi partial pressure). The reactor was sealed and the reaction mixtures were stirred for 20 h at 80 °C. The reactor was then cooled to ambient temperature and carefully vented in a well-ventilated hood. The crude reaction mixtures were concentrated *in vacuo* and then purified by flash column chromatography.

More recently, Yamakawa and Yoshikai developed a cobalt-catalyzed annulation reaction of an α,β-unsaturated imine and an internal alkyne.[104] Dihydropyridine derivatives were produced in good yields under mild conditions (Scheme 5.34). Considering the reaction mechanism, initially a low-valent cobalt species was generated from the precatalysts and underwent nitrogen-assisted oxidative addition of the olefinic C–H bond to form a five-membered cobaltacycle intermediate. Subsequent migratory insertion of the alkyne was followed by reductive elimination to afford an azatriene intermediate, which readily underwent 6π electrocyclization to furnish the dihydropyridine product.

Additionally, a mild methodology to obtain thiopropenoylstannanes was established in 2004.[105] Reaction of stannylated allenes with bis-(trimethylsilyl) sulfide–CoCl$_2$ · 6H$_2$O afforded the first example of the synthesis

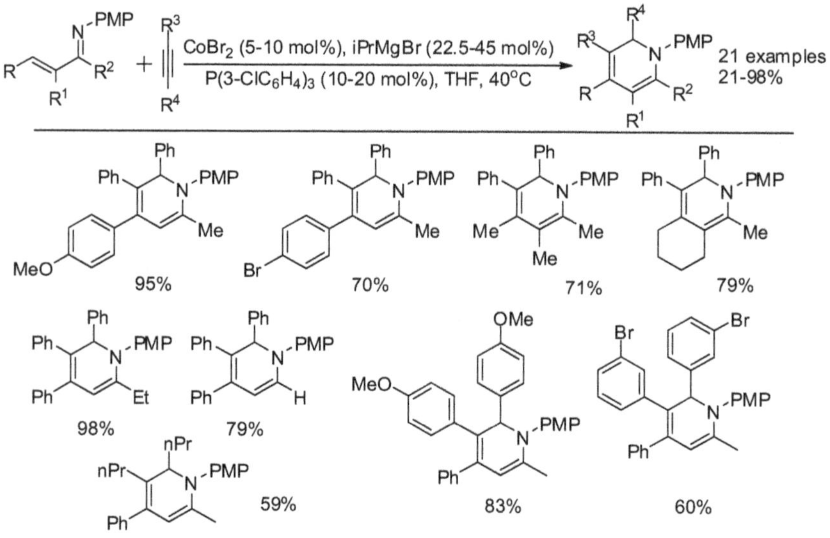

Scheme 5.34 Cobalt-catalyzed synthesis of dihydropyridines.

Reaction Procedure (Scheme 5.34): In a Schlenk tube equipped with a stirrer bar were placed α,β-unsaturated imine (0.20 mmol), alkyne (0.24 mmol), P(3-ClC$_6$H$_4$)$_3$ (7.3 mg, 0.020 mmol, 10 mol%), CoBr$_2$ (0.067 M solution in THF, 0.15 mL, 0.010 mmol, 5 mol%) and THF (0.47 mL). To the mixture was added iPrMgBr (0.89 M in THF, 51 μL, 0.045 mmol, 22.5 mol%) at room temperature. The resulting mixture was stirred at 40 °C for 3 h and then quenched with water. The aqueous layer was extracted with ethyl acetate (3 × 4 mL). The combined organic layer was dried over Na$_2$SO$_4$ and concentrated under reduced pressure. The crude product was purified by silica gel chromatography to afford the dihydropyridine derivative.

of thioacylstannanes. The so obtained thiopropenoylstannanes behave as efficient thiabutadienes towards different *in situ*-generated thioaldehydes and thioacylsilanes, leading to 2-substituted 4-stannyl-1,3-dithiacyclohex-4-ene derivatives through a regioselective hetero Diels–Alder reaction.

Moreover, cobalt catalysts have also been applied in the synthesis of four-membered heterocycles. In 1999, Alper and co-workers established a dicobalt octacarbonyl-catalyzed carbonylation of alkylaziridines including 2-ethoxycarbonyl- and silylated 2-hydroxymethyl-3-methylaziridines.[106] *N*-Benzyl-TBDMSO-protected hydroxymethylaziridines were successfully carbonylated to β-lactams using $Co_2(CO)_8$ as the catalyst: *trans*-β-lactams were obtained from a *cis*-aziridine in quantitative yields, whereas *cis*-β-lactams were isolated using a *trans*-aziridine. The reaction proceeds through nucleophilic ring opening of the aziridine by the *in situ*-generated tetra-carbonylcobaltate anion and shows high regioselectivity, with preferential CO insertion into the alkyl-bearing ring carbon–nitrogen bond rather than into the *O*-protected hydroxymethyl-bearing ring carbon–nitrogen bond. Later, they reported a cobalt-catalyzed carbonylation of epoxides.[107] In the presence of $PPNCo(CO)_4$ and $BF_3 \cdot Et_2O$, both simple and functionalized epoxides were carbonylated in DME and gave the corresponding β-lactones regioselectively in good to high yields. The carbonylation occurred selectively at the unsubstituted C–O bond of the epoxide ring and the reaction tolerates various functional groups such as alkenyl, halide, hydroxy and alkyl ether.

Coates and co-workers developed a highly active and selective porphyrin-based epoxide carbonylation catalyst for the synthesis of β-lactones.[108–111] $[(OEP)Cr(THF)_2][Co(CO)_4]$ (OEP = octaethylporphyrinato) as a separated ion pair composed of a tetracarbonylcobaltate anion and an octahedral chromium porphyrin complex axially ligated by two THF ligands was used as the bimetallic catalyst. Regarding the carbonylation of epoxides to β-lactones, the catalyst exhibited excellent turnover numbers (up to 10 000) and turn-over frequencies (up to 1670 h^{-1}), with regioselective carbonyl insertion occurring between the oxygen and the sterically less hindered carbon of the epoxide substrate. This catalyst is highly tolerant of non-protic functional groups, carbonylating an array of aliphatic and cycloaliphatic epoxides, and also epoxides with pendant ethers, esters and amides. With careful control of reaction conditions in the carbonylation of glycidyl esters, the exclusive production of either the β- or γ-lactone isomer was achieved. Through analysis of the reaction stereochemistry, a mechanism for the formation of γ-lactone products was proposed. Overall, a broad array of synthetically useful lactones were synthesized in a rapid and selective fashion by catalytic carbonylation using $[(OEP)Cr(THF)_2][Co(CO)_4]$. Later, two new complexes were prepared and applied in the regioselective carbonylation of *trans*-disubstituted epoxides to *cis*-β-lactones. In 2011, Ibrahim and co-workers reported a highly active catalytic system for the carbonylation of *meso*- and terminal epoxides to β-lactones (Scheme 5.35).[112] The active catalyst, analogous to Coates' catalyst, was generated *in situ* from commercially available (TPP)CrCl and $Co_2(CO)_8$. This practical system circumvents the

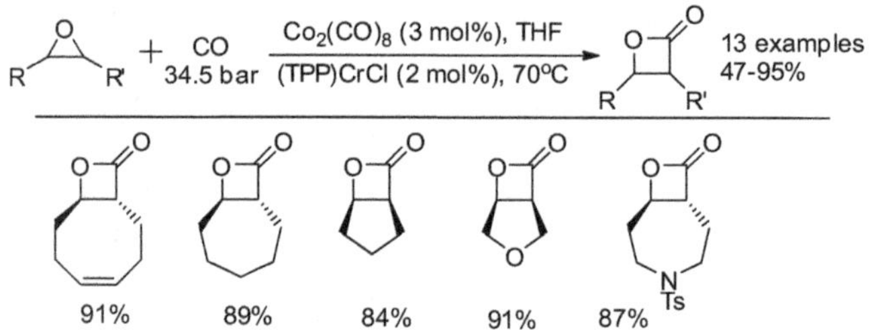

Scheme 5.35   Cobalt-catalyzed synthesis of β-lactones.

preparation of air-sensitive cobaltate salts, operates at low catalyst loadings and allows the carbonylation of functionalized, sterically demanding and heterocyclic *meso*-epoxides

5.3   Summary

The main contributions to the cobalt-catalyzed synthesis of heterocycles have been summarized and discussed. The main applications of cobalt catalysts in organic synthesis are in $[2+2+2]$, $[4+2]$ and $[2+2+1]$ cycloaddition reactions. The effects of ligands on cobalt catalysts are non-significant. More efforts are still needed to improve the catalyst efficiency.

References

1. H. Bhandal, G. Pattenden and J. J. Russell, *Tetrahedron Lett.*, 1986, **27**, 2299–2302.
2. S. Inoki and T. Mukaiyama, *Chem. Lett.*, 1990, 67–70.
3. B. M. Perez, D. Schuch and J. Hartung, *Org. Biomol. Chem.*, 2008, **6**, 3532–3541.
4. B. M. Perez and J. Hartung, *Tetrahedron Lett.*, 2009, **50**, 960–962.
5. D. Schuch, P. Fries, M. Doenges, B. M. Perez and J. Hartung, *J. Am. Chem. Soc.*, 2009, **131**, 12918–12920.
6. Z. M. Wang, S. K. Tian and M. Shi, *Tetrahedron: Asymmetry*, 1999, **10**, 667–670.
7. Z. M. Wang, S. K. Tian and M. Shi, *Tetrahedron Lett.*, 1999, **40**, 977–980.
8. Z. M. Wang, S. K. Tian and M. Shi, *Eur. J. Org. Chem.*, 2000, 349–356.
9. P. Fries, D. Halter, A. Kleinschek and J. Hartung, *J. Am. Chem. Soc.*, 2011, **133**, 3906–3912.
10. C. Palmer, N. A. Morra, A. C. Stevens, B. Bajtos, B. P. Machin and B. L. Pagenkopf, *Org. Lett.*, 2009, **11**, 5614–5617.
11. A. Kubas, J. Hartung and K. Fink, *Dalton Trans.*, 2011, **40**, 11289–11295.
12. (a) M. Okabe, M. Abe and M. Tada, *J. Org. Chem.*, 1982, **47**, 1775–1777; (b) M. Okabe and M. Tada, *J. Org. Chem.*, 1982, **47**, 5382–5384.
13. M. Ladlow and G. Pattenden, *Tetrahedron Lett.*, 1984, **25**, 4317–4320.

14. K. Wakabayashi, H. Yorimitsu and K. Oshima, *J. Am. Chem. Soc.*, 2001, **123**, 5374–5375.
15. T. Fujioka, T. Nakamura, H. Yorimitsu and K. Oshima, *Org. Lett.*, 2002, **4**, 2257–2259.
16. H. Someya, A. Kondoh, A. Sato, H. Ohmiya, H. Yorimitsu and K. Oshima, *Synlett*, 2006, 3061–3064.
17. L. Doszczak and R. Tacke, *Organometallics*, 2007, **26**, 5722–5723.
18. X. Cui, X. Xu, L. Wojtas, M. M. Kim and X. P. Zhang, *J. Am. Chem. Soc.*, 2012, **134**, 19981–19984.
19. R. L. Paddock, Y. Hiyama, J. M. McKay and S. T. Nguyen, *Tetrahedron Lett.*, 2004, **45**, 2023–2026.
20. H. T. Chang, M. Jeganmohan and C. H. Cheng, *Chem. Commun.*, 2005, 4955–4957.
21. H. T. Chang, M. Jeganmohan and C. H. Cheng, *Chem. Eur. J.*, 2007, **13**, 4356–4363.
22. (a) C. Le Floch, C. Bughin, E. Le Gall, E. Leonel and T. Martens, *Tetrahedron Lett.*, 2009, **50**, 5456–5458; (b) C. Le Floch, E. Le Gall, E. Leonel, J. Koubaa and T. Martens, *Eur. J. Org. Chem.*, 2010, 5279–5286.
23. J. H. Park, E. Kim and Y. K. Chung, *Org. Lett.*, 2008, **10**, 4719–4721.
24. S. H. Hong, J. W. Kim, D. S. Choi, Y. K. Chung and S. G. Lee, *Chem. Commun.*, 1999, 2099–2100.
25. S. U. Son, S. J. Paik, S. I. Lee and Y. K. Chung, *J. Chem. Soc., Perkin Trans.*, 2000, **1**, 141–143.
26. A. B. Mohamed and J. R. Green, *Chem. Commun.*, 2003, 2936–2937.
27. N. Saino, D. Kogure and S. Okamoto, *Org. Lett.*, 2005, 7, 3065–3067.
28. A. Goswami, T. Ito and S. Okamoto, *Adv. Synth. Catal.*, 2007, **349**, 2368–2374.
29. P. Novak, R. Pohl, M. Kotora and M. Hocek, *Org. Lett.*, 2006, **8**, 2051–2054.
30. T. Sugihara, A. Wakabayashi, Y. Nagai, H. Takao, H. Imagawa and M. Nishizawa, *Chem. Commun.*, 2002, 576–577.
31. (a) A. Geny, N. Agenet, L. Lannazzo, M. Malacria, C. Aubert and V. Gandon, *Angew. Chem. Int. Ed.*, 2009, **48**, 1810–1813; (b) P. Garcia, Y. Evanno, P. George, M. Sevrin, G. Ricci, M. Malacria, C. Aubert and V. Gandon, *Chem. Eur. J.*, 2012, **18**, 4337–4344.
32. I. Thiel, A. Spannenberg and M. Hapke, *ChemCatChem*, 2013, **5**, 2865–2868.
33. H. Someya, H. Ohmiya, H. Yorimitsu and K. Oshima, *Org. Lett.*, 2007, **9**, 1565–1567.
34. (a) T. Mita, N. Ohtsuki, T. Ikeno and T. Yamada, *Org. Lett.*, 2002, **4**, 2457–2460; (b) N. Ohtsuki, S. Kezuka, Y. Kogami, T. Mita, T. Ashizawa, T. Ikeno and T. Yamada, *Synthesis*, 2003, 1462–1466.
35. W. Li, P. Jia, B. Han, D. Li and W. Yu, *Tetrahedron*, 2013, **69**, 3274–3280.
36. D. Roberto and H. Alper, *J. Am. Chem. Soc.*, 1989, **111**, 7539–7543.
37. Y. C. Wong, K. Parthasarathy and C. H. Cheng, *J. Am. Chem. Soc.*, 2009, **131**, 18252–18253.

38. A. J. Clark, D. I. Davies, K. Jones and C. Millbanks, *J. Chem. Soc., Chem. Commun.*, 1994, 41–42.

39. (a) S. Horiie and S. Murahashi, *Bull. Chem. Soc. Jpn.*, 1960, **33**, 247–251; (b) R. F. Heck, *J. Am. Chem. Soc.*, 1967, **89**, 313–317; (c) M. I. Bruce, B. L. Goodall and F. G. A. Stone, *J. Chem. Soc., Dalton Trans.*, 1975, 1651–1655; (d) S. Murahashi and S. Horiie, *J. Am. Chem. Soc.*, 1956, **78**, 4816–4817.

40. M. A. Chari, D. Shobha and T. Sasaki, *Tetrahedron Lett.*, 2011, **52**, 5575–5580.

41. H. Sharghi, M. Aberi and M. M. Doroodmand, *Adv. Synth. Catal.*, 2008, **350**, 2380–2390.

42. T. H. Zhu, S. Y. Wang, G. N. Wang and S. J. Ji, *Chem. Eur. J.*, 2013, **19**, 5850–5853.

43. P. Saha, M. A. Ali, P. Ghosh and T. Punniyamurthy, *Org. Biomol. Chem.*, 2010, **8**, 5692–5699.

44. C. Feng, J. H. Su, Y. Yan, F. Guo and Z. Wang, *Org. Biomol. Chem.*, 2013, **11**, 6691–6694.

45. J. V. Ruppel, R. M. Kamble and X. P. Zhang, *Org. Lett.*, 2007, **9**, 4889–4892.

46. H. Lu, J. Tao, J. E. Jones, L. Wojitas and X. P. Zhang, *Org. Lett.*, 2010, **12**, 1248–1251.

47. Y. Wakatsuki and H. Yamazaki, *Tetrahedron Lett.*, 1973, **36**, 3383–3384.

48. Y. Wakatsuki and H. Yamazaki, *J. Chem. Soc., Dalton Trans.*, 1978, 1278–1282.

49. A. Naiman and K. P. C. Vollhardt, *Angew. Chem. Int. Ed. Engl.*, 1977, **16**, 708–709.

50. D. J. Brien, A. Naiman and K. P. C. Vollhardt, *J. Chem. Soc., Chem. Commun.*, 1982, 133–134.

51. C. A. Parnell and K. P. C. Vollhardt, *Tetrahedron*, 1985, **41**, 5791–5796.

52. R. E. Geiger, M. Lalonde, H. Stoller and K. Schleich, *Helv. Chim. Acta*, 1984, **67**, 1274–1282.

53. R. L. Hillard III, C. A. Parnell and K. P. C. Vollhardt, *Tetrahedron*, 1983, **39**, 905–911.

54. H. Boennemann, W. Brijoux, R. Brinkmann, W. Meurers, R. Mynott, W. von Philipsborn and T. Egolf, *J. Organomet. Chem.*, 1984, **272**, 231–249.

55. W. Schulz, H. Pracejus and G. Oehme, *Tetrahedron Lett.*, 1989, **30**, 1229–1232.

56. F. Karabet, B. Heller, K. Kortus and G. Oehme, *Appl. Organomet. Chem.*, 1995, **9**, 651–656.

57. G. Vitulli, S. Bertozzi, R. Lazzaroni and P. Salvadori, *J. Organomet. Chem.*, 1986, **307**, C35–C37.

58. B. Heller, D. Heller, P. Wagler and G. Oehme, *J. Mol. Catal. A: Chem.*, 1998, **136**, 219–233.

59. A. Gutnov, V. Abaev, D. Redkin, C. Fischer, W. Bonrath and B. Heller, *Synlett*, 2005, 1188–1190.

60. A. Gutnov, B. Heller, C. Fischer, H. J. Drexler, A. Spannenberg, B. Sundermann and C. Sundermann, *Angew. Chem. Int. Ed.*, 2004, **43**, 3795–3797.

61. M. Hapke, K. Kral, C. Fischer, A. Spannenberg, A. Gutnov, D. Redkin and B. Heller, *J. Org. Chem.*, 2010, **75**, 3993–4003.

62. I. Thiel, H. Jiao, A. Spannenberg and M. Hapke, *Chem. Eur. J.*, 2013, **19**, 2548–2554.

63. I. Thiel, M. Lamac, H. Jiao, A. Spannenberg and M. Hapke, *Organometallics*, 2013, **32**, 3415–3418.

64. J. A. Varela, L. Castedo and C. Saa, *J. Org. Chem.*, 1997, **62**, 4189–4192.

65. J. A. Varela, L. Castedo and C. Saa, *J. Am. Chem. Soc.*, 1998, **120**, 12147–12148.

66. J. A. Varela, L. Castedo and C. Saa, *Org. Lett.*, 1999, **1**, 2141–2143.

67. A. W. Fatland and B. E. Eaton, *Org. Lett.*, 2000, **2**, 3131–3133.

68. C. Braendli and T. R. Ward, *J. Comb. Chem.*, 2000, **2**, 42–47.

69. R. S. Senaiar, D. D. Young and A. Deiters, *Chem. Commun.*, 2006, 1313–1315.

70. A. F. Moretto, H. C. Zhang and B. E. Maryanoff, *J. Am. Chem. Soc.*, 2001, **123**, 3157–3158.

71. L. V. R. Bonaga, H. C. Zhang, A. F. Moretto, H. Ye, D. A. Gauthier, J. Li, G. C. Leo and B. E. Maryanoff, *J. Am. Chem. Soc.*, 2005, **127**, 3473–3485.

72. L. V. R. Bonaga, H. C. Zhang and B. E. Maryanoff, *Chem. Commun.*, 2004, 2394–2395.

73. U. Groth, T. Huhn, C. Kesenheimer and A. Kalogerakis, *Synlett*, 2005, 1758–1760.

74. Y. Zhou, J. A. Porco Jr and J. K. Snyder, *Org. Lett.*, 2007, **9**, 393–396.

75. (a) R. Hrdina, I. G. Stara, L. Dufkova, S. Mitchel, I. Cisarova and M. Kotora, *Tetrahedron*, 2006, **62**, 968–976; (b) P. Sehnal, Z. Krausova, F. Teply, I. G. Stara, I. Stary, L. Rulisek, D. Saman and I. Cisarova, *J. Org. Chem.*, 2008, **73**, 2074–2082.

76. Y. Miclo, P. Garcia, Y. Evanno, P. George, M. Sevrin, M. Malacria, V. Gandon and C. Aubert, *Synlett*, 2010, 2314–2318.

77. K. E. Augustin and H. J. Schaefer, *Eur. J. Lipid Sci. Techanol.*, 2011, **113**, 72–82.

78. (a) C. Yuan, C. T. Chang, A. Axelrod and D. Siegel, *J. Am. Chem. Soc.*, 2010, **132**, 5924–5925; (b) M. J. Eichberg, R. L. Dorta, D. B. Grotjahn, K. Lamottke, M. Schmidt and K. P. C. Vollhardt, *J. Am. Chem. Soc.*, 2001, **123**, 9324–9337.

79. M. W. Buettner, J. B. Naetscher, C. Burschka and R. Tacke, *Organometallics*, 2007, **26**, 4835–4838.

80. N. Agenet, J. H. Mirebeau, M. Petit, R. Thouvenot, V. Gandon, M. Malacria and C. Aubert, *Organometallics*, 2007, **26**, 819–830.

81. D. D. Young and A. Deiters, *Angew. Chem. Int. Ed.*, 2007, **46**, 5187–5190.

82. P. Turek, M. Hocek, R. Pohl, B. Klepetarova and M. Kotora, *Eur. J. Org. Chem.*, 2008, 3335–3343.

83. P. Garcia, S. Moulin, Y. Miclo, D. Leboeuf, V. Gandon, C. Aubert and M. Malacria, *Chem. Eur. J.*, 2009, **15**, 2129–2139.

84. P. Garcia, Y. Evanno, P. George, M. Sevrin, G. Ricci, M. Malacria, C. Aubert and V. Gandon, *Org. Lett.*, 2011, **13**, 2030–2033.

85. P. Garcia, Y. Evanno, P. George, M. Sevrin, G. Ricci, M. Malacria, C. Aubert and V. Gandon, *Chem. Eur. J.*, 2012, **18**, 4337–4344.

86. H. T. Chang, M. Jeganmohan and C. H. Cheng, *Org. Lett.*, 2007, **9**, 505–508.

87. K. Kase, A. Goswami, K. Ohtaki, E. Tanbe, N. Saino and S. Okamoto, *Org. Lett.*, 2007, **9**, 931–934.

88. A. Goswami, K. Ohtaki, K. Kase, T. Ito and S. Okamoto, *Adv. Synth. Catal.*, 2008, **350**, 143–152.

89. Y. K. Sugiyama and S. Okamoto, *Synthesis*, 2011, 2247–2254.

90. G. Dazinger, M. Torres-Rodrigues, K. Kirchner, M. J. Calhorda and P. J. Costa, *J. Organomet. Chem.*, 2006, **691**, 4434–4445.

91. A. R. A. Dahy, K. Yamada and N. Koga, *Organometallics*, 2009, **28**, 3636–3649.

92. A. A. Dahy and N. Koga, *J. Organomet. Chem.*, 2010, **695**, 2240–2250.

93. A. M. Rodriguez, C. Cebrian, P. Prieto, J. I. Garcia, A. de la Hoz and A. Diaz-Ortiz, *Chem. Eur. J.*, 2012, **18**, 6217–6224.

94. J. Jacob, C. M. Cavalier and W. D. Jones, *J. Mol. Catal. A: Chem.*, 2002, **182–183**, 565–570.

95. R. Gleiter and V. Schehlmann, *Tetrahedron Lett.*, 1989, **30**, 2893–2896.

96. (a) Y. J. Hu, X. D. Huang, Z. J. Yao and Y. L. Wu, *J. Org. Chem.*, 1998, **63**, 2456–2461; (b) L. S. Li, Y. Wu, Y. J. Hu, L. J. Xia and Y. L. Wu, *Tetrahedron: Asymmetry*, 1998, **9**, 2271–2277.

97. M. Malinowskaa, P. Kwiatkowski and J. Jurczak, *Tetrahedron Lett.*, 2004, **45**, 7693–7696.

98. P. Kwiatkowski, M. Asztemborska and J. Jurczak, *Tetrahedron: Asymmetry*, 2004, **15**, 3189–3194.

99. P. Kwiatkowski, W. Chaladaj, M. Malinowska, M. Asztemborska and J. Jurczak, *Tetrahedron: Asymmetry*, 2005, **16**, 2959–2964.

100. Z. Dong, X. Liu, J. Feng, M. Wang, L. Lin and X. Feng, *Eur. J. Org. Chem.*, 2011, 137–142.

101. L. V. R. Bonaga, H. C. Zhang, D. A. Gauthier, I. Reddy and B. E. Maryanoff, *Org. Lett.*, 2003, **5**, 4537–4540.

102. L. Lv, X. Wang, Y. Zhu, X. Liu, X. Huang and Y. Wang, *Organometallics*, 2013, **32**, 3837–3849.

103. M. Mulzer and G. W. Coates, *Org. Lett.*, 2011, **13**, 1426–1428.

104. T. Yamakawa and N. Yoshikai, *Org. Lett.*, 2013, **15**, 196–199.

105. A. Degl'Innocenti, A. Capperucci, T. Nocentini, S. Biondi, V. Fratini, G. Castagnoli and I. Malesci, *Synlett*, 2004, 2159–2162.

106. (a) M. E. Piotti and H. Alper, *J. Am. Chem. Soc.*, 1996, **118**, 111–116; (b) P. Davoli, I. Moretti, F. Prati and H. Alper, *J. Org. Chem.*, 1999, **64**, 518–521.

107. J. T. Lee, P. J. Thomas and H. Alper, *J. Org. Chem.*, 2001, **66**, 5424–5426.

108. J. A. R. Schmidt, V. Mahadevan, Y. D. Y. L. Getzler and G. W. Coates, *Org. Lett.*, 2004, **6**, 373–376.

109. J. A. R. Schmidt, E. B. Lobkovsky and G. W. Coates, *J. Am. Chem. Soc.*, 2005, **127**, 11426–11435.
110. Y. D. Y. L. Getzler, V. Mahadevan, E. B. Lobkovsky and G. W. Coates, *J. Am. Chem. Soc.*, 2002, **124**, 1174–1175.
111. M. Mulzer, B. T. Whiting and G. W. Coatess, *J. Am. Chem. Soc.*, 2013, **135**, 10930–10933.
112. P. Ganji, D. J. Doyle and H. Ibrahim, *Org. Lett.*, 2011, **13**, 3142–3145.

CHAPTER 6

Manganese-Catalyzed Heterocycle Synthesis

Manganese is a silvery gray metal with properties including hard, very brittle, difficult to fuse and easy to oxidize. The most common oxidation states of manganese are +2, +3, +4, +6 and +7, although oxidation states from −3 to +7 are observed. In addition to their vital biological applications, manganese salts have been used as powerful oxidants in chemistry for many years. More recently, with the increasing adoption of 'Green Chemistry,' manganese-catalyzed organic syntheses have been explored. In this chapter, the main applications of manganese catalysts in the synthesis of heterocyclic compounds are discussed.

6.1   Three-Membered Heterocycles

6.1.1   Manganese-Catalyzed Synthesis of Epoxides

Owing to the importance of epoxides and the ready availability of alkenes, epoxidation reactions of alkenes have been extensively explored in recent decades. Among all the catalysts reported, manganese complexes are certainly among the most outstanding, and especially manganese–porphyrin complexes have shown excellent activity. In 1984, Meunier *et al.* reported that using a manganese–porphyrin complex as the catalyst and NaOCl as the oxygen donor in the epoxidation of alkenes, epoxides were produced with high selectivity.[1] In their systemic study, they found the reaction rate, chemoselectivity and stereoselectivity are greatly increased by the addition of pyridine, which may behave as an axial ligand of the catalyst. Di-, tri- and tetrasubstituted alkenes were selectively converted (Scheme 6.1). In the case of terminal aliphatic alkenes, they succeeded in meeting this challenge by performing the reaction in a biphasic system.[2] In 1999, $NaIO_4$ was applied in

RSC Catalysis Series No. 16
Economic Synthesis of Heterocycles: Zinc, Iron, Copper, Cobalt, Manganese and Nickel Catalysts
By Xiao-Feng Wu and Matthias Beller
© Wu and Beller, 2014
Published by the Royal Society of Chemistry, www.rsc.org

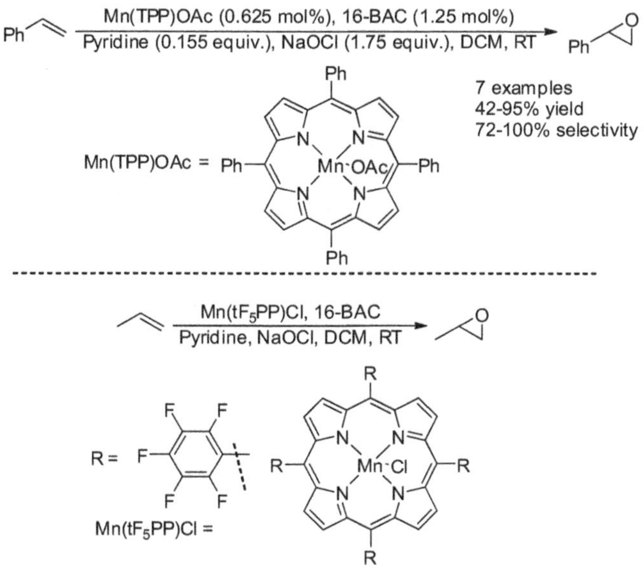

Scheme 6.1 depicts two manganese-catalyzed epoxidation reactions. The first shows Ph-CH=CH₂ converting to the corresponding epoxide:

Ph⌒ →[Mn(TPP)OAc (0.625 mol%), 16-BAC (1.25 mol%)] / [Pyridine (0.155 equiv.), NaOCl (1.75 equiv.), DCM, RT] → Ph epoxide

7 examples
42–95% yield
72–100% selectivity

Mn(TPP)OAc = Ph—[porphyrin]—Ph with Mn·OAc center, Ph groups at meso positions

The second reaction:

→[Mn(tF₅PP)Cl, 16-BAC] / [Pyridine, NaOCl, DCM, RT] → epoxide

R = F—[C₆F₄]— (pentafluorophenyl)

Mn(tF₅PP)Cl = [porphyrin with R groups and Mn·Cl center]

Scheme 6.1 Mn(TPP)OAc-catalyzed epoxidation of alkenes.

the epoxidation of alkenes with Mn(TPP)OAc as the catalyst at room temperature. Imidazole was needed as both base and assistant ligand.[3]

Li and co-workers, using MeCN–H_2O (1:1 v/v) as the medium, tested several oxidants ($KHSO_5$, NaClO, $NaIO_4$, H_2O_2) using styrene as a model. They concluded that $KHSO_5$ and NaClO are suitable oxidants for the manganese-porphyrin-catalyzed epoxidation of styrene and low activity was observed with H_2O_2.[4] More recently, a methodology for the epoxidation of different alkenes with $NaIO_4$ was reported, using Mn(TPP)Cl or Mn(Br₈TPP)Cl in the presence of [bmim][BF₄] in DCE.[5] The advantages of this method are the relatively mild reaction conditions, excellent yield and selectivity of epoxides. Further, introducing electron-withdrawing substituents such as bromine increased the robustness of the catalyst, which could be recovered and reused. In 2010, a model study on the epoxidation of styrene with *m*-CPBA in the presence of Mn(TF₅PP)Cl was reported.[6] The yield can be improved to 98–100% and with excellent selectivity (100%) by using *n*-Bu₄NOAc or *n*-Bu₄NBr as co-catalyst, and the reaction is completed in less than 5 min. In a comparison of the co-catalytic activity of *n*Bu₄NOAc, *n*Bu₄NBr and imidazole in the epoxidation of *cis*-stilbene and styrene, imidazole gave lower yields (26–41%). *m*-CPBA as oxidant was later combined with manganese-salen to epoxidize unfunctionalized alkenes.[7] More recently, this catalyst was reported to be immobilized on mesoporous materials and was applied in the epoxidation of styrene.[8] Good selectivity and yields can be achieved and the catalyst can be reused several times.

In 2004, studies on the highly selective epoxidation (>95%) of unfunctionalized alkenes using tetrabutylammonium periodate (*n*Bu₄NIO₄) as oxidant with six different phenyl-substituted manganese(III) *meso*-tetra-phenylporphyrins [Mn(TPP)OAc] and imidazole as the catalyst system in CH_2Cl_2 were described.[9] Electron-withdrawing and bulky substituents on the phenyl groups lowered the catalytic activity of the corresponding manganese catalyst. Less bulky alkenes with electron-rich double bonds showed greater reactivity in the epoxidation. The co-catalytic activities of four different classes of axial nitrogen donors were compared in the presence of various catalysts. In general, no direct correlation was found between the co-catalytic activities and the pK_a values of the nitrogen donors. The conclusions were as follows: (i) Mn(TPP) with electron-withdrawing substituents on the phenyl groups are much less effective catalysts than those with electron-donating substituents; (ii) electron-rich alkenes display higher reactivity than electron-poor alkenes; (iii) nitrogen donor axial ligands with π-donor capability are much more effective co-catalysts than pure δ-donor amines and no definite general correlation exists between the co-catalytic activities and pK_a values of nitrogenous bases; (iv) pyridines with methyl or amino groups in the 2- or 2,6-positions demonstrate exceptionally high co-catalytic activities in the presence of $Mn(TF_5PP)OAc$.

Later, the same group applied $Mn(TF_5PP)OAc$ as the catalyst and tetra-butylammonium monopersulfate (nBu_4NHSO_5) for the epoxidation of various alkenes.[10] Low to high yields (29–100%) and good to excellent selectivities (75–100%) were achieved in the presence of tetra-butylammonium acetate or fluoride or imidazole as co-catalysts in CH_2Cl_2, in less than 10 min at room temperature. In a study of the effects of nitrogen donors on the epoxidation of cyclooctene, they found that the co-catalytic effects of the nitrogen donors were critically dependent upon the steric hindrances and electronic structures of both the nitrogen donors and the manganese catalysts. No direct general correspondence was found between the co-catalytic effects of the nitrogen donors and their pK_a (BH^+) values. While the strong π-donating N–H imidazoles displayed higher co-catalytic activities than the weak π-donor pyridines, in the presence of MnTPP(OAc) an inverted order was observed for the $MnTF_5PP(OAc)$ complex.[11]

In 2011, Mohajer and co-workers reported their discovery of the effect of water on alkene epoxidation.[12] Cyclooctene was chosen as the substrate and IO_4^- as the oxidant in the presence of excess imidazoles, both in dry CH_2Cl_2 and in CH_2Cl_2 saturated with water. The reaction rates of the electron-deficient $Mn(TF_5PP)OAc$ were a factor 24 lower than those of Mn(TPP)OAc; however, the reaction rates of $Mn(TF_5PP)OAc$ increased 15–30-fold in the presence of water, whereas those of Mn(TPP)OAc did so by a factor of 2–3. The most striking catalytic enhancement caused by the addition of water was observed with 2-methylimidazole and $Mn(TF_5PP)OAc$. As deprotonation of imidazoles may play a significant role in the presence of water, they found that the manganese(III)-*meso*-tetrakis(phenyl-4-sulfonato)porphyrin complex [Mn(TPPS)] decreased the NH proton pK_a value of the axially coordinated imidazole from 14.2 to 9.5. It was concluded that the imidazole ligand is partially deprotonated in the presence of water, which permits the solvation of imidazolium ions that are formed simultaneously. The imidazolate form of the co-catalyst is a much stronger donor than the imidazole itself, providing electron density to Mn(III) and thus promoting oxygen transfer. The failure of *N*-methylimidazole to increase the reaction rates upon addition of water supports this hypothesis. Mn(III)–porphyrins in combination with imidazoles and water constitute a functional biomimetic model of peroxidases.

Zakavi and Ebrahimi investigated substitution effects on the activity of Mn(TPP)OAc complexes.[13] In the competitive oxidation of *cis*- and *trans*-stilbene, the results strongly suggested that a high-valent manganese oxo complex was involved as the reactive intermediate responsible for oxygen atom transfer. A complex order of the catalytic activity for the manganese–porphyrins was observed in the oxidation of various alkenes with PhIO, with the electron-rich manganese–porphyrins showing higher catalytic efficiencies than the electron-deficient compounds in the oxidation of most of the alkenes used. Different orders of catalytic performance of the manganese–porphyrins used in the oxidation of alkenes of various types cannot be simply explained by steric and electronic effects of the substituents attached to the double bond or the oxidative stability of the manganese–porphyrins. Subsequently, the catalytic activity of Mn(III) and Fe(III) complexes of *meso*-tetra(*n*-propyl)porphyrin, MnT(*n*Pr)P(X) and FeT(*n*Pr)P(X) (X = Cl, SCN, OAc), in the oxidation of alkenes with tetra-*n*-butylammonium periodate at room temperature was studied.[14] The influence of different parameters, including the molar ratio of catalyst to imidazole, type of counter ion (X) and oxidative stability of metalloporphyrins, on the efficiency of the catalysts was investigated. The results for the competitive oxidation of *cis*- and *trans*-stilbene suggest the presence of a high-valent manganese–oxo structure as the predominant oxidant species in equilibrium with a six-coordinate complex, MnT(*n*Pr)P(ImH)(IO$_4$) in the case of MnT(*n*Pr)P(OAc). An unusual preference for *trans*-stilbene over *cis*-stilbene was observed in the reaction catalyzed by FeT(*n*Pr)P(OAc). The reaction indicated a significant *cis*- to *trans*-isomerization (81%) in the oxidation of *cis*-stilbene catalyzed by FeT(*n*Pr)P(OAc), which may explain the observed unusual *cis*- to *trans*-stilbene oxide ratio. Whereas oxidation of cyclooctene and styrene led to the exclusive formation of the corresponding epoxides, oxidation of cyclohexene gave 2-cyclohexen-1-ol and cyclohexene oxide as the products. However, the results of this study clearly demonstrate the key role played by the group substituted at the *meso* positions of metalloporphyrins on their catalytic activity, apart from the electron-donating or electron-withdrawing properties of the substituents.

In 2012, a study on the influence of anionic co-catalysts was reported by the same group.[15] The epoxidation of *cis*-stilbene by *n*Bu$_4$NHSO$_5$ was studied in the presence of two manganese–porphyrin complexes, [MnTF$_5$P(Cl) and MnTPP(Cl)], and various *n*Bu$_4$NX (X = OAc, F, Cl, Br, OCN, NO$_3$, BF$_4$). In general, a direct correlation was found between the stereoselectivity of the epoxidation reaction and the nucleophilic properties of these anionic co-catalysts. Also, the epoxidation of naphthalene was carried out with high yield and good selectivity using nBu_4NHSO_5 in the presence of $MnTF_5P(Cl)$ in association with nBu_4NOAc or nBu_4NF as co-catalyst. The coordinating strength of an anionic co-catalyst could have a great influence on the yields and selectivities of oxidation reactions of alkenes with nBu_4NHSO_5 catalyzed by manganese–porphyrins. The hardest anionic axial ligand bonds to the metal center more strongly than others, leading to higher conversion and greater stereoselectivity in the reaction. The importance of the effects of anionic co-catalysts is more pronounced when their relative activities are examined in the presence of manganese–porphyrins with electron-with-drawing substituents on the porphyrin periphery.

H_2O_2 as an ideal oxidant has also been tested and applied in the man-ganese-catalyzed epoxidation of alkenes. The effects of bases and ligands were tested by Johnstone's group in 1991.[16] Using a manganese–porphyrin catalyst, they found that tertiary amine N-oxides can be used as stable lig-ands in place of the easily destroyed imidazole used previously. Further, the use of imidazole as base can be avoided through the use of tetra-butylammonium hydroxide or other bases such as sodium carbonate or sodium acetate. Later, the combination of chloro[tetra(2,6-dichlor-ophenyl)porphyrinato]manganese(III) and ammonium acetate was found to show superior reactivity compared with previous systems.[17]

An interesting investigation of Mn(III)-catalyzed epoxidation of naph-thalenes using H_2O_2 as oxidant was reported in 2004.[18] In the presence of Mn(III)–porphyrins [Mn(TDCPP)Cl, Mn(βNO_2TDCPP)Cl, Mn(TPFPP)Cl] as catalysts, naphthalene and anthracene afforded the *anti*-1,2:3,4-arene dioxides whereas with phenanthrene the 9,10-oxide was obtained (Scheme 6.2). Later, this catalytic system was applied in the oxidation of Δ^4- and Δ^5-steroids.[19] In 2008, the same group prepared a chloro[5,10,15,20-(1,3-dimethylimidazolium-2-yl)porphyrinatomanganese(III)] tetraiodide and evaluated it in the epoxidation of alkenes under homogeneous conditions, using H_2O_2 as oxidant. The catalytic behavior of this metalloporphyrin is

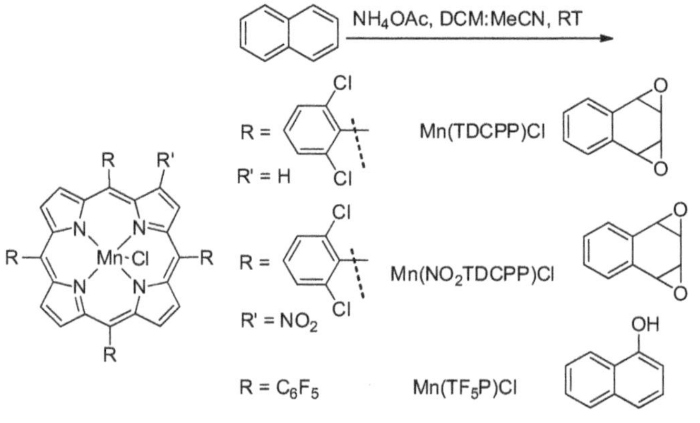

Scheme 6.2 Mn(TPP)Cl-catalyzed epoxidation of naphthalenes.

strongly dependent on the co-catalyst used; high conversions were found for all alkenes studied, affording the corresponding epoxides as the major products with acetic acid or benzoic acid as additive.[20]

In 2010, an Mn(III)-porphyrin-based catalytic system was explored for alkene epoxidations under mild reaction conditions using sodium bicarbonate–hydrogen peroxide as oxidant.[21] The Mn(TPP)OAc–imidazole–NaHCO$_3$ system efficiently catalyzed the epoxidation of alkenes with H$_2$O$_2$. Cyclic alkenes were converted with excellent yields (80–100%) and selectivities (87–100%) that were much better than those observed in the absence of bicarbonate. In the presence of an excess of substrate, a turnover number (TON) of 4286 was obtained after 2 h with the Mn(TPP)OAc–imidazole–NaHCO$_3$–H$_2$O$_2$ system. The bicarbonate-activated oxidation system is a simple, inexpensive and relatively non-toxic alternative to other oxidants and peroxy acids and it can be used in a variety of oxidations where a mild, pH-neutral oxidant is required. Based on NMR evidence, peroxymonocarbonate ion (HCO$_4^-$) formed by a labile pre-equilibrium reaction between bicarbonate ion and H$_2$O$_2$ was proposed as the active oxidant in the catalytic pathway.

Solati *et al.* prepared a chloro(tetramesitylporphyrinato)manganese(III) complex and applied it in the epoxidation of *cis*-stilbene.[22] Some nitrogen donors, acetate and bromide ions were used as co-catalysts in the epoxidation reaction of *cis*-stilbene using *m*-chloroperbenzoic acid (*m*-CPBA) as the oxidant. Cyclic voltammetry was used for comparison of the coordination abilities of these axial ligands. The stability of the manganese–porphyrin complex and the lifetime of the manganese–oxo intermediate species were compared using UV–visible spectroscopy in the presence of these axial ligands. It was shown that an increasing coordination ability of these axial ligands has a positive effect on the stability of the catalyst and also on the stability of the manganese–oxo intermediate species but a negative effect on the catalytic activity of the manganese–porphyrin complex.

Ji's group studied the effect of cobalt acetate on the aerobic epoxidation of cyclohexene with manganese–porphyrin (MnTPPCl) as the catalyst.[23] Using molecular oxygen as the oxidant, significant improvements in the conversion of cyclohexene (from 28 to 72%) and the selectivity toward epoxide were observed, whereas the formation of 2-cyclohexen-1-ol was restricted in the presence of cobalt acetate as co-catalyst. The effect of the amount of catalyst, reaction pressure and reaction temperature on the aerobic oxidation of cyclohexene was also investigated. A co-catalytic mechanism involving cobalt oxide ([CoO$_2$]$^+$) as the active species for enhancing the cyclohexene activity and selectivity toward epoxide was proposed.

With the advantages of heterogeneous catalysts, attempts to prepare supported manganese–porphyrin catalysts were made by different groups. de Miguel's group supported the catalysts on Merrifield and Argogel resins and applied them in alkene epoxidations. Furthermore, recyclability studies have shown that the Merrifield-supported catalyst can be reused three times with minimum loss of activity.[24] In 2002, a carboxymethylated cross-linked polystyrene resin [poly(4-styrylmethylacyl chloride) (PSA)] support was used to attach manganese(III) tetrakis(4-aminophenyl)porphyrin covalently and make this catalyst heterogenized. This catalyst was applied in alkene epoxidation and alkane hydroxylation with sodium periodate as the oxidant and found to be efficient. This new hydrogenized catalyst shows of high stability and reusability.[25]

In 2006, Liu and Nocera described a simple and versatile method for the catalytic epoxidation of alkenes with aqueous H_2O_2 using manganese–salophen catalysts.[26] Low catalyst loadings, short reaction times and a simple reaction setup (*e.g.* no pH buffer is required) are salient features of the system, which combines the benefits of H_2O_2 as an oxidant with the versatility and modularity of salen-based catalysts. A wide range of unfunctionalized and functionalized alkenes can be epoxidized with high TON (Scheme 6.3). More recently, a new manganese(III) complex [Mn(saldien)(N$_3$)] was anchored on mesoporous SBA-15 to produce a stable, active and selective heterogeneous catalyst, SBA-15-[Mn(saldien)(N$_3$)] [saldien = *N,N*-bis(salicylidene)diethylenetriamine].[27]

Recently, Neier and co-workers synthesized a novel manganese non-porphyrin catalyst, using the conformationally rigid and saturated calix[4]pyrrolidine as ligand (Scheme 6.4).[28] The Mn(II) cationic complex, bearing acetate ions as axial ligand and as counter ion, displayed interesting catalytic properties. Epoxidation of simple alkenes using hydrogen peroxide as terminal oxidant gave reasonable to good yields of product. The catalyst displays good robustness in acidic media. The addition of potential apical ligands led to an unusual reactivity pattern. Most of the ligands known to

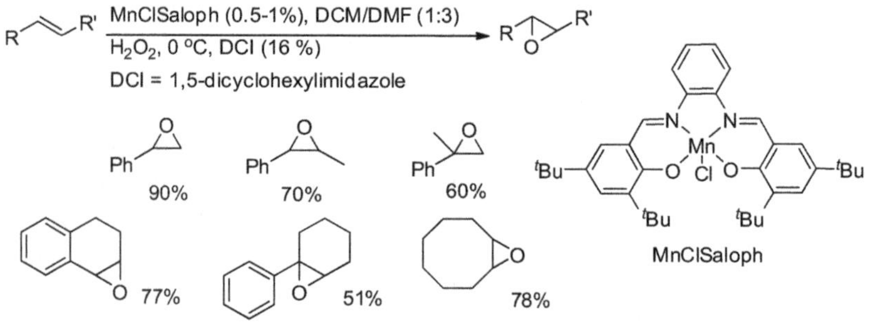

Scheme 6.3 MnClSaloph-catalyzed epoxidation of alkenes.

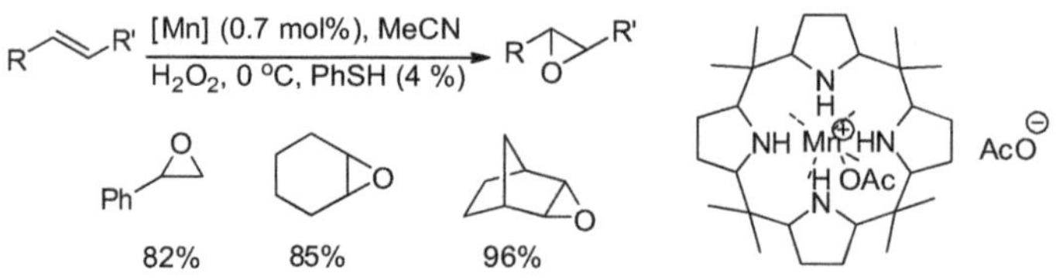

Scheme 6.4 Mn(II) tetraaza-catalyzed epoxidation of alkenes.

accelerate epoxidation were not effective, whereas sulfur-containing ligands showed interesting effects. Addition of thiophenol afforded the most significant increase in the reactivity of the catalytic system, with the formation of the epoxide products in high yields. The saturated tetraaza ligand and the absence of a π-electron donor system to stabilize the high-valent Mn(V)=O led to the hypothesis of an Mn(IV) hydroperoxide adduct as the active intermediate in epoxidation. Moreover, this novel manganese catalyst is the first non-porphyrin Mn complex to epoxidize the unsaturated Δ^5-steroid 3β-acetoxy-5-cholestene with a reasonable yield and β-selectivity.

In 2003, Chan and co-workers reported a manganese/bicarbonate-catalyzed epoxidation of lipophilic alkenes using hydrogen peroxide as the oxidant.[29] The reaction was carried out in an ionic liquid at room temperature and the desired epoxides were formed in good yields (Scheme 6.5). Tetramethylammonium hydrogencarbonate can promote the reaction efficiently, but no activity was observed with NaHCO$_3$ as promoter. The ionic liquid can be reused in up to at least 10 cycles without any diminished capacity to act as the medium for the reaction. With the recycling of the ionic liquid, the reaction can be considered as a cheap, catalytic, scalable and environmentally benign method for alkene epoxidations.

Liu and co-workers prepared a series of metal oxides (MnO, Mn$_3$O$_4$, Mn$_2$O$_3$, MnO$_2$) and applied them in the epoxidation of alkenes using H$_2$O$_2$ as oxidant in DMF at 0–25 °C.[30] MnO exhibited the highest activity for the selective epoxidation of styrene to styrene oxide with 4 equiv. of 30% H$_2$O$_2$. The oxidants and the concentration of hydrogencarbonate had a great influence on the epoxidation reaction. The MnO catalyst displayed higher activity for the catalytic epoxidation of aromatic alkenes (styrene, α-methylstyrene, indene and 1-phenylcyclohexene) and norbornene, mild activity for carbocyclic alkenes and no activity for 1-octene. The MnO catalyst could be reused four times without an obvious decrease in activity.

In 2011, a new biomimetic ligand bearing two imine groups and three imidazole residues was prepared (Scheme 6.6).[31] The Mn(II)–trisimidazole

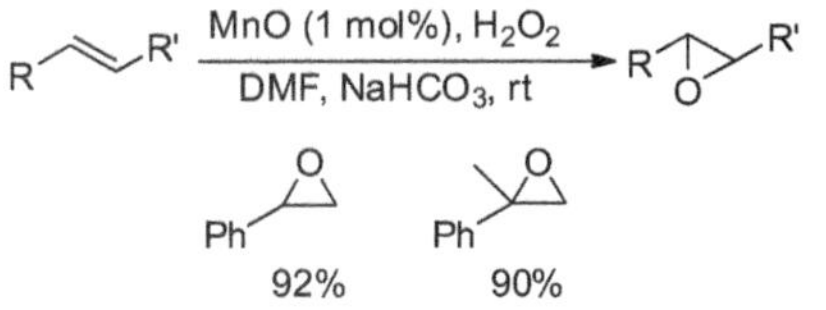

Scheme 6.5 MnO-catalyzed epoxidation of alkenes.

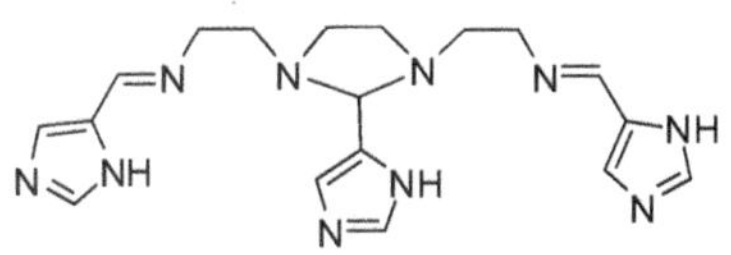

Scheme 6.6 Three-imidazole ligand.

system obtained was evaluated as an oxidation catalyst for the epoxidation of simple alkenes with H_2O_2. The results indicated that this homogeneous system is more reactive than the imidazole-based acetamide or acetylacetone-based Mn(II)–Schiff base complexes reported previously. This is due to the incorporation of a biomimetic ligand that combines structural features such as Schiff base imine groups and imidazole rings. Thus the present Mn(II)–L_{3imid} system represents an optimized form of the previous biomimetic Mn(II) systems. Importantly, the Mn(II)–L_{3imid} catalyst is also ammonium acetate dependent and can activate H_2O_2, favoring productive alkene epoxidations to a remarkable extent with high selectivity.

In 2012, a practical, fast and readily implemented method for the selective epoxidation of electron-rich alkenes with H_2O_2 and an *in situ*-prepared catalyst system was established by Browne and co-workers (Scheme 6.7).[32] Advantageous features include their TONs (up to 300 000) and tolerance towards other oxidation-sensitive functional groups. Additionally, the mild conditions and solvent scope make this system highly competitive with stoichiometric oxidants such as *m*-CPBA. The system is especially suited to the epoxidation of electron-rich alkenes and showed good to excellent selectivity in the epoxidation of dienes and bifunctional substrates. The method can gave exceptional selectivity and activity in the *cis*-dihydroxylation of electron-deficient alkenes. The preliminary mechanistic study was focused on the role of butanedione and indicated that further optimization of the system should focus on overcoming the oxidation of the ketone as a competing reaction.

In 2011, an efficient and highly selective heterogeneous catalyst was developed by immobilization of a manganese complex on an inorganic support to yield (silica gel)–O_2(EtO)Si–L^1–Mn(HL^2) [L^1 = modified salicylaldiminato and HL^2 = (*E*)-*N'*-(2-hydroxy-3-methoxybenzylidene)benzohydrazide].[33] Mn(II) was anchored on the surface of functionalized silica by means of *N,O*-coordination to the covalently Si–O-bound modified salicylaldiminato Schiff base ligand. The prepared (silica gel)–O_2(EtO)Si–L^1–Mn(HL^2) material was characterized by elemental and thermogravimetric analyses (TGA and DTA) and UV–visible and FT-IR spectroscopy. This new material was demonstrated to be a very active catalyst in clean epoxidation reactions using a combined oxidant of aqueous hydrogen peroxide and acetonitrile in the presence of aqueous sodium hydrogencarbonate. The effects of reaction parameters such as solvent, $NaHCO_3$ and oxidant on the epoxidation of *cis*-cyclooctene were investigated. Cycloalkenes were oxidized efficiently to their corresponding epoxides with 87–100% selectivity in the presence of this catalyst. This catalytic system also showed good activity in the epoxidation of linear

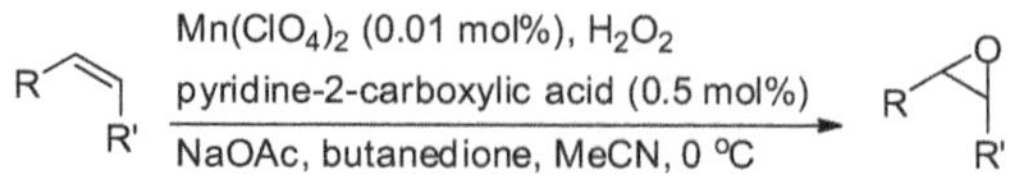

Scheme 6.7   Mn(ClO_4)$_2$-catalyzed epoxidation of electron-rich alkenes.

alkenes. The results showed that it is a robust and stable heterogeneous catalyst that can be recovered quantitatively by simple filtration and reused multiple times without loss of activity. Subsequently, in 2012, nano-sized particles of manganese oxides were prepared by a very simple and cheap process using a decomposing aqueous solution of manganese nitrate at 100 °C.[34] Scanning electron microscopy, transmission electron microscopy and X-ray diffraction (XRD) were used to characterize the phase and the morphology of the manganese oxide. This nano-sized manganese oxide showed efficient catalytic activity toward water oxidation and the epoxidation of alkenes in the presence of cerium(IV) ammonium nitrate and hydrogen peroxide, respectively. In the same year, the incorporation of calcium(II), zinc(II) and aluminum(III) to manganese oxides was found to improve greatly the activity of manganese oxide towards the epoxidation of alkenes in the presence of anhydrous *tert*-butyl hydroperoxide as an oxidant.[35]

Also in 2012, a manganese-based hybrid mesoporous material was synthesized by covalent grafting of $[Mn(II)(1)_2](OAc)_2$ {$1 = [3$-$(2$-pyridyl)pyrazol-1-yl]acetic acid amide$} onto the surface of SBA-15 and characterized by means of XRD, N_2 adsorption–desorption and FT-IR, Raman, EPR and UV–visible spectroscopic techniques.[36] Catalytic tests showed that this catalyst could act as an efficient heterogeneous catalyst for the epoxidation of a wide range of alkenes (including terminal alkenes) under mild reaction conditions when peracids (*e.g.* *m*-CPBA) are used as oxidants. Moreover, the catalytic performance of this catalyst was solvent dependent; it exhibited higher catalytic activity and selectivity to epoxides when the reaction was carried out in an aprotic solvent such as CH_3CN. UV–visible and electrochemical measurements revealed that high-valent Mn species were easily formed during the reaction course when *m*-CPBA was used as oxidant and CH_3CN as solvent, this probably being the main reason for the high activity of this catalyst and its selectivity toward epoxide formation.

In 2013, the functionalization of the surface of SiO_2 particles with attached $[Mn_4O_2(CH_3COO)_7(bipy)_2](ClO_4) \cdot 3H_2O$ and the stacking of SiO_2-matrix overlayers were successfully used to prepare a new and robust heterogeneous manganese catalyst that was applied in the selective epoxidation of alkenes.[37] The catalyst was chemically attached to an SiO_2 particle surface and the attached structure was characterized by FT-IR, UV–visible, Raman, XPS and Mn K-edge XAFS spectroscopic methods. In the same year, a Schiff base ligand was synthesized by the condensation of salicylaldehyde with L-tyrosine.[38] Interaction of this ligand with Mn(II)-exchanged zeolite Y led to encapsulation of the ligand within the zeolite and complexation of the metal. The encapsulated complex was characterized by spectroscopic studies and chemical analyses. This material serves as a catalyst for the oxidation of cyclohexene to cyclohexene epoxide and 2-cyclohexen-1-ol using H_2O_2 as oxidant. The reaction conditions were optimized for solvent, temperature and amounts of oxidant and catalyst. The catalyst showed high activity and selectivity toward the production of cyclohexene epoxide in acetonitrile at 60 °C with an $[H_2O_2]:[C_6H_{10}]$ molar ratio of 2.5. Comparison of the encapsulated catalyst with the corresponding homogeneous catalyst showed that the heterogeneous catalyst had higher activity and selectivity than the homogeneous catalyst.

In addition to the Mn(TPP)OAc- and supported complex-catalyzed epoxidation of alkenes, asymmetric epoxidations have also been realized. In 1998, an asymmetric epoxidation of unfunctionalized alkenes was reported using chiral (salen)Mn(III) complexes together with a carboxylate salt co-catalyst in the presence of either aqueous H_2O_2 or anhydrous urea–H_2O_2 adduct as oxidant.[39] Several simple soluble salts (acetates, formates, benzoates) were studied, all giving good yields of epoxides with moderate to excellent enantioselectivity (Scheme 6.8). For example, 1,1-diphenyl-1-propene was converted into the corresponding epoxide in 84% yield with 96% *ee*. This catalyst was later immobilized and can be reused several times without losing activity.[40] More recently, this catalyst was further modified; for example, ionic liquids were applied as a solvent[41] and sulfonated and silica gel-supported,[42] phosphate-immobilized,[43] crystalline aluminum oligo-styrenylphosphonate hydrogenphosphate-supported[44] and methacrylic terpolymer-supported[45] systems were developed. Good yields and *ee*s were observed in all cases and the catalysts could be easily reused.

More recently, Ruffo *et al.* described the design and preparation of a series of *N,N',O,O'*-chiral ligands (elpaN-salen-H_2), which represent a subset of the elpaN-type library of ligands based on β-1,2-D-glucodiamine (Scheme 6.9).[46] The corresponding Mn(III) complexes were examined in the asymmetric epoxidation of styrenes with satisfactory results using *m*-CPBA as oxidant,

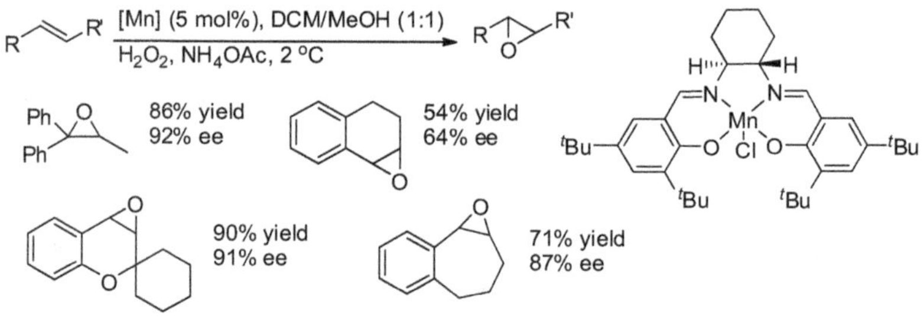

Scheme 6.8   Mn–salen-catalyzed asymmetric epoxidation of alkenes.

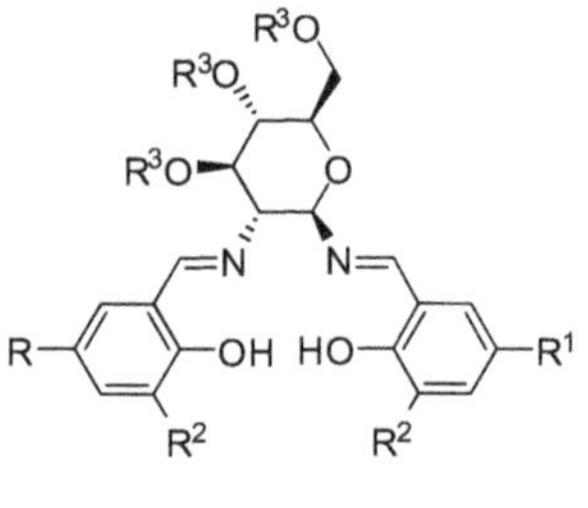

Scheme 6.9   General formula of elpaN-salen-H_2 ligands.

e.g. cis-β-methylstyrene was functionalized with *ee* up to 88%. This work substantiates the original idea of the possibility of preparing libraries of pseudo-enantiomeric ligands based on D-glucose by simply switching the positions of coordination from C2 and C3 (Naple-type) to C1 and C2 (elpaN-type).

Additionally, the previously mentioned manganese–porphyrins were also studied in the asymmetric epoxidation of alkenes.[47] A selection of alkenes were epoxidized with iodosylbenzene, catalyzed by three related iron(III) tetraarylporphyrins, **1***, **2*** and **3***, **1*** with four 2,6-di(1-phenylbutoxy)phenyl groups, **2*** with one pentafluorophenyl and three 2,6-di(1-phenylbutoxy)-phenyl groups and **3*** with two pentafluorophenyl and two 2,6-di(1-phenyl-butoxy)phenyl groups. Catalyst **1*** is very sterically hindered and prone to self-oxidation, which makes it a relatively poor epoxidation catalyst. Introducing the smaller pentafluorophenyl groups in place of 2,6-di(1-phenyl-butoxy)phenyl increases the catalyst reactivity, stability and selectivity. This change allows easier access of the substrates to the active oxidant and also, by decreasing the electron density on the porphyrin ligand, increases the reactivity of the oxoiron intermediate and its stability towards self-oxidation. A family of five homochiral catalysts, **1**, **2** and **3** [the analogs of **1***, **2*** and **3***, prepared from (*R,R*)-2,6-di(1-phenylbutoxy)benzaldehyde], **4**, with three pentafluorophenyl and one (*R,R*)-2,6-di(1-phenylbutoxy)phenyl group, and **5**, the manganese(III) analog of **3**, were used to epoxidize three prochiral alkenes. All the reactions gave low enantioselectivities. Using styrene as the substrate, (*S*)-styrene epoxide was the major enantiomer obtained with all the catalysts except **1**, which led to the (*R*)-styrene epoxide being preferred. In contrast, *cis*-hept-2-ene and 2-methylbut-2-ene gave the same major epoxide enantiomer with all the catalysts. The dependence of the *ee* values on the catalyst and substrate structure, temperature and solvent was examined and discussed.

Che, Wong and co-workers developed general and efficient methods for the highly diastereoselective epoxidation of allylically substituted cycloalkenes by sterically bulky Mn– and Ru–porphyrin catalysts.[48] Highly diastereoselective epoxidations of allyl-substituted cycloalkenes, including allylic alcohols, esters and amines, can be achieved using the sterically bulky metalloporphyrins [Mn(TDCPP)Cl] and [Ru(TDCPP)CO] as catalysts. The '[Mn] + H_2O_2' and '[Ru] + 2,6-Cl_2pyNO' protocols afforded *trans*-epoxides selectively in good yields (up to 99%) with up to >99 : 1 *trans* selectivity. Soon afterwards, they found that by using [Mn(TDCPP)Cl] as a catalyst and Oxone–H_2O_2 as an oxidant, allyl-substituted alkenes could be efficiently *erythro*-selectively epoxidized.[49] Up to 9 : 1 *erythro* selectivities for terminal allylic alkenes could be achieved, which are significantly higher than those achieved using *m*-CPBA as an oxidant. In addition, the synthetic utilities of this epoxidation method were highlighted in the stereoselective synthesis of key anti-HIV drug intermediates and the epoxidation of glycals (Scheme 6.10). In 2012, a sulfonated manganese–porphyrin complex was prepared and applied in the epoxidation of alkenes and alkanes in water and methanol solution. Moderate *ee*s were observed.[50]

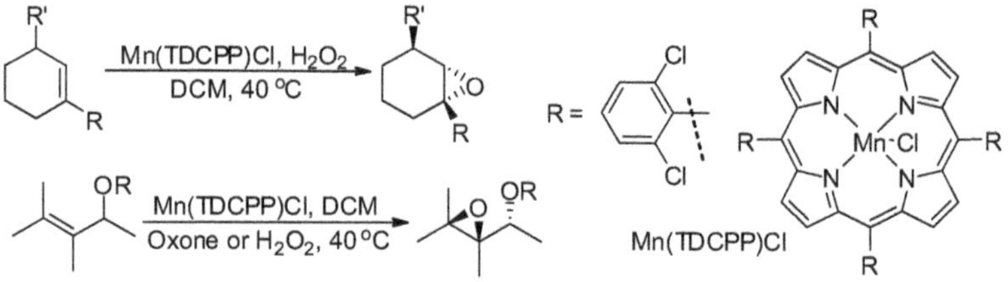

Scheme 6.10   [Mn(TDCPP)Cl]-catalyzed asymmetric epoxidation of alkenes.

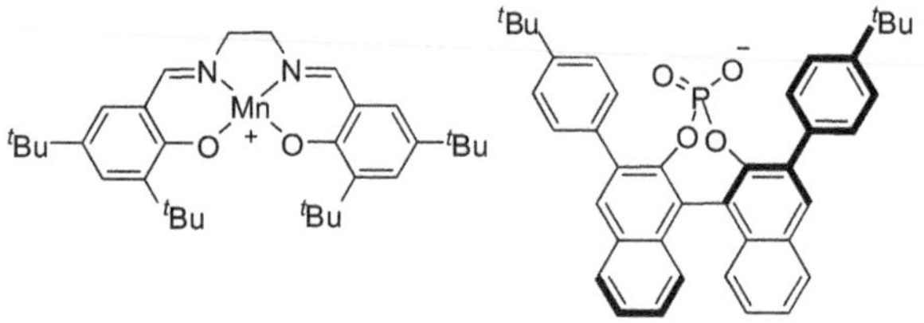

Scheme 6.11   Manganese–salen–phosphate complex.

R' = nPr, 66% yield; 90% ee
R' = iPr, 53% yield; 94% ee

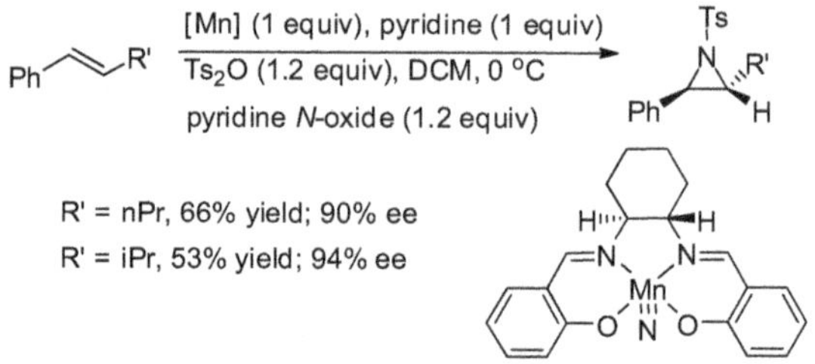

Scheme 6.12   Mn-mediated asymmetric aziridination of styrenes.

In 2010, Liao and List reported a manganese–salen–phosphate complex-catalyzed asymmetric epoxidation of alkenes.[51] Based on the concept of asymmetric counteranion-directed catalysis, this ion-pair epoxidation catalyst, which consists of an achiral Mn–salen complex and a chiral phosphate counteranion, represents another powerful catalyst system (Scheme 6.11). Numerous alkenes were converted into the corresponding epoxides in an asymmetric manner with excellent yields and selectivities. All the reactions were carried out in benzene at room temperature with PhIO as the oxidant.

The aziridination of styrene derivatives can be considered as analogous to the transformation of epoxidation. In 1998, Komatsu and co-workers reported a chiral nitridomanganese complex-mediated aziridination of styrenes.[52] The desired products were produced in good yields and with good enantioselectivity (Scheme 6.12). In addition, this methodology resulted in stereospecific aziridination. They also found that additives play an important role in the reaction; Ts$_2$O was the most effective reagent in the activation of the manganese complex and the use of pyridine *N*-oxide was necessary to obtain high enantioselectivity.

6.2 Five-Membered Heterocycles

6.2.1 Manganese-Catalyzed Synthesis of Lactones

In 1968, a manganese(III) acetate-mediated synthesis of γ-lactones from alkenes and acetic acid was reported by Bush and Finkbeiner (Scheme 6.13).[53] The reaction was performed in refluxing acetic acid and using acetic anhydride as additive. Good yields of the corresponding lactones were produced calculated based on Mn(OAc)₃. Around the same time, Heiba *et al.* reported the same procedure with KOAc as additive.[54] They found that not only can Mn(OAc)₃ initiate the reaction, but also MnO₂, Mn₂O₃, Ce(OAc)₄, Ce(NH₄)₂(NO₃)₆ and NH₄VO₃ can be applied as promoters.[55]

The intramolecular version of Mn(OAc)₃-mediated lactone synthesis has also been explored. In 1984, this method was applied by Corey and Kang for the preparation of polycyclic γ-lactones *via* double annulation.[56] Starting from the corresponding substrates, good yields of the desired lactones were isolated (Scheme 6.14). In addition to acetic acid, it was subsequently found that cyanoacetic and melonic acids could also be used.[57] By applying Mn₃O(OAc)₇ as promoter, the desired lactones were obtained in good yields.

Also in the mid-1980s, Fristad and co-workers reported a series of studies on the Mn(OAc)₃-mediated synthesis of lactones from alkenes and activated carboxylic acids.[58–62] They described for the first time the reaction of malonic acid with alkenes, giving spiro-fused lactones in moderate to excellent yields (Scheme 6.15). In their mechanism study, the effect of acetic anhydride elucidated and it was found to be a result of it being oxidized itself. The fate of the oxidized acetic acid and anhydride in the absence of a suitable acceptor molecule was identified. The linear relationship between the logarithm of the relative reactivity of substituted acetic acids and their C–H acidity (pK_a of the corresponding ester) demonstrated the importance of substrate acidity in this reaction. In 1991, Snider *et al.* studied the effects

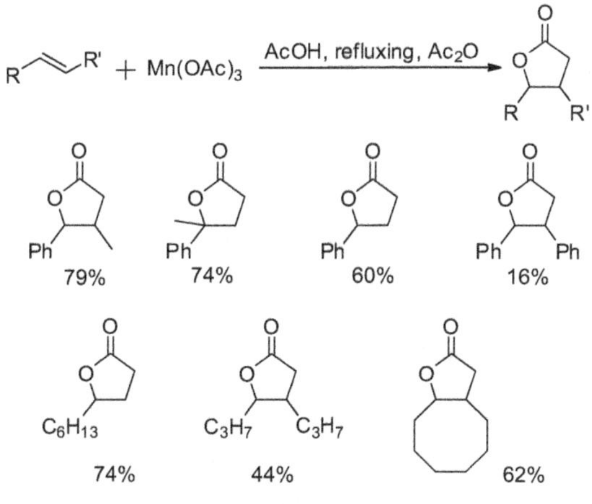

Scheme 6.13 Mn(OAc)₃-promoted synthesis of lactones.

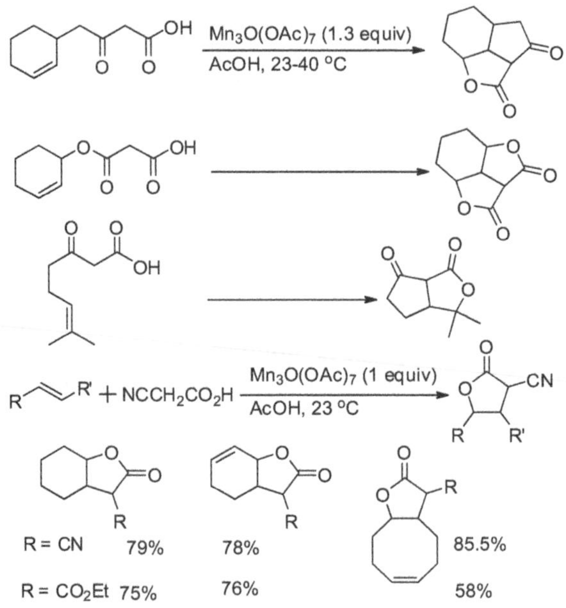

Scheme 6.14   Mn$_3$O(OAc)$_7$-promoted synthesis of lactones.

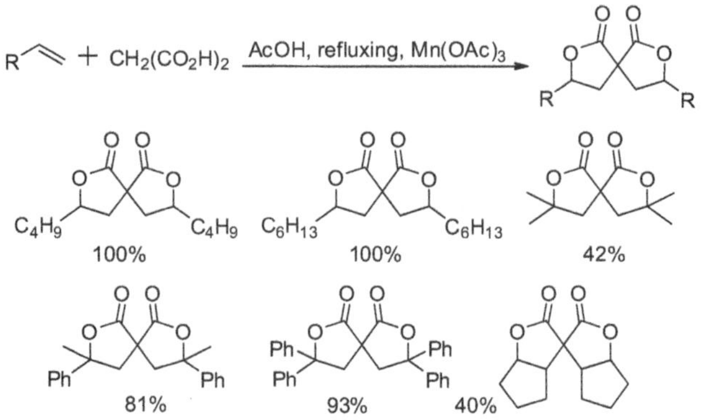

Scheme 6.15   Mn(OAc)$_3$-promoted synthesis of spiro-fused lactones.

of solvents on manganese-promoted cyclization reactions.[63] They found that ethanol can be applied as an alternative solvent in the intramolecular oxidative free-radical cyclization reaction. In some cases, higher yields can be achieved in ethanol than acetic acid.

In 2005, Burton's group reported the synthesis of fused tricyclic γ-lactones using manganese(III) acetate as the promoter (Scheme 6.16).[64] Starting from cyclic alkenes bearing a carboxylic acid and a malonate group, good yields of the desired product were obtained using Cu(OTf)$_2$ or Cu(BF$_4$)$_2$ as additive. Notably, MeCN was applied as the solvent instead the previously needed

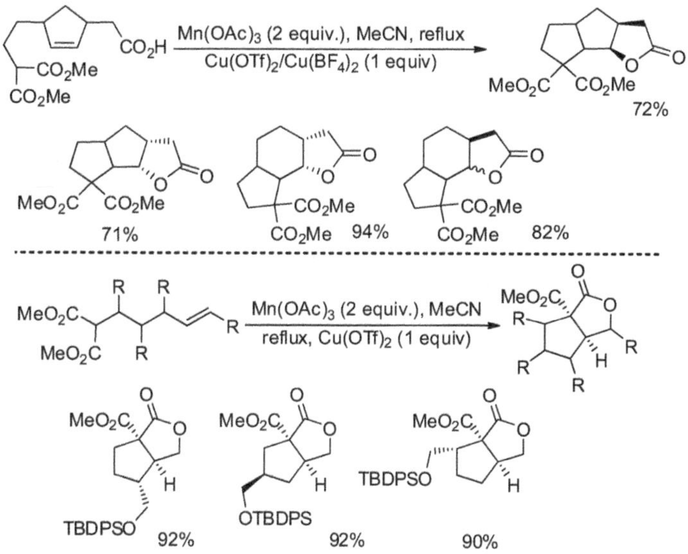

Scheme 6.16 Mn(OAc)₃-promoted synthesis of tricyclic γ-lactones.

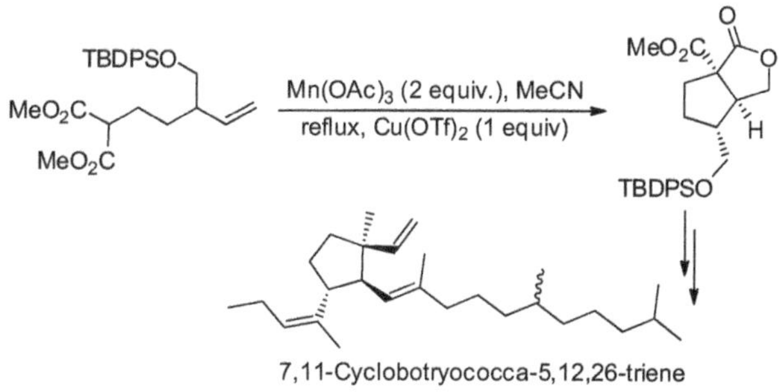

Figure 6.1 Total synthesis of 7,11-cyclobotryococca-5,12,26-triene.

acetic acid. Later, they found that this methodology can be extended to 4-pentenylmalonates.[65,66] Excellent yields of [3.3.0]-bicyclic γ-lactones were produced in the presence of Mn(OAc)₃ and Cu(OTf)₂. The successful application of this procedure in total synthesis was reported by the same group in 2010 (Figure 6.1).[67] The hydrocarbon natural product 7,11-cyclobotryococca-5,12,26-triene was efficiently prepared, the reaction featuring an oxidative radical cyclization for the diastereoselective synthesis of a [3.3.0]-bicyclic γ-lactone as a key step, along with copper-catalyzed coupling reactions for the stereo-controlled synthesis of the two trisubstituted alkenes.

In 2011, a regio- and stereoselective one-pot synthesis of carbohydrate-based butyrolactones was developed (Scheme 6.17).[68] Use manganese(III) acetate as the promoter, diverse arrays of [4.3.0] bicyclic carbohydrate-based γ-lactone building blocks were directly synthesized from glycals.

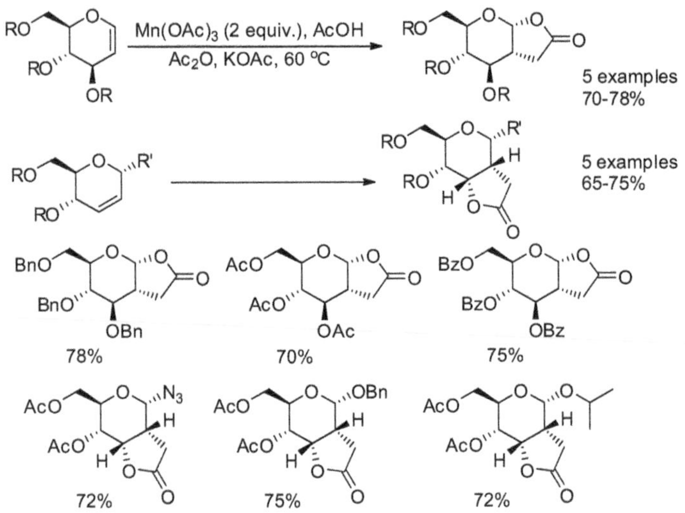

Scheme 6.17 Mn(OAc)$_3$-promoted cyclization of carbohydrate-based butyro-lactones.

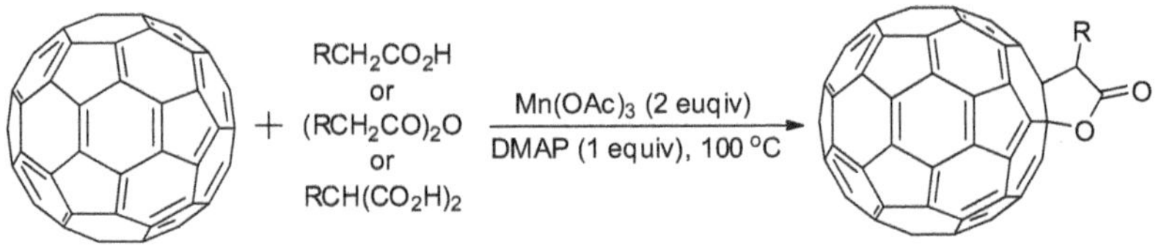

Scheme 6.18 Mn(OAc)$_3$-promoted synthesis of [60]fullerene-fused lactones.

A mechanism to explain the high regio- and stereoselectivity was proposed. In general, all the reactions proceeded via a radical reaction mechanism with the formation of a C–C bond, followed by a lactonization step. The regio-selectivity was determined at the level of formation of the C–C bond. Possibly, the key interactions are between the SOMO of the electron-deficient acetoxy radical and the HOMO of the glucal double bond, which determines the site of initial attack of the acetoxy radical. The high stereoselectivities of the lactonization may be rationalized by considering the ground-state con-formational distribution of glycals.

Wang and co-workers applied Mn(OAc)$_3$-mediated lactone synthesis in the preparation of [60]fullerene-fused lactones in 2006 (Scheme 6.18).[69] Mod-erate to good yields of the desired lactones were produced by reacting [60]fullerene with carboxylic acids, carboxylic anhydrides or malonic acids. Reductive ring opening of the lactones with Grignard reagents was also observed. It should be noted that DMAP played a crucial role in the suc-cessful synthesis of lactones from both carboxylic acids and anhydrides. No lactones were obtained with carboxylic anhydrides as the reagents in the absence of DMAP. As for carboxylic acids, by-products were formed without the addition of DMAP. Other bases such as pyridine, triethylamine and triethylenediamine were examined and found to be inferior to DMAP.

Sung and Wang reported an Mn(III)-based oxidative free-radical cyclizations of substituted allyl α-methyl-β-keto esters to lactones in 2003.[70] They performed detailed experimental studies and DFT calculations and the results showed that the *cis* cyclization is easier than the corresponding *trans* cyclization, but the *cis* radicals generated are not necessarily more stable than the corresponding *trans* radicals generated after the cyclizations. The free-radical cyclizations of substrates in the presence of Mn(OAc)₃ in acetic acid or acetonitrile are all reversible and operate under thermodynamic control; the stereoselectivity of the cyclizations depends on the relative stability of the cyclization-generated radicals. Therefore, the oxidative free-radical cyclization of allyl α-methyl-β-keto esters with Mn(OAc)₃ gives a *cis* compound as the major product, whereas the same oxidative free-radical cyclizations of substituted allyl α-methyl-β-keto esters with Mn(OAc)₃ produce *trans* products as major products.

More recently, Chen and co-workers described the construction of the [4 + 2] and [3 + 2] core skeletons of dimeric pyrrole–imidazole alkaloids and also applied them in the asymmetric synthesis of ageliferins (Figure 6.2).[71] By this oxidative method, the core skeletons of two classes of pyrrole–imidazole dimers were successfully prepared. Using this method, they successfully synthesized ageliferin, bromoageliferin and dibromoageliferin in their natural enantiomeric form. Through computational and experimental studies, they demonstrated that the biogenic dimerization of oroidin may proceed through a radical mechanism. Activation of oroidin by SET oxidation is likely used by enzymes to promote the otherwise difficult cycloaddition reactions. They further designed a biomimetic synthesis of the [4 + 2] dimer of oroidin using a manganese(III)-promoted radical tandem cyclization reaction.

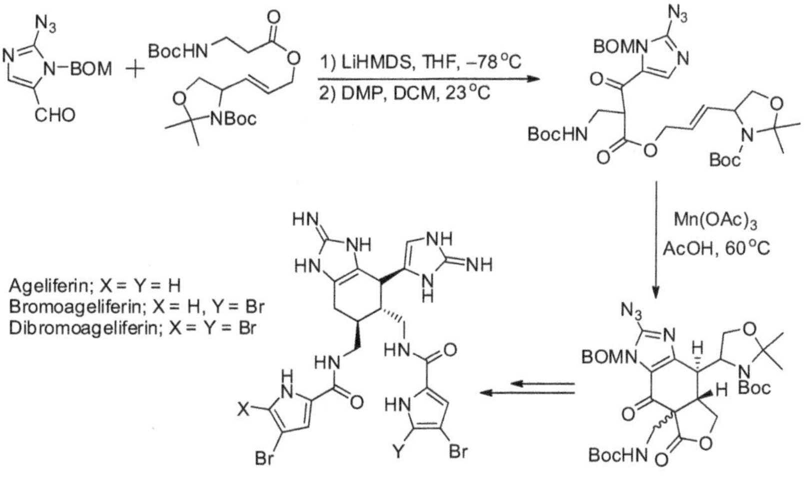

Figure 6.2 Total synthesis of ageliferins.

6.2.2   Manganese-Catalyzed Synthesis of Furans

In 1974, Heiba and Dessau reported a manganese-promoted cyclization of 1,3-diketones and alkenes to dihydrofurans.[72] The desired dihydrofurans were produced in good yields using acetic acid as solvent (Scheme 6.19). The reaction was proposed to proceed through a free-radical pathway.

In 1996, Nishino *et al.* investigated the reaction of alkenes with manganese(III) acetate dihydrate in the presence of diketene and elucidated the acetonylation pathway *via* dehydration of an equilibrium mixture of 5-hydroxy-2-pentanones and 2-tetrahydrofuranols.[73] When alcohols were added to the reaction mixture, the pentanones were trapped as 2-alkoxyte-trahydrofurans. Addition of amines resulted in the formation of 2-amino-4,5-dihydrofurans, when acetonitrile was used as the solvent. This was the first example of the introduction of an amino group into the 2-position of the 4,5-dihydrofuran ring using manganese(III) oxidation. Although most products were derived from conjugated enolate, minor products were also isolated from the unstable unconjugated enolate. Tetrahydrofurylideneacetates de-rived from the unstable unconjugated enolate were also obtained as minor products.

In 2004, the same group presented a novel Mn(III)-based procedure for the synthesis of dihydropyrans and 2,8-dioxabicyclo[3.3.0]oct-3-enes using the reaction of alkenes with 2-(2-oxoethyl)malonates and 3-acetyl-1,4-pentane-diones, respectively (Scheme 6.20).[74] This is an interesting example of the use of the tricarbonyl system for Mn(III)-based oxidative cyclization. These routes rely on the nucleophilic character of the carbonyl oxygen atoms of the malonates and pentanediones used to obtain the products by a cycloaddi-tion reaction or cycloaddition–tandem cyclization reactions.

Later, they reported a manganese(III)-induced oxidative cyclization of 3-(2-oxoethyl)piperidine-2,4-diones with 1,1-diarylethenes in acetic acid at reflux temperature to produce 3-aza-7,12-dioxatricyclo[4.3.0^{1,6}]dodec-8-en-2-ones, simply called azadioxa[4.3.3]propellanes, in excellent yields.[75,76] A similar oxidation of 2-(2-oxoethyl)cycloalkane-1,3-diones gave the corresponding [4.3.3]-, [5.3.3]- and [6.3.3]-propellanes. The oxidation of 3-oxopropyl-sub-stituted cycloalkane-1,3-diones also afforded the corresponding propellanes along with the 3-oxopropyl-substituted bicyclic intermediates. The bicyclic intermediates were definitely converted into the corresponding propellanes

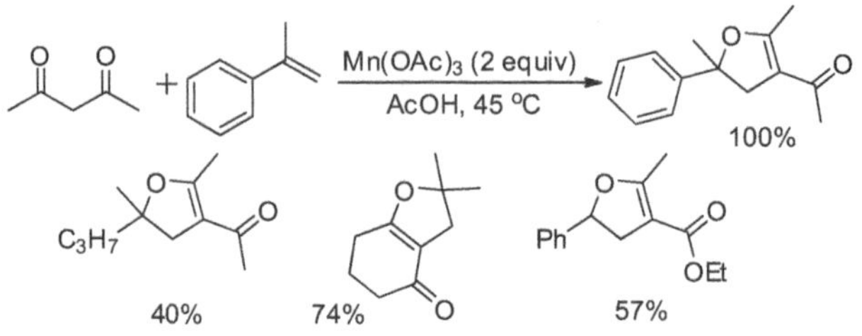

Scheme 6.19   Mn(OAc)$_3$-promoted synthesis of dihydrofurans.

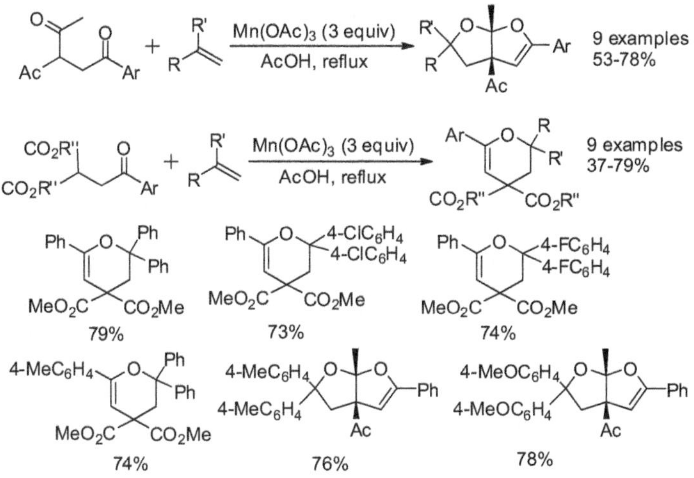

Scheme 6.20 Mn(OAc)₃-promoted synthesis of dihydropyrans.

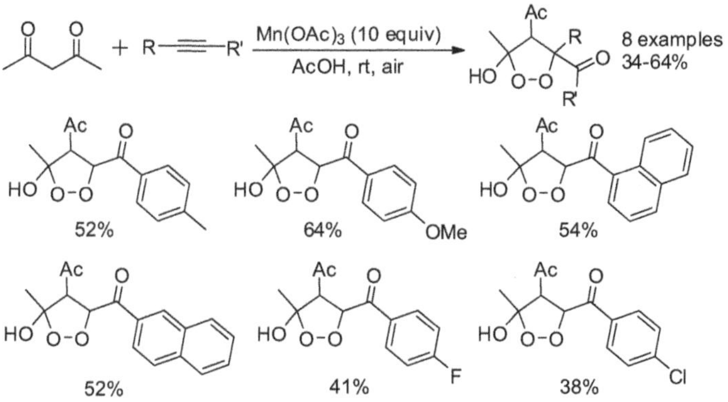

Scheme 6.21 Mn(OAc)₃-promoted synthesis of 1,2-dioxolanes.

in the presence of a Lewis acid. The structure determination and the reaction pathway were also described. In 2009, they reported the manganese(III)-based aerobic oxidation of arylacetylenes with 2,4-pentanedione at ambient temperature.[77] The reaction unexpectedly gave the 1,2-dioxolane derivatives in moderate yields together with a small amount of the oxiranes (Scheme 6.21). The 1,2-dioxolanes underwent silica gel-assisted contraction to give the oxiranes quantitatively.

Additionally, they found that manganese(III) acetate is an excellent catalyst for aerobic peroxidation.[78] Quinolinones bearing no substituent at the C3 position reacted with ethenes under aerobic oxidation conditions to produce bis(hydroperoxyethyl)quinolinediones together with [4.4.3]propellane-type cyclic peroxides. However, it was reported that the reaction of ethenes with quinolinones substituted at the C3 position gave the desired 1,2-dioxane-fused quinolinones and/or hydroperoxyethyl derivatives.

In 2005, manganese(III)-catalyzed aerobic oxidation of 2,4-piperidinediones was reported by the same group.[79] The reaction was performed in the presence of alkenes at room temperature, producing 1-hydroxy-8-aza-2,3-dioxabicyclo[4.4.0]decan-7-ones in excellent yields. On the other hand, 6-acetoxy-3-aza-7-oxabicyclo[4.3.0]nonan-2-ones were obtained by the oxidation of 2,4-piperidinedione-3-carboxylates with manganese(III) acetate in the presence of alkenes at elevated temperature under an argon atmosphere (Scheme 6.22). A similar oxidation using decarboxylated 2,4-piperidinediones produced the 2,3,6,7-tetrahydrofuro[3,2-c]pyridin-4(5H)-ones and/or 2,3,6,7-tetrahydrofuro[2,3-b]pyridin-4(5H)-ones in good yields. The determination of the structure and the decomposition reaction of the azabicyclic peroxides in acetic acid or acetic anhydride and the reaction pathway were also described.

In 2000, Brun's group reported the asymmetric radical synthesis of 2,3-dihydrofurans by applying Mn(OAc)₃ as the promoter.[80] trans-Disubstituted 2,3-dihydrofurans, with dr ranging from 2:1 to 9:1, depending on the substituent of the chiral auxiliary, were produced from alkyl acetoacetates and p-methoxycinnamoyloxazolidinones (Scheme 6.23). After chromatographic separation of the two diastereomers, the oxazolidinone can be

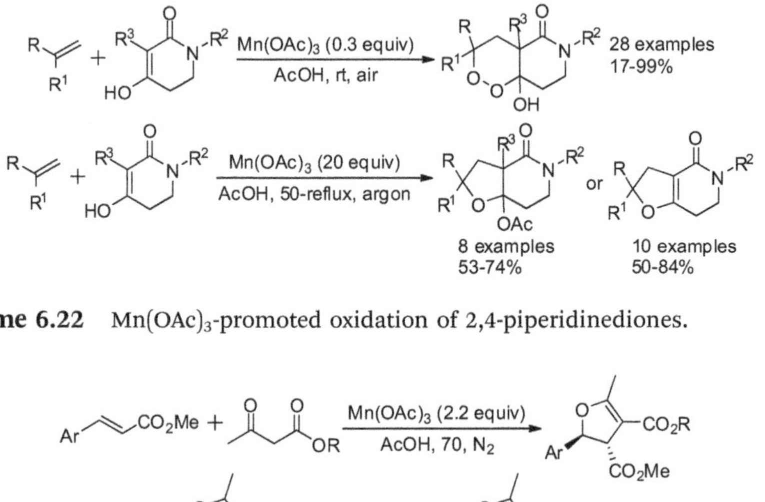

Scheme 6.22   Mn(OAc)₃-promoted oxidation of 2,4-piperidinediones.

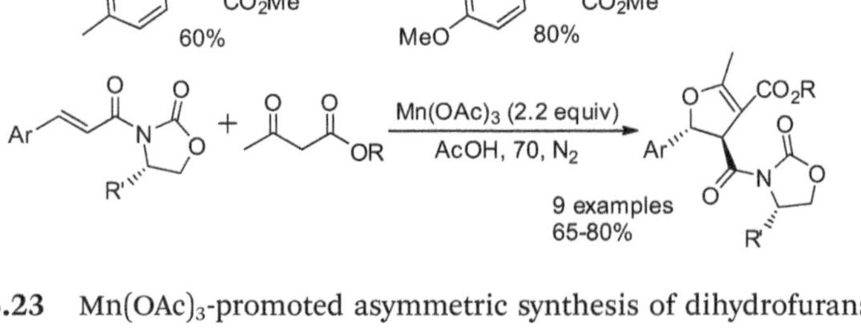

Scheme 6.23   Mn(OAc)₃-promoted asymmetric synthesis of dihydrofurans.

removed to afford enantiopure dihydrofuranyl esters in good overall yield. Later, this procedure was applied to the controlled preparation of functionalized mixed thiophene–furan oligomers.[81] Remarkably, this methodology was successfully used in the total synthesis of (+)-phyltetralin in 2003 (Figure 6.3).[82]

In 2005, Huang and Shi reported a manganese(III)-mediated oxidative annulation of methylenecyclopropanes with 1,3-dicarbonyl compounds.[83] The reaction was carried out in acetic acid and produced 4,5-dihydrofuran derivatives as [3 + 2] annulation products in moderate to good yields under mild conditions (Scheme 6.24). The formation of a stable and long-lived cyclopropylcarbinyl cation intermediate corresponds to this unpredictable result. More recently, the same group found that Mn(OAc)$_3$ can also promote the oxidative annulation of vinylidenecyclopropanes with 1,3-dicarbonyl

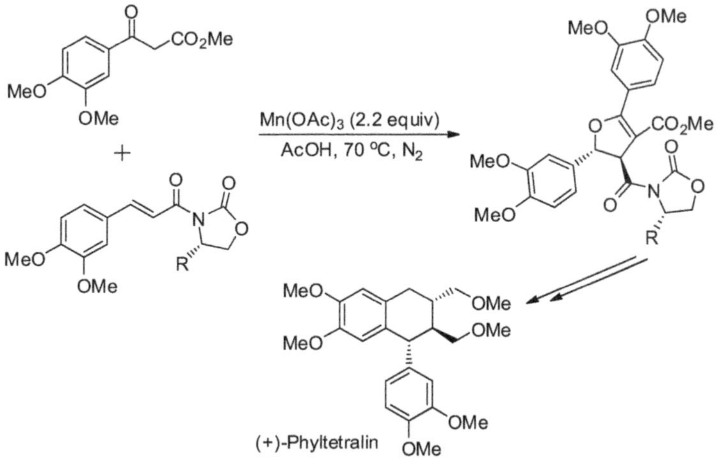

Figure 6.3 Total synthesis of (+)-phyltetralin.

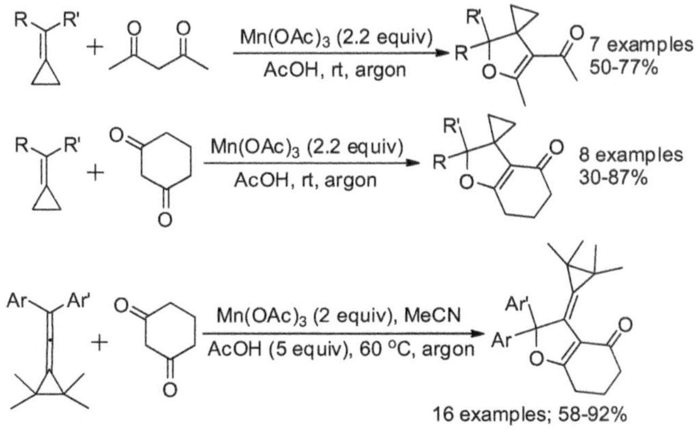

Scheme 6.24 Mn(OAc)$_3$-promoted annulation of methylenecyclopropanes.

compounds. The corresponding functionalized dihydrofuran derivatives were produced in moderate to good yields in acetonitrile. Notably, acetic acid was only needed as an additive in this new procedure rather than as a solvent.[84] Oxidative additions to benzonorbornadiene and oxabenzonorbornadiene were also reported.[85]

Wang *et al.* developed the synthesis of *trans*-2-acyl-3-aryl/alkyl-2,3,6,7-tetrahydro-4(5*H*)-benzofuranone derivatives through the radical reactions of 1-(pyridin-2-yl)enones with 1,3-cyclohexanediones mediated by Mn(OAc)$_3$ · 2H$_2$O (Scheme 6.25).[86] With the assistance of ball-milling, this methodology can be performed *via* solvent-free radical reactions, which were very efficient and afforded the unexpected products in yields as high as 91%. More importantly, the reversed regioselectivity of 1-(pyridin-2-yl)enones was revealed for the first time. The cyclization reactions gave good to excellent yields, and also extremely high diastereoselectivity and unexpected regioselectivity. Mn(OAc)$_3$ · 2H$_2$O played the roles of both an oxidant and a Lewis

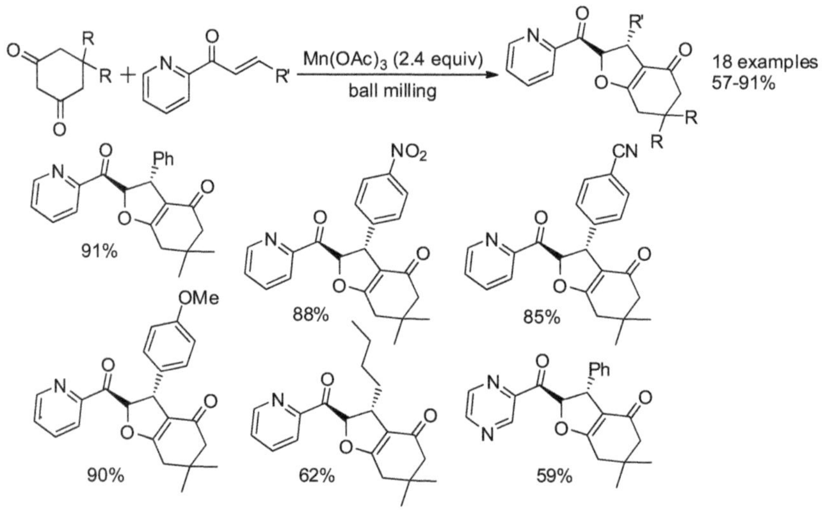

Scheme 6.25   Mn(OAc)$_3$-promoted synthesis of tetrahydro-4(5*H*)-benzofuranone derivatives.

Reaction Procedure (Scheme 6.25): Two stainless-steel jars, each containing a mixture of 1,3-cyclohexanedione (0.10 mmol), enone (0.12 mmol), Mn(OAc)$_3$ · 2H$_2$O (64.3 mg, 0.24 mmol), and a stainless-steel ball, were milled vigorously at a rate of 1800 rpm at room temperature for 1 h. The resulting mixtures were extracted with ethyl acetate, and the combined solution was evaporated to remove the solvent *in vacuo*. The residue was separated by flash column chromatography on silica gel (eluent: 2.5:1 petroleum ether–ethyl acetate) to afford 6,6-dimethyl-3-phenyl-2-picolinoyl-2,3,6,7-tetrahydrobenzofuran-4(5*H*)-one.

acid and thus altered the reaction pathway of 1-(pyridin-2-yl)enones. Additionally, the ionic liquid 1-butyl-3-methylimidazolium tetrafluoroborate ([bmim][BF$_4$]) as an unusual solvent was also applied in manganese(III)-mediated radical reactions.[87] This modification led to the efficient formation of carbon–carbon bonds and can avoid the use of AcOH as solvent. In the presence of Mn(OAc)$_3$ (2.1 equiv.) in [bmim][BF$_4$] and DCM (1 : 4 v/v), moderate yields of the desired products were obtained. Notably, the Mn(OAc)$_3$ can be recovered (on precipitation) by addition of further organic solvent to the reaction mixture, which is not dependent on the nature of the ionic liquid. After filtration, the manganese acetate was reacted with potassium permanganate to reoxidize the manganese(II) back to manganese(III).

In 2007, Yilmaz and co-workers reported the synthesis of 4-cyano-2,3-dihydrofuran-3-carboxamides *via* oxidative cyclization of 3-oxopropanenitriles with unsaturated amides using manganese(III) acetate as the promoter.[88] Treatment of 3-oxopropanenitriles with (2*E*)-3-(5-methyl-2-furyl)acrylamide and (2*E*)-3-(2-thienyl)acrylamide gave 2-(5-methyl-2-furyl)- and 2-(2-thienyl)-substituted 4-cyano-2,3-dihydrofuran-3-carboxamides in moderate yields, respectively. However, (2*E*)-3-(2-furyl)acrylamide and (2*E*)-3-phenylacrylamide did not give any product under the same conditions. On the other hand, reaction of a dienamide such as (2*E*,4*E*)-5-phenylpenta-2,4-dienamide with 3-oxopropanenitriles gave diastereomeric mixtures of 2-(2-vinylphenyl)-4-cyano-2,3-dihydrofuran-3-carboxamides. Later, they and other groups reported the cyclization of 1,3-dicarbonyl compounds with 2-thienyl- and 2-furyl-substituted alkenes,[89] internal alkenes[90–92] and dienes (Table 6.1).[93] Mechanisms were also proposed for the formation of all of these compounds.

Table 6.1 Mn(OAc)$_3$-mediated cyclization of alkenes with carbonyl compounds.

Entry	Substrate 1	Substrate 2	Product	Yield (%)
1				56
2				64
3				55

Table 6.1 *(Continued)*

Entry	Substrate 1	Substrate 2	Product	Yield (%)
4				53
5				49
6				42
7				47
8				88
9				85
10				87
11				79
12				90

Table 6.1 (*Continued*)

Entry	Substrate 1	Substrate 2	Product	Yield (%)
13	Ph–CO–CH₂–CO–CF₃	Ph, thienyl alkene	thiophene-furan product with Ph, CF₃	85
14	thienyl–CO–CH₂–CO–CF₃	Ph-cyclohexene	fused bicyclic furan product with Ph, thienyl, CF₃	83
15	thienyl–CO–CH₂–CO–CF₃	Ph,Ph-diene	dihydrofuran product with Ph, thienyl, CF₃	84
16	thienyl–CO–CH₂–CO–C₃F₇	Ph,Ph-diene	dihydrofuran product with Ph, thienyl, C₃F₇	88
17	thienyl–CO–CH₂–CO–CF₃	Ph-diene	dihydrofuran product with thienyl, CF₃	72
18	thienyl–CO–CH₂–CO–C₃F₇	Ph-diene	dihydrofuran product with thienyl, C₃F₇	74
19	thienyl–CO–CH₂–CO–CF₃	Ph,H-diene	dihydrofuran product with thienyl, CF₃	54
20	thienyl–CO–CH₂–CO–C₃F₇	Ph,H-diene	dihydrofuran product with thienyl, C₃F₇	57

Vanelle and co-workers reported a microwave-assisted synthesis of 5-(4-nitrophenyl)-2-phenyl-4-(phenylsulfonyl)-2,3-dihydrofuran in 2009 (Scheme 6.26).[94] The procedure was based on manganese(III) acetate-initiated oxidative cyclization of 1-(4-nitrophenyl)-2-(phenylsulfonyl)ethanone

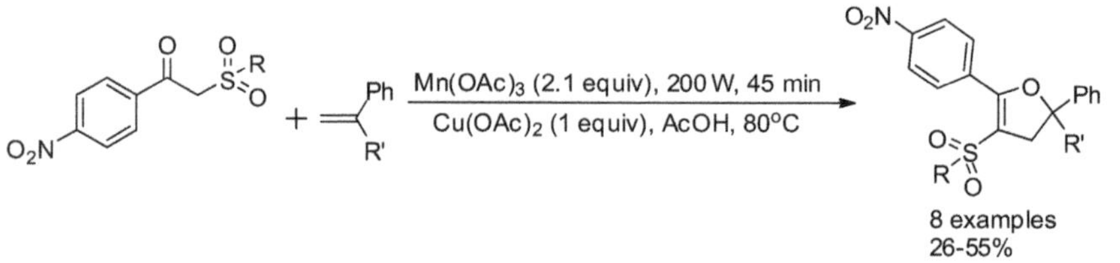

Scheme 6.26 Mn(OAc)$_3$-promoted cyclization of 1-(4-nitrophenyl)-2-(phenylsulfo-nyl)ethanone and vinylbenzene.

and vinylbenzene. This new protocol was applied to four sulfone derivatives, using vinylbenzene and diphenylethene, affording a series of 2,3-dihy-drofurans in moderate to good yields (26–55%). Similar methodology, ap-plied to allylbenzene, surprisingly led to dehydronaphthalene derivatives in moderate yields.

In 1997, Oshima and co-workers reported a radical cyclization of allyl 2-iodophenyl ether, *N,N*-diallyl-2-iodoaniline and 2-iodoethanal acetal using trialkylmanganate(II).[95] Treatment of allyl 2-halophenyl ethers with tribu-tylmanganate (*n*Bu$_3$MnLi or *n*Bu$_3$MnMgBr) provided dihydrobenzofuran derivatives in good yields. Indoline derivatives were also produced effectively starting from 2-iodoaniline compounds. The reaction are suggested to pro-ceed by the following sequences: (1) formation of a radical by treatment of iodophenol or iodoaniline derivatives with tributylmanganate(II), (2) radical cyclization and (3) recombination of radical and manganese species giving an alkylmanganese(II) compound. The reaction proved to proceed in the presence of a catalytic amount of manganese(II) chloride under atmospheric oxygen. The manganese-catalyzed radical cyclization could also be applied to 2-iodoethanal acetal, in which case the presence of oxygen was not neces-sary. With the knowledge that Mn(0) could be an active catalyst and play a critical role in manganese-catalyzed cyclization reactions, they subsequently developed a new strategy for the preparation of active Mn(0) and applied it in radical cyclization reactions.[96] They reduced Li$_2$MnCl$_4$ with magnesium in THF to afford a fairly active manganese species that readily initiated the radical cyclization of 2-iodoethanal allylic acetals at room temperature. The corresponding 2-bromoethanal acetals also provided the same cyclized products upon treatment with the activated manganese reagent at reflux in THF. This reagent can also be used to induce tandem radical cyclizations and the 5-*exo*/6-*endo* and 5-*exo*/6-*exo* modes are both available to give the products with *trans*-stereochemistry with regard to the ring junction. Fur-ther, the intramolecular type of sequential generation and utilization of radical and anionic species with this reagent have also been studied. Many kinds of 2-haloethanal acetal derivatives with a radical acceptor in a suitable position undergo cyclization reactions smoothly under very mild conditions with this new active Mn(0) reagent (Scheme 6.27 and Table 6.2).

It was demonstrated by Dickschat and Studer that radical can also be generated from an aromatic boronic acid using Mn(OAc)$_3$ as oxidant,

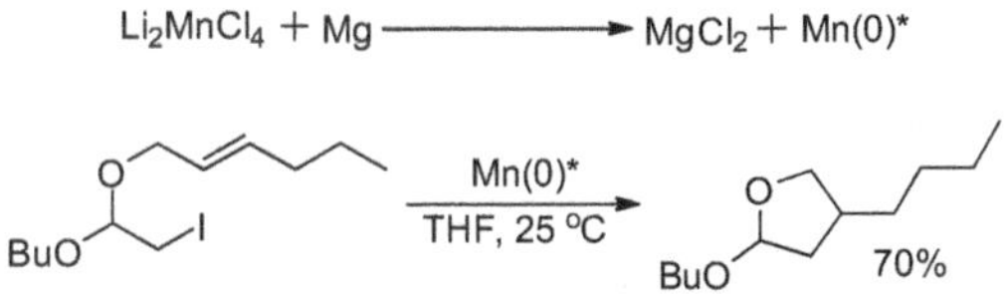

Scheme 6.27 Mn(0)-promoted cyclization of 2-iodoethanal allylic acetals.

Table 6.2 Substrate scope for Mn(0)-promoted cyclization.

Entry	Substrate	Product	Yield (%)
1	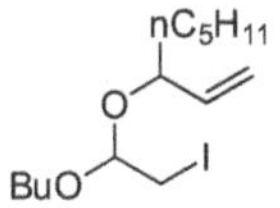	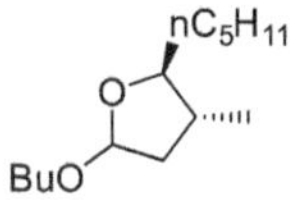	72
2	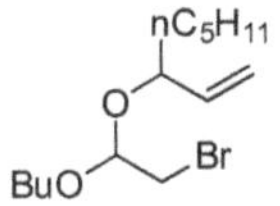	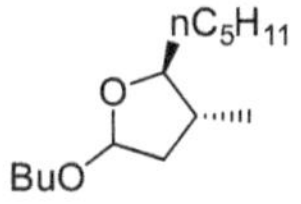	61
3	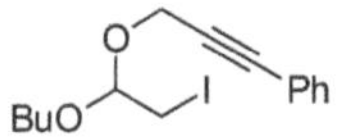	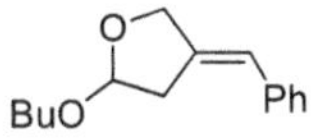	85
4	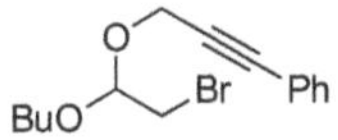	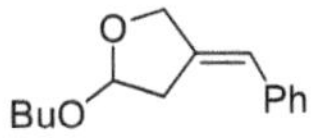	80
5	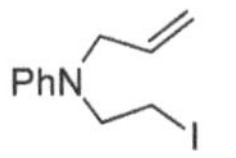	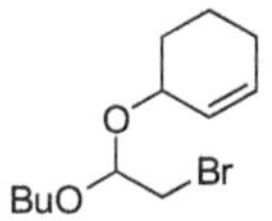	60
6	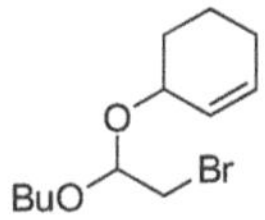	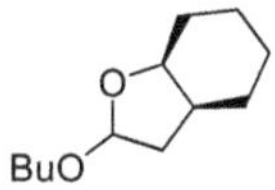	62
7	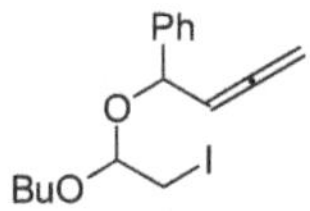	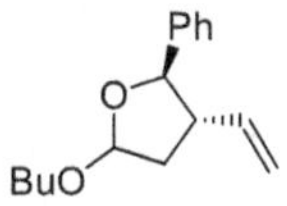	70

followed by cyclization.[97] Hydroxylated dihydrobenzofuran was produced in 28% yield starting from the corresponding arylboronic acid with Mn(OAc)$_3$ in the presence of dioxygen (Scheme 6.28).

In 2004, Burton and co-workers described a manganese(III) acetate-mediated synthesis of oxygen heterocycles.[98] The influence of copper(II) salts [Cu(OTf)$_2$, Cu(BF$_4$)$_2$ and Cu(SbF$_6$)$_2$] on the product distribution was studied. Unsaturated malonates bearing pendant alcohols yield carbocycles tethered to oxygen heterocycles on exposure to manganese(III) acetate and

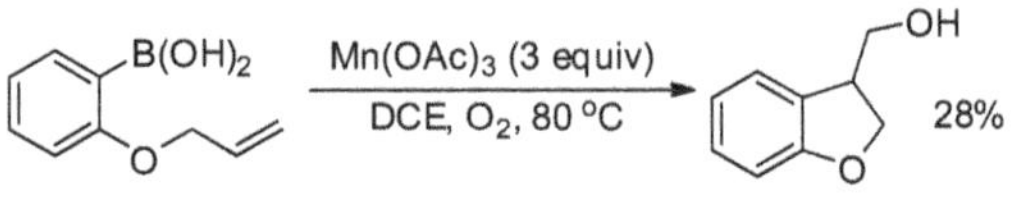

Scheme 6.28 Mn(OAc)$_3$-promoted synthesis of hydroxylated dihydrobenzofuran.

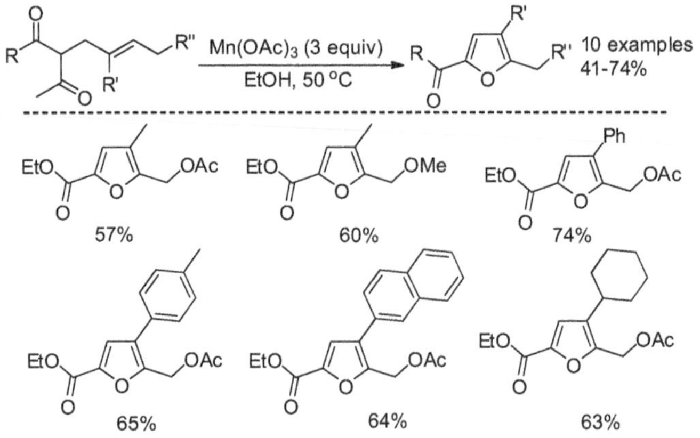

Scheme 6.29 Mn(OAc)$_3$-promoted synthesis of furans.

an appropriate copper(II) salt. The formation of tetrahydrofurans (THFs) may be achieved without copper(II) additives whereas the formation of tetrahydropyrans (THPs) necessitates the presence of copper(II) salts bearing poorly coordinating anions. With both THFs and THPs, the use of copper(II) additives bearing poorly coordinating anions results in products of oxidative substitution in preference to products arising from β-hydride elimination.

More recently, Koo and co-workers studied the mechanism and scope of the Mn(OAc)$_3$-mediated oxidative synthesis of furans.[99] Unless the Mn(III)-produced carbon radical from β-ketocarbonyl compounds undergoes smooth intramolecular addition to alkenes, it traps molecular oxygen in the reaction medium to produce a peroxy radical, which reacts with the neighboring carbonyl group to form 1,2-dioxetane. Thermal decomposition of 1,2-dioxetane completes the oxidation to produce an α-oxo ester. This oxidation seems to be general at 50 °C under aerobic conditions and can be catalytic for Mn(III) in AcOH with ultrasonic irradiation. Thus, the development of a new synthetic method for diversely substituted furans has been accomplished based on a combination of the Mn(III)-initiated domino oxidation of β-ketocarbonyl compounds with a suitable α-allylic substitution (Scheme 6.29). Rigorous exclusion of O$_2$ and a suitable disposition of alkenyl groups relative to the radical center are prerequisites for smooth carbon-radical cyclization, together with the assistance of the Cu(II) salt. Otherwise, oxidation of the α-carbon radicals is the general pathway.

6.2.3 Manganese-Catalyzed Synthesis of Pyrroles and Related Compounds

Chiba, Narasaka and co-workers reported an Mn(III)-catalyzed method for the synthesis of tri- and tetrasubstituted N–H pyrroles in 2008.[100] They applied readily available vinyl azides and 1,3-dicarbonyl compounds as substrates, and polysubstituted N–H pyrroles with a wide variety of substituents were obtained in good yields (Scheme 6.30). This catalytic reaction may be initiated by addition of Mn(III) enolate to vinyl azide via a radical pathway, giving an iminyl radical with release of Mn(II) species and dinitrogen. The resulting iminyl radical undergoes intramolecular addition to a carbonyl group to give an alkoxyl radical. Reduction of this alkoxyl radical by Mn(II) species gives Mn(III) alkoxide. Alternatively, reaction of the iminyl radical

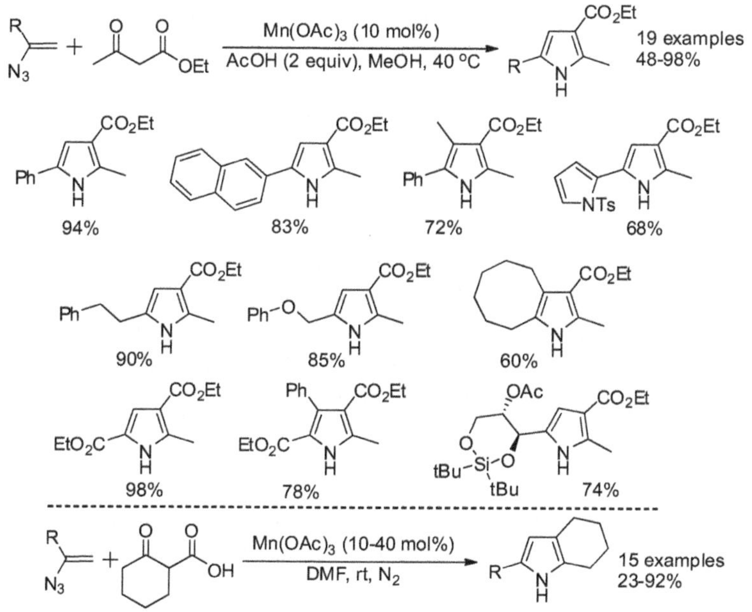

Scheme 6.30 Mn(OAc)$_3$-catalyzed synthesis of pyrroles.

Reaction Procedure (Scheme 6.30): To a solution of α-azidostyrene (145 mg, 1.00 mmol) and ethyl acetoacetate (191 μL, 1.50 mmol) in MeOH (10 mL) were added AcOH (114 μL, 2.00 mmol) and manganese(III) acetate dehydrate (27.6 mg, 0.10 mmol), and the mixture was stirred at 40 °C for 2 h. The reaction mixture was quenched with pH 9 ammonium buffer, then extracted twice with EtOAc. The combined organic extracts were washed with brine, dried over MgSO$_4$ and concentrated. Purification of the crude product by flash column chromatography on silica gel (eluent: 85:15 hexane–ethyl acetate) afforded the pyrrole (207 mg, 0.90 mmol) in 90% yield.

Reaction Procedure: To a solution of α-azidostyrene (52.8 mg, 0.364 mmol) and 2-oxocyclohexanecarboxylic acid (155.2 mg, 1.09 mmol) in DMF (3.6 mL) was added Mn(acac)$_3$ (12.8 mg, 0.0364 mmol) and the mixture was stirred for 5 h. The reaction mixture was quenched with pH 9 ammonium buffer, then extracted with Et$_2$O. The combined organic extracts were washed with brine, dried over MgSO$_4$ and concentrated. Purification of the crude product by flash column chromatography on Florisil (eluent: 97 : 3 hexane–EtOAc) afforded the pure product (59.5 mg, 0.302 mmol) in 83% yield.

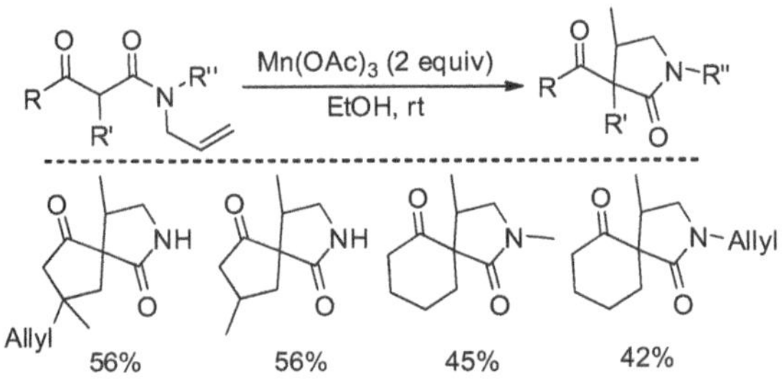

Scheme 6.31   Mn(OAc)$_3$-promoted synthesis of lactams.

with Mn(II) species affords alkylideneaminomanganese(III), nucleophilic attack of which on a carbonyl group gives an addition intermediate and yields a pyrrole after protonation with acetic acid followed by dehydration. In 2012, Swamy's group reported the thermolysis of phosphorus-based vinyl azides under solvent- and catalyst-free conditions and furnished a new route to 1,4-pyrazines.[101] A simple one-pot, Mn(III)-catalyzed photochemical route has been developed for multisubstituted pyrroles starting from allenes and 1,3-dicarbonyls via *in situ*-generated vinyl azides. The utility of new phosphorus-based pyrroles was also demonstrated in the Horner reaction. In 2011, Chiba's group reported another methodology for the synthesis of pyrroles.[102] The reaction started from vinyl azides and β-keto acids and N–H pyrroles with a variety of substituents were obtained in good yields.

In 2000, a methodology for the synthesis of substituted lactams and spirolactams was developed by Cossy *et al. via* Mn(III)-induced radical cyclization of unsaturated β-keto carboxamides (Scheme 6.31).[103] Treatment of the corresponding tertiary enamines under similar reaction conditions and in the presence of K$_2$CO$_3$ afforded the same cyclized products but with inversion of diastereoselectivity. The oxidation of optically pure secondary enamines leads to diastereomeric spirolactams in a ratio of ~3 : 1. The diastereomeric products were readily separated by silica gel chromatography and were obtained in enantiomerically pure form, suitable for use in the synthesis of natural products having a 2-azaspiro framework. Ishibashi and co-workers reported that treatment of *N*-[2-(3,4-dimethoxyphenyl)ethyl]-α-(methylthio)acetamide with Mn(OAc)$_3$ in the presence of Cu(OAc)$_2$ gave tetrahydroindol-2-one, which then cyclized with $Mn(OAc)_3$ to give 4-acetoxyer-ythrinane.[104] A similar reaction of the 3,4-methylenedioxyphenyl congener also gave tetrahydroindol-2-one, which, however, gave only a trace amount of the $Mn(OAc)_3$-mediated cyclization product and afforded the oxidation product. On the basis of these results, formation of 4-acetoxyerythrinane from tetra-hydroindol-2-one was thought to proceed *via* nucleophilic attack of the pyrrole ring on the cation radical, generated by a single-electron transfer reaction of the acetoxy-substituted intermediate. No reaction occurred on treatment of a compound without a methylthio group with $Mn(OAc)_3$–$Cu(OAc)_2$ and with re-covery of the starting material. On the other hand, treatment of the starting material with $Mn(OAc)_3$ using $Cu(OTf)_2$ as an additive in place of $Cu(OAc)_2$ gave another erythrinane. This method was applied to a formal synthesis of 3-demethoxyerythratidinone, a naturally occurring *Erythrina* alkaloid.

In 2009, Nishino's group reported that *N*-propenyl-3-oxobutanamides underwent manganese(III)-induced oxidative intramolecular cyclization in ethanol to produce 3-azabicyclo[3.1.0]hexan-2-ones in good yields.[105] A similar reaction of propenyl 3-oxobutanoates and *S*-propenyl 3-oxobuta-nethioates also gave the corresponding 3-oxa- and 3-thiabicyclo[3.1.0]hexan-2-ones. The reaction details, the structure determination and the reaction pathway were described.

In addition of the cyclization of alkenes, Burton and co-workers reported an intramolecular cyclization of alkyne derivatives (Scheme 6.32).[106] By this manganese(III)-mediated cyclization reaction of alkynoic amidomalonates and protected aminomalonates, (*Z*)-*exo*-alkylidenepyrrolidinones and -pyr-rolidines were produced in a convenient and practical manner. The products can be obtained in good yields with a clear preference for the sterically more demanding *Z*-products, rendering the approach complementary to

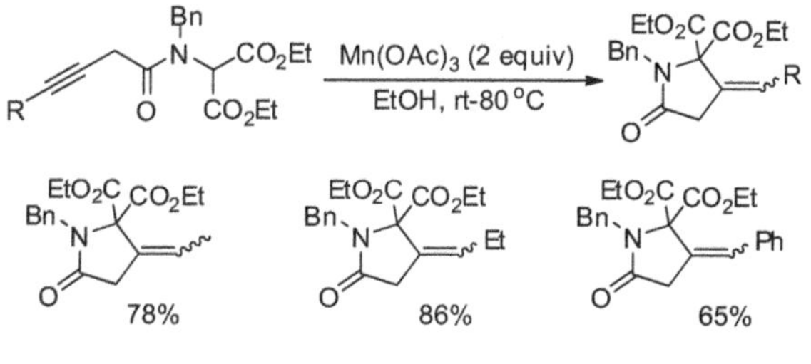

Scheme 6.32 $Mn(OAc)_3$-promoted synthesis of pyrrolidines.

Reaction Procedure (Scheme 6.32): Amidomalonate and manganese(III) acetate (2.0 equiv.) were dissolved in degassed alcohol solvent (20 mL/mmol malonate). The solution was heated to 80 °C and stirred for 15 h, then allowed to cool to room temperature and the solvent was removed under reduced pressure. The resulting residue was suspended in diethyl ether and filtered through a silica pad, eluting with diethyl ether. The solvent was removed under reduced pressure to yield the crude product.

previously developed Conia-ene type reactions. Not only are these cyclizations applicable to the formation of five-membered rings but it is also possible to synthesize piperidinones. Thus, exposure of the alkynyl malonates to manganese(III) acetate in ethanol at 80 °C gave the corresponding piperidinones in 66–71% yields.

In 2008, oxidative free-radical reactions of 2-substituted-1,4-quinone derivatives were described.[107] The electrophilic carbon-centered radical produced by the manganese(III) acetate oxidation of an α-chloro-β-keto ester undergoes efficient addition to the C=C double bond of 5,6-dimethyl-2-(methylamino)-1,4-benzoquinone and this reaction provides a novel method for the synthesis of a spirolactam and indole-2,4,7-trione. It shows high chemoselectivity depending on the migratory aptitude of the substituent on the α-chloro-β-keto ester. An imine radical can be generated from the oxidation of a β-enamino carbonyl compound with an Mn(III) or Ce(IV) salt. With 2-hydroxy-1,4-naphthoquinone, a spirolactam was effectively prepared from a β-enamino carbonyl compound. TBACN–CHCl₃ provide the most effective reaction conditions for the formation of a spirolactam.

The first chiral metallosalen-catalyzed enantioselective intramolecular amidation of sulfamate esters was developed by Che and co-workers in 2005 (Scheme 6.33).[108] The reaction proceeded with moderate to good yields and substrate conversions, with exclusive *cis* selectivity and with moderate to good enantioselectivity. The manganese(III) Schiff base-catalyzed reaction also represents the first step toward the development of a general catalytic system for asymmetric intramolecular C–N bond formation.

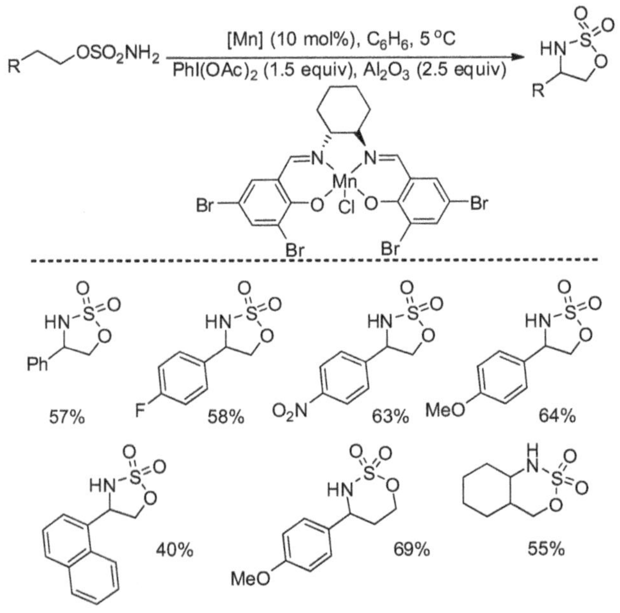

Scheme 6.33 Mn(OAc)₃-catalyzed intramolecular amidation of sulfamate esters.

6.3 Six-Membered Heterocycles

In 2009, Wang and Chiba developed an Mn(III)-mediated divergent synthesis of substituted pyridines and 2-azabicyclo[3.3.1]non-2-en-1-ol derivatives from readily available vinyl azides and cyclopropanols with a range of substituents.[109] In addition, versatile transformations of 2-azabicyclo[3.3.1]non-2-en-1-ol to 2-azabicyclo[3.3.1]nonane or -non-2-ene frameworks were exploited. Vinyl azides were successfully applied as a three-atom unit including one nitrogen to prepare pyridines and δ-lactams by reactions with monocyclic cyclopropanols and also to construct 2-azabicyclo[3.3.1] and 2-azabicyclo[4.3.1] frameworks with bicyclic cyclopropanols, bicyclo[3.1.0]hexan-1-ols and bicycle[4.1.0]heptan-1-ols. These reactions were initiated by a radical addition of β-carbonyl radicals, generated by the one-electron oxidation of cyclopropanols with Mn(III), to vinyl azides to give iminyl radicals, which cyclized with the intramolecular carbonyl groups. In addition, application of this methodology to the synthesis of the quaternary indole alkaloid melinonine-E was accomplished (Scheme 6.34).[110]

In 2012, Chatani, Tobisu and co-workers developed a novel bimolecular coupling of 2-isocyanobiaryls with organoboronic acids by a formal homolytic aromatic substitution (HAS) (Scheme 6.35).[111] This method permits the rapid divergent synthesis of phenanthridine and its π-extended analogs bearing aryl, heteroaryl and alkyl groups at the C6 position from readily accessible starting materials and a promoter. A mechanistic pathway for this manganese(III)-mediated annulation of 2-isocyanobiphenyls with boronic acid was proposed. The reaction of boronic acid with an Mn(III) salt generates an aryl or alkyl radical, which undergoes intermolecular addition to isocyanide to form an imidoyl radical. Intramolecular attack of the imidoyl radical on the pendant aromatic ring subsequently provides a cyclohexadienyl-type radical, which ultimately aromatizes to afford a phenanthridine. Several experiments were performed to provide support for the proposed mechanism. One such experiment involved the addition of 2,2,6,6-tetramethyl-1-piperidinoxyl (TEMPO) to quench the reaction and the putative alkyl radical intermediate was intercepted by TEMPO. In a separate experiment, the use of 5-hexenylboronic acid exclusively afforded a phenanthridine bearing a cyclopentylmethyl group, indicating that a rapid 5-*exo* process preceded the intermolecular addition to isocyanide.

In 2003, a simple method was developed for the diastereoselective synthesis of 2,3-aryl-substituted piperazines by Mercer and Sigman (Scheme 6.36).[112] They used both Mn(0) and a simple Brønsted acid as promoter and provided a convenient protocol for the synthesis of this important class of heterocycles. The method was shown to work for a variety of aryl substrates and can be easily accomplished on a 10 g scale. Additionally, highly substituted piperazines and seven-membered heterocyclic rings can be formed in good yields.

In 2003, Taylor and co-workers developed another procedure for the preparation of piperazines.[113] They applied α-hydroxyketones and 1,2-diamines as substrates and obtained quinoxalines or dihydropyrazines in moderate to

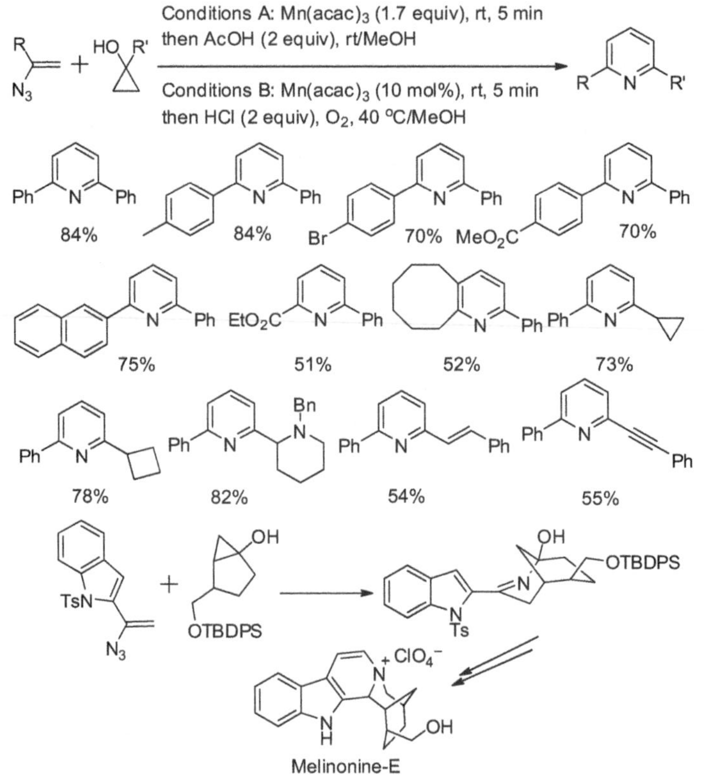

Scheme 6.34 Mn(OAc)$_3$-mediated synthesis of pyridines.

Reaction Procedure (Scheme 6.34): To a solution of α-azidostyrene (43.6 mg, 0.30 mmol) and 1-phenylcyclopropanol (48.4 mg, 0.36 mmol) in MeOH (3.0 mL) was added Mn(acac)$_3$ (10.6 mg, 0.03 mmol) at room temperature under a nitrogen atmosphere. After 5 min, HCl (0.20 mL, 0.60 mmol, 3.0 M in MeOH) was added and the nitrogen balloon was then replaced with an oxygen balloon. The reaction mixture was heated at 40 °C for 1 h and quenched with pH 9 ammonium buffer, then extracted twice with ethyl acetate. The combined organic extracts were washed with brine, dried over MgSO$_4$ and concentrated. Purification of the crude product by flash column chromatography on silica gel (eluent: 98:2 hexane–ethyl acetate) afforded the pure product (55.6 mg, 0.24 mmol) in 80% yield as a pale-yellow solid.

---

good yields (Scheme 6.37). The reaction proceeded *via* MnO$_2$-mediated oxidation followed by *in situ* trapping with aromatic or aliphatic 1,2-diamines. The reactions can be carried out by a one pot procedure, which avoids the need to isolate the highly reactive 1,2-dicarbonyl intermediates.

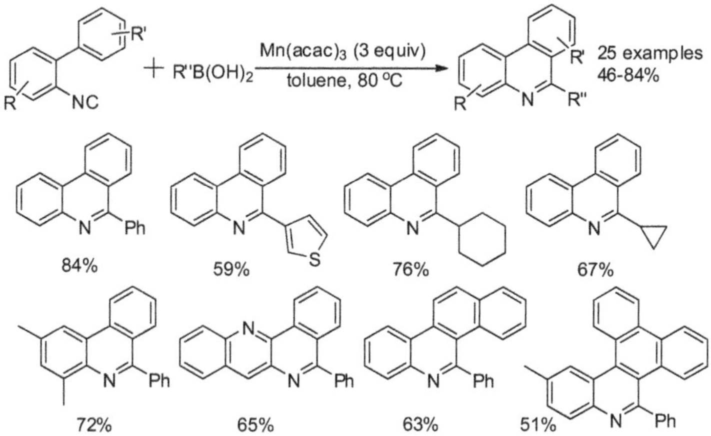

Scheme 6.35 Mn(OAc)$_3$-mediated synthesis of pyridines from isocyanides.

Reaction Procedure (Scheme 6.35): To an oven-dried 10 mL two-necked flask, isocyanide (0.30 mmol), phenylboronic acid (0.45 mmol), Mn(acac)$_3$ (0.90 mmol) and toluene (2.0 mL) were added sequentially under a gentle stream of nitrogen. The mixture was stirred at 80 °C for 1 h under a nitrogen atmosphere. The reaction mixture was then cooled to room temperature and purified by column chromatography on silica gel to afford the desired product.

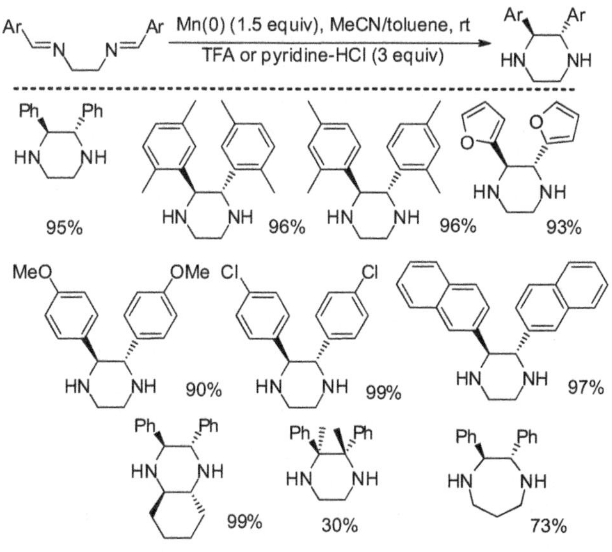

Scheme 6.36 Mn(0)-mediated synthesis of piperazines.

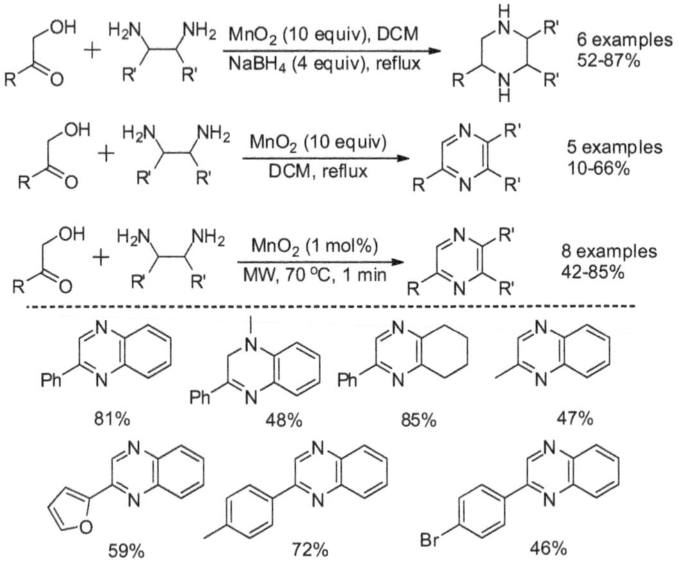

Scheme 6.37 MnO$_2$-catalyzed synthesis of piperazines and quinoxalines.

Reaction Procedure: General Procedure for Quinoxalines and Dihydropyrazines (Scheme 6.37): To a mixture of α-hydroxyketone (0.50 mmol), 1,2-diamine (1.00 mmol) and powdered 4 Å molecular sieves (0.50 g) in dry CH$_2$Cl$_2$ (25 mL) was added activated MnO$_2$ (0.435 g, 5.00 mmol) and the mixture was heated to reflux. With ethylenediamine, 2.0 M HCl in Et$_2$O (1 equiv. w.r.t. amine) was also added to suppress the formation of amide by-product. After complete reaction, the mixture was cooled to r.t., filtered through Celite and the residue washed well with CH$_2$Cl$_2$. Concentration and purification of the crude product by flash column chromatography on silica gel for quinoxalines or deactivated, neutral alumina for dihydropyrazines gave the desired product.

Reaction Procedure: General Procedure for Piperazines: As described above, but with diamine reduced to 0.60 mmol and NaBH$_4$ (0.076 g, 2.00 mmol) included. After complete consumption of substrate, MeOH (6 mL) was added at r.t. and the mixture was stirred for 20 h. Work-up and purification by acid–base extraction gave the desired product.

Reaction Procedure for 2-Phenylquinoxaline: To a 10 mL glass vial were added α-hydroxyacetophenone (0.133 g, 0.97 mmol), 1,2-diaminobenzene (0.211 g, 1.95 mmol), 4 Å molecular sieves (1.0 g), and activated MnO$_2$ (Aldrich, product No. 21764-6; 1 mg). The vessel was then heated at 70 °C under microwave irradiation using a Discover Synthesizer (monomode microwave cavity at 2.45 GHz; temperature control by automated adjustment of irradiation power in the range 0–300 W). After 1 min, the vial was cooled to r.t. The solution was filtered and concentrated under reduced pressure and purified by flash chromatography on silica (eluent: hexane–diethyl ether) to give 163 mg (81% yield) of 2-phenylquinoxaline.

Modifications of the procedure allowed the selective formation of pyrazines and piperazines. Later, Chung and co-workers showed that quinoxalines can be prepared from the same substrates using a catalytic amount of MnO_2 with microwave irradiation.[114]

More recently, an $Mn(acac)_3$-catalyzed aerobic dehydrogenative cyclization of indole derivatives was developed.[115] This catalytic system can convert from two C–H bonds (C2–H of indole and α-C–H bond of malonate ester) to a C–C bond as part of a polycyclic indole skeleton with a synthetically useful yield. In this method, aerobic oxygen can be used as the stoichiometric oxidant and water is the sole stoichiometric side-product. This was the first example of base metal-catalyzed aerobic dehydrogenative cyclization producing ring-fused indole skeletons. In 2009, Nishino's group developed a manganese(III)-mediated synthesis of 4,4-bis(ethoxycarbonyl)-3,4-dihydro-2(1*H*)-quinolinones from the corresponding 2-[2-(*N*-arylamino)-2-oxoethyl]malonates.[116] It was found that the malonates having not only electron-releasing but also electron-withdrawing groups on the aromatic ring of the aniline moiety efficiently underwent oxidative cyclization to give the corresponding dihydroquinolinones. Furthermore, it was also possible to introduce an acetoxymethyl functionality into the dihydroquinolinone skeleton when an excess amount of the oxidant was used. The dihydroquinolinones could then be converted into the corresponding quinoline derivatives. In addition, the oxidative cyclization of the *ortho*-substituted malonates having a relatively lower ionization potential led to spirolactams. The 6-*endo*/5-*exo* selectivity of the cyclization was interpreted as being due to the difference in activation energy between the 6-*endo* and 5-*exo* cyclizations.

In 2000, Ishii and co-workers reported an example of manganese-catalyzed oxidative synthesis of a peroxide from styrene and cyclohexanone.[117] In the presence of 0.5 mol% of $Mn(OAc)_2$ in AcOH under pressure of O_2, the corresponding peroxide was isolated in 70% yield (Scheme 6.38).

In 2008, Nishino's group studied the cyclization of 1,1-diarylethenes and cyclic 1,3-dicarbonyl compounds.[118,119] Azatrioxa- and trioxa[4.4.3]-propellanes were produced selectively by manganese(III)-catalyzed aerobic oxidation. The oxopropyl-substituted diketones did not afford the desired propellanes, but rather the dioxabicyclic intermediates. The conversion of these acid-sensitive dioxabicyclic intermediates into the corresponding endoperoxypropellanes was achieved by Lewis acid-induced intramolecular cyclization. Additionally, the aerobic oxidation of tetronic acid in the presence of 1,1-disubstituted alkenes afforded hydroperoxyethyl peroxylactones, and a similar reaction using 3-alkyl-substituted tetronic acids gave stable crystalline peroxylactones in good to excellent yields. The oxidation using a

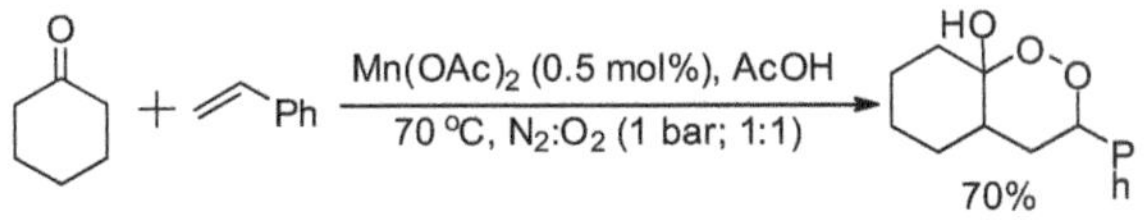

Scheme 6.38 $Mn(OAc)_2$-catalyzed synthesis of a peroxide.

stoichiometric amount of manganese(III) acetate did not give the bicyclic lactone but the ethenyl- and/or ethyltetronic acid derivatives.

In 2011, the biomimetic oxidation of lapachol using aqueous hydrogen peroxide as oxidant and chloro[5,10,15,20-tetrakis(2,6-dichloro-phenyl)porphyrinatomanganese(III)] (Mn-Porph) as catalyst was described.[120] *o*-Naphthoquinones were obtained from *m*-CPBA, whereas *p*-naphthoquinones were highly favored when using Mn-Porph and H_2O_2.

In 2011, Koo and co-workers developed a protocol of the Mn(III)-promoted tandem oxidation and cyclization of terpenoids containing a β-keto ester unit (Scheme 6.39).[121] The carbon radical center of the β-keto ester undergoes oxidation and subsequent intramolecular hetero-Diels–Alder reaction with the terpenoid chain to give bicyclic dihydropyrans. The dialkyl substitution pattern at the terminal sp^2 carbon of the terpenoids is a structural requirement for the intramolecular hetero-Diels–Alder reaction of inverse electronic demand. Only the oxidation product, δ-hydroxy-β,γ-unsaturated-α-oxo esters, prevails for the terpenoid chains with monoalkyl or no substitution at the terminal sp^2 carbon. This tandem sequence is highly effective and various polycyclic dihydropyrans were synthesized in high yields and stereoselectivities by a single reaction from β-keto esters of terpenoids.

Wang, Zha and co-workers developed a new catalytic system for the one-pot synthesis of 3,4-disubstituted coumarins by using Mn_3O_4 nanoparticles as a heterogeneous catalyst.[122] A variety of 3,4-disubstituted coumarin derivatives were synthesized from substituted 2-(hydroxymethyl)phenols and β-keto esters in good yields (Scheme 6.40). The advantages of this Mn_3O_4 nanoparticle system include low catalyst loading, high activity and good recyclability.

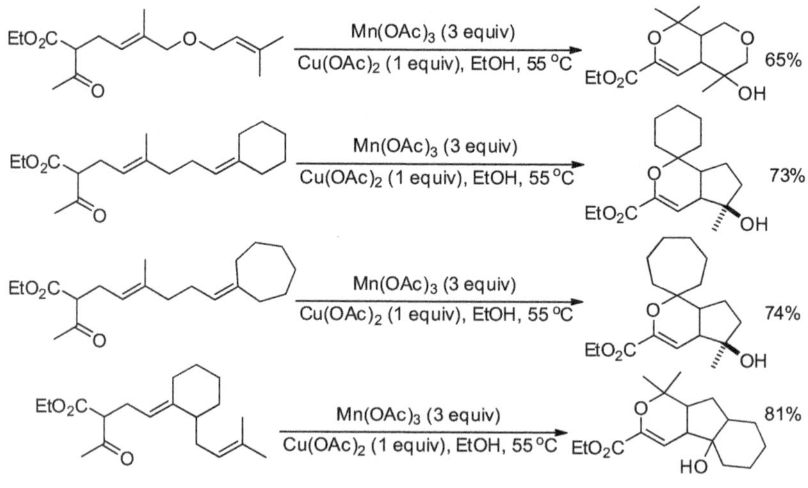

Scheme 6.39   Mn(OAc)$_3$-promoted cyclization of terpenoids.

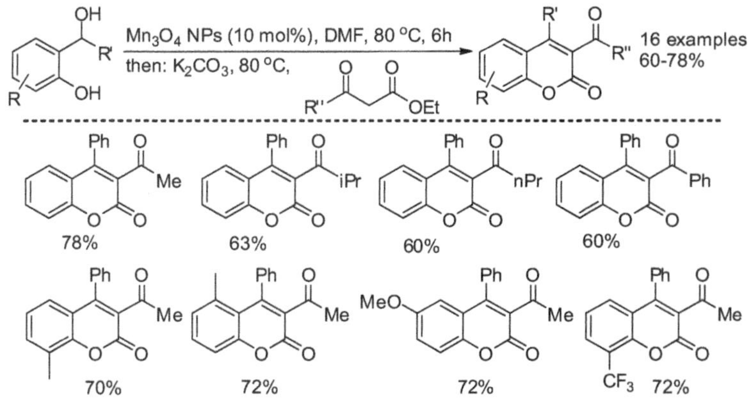

Scheme 6.40 Mn_3O_4-catalyzed synthesis of coumarins.

6.4 Other Heterocycles

In addition to the above-mentioned manganese-catalyzed radical reactions, manganese catalysts have also been applied in the preparation of four-membered heterocycles. In 2000, a manganese-mediated diastereoselective 4-*exo-trig* cyclization of enamides to β-lactams was developed.[123] Various desired products were produced in moderate to good yields (Scheme 6.41). The effect of chiral substituents on the enamide nitrogen atom on the diastereoselection of the Mn(III)-mediated 4-*exo-trig* cyclization to β-lactams was studied. A significant level of diastereoselectivity was achieved when an amino acid ester moiety was included in the enamidic skeleton.

Nishino and co-workers developed a new approach to the synthesis of dibenz[*b,f*]oxepincarboxylates *via* Mn(OAc)$_3$-mediated oxidative intramolecular rearrangement.[124] They also proposed a reaction mechanism for the formation of the product involving 1,2-aryl radical rearrangement and subsequent decarboxylation. The oxidation of monoalkyl 2-(9*H*-xanthenyl)malonates with Mn(OAc)$_3$ gave the 9- or 10-dibenz[*b,f*]oxepincarboxylates in good yields (Scheme 6.42). The reaction proceeds with high regioselectivity except in the case of (1-methoxyxanthenyl)malonate (R = 1-MeO, R' = Me), which gave two regioisomers. It was proposed that the process for the formation of dibenz[*b,f*]oxepincarboxylates must include a 1,2-aryl radical rearrangement followed by oxidative decarboxylation.

Macrocyclic compounds reveal supramolecular behavior such as molecular recognition, anion binding, metal ion transport, enzymatic catalysis and chemical switching, hence considerable attention has been devoted to this topic and manganese salts along with other catalysts have been explored in this area. Based on their developed methodologies, Nishino and co-workers reported a series of studies on the preparation of macrocycles using manganese(III) acetate as promoter.[125] Various size of rings

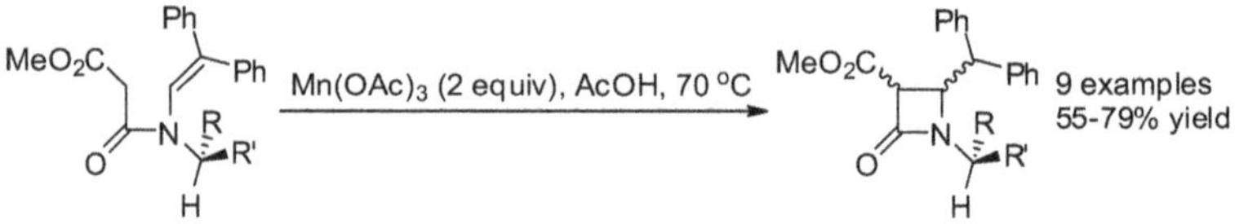

Scheme 6.41   Mn(OAc)$_3$-promoted synthesis of β-lactams.

Reaction Procedure (Scheme 6.41): To a solution of enamide (1 mmol) in glacial acetic acid (5 mL), Mn(OAc)$_3$ · 2H$_2$O was added (536 mg, 2 mmol) under an argon atmosphere. The reaction mixture was stirred at 70 °C until the brown color disappeared; then it was cooled to room temperature and poured into water (50 mL). The resulting mixture was extracted with CH$_2$Cl$_2$ and the organic phase was washed with saturated NaHCO$_3$ solution, then with water, and finally dried over Na$_2$SO$_4$. Removal of the solvent under reduced pressure and chromatographic separation on a silica gel column (eluent: petroleum ether–Et$_2$O) afforded pure β-lactam.

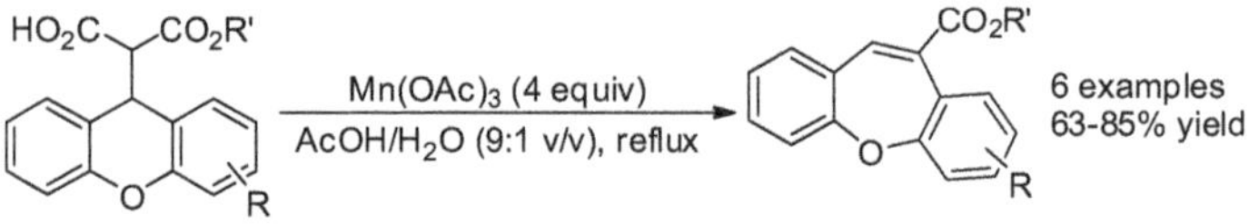

Scheme 6.42   Mn(OAc)$_3$-promoted synthesis of dibenz[*b,f*]oxepincarboxylates.

Reaction Procedure (Scheme 6.42): To a heated solution of monoalkyl 2-(9*H*-xanthenyl)malonate (0.5 mmol) in glacial acetic acid (10 mL) in the presence and absence of an additive was added Mn(OAc)$_3$ (2 mmol) just before refluxing. The reaction was stopped when the dark-brown color of the solution turned clear red. The reaction mixture was then cooled to room temperature and the solvent was removed *in vacuo*. The residue was triturated with 2 M HCl (15 mL) followed by extraction with CHCl$_3$ (3×10 mL). The combined extracts were washed with a saturated aqueous solution of NaHCO$_3$ (2×15 mL) and water (2×10 mL). The organic layer was dried over MgSO$_4$ and again concentrated to dryness. The crude products were separated by silica gel TLC (Wako B-10) with CHCl$_3$ as eluent to give dibenz[*b,f*]oxepincarboxylate. 9-Xanthenone and xanthene were also isolated in some cases.

---

were produced, including macrodiolide and 13-methyl-11,11-diphenyl-3,8,12-trioxabicyclo[8.3.0]tridec-13-en-2-one (Figure 6.4).

Pyrrole–imidazole alkaloids are a family of highly nitrogenated and halogenated natural products that possess unique molecular skeletons and

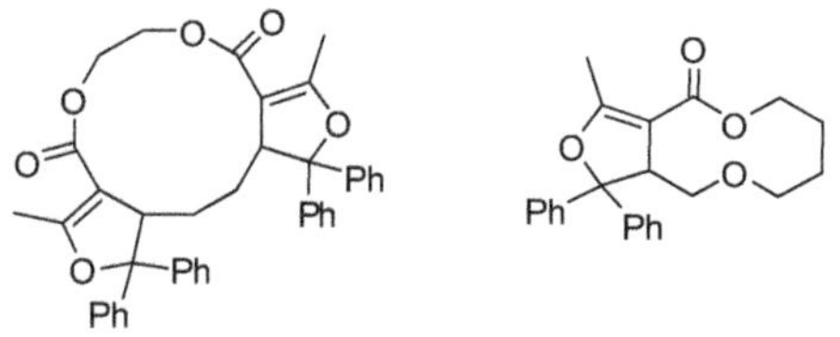

Figure 6.4 Macrocyclic compounds produced by Mn(OAc)₃-promoted synthesis.

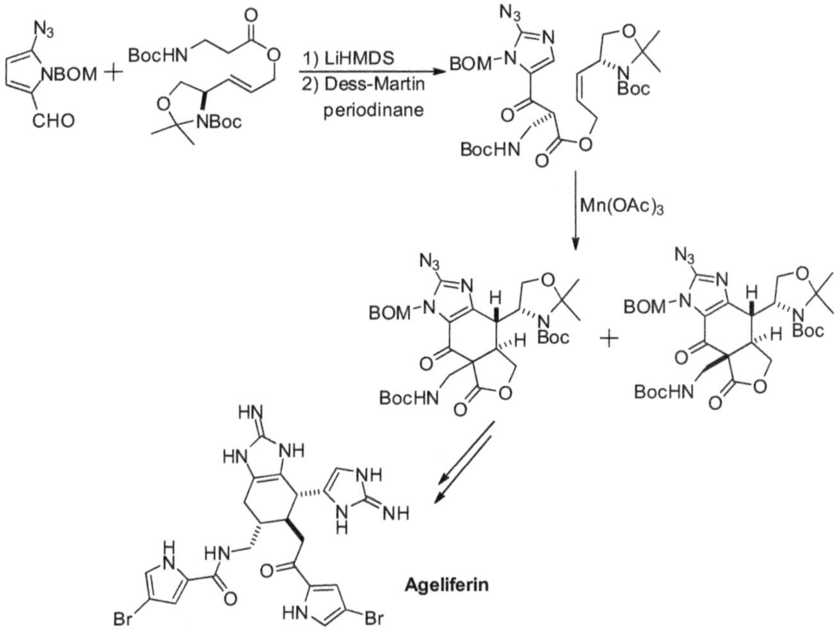

Figure 6.5 Mn(OAc)₃-promoted total synthesis of ageliferin.

significant biological activities. Ageliferin is a dimeric member found in many *Agelas* and *Stylissa* sponges. In 2011, an Mn(III)-mediated oxidative radical cyclization reaction was used as the key step to construct the core skeleton of this pyrrole–imidazole dimer (Figure 6.5).[126] This approach resembles biogenic [4 + 2] dimerization in an intramolecular fashion. Chen and co-workers also reported the total synthesis of palau'amine (Figure 6.6).[127]

Among the vast numbers of indole and pyrrole alkaloids there exist a small number that bear a six-membered ring fused to the 1,2-positions and which have an all-carbon quaternary center attachment at the heterocyclic 2-position. In 2006, Magolan and Kerr demonstrated that the intramolecular cyclization of a malonic radical, generated with Mn(OAc)₃, onto the 2-position of indoles and pyrroles is an effective method for

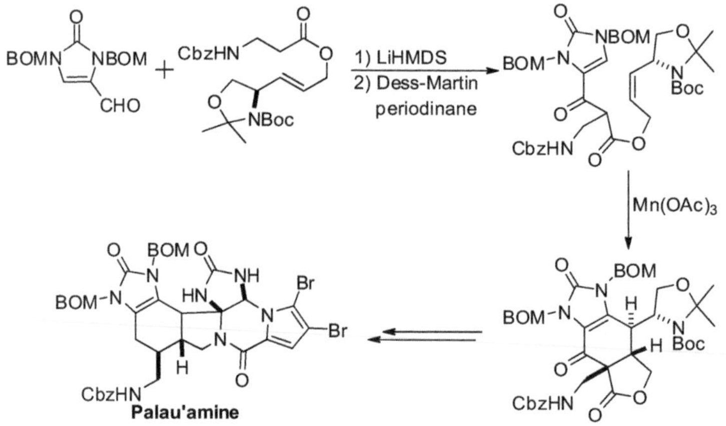

Figure 6.6 Mn(OAc)$_3$-promoted total synthesis of palau'amine.

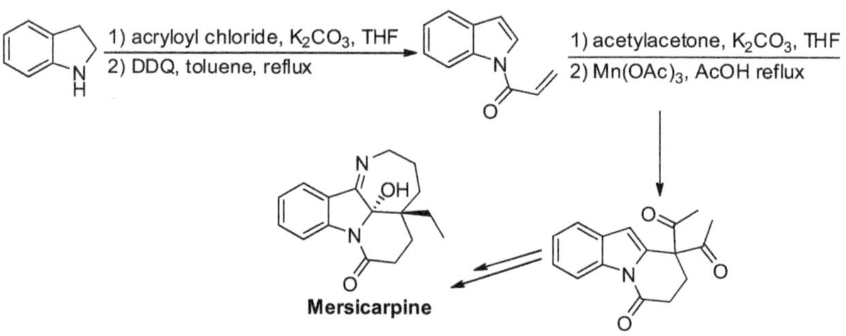

Figure 6.7 Mn(OAc)$_3$-promoted total synthesis of mersicarpine.

the generation of annulated heterocycles.[128] Indoline substrates can also be employed as they are oxidized to indoles under the reaction conditions. The application of this method to the synthesis of the tetracyclic core of the natural product tronocarpine was also illustrated. Later, they applied this methodology in the total synthesis of mersicarpine (Figure 6.7).[129]

The fungal metabolite (+)-fusarisetin A [(+)-**1**], produced by the soil fungus *Fusarium* sp. FN080326, potently inhibits acinar morphogenesis, cell migration and cell invasion in MDAMB-231 cells. Tests of cell growth and cell death using the same cell line showed that (+)-**1** does not exhibit significant cytotoxicity. These findings suggest that (+)-**1** holds promise as a valuable anticancer agent. In 2012, Gao and co-workers accomplished the first asymmetric synthesis of (+)-**1** (Figure 6.8).[130] This strategy relies on a biosynthetic pathway involving (1) a one-pot IMDA/Roskamp reaction to

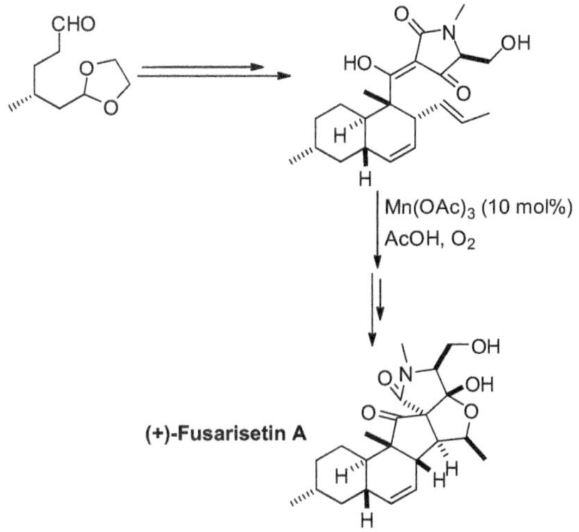

Figure 6.8 Mn(OAc)$_3$-promoted total synthesis of (+)-fusarisetin A.

construct the transdecalin skeleton and (2) biosynthetic oxidation of equisetin promoted by Mn(III)–O$_2$.

6.5   Summary

The main contributions of manganese-catalyzed heterocycle synthesis have been summarized and discussed. Manganese salts are used typically as radical initiators and also act as a Lewis acid. For future work, efforts should be made to reduce the manganese promoter to catalytic amounts with a 'green' extra oxidant. The possibility of applying manganese catalysts in cross-coupling reactions still needs to be explored.

References

1. B. Meunier, E. Guilmet, M. De Carvalho and R. Poilblanc, *J. Am. Chem. Soc.*, 1984, **106**, 6668–6676.
2. B. De Poorter and B. Meunier, *J. Chem. Soc., Perkin Trans.*, 1985, **2**, 1735–1740.
3. D. Mohajer, R. Tayebee and H. Goudarziafshar, *J. Chem. Res. (S)*, 1999, 168–169.
4. J. Y. Liu, X. F. Li, Y. Z. Li, W. B. Chang and A. J. Huang, *J. Mol. Catal. A: Chem.*, 2002, **187**, 163–167.
5. S. Tangestaninejad, M. Moghadam, V. Mirkhani, I. Mohammadpoor-Baltork and R. Hajian, *Inorg. Chem. Commun.*, 2010, **13**, 1501–1503.
6. Z. Solati, M. Hashemi and L. Ebrahimi, *Catal. Lett.*, 2011, **141**, 163–167.

7. X. Huang, X. Fu, X. Wu and Z. Jia, *Tetrahedron Lett.*, 2013, **54**, 4041–4044.

8. (a) L. Ma, F. Su, W. Guo, S. Q. Zhang, Y. Guo and J. Hu, *Micropor. Mesopor. Mater.*, 2013, **169**, 16–24; (b) L. Ma, F. Su, X. Zhang, D. Song, Y. Guo and J. Hu, *Micropor. Mesopor. Mater.*, 2014, **184**, 37–46.

9. D. Mohajer, G. Karimipour and M. Bagherzadeh, *New J. Chem.*, 2004, **28**, 740–747.

10. D. Mohajer and Z. Solati, *Tetrahedron Lett.*, 2006, **47**, 7007–7010.

11. D. Mohajer and L. Sadeghian, *J. Mol. Catal. A: Chem.*, 2007, **272**, 191–197.

12. L. Mahmoudi, D. Mohajer, R. Kissner and W. H. Koppenol, *Dalton Trans.*, 2011, **40**, 8695–8700.

13. S. Zakavi and L. Ebrahimi, *Polyhedron*, 2011, **30**, 1732–1738.

14. S. Zakavi, S. Talebzadeh and S. Rayati, *Polyhedron*, 2012, **31**, 368–372.

15. Z. Solati, M. Hashemi, A. Keshavarzi and E. Rafiee, *J. Porphyrins Phthalocyanines*, 2012, **16**, 149–153.

16. A. M. d'A. R. Gonsalves, R. A. W. Johnstone, M. M. Pereira and J. Shaw, *J. Chem. Soc., Perkin Trans.*, 1991, **1**, 645–649.

17. A. Thellend, P. Battioni and D. Mansuy, *J. Chem. Soc., Chem. Commun.*, 1994, 1035–1036.

18. S. L. H. Rebelo, M. M. Q. Simoes, M. G. P. M. S. Neves, A. M. S. Silva and J. A. S. Cavaleiro, *Chem. Commun.*, 2004, 608–609.

19. S. L. H. Rebelo, M. M. Q. Simoes, M. G. P. M. S. Neves, A. M. S. Silva, J. A. S. Cavaleiro, A. F. Peixoto, M. M. Pereira, M. R. Silva, J. A. Paixao and A. M. Beja, *Eur. J. Org. Chem.*, 2004, 4778–4787.

20. R. De Paula, M. M. Q. Simoes, M. G. P. M. S. Neves and J. A. S. Cavaleiro, *Catal. Commun.*, 2008, **10**, 57–60.

21. H. H. Monfared, V. Aghapoor, M. Ghorbanloo and P. Mayer, *Appl. Catal. A: Gen.*, 2010, **372**, 209–216.

22. Z. Solati, M. Hashemi, S. Hashemnia, E. Shahsevani and Z. Karmand, *J. Mol. Catal. A: Chem.*, 2013, **374–375**, 27–31.

23. Y. Li, X. T. Zhou and H. B. Ji, *Catal. Commun.*, 2012, **27**, 169–173.

24. E. Brulé and Y. R. de Miguel, *Tetrahedron Lett.*, 2002, **43**, 8555–8558.

25. S. Tangestaninejad, M. H. Habibi, V. Mirkhani and M. Moghadam, *Synth. Commun.*, 2002, **32**, 3331–3337.

26. S. Y. Liu and D. G. Nocera, *Tetrahedron Lett.*, 2006, **47**, 1923–1926.

27. S. Alavi, H. Hosseini-Monfared and M. Siczek, *J. Mol. Catal. A: Chem.*, 2013, **377**, 16–28.

28. F. Bruyneel, C. Letondor, B. Bastürk, A. Gualandi, A. Pordea, H. Stoeckli-Evans and R. Neier, *Adv. Synth. Catal.*, 2012, **354**, 428–440.

29. K. H. Tong, K. Y. Wong and T. H. Chan, *Org. Lett.*, 2003, **5**, 3423–3425.

30. B. Qi, L. L. Lou, K. Yu, W. Bian and S. Liu, *Catal. Commun.*, 2011, **15**, 52–55.

31. A. Stamatis, C. Vartzouma and M. Louloudi, *Catal. Commun.*, 2011, **12**, 475–479.

32. J. J. Dong, P. Saisaha, T. G. Meinds, P. L. Alsters, E. G. Ijpeij, R. P. van Summeren, B. Mao, M. Fananas-Mastral, J. W. de Boer, R. Hage, B. L. Feringa and W. R. Browne, *ACS Catal.*, 2012, **2**, 1087–1096.

33. M. Ghorbanloo, H. H. Monfared and C. Janiak, *J. Mol. Catal. A: Chem.*, 2011, **345**, 12–20.

34. M. M. Najafpour, F. Rahimi, M. Amini, S. Nayeri and M. Bagherzadeh, *Dalton Trans.*, 2012, **41**, 11026–11031.

35. M. Amini, M. M. Najafpour, S. Nayeri, B. Pashaei and M. Bagherzadeh, *RSC Adv.*, 2012, **2**, 3654–3657.

36. J. Tang, Y. Zu, W. Huo, L. Wang, J. Wang, M. Jia, W. Zhang and W. R. Thiel, *J. Mol. Catal. A: Chem.*, 2012, **355**, 201–209.

37. S. Muratsugu, Z. Wenig and M. Tada, *ACS Catal.*, 2013, **3**, 2020–2030.

38. M. Ghorbanloo, S. Rahmani and H. Yahiro, *Transition Met. Chem.*, 2013, **38**, 725–732.

39. P. Oietikäinen, *Tetrahedron*, 1998, **54**, 4319–4326.

40. S. Wie, Y. Tang, G. Xu, X. Tang, Y. Ling, R. Li and Y. Sun, *React. Kinet. Catal. Lett.*, 2009, **97**, 329–333.

41. J. Teixeira, A. R. Silva, L. C. Branco, C. A. M. Afonso and C. Freire, *Inorg. Chim. Acta*, 2010, **363**, 3321–3329.

42. S. Wie, Y. Tang, X. Xu, G. Xu, Y. Yu, Y. Sun and Y. Zheng, *Appl. Organomet. Chem.*, 2011, **25**, 146–153.

43. X. Hu, X. Fu, J. Xu and C. Wang, *J. Organomet. Chem.*, 2011, **696**, 2797–2804.

44. X. Huang, X. Fu, Z. Jia, Q. Miao and G. Wang, *Catal. Sci. Technol.*, 2013, **3**, 415–424.

45. K. Matkiewicz, A. Bukowska and W. Bukowski, *J. Inorg. Organomet. Polym.*, 2012, **22**, 332–341.

46. F. Ruffo, A. Bismuto, A. Carpentieri, M. E. Cucciolito, M. Lega and A. Tuzi, *Inorg. Chim. Acta*, 2013, **405**, 288–294.

47. J. R. L. Smith and G. Reginato, *Org. Biomol. Chem.*, 2003, **1**, 2543–2549.

48. W. K. Chan, P. Liu, W. Y. Yu, M. K. Wong and C. M. Che, *Org. Lett.*, 2004, **6**, 1597–1599.

49. W. K. Chan, M. K. Wong and C. M. Che, *J. Org. Chem.*, 2005, **70**, 4226–4232.

50. H. Srour, P. Le Maux and G. Simonneaux, *Inorg. Chem.*, 2012, **51**, 5850–5856.

51. S. Liao and B. List, *Angew. Chem. Int. Ed.*, 2010, **49**, 628–631.

52. S. Minakata, T. Ando, M. Nishimura, I. Ryu and M. Komastu, *Angew. Chem. Int. Ed.*, 1998, **37**, 3392–3394.

53. J. B. Bush and H. Finkbeiner, *J. Am. Chem. Soc.*, 1968, **90**, 5903–5905.

54. E. I. Heiba, R. M. Dessau and W. J. Koehl Jr, *J. Am. Chem. Soc.*, 1968, **90**, 5905–5906.

55. E. I. Heiba, R. M. Dessau and P. G. Rodewald, *J. Am. Chem. Soc.*, 1974, **96**, 7977–7981.

56. E. J. Corey and M. C. Kang, *J. Am. Chem. Soc.*, 1984, **106**, 5384–5385.

57. E. J. Corey and A. W. Gross, *Tetrahedron Lett.*, 1985, **26**, 4291–4294.

58. A. B. Ernst and W. E. Fristad, *Tetrahedron Lett.*, 1985, **26**, 3761–3764.

59. W. E. Fristad and J. R. Peterson, *J. Org. Chem.*, 1985, **50**, 10–18.

60. W. E. Fristad and S. S. Hershberger, *J. Org. Chem.*, 1985, **50**, 1026–1031.

61. W. E. Fristad, J. R. Peterson and A. B. Ernst, *J. Org. Chem.*, 1985, **50**, 3143–3148.
62. W. E. Fristad, J. R. Peterson, A. B. Ernst and G. B. Urbi, *Tetrahedron*, 1986, **42**, 3429–3442.
63. B. B. Snider, J. E. Merritt, M. A. Dombroski and B. O. Buckman, *J. Org. Chem.*, 1991, **56**, 5544–5553.
64. D. G. Hulcoop and J. W. Burton, *Chem. Commun.*, 2005, 4687–4689.
65. L. H. Powell, P. H. Docherty, D. G. Hulcoop, P. D. Kemmitt and J. W. Burton, *Chem. Commun.*, 2008, 2559–2561.
66. A. W. J. Logan, J. S. Parker, M. S. Hallside and J. W. Burton, *Org. Lett.*, 2012, **14**, 2940–2943.
67. J. J. Davies, T. M. Krulle and J. W. Burton, *Org. Lett.*, 2010, **12**, 2738–2741.
68. S. K. Yousuf, D. Mukherjee, L. Mallikharjunrao and S. C. Taneja, *Org. Lett.*, 2011, **13**, 576–579.
69. (a) G. Wang, F. Li and T. Zhang, *Org. Lett.*, 2006, **8**, 1355–1358; (b) F. Li, T. Liu, Y. Huang and G. Wang, *J. Org. Chem.*, 2009, **74**, 7743–7749.
70. K. Sung and Y. Y. Wang, *J. Org. Chem.*, 2003, **68**, 2771–2778.
71. X. Wang, X. Wang, X. Tan, J. Lu, K. W. Cormier, Z. Ma and C. Chen, *J. Am. Chem. Soc.*, 2012, **134**, 18834–18842.
72. E. I. Heiba and R. M. Dessau, *J. Org. Chem.*, 1974, **39**, 3456–3457.
73. H. Nishino, V. H. Nguyen, S. Yoshinaga and K. Kurosawa, *J. Org. Chem.*, 1996, **61**, 8264–8271.
74. V. H. Nguyen and H. Nishino, *Tetrahedron Lett.*, 2004, **45**, 3373–3377.
75. K. Asahi and H. Nishino, *Tetrahedron Lett.*, 2006, **47**, 7259–7262.
76. K. Asahi and H. Nishino, *Tetrahedron*, 2008, **64**, 1620–1634.
77. T. Tsubusaki and H. Nishino, *Tetrahedron*, 2009, **65**, 3745–3752.
78. R. Kumabe and H. Nishino, *Tetrahedron Lett.*, 2004, **45**, 703–706.
79. K. Asahi and H. Nishino, *Tetrahedron*, 2005, **61**, 11107–11124.
80. F. Garzino, A. Méou and P. Brun, *Tetrahedron Lett.*, 2000, **41**, 9803–9807.
81. F. Garzino, A. Méou and P. Brun, *Helv. Chim. Acta*, 2002, **85**, 1989–1998.
82. F. Garzino, A. Méou and P. Brun, *Eur. J. Org. Chem.*, 2003, 1410–1414.
83. J. W. Huang and M. Shi, *J. Org. Chem.*, 2005, **70**, 3859–3863.
84. W. Yuan, Y. Wie and M. Shi, *Tetrahedron*, 2011, **67**, 7139–7142.
85. R. Caliskan, T. Pekel, W. H. Watson and M. Balci, *Tetrahedron Lett.*, 2005, **46**, 6227–6230.
86. G. W. Wang, Y. W. Dong, P. Wu, T. T. Yuan and Y. B. Shen, *J. Org. Chem.*, 2008, **73**, 7088–7095.
87. G. Bar, A. F. Parsons and C. B. Thomas, *Chem. Commun.*, 2001, 1350–1351.
88. E. V. Burgaz, M. Yilmaz, A. T. Pekel and A. Öktemer, *Tetrahedron*, 2007, **63**, 7229–7239.
89. M. Yilmaz, *Tetrahedron*, 2011, **67**, 8255–8263.
90. M. Yilmaz and A. T. Pekel, *J. Fluorine Chem.*, 2011, **132**, 628–635.

91. E. Bicer, M. Yilmaz, M. Karatas and A. T. Pekel, *Helv. Chim. Acta*, 2012, **95**, 795–804.
92. R. Caliskan, M. F. Ali, E. Sahin, W. H. Watson and M. Balci, *J. Org. Chem.*, 2007, **72**, 3353–3359.
93. M. Yilmaz, E. V. B. Yilmaz and A. T. Pekel, *Helv. Chim. Acta*, 2011, **94**, 2027–2038.
94. C. Curti, M. D. Crozet and P. Vanelle, *Tetrahedron*, 2009, **65**, 200–205.
95. R. Inoue, J. Nakao, H. Shinokubo and K. Oshima, *Bull. Chem. Soc. Jpn.*, 1997, **70**, 2039–2049.
96. (a) J. Tang, H. Shinokubo and K. Oshima, *Synlett*, 1998, 1075–1076; (b) J. Tang, H. Shinokubo and K. Oshima, *Tetrahedron*, 1999, **55**, 1893–1904.
97. A. Dickschat and A. Studer, *Org. Lett.*, 2010, **12**, 3972–3974.
98. D. G. Hulcoop, H. M. Sheldrake and J. W. Burton, *Org. Biomol. Chem.*, 2004, **2**, 965–967.
99. C. Wang, Z. Li, Y. Ju and S. Koo, *Eur. J. Org. Chem.*, 2012, 6976–6985.
100. Y. F. Wang, K. K. Toh, S. Chiba and L. Narasaka, *Org. Lett.*, 2008, **10**, 5019–5022.
101. K. V. Sajna and K. C. K. Swamy, *J. Org. Chem.*, 2012, **77**, 8712–8722.
102. E. P. J. Ng, Y. F. Wang and S. Chiba, *Synlett*, 2011, 783–786.
103. J. Cossy, A. Bouzide and C. Leblanc, *J. Org. Chem.*, 2000, **65**, 7257–7265.
104. S. Chikaoka, A. Toyao, M. Ogasawara, O. Tamura and H. Ishibashi, *J. Org. Chem.*, 2003, **68**, 312–318.
105. K. Asahi and H. Nishino, *Synthesis*, 2009, 409–423.
106. H. A. Keane, W. Hess and J. W. Burton, *Chem. Commun.*, 2012, **48**, 6496–6498.
107. A. I. Tsai and C. P. Chuang, *Tetrahedron*, 2008, **64**, 5098–5102.
108. J. Zhang, P. W. H. Chan and C. M. Che, *Tetrahedron Lett.*, 2005, **46**, 5403–5408.
109. Y. F. Wang and S. Chiba, *J. Am. Chem. Soc.*, 2009, **131**, 12570–12572.
110. Y. F. Wang, K. K. Toh, E. P. J. Ng and S. Chiba, *J. Am. Chem. Soc.*, 2011, **133**, 6411–6421.
111. M. Tobisu, K. Koh, T. Furukawa and N. Chatani, *Angew. Chem. Int. Ed.*, 2012, **51**, 11363–11366.
112. G. J. Mercer and M. S. Sigman, *Org. Lett.*, 2003, **5**, 1591–1594.
113. S. A. Raw, C. D. Wilfred and R. J. K. Taylor, *Chem. Commun.*, 2003, 2286–2287.
114. S. Y. Kim, K. H. Park and Y. K. Chung, *Chem. Commun.*, 2005, 1321–1323.
115. K. Oisaki, J. Abe and M. Kanai, *Org. Biomol. Chem.*, 2013, **11**, 4569–4572.
116. T. Tsubusaki and H. Nishino, *Tetrahedron*, 2009, **65**, 9448–6459.
117. T. Iwahama, S. Sakaguchi and Y. Ishii, *Chem. Commun.*, 2000, 2317–2318.
118. K. Asahi and H. Nishino, *Eur. J. Org. Chem.*, 2008, 2404–2416.
119. M. A. Haque and H. Nishino, *Heterocycles*, 2011, **83**, 1783–1805.
120. S. M. G. Pires, R. De Paula, M. M. Q. Simoes, A. M. S. Silva, M. R. M. Domingues, I. C. M. S. Santos, M. D. Vargas, V. F. Ferreira, M. G. P. M. S. Neves and J. A. S. Cavaleiro, *RSC Adv.*, 2011, **1**, 1195–1199.

121. Z. Li, H. Jung, M. Park, M. S. Lah and S. Koo, *Adv. Synth. Catal.*, 2011, **353**, 1913–1917.
122. H. Sun, Y. Zhang, F. Guo, Y. Yan, C. Wan, Z. Zha and Z. Wang, *Eur. J. Org. Chem.*, 2012, 480–483.
123. A. D'Annibale, D. Nanni, C. Trogolo and F. Umani, *Org. Lett.*, 2000, **2**, 401–402.
124. Z. Cong, T. Miki, O. Urakawa and H. Nishino, *J. Org. Chem.*, 2009, **74**, 3978–3981.
125. (a) T. Yoshinaga, H. Nishino and K. Kurosawa, *Tetrahedron Lett.*, 1998, **39**, 9197–9200; (b) S. Jogo, H. Nishino, M. Yasutake and T. Shinmyozu, *Tetrahedron Lett.*, 2002, **43**, 9031–9034; (c) Y. Ito, T. Yoshinaga and H. Nishino, *Tetrahedron*, 2010, **66**, 2683–2694; (d) Y. Ito, Y. Tomiyasu, T. Kawanabe, K. Uemura, Y. Ushimizu and H. Nishino, *Org. Biomol. Chem.*, 2011, **9**, 1491–1507.
126. X. Wang, Z. Ma, J. Lu, X. Tan and C. Chen, *J. Am. Chem. Soc.*, 2011, **133**, 15350–15353.
127. Z. Ma, J. Lu, X. Wang and C. Chen, *Chem. Commun.*, 2011, **47**, 427–429.
128. J. Magolan and M. A. Kerr, *Org. Lett.*, 2006, **8**, 4561–4564.
129. J. Magolan, C. A. Carson and M. A. Kerr, *Org. Lett.*, 2008, **10**, 1437–1440.
130. J. Yin, C. Wang, L. Kong, S. Cai and S. Gao, *Angew. Chem. Int. Ed.*, 2012, **51**, 7786–7789.

Nickel-Catalyzed Heterocycle Synthesis

Nickel has two electron configurations, $[Ar]4s^2 3d^8$ and $[Ar]4s^1 3d^9$, with very close energies, and was first isolated and classified as a chemical element in 1751 by Axel Fredrik Cronstedt. Nickel catalysts have been widely explored in carbonylation reactions, coupling reactions and many other types of catalytic transformations. More recently, nickel-catalyzed C–O bond activation has seen important achievements and nickel showed superior activity, even better than palladium. In this chapter, the main applications of nickel catalysts in heterocycle synthesis are discussed.

7.1 Five-Membered Heterocycles

Radical cyclization constitutes one of the important methodologies for the construction of heterocyclic compounds. In the reported procedures, most of the methodologies need Bu$_3$SnH as a radical initiator. In the development of new methodologies, electrochemical reactions have also been developed and applied. Since the early 2000s, Medeiros's group has carried out a significant amount of work on electrochemical cyclization using nickel complexes as electron transfer mediators.[1] Various nickel complexes have been prepared for the radical cyclization of propargyl derivatives and unsaturated 2-bromophenyl ethers (Scheme 7.1).

In contrast, nickel-catalyzed cyclization of alkynes combined with addition reactions in a domino process was reported in 2011 (Scheme 7.2).[2] Carbonickelations of alkynes and functionalization of the resulting vinylnickel moiety were performed efficiently in a nickel-catalyzed domino cyclization–condensation process. This reaction proceed only by *exo-dig* cyclization and constitutes a one-pot synthesis of substituted dihydro-benzofurans, chromans, isochromans, indoles and indanes. These valuable

RSC Catalysis Series No. 16
Economic Synthesis of Heterocycles: Zinc, Iron, Copper, Cobalt, Manganese and Nickel Catalysts
By Xiao-Feng Wu and Matthias Beller
© Wu and Beller, 2014
Published by the Royal Society of Chemistry, www.rsc.org

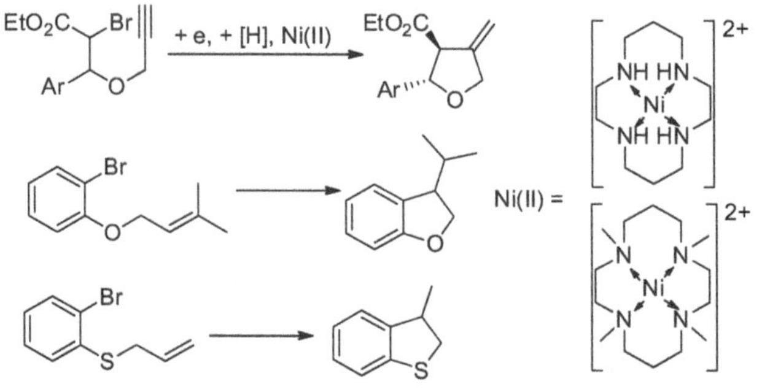

Scheme 7.1   Electrochemical cyclization of propargyl derivatives and unsaturated 2-bromophenyl ethers.

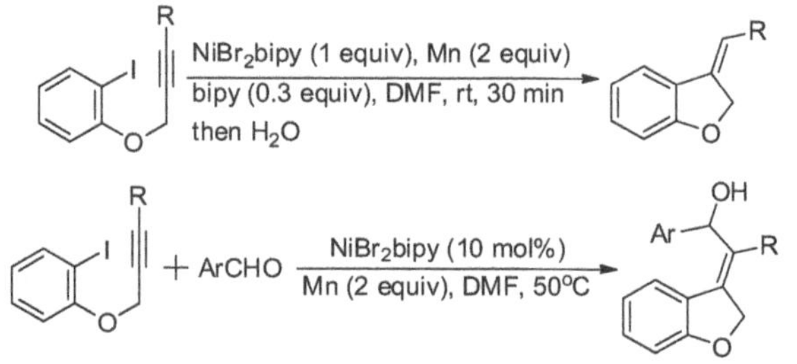

Scheme 7.2   Nickel-catalyzed cyclization of alkynes.

products were generally obtained in good yields and high stereoselectivity, and have been shown to be useful synthons for rapid access to functionalized polycyclic skeletons. The tandem aspect of this reaction has been shown to apply to a significant variety of activated electrophiles, and products that bear diverse functionalities can be obtained in generally good yields, albeit limited by purification issues with these delicate vinylidenic heterocycles. When the substrate bears an aldehyde group, either on the side chain or on the aromatic nucleus, intramolecular trapping of the vinylnickel intermediate takes place, which affords the expected tricyclic aldol products. In one case, dehydration of this latter function provided a diene that could be employed in a Diels–Alder cycloaddition with a classical dienophile, despite the aromatic character of one of the double bonds. This supplementary step gives access to functionalized tetracyclic structures of interest from the perspective of natural product synthesis.

   In 2011, Zhang and co-workers reported a nickel(II)-catalyzed diastereoselective [3 + 2] cycloaddition of *N*-tosylaziridines and aldehydes *via* selective carbon–carbon bond cleavage (Scheme 7.3).[3] They used Ni(ClO4)2 as catalyst and the cycloaddition reaction proceeded with high diastereoselectivity and regioselectivity, leading to highly substituted 1,3-oxazolidines. Notably, this novel reaction can be easily expanded to the gram scale. The reaction undergoes a different pathway under classical thermal conditions.

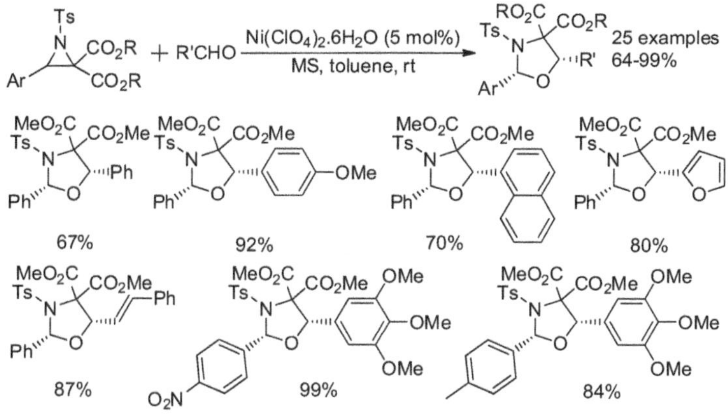

Scheme 7.3   Nickel-catalyzed synthesis of 1,3-oxazolidines.

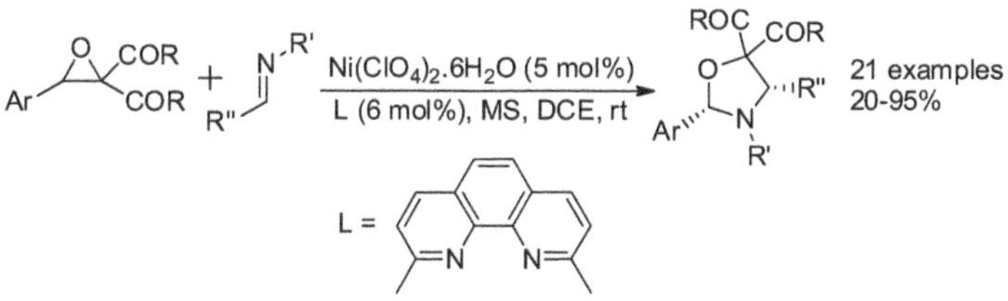

Scheme 7.4   Nickel-catalyzed synthesis of 2,4-oxazolidines.

Moderate enantioselectivity can be achieved by application of Pybox as the chiral ligand, which may be improved by further modification.

More recently, the same group developed an efficient formal $[3+2]$ cycloaddition of imines and carbonyl ylides generated *in situ* from the Lewis acid-catalyzed C–C bond cleavage of donor–acceptor oxiranes under mild conditions (Scheme 7.4).[4] This procedure provides highly substituted 2,4-*trans*-oxazolidines in high yields with moderate to high *dr*. Addition of a ligand could tune the selectivity and the catalytic activity of the Lewis acid. Control experiments indicated that two activating groups in an appropriate orientation to bind to the Lewis acid simultaneously are crucial for achieving chemoselective C–C bond cleavage of the oxirane ring. This transformation can also be realized under catalyst-free conditions, but high temperatures or microwave heating are required.[5]

In 2012, they reported a method for the Lewis acid-catalyzed chemo-divergent cycloaddition of aryl oxiranyldicarboxylates with aldehydes (Scheme 7.5).[6] In this methodology, the C–C or C–O bond cleavage of oxiranes can be controlled by the use of an $Ni(ClO_4)_2$ or $Sn(OTf)_2$ catalyst, respectively. Density functional theory (DFT) calculations supported the proposed mechanism of these two reaction pathways. These chemodivergent transformations may be used as a probe to classify the Lewis acids into subgroups, which will make a fundamental contribution to this field.

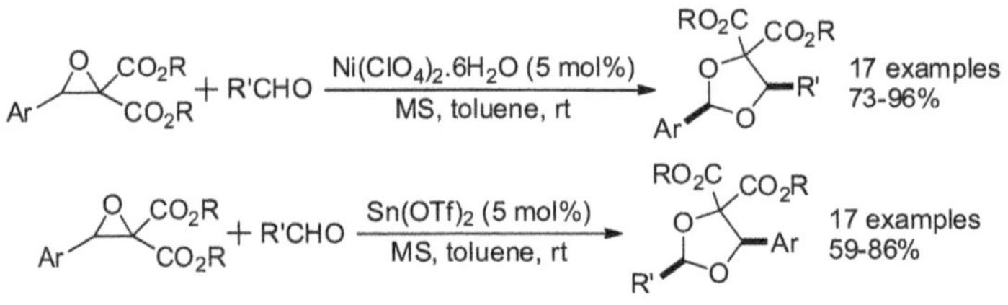

Scheme 7.5   Lewis acid-catalyzed ring opening of oxiranes.

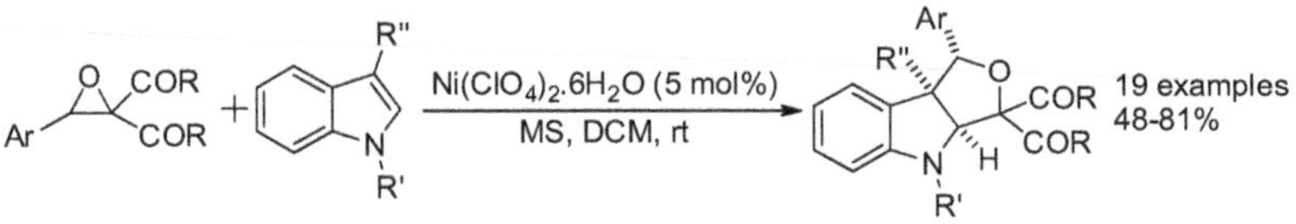

Scheme 7.6   Ni(ClO$_4$)$_2$-catalyzed synthesis of 1*H*-furo[3,4-*b*]indoles.

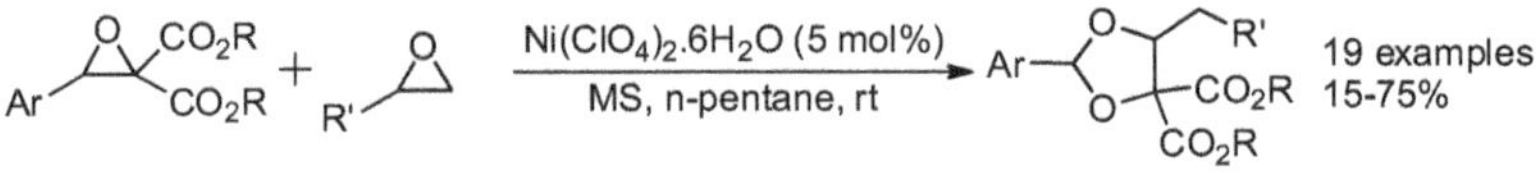

Scheme 7.7   Ni(ClO$_4$)$_2$-catalyzed synthesis of 1,3-dioxolanes.

Ni(ClO$_4$)$_2 \cdot$ 6H$_2$O-catalyzed regio- and diastereoselective [3 + 2] annulations of aryl oxiranyl dicarboxylates and indoles *via* selective C–C bond cleavage of oxiranes were reported in 2012 by Zhang, Wu and co-workers.[7] The cycloadditions proceeded smoothly with high regio- and diastereoselectivity under mild conditions, leading to 1*H*-furo[3,4-*b*]indoles in good to excellent yields (Scheme 7.6). It was proved that enantioselectivity could be achieved when a chiral ligand was used in the reaction.

More recently, it was found that two different functionalized oxiranes can be converted into highly substituted 1,3-dioxolanes (Scheme 7.7).[8] In detail, the reaction follows sequential Meinwald rearrangement of a terminal oxirane through C–O bond cleavage and cycloaddition with a donor–acceptor oxirane through C–C bond cleavage with the assistance of nickel perchlorate. The same catalyst mediates the ring opening of the oxirane moiety in two different ways. This method provides an alternative route for the synthesis of highly substituted 1,3-dioxolanes by using oxiranes instead of aldehydes.

In 2002, Kimura, Tamaru and co-workers developed a nickel-catalyzed conjugate addition reaction of [Me$_2$Zn] and carbonyl compounds with 1,ω-dienynes, which afforded cycloalkanes and their heterocyclic analogs in good yields (Scheme 7.8).[9] The products are characterized by the stereo-defined exocyclic tri- and tetrasubstituted double bonds (100% purity) and also by the remarkably high 1,5-diastereomeric purity (>97% in most cases) with respect to C2 of cycloalkane rings and C4 of *trans*-4-hydroxy-1-butenyl side chains.

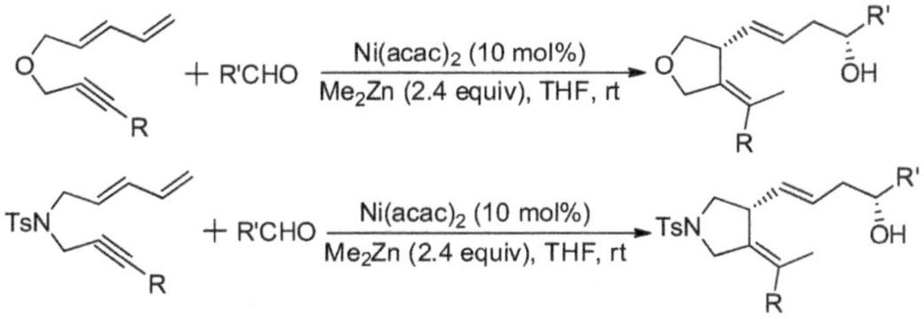

Scheme 7.8 Ni(acac)$_2$-catalyzed cyclization of 1,ω-dienynes.

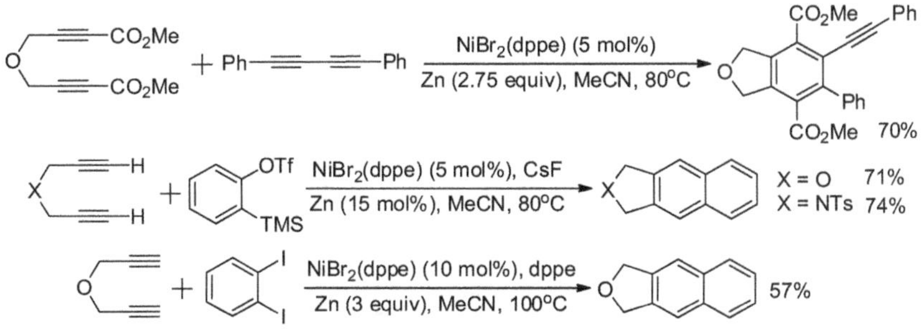

Scheme 7.9 NiBr$_2$(dppe)–Zn-catalyzed cyclization of alkynes.

A nickel-catalyzed highly regio- and chemoselective co-cycloaddition of non-conjugated diynes with 1,3-diynes was reported by Cheng and co-workers in 2002.[10] This method provides a novel pathway for the preparation of polysubstituted arylalkynes. Later, they developed a new methodology for the [2 + 2 + 2] co-cyclotrimerization of arynes with diyne using NiBr$_2$(dppe)–Zn as catalyst.[11] They proved that diynes can undergo cycloaddition with benzynes with excellent tolerance of functional groups and fused-ring sizes to furnish naphthalene derivatives in moderate to good yields. In 2008, the same group found that *o*-dihaloarenes can act as aryne precursors and react with acetylenes and nitriles.[12] The transformation was catalyzed by NiBr$_2$(dppe)–dppe–Zn and gave the desired substituted naphthalene, phenanthridine or triphenylene derivatives in moderate to excellent yields with good tolerance of functional groups. Later, a system without the use of zinc metal as reductant was reported by Iwayama and Sato[13] (Scheme 7.9).

6-Arylpurine derivatives have been found to exhibit diverse types of biological activity: some substituted 6-arylpurine bases are antagonists of corticotrophin-releasing hormone or possess antimycobacterial and anti-bacterial activity, and 6-arylpurine ribonucleosides are potent cytostatics. With this back ground, a novel approach to 6-arylpurines based on [2 + 2 + 2] co-cyclotrimerization of 6-alkynylpurines with various α,ω-diynes was described in 2003.[14] This co-cyclotrimerization was catalyzed by Ni–phosphine catalysts and the corresponding Co–phosphine catalysts can give even better yields (Scheme 7.10).

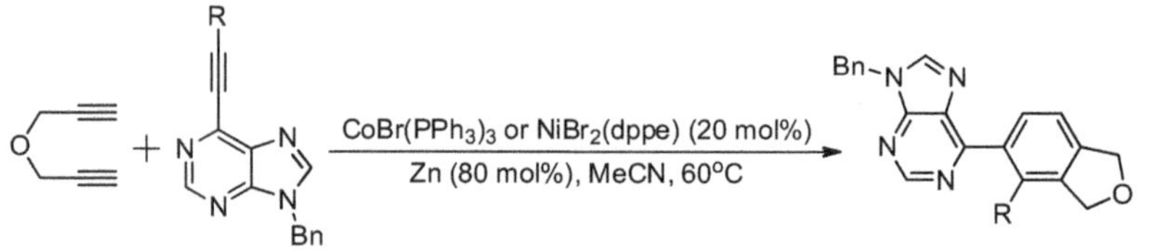

Scheme 7.10   NiBr$_2$(dppe)–Zn-catalyzed synthesis of 6-arylpurines.

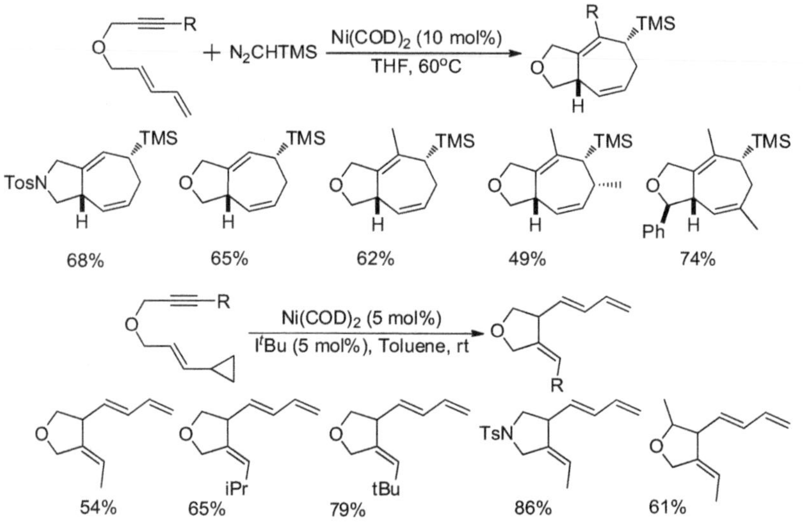

Scheme 7.11   Nickel-catalyzed synthesis of furan derivatives.

In 2004, a novel, nickel-catalyzed $[4+2+1]$ cycloaddition of a diazo-alkane, diene and alkyne was developed by Ni and Montgomery.[15,16] This methodology provides a preparatively useful and chemoselective entry to seven-membered rings. Around the same time, Zuo and Louie discovered a variety of conditions based on Ni–NHC systems for the rearrangement of cyclopropylenynes to afford cyclopentane- and cycloheptene-based hetero-cycles (Scheme 7.11).[17]

Recently, Houk's group performed DFT calculations on this topic and evaluated the theoretical aspects of the reaction mechanism.[18] They found that the preferred catalytic cycle involves oxidative cyclization to form a metallacyclopentene intermediate, followed by cyclopropane cleavage to yield a metallacyclooctadiene intermediate. Subsequent direct C–C reductive elimination leads to the cycloheptadiene product, whereas β-hydride elim-ination and C–H reductive elimination lead to the homo-ene product. The selectivity was controlled by the shape and orientation of the NHC ligand. With the SIPr ligand, larger terminal alkyne substituents destabilize the β-hydride elimination transition state, leading to the $[5+2]$ cycloaddition product. This was attributed to the steric repulsions with the *i*Pr groups located perpendicular to the imidazolidine ring. With the I*t*Bu ligand, the β-hydride elimination transition state leading to the homo-ene product is preferred.

In 2006, Tamaru and co-workers demonstrated that Ni(acac)$_2$-catalyzed reactions of Me$_2$Zn, alkynes, dienes (of 1,3-dien-8-ynes and 1,3-dien-9-ynes), aldehydes and anisidine furnish cyclic dienylamines (Scheme 7.12).[19] Despite the low reactivity, aldimines generated *in situ* showed comparable reactivity to aldehydes under the nickel catalysis and even showed better performance than aldehydes regarding the yields and stereoselectivity, providing only single diastereomers. Lactols failed in this reaction, but the corresponding lactamines successfully underwent the five-component connection reaction to give cyclic dienylamino alcohols in acceptable yields. The authors reported the conjugate addition reaction of Me$_2$Zn with aldehydes across 1,3-dien-8-ynes and 1,3-dien-9-ynes **1** ($n=1$, 2) that furnished cyclic dienyl alcohols with high 1,5-*syn* stereoselectivity and in good yields.[20]

In 2011, Louie and co-workers successfully incorporated ketenes in [2 + 2 + 2] cycloaddition reactions with diynes.[21] A variety of 2,4-cyclohexadienones were formed in moderate to excellent yields (Scheme 7.13). An enantiopure cyclohexadienone product was obtained when (*R*)-BINAP was used as the ligand.

Additionally, in 2004, Cheng and co-workers demonstrated a nickel-catalyzed cyclization of *o*-bromobenzoate with aldehydes to afford phthalides.[22] In the presence of [NiBr$_2$(dppe)] and zinc powder in THF (24 h, reflux

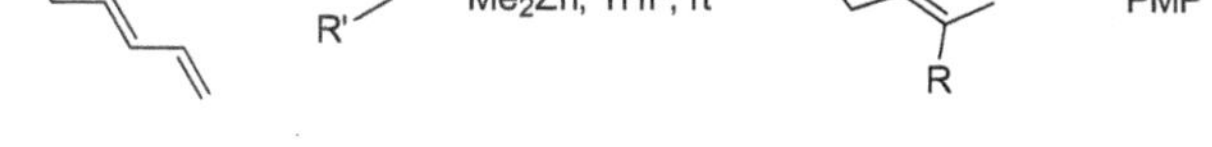

Scheme 7.12 Ni(acac)$_2$-catalyzed reaction of diynes with aldehydes and imines.

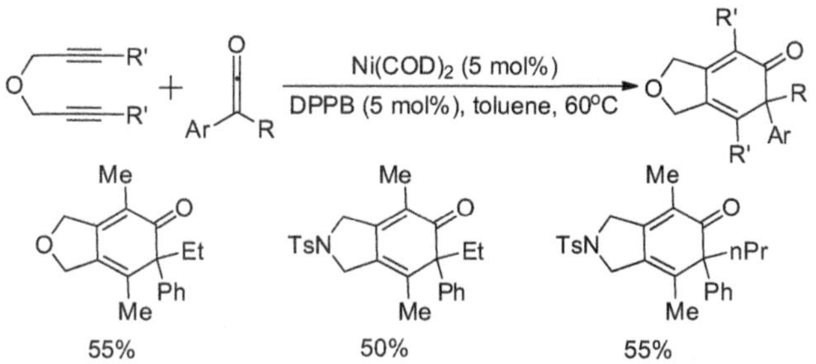

Scheme 7.13 Ni(COD)$_2$-catalyzed synthesis of 2,4-cyclohexadienones.

temperature), a series of substituted aromatic and aliphatic aldehydes also underwent cyclization with *o*-bromobenzoate, producing the corresponding phthalide derivatives in moderate to excellent yields and with high chemoselectivity (Scheme 7.14). A possible reason for the enhanced catalytic activity of bidentate phosphine ligands compared with those that are monodentate is the formation of catalytic intermediates with *cis* structures. The *cis* arrangement facilitates the insertion of the coordinated aldehyde into the nickel–carbon bond. For a nickel catalyst with monodentate ligands, such as triphenylphosphine, the oxidative addition of an aryl halide to nickel(0) generally gives *trans*-[NiL(Ar)X] because of steric repulsion of the two bulky phosphine ligands. The substituents of the *trans* structure are inappropriately positioned for aryl migration to the carbonyl carbon of the coordinated aldehyde, hence greatly reducing the yield of the product.

In 2006, Bowman and Johnson developed a mild Ni(0)-catalyzed rearrangement of 1-acyl-2-vinylcyclopropanes to dihydrofurans.[23] The room-temperature isomerizations afforded dihydrofuran products in yields regularly greater than 90% (Scheme 7.15). A highly substituted, stereochemically defined cyclopropane was employed in the rearrangement to evaluate the

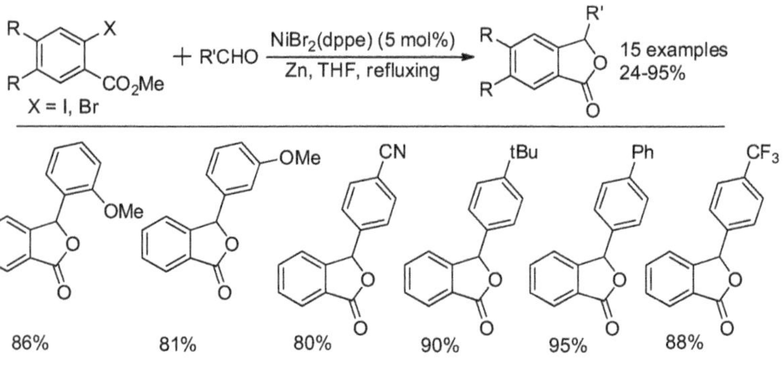

Scheme 7.14   Nickel-catalyzed synthesis of phthalides.

Reaction Procedure (Scheme 7.14): A round-bottomed side-arm flask (25 mL) fitted with a reflux condenser containing [NiBr$_2$(dppe)] (0.050 mmol, 5.0 mol%) and zinc powder (2.75 mmol) was evacuated and purged three times with nitrogen. Freshly distilled THF (2.0 mL), *o*-bromobenzoate (1.50 mmol) and aldehyde (1.00 mmol) were added sequentially to the system and the reaction mixture was stirred under reflux conditions for 24 h. The reaction mixture was cooled to r.t., diluted with dichloromethane and then stirred in air for 15 min. The mixture was filtered through a short Celite and silica-gel pad and washed with dichloromethane several times. The filtrate was concentrated and the residue purified on a silica-gel column using hexanes–ethyl acetate as the eluent to afford the cyclization product.

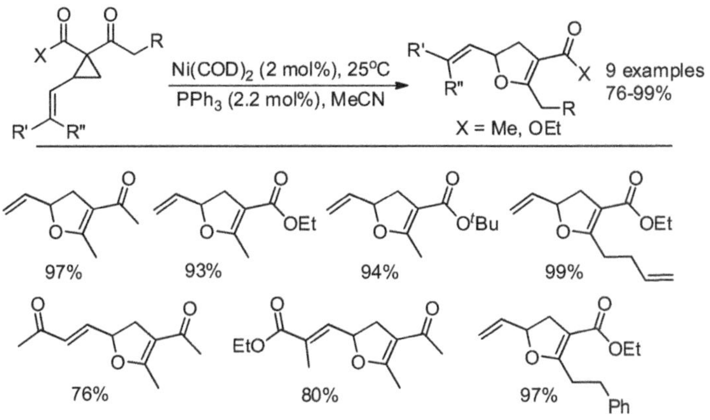

Scheme 7.15 Nickel-catalyzed rearrangements of 1-acyl-2-vinylcyclopropanes.

Reaction Procedure (Scheme 7.15): In an inert-atmosphere glove-box, a dry shell vial with a magnetic stirrer bar was charged with Ni(COD)$_2$, ligand and dry CH$_3$CN. The metal–ligand mixture was stirred vigorously for 1 h. The vinylcyclopropane was added as a solution in CH$_3$CN to the vial, which was sealed with a septum and then removed from the glovebox. The reaction mixture was stirred under argon for the required time at 25 °C, then filtered through a small plug of silica, eluting with Et$_2$O, and the solvent was removed with a rotary evaporator. The product was purified as needed by flash chromatography to afford the dihydrofuran product. TLC visualization was performed with a UV lamp or KMnO$_4$ solution.

reaction mechanism. Product analysis indicated that the overall reaction proceeds with retention of configuration at the vinyl-bearing stereogenic center. This method is noteworthy for low catalyst loadings and short reaction times and is tolerant of both functional and structural changes made to the cyclopropane.

Nickel catalysis has been applied in the electrochemical synthesis of cyclic carbonates. Epoxides was applied as substrates and the desired products were produced in good yields under a carbon dioxide atmosphere.[24] In 2003, Xia and co-workers developed a nickel-mediated reaction of epoxides with CO$_2$ under solvent-free conditions.[25] Ni(PPh$_3$)$_2$Cl$_2$–PPh$_3$–Zn with nBu$_4$NBr is a highly efficient catalyst system in the cycloaddition of carbon dioxide to epoxides under mild reaction conditions. It is an air-stable, easily synthesized, cheap, extremely robust and environmentally benign catalyst system, which is free of co-solvent and can tolerate multiple substrates. Additionally, Mori and co-workers developed a nickel-mediated sequential addition of carbon dioxide and arylaldehydes to terminal allenes in 2003.[26] The reaction proceeded in a diastereoselective manner to afford α-methylene-γ-hydroxycarboxylic acids,

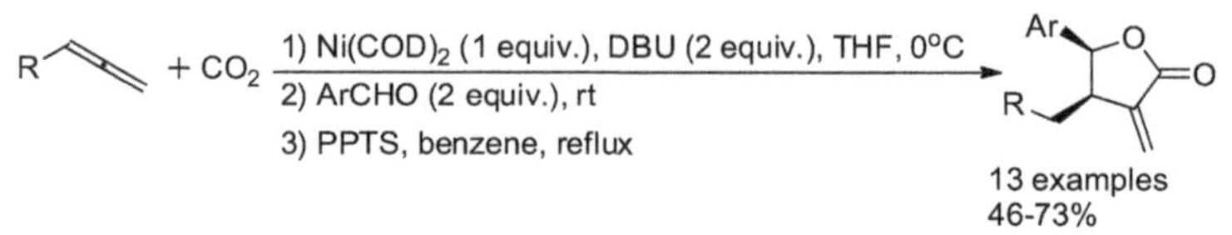

Scheme 7.16   Nickel-catalyzed synthesis of lactones from allenes.

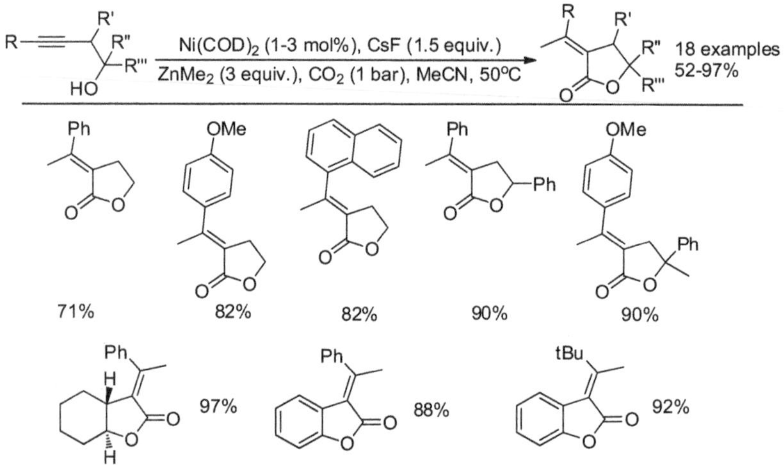

Scheme 7.17   Nickel-catalyzed synthesis of lactones from homopropargylic alcohols.

which allowed the stereoselective preparation of *cis*-β,γ-disubstituted α-methylene-γ-lactones in the presence of pyridinium *p*-toluenensulfonate (Scheme 7.16).

   In 2011, Li and Ma successfully developed the first example of the nickel-catalyzed highly regio- and stereoselective alkylative carboxylation of alkynes.[27] With CO_2 and just 1–2 mol% of Ni(COD)$_2$ as catalyst, the desired lactones were isolated in good yields (Scheme 7.17). Considering the unique character of the directing OH group, the authors reasoned that the deprotonation of this functionality with the zinc reagent makes the interaction between the oxygen atom and Zn much stronger than that of tosylamide with Zn; hence the OH group acts as a much better activating/directing group to enhance the yield of CO_2 fixation and also the regioselectivity. Considering the role of CsF, they reasoned that it increased the reactivity of the alkenyl zinc intermediate toward CO_2. Because of the ready availability of various homopropargylic alcohols, excellent catalytic activity, high regio- and stereoselectivity and good compatibility of various functional groups, this transformation will be a useful and practical method for the highly selective synthesis of natural and unnatural lactones with synthetic or biological potential, which opens up new and efficient means for CO_2 activation. However, it should be noted that such non-methyl alkyl- or arylative products are still a challenge in preparation.

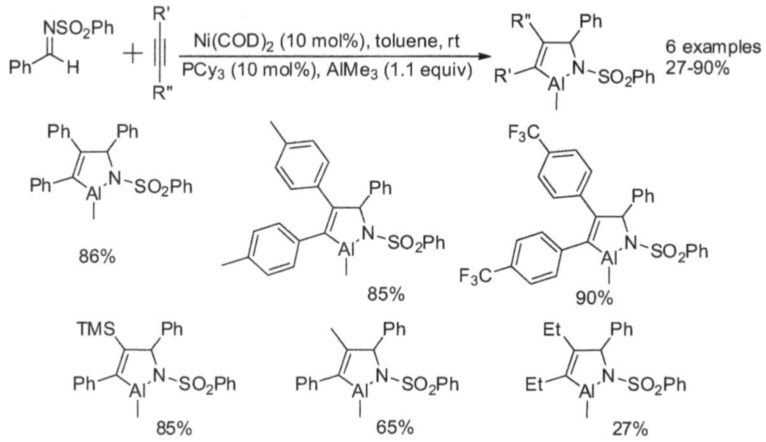

Scheme 7.18 Nickel-catalyzed activation of nitriles.

In 2009, Ogoshi and co-workers demonstrated a nickel-catalyzed three-component cyclocondensation of imines, alkynes and AlMe$_3$ to yield unique azaaluminacyclopentenes (Scheme 7.18).[28] Concerning the reaction mechanism, nickelacycle, generated by the oxidative cyclization of an alkyne and an imine, was a key intermediate in the cyclocondensation reaction and also in the three-component coupling reaction with ZnMe$_2$. Later, the same group reported the nickel(0)-catalyzed formation of oxaaluminacyclopentenes.[29] The reactions proceeded *via* an oxanickelacyclopentene as the key intermediate and used Me$_2$AlOTf as additive and aldehyde and alkyne as substrates. More recently, the use of Me$_2$AlCl as an additive was found to allow the oxidative addition of the Ar–CN bond in 2-(2-methylallyl)benzonitrile on nickel(0) in the presence of PnBu$_3$, giving a *trans*-arylnickelcyanide complex.[30] In contrast, in the presence of PCy$_3$, intramolecular oxidative cyclization on nickel(0) took place to afford a nickeladihydropyrrole. Without the addition of Me$_2$AlCl, the quantitative generation of an $\eta^2 : \eta^2$-5-ene-nitrile–Ni(0) species, which was definitely converted to the nickeladihydropyrrole after treatment with Me$_2$AlCl, was observed. In addition, TfOH also promoted the oxidative cyclization of the $\eta^2 : \eta^2$-5-ene-nitrile complex to yield the corresponding five-membered aza-nickelacycle. A similar intramolecular oxidative cyclization occurred when 2-allylbenzonitrile was used in the presence of a Lewis acid, such as Me$_2$AlCl, Me$_2$AlOTf and Me$_3$SiOTf, or of TfOH to give the corresponding nickeladihydropyrroles in quantitative yield. The molecular structures of a series of nickeladihydropyrroles were unambiguously determined by means of X-ray crystallography. The nickeladihydropyrrole derived from (2-methylallyl)benzonitrile and TfOH was found to react with HSiMe$_2$Ph at 80 °C to furnish a silanamine derivative. The reaction was expanded to an Ni(0)–PCy$_3$–TfOH-catalyzed coupling reaction of 5-ene-nitriles and HSiMe$_2$Ph, yielding the corresponding silanamine in 84% yield.

In 2001, Lozanov and Montgomery reported a nickel-catalyzed cyclization of silicon-tethered ynals (Scheme 7.19).[31] The silicon heterocycles obtained can be converted into allylic alcohols that possess a stereo-defined alkene unit *via* a cleavage process that involves stereospecific protodesilylation of the vinylsilane functionality. Significantly, a high degree of selectivity in the formation of trisubstituted alkenes was observed, thus avoiding the need for the tedious separation of alkene stereoisomers.

In 1994, Mori and co-workers reported a nickel(0)-catalyzed asymmetric [2 + 2 + 2] cocyclization to isoindoline and isoquinoline derivatives (Scheme 7.20),[32,33] and modest enantioselectivity was observed. It may be possible to develop a conceptually new methodology for the construction of benzylic chiral carbon centers, which are important for the synthesis of biologically active substances.

A procedure for the construction of pyrrolizidine and indolizidine skeletons was successfully realized by applying nickel-catalyzed cyclization of a 1,3-diene and aldehyde in a chain.[34,35] A formal total synthesis of an *Elaeocarpus* alkaloid, (–)-elaeokanine C, in the naturally occurring form was achieved using this cyclization (Scheme 7.21).

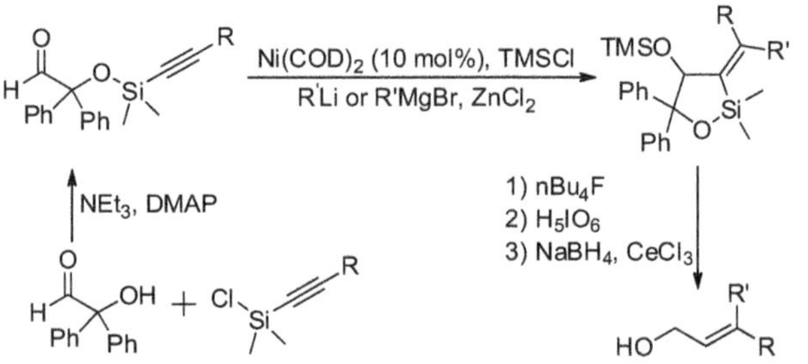

Scheme 7.19 Nickel-catalyzed cyclization of silicon-tethered ynals.

Reaction Procedure (Scheme 7.19): A 0.25 M CH_2Cl_2 solution of 2,2-diphenyl-2-hydroxyethanal (1.0 equiv.) was cooled to 0 °C and Et_3N (2.5 equiv.), DMAP (10 mol%) and chloroalkynylsilane were added sequentially. The resulting suspension was stirred at 0 °C for 15 min and standard extraction and SiO_2 chromatography afforded pure ynals. For the cyclization of ynals, the appropriate organolithium or organomagnesium (4.0 equiv.) was added by syringe to a 0.6 M solution of $ZnCl_2$ (2.5 equiv.) in THF at 0 °C. After 15 min, a 0.025 M THF solution of $Ni(COD)_2$ (10 mol%) was transferred by cannula to the mixture and the resulting solution was immediately transferred to a mixture of ynal (1.0 equiv.) and Me_3SiCl (1.25 equiv.) at 0 °C. After consumption of the ynal as judged by TLC (typically 1 h), a standard extractive work-up afforded product. The crude product was dissolved in THF (0.1 M) at 0 °C and the solution was treated with nBu_4NF (6.0 equiv.) for 0.25–1.0 h at 25 °C. Standard extraction and SiO_2 chromatography afforded the pure diol.

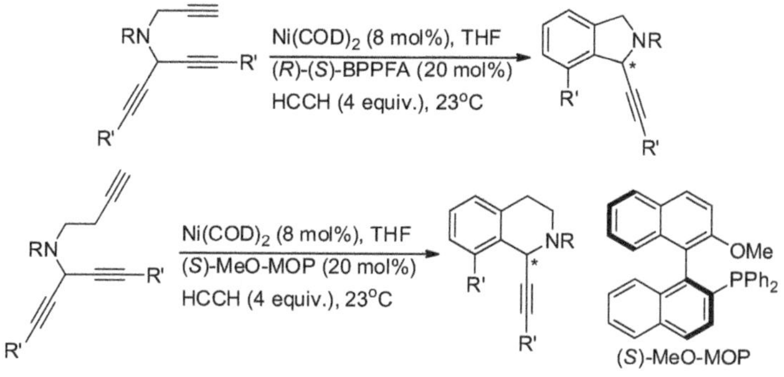

Scheme 7.20 Nickel-catalyzed synthesis of isoindolines and isoquinolines.

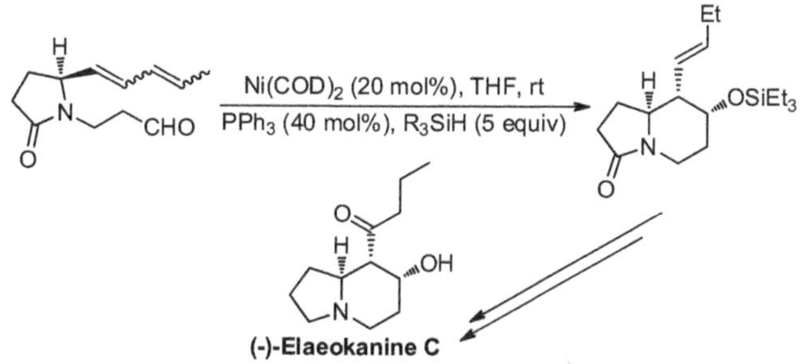

Scheme 7.21 Nickel-catalyzed cyclization of a 1,3-diene and aldehyde.

The same group reported a nickel(0)-catalyzed asymmetric cyclization of a 1,3-diene and tethered aldehyde in 2000.[36,37] If the silane was replaced with $PhF_2SiSiMe_3$ or $Me_3SiSnBu_3$, the corresponding cyclized product with an allylsilyl or an allylstannyl unit in the side chain was formed in good yield. The cyclized product obtained from the reaction in the presence of $Me3SiSnBu_3$ had reactivity as an allylstannane derivative and the coupling reaction with benzaldehyde proceeded in a diastereoselective manner. When the silastannylative cyclization was carried out in the presence of a chiral monodentate phosphine ligand, the cyclized product was produced in an optically active form with modest *ee*. Five-membered heterocycles were produced in moderate yields. In 2002, they developed a nickel-catalyzed ring-closing carboxylation of bis-1,3-dienes.[38] The reaction proceeds *via* insertion of CO_2 into a bis-π-allylnickel intermediate followed by a transmetalation process of the resulting cyclic nickel carboxylate with an organozinc reagent. This reaction can be carried out easily under mild conditions and the yields and regio- and stereoselectivities are generally high (Scheme 7.22). In 2005, they reported a nickel-mediated carboxylative cyclization of enynes.[39,40] The transformation proceeds *via* insertion of CO_2 into a nickelacyclopentene intermediate. The effects of substituents were investigated in detail and it was found that an electron-withdrawing group on the alkene was necessary

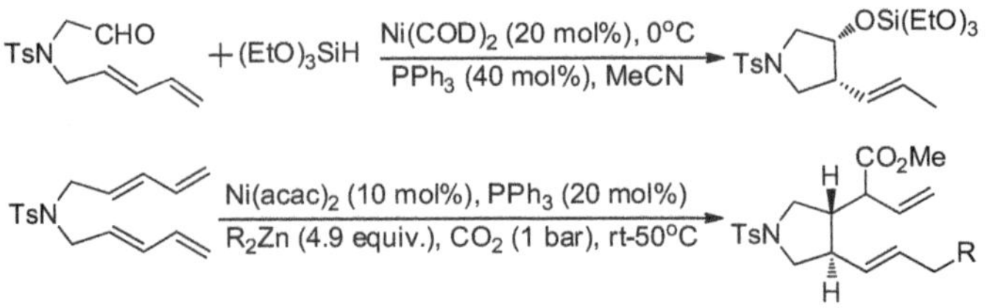

Scheme 7.22   Nickel-catalyzed cyclization of 1,3-dienes.

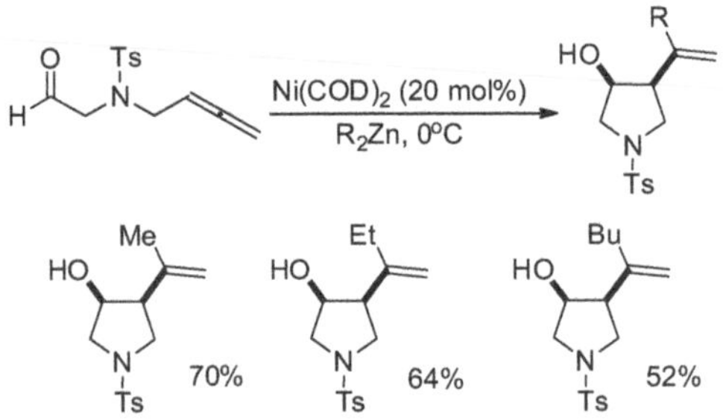

Scheme 7.23   Nickel-catalyzed cyclization of allenyl aldehydes.

Reaction Procedure (Scheme 7.23): To a 0.1 M solution of Ni(COD)$_2$ (0.2 equiv.) in THF was added dropwise dialkylzinc (3.5 equiv.) at 0 °C. The resulting mixture was immediately transferred to a 0.05 M solution of dienal in THF at 0 °C. After consumption of starting material as monitored by TLC (typically 15–30 min at 0 °C), the reaction mixture was quenched with pH 8 NH$_4$Cl–NH$_4$OH buffer at 0 °C and extracted twice with Et$_2$O. The combined organic layers were washed with brine, dried over MgSO$_4$, filtered and concentrated and the residue was purified by column chromatography on silica gel.

for oxidative cyclization of enynes, which was an initial key step of this carboxylative cyclization. The utility of this novel method was further demonstrated by application to the synthesis of various carboxylic acid derivatives having five- and six-membered ring skeletons.

Montgomery and Song developed the direct cyclization of allenylaldehydes with organozincs in the presence of Ni(COD)$_2$ as catalyst (Scheme 7.23).[41] The reaction provides synthetically versatile homoallylic alcohols as the terminal product. Both monosubstituted and 1,3-disubstituted allenes participate in the process, with the latter allowing the preparation of stereochemically defined trisubstituted alkenes. Concerning the mechanism of the alkylative cyclizations, several potential pathways were proposed. Mechanisms that seem most likely involve either (a) the formation of a nickel metallacycle followed by organozinc transmetalation or (b) carbozincation or carbonickelation of the allene followed by addition to the aldehyde. Although simple allenes lacking

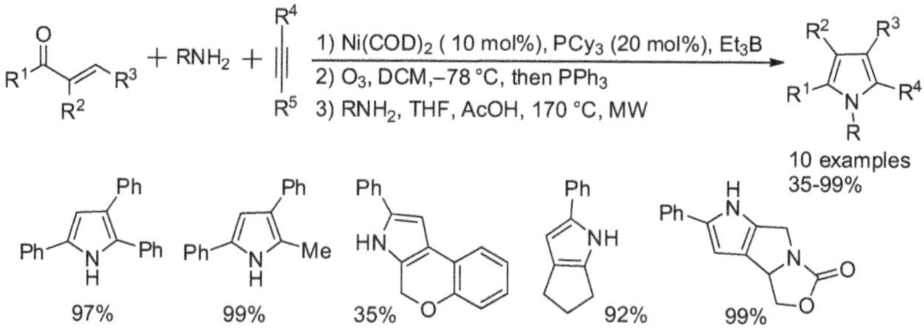

Scheme 7.24 Nickel-catalyzed synthesis of pyrroles.

an aldehyde undergo rapid oligomerization when exposed to Ni(COD)$_2$, no evidence for allene carbometalation was noted when simple allenes were treated with organozincs and Ni(COD)$_2$ under the standard conditions. Based on their experience, the authors favored the metallacycle-based mechanism.

Montgomery's group reported the reductive coupling of enones or enals with alkynes, followed by alkene oxidative cleavage and Paal–Knorr cyclization to give a variety of pyrroles (Scheme 7.24).[42] A number of limitations of alternative entries to the requisite 1,4-dicarbonyl intermediate are avoided. Classes of pyrroles that are accessible by this approach include 2,3-, 2,4-, 1,2,3-, 1,2,4-, 2,3,5- and 1,2,3,5-substituted monocyclic pyrroles and also a number of fused-ring polycyclic derivatives. Some of the classes of pyrroles that may be obtained by this method are difficult to access by alternative approaches and the reductive coupling approach provides complementary characteristics to alternative procedures, including enolate heterodimerizations and Stetter reactions.

A nickel(0) catalyst-mediated cyclization of *N*-benzoylaminals in the presence of a stoichiometric Lewis acid was described by Watson and co-workers in 2011.[43] This method permits the preparation of a variety of isoindolinones with substitution on the benzoyl fragment and C3 carbon. This reaction likely proceeds *via* an α-amidoalkylnickel(II) intermediate, which then may cyclize *via* either an electrophilic aromatic substitution or an insertion pathway. Various isoindolinones were produced in moderate to good yields (Scheme 7.25).

In 2011, a nickel-catalyzed [6 − 3 + 2] cycloaddition was developed.[44] This procedure was also applied for divergent syntheses of indoles from readily available anthranilic acid derivatives with alkynes (Scheme 7.26). Regarding the reaction mechanism, it is reasonable to consider that the catalytic cycle of this reaction should consist of the oxidative addition of an ester CO–O bond to an Ni(0) complex. Decarbonylation and coordination of alkyne take place, during which the steric repulsive interaction is minimal between the bulkier R^L and the PPr$_3$ ligand on the nickel, to give a nickel(II) intermediate. The alkyne would then insert into the C–Ni bond to give a nickelacycle. With its eight-membered ring strain, the Ni–O bond can undergo a facile 1,3-acyl migration to give a thermodynamically more stable six-membered

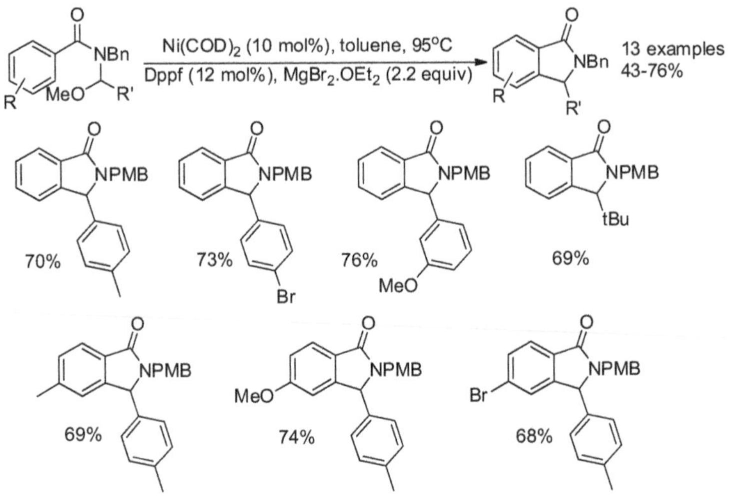

Scheme 7.25   Nickel-catalyzed synthesis of isoindolinones.

Reaction Procedure (Scheme 7.25): In a N_2-atmosphere glove-box, Ni(COD)$_2$ (3.8 mg, 0.0138 mmol, 10 mol%) was weighed into a 1 dram vial equipped with a magnetic stirrer bar. Dppf (8.9 mg, 0.0166 mmol, 12 mol%), aminal (0.138 mmol, 1.0 equiv.), MgBr$_2$·OEt$_2$ (78.6 mg, 0.304 mmol, 2.2 equiv.) and then PhMe (500 μL, 0.28 M) were added. The vial was capped with a Teflon-lined cap and heated in an aluminum heating block at 95 °C. After cooling to room temperature, the crude material was directly purified by silica gel chromatography (eluent: 10–20% EtOAc–hexanes) to furnish the isoindolinone.

---

nickelacycle. Subsequent reductive elimination gives the desired product and regenerates the starting Ni(0) complex.

The same group developed another procedure for the synthesis of indoles based on nickel-catalyzed heteroannulation of *o*-haloanilines with alkynes (Scheme 7.27).[45] During the investigation of the reaction, it was found that IPr [1,3-bis(2,6-diisopropylphenyl)imidazol-2-ylidene] is an effective ligand for the reaction. This nickel-catalyzed heteroannulation of *o*-haloanilines with alkynes provides the same regioisomer as in palladium-catalyzed Larock heteroannulations.

Pyrrolidines are important structural units in organic chemistry and are frequently found in primary and secondary metabolites and also in other biomolecules and synthetic pharmaceuticals. In 2013, a nickel-catalyzed [3 + 2] cycloaddition of vinylcyclopropanes with imines was reported.[46] Pyrrolidine derivatives were obtained with high diastereoselectivities in good yields under mild conditions (Scheme 7.28). In this nickel catalytic system, the coordination of a tosyl substituent to nickel effectively promoted diastereoselective cycloaddition.

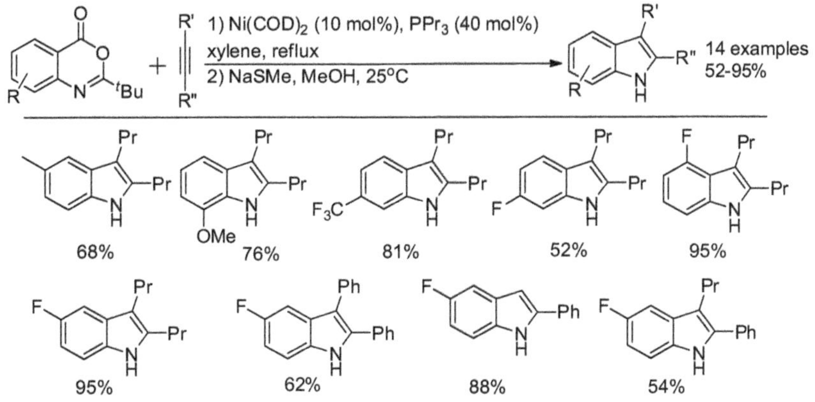

Scheme 7.26 Nickel-catalyzed synthesis of indoles.

Reaction Procedure (Scheme 7.26): The reaction was performed in a 20 mL round-bottomed flask equipped with a Teflon-coated magnetic stirrer bar and Dimroth reflux condenser. The top of condenser was connected with a balloon filled with argon ($\sim$1 atm). Benzoxazinones (0.5 mmol) and alkyne (1.0 mmol) were added to a solution of bis(1,5-dicyclooctadiene)nickel (14 mg, 0.05 mmol) and tripropylphosphine (32 mg, 0.20 mmol) in xylene (2 mL) in a dry box. The flask was taken outside the dry box and heated at 160 °C for 24 h under an argon atmosphere. The resulting reaction mixture was cooled to ambient temperature, filtered through a silica gel pad and concentrated *in vacuo*. The residue was treated with NaSMe (70 mg, 1 mmol) in MeOH (2 mL) for 4 h at ambient temperature. The resulting reaction mixture was diluted with CH_2Cl_2 (50 mL) and washed with water (10 mL). The organic layer was separated, dried over $MgSO_4$, filtered and concentrated *in vacuo*. The residue was purified by flash silica gel column chromatography (eluent: 20:1 hexane–ethyl acetate) to give the corresponding indole.

The hydantoin (imidazolidine-2,4-dione) skeleton is an important structural motif found in a number of pharmaceutically active compounds, such as phenytoin and fosphenytoin, which are used for the treatment of epilepsy. In addition, substituted hydantoins act as valuable intermediates for the synthesis of enantiomerically pure amino acids through dynamic kinetic resolution using hydantoinase biocatalysis. In 2011, Murakami and co-workers developed a new synthetic route to 1,3,5-trisubstituted hydantoins starting from acrylates and isocyanates.[47] Ni(0)–SIPr [SIPr = 1,3-bis(2,6-diisopropylphenyl)-4,5-dihydroimidazol-2-ylidene] was used as the catalyst and 1,3,5-trisubstituted hydantoins were isolated in good yields (Scheme 7.29). A mechanism was proposed: initially, oxidative cyclization on nickel(0) occurs with a heteropair of substrates to give a five-membered ring

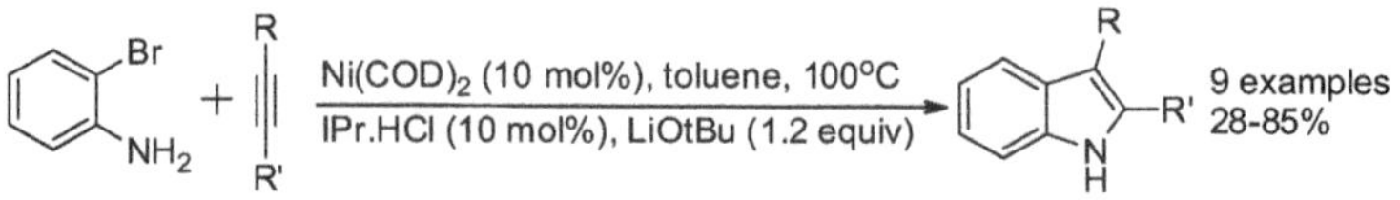

Scheme 7.27   Nickel-catalyzed synthesis of indoles from 2-bromoaniline.

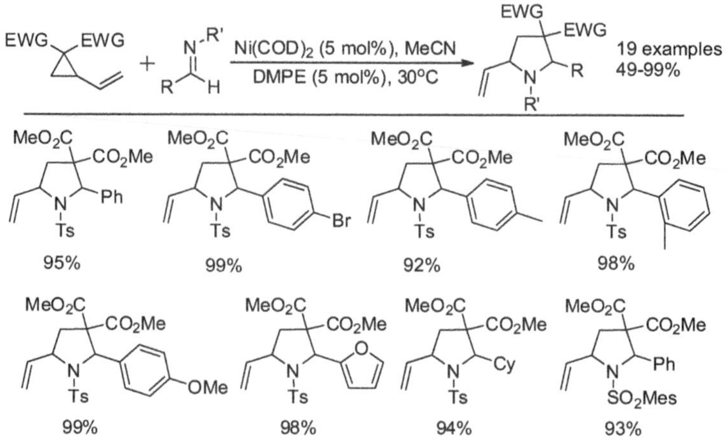

Scheme 7.28   Nickel-catalyzed synthesis of pyrrolidines.

Reaction Procedure (Scheme 7.28): The reaction was performed in a 15 mL sealed tube equipped with a Teflon-coated magnetic stirrer bar. A vinylcyclopropane (46.0 mg, 0.25 mmol), CH_3CN (0.5 mL) and an imine (0.30 mmol) were added to a solution of bis(1,5-dicyclooctadiene)nickel (3.4 mg, 0.0125 mmol) and bis(dimethylphosphine)ethane (0.0021 mL, 0.0125 mmol) in CH_3CN (0.5 mL) in a dry box. The flask was taken outside the dry box and heated at 30 °C for the indicated time under an argon atmosphere. The resulting reaction mixture was subsequently diluted with hexane–EtOAc (1 : 1), filtered through a silica gel pad and concentrated *in vacuo*. The residue was purified by flash silica gel column chromatography using hexane–EtOAc (3 : 1 or 2 : 1) as an eluent to give the corresponding pyrrolidine.

---

azanickelacycle, with which the ester group is bound to the carbon R (Scheme 7.28) to nickel. They assumed that a partial negative charge that develops on the R carbon is stabilized by the ester substituent. Subsequent β-hydride elimination and reductive elimination result in the formation of *N*-substituted fumaramates. Next, the amide group of the fumaramates is deprotonated by the NHC ligand (SIPr). The resulting amide anion adds to another molecule of isocyanate to afford an anionic intermediate, which subsequently undergoes ring closure by conjugate addition in a 5-*exo* mode to furnish the hydantoins.

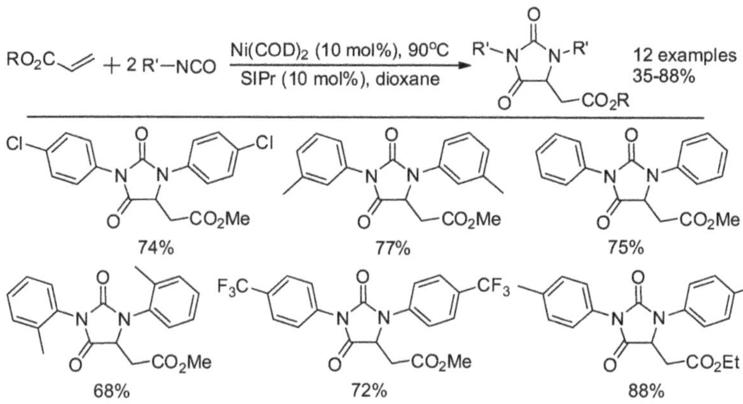

Scheme 7.29   Nickel-catalyzed synthesis of 1,3,5-trisubstituted hydantoins.

Reaction Procedure (Scheme 7.29): In an N_2-filled glove-box, Ni(COD)$_2$ (11.1 mg, 40 μmol, 10 mol%), SIPr (15.7 mg, 40 μmol, 10 mol%) and 1,4-dioxane (2 mL) were placed in an oven-dried 4 mL vial equipped with a stirrer bar. The reaction mixture was stirred for 20 min, then methyl acrylate (34.6 mg, 0.40 mmol, 1.0 equiv.) and *p*-tolyl isocyanate (151.3 μL, 1.2 mmol, 3.0 equiv.) were added *via* syringe. The vial was capped with a Teflon film and the reaction mixture was taken outside the glove-box. After being heated at 90 °C for 18 h, the reaction mixture was cooled to room temperature and stirred for 30 min in open air. The resulting mixture was passed through a pad of Florisil and eluted with ethyl acetate. The filtrate was concentrated under reduced pressure. The residue was purified by gel permeation chromatography (CHCl$_3$), then preparative TLC (eluent: 4 : 1 hexane–ethyl acetate) to give the desired product.

In 2000, a simple catalytic hydrocyanation procedure under exceptionally mild conditions was described by Rosas *et al.*[48] The proposed oxidation–reduction mechanism with the Ni(CN)$_2$–CO–KCN system in aqueous alkaline medium was the first reported example confirmed by IR spectroscopic studies, which extends the scope of this catalytic system to carbonylation and hydrocyanation reactions. It was claimed that the [Ni(CN)$_4$]$^{4-}$ ion is the active species in the hydrocyanation reaction, which was obtained by replacement of CO ligands of [Ni(CN)$_2$(CO)$_2$]$^{2-}$ by CN$^-$. In addition, this study provides an interesting example of some catalyzed organic reactions in aqueous media. 5-Hydroxy-3-pyrrolin-2-ones were regioselectively synthesized in a good yield under very mild conditions (Scheme 7.30).

A simple, efficient and rapid protocol for the preparation of 2-arylbenzimidazoles, 2-arylbenzothiazoles and azomethines from 1,2-diamines or 2-aminophenols using PVP-stabilized Ni nanoparticles dispersed in ethylene

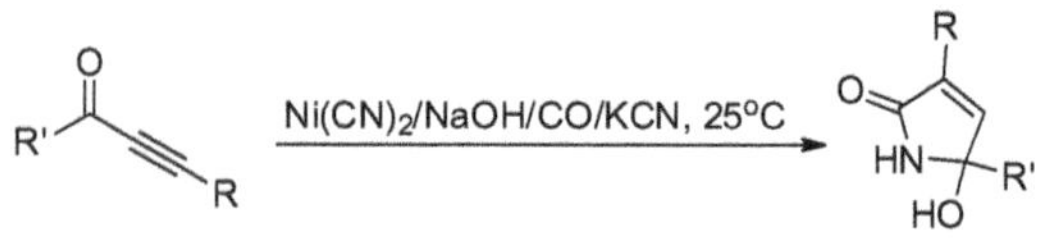

Scheme 7.30   Nickel-catalyzed synthesis of pyrrolinones.

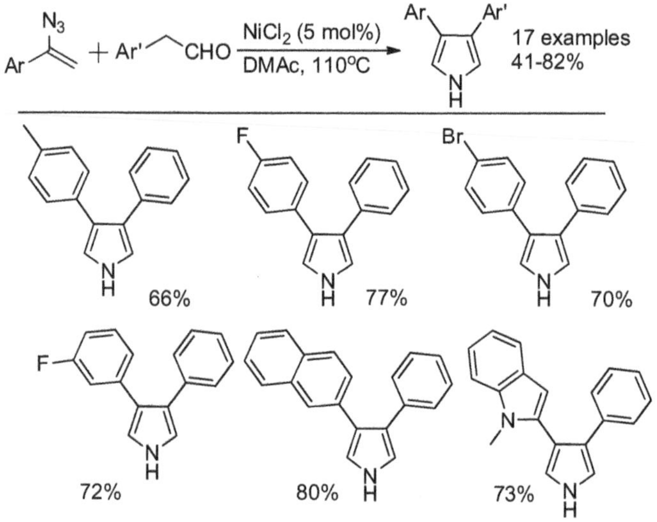

Scheme 7.31   Nickel-catalyzed synthesis of pyrroles.

Reaction Procedure (Scheme 7.31): A mixture of 1-(1-azidovinyl)benzene (0.6 mmol, 87 mg), 2-phenylacetaldehyde (0.4 mmol, 49 mg), NiCl₂ (2.6 mg, 0.02 mmol), DMAc (1.5 mL) was heated with stirring at 110 °C under an argon atmosphere. After 4 h, water (5 mL) was added and the mixture was extracted with ethyl acetate. The extract was dried with anhydrous magnesium sulfate, concentrated and purified by flash chromatography on silica gel (eluent: 8 : 1 petroleum ether–ethyl acetate) to afford the pure product.

glycol as the catalyst at room temperature was reported in 2012.[49] The methodology offers the competitive advantages of recyclability of the catalyst without significant loss of catalytic activity, lower catalyst loadings and excellent yields in short reaction times.

A novel and efficient copper- or nickel-catalyzed highly selective denitrogenative annulation of vinyl azides with arylacetaldehydes was developed by Jiao and co-workers in 2012 (Scheme 7.31).[50] It was reported that 2,4- and 3,4-diaryl-substituted pyrroles, which are difficult to synthesize by previous methods, can be prepared highly regioselectively by this protocol simply by switching the selection of the transition metal catalysts. Compared with the previous acidic or basic conditions for polysubstituted pyrrole synthesis, the present reaction conditions are mild, neutral and very simple without any additives. Concerning the reaction mechanism for Ni-catalyzed 3,4-disubstituted pyrrole formation, NiCl$_2$-promoted thermal denitrogenative decomposition of the vinyl azide generates 2H-azirines, which could not be produced in the presence of a Cu catalyst. Subsequent nucleophilic attack by the enol tautomers of the phenylacetaldehydes affords the intermediates, intramolecular ring opening of which yields five-membered species that produce the 2H-pyrroles after β-OH elimination and Ni complexes, which react with aldehydes to regenerate enol intermediates. Finally 3,4-disubstituted pyrroles are achieved *via* tautomerization of the intermediates.

In 2005, Hsieh and Cheng developed a new methodology for the cyclization of isocyanates with iodoesters and haloarenes using nickel complexes as catalysts and giving imide and amide derivatives as the products (Scheme 7.32).[51] This was the first report that isocyanates can undergo cyclization with 1,3-iodoesters with good tolerance of functional groups. This catalytic reaction was probably initiated by the reduction of Ni(II) species to Ni(0) species by zinc powder. The iodo compound then undergoes oxidative addition to the Ni(0) complex to form an intermediate. Insertion of an isocyanate molecule into the intermediate and subsequent imidation lead to product and the regeneration of the Ni(II) species.

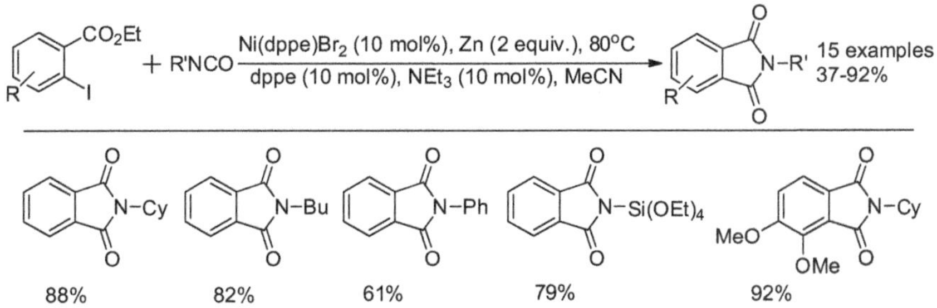

Scheme 7.32 Nickel-catalyzed synthesis of phthalimides.

Reaction Procedure (Scheme 7.32): In a screw-capped vial were placed NiBr$_2$(dppe) (62 mg, 0.10 mmol), dppe (40 mg, 0.10 mmol) and Zn (128 mg, 2.0 mmol) and the vial was sealed with a septum and flushed several times with nitrogen. Iodobenzoate (1.0 mmol), isocyanate (5.0 mmol), triethylamine (10 mg, 0.10 mmol) and acetonitrile (2.0 mL) were injected into the reaction mixture *via* a syringe (Solid isocyanates and dimethoxyiodobenzoate could be added to the vial immediately after the catalyst.). The septum was removed and the vial was quickly sealed with a screw cap under nitrogen. The reaction mixture was stirred at 80 °C for 36 h. The crude reaction mixture was diluted with CH$_2$Cl$_2$, filtered through a thin Celite pad and concentrated *in vacuo*. The residue was chromatographed on a silica gel column using hexane–ethyl acetate as the eluent to give the pure product.

Later, Cheng's group demonstrated a simple and convenient nickel-catalyzed ene–imine reductive coupling reaction for the synthesis of γ-amino derivatives using zinc powder as the reducing agent and water as the proton source (Scheme 7.33).[52] The coupling reaction occurs at the β-carbon atom of the conjugated alkene, in contrast to the known coupling reactions at the α-carbon atom. The catalytic reaction is a simple and efficient method for the synthesis of GABA derivatives. The active nickel catalyst requires bipyridine or phenanthroline instead of the commonly used phosphine ligands. The reaction was initiated by the reduction of Ni(II) to Ni(0) by zinc powder. Coordination of the imine and acrylate to the nickel center results in the formation of an azanickelacycle, which undergoes hydrolysis to give the γ-amino derivative and an Ni(II) species. The nickel(II) species was further reduced by zinc to regenerate the Ni(0) species for the next cycle.

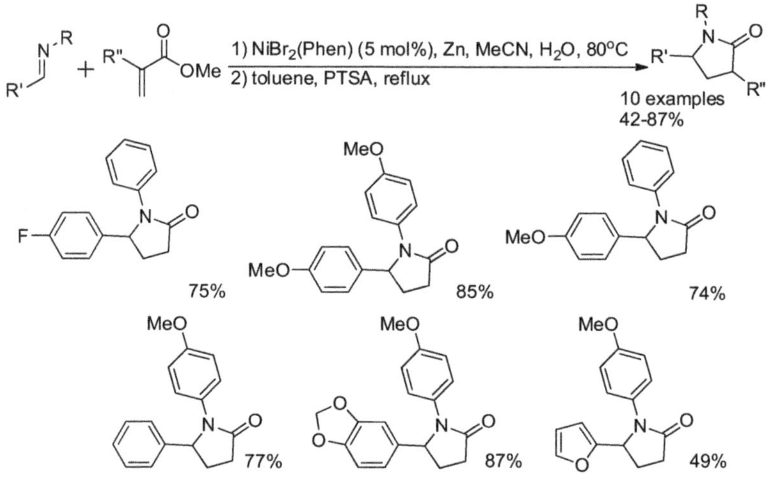

Scheme 7.33   Nickel-catalyzed synthesis of pyrrolidinones from imines.

Reaction Procedure (Scheme 7.33): A screw-capped sealed tube initially fitted with a septum (15 mL) containing [Ni(phen)Br$_2$] (0.01250 mmol) and zinc powder (0.75 mmol) was evacuated and purged three times with nitrogen. Freshly distilled acetonitrile (1.0 mL), imine (0.25 mmol), ethyl acrylate (0.75 mmol) and water (0.50 mmol) were added and the reaction mixture was stirred at 80 °C for 20 h. After filtration of the zinc, the acetonitrile was removed under vacuum, toluene (2.0 mL) was added along with PTSA (0.0250 mmol) and the mixture was kept at 120 °C for 4 h. After completion of the reaction, the mixture was cooled and diluted with dichloromethane. Et$_3$N (0.05 mmol) was added and the mixture was filtered and then concentrated. Separation on a column of silica gel using hexane–ethyl acetate as the eluent gave the pure diphenylpyrrolidinone product.

Five-membered azanickelacycles formed from imines and alkynes were proposed as key catalytic intermediates. The formation of a nickelacycle by the addition of two π components is generally regioselective, with the carbon atom having an electron-withdrawing functionality near the metal center. This mechanism explains the regioselectivity of the present reductive coupling product. Alternatively, the acrylate first interacts with Ni(0) to give an oxy-π-allylnickel(II) intermediate. A β-coupling reaction of the oxy-π-allyl group with the imine followed by hydrolysis would lead to the final γ-amino derivative. This pathway cannot be totally excluded.

Wender and Christy reported nickel(0)-catalyzed [2+2+2+2] cyclo-additions of terminal diynes for the synthesis of substituted cyclooctate-traenes in 2007 (Scheme 7.34).[53] The catalysts and conditions that allow for the efficient and selective conversion of a wide range of 1,6- and 1,7-diynes to symmetrical 1,2,5,6-tetrasubstituted cyclooctatetraenes were identified. In all cases, the [2+2+2+2] cycloaddition was favored over the [2+2+2] process and, in all but one case, a single alkene positional isomer was obtained. A hitherto unexplored cross-cyclotetramerization was also reported, providing access to differentially substituted cyclooctatetraenes. This methodology provides facile access to cyclooctatetraenes and to other cyclooctane deriva-tives, the latter through selective partial hydrogenations or alkene additions.

In 2002, Kang and Yoon described a new procedure for the cyclization/ alkylation and arylation reactions of allenyl aldehydes and ketones with organozinc reagents.[54] Ni(COD)$_2$ was used as the catalyst and *cis*-fused homoallylic cyclopentanols were produced in good yields (Scheme 7.35). It was suggested that the oxidative cyclization of the Ni(0)–allenylaldehyde

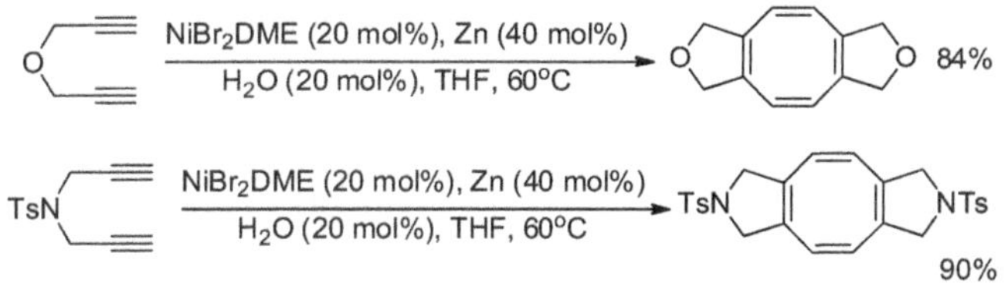

Scheme 7.34 Nickel-catalyzed synthesis of cyclooctatetraenes.

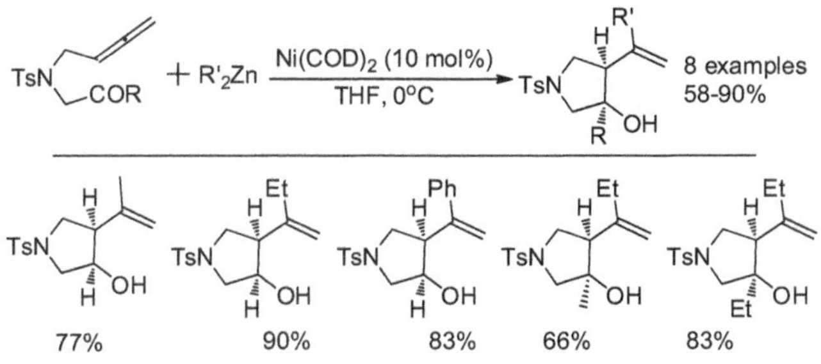

Scheme 7.35 Nickel-catalyzed synthesis of homoallylic cyclopentanols.

complex occurs to form an oxametallacycle in the first stage. Transmetalation between the organozinc reagent and oxametallacycle, involving nickel–oxygen bond cleavage, then gives the intermediate, which upon reductive elimination affords the zinc alkoxide precursor of the cyclopentanol.

A nickel-catalyzed denitrogenative alkyne insertion reaction of N-sulfonyl-1,2,3-triazoles was developed in 2009 by Murakami and co-workers.[55] Substituted pyrroles were obtained when N-sulfonyl-1,2,3-triazoles reacted with alkynes in the presence of a nickel(0)–phosphine catalyst with the extrusion of molecular nitrogen (Scheme 7.36). In detail, the triazole moiety isomerized to an α-imino diazo species and denitrogenative addition to nickel(0) was followed by insertion of alkynes and reductive elimination.

Wan and co-workers recently developed an Ni-catalyzed [3 + 2] cycloaddition of methyleneaziridines with diynes to give pyrroles with a free alkyne unit under mild conditions (Scheme 7.37).[56] This procedure includes a significant indirect C–C bond activation of three-membered azacycles and represents an alternative strategy for the ring enlargement of methyleneaziridines compared with the previous C–N bond cleavage. The excellent regioselectivity and the free alkyne unit in the products are advantageous for further derivatization. The success of C–C bond cleavage of the methyleneaziridines makes it possible to promote metal-catalyzed cycloaddition of

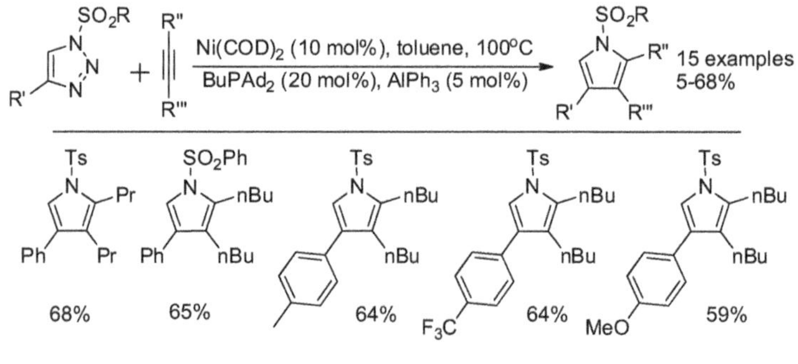

Scheme 7.36   Nickel-catalyzed synthesis of pyrroles from triazoles.

Reaction Procedure (Scheme 7.36): In a glove-box, triazoles (0.20 mmol) and AlPh$_3$ (2.6 mg, 10 mmol) were charged into an oven-dried 4 mL vial equipped with a stirrer bar. A solution of Ni(COD)$_2$ (5.5 mg, 20 mmol) and P(nBu)Ad$_2$ (14.3 mg, 40 mmol) in toluene (2 mL) and alkyne (0.40 mmol) were added. The vial was then capped with Teflon film and removed from the glove-box. The reaction mixture was heated at 100 °C for 12 h, then cooled to room temperature and stirred in the open air for 30 min. The resulting mixture was passed through a pad of Florisil and eluted with ethyl acetate. The filtrate was concentrated under reduced pressure. The residue was purified by preparative TLC (eluent: hexane–dichloromethane) to give the pure product.

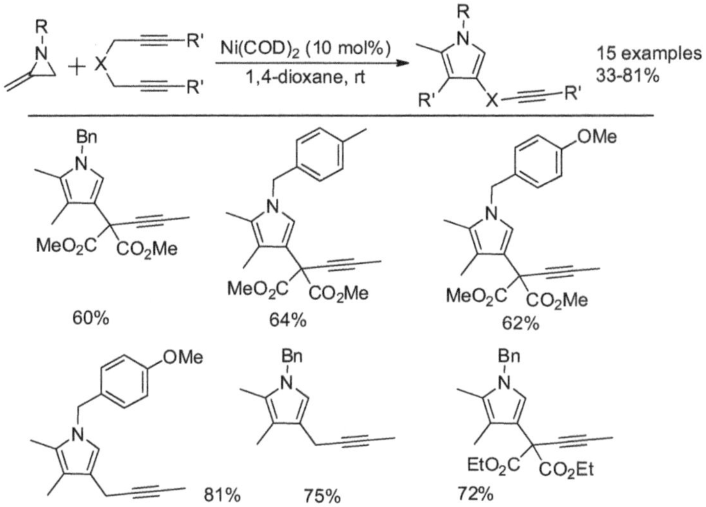

Scheme 7.37 Nickel-catalyzed synthesis of pyrroles from diynes.

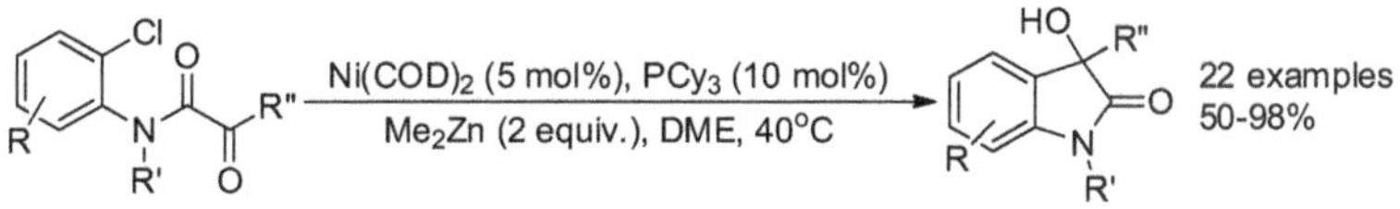

Scheme 7.38 Nickel-catalyzed synthesis of 3-hydroxyoxindoles.

Reaction Procedure (Scheme 7.38): [Ni(COD)$_2$] (2.8 mg, 0.01 mmol) and PCy$_3$ (5.7 mg, 0.02 mmol) were placed in a dried Schlenk tube under N$_2$ and then freshly distilled DME (2 mL) was introduced *via* a syringe. The resulting mixture was stirred at room temperature for 1 h before addition of the substrate (56 mg, 0.2 mmol) under N$_2$, followed by treatment with a solution of Me$_2$Zn in toluene (0.2 mL, 0.4 mmol, 2 M). The reaction mixture was then stirred at 40 °C for 12 h (monitored by TLC). After cooling to room temperature, the mixture was quenched with saturated NH$_4$Cl (5 mL). Extraction with diethyl ether (2×10 mL), drying of the combined organic phases over MgSO$_4$ and final evaporation of the diethyl ether afforded a crude product that was then purified by flash chromatography to give the oxindole (46.5 mg, 95%).

this three-membered azacycle with other substrates to give heterocycles with various structures.

In 2011, Gao and co-workers reported an efficient process for the direct nucleophilic addition of aryl and vinyl chlorides to ketoamides.[57] 3-Hydroxyoxindoles and pyrrolidinones were successfully constructed under mild reaction conditions in good yields (Scheme 7.38). This method was then successfully extended to the analogous bromides. Concerning the reaction mechanism, nucleophilic addition occurs after the generation of the nucleophilic aryl–nickel intermediate. An oxidative addition gives the adduct, followed by metathesis with Me$_2$Zn, to afford the product precursor and the Ni(II) species, which is reduced to Ni(0) in the presence of Me$_2$Zn to complete the catalytic cycle.

A nickel-catalyzed aromatic C–H alkylation with tertiary or secondary alkyl–Br bonds for the construction of indolones was demonstrated by Lei and co-workers more recently.[58] Various functional groups were well tolerated. Moreover, the challenging secondary alkyl bromides were well introduced in this transformation (Scheme 7.39). Radical trapping and photocatalysis conditions indicated that most likely this aromatic C–H alkylation is a radical process. In detail, first, the active Ni(I) species was generated by the reaction of alkyl halide with Ni(PPh$_3$)$_4$. The catalytic cycle starts from the further reaction of the Ni(I) species with alkyl halide *via* a SET process. The radical species and an Ni(II) species are generated. Then, intramolecular radical addition of the carbon radical to the aromatic ring affords an intermediate, which is further oxidized by the Ni(II) species to afford the final product. Meanwhile, the Ni(II) species was reduced to re-generate the active Ni(I) species.

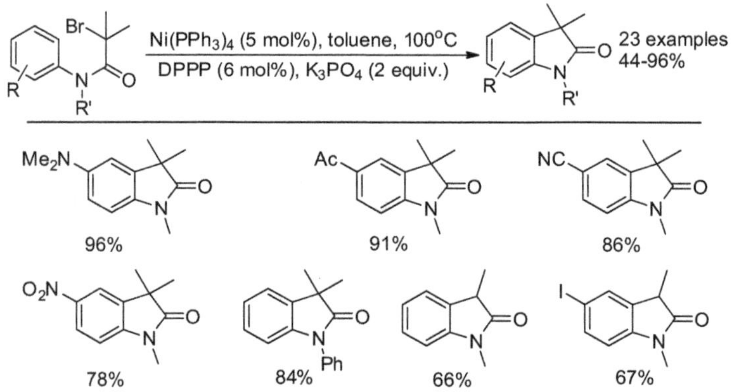

Scheme 7.39   Nickel-catalyzed synthesis of indolones *via* C–H activation.

Reaction Procedure (Scheme 7.39): In a glove-box, Ni(PPh$_3$)$_4$ (27.7 mg, 0.025 mmol), dppp (12.4 mg, 0.03 mmol) and K$_3$PO$_4$ (212 mg, 1.0 mmol) were placed in a Schlenk tube. The tube was then sealed with a septum and taken out of the glove-box and toluene (2.0 mL) was injected *via* a syringe. After stirring for several minutes, α-bromoamide (0.5 mmol) was added to the reaction tube. The reaction mixture was heated to 100 °C and stirring was continued for 24 h. After the mixture had cooled to room temperature, the pure product was obtained by flash column chromatography on silica gel (eluent: 20 : 1 petroleum ether–ethyl acetate).

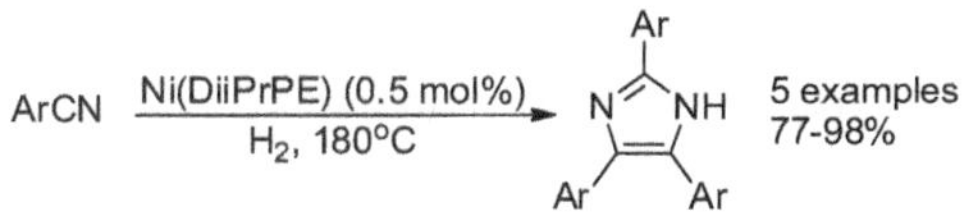

$$\text{ArCN} \xrightarrow[\text{H}_2,\ 180^\circ\text{C}]{\text{Ni(DiiPrPE) (0.5 mol\%)}}$$

Ar

N⁀NH

Ar⁀Ar

5 examples
77-98%

Scheme 7.40 Nickel-catalyzed synthesis of imidazoles from nitriles.

A simple and efficient method for the conversion of isoxazoles to the corresponding pyrazoles was developed in 2007.[59] By treating the isoxazoles with hydrazine in methanol in the presence of Raney nickel, the corresponding pyrazoles were formed in moderate to excellent yields. Although this straightforward method is restricted to certain types of substituents in the heterocyclic ring, the target product can be prepared in special cases with the use of a modified procedure.

A one-pot cyclization of aromatic nitriles to yield 2,4,5-trisubstituted imidazoles was described in 2011 by Garcia *et al.*[60] The desired products were formed in high yields using single-site nickel catalysts (Scheme 7.40). This synthetic method can be used to prepare a variety of valuable imidazoles from cheap raw materials such as *para*-substituted benzonitriles and *N*-heterocycles such as 4-cyanopyridine under neat conditions and at relatively low H_2 pressure. The generation of by-products such as triazines can be substantially or totally inhibited by increasing the pressure of H_2 in the system. A mechanistic proposal for this transformation, based on the findings for the catalytic reduction of nitriles using H_2 and a variety of solvents, involved the formation of a Schiff base as the main product. To a much lesser extent, the Schiff base underwent further reduction to *N*-dibenzylamine. The synthesis of the Schiff base constituted a definite proof of a catalytic tandem hydrogenation–condensation process during which NH_3 is also extruded. The Schiff base was envisioned to act as a key organic intermediate towards the final cyclization product, *via* a series of basic reactions.

7.2 Six-Membered Heterocycles

In 2005, Louie and co-workers developed a mild and efficient method for preparing a wide range of pyridines from alkynes and nitriles (Scheme 7.41).[61] Both intramolecular and intermolecular reactions were catalyzed by a combination of an Ni(0) precursor and an imidazolylidene ligand. Furthermore, cycloaddition of an asymmetric diyne afforded a single pyridine regioisomer. owing to the difficulty in managing an Ni(0) complex, a convenient method for preparing pyridines from air-stable, commercially available catalyst precursors was described in 2006.[62] The addition of *n*BuLi to Ni(acac)$_2$ and an NHC salt (such as IPr · HCl or SIPr · HCl) rapidly generates an active Ni(0)–NHC catalyst for the cycloaddition of diynes and nitriles that affords pyridines without a decrease in yield. The *in situ* method also converts diynes and carbon dioxide to the corresponding pyrones. When diynes and cyanamides were applied as substrates, a variety of bicyclic *N,N*-disubstituted 2-aminopyridines were prepared from by nickel-catalyzed

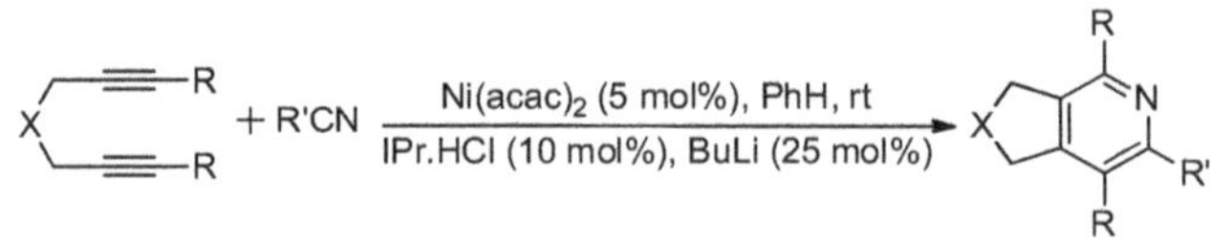

Scheme 7.41   Nickel-catalyzed synthesis of pyridines.

[2 + 2 + 2] cycloaddition reactions.[63] The reactions proceeded at room temperature with a low catalyst loading to afford 2-aminopyridines in good to excellent yields. The method is amenable to both internal and terminal diynes and proceeds in a regioselective manner. A number of cyanamides with diverse functional group tolerance were used. The intermolecular version employing 3-hexyne and *N*-cyanopyrrolidine also afforded the desired *N,N*-disubstituted 2-aminopyridine in good yields. Later, they found that the reaction of $Ni(COD)_2$, IPr and nitrile affords dimeric $[Ni(IPr)RCN]_2$ in high yields. X-ray analysis revealed that these species display simultaneous η^1- and η^2-nitrile binding modes.[64] These dimers are catalytically competent in the formation of pyridines from the cycloaddition of diynes and nitriles. Kinetic analysis showed the reaction to be first order in $[Ni(IPr)RCN]_2$ and zeroth order in added IPr, nitrile and diyne. Extensive stoichiometric competition studies were performed and selective incorporation of the exogenous, not dimer-bound, nitrile was observed. After cycloaddition, the dimeric state was found to be largely preserved. Nitrile- and ligand-exchange experiments were performed and found to be inoperative in the catalytic cycle. These observations suggest a mechanism in which the catalyst is activated by partial dimer opening followed by binding of exogenous nitrile and subsequent oxidative heterocoupling. More recently, a detailed mechanistic evaluation of the $Ni(IPr)_2$-catalyzed [2 + 2 + 2] cycloaddition of diynes and nitriles was conducted.[65] Through kinetic analysis of these reactions, the observed regioselectivities of the products and stoichiometric reactions, $Ni(IPr)_2$-catalyzed cycloadditions of diynes and nitriles appear to proceed by a heterooxidative coupling mechanism, in contrast to reactions with other common cycloaddition catalysts. The reaction profiles demonstrated a strong dependence on nitrile, resulting in variable nitrile-dependent resting states. The strong coordination and considerable steric bulk of the carbene ligands facilitate selective initial binding of nitrile, thereby forcing a heterocoupling pathway. *In situ* IR data suggest that the initial binding of the nitrile resides in a rare, η^1-bound conformation. Following nitrile coordination, a rate-determining hapticity shift of the nitrile and subsequent loss of carbene occur. Alkyne coordination then leads to heterooxidative coupling, insertion of the pendant alkyne and reductive elimination to afford pyridine products.

   In 2011, Louie and co-workers discovered an Ni–phosphine system that promotes both oxidative coupling between an alkyne and a nitrile and also C–N bond-forming reductive elimination (Scheme 7.42).[66] The combination of catalytic amounts of Xantphos and Ni(0) produced a variety of pyridines from unactivated nitriles and diynes under mild and efficient reaction conditions.

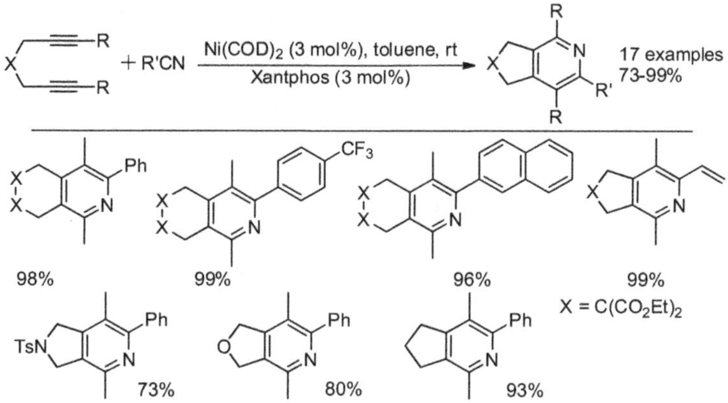

Scheme 7.42 Nickel-catalyzed synthesis of pyridines from nitriles.

Reaction Procedure (Scheme 7.42): In a nitrogen-filled glove-box, diyne (1 equiv., 0.1 M) and nitrile (1.5 equiv.) were added to an oven-dried screw-capped vial equipped with a magnetic stirrer bar. In a separate vial, [Ni(COD)$_2$] and Xantphos were weighed out (in a 1:1 molar ratio) and dissolved in toluene. The catalyst (3 mol%) solution was added to the reaction mixture. The vial was sealed and taken out of the glove-box and the reaction mixture was stirred at r.t. for 3 h. The resulting mixture was concentrated and purified by flash column chromatography on silica gel using first 15%, then 30% and finally 50% EtOAc–hexanes as eluent.

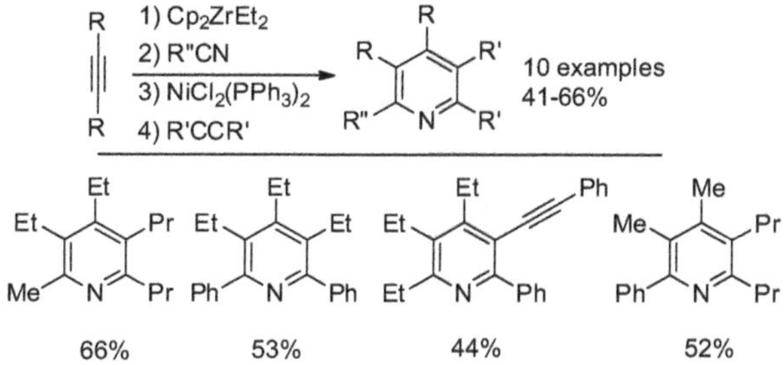

Scheme 7.43 Nickel-catalyzed synthesis of penta-substituted pyridines.

In 2000, Takahashi *et al.* reported a nickel-catalyzed synthesis of substituted pyridines from two different alkynes and a nitrile.[67] The reaction proceeds through azazirconacyclopentadienes as intermediates. Various penta-substituted pyridines were isolated in moderate to excellent yields (Scheme 7.43).

A novel nickel-catalyzed dehydrogenative [4 + 2] cycloaddition of 1,3-dienes with nitriles was developed by Ogoshi and co-workers in 2011.[68] A variety of pyridines were prepared in good yields using Ni(0) complexes

(Scheme 7.44). Moreover, the reaction can be applied to both di- and tricyano compounds. These polypyridine derivatives were shown to be potentially useful ligands in transition metal complexes. Moreover, the formation of the expected intermediate and its molecular structure were revealed by crystallography. Concerning the reaction mechanism, the oxidative cyclization of a nitrile and a diene with Ni(0) occurred to give an η^3-azaallylnickel intermediate that underwent isomerization to yield the tautomer, which reacted reversibly with a nitrile to form a diazanickelacycle species. The diazanickelacycle generated by the reaction of 2,3-dimethyl-1,3-butadiene with benzonitrile could be isolated and its structure determined by X-ray diffraction. Reductive elimination from the diazanickelacycle species gave rise to 1,2-dihydropyridines followed by isomerization and dehydrogenation to yielded a pyridine. The generated hydrogen molecule was consumed by nickel-catalyzed hydrogenation of 1,3-butadiene under the reaction conditions. The large difference in reactivity between PCy_3 and IPr is mostly caused by the difference in the stability of the diazanickelacycle.

In 2005, Korivi and Cheng demonstrated a very convenient and highly regioselective synthetic approach for the preparation of substituted isoquinolines (Scheme 7.45a).[69] This nickel-catalyzed annulation was much more efficient than the known palladium-catalyzed reaction in terms of the catalytic reaction rate and the scope of the alkyne substrates. The method tolerates a wide variety of functional groups and utilizes inexpensive

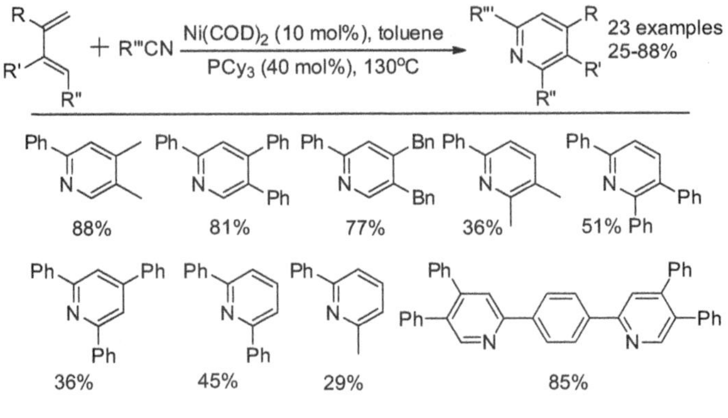

Scheme 7.44 Nickel-catalyzed synthesis of pyridine-1,3-dienes.

Reaction Procedure (Scheme 7.44): To a solution of Ni(COD)$_2$ (4.9 mg, 0.02 mmol), PCy$_3$ (22.4 mg, 0.08 mmol) and benzonitrile (22.5 mg, 0.22 mmol) in toluene-d_8 (0.5 mL) was added 2,3-dimethyl-1,3-butadiene (45.4 mg, 0.55 mmol). The resulting mixture was transferred into a J-Young NMR tube. The tube was tightly sealed and thermostated at 130 °C. The reaction mixture was cooled to room temperature at a giving time and monitored by ^{1}H and ^{31}P NMR spectroscopy.

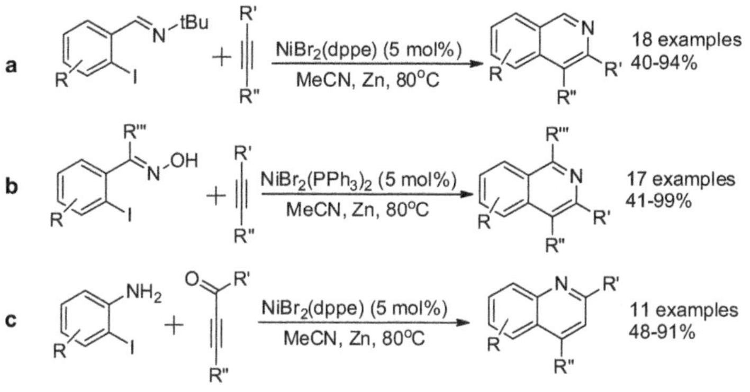

Scheme 7.45 Nickel-catalyzed synthesis of quinolines.

catalysts. In addition, the unusual regioselectivity indicated two pathways of the alkyne insertion into the nickelacycle intermediate. Regarding the reaction mechanism, the reaction likely starts with reduction of Ni(II) to Ni(0) by zinc powder. The oxidative addition of 2-iodobenzaldimine to Ni(0) leads to the formation of a five-membered ring nickelacycle. Coordinative insertion of alkyne into the nickelacycle, followed by reductive elimination, gave an iminium cation and regeneration of the Ni(0) catalyst. The tertiary butyl group on the iminium ion was then removed by the attack of iodide ion produced during the reaction to give the final product. In 2012, the same group developed another convenient method for the synthesis of highly substituted isoquinolines and isoquinolinium salts by the nickel-catalyzed cyclization of o-haloketoximes and -ketimines, respectively, with alkynes as the partner.[70] The reaction of o-haloketoximes and various alkynes in the presence of [Ni(PPh$_3$)$_2$Br$_2$] and zinc powder in a mixture of acetonitrile and tetrahydrofuran at 80 °C for 15 h gave 1,3,4-trisubstituted isoquinoline products in moderate to excellent yields and with high regioselectivity (Scheme 7.45b). The corresponding isoquinoline N-oxide was found to be the intermediate in the cyclization reaction pathway. In contrast, the reaction of o-haloketimines and alkynes under similar catalytic conditions in tetrahydrofuran at 70 °C for 2 h gave 1,2,3,4-tetrasubstituted isoquinolinium salts in good to excellent yields. They also developed[71] a synthetic approach for the preparation of 2,4-disubstituted quinolines.[71] This nickel-catalyzed reaction is interesting, since most of the synthetic methods available are for the preparation of 2,3-disubstituted quinolines. The method tolerates a range of functional groups and utilizes inexpensive catalysts and readily available reagents (Scheme 7.45c). They prepared halogen derivatives which are of particular interest in medicinal chemistry. The catalytic cycle was likely initiated by the reduction of Ni(II) to Ni(0) by zinc metal powder. The oxidative addition of 2-iodoaniline to Ni(0) species affords an *ortho*-metalated aniline–nickel complex. Coordination of alkyne and insertion of this alkyne into an Ni–C bond generates another intermediate. Subsequent protonation with water yields an amino chalcone and an Ni(II) species. Further reduction of the Ni(II) species by zinc regenerates the catalyst Ni(0) along with a Zn(II) derivative. The keto group and the *o*-amino group in the obtained amino chalcone are expected to be *trans* to each other after protonation of the intermediate. Thus, *trans* to *cis* isomerization to yield a *cis*-amino chalcone should occur. Further condensation between the keto and amino groups yields the final product.

Driver and co-workers developed a synthetic strategy to access 2,6-disubstituted pyridines from triazolopyridines through a regioselective Ni-catalyzed alkenylation of the C7–H bond.[72] The N_2 fragment embedded in the resulting C–H functionalized triazolopyridine can be readily excised using acidic or oxidative conditions without destruction of the C7-alkene substituent. This nickel alkenylation method employs the cheap and air-stable triphenylphosphine ligand and a slight excess of the acetylene. It is tolerant of a range of functionality on either the triazolopyridine or alkyne without reducing the reaction yield (Scheme 7.46). Importantly, alkyne insertion into the C7–H triazolopyridine bond is regioselective.

A one-pot catalytic synthesis of substituted 1,8-naphthyridines and 2*H*-pyrano[3,2-*g*]quinolin-2-ones by the reaction of α-ketoalkynes with 6-aminonicotinamide and 7-amino-4-methylcoumarin, respectively, in water, using a homogeneous nickel catalyst under very mild reaction conditions, was described in 2003 by Rosas *et al.*[73] The desired products were isolated in moderate to good yields. In the absence of this catalytic system, very low yields are obtained even after long reaction times.

Cheng and co-workers demonstrated for the first time a general method for the synthesis of *N*-aryl- and *N*-alkylisoquinolinium salts in 2009.[74] A vast number of functionalities can be introduced into the salts (Scheme 7.47). As isoquinolinone is an important core structure for a wide range of natural

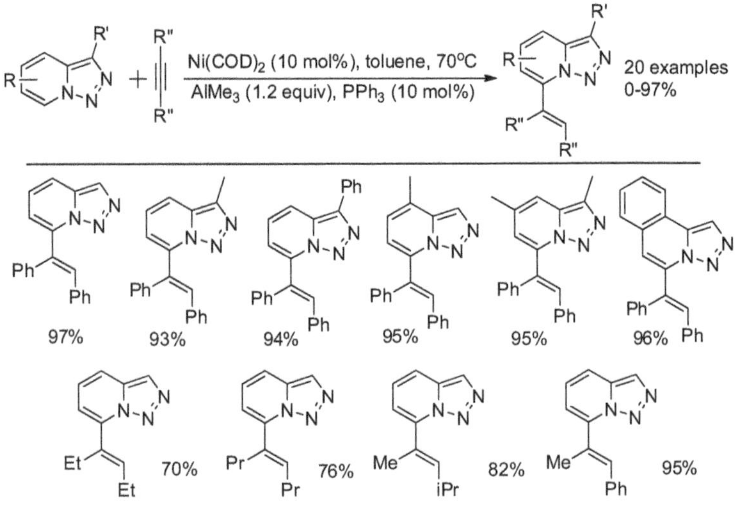

Scheme 7.46 Nickel-catalyzed synthesis of pyridines from triazolopyridines.

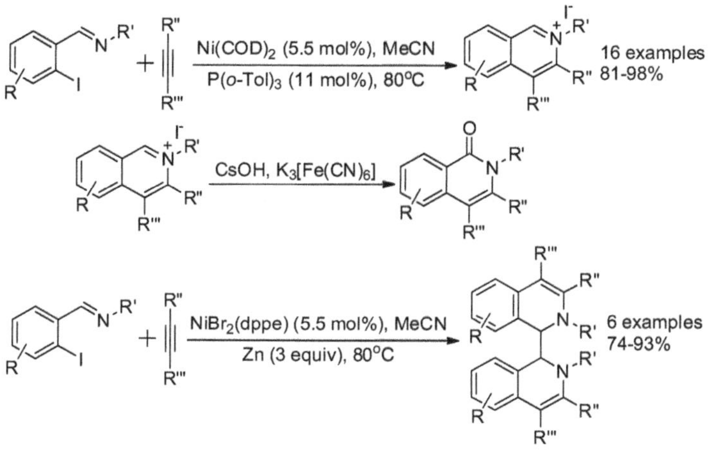

Scheme 7.47 Nickel-catalyzed synthesis of isoquinolinium salts.

Reaction Procedure (Scheme 7.47): A screw-capped sealed tube fitted with a septum containing 2-iodobenzaldehyde (0.20 mmol) and *p*-toluidine (0.20 mmol) was evacuated and purged three times with nitrogen. The tube was charged with [Ni(COD)$_2$] (0.011 mmol, 3.0 mg) and P(*o*-Tol)$_3$ (0.023 mmol, 7.0 mg) inside a glove-box. The tube was then kept under an atmosphere of nitrogen on a dual-manifold Schlenk line. Alkyne (0.25 mmol) was dissolved in acetonitrile (3.0 mL) and added to the stirred mixture *via* a syringe. The septum was quickly exchanged for a screw-cap and the reaction mixture was stirred at 80 °C for 1.5 h. At the end of the reaction, the reaction mixture was diluted with CH$_2$Cl$_2$ and then filtered through a silica-gel pad using methanol as the eluent ($\sim$20 mL). The combined filtrate was concentrated *in vacuo* and the residue was carefully washed with ethyl acetate and hexane to afford the desired pure product.

products. The application of this methodology was illustrated by the facile conversion of the salts into other structures such as isoquinolinones and reductive dimerization products.

In 2011, a nickel-catalyzed [2 + 2 + 2] cycloaddition of an imine bearing an *N*-pyridyl group and two alkyne units to 1,2-dihydropyridines was reported by Yoshikai and co-workers.[75] This reaction represents a rare example of metal-catalyzed cycloaddition reactions involving imines as substrates and the desired products were formed in good yields in general (Scheme 7.48). The directing effect of the pyridyl group was supported by DFT calculations.

In 2008, a new nickel-catalyzed reaction of alkynes with *N*-arylphthali-mides to provide isoquinolones was developed by Kurahashi and co-workers.[76] It was demonstrated that amide C–N bonds are susceptible to nucleophilic attack by the Ni(0) complex, which allows intermolecular addition to alkynes *via* decarbonylation. Various desired products were formed in good yields (Scheme 7.49a). More recently, they reported an unprecedented reaction pattern that provides isoindolinones with a nickel catalyst and MAD [methylaluminum bis(2,6-di-*tert*-butyl-4-methylphenoxide)] co-catalyst from phthalimides and trimethylsilyl-substituted alkynes (Scheme 7.49b).[77] This reaction represents the first example of alkylidenation with alkynes involving decarbonylation. In view of the reaction mechanism of the previously reported nickel-catalyzed reaction of phthalimides with alkynes, the catalytic cycle of the present reaction could be considered to involve the oxidative addition of an imide CO–N bond to an Ni(0) complex

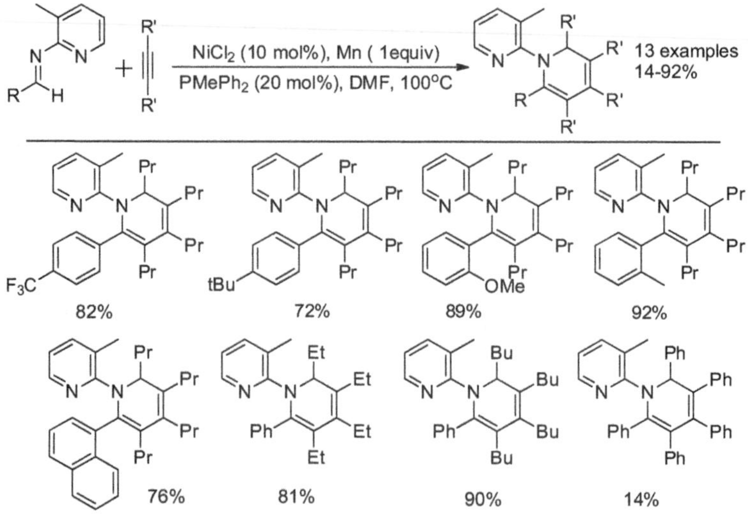

Scheme 7.48   Nickel-catalyzed synthesis of 1,2-dihydropyridines.

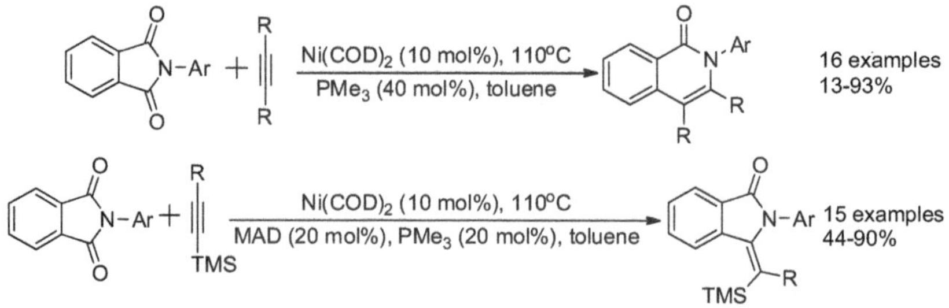

Scheme 7.49   Nickel-catalyzed synthesis of isoquinolones and isoindolinones.

Reaction Procedure (Scheme 7.49): The reaction was performed in a 20 mL round-bottomed flask equipped with a Teflon-coated magnetic stirrer bar and Dimroth reflux condenser. The top of condenser was connected with a balloon filled with argon (~1 atm). An *N*-arylphthalimide (0.5 mmol) and an alkyne (0.75 mmol) were added to a solution of bis(1,5-dicyclooctadiene)nickel (14 mg, 0.05 mmol) and trimethylphosphine (15 mg, 0.20 mmol) in toluene (2 mL) in a dry box. The flask was taken outside the dry box and heated at 110 °C for the indicated time under an argon atmosphere. The resulting mixture was cooled to ambient temperature, filtered through a silica gel pad and concentrated *in vacuo*. The residue was purified by flash silica gel column chromatography (20 g, 15 × 2 cm i.d., 5 : 1 hexane–ethyl acetate as eluent) to give the corresponding carboamination products.

Reaction Procedure: The reaction was performed in a 20 mL round-bottomed flask equipped with a Teflon-coated magnetic stirrer bar and Dimroth reflux condenser. The top of condenser was connected with a balloon filled with argon (~1 atm). An *N*-arylphthalimide (0.2 mmol) and an alkyne (0.3 mmol) were added to a solution of bis(1,5-cyclooctadiene)-nickel (5.5 mg, 0.02 mmol), trimethylphosphine (3 mg, 0.4 mmol) and methylaluminum bis(2,6-di-*tert*-butyl-4-methylphenoxide) (MAD) (0.04 mmol) in toluene (2 mL) in a dry box. The flask was taken outside the dry box and refluxed (110 °C, oil bath at 140 °C) for the indicated time under an argon atmosphere. The resulting mixture was cooled to ambient temperature, filtered through a silica gel pad and concentrated *in vacuo*. The residue was purified by flash silica gel column chromatography (eluent: 5 : 1 hexane–ethyl acetate) to give the corresponding carboamination products.

in association with MAD. Subsequent decarbonylation and coordination of alkyne takes place to afford a five-membered nickelacycle. MAD, which is coordinated to carbonyl oxygen, would then promote the formation of acyclic cationic nickel; the electron-rich phosphine ligand PMe_3 may also promote the formation of acyclic cationic nickel by stabilizing the cationic nature of the nickel center. The nucleophilic addition of alkyne takes place to afford a cationic nickel–vinylidene complex *via* a [1,2]-shift of the silyl group on the alkyne. The insertion of a vinylidene moiety in the C–Ni bond affords a six-membered nickelacycle which undergoes reductive elimination to furnish isoindolinone and regenerates the starting Ni(0).

Nakao, Hiyama and co-workers reported a nickel-catalyzed dehydrogenative [4 + 2] cycloaddition of formamides with alkynes in 2011 (Scheme 7.50).[78] They demonstrated that *N*,*N*-bis(1-arylalkyl)formamides undergo an unprecedented dehydrogenative [4 + 2] cycloaddition reaction with alkynes *via* nickel–$AlMe_3$ cooperative catalysis through double functionalization of otherwise unreactive $C(sp^2)$–H and $C(sp^3)$–H bonds to give highly substituted dihydropyridone derivatives, which can serve as versatile synthetic precursors for nitrogen-containing six-membered heterocycles. Later, Anand and Sunoj performed a computation study on this topic in order to elucidate the reaction mechanism.[79] They showed, using the computed Gibbs free energies, that the origin of the cooperativity offered by $AlMe_3$ in the Ni(0)-catalyzed dehydrogenative cycloaddition between alkynes and formamides is the significant stabilization of the vital transition states

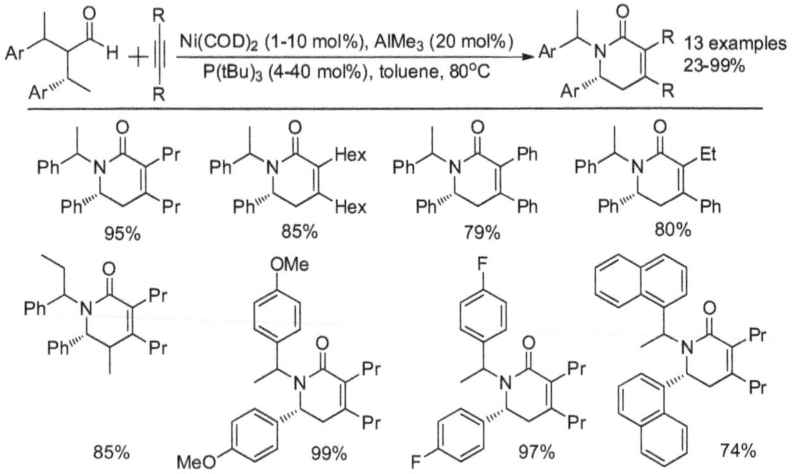

Scheme 7.50 Nickel-catalyzed synthesis of dihydropyridones.

Reaction Procedure (Scheme 7.50): An alkyne (2.2–6.6 mmol) was added to a solution of a formamide (1.00 mmol), Ni(COD)$_2$ (2.8–28 mg, 10–100 μmol), P(*t*Bu)$_3$ (8.1–81 mg, 40–400 μmol), a 1.08 M solution of AlMe$_3$ in hexane (0.19 mL, 0.20 mmol) and dodecane (internal standard, 57 mg, 0.33 mmol) in toluene prepared in a 15 mL vial under an argon atmosphere in a dry box. The vial was sealed with a screw-cap, taken outside the dry box and heated as specified. The resulting mixture was filtered through a silica gel pad, concentrated *in vacuo* and purified by flash column chromatography on silica gel to give the corresponding cycloadducts. Isomers were further separated by preparative recycling silica gel chromatography.

---

compared with the non-cooperative pathways. The rate-limiting oxidative insertion was found to be 14 kcal mol^{-1} lower in energy with Lewis acid coordination than in its absence. The modulation of electronic effects due to the closer geometric proximity of the coordinated Lewis acid to the formyl group renders the preferred order of dual C–H activation formyl C(sp^2)–H first followed by benzyl methyl C(sp^3)–H.

3,4-Dihydroisoquinolin-1(2*H*)-ones are found in a wide variety of plant alkaloids and bioactive compounds. Murakami and co-workers reported an enantioselective synthesis of 3,4-dihydroisoquinolin-1(2*H*)-ones in 2010.[80] The procedure involves nickel-catalyzed denitrogenative annulation of 1,2,3-benzotriazin-4(3*H*)-ones with allenes and the desired products were formed in a regio- and enantioselective manner in moderate to excellent yields (Scheme 7.51). Enantioselectivity is achieved by applying a chiral ligand (such as *i*Pr-FOXAP).

In 2011, Chatani's group described a nickel-catalyzed regioselective oxidative cycloaddition of aromatic amides to alkynes.[81] As the first example of

Ni-catalyzed transformation of *ortho* C–H bonds, their procedure takes advantage of chelation assistance by a 2-pyridinylmethylamine moiety. The corresponding products were isolated in good yields (Scheme 7.52). The directing group, a 2-pyridinylmethylamine, can be easily removed by treatment of the product with lithium diisopropylamide (LDA) and then bubbling O_2 followed by hydrolysis to give the NH-isoquinolone in good yield.

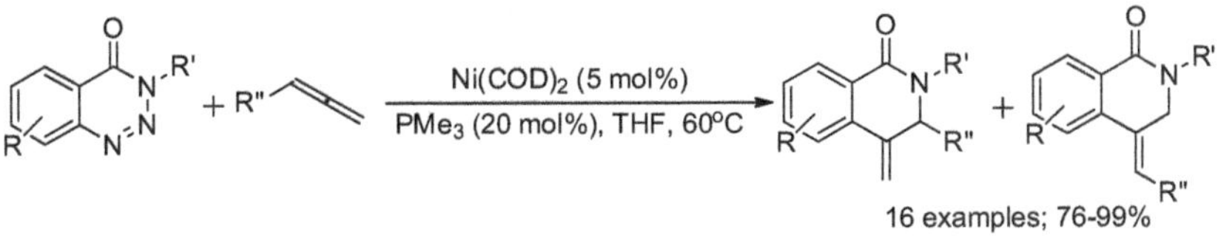

Scheme 7.51 Nickel-catalyzed synthesis of 3,4-dihydroisoquinolin-1(2*H*)-ones.

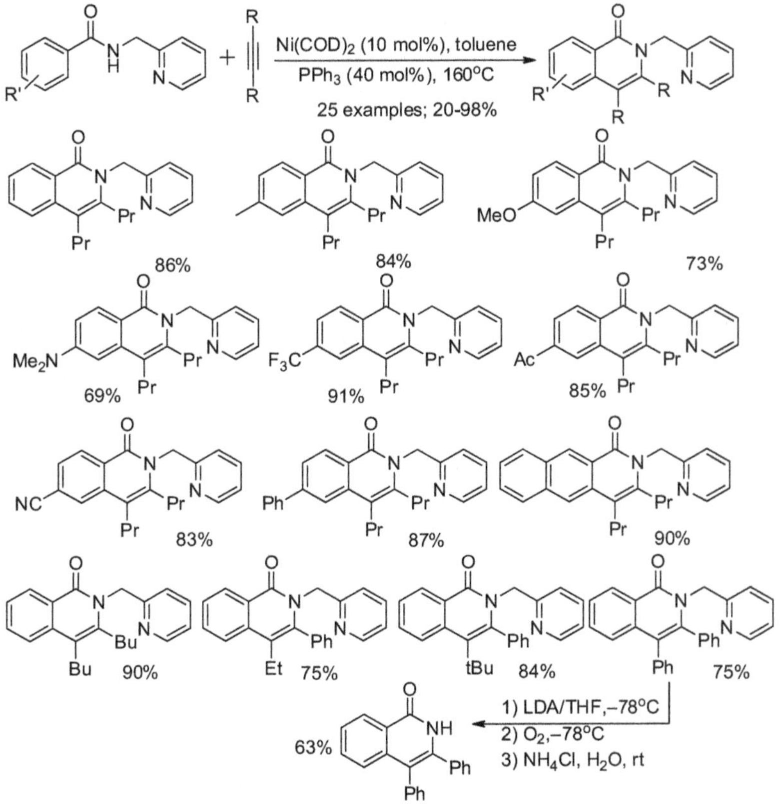

Scheme 7.52 Nickel-catalyzed synthesis of isoquinolones *via* C–H activation.

Reaction Procedure (Scheme 7.52): An oven-dried 5 mL screw-capped vial was charged with Ni(COD)$_2$ (13.8 mg, 0.05 mmol), *N*-(pyridin-2-ylmethyl)benzamide (0.5 mmol), PPh$_3$ (52 mg, 0.2 mmol), 4-octyne (165 mg,

1.5 mmol) and toluene (2 mL) in a glove-box filled with N_2. After closing the cap, the vessel was heated in an oil-bath at 160 °C for 6 h, followed by cooling. After transferring the contents to a round-bottomed flask with EtOAc, the isomeric composition was determined by NMR and GC analysis. The volatiles were removed *in vacuo* and the residue was subjected to flash column chromatography on silica-gel (eluent: $1:0 \rightarrow 4:1 \rightarrow 0:1$ hexane–ethyl acetate) to give the pure product (138 mg, 86% yield) as a white solid.

Interestingly, in 2008, Murakami and co-workers found that 1,2,3-benzotriazin-4(3H)-ones can react with internal and terminal alkynes in the presence of a nickel(0)–phosphine catalyst.[82] A wide variety of alkyne substrates including borylalkynes were regioselectively incorporated into 1,2,3-benzotriazin-4(3H)-ones with loss of a dinitrogen molecule. Various substituted 1(2H)-isoquinolones were produced in high yield (Scheme 7.53a). The reaction was initiated by insertion of nickel(0) into the N–N linkage of the triazinone, which prompted extrusion of a molecular dinitrogen to give an azanickelacycle. Subsequent insertion of the alkyne into the Ni–C bond led to the seven-membered ring nickelacycle. Finally, reductive elimination afforded the desired product and regenerated the nickel(0) catalyst. Subsequently, Cheng's group demonstrated an easy and convenient nickel-catalyzed annulation of substituted 2-halobenzamides with alkynes to give the corresponding isoquinolinone (Scheme 7.53b).[83] This protocol was successfully applied to the total synthesis of oxyavicine in excellent yield. Regarding the reaction mechanism, the oxidative addition of 2-iodobenzamide to Ni(0) in the presence of Et_3N leads to the formation of a five-membered ring nickelacycle. Coordinative insertion of an alkyne into the nickelacycle gives seven-membered ring nickelacycle intermediates. Reductive elimination then provides the final isoquinolinone and regenerates the Ni(0) catalyst for the next catalytic cycle. There are two possible pathways for the insertion of a coordinated alkyne into a nickelacycle: the authors proposed that a carbon–carbon triple bond can insert into the carbon–nickel bond or the nitrogen–nickel linkage, depending on the nature of the alkyne.

In 2010, a decarbonylative cycloaddition reaction of phthalimides with 1,3-dienes was reported.[84] It was found that the cycloaddition proceeded regioselectively with respect to the 1,3-dienes. Moreover, regioselective cycloadditions of an unsymmetrically functionalized phthalimide with 1,3-dienes were also achieved. These cycloadditions displayed excellent regio- and chemoselectivity in the presence of various functional groups, which may open the way for a facile divergent synthesis of functionalized isoquinolones (Scheme 7.54a). Regarding the reaction mechanism, it was proposed that the catalytic cycle consists of the oxidative addition of an amide CO–N bond to an Ni(0) complex, with subsequent decarbonylation and coordination of the bidentate diene giving a nickel(II) intermediate. The diene then inserts into the C–Ni bond to give a more stable acyclic π-allylnickel

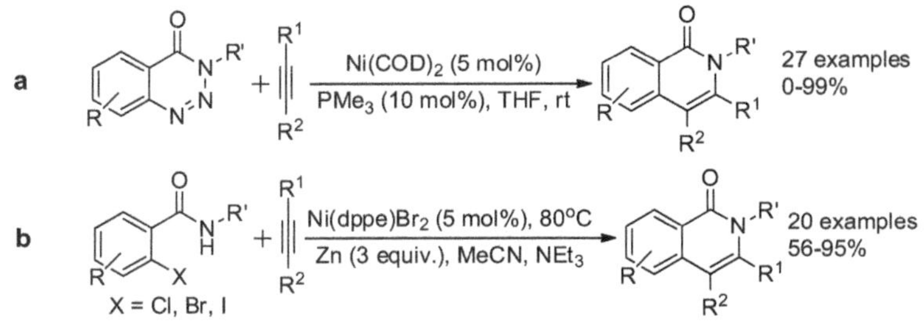

Scheme 7.53 Nickel-catalyzed synthesis of isoquinolinones.

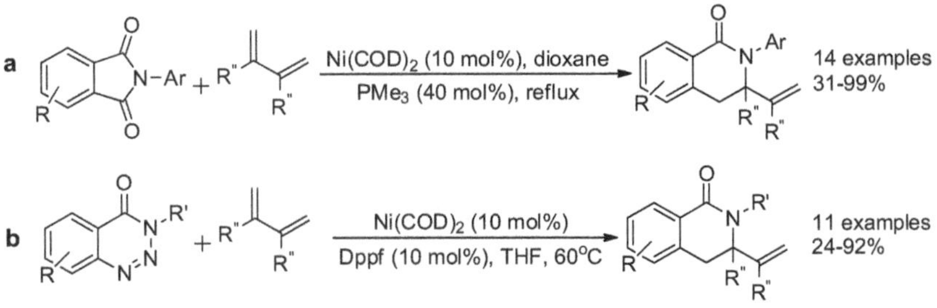

Scheme 7.54 Nickel-catalyzed synthesis of 3,4-dihydroisoquinolin-1(2*H*)-ones.

intermediate. Nucleophilic addition of a nitrogen atom to the π-allylnickel at the more substituted carbon takes place to afford the pure product and regenerate the starting Ni(0) complex. Around the same time, Murakami and co-workers found that 1,2,3-benzotriazin-4(3*H*)-ones can react with 1,3-dienes in the presence of a nickel(0)–phosphine complex to give a variety of 3,4-dihydroisoquinolin-1(2*H*)-ones (Scheme 7.54b).[85] In detail, the oxidative insertion of nickel(0) into the triazinone moiety prompts extrusion of dinitrogen to give a five-membered ring azanickelacyclic intermediate. Subsequent insertion of 1,3-dienes into the Ni–C bond followed by allylic amidation affords 3,4-dihydroisoquinolin-1(2*H*)-ones. Alkenes also undergo insertion into the five-membered ring azanickelacyclic intermediate and subsequent reductive elimination gives 3-substituted 3,4-dihydroisoquinolin-1(2*H*)-ones along with regenerated nickel(0).

As early as in 1982, Hoberg and Oster reported a nickel-catalyzed [2 + 2 + 2'] cycloaddition of two alkynes with isocyanates.[86] 2-Oxo-1,2-dihydropyridines were produced in good yields (Scheme 7.55a). Louie's group found that the reaction can be performed at room temperature by using an NHC-carbene (SIPr) as ligand.[87] Alternatively, the combination of Ni(0) and PEt$_3$ can also give regioisomeric mixtures of pyridones by cycloaddition of asymmetrical alkynes and isocyanates. Regioselectivity was highly dependent on the size of the terminal groups. In some cases, electronic factors override steric interactions to afford ultimately a single pyridone regioisomer. NMR analysis indicated that the two regioisomers possessed a large substituent at the 3-position on the pyridone ring. Under similar

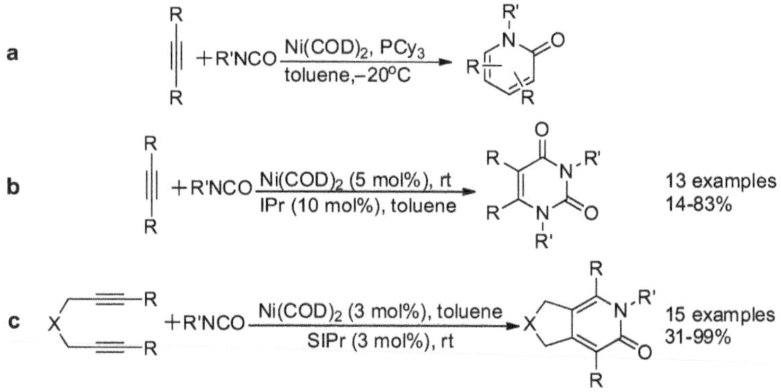

Scheme 7.55 Nickel-catalyzed synthesis of 2-pyridones.

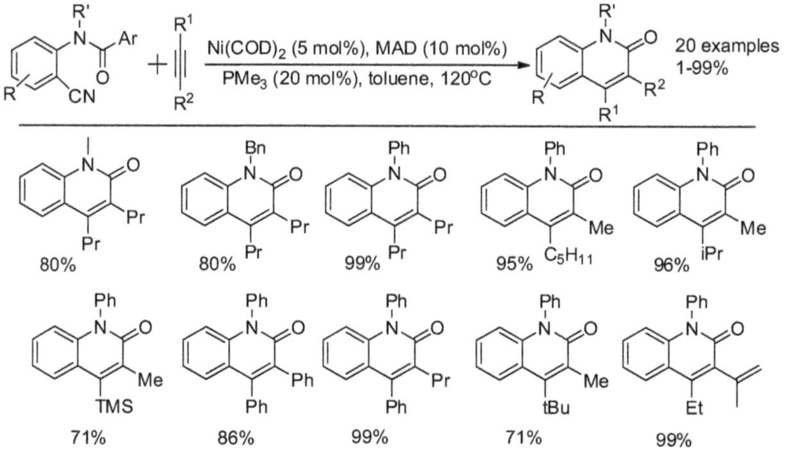

Scheme 7.56 Nickel-catalyzed synthesis of quinolones from *o*-cyanophenylbenza-
             mides and alkynes.

conditions, pyrimidinediones can be produced by cycloaddition of one alkyne and two isocyanates.[88] The key to the success of this protocol is the use
of unsymmetrically substituted alkynes, which favors the formation of pyrimidinediones over pyridones. A variety of pyrimidinediones were prepared
(Scheme 7.55b). A one-pot cycloaddition and Stille coupling were reported for
tributyl(1-propynyl)tin. In 2004, Louie and co-workers succeeded in devising a
mild and efficient method for preparing a wide range of 2-pyridones from
diynes and isocyanates (Scheme 7.55c).[89] Both intramolecular and intermolecular reactions were catalyzed by a combination of an Ni(0) precursor
and an NHC ligand under mild reaction conditions.

   More recently, a nickel-catalyzed synthesis of quinolones was developed by
Nakai *et al.*[90] The procedure used *o*-cyanophenylbenzamides and alkynes as
substrates and the desired quinolones were isolated in good to excellent
yields (Scheme 7.56). The reaction involves elimination of a nitrile group by

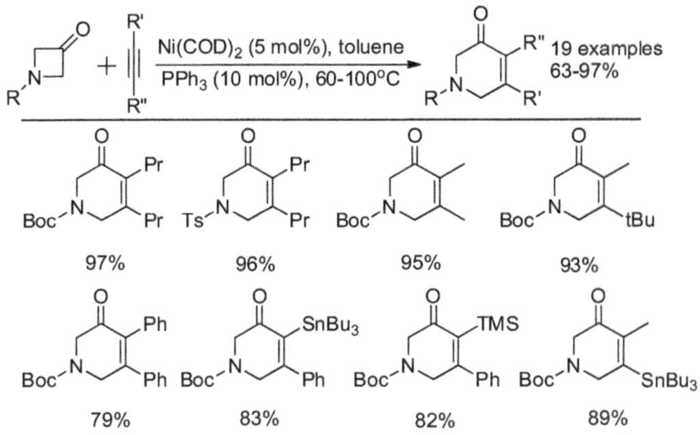

Scheme 7.57 Nickel-catalyzed synthesis of 3-piperidones.

cleavage of the two independent aryl–cyano and aryl–carbonyl C–C bonds of the amides.

In 2012, Kumar and Louie developed a nickel-catalyzed-method for the [4 + 2] cycloaddition reaction of azetidinones and alkynes.[91] This reaction included an interesting C–C bond cleavage that ultimately afforded 3-piperidone products. The reaction conditions are both mild and practical and afford the *N*-heterocycles in excellent yields (Scheme 7.57). Importantly, minimal side products such as C–H activation products were formed. The authors suggested that the reaction is initiated with oxidative coupling between the alkyne and the carbonyl of the azetidinone. With interest in this reaction, Lin and Li carried out a theoretical study.[92] With the aid of DFT calculations, they examined the mechanism of the Ni-catalyzed cycloaddition of 3-azetidinone with alkynes. The results did not support the originally proposed mechanism, which involved a ring expansion through β-carbon elimination. Instead, their calculations supported a mechanism that involves oxidative addition of 1-Boc-3-azetidinone to an Ni(0) center to form an intermediate having both Ni–C(O) and Ni–C(sp³) bonds, then alkyne insertion into either the Ni–C(O) or Ni–C(sp³) bond depending on the alkyne substrate being studied, and finally reductive elimination to give the cycloaddition products. The nature of the insertion of alkynes into an Ni–C bond was also discussed. When an alkyne substrate inserts into an Ni–C(O) bond, they found that the alkyne acts as a nucleophile to attack the electron-deficient, metal-bonded C=O carbon center. When an alkyne substrate inserts into an Ni–C(sp³) bond, the alkyne acts as an electrophile to interact with the Ni–C(sp³) bond. The regioselectivities observed experimentally in the cycloaddition reactions of 1-Boc-3-azetidinone with MeC≡C*t*Bu, MeC≡CTMS, PhC≡CTMT and MeC≡CTMT were also investigated theoretically. The theoretical calculation results showed that the regioselectivity was determined in the step involving alkyne insertion into an intermediate having both Ni–C(O) and Ni–C(sp³) bonds. It was suggested that in the insertions of both MeC≡C*t*Bu and

MeC≡CTMS the sterically bulky substituents *t*Bu and TMS prevent the *t*Bu-
and TMS-substituted carbon from coupling with an Ni-bonded carbon. In
both cases, they found that the methyl-substituted carbons couple with one
of the Ni-bonded carbons. The methyl-substituted carbon of MeC≡C*t*Bu is
π-electron rich, whereas the methyl-substituted carbon of MeC≡CTMS is π-
electron deficient, due to the π-accepting properties of TMS. As a result, the
methyl-substituted carbon of MeC≡C*t*Bu couples with the Ni–C(O) carbon
whereas the methyl-substituted carbon of MeC≡CTMS couples with the
Ni–C(sp^3) carbon, giving rise to the completely different regioselectivities
observed experimentally. The authors also examined the C–C coupling pathways for the reactions of PhC≡CTMT and MeC≡CTMT and found that the
carbonyl carbon prefers to couple with the stannyl-substituted carbon. The
results were in agreement with the experimental observations for stannyl-
phenylalkynes, but not for stannylmethylalkynes.

A new nickel-catalyzed reaction of alkynes with isatoic anhydrides to provide quinolones was reported in 2009 (Scheme 7.58).[93] It was shown that
carbamates are susceptible to oxidative addition of an Ni(0) complex, which
allows intermolecular addition to alkynes *via* decarboxylation. In detail,
the catalytic cycle of the present reaction is considered to involve oxidative
addition of an anhydride O–CO bond to nickel, then subsequent decarboxylation and coordination of alkyne to afford an Ni(II) intermediate. The alkyne
then inserts into the acyl–nickel bond to give a nickelacycle, which undergoes
reductive elimination to give final product and regenerate the starting Ni(0).
The regioselectivity of the reaction can be rationalized in terms of the dir-
ection of alkyne insertion, in which the repulsive steric interaction is minimal
between the bulkier R^L (the substitution on alkyne) and the PCy$_3$ ligand on
the nickel to give the nickel(II) intermediate. Thus, the reaction of substrates
using less sterically hindered ligands, such as PPh$_3$ and PMe$_3$, affords the
product with lower regioselectivity, while the reaction using a more sterically
hindered *N*-heterocarbene ligand shows higher regioselectivity.

More recently, Yang and co-workers developed a nickel-catalyzed
decarboxylative cycloaddition of alkenes with isatoic anhydrides. This procedure provided a new method for the synthesis of quinolone derivatives
(Scheme 7.59).[94] Nickel(0) catalyst was generated *in situ* from NiCl$_2$(PMe$_3$)$_2$
and activated Zn powder, which was found to be very important to ensure the
success of the reaction. Isatoic anhydrides were reacted with various nor-
bornenes to afford novel quinolone structures. This protocol is simple and
easy to handle.

Interestingly, Devaraj and co-workers discovered that metal salts such as
nickel and zinc triflates can be used to catalyze the one-pot synthesis of
1,2,4,5-tetrazines from unactivated nitriles (Scheme 7.60).[95] This methodology conveniently provides a range of symmetric and asymmetric tetrazines
directly from commercially available nitriles. In this reaction, the metal acts
as a Lewis acid by coordinating to the nitrile and promoting nucleophilic
addition by hydrazine. It is also plausible that the metal binds both the
nitrile and hydrazine, promoting synthesis of the amidrazone intermediate.

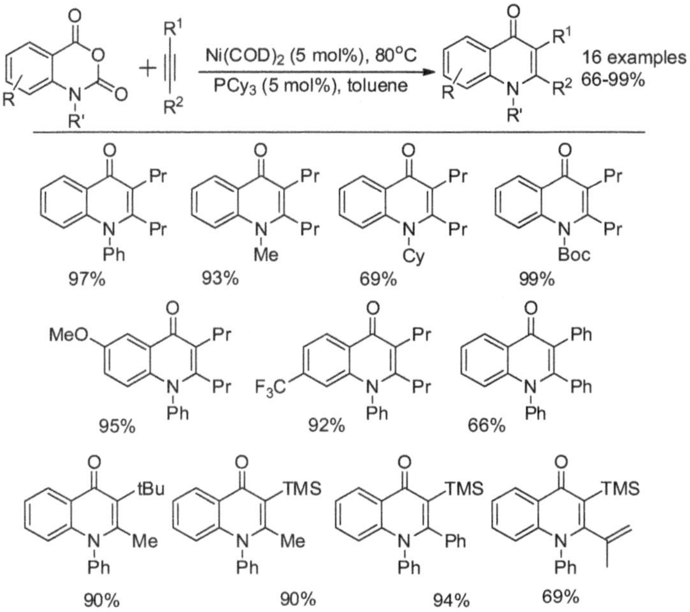

Scheme 7.58 Nickel-catalyzed synthesis of quinolones from isatoic anhydrides.

> **Reaction Procedure** (Scheme 7.58): The reaction was performed in a 15 mL sealed tube equipped with a Teflon-coated magnetic stirrer. An isatoic anhydride (0.5 mmol) and an alkyne (0.6 mmol) were added to a solution of bis(1,5-dicyclooctadiene)nickel (7 mg, 0.025 mmol) and tricyclohexylphosphine (7 mg, 0.25 mmol) in toluene (1.5 mL) in a dry box. The flask was taken out of the dry box and heated at 80 °C for the indicated time under an argon atmosphere. The resulting mixture was cooled to ambient temperature, filtered through a silica gel pad and concentrated *in vacuo*. The residue was purified by flash silica gel column chromatography (eluent: 5 : 1 hexane–ethyl acetate) to give the pure 4-quinolone.

A three-component reaction of 2-alkynylbenzaldehydes, a sulfonohydrazide and dimethyl cyclopropane-1,1-dicarboxylate was developed by Wu and co-workers.[96] They used silver(I) triflate and nickel(II) perchlorate hexahydrate as a bimetallic catalyst and 2,3,4,11b-tetrahydro-1*H*-pyridazino[6,1-*a*]isoquinolines were formed in moderate to good yields. A possible reaction pathway that proceeds *via* a tandem 6-*endo*-cyclization and [3 + 3] cycloaddition sequence was proposed. Subsequently, Sharma's group found that the reaction between 7-amino-4-methylcoumarin and several ferrocenyl-α-ketoalkynes in the presence of a nickel catalytic system affords ferrocenyl-β-enaminonecoumarins and ferrocenylpyrano[3,2-*g*]quinolin-2-ones.[97] It was noted that the substitution at C8 depends on the nature of the α-ketoalkyne used. Using a ketoalkyne with a weak

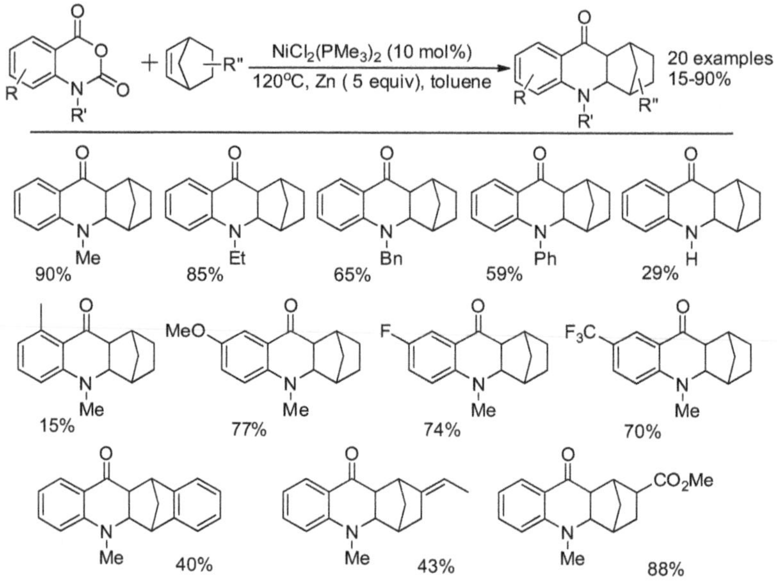

Scheme 7.59 Nickel-catalyzed synthesis of 2,3-dihydro-4-quinolones from alkenes.

Reaction Procedure (Scheme 7.59): An oven-dried 20 mL screw-capped vial was charged with NiCl$_2$(PMe$_3$)$_2$ (10%, 14.1 mg), activated Zn dust (5.0 equiv., 162 mg) and isatoic anhydride (0.5 mmol) under a gentle stream of argon. Toluene (5 mL) was added to the vial *via* a syringe, followed by norbornene (94.1 mg). The vessel was heated in an oil-bath at 100 °C for 12 h followed by cooling. The contents were subjected to flash chromatography to give 10-methyl-1,3,4,4a,9a,10-hexahydro-1,4-methanoacridin-9(2*H*)-one (92%) as a colorless liquid.

electron-donating group, *e.g.* an alkyl or phenyl group, the substitution of this group will occur at C6 of the pyrano[3,2-*g*]quinolinone, and with a ferrocenyl group with a higher electron-donating capacity, the substitution will be at the C8 of the heterocycle.

In 2009, Louie's group described a mild and general route for preparing dienamides.[98] Nickel–imidazolylidene complexes were used to mediate cycloadditive coupling between enynes and isocyanates (Scheme 7.61a). Dienamides were prepared in excellent yields and with good $E:Z$ selectivity. These dienamides can be further manipulated through oxidative cyclization methods. When a terminal enyne is employed, cyclization affords a lactam rather than a dienamide. A nickel catalyst was also applied in the activation of the C–Cl bond and coupling with NH.[99] The use of an *in situ*-generated Ni(0) catalyst associated with 2,2′-bipyridine or *N,N′*-bis(2,6-diisopropylphenyl)dihydroimidazol-2-ylidene (SIPr) as a ligand and NaO*t*Bu as the base for the intramolecular coupling of aryl chlorides with amines was successful developed

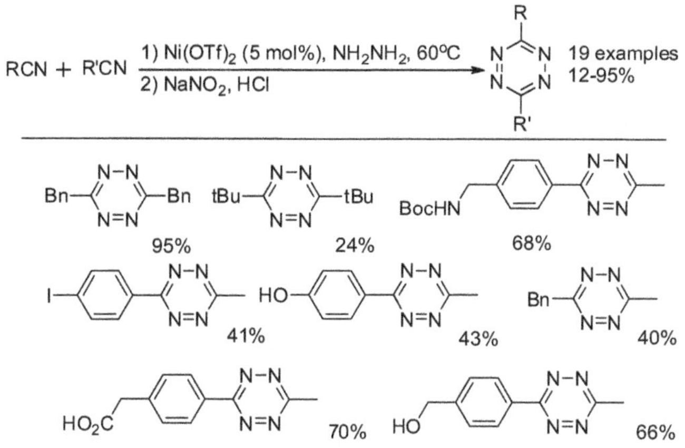

Scheme 7.60 Nickel-catalyzed synthesis of tetrazines.

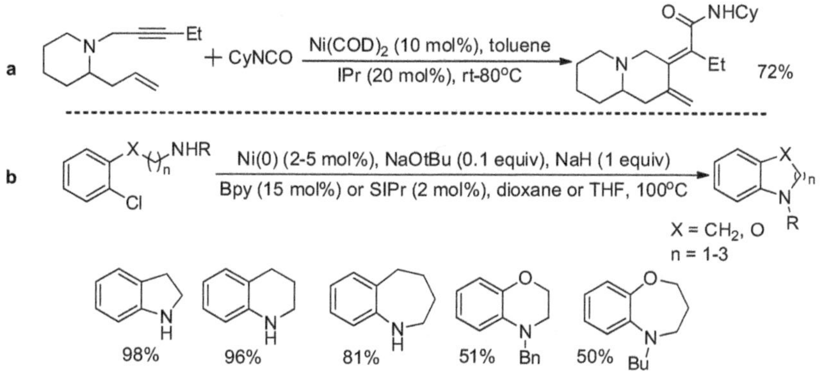

Scheme 7.61 Nickel-catalyzed C–N bond formation.

(Scheme 7.61b). Five-, six- and seven-membered rings can be prepared by this procedure. The complexes show good catalytic activity for the synthesis of indoles, quinolines, benzazepines, benzoxazines and benzoxazepines.

In 1999, Tang and Montgomery found that Ni(COD)$_2$–PBu$_3$ is a highly effective catalyst system for the triethylsilane-mediated reductive cyclization of ynals (Scheme 7.62).[100] The method was demonstrated to be efficient and highly stereoselective in the preparation of a variety of quinolizidine, indolizidine and pyrrolizidine alkaloids. The cyclization method allows the direct introduction of an allylic alcohol moiety with completely stereoselective introduction of an exocyclic double bond and highly diastereoselective alcohol introduction relative to pre-existing chirality. A total synthesis of (+)-allopumiliotoxin 267A was carried out, which highlights the utility of nickel-catalyzed ynal cyclizations in complex synthetic strategies. Soon afterwards, they also achieved the total syntheses of (+)-allopumiliotoxin 339A and (+)-allopumiliotoxin 339B by utilizing this ynal cyclization as the key step.[101]

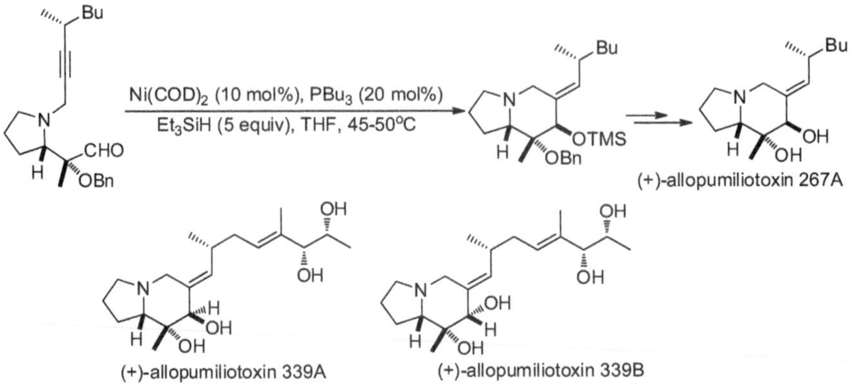

Scheme 7.62   Application of nickel catalysis in total synthesis.

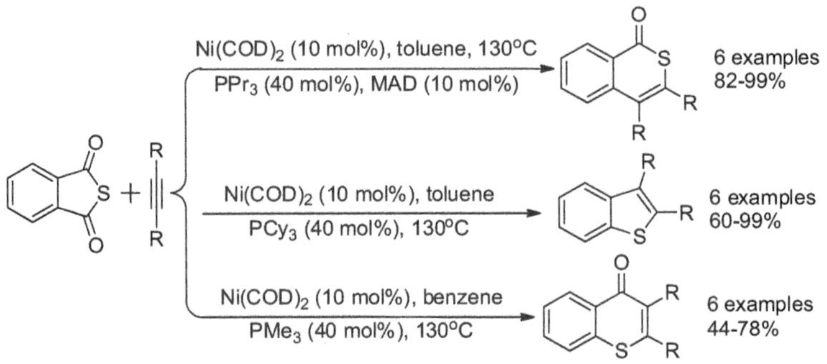

Scheme 7.63   Nickel-catalyzed cycloadditions of thiophthalic anhydride with alkynes.

These syntheses provide short and efficient entries to the allopumiliotoxins. Additionally, the total synthesis of erythrocarine was achieved in 2003 by Mori and co-workers.[102] They used nickel as the promoter under a carbon dioxide atmosphere and various heterocyclic compounds were formed in good yields.

Nickel-catalyzed cycloadditions of thiophthalic anhydride with alkynes to afford sulfur-containing heterocyclic compounds were developed in 2011.[103] It was demonstrated that the nickel-catalyzed reaction gave three types of compounds selectively, depending on the reaction conditions employed (Scheme 7.63). The use of Ni(0)–PPr$_3$ catalyst in combination with a Lewis acid afforded thioisocoumarins. In contrast, the use of Ni(0)–PCy$_3$ catalyst in the reaction afforded benzothiophenes, whereas the use of Ni(0)–PMe$_3$ catalyst furnished thiochromones.

In 2010, a nickel-catalyzed regio- and enantioselective annulation reaction of 1,2,3,4-benzothiatriazine-1,1(2H)-dioxides with allenes was developed by Murakami and co-workers.[104] They demonstrated that a highly reactive azanickelacycle can be generated from 1,2,3,4-benzothiatriazine-1,1(2H)-di-oxides in this procedure through extrusion of N$_2$. The azanickelacycle incorporates a variety of allenes in a regio- and enantioselective manner, providing a new synthetic route to substituted 3,4-dihydro-1,2-benzothiazine-1,1(2*H*)-dioxides, the biological activities of which are of much interest (Scheme 7.64).

In addition to nitrogen-containing heterocycles, nickel catalysts have also been applied in the preparation of oxygen-containing heterocyclic compounds. In 2008, a nickel-catalyzed decarbonylative addition of anhydrides to alkynes was reported,[105] various isocoumarins being isolated in good yields (Scheme 7.65). The reaction mechanism is considered to involve oxidative addition of an anhydride O–CO bond to nickel. Subsequent decarbonylation and coordination of alkyne take place, in which the steric repulsive interaction is minimal between the bulkier R′ and the PMe₃ ligand on the nickel, to give a nickel(II) intermediate. The alkyne then inserts into

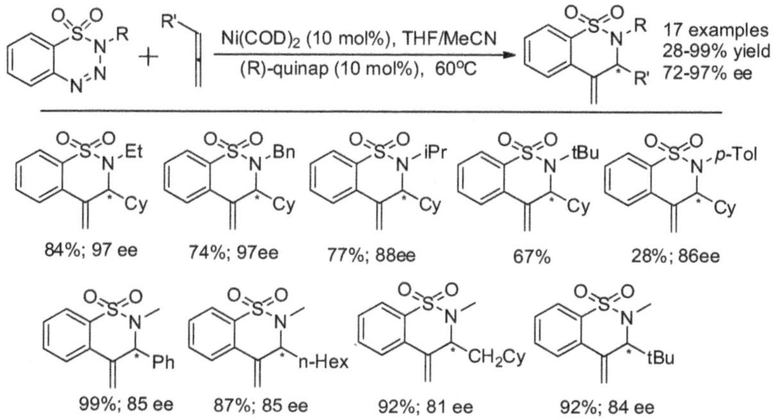

Scheme 7.64 Nickel-catalyzed synthesis of 3,4-dihydro-1,2-benzothiazine-1,1(2*H*)-dioxides.

Reaction Procedure (Scheme 7.64): In an N₂-filled glove-box, a 1,2,3,4-benzothiatriazine-1,1(2*H*)-dioxide (39.7 mg, 0.20 mmol), [Ni(COD)₂] (5.6 mg, 0.02 mmol), (*R*)-quinap (8.8 mg, 0.02 mmol), 1,4-dioxane (2 mL) and an alkyne (58 mL, 0.40 mmol) were placed at room temperature in an oven-dried 4 mL vial containing a magnetic stirrer bar. The vial was sealed with a Teflon cap and taken out of the glove-box. After being heated at 60 °C for 6 h, the reaction mixture was cooled to room temperature and stirred for 1 h in the open air. The resulting mixture was passed through a pad of Florisil and eluted with ethyl acetate, then the filtrate was concentrated under reduced pressure. The residue was purified by preparative TLC (eluent: 5 : 1 hexane–ethyl acetate) to give an isomeric mixture of products. The enantiomeric excess of the major isomer was determined by HPLC using a Chiralcel OD-H column.

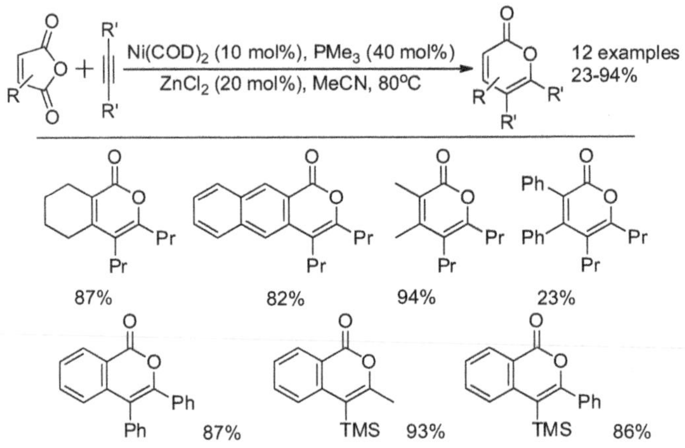

Scheme 7.65   Nickel-catalyzed synthesis of isocoumarins from anhydrides.

the aryl–nickel bond to give a nickelacycle, which undergoes reductive elimination to give the desired product and regenerates the starting nickel(0). *In situ* IR spectroscopic analysis demonstrated that the stoichiometric reaction of Ni(COD)$_2$–PMe$_3$ with substrates without ZnCl$_2$ resulted in gradual consumption of the anhydrides without the formation of a product. Importantly, constant generation of alkynes was observed simultaneously on addition of ZnCl$_2$. These results imply that reductive elimination is specifically promoted by the addition of ZnCl$_2$. The effect of ZnCl$_2$ is likely to result from the coordination of a Lewis acid to a carbonyl group, which may generate an electron-poor alkenylnickel through a conjugated system.

In 2011, the same group reported the decarbonylative cycloaddition of phthalic anhydride with allene.[106] δ-Lactones were produced in a single step. The reaction represents an unprecedented insertion reaction of a carbon–carbon double bond into a carbon–oxygen bond. They also demonstrated that δ-thiolactone can be prepared *via* decarbonylative cycloaddition of thiophthalic anhydride with allene. Moreover, asymmetric insertion reactions of a carbon–carbon double bond into a carbon–oxygen bond and into a carbon–sulfur bond were also successfully demonstrated with chiral phosphine ligands to provide δ-lactones and δ-thiolactones.

In 2009, they reported a nickel-catalyzed cycloaddition of salicylic acid ketals with alkynes *via* elimination of ketones,[107] which opened the way to the divergent synthesis of chromones (Scheme 7.66). It was shown that ketones are capable of β-elimination, which allows the formation of the key oxa-nickelacycle intermediate. With regard to the reaction mechanism, it is reasonable to consider that the catalytic cycle of the reaction consists of the oxidative addition of an ester CO–O bond to an Ni(0) complex. Subsequent elimination of benzophenone and coordination of alkyne take place, in which the steric repulsive interaction is minimal between the bulkier R' and the PCy$_3$ ligand on the nickel, to give an nickel(II) intermediate. The alkyne then inserts into the C–Ni bond to give a nickelacycle, which undergoes

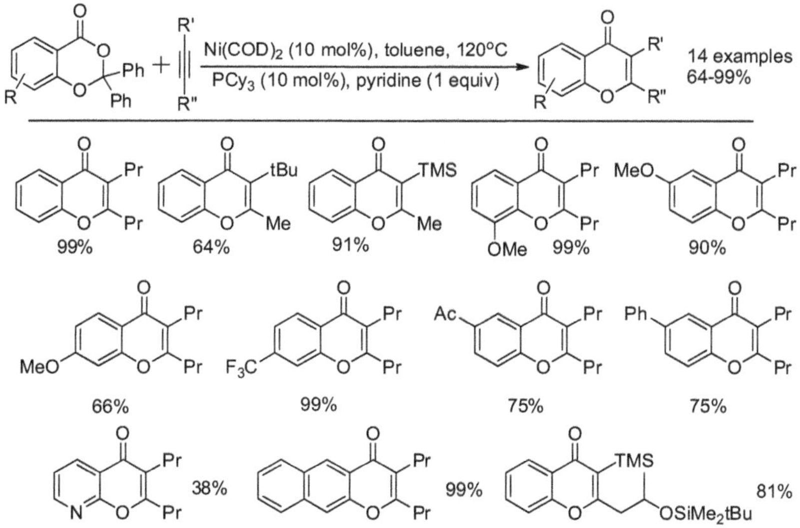

Scheme 7.66 Nickel-catalyzed synthesis of chromones.

> **Reaction Procedure** (Scheme 7.66): The reaction was performed in a 15 mL sealed tube equipped with a Teflon-coated magnetic stirrer. A salicylic acid ketal (0.5 mmol) and an alkyne (1.0 mmol) were added to a solution of bis(1,5-dicyclooctadiene)nickel (14 mg, 0.05 mmol), tricyclohexylphosphine (14 mg, 0.05 mmol) and pyridine (0.5 mmol) in toluene (1.5 mL) in a dry box. The flask was taken out of the dry box and heated at 120 °C for the indicated time under an argon atmosphere. The resulting reaction mixture was cooled to ambient temperature, filtered through a silica gel pad and concentrated *in vacuo*. The residue was purified by flash silica gel column chromatography (20 g, 15×2 cm i.d., 5:1 hexane–ethyl acetate as eluent) to give the corresponding chromone.

reductive elimination to give the final product and regenerates the starting Ni(0) complex.

More recently, they developed another interesting nickel-catalyzed cycloaddition of *o*-arylcarboxybenzonitriles and alkynes *via* cleavage of two carbon–carbon σ-bonds.[108] This is an unprecedented type of cycloaddition; *o*-arylcarboxybenzonitrile reacted with alkynes to afford coumarins over a nickel catalyst. As the authors demonstrated, this reaction represents the first example of intermolecular cycloaddition involving the cleavage of two carbon–carbon σ-bonds of two independent C–CN and C–CO bonds. The desired coumarins were formed in excellent yields with broad functional group tolerance (Scheme 7.67). The eliminated aromatic nitriles can also be detected.

In 2002, Louie and co-workers developed a mild and efficient method for the preparation of pyrones from diynes and CO_2.[109,110] The reaction employs

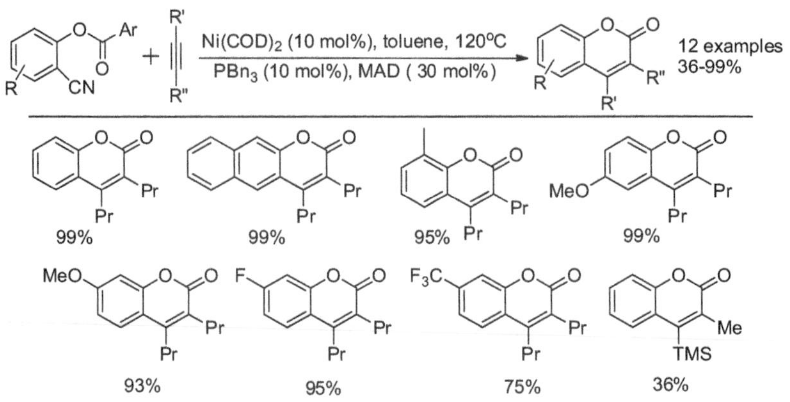

Scheme 7.67 Nickel-catalyzed synthesis of coumarins from alkynes.

catalytic amounts of Ni(0) and IPr ligand, CO_2 at atmospheric pressure and mild reaction conditions, and good yields of the desired pyrones were formed (Scheme 7.68). The mechanism was believed to involve an initial [2 + 2] cycloaddition of CO_2 and a single alkynyl unit of the substrate. Subsequent insertion of the second pendant alkynyl unit followed by a carbon–oxygen bond-forming reductive elimination would then release the pyrone product and regenerate the catalyst (NiLn). The proposed mechanism was supported by the observation that no conversion occurred with sterically hindered diynes. Interestingly, the proposed mechanism suggested that asymmetric diynes that possess at least one unhindered alkyne may undergo a regioselective cyclization. This was tested by analyzing the cycloaddition product of the asymmetric diyne with CO_2, and only one pyrone regioisomer was observed and isolated in 83% yield. In their continuing studies, they found that regioselectivity was highly dependent on the ligand employed and the size of the terminal groups on the diyne. When one alkynyl unit on the diyne was substituted with a TMS group (and the other with a relatively small methyl group), the IMes ligand afforded only one pyrone regioisomer. Improved performance was obtained with the IPr ligand as high regioselectivity was observed with diynes that contained a terminal *i*Pr, *t*Bu or TMS substituent. X-ray crystallographic analysis indicated that in the predominant regioisomer the relatively large substituent was located at the 3-position on the pyrone ring.

The coumarins are an important class of naturally occurring compounds, many of which exhibit useful biological activity. Among them, benzo-annulated coumarin derivatives are known as electron-transporting emitters. With this background, a nickel-catalyzed highly regio- and stereoselective cyclization of oxanorbornenes with alkyl propiolates to give benzocoumarin derivatives was developed in 2001 by Cheng and co-workers.[111] Using this novel nickel-catalyzed cyclization of tricyclic alkenes with propiolates, benzocoumarins or tetrahydrocoumarins were produced in good yields in a one-pot manner (Scheme 7.69). The mechanism of benzocoumarin formation is

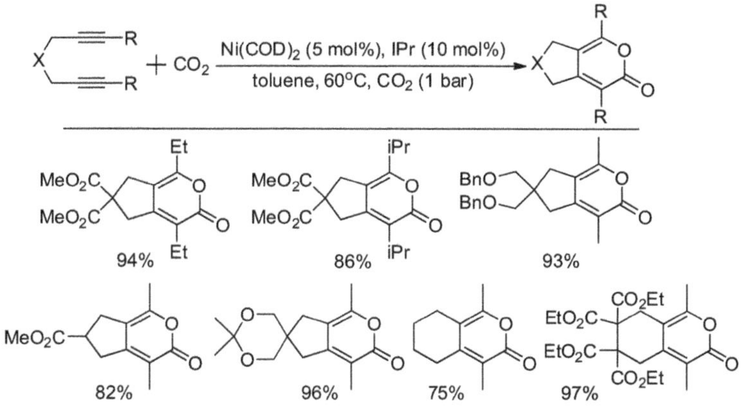

Scheme 7.68 Nickel-catalyzed synthesis of pyrones from diynes.

Reaction Procedure (Scheme 7.68): A solution of $Ni(COD)_2$ and IPr or IMes is prepared and allowed to equilibrate for at least 6 h. It was previously discovered that $Ni(COD)_2$ and IPr exist in equilibrium with the catalyst, $Ni(IPr)_2$ and COD and that at least 6 h is necessary to reach equilibrium. An oven-dried two-necked round-bottomed flask equipped with a magnetic stirrer bar, septum, gas adapter and balloon was evacuated and filled with CO_2. A solution of diyne was added and the flask was immersed in a 60 °C oil-bath. To the stirred solution, the equilibrated solution of $Ni(COD)_2$ and IPr or IMes was added. The dark greenish black reaction mixture was heated for 30 min (or until complete consumption of starting material was observed as judged by GC or TLC). The mixture was then cooled to ambient temperature, concentrated and purified by flash chromatography on SiO_2.

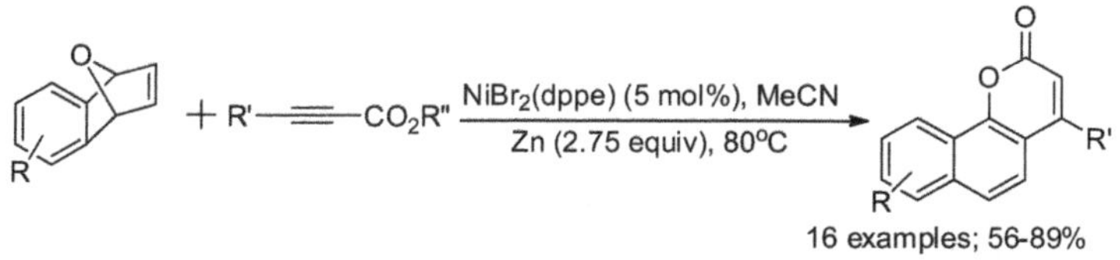

Scheme 7.69 Nickel-catalyzed synthesis of coumarins from oxanorbornenes and propiolates.

Reaction Procedure (Scheme 7.69): Freshly distilled CH_3CN (3.0 mL) and propiolates (1.2–2.0 mmol) were added to oxanorbornenes (1.00 mmol), [$NiBr_2(dppe)$] (0.05 mmol) and zinc powder (0.180 g, 2.75 mmol) under nitrogen. The reaction mixture was heated with stirring at 80 °C for 12 h and then cooled and stirred in air for 15 min. The mixture was filtered through Celite and silica gel and washed with dichloromethane. The

filtrate was concentrated and the residue was purified on a silica gel
column with hexane–ethyl acetate as eluent to afford the desired cyclization product.

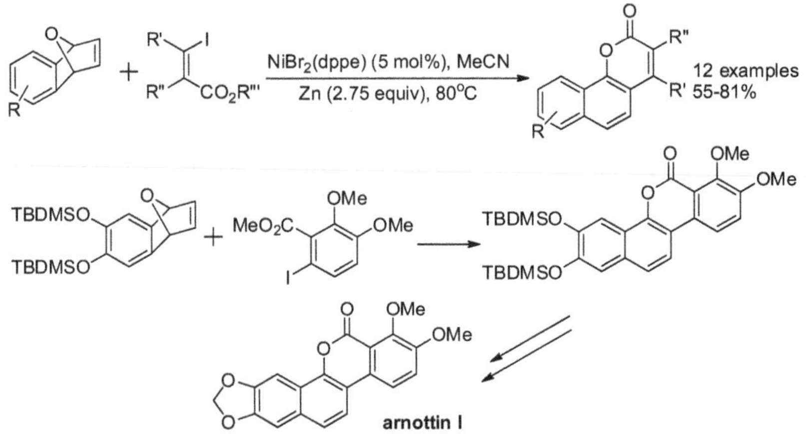

Scheme 7.70   Total synthesis of arnottin I.

intriguing in view of the extensive bond-formation and bond-breaking processes required. The reduction of [NiBr$_2$(dppe)] by zinc likely initiates the
catalytic reaction. Coordination of both oxanorbornenes and propiolate to
Ni(0) followed by cyclometalation forms a nickelacyclopentene intermediate.
Subsequent β-oxy elimination and protonation give the intermediate organic
product and an Ni(II) species. The latter is reduced by Zn to regenerate the
Ni(0) catalyst, while the organic product undergoes *cis–trans* isomerization,
dehydrogenation and lactonization to give the final benzocoumarin product.
In the same reaction system, adding 1.5 equiv. of water provides another
novel nickel-catalyzed procedure for the production of functionalized
cyclohexenols and 1,2-dihydroarenes.[112] Using bicyclic alkenes and alkynes
as substrates, the desired products were formed in fair to excellent yields
with complete regio- and stereoselectivity.

Cheng and co-workers also described the cyclization of oxabicyclic alkenes
with β-iodo-(Z)-propenoates and *o*-iodobenzoate using nickel complexes as
the catalyst.[113] This procedure gives a simple and efficient route to annulated coumarins (Scheme 7.70). In the presence of Ni(dppe)Br$_2$ and Zn
powder in acetonitrile at 80 °C, oxabicyclic alkenes undergo cyclization with
o-iodobenzoate and with β-iodo-(Z)-propenoates to give the benzocoumarin
derivatives in moderate to good yields. Regarding the reaction pathway, the
reduction of Ni(II) by zinc to Ni(0) likely initiates the catalytic cycle. Oxidative
addition of iodo compounds to nickel(0) yields a nickel(II) intermediate.
Coordination of 7-oxabenzonorbornadiene and subsequent insertion followed by β-oxy elimination give a nickel alkoxide, which undergoes transmetalation with ZnX$_2$, followed by lactonization and dehydrogenation to give

the final coumarin product and Ni(II) species. The latter is reduced by Zn to regenerate the Ni(0) catalyst for the catalytic cycle. In 2006, the same group succeeded in applying this procedure in total synthesis.[114] By ring-opening addition of methyl 2,3-dimethoxy-6-iodobenzoate to oxabenzonorborna-dienes followed by cyclization in the presence of NiBr$_2$(dppe) and Zn metal powder in acetonitrile at 80 °C, the corresponding benzocoumarin deriva-tives were formed. The natural product arnottin I, which was isolated from *Xanthoxylum arnottianum* Maxim, was obtained in 21% overall yield after six steps starting from catechol.

Recently, Zhu and co-workers reported a novel nickel-catalyzed insertion reaction of *tert*-butyl isocyanide.[115] In this approach, 1,2-bis(diphenylphos-phino)ethane serves as an efficient ligand, thereby allowing the preparation of lactones from (*o*-bromophenyl)phenylethanone derivatives (Scheme 7.71). Additionally, a new nickel-catalyzed [4 + 2] cycloaddition reaction of enones with alkynes to provide polysubstituted pyrans was developed in 2009 by Koyama *et al.* (Scheme 7.72).[116] This was the first report that enones

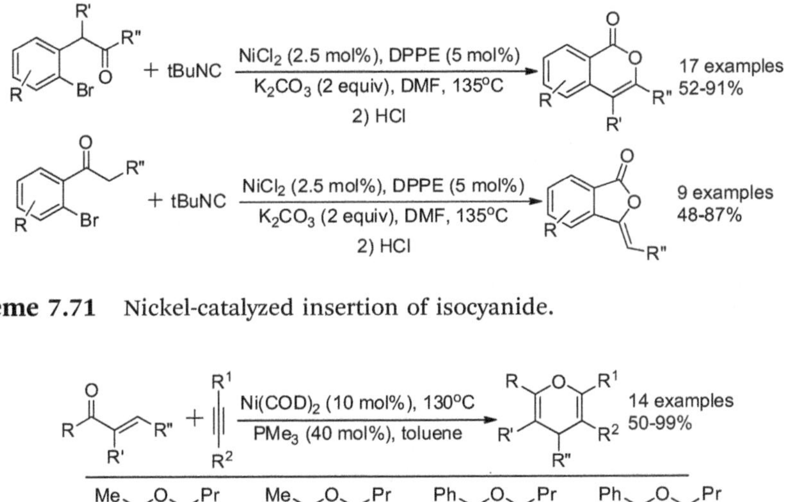

Scheme 7.71 Nickel-catalyzed insertion of isocyanide.

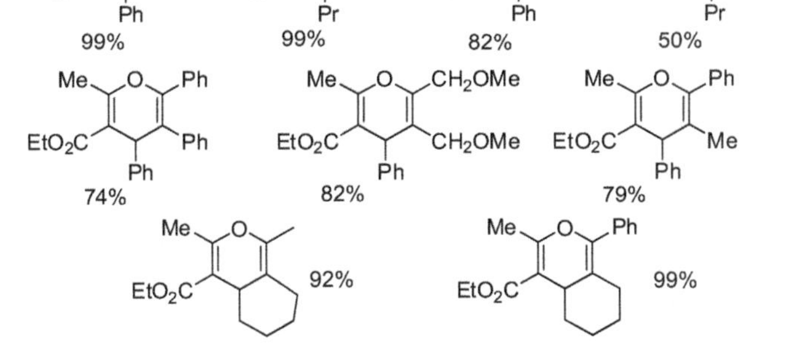

Scheme 7.72 Nickel-catalyzed synthesis of pyrans.

are susceptible to oxidative cyclization of nickel(0); such reactions allow inter- or intramolecular cycloaddition with alkynes. Concerning the reaction mechanism, the formation of polysubstituted pyrans can be rationalized as arising from oxidative cyclization of Ni(0) with an enone to form an oxa-nickelacycle. Subsequent coordination of alkyne takes place to give an intermediate; here, the steric repulsive interaction is minimal between the bulkier R(R1 or R2) and R″ of the oxanickelacycle. Insertion of an alkyne into the C–Ni bond leads to the seven-membered oxanickelacycle, which under-goes reductive elimination to give the final product and regenerates the starting nickel(0) complex.

Moliaro and Jamison developed a nickel-catalyzed reductive coupling of alkynes and epoxides in 2003 (Scheme 7.73a).[117] This catalytic reaction represents the first use of a non-π-based electrophile in a growing class of nickel-catalyzed, multicomponent coupling reactions. It was also the first catalytic method of reductive coupling of alkynes and epoxides that is effective for both intermolecular and intramolecular cases, and mech-anistically distinct from these, possibly involving a nickella(II)oxetane. An-other feature that is unprecedented is the complete selectivity for the usually disfavored *endo* epoxide opening product in alkyne–epoxide reductive cyclizations. Finally, the utility and ease of implementation of this method are direct results of the availability of terminal epoxides in >99% *ee*. In 2011, the same group reported a modified procedure for the reductive coupling of epoxides and alkynes.[118] This new method used air-stable and inexpensive Ni(II) precatalysts (e.g., NiBr$_2 \cdot$3H$_2$O) as the source of Ni(0) and simple al-cohols (e.g., 2-propanol) as the reducing agent, and the desired products were isolated in good yields (Scheme 7.73b). Deuterium-labeling experi-ments were consistent with oxidative addition of an epoxide C–O bond that occurs with inversion of configuration.

In 2011, another reaction procedure for pyran and tetrahydrofuran prep-aration was developed by Kim and Lee (Scheme 7.74).[119] As a mild and convenient nickel-catalyzed method for radical cyclization, they used zinc as the reductant in methanol. An Ni–Pybox catalyst system efficiently promotes the reductive cyclization of various alkyl halides with an effectiveness com-parable or superior to that of organotin hydride methods without the

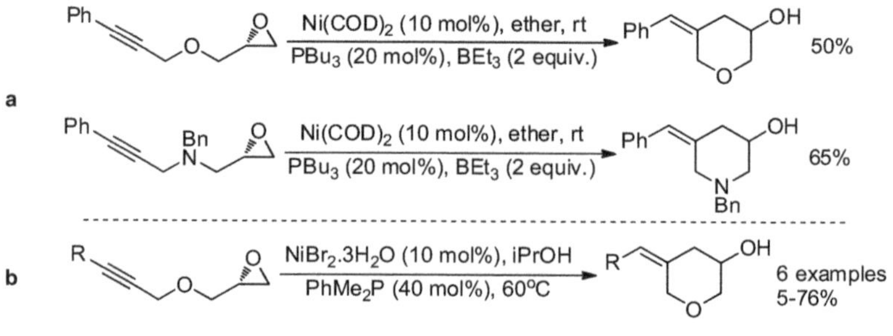

Scheme 7.73   Nickel-catalyzed ring opening of epoxides.

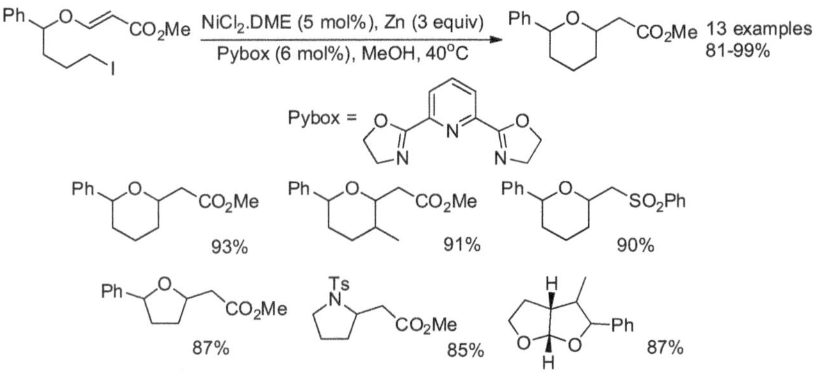

Scheme 7.74 Nickel-catalyzed cyclization of organohalides.

necessity for slow addition. In addition to the remarkable operational simplicity, the high functional group tolerance is also a noteworthy feature of this method. A free-radical mechanism was proposed for this Ni-catalyzed cyclization. After reduction of the Ni(II) precatalyst salt, the resulting Ni(0) complex brings about homolysis of the C–I bond of the substrate to give a carbon-centered radical through a single-electron transfer or direct iodine abstraction process. The radical enters into 6-*exo-trig* addition and subsequent reduction to yield the final product, and the active Ni(0) species may be regenerated through reduction mediated by zinc. This free-radical mechanism, rather than a two-electron pathway involving a carbon–metal σ-complex, is more consistent with the enhanced reactivity of the secondary halide substrate and the 2,5- and 2,6-*cis* diastereoselectivity in the formation of the final products.

Tekevac and Louie described a mild and general route for preparing pyrans and dienones from carbonyls and diynes in 2005.[120] Nickel–imidazolylidene complexes were used to mediate cyclizations between diynes and aldehydes. The reaction of an enyne and an aldehyde afforded a mixture of cyclized products. In addition, a spiropyran was prepared from the cycloaddition of a diyne and cyclohexanone (Scheme 7.75a). Later, they reported a comprehensive study on the nickel-catalyzed cycloaddition of unsaturated hydrocarbons and carbonyls (Scheme 7.75b).[121] Diynes and enynes were used as coupling partners. Carbonyl substrates included both aldehydes and ketones. Reactions of diynes and aldehydes afforded the [3,3] electrocyclic ring-opened tautomers, rather than pyrans, in high yields. The cycloaddition reaction of enynes and aldehydes afforded two distinct products. A new carbon–carbon bond was formed, prior to a competitive β-hydrogen elimination of a nickel alkoxide, between the carbonyl carbon and either one of the carbons of the alkene or the alkyne. The steric hindrance of the enyne greatly affected the chemoselectivity of the cycloaddition of enynes and aldehydes. In some cases, a dihydropyran was also formed. The scope of the cycloaddition reaction was extended to include the coupling of enynes

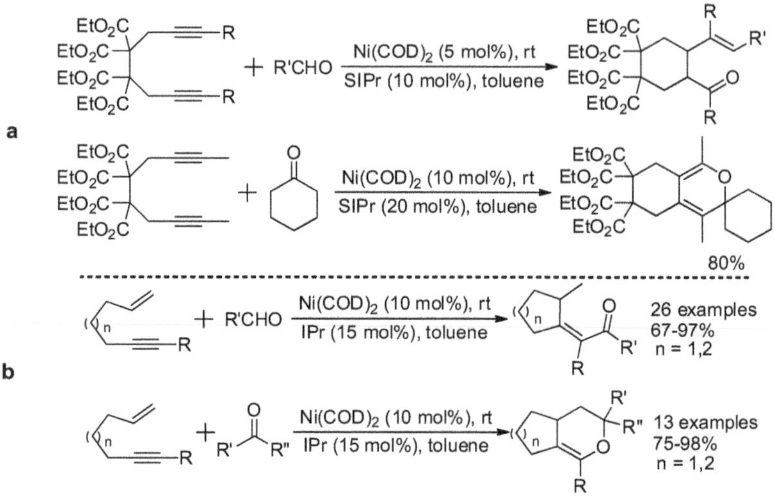

Scheme 7.75   Nickel-catalyzed synthesis of dihydropyrans.

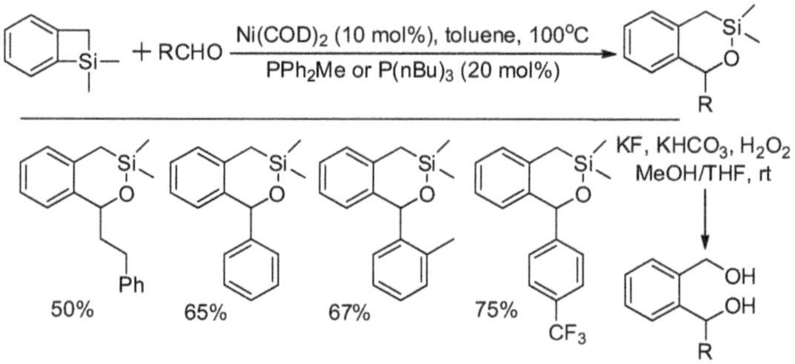

Scheme 7.76   Nickel-catalyzed ring expansion of benzosilacyclobutene.

and ketones. No β-hydrogen elimination was observed in the cycloaddition reaction of enynes and ketones. Instead, C–O bond-forming reductive elimination occurred exclusively to afford dihydropyrans in excellent yields. In all cases, complete chemoselectivity was observed; only dihydropyrans in which the carbonyl carbon forms a carbon–carbon bond with a carbon of the alkene, rather than of the alkyne, were observed. All cycloaddition reactions proceeded at room temperature and employed nickel catalysts bearing the hindered 1,3-bis(2,6-diisopropylphenyl)imidazol-2-ylidene (IPr) or its saturated analog, 1,3-bis(2,6-diisopropylphenyl)-4,5-dihydroimidazolin-2-ylidene (SIPr).

A nickel-catalyzed ring expansion reaction of benzosilacyclobutene with aldehydes was reported by Oshima and co-workers in 2006.[122] Oxasilacyclohexenes were formed in good yields (Scheme 7.76), and could be easily converted into 1,2-bis(hydroxymethyl)benzenes by Tamao–Fleming oxidation.

7.3 Other Heterocycles

In addition to the mentioned five- and six-membered heterocycles, nickel catalysts have also been applied in the preparation of other heterocyclic compounds. In 2008, Louie and co-workers reported a nickel-catalyzed coupling of vinylaziridines and phenyl isocyanates.[123] Under thermal conditions, oxazolidinone products were predominant, and some reactions even afforded a seven-membered ring heterocycle. When Ni(0)–IMes was employed as a catalyst, a wider array of vinylaziridines underwent coupling reactions. The Ni-catalyzed reactions generally afforded vinylimidazolidinones as major products (Scheme 7.77). Subsequently, they reported the rearrangement of vinylaziridines and aziridinylenynes using nickel as catalyst.[124]

In 2002, Rayabarapu and Cheng demonstrated for the first time a bimetal-catalyzed regio- and stereoselective cyclization of 2-iodobenzyl alcohols with alkyl propiolates.[125] This catalytic reaction provides a convenient and unique method for the synthesis of seven-membered ring lactones (Scheme 7.78). The catalytic reaction involves an unusual *E*–*Z* isomerization of a carbon–carbon double bond prior to ring closure. A possible catalytic cycle was suggested. Reduction of Ni(II) to Ni(0) by Zn metal likely initiates the catalytic cycle. Oxidative addition of 2-iodobenzyl alcohol to Ni(0) to generate organonickel(II) species was followed by regioselective propiolate insertion and subsequent protonation to give (*E*)-3-arylacrylate. *E*–*Z* isomerization of the C–C double bond in the arylacrylate and ring closure then occurred to afford the final lactone product.

More recently, Louie and co-workers demonstrated that eight-membered heterocycles can be easily accessed through Ni–IPr-catalyzed coupling of 3-azetidinone (or 3-oxetanone) and diynes (Scheme 7.79a).[126] The

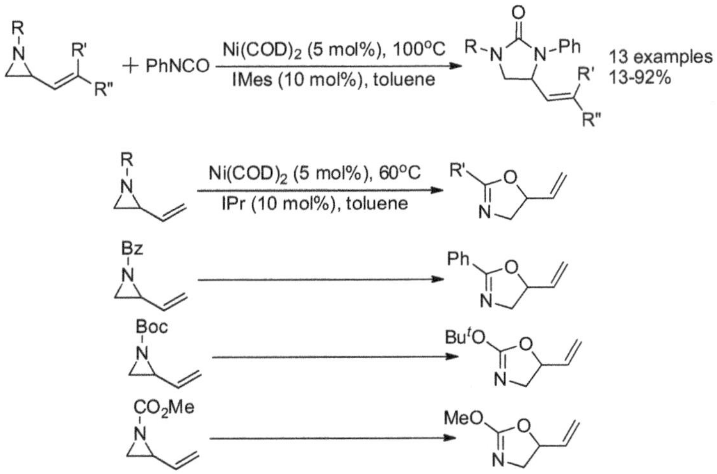

Scheme 7.77 Nickel-catalyzed synthesis of vinyl imidazolidinones.

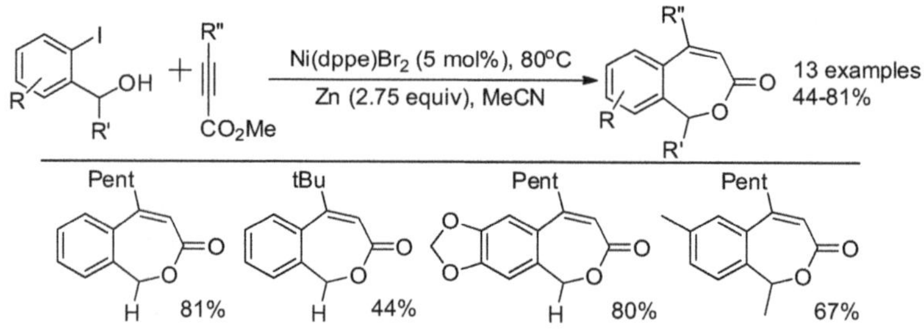

Scheme 7.78 Nickel-catalyzed synthesis of seven-membered ring lactones.

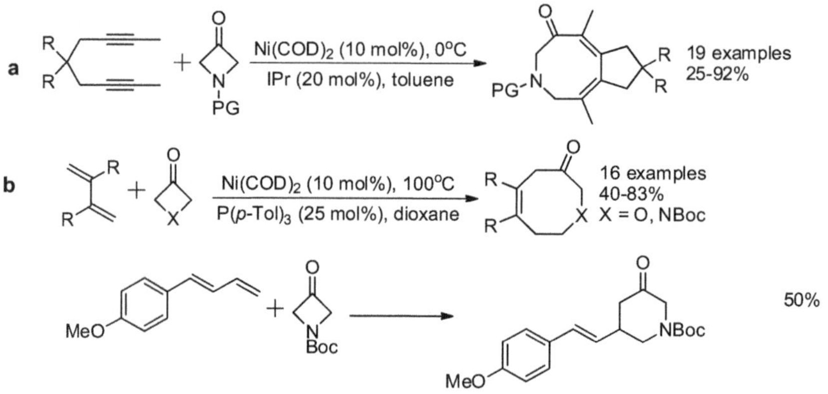

Scheme 7.79 Nickel-catalyzed cyclization of azetidinones.

Reaction Procedure (Scheme 7.79): In a nitrogen-filled glove-box, a scintillation vial equipped with a magnetic stirrer bar was charged with a solution of azetidinone (1 equiv.) and diyne (1.2 equiv.). At room temperature, a solution of the catalyst, which was prepared by stirring a mixture of [Ni(COD)$_2$] and IPr (1:2 molar ratio) in toluene for at least 6 h, was added. The vial containing the reaction mixture was immediately taken out of the glove-box and sealed; then reaction mixture was stirred at 0 °C for 8 h. The solvent was removed under vacuum and the product was purified by silica gel flash column chromatography.

Reaction Procedure: In a nitrogen-filled glove-box, a solution of the catalyst {10 mol%; prepared from [Ni(COD)$_2$] and tri(p-tolyl)phosphane in a 1:2.5 molar ratio in 1,4-dioxane} was added to a solution of the 3-azetidinone (1 equiv., 0.4 M) and the 1,3-diene (2 equiv.) in 1,4-dioxane at room temperature. The resulting reaction mixture was stirred for 24 h at 100 °C and then opened to the air, concentrated *in vacuo* and purified by silica gel flash column chromatography.

decomposition of these constrained heterocycles could be avoided by using specific reaction conditions. This method involves an interesting C–C bond cleavage step, which operates smoothly at 0°C. Typically, such a cycloaddition would begin with oxidative coupling between an alkyne and the carbonyl group. In this case, oxidative coupling would afford a spirocyclic intermediate. Subsequent insertion of the pendant alkyne would give another intermediate, which could then undergo either reductive elimination (to afford products) or β-carbon elimination to afford a metallacycle. Finally, C–C bond-forming reductive elimination occurs to give the dihydroazocine product and the catalyst. Soon afterwards, they developed an Ni–P(*p*-Tol)₃-catalyzed intermolecular cycloaddition of 1,3-dienes and 3-azetidinones/3-oxetanones (Scheme 7.79b).[127] This synthetic method involves C–C activation of the strained four-membered heterocycle to form monocyclic and bicyclic eight-membered heterocyclic products, which are difficult to access by conventional methods. Interestingly, the use of a diene conjugated with a benzene ring led to the formation of a piperidinone rather than an eight-membered heterocycle.

7.4 Summary

The main contributions to the nickel-catalyzed synthesis of heterocyclic compounds have been presented and discussed. In general, the procedures still need a relatively high loading of catalyst and unstable nickel(0) as catalyst precursor. In the future, the use of more stable nickel salts and *in situ* reduction should be encouraged. The reaction efficiency still needs to be improved.

References

1. (a) E. Dunach, A. P. Esteves, A. M. Freitas, M. A. Lemos, M. J. Medeiros and S. Olivero, *Pure Appl. Chem.*, 2001, **73**, 1941–1945; (b) A. P. Esteves, D. M. Goken, L. J. Klein, M. A. Lemos, M. J. Medeiros and D. G. Peters, *J. Org. Chem.*, 2003, **68**, 1024–1029; (c) E. Dunach, A. P. Esteves, M. J. Medeiros and S. Olivero, *Tetrahedron Lett.*, 2004, **45**, 7935–7937; (d) E. Dunach, A. P. Esteves, M. J. Medeiros and S. Olivero, *New J. Chem.*, 2005, **29**, 633–636; (e) E. Dunach, A. P. Esteves, M. J. Medeiros and S. Olivero, *Green Chem.*, 2006, **8**, 380–385; (f) A. P. Esteves, E. C. Ferreira and M. J. Medeiros, *Tetrahedron*, 2007, **63**, 3006–3009; (g) E. Dunach and M. J. Medeiros, *Electrochim. Acta*, 2008, **53**, 4470–4477; (h) A. P. Esteves, C. S. Neves, M. J. Medeiros and D. Pletcher, *J. Electroanal. Chem.*, 2008, **614**, 131–138; (i) M. J. Medeiros, C. S. S. Neves, A. R. Pereira and E. Dunach, *Electrochim. Acta*, 2011, **56**, 4498–4503; (j) E. Dunach, A. P. Esteves, M. J. Medeiros, D. Pletcher and S. Olivero, *J. Electroanal. Chem.*, 2004, **566**, 39–45; (k) J. Pelletier, S. Olivero and E. Dunach, *Synth. Commun.*, 2004, **34**, 3343–3348.

2. (a) M. Durandetti, L. Hardou, R. Lhermet, M. Rouen and J. Maddaluno, *Chem. Eur. J.*, 2011, **17**, 12773–12783; (b) R. Lhermet, M. Durandetti and J. Maddaluno, *Beilstein J. Org. Chem.*, 2013, **9**, 710–716.
3. X. Wu, L. Li and J. Zhang, *Chem. Commun.*, 2011, **47**, 7824–7826.
4. J. Zhang, Y. Xiao and J. Zhang, *Adv. Synth. Catal.*, 2013, **355**, 2793–2797.
5. G. Bentabed-Ababsa, A. Derdour, T. Roisnel, J. A. Sáez, P. Pérez, E. Chamorro, L. R. Domingo and F. Mongin, *J. Org. Chem.*, 2009, **74**, 2120–2133.
6. Z. Chen, Z. Tian, J. Zhang, J. Ma and J. Zhang, *Chem. Eur. J.*, 2012, **18**, 8591–8595.
7. J. Zhang, Z. Chen, H. H. Wu and J. Zhang, *Chem. Commun.*, 2012, **48**, 1817–1819.
8. Z. Chen, Y. Xiao and J. Zhang, *Eur. J. Org. Chem.*, 2013, 4748–4751.
9. A. Ezoe, M. Kimura, T. Inoue, M. Mori and Y. Tamaru, *Angew. Chem. Int. Ed.*, 2002, **41**, 2784–2786.
10. A. Jeevanandam, R. P. Korivi, I. W. Huang and C. H. Cheng, *Org. Lett.*, 2002, **4**, 807–810.
11. J. C. Hsieh and C. H. Cheng, *Chem. Commun.*, 2005, 2459–2461.
12. J. C. Hsieh and C. H. Cheng, *Chem. Commun.*, 2008, 2992–2994.
13. T. Iwayama and Y. Sato, *Chem. Commun.*, 2009, 5245–5247.
14. (a) P. Turek, M. Kotora, M. Hocek and I. Cisarova, *Tetrahedron Lett.*, 2003, **44**, 785–788; (b) P. Turek, M. Kotora, I. Tislerova, M. Hocek, I. Votruba and I. Cisarova, *J. Org. Chem.*, 2004, **69**, 9224–9233.
15. Y. Ni and J. Montgomery, *J. Am. Chem. Soc.*, 2004, **126**, 11162–11163.
16. Y. Ni and J. Montgomery, *J. Am. Chem. Soc.*, 2006, **128**, 2609–2614.
17. G. Zuo and J. Louie, *J. Am. Chem. Soc.*, 2005, **127**, 5798–5799.
18. X. Hong, P. Liu and K. N. Houk, *J. Am. Chem. Soc.*, 2013, **135**, 1456–1462.
19. M. Kimura, M. Mori, N. Mukai, K. Kojima and Y. Tamaru, *Chem. Commun.*, 2006, 2813–2815.
20. (a) M. Kimura, A. Ezoe, M. Mori and Y. Tamaru, *J. Am. Chem. Soc.*, 2005, **127**, 201–209; (b) A. Ezoe, M. Kimura, T. Inoue, M. Mori and Y. Tamaru, *Angew. Chem. Int. Ed.*, 2002, **41**, 2784–2786.
21. P. Kumar, D. M. Troast, R. Cella and J. Louie, *J. Am. Chem. Soc.*, 2011, **133**, 7719–7721.
22. D. K. Rayabarapu, H. T. Chang and C. H. Cheng, *Chem. Eur. J.*, 2004, **10**, 2991–2996.
23. R. K. Bowman and J. S. Johnson, *Org. Lett.*, 2006, **8**, 573–576.
24. P. Tascedda and E. Dunach, *J. Chem. Soc. Chem. Commun.*, 1995, 43–44.
25. F. Li, C. Xia, L. Xu, W. Sun and G. Chen, *Chem. Commun.*, 2003, 2042–2043.
26. M. Takimoto, M. Kawamura and M. Mori, *Org. Lett.*, 2003, **5**, 2599–2601.
27. S. Li and S. Ma, *Org. Lett.*, 2011, **13**, 6046–6049.
28. M. Ohashi, O. Kishizaki, H. Ikeda and S. Ogoshi, *J. Am. Chem. Soc.*, 2009, **131**, 9160–9161.

29. M. Ohashi, H. Saijo, T. Arai and S. Ogoshi, *Organometallics*, 2010, **29**, 6534–6540.
30. M. Ohashi, M. Ikawa and S. Ogoshi, *Organometallics*, 2011, **30**, 2765–2774.
31. M. Lozanov and J. Montgomery, *Tetrahedron Lett.*, 2001, **42**, 3259–3261.
32. Y. Sato, T. Nishimata and M. Mori, *J. Org. Chem.*, 1994, **59**, 6133–6135.
33. Y. Sato, T. Nishimata and M. Mori, *Heterocycles*, 1997, **44**, 443–457.
34. Y. Sato, N. Saito and M. Mori, *Tetrahedron Lett.*, 1997, **38**, 3931–3934.
35. Y. Sato, N. Saito and M. Mori, *Tetrahedron*, 1998, **54**, 1153–1168.
36. Y. Sato, N. Saito and M. Mori, *J. Am. Chem. Soc.*, 2000, **122**, 2371–2372.
37. N. Saito, M. Mori and Y. Sato, *J. Organomet. Chem.*, 2007, **692**, 460–471.
38. M. Takimoto and M. Mori, *J. Am. Chem. Soc.*, 2002, **124**, 10008–10009.
39. M. Takimoto, T. Mizuno, Y. Sato and M. Mori, *Tetrahedron Lett.*, 2005, **46**, 5173–5176.
40. M. Takimoto, T. Mizuno, M. Mori and Y. Sato, *Tetrahedron*, 2006, **62**, 7589–7597.
41. J. Montgomery and M. Song, *Org. Lett.*, 2002, **4**, 4009–4011.
42. B. B. Thompson and J. Montgomery, *Org. Lett.*, 2011, **13**, 3289–3291.
43. D. M. Shacklady-McAtee, S. Dasgupta and M. P. Watson, *Org. Lett.*, 2011, **13**, 3490–3493.
44. N. Maizuru, T. Inami, T. Kurahashi and S. Matsubara, *Org. Lett.*, 2011, **13**, 1206–1209.
45. Y. Yoshida, T. Kurahashi and S. Matsubara, *Chem. Lett.*, 2011, **40**, 1067–1068.
46. R. Tombe, T. Kurahashi and S. Matsubara, *Org. Lett.*, 2013, **15**, 1791–1793.
47. T. Miura, Y. Mikano and M. Murakami, *Org. Lett.*, 2011, **13**, 3560–3563.
48. N. Rosas, A. Cabrera, P. Sharma, J. L. Arias, J. L. Garcia and H. Arzoumanian, *J. Mol. Catal. A: Chem.*, 2000, **156**, 103–112.
49. J. M. Khurana, Sneha and K. Vij, *Synth. Commun.*, 2012, **42**, 2606–2616.
50. F. Chen, T. Shen, Y. Cui and N. Jiao, *Org. Lett.*, 2012, **14**, 4926–4929.
51. J. C. Hsieh and C. H. Cheng, *Chem. Commun.*, 2005, 4554–4556.
52. C. H. Yeh, R. P. Korivi and C. H. Cheng, *Angew. Chem. Int. Ed.*, 2008, **47**, 4892–4895.
53. P. A. Wender and J. P. Christy, *J. Am. Chem. Soc.*, 2007, **129**, 13402–13403.
54. S. K. Kang and S. K. Yoon, *Chem. Commun.*, 2002, 2634–2635.
55. T. Miura, M. Yamauchi and M. Murakami, *Chem. Commun.*, 2009, 1470–1471.
56. B. Pan, C. Wang, D. Wang, F. Wu and B. Wan, *Chem. Commun.*, 2013, **49**, 5073–5075.
57. J. X. Hu, H. Wu, C. Y. Li, W. J. Sheng, Y. X. Jia and J. R. Gao, *Chem. Eur. J.*, 2011, **17**, 5234–5237.
58. C. Liu, D. Liu, W. Zhang, L. Zhou and A. Lei, *Org. Lett.*, 2013, **15**, 6166–6169.

59. S. I. Sviridov, A. A. Vasil'ev and S. V. Shorshnev, *Tetrahedron*, 2007, **63**, 12195–12201.
60. J. J. Garcia, P. Zerecero-Silva, G. Reyes-Rios, M. G. Crestani, A. Arevalo and R. Barrios-Francisco, *Chem. Commun.*, 2011, **47**, 10121–10123.
61. M. M. McCormick, H. A. Duong, G. Zuo and J. Louie, *J. Am. Chem. Soc.*, 2005, **127**, 5030–5031.
62. T. N. Tekavec, G. Zuo, K. Simon and J. Louie, *J. Org. Chem.*, 2006, **71**, 5834–5836.
63. R. M. Stolley, M. T. Maczka and J. Louie, *Eur. J. Org. Chem.*, 2011, 3815–3824.
64. R. M. Stolley, H. A. Duong, D. R. Thomas and J. Louie, *J. Am. Chem. Soc.*, 2012, **134**, 15154–15162.
65. R. M. Stolley, H. A. Duong and J. Louie, *Organometallics*, 2013, **32**, 4952–4960.
66. P. Kumar, S. Prescher and J. Louie, *Angew. Chem. Int. Ed.*, 2011, **50**, 10694–10698.
67. T. Takahashi, F. Y. Tsai and M. Kotora, *J. Am. Chem. Soc.*, 2000, **122**, 4994–4995.
68. M. Ohashi, I. Takeda, M. Ikawa and S. Ogoshi, *J. Am. Chem. Soc.*, 2011, **133**, 18018–18021.
69. R. P. Korivi and C. H. Cheng, *Org. Lett.*, 2005, **7**, 5179–5182.
70. W. C. Shih, C. C. Teng, K. Parthasarathy and C. H. Cheng, *Chem. Asian J.*, 2012, **7**, 306–313.
71. R. P. Korivi and C. H. Cheng, *J. Org. Chem.*, 2006, **71**, 7079–7082.
72. S. Liu, J. Sawicki and T. G. Driver, *Org. Lett.*, 2012, **14**, 3744–3747.
73. N. Rosas, P. Sharma, A. Cabrera, G. Penieres, J. L. Garcia and L. A. Maldonado, *Heterocycles*, 2003, **60**, 2631–2636.
74. R. P. Korivi, Y. C. Wu and C. H. Cheng, *Chem. Eur. J.*, 2009, **15**, 10727–10731.
75. L. Adak, W. C. Chan and N. Yoshikai, *Chem. Asian J.*, 2011, **6**, 359–362.
76. Y. Kaijita, S. Matsubara and T. Kurahashi, *J. Am. Chem. Soc.*, 2008, **130**, 6058–6059.
77. T. Shiba, T. Kurahashi and S. Matsubara, *J. Am. Chem. Soc.*, 2013, **135**, 13636–13639.
78. Y. Nakao, E. Morita, H. Idei and T. Hiyama, *J. Am. Chem. Soc.*, 2011, **133**, 3264–3267.
79. M. Anand and R. B. Sunoj, *Org. Lett.*, 2012, **14**, 4584–4587.
80. M. Yamauchi, M. Morimoto, T. Miura and M. Murakami, *J. Am. Chem. Soc.*, 2010, **132**, 54–55.
81. H. Shiota, Y. Ano, Y. Aihara, Y. Fukumoto and N. Chatani, *J. Am. Chem. Soc.*, 2011, **133**, 14952–14955.
82. T. Miura, M. Yamauchi and M. Murakami, *Org. Lett.*, 2008, **10**, 3085–3088.
83. C. C. Liu, K. Parthasarathy and C. H. Cheng, *Org. Lett.*, 2010, **12**, 3518–3521.

84. K. Fujiwara, T. Kurahashi and S. Matsubara, *Org. Lett.*, 2010, **12**, 4548–4551.
85. T. Miura, M. Morimoto, M. Yamauchi and M. Murakami, *J. Org. Chem.*, 2010, **75**, 5359–5362.
86. H. Hoberg and B. W. Oster, *Synthesis*, 1982, 324–325.
87. H. A. Duong and J. Louie, *J. Organomet. Chem.*, 2005, **690**, 5098–5104.
88. H. A. Duong and J. Louie, *Tetrahedron*, 2006, **62**, 7552–7559.
89. H. A. Duong, M. J. Cross and J. Louie, *J. Am. Chem. Soc.*, 2004, **126**, 11438–11439.
90. K. Nakai, T. Kurahashi and S. Matsubara, *Org. Lett.*, 2013, **15**, 856–859.
91. P. Kumar and J. Louie, *Org. Lett.*, 2012, **14**, 2026–2029.
92. Y. Li and Z. Lin, *Organometallics*, 2013, **32**, 3003–3011.
93. Y. Yoshino, T. Kurahashi and S. Matsubara, *J. Am. Chem. Soc.*, 2009, **131**, 7494–7495.
94. M. Sun, Y. N. Ma, Y. M. Li, Q. P. Tian and S. D. Yang, *Tetrahedron Lett.*, 2013, **54**, 5091–5095.
95. J. Yang, M. R. Karver, W. Li, S. Sahu and N. K. Devaraj, *Angew. Chem. Int. Ed.*, 2012, **51**, 5222–5225.
96. X. Yu, G. Qiu, J. Liu and J. Wu, *Synthesis*, 2011, 2268–2274.
97. I. Arellano, P. Sharma, L. Rubio-Perez, A. Cabrera, N. Rosas and A. Toscano, *J. Organomet. Chem.*, 2012, **700**, 29–35.
98. B. R. D'Souza and J. Louie, *Org. Lett.*, 2009, **11**, 4168–4171.
99. R. Omar-Amrani, A. Thomas, E. Brenner, R. Schneider and Y. Fort, *Org. Lett.*, 2003, **5**, 2311–2314.
100. X. Q. Tang and J. Montgomery, *J. Am. Chem. Soc.*, 1999, **121**, 6098–6099.
101. X. Q. Tang and J. Montgomery, *J. Am. Chem. Soc.*, 2000, **122**, 6950–6954.
102. K. Shimizu, M. Takimoto and M. Mori, *Org. Lett.*, 2003, **5**, 2323–2325.
103. T. Inami, Y. Baba, T. Kurahashi and S. Matsubara, *Org. Lett.*, 2011, **13**, 1912–1915.
104. T. Miura, M. Yamauchi, A. Kosaka and M. Murakami, *Angew. Chem. Int. Ed.*, 2010, **49**, 4955–4957.
105. Y. Kajita, T. Kurahashi and S. Matsubara, *J. Am. Chem. Soc.*, 2008, **130**, 17226–17227.
106. Y. Ochi, T. Kurahashi and S. Matsubara, *Org. Lett.*, 2011, **13**, 1374–1377.
107. A. Ooguri, K. Nakai, T. Kurahashi and S. Matsubara, *J. Am. Chem. Soc.*, 2009, **131**, 13194–13195.
108. K. Nakai, T. Kurahashi and S. Matsubara, *J. Am. Chem. Soc.*, 2011, **133**, 11066–11068.
109. J. Louie, J. E. Gibby, M. V. Farnworth and T. N. Tekavec, *J. Am. Chem. Soc.*, 2002, **124**, 15188–15189.
110. T. N. Tekavec, A. M. Arif and J. Louie, *Tetrahedron*, 2004, **60**, 7431–7437.
111. D. K. Rayabarapu, T. Sambaiah and C. H. Cheng, *Angew. Chem. Int. Ed.*, 2001, **40**, 1286–1288.
112. D. K. Rayabarapu and C. H. Cheng, *Chem. Eur. J.*, 2003, **9**, 3164–3169.

113. D. K. Rayabarapu, P. Shukla and C. H. Cheng, *Org. Lett.*, 2003, **5**, 4903–4906.
114. S. Madan and C. H. Cheng, *J. Org. Chem.*, 2006, **71**, 8312–8315.
115. X. D. Fei, T. Tang, Z. Y. Ge and Y. M. Zhu, *Synth. Commun.*, 2013, **43**, 3262–3271.
116. I. Koyama, T. Kurahashi and S. Matsubara, *J. Am. Chem. Soc.*, 2009, **131**, 1350–1351.
117. C. Molinaro and T. F. Jamison, *J. Am. Chem. Soc.*, 2003, **125**, 8076–8077.
118. M. G. Beaver and T. F. Jamison, *Org. Lett.*, 2011, **13**, 4140–4143.
119. H. Kim and C. Lee, *Org. Lett.*, 2011, **13**, 2050–2053.
120. T. N. Tekevac and J. Louie, *Org. Lett.*, 2005, **7**, 4037–4039.
121. T. N. Tekevac and J. Louie, *J. Org. Chem.*, 2008, **73**, 2641–2648.
122. K. Hirano, H. Yorimitsu and K. Oshima, *Org. Lett.*, 2006, **8**, 483–485.
123. K. Zhang, P. R. Chopade and J. Louie, *Tetrahedron Lett.*, 2008, **49**, 4306–4309.
124. G. Zuo, K. Zhang and J. Louie, *Tetrahedron Lett.*, 2008, **49**, 6797–6799.
125. D. K. Rayabarapu and C. H. Cheng, *J. Am. Chem. Soc.*, 2002, **124**, 5630–5631.
126. P. Kumar, K. Zhang and J. Louie, *Angew. Chem. Int. Ed.*, 2012, **51**, 8602–8606.
127. A. Thakur, M. E. Facer and J. Louie, *Angew. Chem. Int. Ed.*, 2013, **52**, 12161–12165.

CHAPTER 8
Outlook

In the last six chapters, we discussed in detail the applications of Zn, Fe, Cu, Co, Mn and Ni catalysts in the synthesis of heterocycles. The low cost and biocompatibility of these catalysts and the great importance of heterocycles make this area of high interest. Zinc catalysts are mainly used as Lewis acids, and they are abilities in coupling reactions are still largely unexplored. In comparison, the power of copper catalysts in cross-coupling is impressive, even comparable to that of palladium. Iron catalysts lie between zinc and copper, and a large percentage of iron-catalyzed heterocycle syntheses are based on their Lewis acid properties, but coupling reactions have also been reported. In the area of oxidative transformations, copper and manganese showed excellent activities compared with the others. Regarding C–O bond activation, nickel catalysts are much better than palladium and other noble metals.

In the future, improving the catalytic abilities of zinc and iron still needs more efforts. In the case of copper catalysts, their reaction efficiency still has room for improvement. Well-defined complexes as catalysts are always interesting from the mechanistic understanding point of view. During the methodology design, hazardous reagents should be avoided as much as possible and 'green' solvents should be considered instead of classical solvents. Concerning the concepts of 'atom efficiency' and 'bond-forming economy', the reagents used must have the optimal properties with respect to the bonds formed.

A major issue in Zn-, Fe-, Cu-, Co-, Mn-catalyzed syntheses of heterocycles is that the reactions all need preactivation of the substrates applied. Concerning C–H activation, regarding the reaction efficiency, relatively high catalyst loadings (5–10 mol%) are still required in most cases. Reliable ligands need to be developed to stabilize the active species and further increase the reaction rate. Overall, more efforts are required to make this chemistry more satisfactory – we need a breakthrough!

RSC Catalysis Series No. 16
Economic Synthesis of Heterocycles: Zinc, Iron, Copper, Cobalt, Manganese and Nickel Catalysts
By Xiao-Feng Wu and Matthias Beller
© Wu and Beller, 2014
Published by the Royal Society of Chemistry, www.rsc.org

Subject Index

Note: Page numbers in *italic* refer to schemes and tables.